The Oil Palm

TROPICAL AGRICULTURE SERIES

The Tropical Agriculture Series, of which this volume
forms part, is published under the editorship of
D. Rhind, CMG, OBE, BSc, FLS, FIBiol.

ALREADY PUBLISHED
Tobacco *B. C. Akehurst*
Tropical Pasture and Fodder Plants
A. V. Bogdan
Coconuts *R. Child*
Yams *D. G. Coursey*
Sorghum *H. Doggett*
Tea *T. Eden*
Rice *D. H. Grist*
Termites *W. V. Harris*
The Oil Palm *C. W. S. Hartley*
Tropical Farming Economics
M. R. Haswell
Sisal *G. W. Lock*
Cattle Production in the Tropics Volume I
W. J. A. Payne
Cotton *A. N. Prentice*
Bananas *N. W. Simmonds*
Tropical Pulses *J. Smartt*
Agriculture in the Tropics
C. C. Webster and P. N. Wilson
An Introduction to Animal Husbandry in the Tropics
G. Williamson and W. J. A. Payne
Cocoa *G. A. R. Wood*

The Oil Palm

(*Elaeis guineensis* Jacq.)

Second edition

C. W. S. Hartley

CBE, MA, Dip. Agr. Sci., AICTA, FIBiol
Formerly Director,
West African Institute for Oil Palm Research

Longman
London and New York

Longman Group Limited London

Associated companies, branches and representatives throughout the world

Published in the United States of America by Longman Inc., New York

First edition © C. W. S. Hartley, 1967
This edition © Longman Group Limited, 1977

First published 1967
New impression (revised) 1970
Second edition 1977
New impression 1979

Library of Congress Cataloging in Publication Data

Hartley, Charles William Stewart, 1911–
   The oil palm (Elaeis guineensis Jacq.)

   (Tropical agriculture series)
   Includes bibliographical references and index.
   1. Oil-palm.   I. Title.
SB299.P3H3   633'.85   76–23180
ISBN 0 582 46809 4

Set in IBM Journal 10 on 11pt
and printed in Great Britain by
Lowe and Brydone (Printers) Ltd,
Thetford, Norfolk

Preface to the first edition

During recent years there has been a considerable expansion of oil palm acreages in tropical Asia, Africa and America, and interest in the crop has been steadily increasing. At the same time there has been much improvement in the cultivation of the palm following research carried out by research institutes and plantation companies. Of particular value has been the interchange of information between the great producing regions of Africa and Asia and the realization that work done in one continent is often of great import to producers in another. Nevertheless, much of the work carried out has not been adequately published and original papers are sometimes difficult to obtain. It has therefore been my aim to provide in this book a comprehensive account of the oil palm as a plant, of the industry from its early beginnings to its present stage of development, and of the work carried out in all regions to improve cultivation, production and the extraction of the products. In so doing I have tried to interpret the difficulties that have been encountered in various parts of the world, to trace, historically and critically, the reasons underlying certain practices, and to draw attention to the experimental bases, where such exist, for present procedures.

I have been greatly assisted in the compilation of this book by the ready assistance I have received from many quarters. In the first place I have to thank the Managing Committee of the West African Institute for Oil Palm Research (now N.I.F.O.R.) for assistance given to me and for permission to make use of material being the property of the Institute. Members of the research staff of the Institute, past and present, have contributed much to this book through their work. In particular I would like to acknowledge the help I have had during compilation from Mr G. Blaak, Mr T. Menendez, Mr S. C. Nwanze, Mr A. R. Rees, Mr J. S. Robertson, Mr R. D. Sheldrick, Mr J. M. A. Sly, Dr L. D. Sparnaaij, Dr P. B. H. Tinker and Mr A. C. Zeven; and I am especially grateful to Mr Robertson and Dr Tinker for reading and commenting upon parts of the text and to Mr Rees for answering many queries on problems of germination and physiology generally.

Much assistance and data have also been generously given to me by the principal oil palm plantation companies and their research organizations, and for these I am very glad to be able to thank Mr D. L. Martin, Mr S. de Blank and Mr A. H. Green of Unilever Plantations Group, Mr B. S. Gray, Director of Research, and the headquarters staff of Messrs Harrisons and Crosfield Ltd, and Mr R. A. Bull, Director of Research (Oil Palms), Chemara Plantations Ltd. Discussions over the years with these veterans and stalwarts of the oil palm industry, and with many of their colleagues, have been of inestimable value to me. Dr J. J. Hardon, Oil Palm Geneticist, was kind enough to read and comment upon part of the text, and Mr B. J. Wood provided me with information on, and photographs of, Malaysian insect pests. My thanks are also due to Dunlop Plantations Ltd and Dunlop Malayan Estates Ltd for assistance in many ways, and to managers of oil palm estates in Malaysia, Africa and America, too numerous for separate mention but whose observations have often been of particular moment.

To the Department of Agriculture, Malaya, which first introduced me to the oil palm, and to Dr Ng Siew Kee, my thanks are due for the Malayan soils data included in the Tables in Chapter 3 and for the data in Chapters 5 and 11 of certain field experiments. I would also like to thank the Director of Agriculture and his staff for many helpful discussions in Malaya in recent years.

In dealing with the oil palm in Sumatra my work was much facilitated by discussions and correspondence with workers conversant with the industry in that island. In particular I wish to thank Dr J. J. Duyverman and Mr J. Werkhoven of the Royal Tropical Institute, Amsterdam, Mr A. Kortleve of H.V.A. International, N.V., Mr F. Pronk, previously of A.V.R.O.S., and Mr J. J. Olie and Mr M. J. van der Linde of Gebr. Stork and Co; the latter kindly provided me with drawings and photographs and much information on processing plants.

My task has also been assisted by helpful discussion with research workers of the Institut de Recherches pour les Huiles et Oléagineux, Paris, and I have to thank M. Carrière de Belgarric, Director General, Dr P. Prevot and M. M. Ollagnier for their friendly cooperation and for putting me in touch with their staff, both in Africa and America.

I have to thank the Ministry of Overseas Development for arrangements made for me to visit areas of oil palm development in a number of countries in South and Central America, and I am also grateful to the British Embassies in these countries for the very real assistance which they gave me. To Dr V. M. Patiño of Cali, Colombia, my thanks are due for the supply of information on planting material and on introductions into Latin America, and on the American oil palm. Useful information from the American continent was also supplied to me by the United Fruit Company and, on insect pests, by Mr F. P. Arens of the F.A.O., Ecuador.

I should like particularly to thank Mr D. Rhind, CMG, for the many helpful comments he made during the final preparation of the chapters, and Mr E. O. Pearson, OBE, and his staff at the Commonwealth Institute of Entomology for checking the names of insect pests and supplying information and references.

The writing of this book has been made possible by the warm hospitality I have received from the Commonwealth Forestry Institute, Oxford, and I am especially grateful to Dr T. W. Tinsley, who welcomed me into his Section, to Professor M. V. Laurie for permission to work at the Institute, and to the Librarian, Mr E. F. Hemmings, and his staff for their unfailing help. Lastly, I have to thank my wife and children for some tedious work willingly done on data which I have used in this book.

C. W. S. Hartley

Three Gables,
Amberley,
near Stroud, Glos.
5 October 1966

Preface to the second edition

For this new edition the text has been extensively revised. It is now ten years since the manuscript of the first edition was completed and since that time there has been great progress in research and an unprecedented enlargement of planted areas in Asia, Africa and America. Chapter 1 has been brought up to date and the widely different development methods being employed are discussed in this chapter and in Chapter 8.

In the revision of Chapters 3 and 4 particular account has been taken of the progress made in relating climate and soil to yield and of the work on growth analysis which has given a better understanding of the palm's performance under varying environmental and cultural circumstances. The part of Chapter 3 which deals with the African palm groves has been severely reduced as these areas are of dwindling importance in the total supply of oil palm products.

In Chapter 5 the recent work on heritability and on interspecific hybridization with the American oil palm is now incorporated. Methods of breeding currently employed are compared, while the latest prospections for new material and the research on growth factors in relation to selection and breeding are described.

Chapters 6 to 10 have been revised to take account of the progress made in cultural practices while Chapter 11 has been extensively rewritten and rearranged to allow for the substantial body of new data on many aspects of the nutrition of the palm. Chapter 13, on diseases and pests, has been brought up to date and expanded. In Chapter 14 the new work on oil quality is discussed and information is given on the oils of the American oil palm and the interspecific hybrid; reference is also made to new developments in both large and small mills.

Once again I am indebted to many research workers and organizations for assistance. In particular I wish to thank Mr A. H. Green and the Unilever Plantations Group for permission to quote from their Annual Reviews of Research, Messrs Harrisons and Crosfield for permission to make use of material from their Oil Palm Research Station Annual Reports, the Director of N.I.F.O.R., Nigeria, for supplying data for updating some experimental results, Dr B. S. Gray for information on Indonesian developments, the Department of Botany of the University of

Birmingham for permission to quote from the thesis of Mr N. Rajanaidu, Dr J. A. Cornelius of the Tropical Products Institute, London, for oil analysis data, Mr J. J. McNerney and the Commonwealth Secretariat, and Oil World Publications, Hamburg, for export and other statistical data, and Dr R. H. V. Corley of the Oil Palm Physiology Unit, M.A.R.D.I., and Unipamol Malaysia Ltd, for discussion and correspondence on physiology. I have also again been much helped by discussion with individual planters and members of research organizations in the many countries I have visited over the last ten years, and special mention should be made of the Institut de Recherches pour les Huiles et Oléagineux, Paris, whose publications, *Oléagineux* and Rapports Annuels, continue to be invaluable sources of information.

For this new edition, all data have been converted to the metric system. A conversion table has been provided at the end of Chapter 10 on p. 490. Fifteen new plates and eleven new text figures have been provided.

C. W. S. Hartley

5 October 1976

Contents

List of plates

COLOUR PLATES (between pp. 78 and 79)

Acknowledgements

We are indebted to the following for permission to reproduce copyright material:

The Bailey Hortorium Mann Library, Cornell University, for Figs 2.1 and 2.5, reproduced from *Principes*; the Librarie générale de l'Enseignment, Paris, for part of Fig. 2.3, reproduced from the *Revue générale de botanique*, vol. 62, 1955; the Director of the Institut National pour l'Etude Agronomique du Congo, for Figs 2.10, 2.11 and 5.14, reproduced from the publications of INEAC; The Clarendon Press for Figs 4.2, 4.3, 4.5, 4.6 and 11.7, reproduced from the *Annals of Botany*, the *Journal of Experimental Botany* and the *Journal of Soil Science* respectively; the Editor, *Plant and Soil*, for Fig. 11.1; the Chief Librarian of the Bibliothèque des Archives Africaines, Brussels, for Figs 5.1, 10.10 and 10.11, reproduced in the *Bulletin Agricole du Congo Belge*; the Director of *Oléagineux* for Figs 8.2, 9.4 and 10.9; the Rubber Research Institute of Malaysia for Figs 10.1 to 10.8; the Tropical Products Institute for Figs 11.3 and 11.4 by R. Ochs, and Fig. 14.2 by M. Loncin and B. Jacobsberg, reproduced from papers presented at the Tropical Products Institute Conference in London in 1965; the Cambridge University Press for Figs 4.7 and 4.8 reproduced from the *Journal of Experimental Agriculture*; the Incorporated Society of Planters, Malaysia, for Figs 7.2, 9.5, 9.6, 10.10 and 11.10.

Cambridge University Press for five tables from *Experimental Agriculture*; Centre for Agricultural Publishing & Documentation for tables from *Agricultural Research Report* No. 823 by Van der Vossen, 1974; Ghana Journal of Agricultural Science for tables from an article by Van der Vossen in *Ghana Journal of Agricultural Science*, vol. 3, 1970; The author for a table by Dr J. J. Hardon in *Euphytica*, 1969, vol. 18; The Incorporated Society of Planters for a table by R. H. Corley in *Advances in Oil Palm Cultivation*, 1973, a table by K. S. Tan and C. K. Hew *et al.* incorporating material from *Advances in Oil Palm Cultivation*, 1973, a table by S. H. Warriar and C. J. Piggott entitled 'Rehabilitation of Oil Palms by Corrective Manuring through Leaf Analysis' and a table from an article by B. Bek-Bielson in *Quality and Marketing of Oil Palm Products*, 1969; S.E.T.C.O., Paris, for a table by J. Meunier incorporating material from

Oléagineux, vol. 24, 1969, a table incorporating material by B. Tailliez and J. Olivin from *Oléagineux*, vol. 26, 1971, a table incorporating material by A. Bachy from *Oléagineux*, vol. 24, 1969 and a table incorporating material by M. Naudet and H. Faulkner from *Oléagineux*, vol. 30, 1975.

We are grateful to the following for permission to reproduce copyright material upon which tables have been based:
The Director of Agriculture, Malaysia, for data from *Malayan Agricultural Journals*, 35, 12, 1952: 41, 131, 1958, and 20, 16, 1932; Direction Générale de l'Administration Bibliotheque africaine for data by R. Pichel published in *Bulletin Agricole du Congo Belge*, 48, 67; Butterworth and Co. (Publishers) Ltd for data from *Tropical Agriculture*, 39, 271, 1962; The Clarendon Press for data from *Annals of Botany*, Vols 26 and 27, *Journal of Experimental Botany*, Vols 7, 9 and 12 and *Empire Journal of Experimental Agriculture*, Vol. 24; the Controller of Her Majesty's Stationery Office for data from *Soils of the Semporna Peninsula* by T. R. Paton; Institut National pour l'Etude Agronomique du Congo for data from *Publs. de l'INEAC Serie Tech*. 66, 1961 and *Bulletin Agricole du Congo Belge*, 24, 1933 and *Bulletin d'Information de l'INEAC*, 6, 351; the Director of *Oléagineux* for data by C. de Berchoux and J. P. Gascon published in issues 20, 1, 1965 and 18, 713, 1963; the Editor of *The Planter* for data by B. S. Gray published in issue 42, 16, 1966; the Tropical Products Institute for data on the 'Colour of samples of palm kernels received in the United Kingdom' by J. A. Cornelius, and the Managing Committee of the West African Institute for Oil Palm Research, Nigeria.

We are also indebted to the Commonwealth Economic Committee for supplying acreage and export statistics ahead of the annual *Vegetable Oils and Oilseeds*, and to the University of Birmingham for access to Mr N. H. Stilliard's unpublished thesis *The rise and development of the legitimate trade in palm oil with West Africa*. Many of the photographs have been supplied from the publications or collection of the West African (now Nigerian) Institute for Oil Palm Research. For the supply of certain other photographs we are indebted to Mr J. M. A. Sly, Mr T. Menendez, Mr B. J. Wood, Mr P. F. Arens, Gebr. Stork and Co. of Amsterdam, Société pour l'Equipement des Industries Chimiques of Paris, the Four Wheel Drive Auto Company, Clintonville, U.S.A., Unilever Ltd, London and Ing. Sanchez Pates of Colombia.

Chapter 1

The origin and development of the oil palm industry

At the present time, the oil palm (*Elaeis guineensis* Jacq.) exists in a wild, semi-wild and cultivated state in the three land areas of the equatorial tropics: in Africa, in South-east Asia and in America. Of all oil-bearing plants it is the highest yielding, even the poorer plantations of Africa out-yielding the best fields of coconuts, a crop which the oil palm has recently overtaken in the export field. Yet the plant and its products are virtually unknown to the vast majority of the world's population and, except in Africa, the people of the tropical regions have but little acquaintance with them.

The reasons for this anomalous situation are to be found in the curious history of the palm. Until recent centuries it appears to have been confined to West and Central Africa, a region inhospitable even to the most adventurous of early traders. The palm had to await the slave trade, starting after the early Portuguese voyages of the fifteenth century, before it could escape to another tropical land mass. It was not suited, like the coconut, to the foreshore, but it soon established itself meagrely behind the coast line in Brazil where it was little used except by the transported Africans who knew its value, and who, no doubt, had brought it with them.

In Africa it remained a domestic plant, supplying a need for oil and vitamin A in the diet, and it was not until the end of the eighteenth and the beginning of the nineteenth centuries that it entered world trade. The fruit of the oil palm is a drupe, the outer pulp of which provides the palm oil of commerce. Within the pulp or mesocarp lies a hard-shelled nut containing the palm kernel, later to provide two further commercial products, palm kernel oil (rather similar in composition to coconut oil) and the residual livestock food, palm kernel cake. The nuts were discarded in the crude preparation of palm oil and the traders of those times did not in any case watch the process. For the most part they stood offshore in their ships and traded for slaves with the great chiefs of the west coast. When the slave trade was suppressed by the British, however, a substitute had to be found and thus palm oil acquired the unique distinction of being the commodity which replaced the traffic in slaves.

1

The origin of the oil palm

There is fossil, historical and linguistic evidence for an African origin of the oil palm. It has also been suggested on the evidence of an analysis by Friedel that fat found in a jar in a tomb at Abydos (*c*. 3000 B.C.) may have been palm oil.[1]*

Fossil evidence has only recently come to light. Seward[2] found no oil palm remains in the upper cretaceous layers in eastern Nigeria but Zeven[3] has reported fossil pollen from Miocene and younger layers in the Niger Delta as being similar to pollen of the oil palm as it grows today. That such pollen is found throughout these layers is strong evidence that the palm has been maintaining an existence in West Africa from very early times. Cook[4] suggested a Brazilian origin for the oil palm; two of his grounds for this contention were that the palm grew 'spontaneously' in the coastal areas of that country and that all allied genera have an American origin. Zeven suggested that both *Elaeis guineensis* and another Cocoid palm, *Jubaeopsis caffra*, originated on the African side of the Tertiary land bridges which are believed to have lain between Africa and America, and that they became separated from other members of the tribe by the ocean. Corner,[5] however, considers that there are botanical and distributional grounds for discarding this theory and suggested that pre-Columbean transportation of *Elaeis* to Africa is probable.

The historical record of the oil palm is meagre and, in many respects, vague; only recently have efforts been made to relate such records as exist to the main landmarks of exploration.[6,7,8] Colombus discovered South America in 1498 and Brazil was discovered by both the Portuguese and Spanish in 1500, but no real interest was taken in the country until the middle of the sixteenth century. If well-authenticated records of the oil palm in West Africa before or around these times can be established, then a Brazilian origin becomes less likely though the possibility of pre-Columbean transportation cannot be ignored.

Portuguese exploration of the Guinea coast started in 1434, Dutch and English exploration some 150 years later. There is no mention of the oil palm by the earlier Arab explorers or by Marco Polo. 'Palm trees' in abundance are mentioned in the account of Diogo Gomes's voyage of 1456 or 1457, but the account of the voyages of Ca' da Mosto (1435–60)[9] gives the first mention of a palm which strongly suggests the oil palm, namely:

> There is to be found in this country a species of tree bearing red nuts with black eyes, in great quantities, but they are small.

Of an oil used with food it was recorded that this

> has three properties, the scent of violets, the taste of our olive oil,

---

* Superior figures refer to the References at the end of each chapter.

and a colour which tinges the food like saffron, but is more attractive.

In the description given by Duarte Pacheco Pereira of his voyage of 1506–8 (Esmeraldo de Situ Orbis) mention is made of palm groves north of, and on an island off, the coast of Liberia, and of trade in palm oil (azeite de palma) near the Forcados river in Nigeria. Later Portuguese, Dutch and English accounts of voyages refer to palm wine as well as to the oil. At the end of the sixteenth century ten Broecke described two types of palm wine, one presumed to be from the oil palm, the other, from its name which has survived in south-eastern Nigeria to this day, being wine from the *Raphia* palm.

None of these descriptions can be considered in any way complete, but the earlier accounts of palm groves and trade in palm oil, and the pre-Columbean description of the oil, make it certain that the palm could not have been brought from America by the Portuguese as claimed by Cook. In addition, the descriptions in the earlier herbals, although written after the discovery of America, are sufficiently early to make it likely that the plant described was native to West Africa and it is clear that the writers thought they were describing a plant indigenous to the Guinea coast. In certain cases specific reference is made to importation or to practices implying importation. For instance, Clusius (1605) stated that the palm was found on the Guinea coast and that the 'fruit, after addition of some flour of a certain root, was used by the Portuguese from San Thomé to feed their slaves during the whole journey' to America. Sloane (1696) reported that oil palms in Jamaica came from Guinea. Miller, in his *Gardeners Dictionary* of 1768, stated that oil palm fruits had been carried from Africa to America by the Negroes. Finally, there are no early descriptions of the oil palm from Brazil.

Linguistic evidence[8] also strongly supports a West African origin for the oil palm. Names for introduced plants are usually a corruption of a foreign name or have two parts, the one referring to a local plant which in some way resembles the introduced plant, the other signifying its possible geographical origin or the person or persons who introduced it. Thus the coconut in many West African languages is 'white man's nut' or 'Portuguese oil palm, nut or kernel', or may be a corruption of coconut. All the vernacular names for the oil palm, however, are short and can only be directly translated to mean oil palm. Moreover the Negro names for the oil palm in Surinam suggest that they are a corruption of the Yoruba, Fanti-Twi and Kikongo names of Africa; and the Brazilian name *dende* may be derived from the Kimbundu word *ndende* of Angola.

Habitat

Well before much of the above evidence was available, Chevalier[10] was a strong champion of the African origin of the palm, and he suggested that its natural habitats were in *galeries forestière* or forest outliers either near

Raphia or in associations of *Elaeis* and *Raphia*. The difficulty in determining the natural habitat of the oil palm lies in the fact that while it does
not grow in the primeval forest it begins to flourish wherever man has
cleared a part of the forest. It requires a relatively open area to grow and
reproduce itself, and it thrives best when soil moisture is well maintained.
It is natural to suppose therefore that the palm would find a place on the
forest fringe near to rivers. Apparently wild associations of the oil palm
and the *Raphia* palm have been noted bordering rivers in Zaire and on
uncultivated islands, subject to annual flooding, in the Zaire and Ubangi
rivers.[11, 10] Chevalier claimed that river habitats of this kind could be found
right across Africa from Senegal to Angola and even to Mozambique.
Similar habitats have now been assumed by 'escapes' in the Far East, both
in Sumatra and in Sabah on the island of Borneo.

The fresh-water swamp has also been suggested as a natural habitat for
the oil palm.[12] Palms growing in such swamps adjoining rivers have been
seen in several parts of southern Nigeria. The palms are often to be found
in standing, though not stagnant, water for many months of the year.
However, these areas are always near to man's habitation, and it is not in
any case possible to differentiate clearly between swamp habitats and the
riverside habitats already mentioned. The differences are probably only
ones of depth of penetration from the riverside, this depending to a large
degree on the extent of the river flooding in the rainy season. As a
producer, the oil palm will not tolerate permanently high water tables in
impervious soils, but it appears in its natural home to be tolerant of
fluctuating water tables and moving water in sandy or silty riverine soils.
In short, it seems most probable that the oil palm grows naturally near to
rivers where the palms will be subject to less competition from the forest
flora, where more light will therefore penetrate and where moisture is
plentiful but not yet excessive.

The oil palm in its African setting

Some authors have attempted to distinguish between *palmeraies* (groves)
naturelles or *spontanées* and *palmeraies subspontanées*, the supposition
being that a 'spontaneous' grove or group of palms is one which has grown
up without the intervention of man, while the 'subspontaneous' or semi-
wild grove is one which owes its existence to the alteration of the natural
vegetation by man in such a way that conditions suitable for unaided
germination and development of oil palms have arisen. Chevalier[10] even
went so far as to give '*Elaeis bien spontanés*' a separate specific name, *E.
ubanghensis*, claiming that this was distinguishable from the domesticated
grove palm by the small bunches and fruit and the very thin mesocarp.
While it is not difficult, in populated areas, to attribute the presence of
palm groves to man's activities, in many isolated areas where oil palms are
found it is virtually impossible to say with certainty whether the palms are

truly wild or have found their position through the intervention, sometime in the past, of man. Even in an area which is under water for a considerable period of the year and which therefore gives the appearance of a natural habitat, the flooding may be of recent origin due to the progressive silting up or alteration in course of rivers during a comparatively short historical period of time. It is largely speculative and unrewarding, therefore, to consider the presence and spread of the oil palm in Africa in terms of wild or semi-wild groves; moreover the presence of the oil palm in certain places is undoubtedly due to the intentional introduction of seed from one country into another.

The actual physical spread of the oil palm by seed may be through the agency of gravity and water, of animals, or of man. Movement by gravity and water must be of limited occurrence especially as the fruit does not float, but some groups of palms on the banks of rivers or streams may have arisen through the transport of fruit, perhaps on floating material, which has rolled down a slope into a stream. Animals considered to be responsible for the spread of seed are the common mammals, rodents and a few birds of tropical Africa. The main agency is thought to be man who has, for the most part, spent a wandering life in Africa and has been in the habit of carrying bunches from palm to homestead, dropping or throwing away odd fruit on the way.

Geographical distribution in Africa

The distribution before the First World War was summarized by Schad[13] and this has been brought up to date by Zeven.[14] On the west side of the continent the oil palm occurs at 16°N near St Louis in Senegal. Inland it has been recorded on the upper Niger at 13°N near Bamako. Though there are planted palms around Dakar and in the Gambia, no extensive areas of palms are found in the countries of the coast until the Fouta Djallon district of Guinea is reached at the 10°−11°N parallel. From here the real palm belt of Africa runs through the southern latitudes of Sierra Leone, Liberia, the Ivory Coast, Ghana, Togoland, Dahomey, Nigeria, Cameroon and into the equatorial regions of the Republics of Congo and Zaire. The northern limit of this belt varies with the isohets and with topography. On the coast there is a small area surrounding Accra where rainfall falls to 65 mm per annum and the presence of grass savannah excludes the oil palm.

Except in the heavy summer rain areas of Guinea and Sierra Leone the oil palm only spreads beyond about 7°N into favoured river valleys under escarpments (e.g. west of the Jos plateau in Nigeria) or where underground water supplies are available. Beyond this latitude the oil palm gradually gives place to the palmyra palm (*Borassus aethiopicum*). However, small groups of palms occur in relatively dry areas in the Central African Republic and south of the Sudan.

In Central Africa the main area of spread of the oil palm lies between

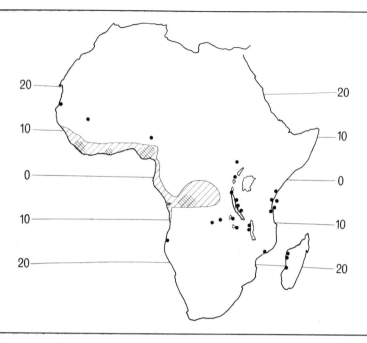

Fig. 1.1   An impression of the distribution of the oil palm in Africa. The shaded areas represent the 'palm belt' where the main groves are situated and where the palm is ubiquitous. The black dots represent the position or boundaries of isolated areas of colonization. The areas of heaviest population and production are double hatched. (Adapted from Zeven and earlier authors.)

3°N and 7°S and covers huge areas of Zaire together with parts of the Republic of Congo and Angola. The groves are most concentrated in the Kwilu and Kwango districts between 4° and 7°S and in Cabinda in Angola. In the coastal districts of Angola the climate is too dry though palms are found as far south as Mocamedes (15°S). Groves in the north are contiguous with those of southern Zaire, and plantations have been established south of the Rio Cuanza.

The spread of the oil palm into East Africa is believed to have been from Zaire. Dense stands of palms are found in the Semliki valley on the Zaire–Uganda border, in the Ruzizi plain between Lakes Kivu and Tanganyika and in a belt about 19 km wide on the eastern shore of Lake Tanganyika. Most of East and South-east Africa is far too dry for the oil palm and it appears only at altitudes below 1,067 m near lakes or water courses where a reasonable rainfall is assured. It exists on the west shore of Lake Malawi as far as 13°S and is also to be found on Lake Bangweulu and Lake Mweru in Zambia and scattered about in the south-eastern regions of Zaire at about 10°S wherever moisture conditions are favourable.

These areas are separated from the coast by belts of very dry territory and it seems not unlikely that the oil palm was brought to the coast during the long period of the Arab slave trade. Today, oil palms are to be found sporadically from the Tana river in Kenya to Dar es Salaam in Tanzania and on the islands of Zanzibar and Pemba (between 3° and 7°S).

The oil palm is also found on the island of Madagascar (Malagasy Republic) on a stretch of the west coast from south of Cap St André to the basins of the Manambolo and Tsiribihina rivers and as far south as 21° of latitude where it is known as Tsingilo.[15] Climatically, the east coast of Madagascar seems more suited to the oil palm and although at one time it was thought to be indigenous on the west coast it seems more probable that it was brought across the sea by Africans of continental Africa when they entered the island and that its distribution is largely due to chance introduction.

From before the beginning of the century there have been many reports of the planting of palms in parts of Africa where the plant was hitherto unknown. These include planting on Government stations in most of the East African countries and in Angola.

The oil palm is a plant of the lowlands and in the great areas of productive groves it is found from sea level to about 300 m. However, when mountainous country adjoins these areas or where the oil palm has been transported to high tablelands having sufficient rainfall, it has been able to survive at much greater heights. The most interesting mountainous habitat is the Cameroon mountain which is not far removed from the greatest grove area of Africa in eastern Nigeria. Here palms occur at over 1,300 m. In Guinea, the palm grows at 1,000 m in the mountains of the Fouta Djallon, and in East Cameroon it is found in abundance at 1,000 m in two favoured localities.[16] Most of the East African inland localities are at a height of around 1,000 m. Nothing is known of the relation of altitude to bunch production in Africa, though production is said to be low in the high-altitude areas of Cameroon.

Within 7° of the equator and in some exceptional areas at higher latitudes, wherever a large population is to be found, the typical palm groves of West and West Central Africa have arisen. Dense groves are found most continuously and extensively in eastern Nigeria, but the palm is so widespread that it is true to say that it would be difficult to stand anywhere in the palm belt except in the heart of a forest reserve and not be able to see a few palms. The dominant position of Nigeria in the export of oil and kernels from Africa was due to a rapid increase in population taking place near to the coast in a region with fair communications and suitable climatic conditions.

In areas of highest population the groves are found at their densest, but in areas of relatively low population, such as in parts of western Nigeria, 'palm bush' country has developed where the density is rarely more than thirty adult palms to the hectare; here they grow in high bush on land which is only farmed at wide intervals of time. In the denser groves of eastern Nigeria the palms are so close to each other that satisfactory farming is difficult; nevertheless the pressure of population has been such that not only shade-loving plants such as the cocoyam are grown, but the ordinary staple foodcrops of the people — cassava, yams and maize — are widely planted though they yield poor crops.

In Zaire the groves are to be found mainly in the south of the country. In some areas dense groves comparable to those of eastern Nigeria exist, but in general the groves are less uniform and occur in long narrow belts or in patches alternating with forest or derived savannah.

Before the establishment of plantations in the Far East and, later, in Zaire, the groves of Africa formed the sole source of supply of palm oil and kernels. It has been said that when the palm is considered in its West African setting it is first and foremost a food crop. Large quantities of palm oil are consumed locally in all West African countries and, while it is impossible to obtain exact figures, there is good reason to believe that, even in Nigeria, more palm oil is consumed within the country than was ever exported. In countries with smaller populations and lower production, e.g. Sierra Leone, Ghana and Dahomey, the bulk of the palm oil is consumed locally or moves to adjoining territories.

In certain circumstances palm kernels are extracted from fruit without the prior extraction of the mesocarp oil; this together with local consumption of oil accounts for the very high export figures for palm kernels compared with those of palm oil. In some parts of West Africa the fruit quality, as judged by the thickness of the mesocarp layer, is so poor that there is little encouragement to extract the oil; this is the case in Sierra Leone and some parts of Ghana, Dahomey and western Nigeria.

Besides providing the main source of palm kernels for export and an important source of palm oil for both export and local consumption, the grove palms can produce large quantities of palm wine for consumption in urban areas and villages. In many parts of West Africa the palms surrounding large towns are regularly used for tapping for wine; the industry is an important one and the wine has dietetic value because of its vitamin B content. The value of an oil palm for wine production depends largely on its position and there is no doubt that where there is a steady market for wine a higher return can be obtained from tapping than from the harvesting of bunches. Little study has been made of the effect of tapping on bunch production, but the two industries are clearly incompatible; the palms should either be used for wine production or for bunch production according to their position and the local market for these products.

The grove palms also provide a number of other products which are of importance in the domestic life of the African peoples. Many of these products, such as leaves for thatching, petioles and rachises for fencing or for protecting the tops of mud walls, may decrease in importance with the increasing use of more durable materials in the building of houses. Until recently little use was made in Africa of the palm kernel, although occasionally the kernel oil was expressed for soap-making and a very crude method of extracting the oil by frying the kernels in a large pan was to be found in Togoland and elsewhere. In Zaire, Nigeria and Dahomey large kernel crushing industries have arisen using kernels from both plantations and groves; by 1972 African countries were exporting over 100,000 tons of palm kernel oil annually.

The ash obtained by burning the refuse which remains after stripping the fruit from the bunches is rich in potash and is sometimes used in soap-making.

Early trade

West Africa was isolated from Europe throughout the middle ages; but when the Ottoman Empire blocked the way to the orient, Portuguese seamen pushed their way round the west coast in a quest for spices, establishing a fort at Elmina in 1482. The discovery of the Americas soon diverted attention from this trade, however, and when interest was revived it was in a trade for slaves to sustain the development of the new countries of the Western world. The slave trade proper survived from 1562 to 1807, and during the whole of that time, though palm oil is mentioned as food for slaves, there was virtually no other commerce on the west coast. Vessels followed the 'triangular passage'; European goods were exchanged for slaves on the coast, slaves were exchanged for West Indian or American goods on the other side of the Atlantic and the ships then returned to Europe.

At the beginning of the nineteenth century the illegal slave trade had to be replaced by what came to be known as the 'legitimate trade' between Europe and the coast, but palm oil hardly figured among the commodities which, at that time and in very small quantities, supplemented the traffic in slaves. Ivory and timber were the most important of these, though gold dust, pepper, rice and gum copal also played a part. Although very small amounts of palm oil had been traded in England in 1588 and 1590, the first import records show that in 1790 less than 130 tons were imported into the United Kingdom and in the year of the abolition of the slave trade (1807) the quantity was no more.

The dominance of the slave trade and the methods of trading afford an explanation for the almost complete lack of knowledge of palm oil in Europe at that time. Until 1804 there was not even an accepted name for the oil in English, the latin form 'oleum palmae' often being used in documents. Some quaint ideas were held by traders about the origin of the oil, some believing that it was drawn off from the roots. Trade restrictions, navigation laws, the small number of traders, lack of access to the interior, local disturbances, disease and the continuance of illegal traffic in slaves all helped to prevent a rapid increase in trade. Imports into the United Kingdom from West Africa reached 1,000 tons in 1810, but thereafter a fluctuating trade continued well into the 1820s. It has been said that the trade was almost established in the legitimate commerce of Africa by

1830, while the developments which took place after that year made its future certain.[17] These developments were the taking of more active measures for the suppression of the slave trade and the active encouragement of the palm oil trade by the British Government.

The main port was Bonny in the Niger Delta, of which the missionary, Hope Waddell, wrote in 1846, 'By its safe and extensive anchorage, its proximity to the sea and connection with the great rivers of Central Africa, Bonny is now the principal seat of the palm oil trade as it was formerly of the slave trade'. By 1830 the oil trade had become larger in the Delta than the trade in men and more than half the oil was handled at Bonny.[17-19]

The merchants' difficulties were not over, however. During the 1840s the expected increase in the trade did not take place. This stagnation was caused by an unexpected increase in the illegal slave trade induced by a demand from Brazil and Cuba. The increase in the number of slaves shipped was considerable and the measures which had to be taken against the trade were temporarily detrimental to the palm oil trade itself. Vessels engaged in the legitimate trade were held on suspicion of being slavers, often on quite trivial grounds. Matters were not brought under control until the 1850s, soon after a British Consul for the Bights of Biafra and Benin had been appointed at Fernando Po. This was followed by the capture of Lagos and the winning over of the traffic in the Bight of Benin to the legitimate trade. Henceforward palm oil was to be exported from the Benin river as well as from Bonny and Calabar in the area known as the Oil Rivers.

The trade was carried forward in an unusual manner. The European trader did not leave his ship; this was through justified fear of disease (as much as one-quarter of the crew often died even when remaining aboard in the rivers) and because the coastal tribes controlled the trade, buying oil from the interior where it was produced and, as middlemen, selling to the traders. The use of vast quantities of cowries, taking the place of modern currency, hindered the completion of transactions in the interior, and poor extraction methods led to arguments over quality. The regulation of the trade and the prevention of a resurgence of slave trading was assisted in the 1840s by the presence of a cruiser squadron in West African waters. The advantages and disadvantages of this protection were much argued at the time and the squadron gave place in the 1850s to the consular system, which was generally considered an improvement. The inauguration of this system gave rise, in its turn, to the idea of having settlements ashore for the trade. The value of British settlements, as opposed to the merchants' own commercial treaties with the chiefs of the coast, was much debated during the 1860s and there were many merchants who were not in favour of British penetration.

The establishment of the legitimate trade in palm oil was made possible by the industrial revolution in Europe which led to a new type of demand for tropical commodities. These were not only required for immediate consumption, but were needed as raw material for industries. 'People began to take washing seriously', and the demand for vegetable oils for soap-making increased rapidly as did the use of both animal and vegetable oils as lubricants for machinery. During the 1830s between 11,000 and 14,000 tons of palm oil per annum were exported from West Africa. A fluctuating trade continued during the 1840s and 1850s but by the end of the 1860s 25,000 to 30,000 tons per annum were being exported from the Delta area alone, with prices varying between £34 and £44 per ton.[18] From then onwards until the First World War, exports steadily increased, over 87,000 tons, valued at nearly £1,900,000, being exported from British Colonial territories in 1911.[20]

This large increase in importation was caused by the oil's wider uses in industry. The early history of its usage was as follows. In the late eighteenth century it was imported for use by wool combers and soap boilers. It continued in use for soap-making throughout the nineteenth century; mixed with tallow and suet to make yellow soaps and, later, bleached for the manufacture of toilet soaps. In the middle of the nineteenth century a London firm patented a process for the treatment of fatty acids as a result of which palm oil became widely used in the manufacture of stearic candles. It was also used as a lubricant in the early days of the railways. Margarine manufacture started in the 1870s and the best quality palm oil was employed for this, though palm kernel oil was, at that time, generally more suitable. In the 1890s a new use for palm oil was found in the tin-plate industry; oxidation of the iron was prevented by a coating of palm oil before the actual plating was carried out.

About 1850 both copra and palm kernels began to take a place in the provision of oil for soap-making. The oil from these two products is very similar. Palm kernel oil is almost colourless and odourless and came to be regarded as better adapted for soap-making than palm oil. Apart from this factor the kernels, after oil expression, left a valuable cattle cake which was increasingly used by farmers in Europe. Exports of palm kernels started in 1832 but production, carried out entirely by African women who extracted the kernels by cracking individual nuts, did not increase rapidly until the latter half of the nineteenth century; by 1905 exports of kernels from British Colonial territories were over 157,000 tons, more than two-and-a-half times the export of palm oil, and by 1911 over 232,000 tons valued at more than £3,400,000 were exported. The rise of this village industry in the British Colonial territories was even more spectacular than that of palm oil; about 75 per cent of the world production came from Nigeria and about 18 per cent from Sierra Leone.

In spite of the increase in the trade in palm oil, in palm kernels and in other oils and fats, the demand at the beginning of the twentieth century proved greater than the supply. The soap and margarine industries were in competition. In 1913 it was stated that as a result of this competition the price of both coconut and palm kernel oil had almost doubled in the previous 10 years.

It was this situation which began to focus attention on the sources of supply and methods of production. Little was known of the native methods of palm oil extraction and many early descriptions were inaccurate or incomplete. Hand cracking of nuts for kernel extraction was slow and tedious and there had been some opposition to the trade by chiefs who dealt in palm oil. There was a general realization that a much larger quantity of palm products would be forthcoming if more efficient methods were employed.

Thus the second decade of the twentieth century saw the beginning of two developments which were to change, during the inter-war period, the pattern of oil palm exports. These were the Lever concessions in the Belgian Congo (now Zaire) and the establishing of the Deli oil palm as a plantation crop in Sumatra and, somewhat later, in Malaysia.

Early in the century, Sir William Lever (Lord Leverhulme), in his search for a further supply of oils and fats, approached the British Colonial Office for rights in Sierra Leone to introduce mills for treatment of palm fruit and to acquire land on which to cultivate the crop. A second application, in 1908, was confined to the first object. After negotiations entailing counter proposals by the Government, a concession in Sierra Leone was given in 1911. In the Gold Coast (now Ghana) and Nigeria, however, the Governors were opposed to such concessions on the grounds that the land belonged to the native peoples and that the Crown made no claim to be able to dispose of it in any way. Some fears were aroused among the peoples themselves and these were voiced in meetings and deputations. Resistance to concessions was also supported by a section of public opinion in England and by a number of prominent merchants trading with West Africa.

The concession system failed in Sierra Leone. The cause of failure was said to be that the price offered compared unfavourably with what could be earned by extracting and selling oil and kernels by native methods. Whether this was so or not, insufficient fruit was brought to the mill for it to be operated successfully, and this might have been partly due to the long distances of carry needed in the relatively sparsely covered Sierra Leone groves. The great grove areas of West Africa therefore continued to be exploited throughout the next 30 to 40 years by crude native methods. Production and exports rose steadily, though not by any means spectacularly, throughout the inter-war period and the quality of the palm oil remained low. The British West African territories continued to dominate the trade in palm kernels, but events elsewhere reduced the relative importance of West Africa in the trade in palm oil.

Having failed to obtain the concessions he required in West Africa, Sir William Lever turned his attention to the great areas of largely unexploited groves in the Congo basin. In 1911 he obtained large concessions of land for the *Huileries du Congo Belge* (H.C.B.) and entered into an agreement with the Belgian Minister for the Colonies by which, in certain defined regions, the company agreed to erect oil mills in return for a guarantee that all fruit harvested would be processed in them. The establishment of this company not only led to a rapid increase of oil and kernel exports, which in 1912 amounted in total to less than 8,000 tons, but also encouraged the exploitation of groves outside the Lever concessions. In 1933, new arrangements were made for concessions of palm grove areas by zones, each concession being limited to 15 years and subject to definite rules regarding methods of harvesting, purchase price, etc., and including an undertaking to plant small estates of selected oil palms.

By 1935, exports from Zaire had reached over 56,000 tons of palm oil and over 64,000 tons of kernels.[21] Half the oil exported and a sixth of the kernels were produced by *Huileries du Congo Belge*. Outside this company the production contained a lower percentage of oil than of kernels, the amount of oil produced by village methods being relatively small and to a great extent locally consumed. The cracking of nuts for the sale of kernels for cash soon became an important village occupation, however, though Nigerian production was never rivalled.

The West African trade in palm kernels took a peculiar course in the early part of the present century. At the outbreak of the First World War, although British colonies were responsible for most of the production, Germany was buying about three-quarters of the kernels exported, and a considerable seed-crushing industry had been set up. During the war West African production had to be directed to the United Kingdom and, indeed, was required for the larger quantities of margarine to be manufactured. New seed-crushing mills were erected and, to maintain the industry after the war, a protective export duty, to be remitted on exports to Commonwealth countries, was imposed in West Africa. This duty soon became unpopular however, and was abolished in 1925.

The development of a plantation industry

The growing of the oil palm in plantations started in the Far East, and, strangely, there was no direct connection between the African groves and the establishment of this new industry. The earliest record of the introduction of oil palms to the East Indies is of four seedlings, two from Bourbon (Réunion) or Mauritius and two from Amsterdam, which were planted in the Botanic Gardens at Buitenzorg, now Bogor, in Java in 1848.

The Deli palm

Unfortunately, the real origin of these Buitenzorg palms remains a mystery. In a report on the Buitenzorg gardens dated 23 March 1850, Teysmann records '*Elaeis guineensis* received from the Hortus Botanicus at Amsterdam and through Mr D. T. Price of Bourbon. This palm produces an oil which is of great trade interest on the Guinea coast'. In 1853 and 1856 Teysmann reported that the introductions of 1848 had flowered after four years. At first it was thought that the plants were either male or female, but their monoecious character was soon discovered and Teysmann's interest clearly lay in determining whether the palms would prove a more useful producer of oil than the ubiquitous coconut palm.[22]

It has not been possible to trace the origin of the Amsterdam plants and confusion over the other two specimens resulted from Teysmann's later mention of 'Bourbon or Mauritius'. As late as 1917 it was stated that the oil palm does not occur wild in Mauritius (nor, presumably, on Réunion), but it is known to have existed in the Botanical Gardens of Mauritius well before 1863. It can only be supposed that the Amsterdam plants came from Africa, and that the Mauritius or Réunion plants came from the same source but were routed through one of these islands.

There are two other puzzling features of these remarkable importations. Hunger reported that not one of the original palms existed in the early 1920s, yet it is widely believed, and very recent reports confirm,[23, 25] that two of the four originals still survive and are over 70 feet in height. Secondly, the progeny of these four palms, transferred to Deli in Sumatra and eventually to give rise, under the term Deli palm, to the industry of the Far East, were uniform and their fruit were of a special constitution in certain respects. Though the fruit is *dura* (thick-shelled) in form, the spikelets of the bunches end in short spikes instead of long spines. The fruit are larger and contain a much higher proportion of mesocarp (around 60 per cent) than the general run of African *dura*; they are paler coloured and the oil percentage of the mesocarp is usually rather lower. Some observers claim to see differences in the colour of the leaves and the shape of the crown. What is most remarkable is that the population of palms descended from the Buitenzorg parents was relatively uniform.

> It has often been stressed by me [wrote Hunger], that the most important point . . . has been the fortunate fact that the Deli oil palms grown for ornamental purposes and which have provided the seed for the first regular plantings have proved to be of such a peculiarly valuable variety. [And he continued.] As soon as *E. guineensis* was grown commercially at Deli, *a priori* selection gave no difficulties as the available ornamental palms were already of a very productive kind and furthermore, from the beginning and relatively speaking, they bred true to type.

These facts cannot but suggest that either all four palms were from the same parent or that only one of the palms was the progenitor of the Deli palms. The former seems the more likely contingency.

In spite of the enthusiasm of Teysmann, the publicity given by the Government of Holland to the usefulness of the plant, and the establishment of plots at Banjar Mas in Java and at Palembang in Sumatra before 1860, no industrial plantation was to be fully established for 50 years. Rutjers[24] attributed this to the lack of enthusiasm of the local authorities in Java and to the doubts of estates both about profitability and the milling methods to be employed. In 1878 a plot was established in the Economic Gardens from seed from the Bogor Botanic Gardens; plants from the same source had been established in the Deli district of Sumatra 3 years earlier. Ornamental avenues began to be established on tobacco estates in Sumatra's East-Coast district from 1884 and further planting from these avenues became widespread. Rutjers[24] considered that the strongest evidence that the avenue palms had a common origin in the Bogor palms is that they were similar to those established in the Economic Gardens and that the source of the latter and of the first-recorded importation into Sumatra was the same.

Plantations in the Far East

The foundation of the industry is generally attributed to M. Adrien Hallet, a Belgian with some knowledge of oil palms in Africa, who planted palms of Deli origin in 1911 in the first large commercial plantations in Sumatra. Hallet's plantings on Sungei Liput (Atjeh) and Pulu Radja (Asahan) Estates are recorded[26] as being contemporary with the establishment of 2,000 palms by a German, K. Schadt, on his Tanah Itam Ulu concession in Deli. Hallet recognised that the avenue palms growing in Deli were not only more productive than palms in Africa, but had a fruit composition superior to the ordinary *dura* palms of the west coast. A potential oil content of 30 per cent in the fruit was recognized right from the start.[27] Within 3 years, 2,600 hectares had been planted, but stagnation followed owing to the First World War and lack of information on likely profits and on extraction methods. In the meantime, however, M. H. Fauconnier, who had been associated with M. Hallet, had established during 1911 and 1912 some palms of Deli origin at Rantau Panjang in the Kuala Selangor district of Malaysia. These palms were in full bearing by 1917 and in that year the first seedlings were planted on an area later to be known as Tennamaram Estate.

The industry grew rapidly in Sumatra, but did not gain its full momentum in the Far East until the 1930s. In 1925 there were 31,600 hectares planted in Sumatra and only 3,348 in Malaysia[25], but by 1938 the areas had risen to 92,300 and 29,196 hectares respectively. With over 120,000 hectares, an industry of considerable importance and capable of producing more oil than was being exported from Africa had, in the space of about 20 years, been established.

Plantations in Africa

Meanwhile some interest in planting the oil palm had emerged in Africa. In both the French territories and the Belgian Congo (Zaire) experimental plantings were started in the early 1920s. The Germans had already planted areas of palms in the Cameroons prior to the First World War and in Nigeria an interesting plot had been established near Calabar between 1912 and 1916.

By the early 1920s the Belgians had decided that plantations were likely to be more profitable than the exploitation of groves and that the rare thin-shelled *tenera* fruit, with its high percentage of oil-bearing mesocarp, would be more valuable than imported thicker-shelled Deli palms.[27] M. Ringoet, who had seen the beginnings of the plantation industry in Sumatra, established plantings at Yangambi, near Stanleyville (now Kisanganu), in 1922 and at a number of other places during the next few years. Open-pollinated selected *tenera* seed was used on commercial plantations as early as 1924[28] and it was soon realized that the *tenera* was a hybrid which would nevertheless give a high proportion of its progeny in the same thin-shelled form. At the same time there seemed no reason to doubt that the *tenera* palm would provide as high a bunch and fruit yield as the *dura* palm. Improved exploitation of the groves, by clearing the undergrowth, providing paths, etc., continued alongside this planting, and by 1939 there were 14,038 hectares of African planted plots. About one-third of the European plantations were not yet in bearing.

In British West Africa a belated attempt was made to make the authorities aware of progress elsewhere and to induce them to encourage similar developments. A committee recommended the Secretary of State for the Colonies in 1923 to encourage commercial firms to introduce oil mills under similar conditions to those imposed in the Belgian Congo and at the same time to allow these firms to establish limited plantations. Recommendations were also made that the groves should be opened up for exploitation and that Africans should plant areas of palms.[29] These recommendations went largely unheeded and no significant progress was made during the inter-war period when enthusiasm was high in Sumatra, Malaysia and Zaire. Twenty years were to pass by before mills were introduced and an appreciable improvement in oil quality could take place. Nigeria, the foremost exporter until 1934, took little or no part in the expanding world trade in palm oil. Attention was confined to improving the groves, the planting of small areas by the local population and the introduction of small hand-operated presses. Experimentation was inadequate and the idea of replacing groves by small plantations of unselected material was largely abandoned in Nigeria in the late 1930s in favour of planting small plots on forest land. This policy was slightly more successful: 3,727 hectares owned by 5,530 farmers were planted by the end of 1938, but land tenure systems impeded further progress.[30, 31] There was considerable extension in the use of hand presses, but this was not accompanied by an appreciable improvement in oil quality. Useful small plantings were established on Agricultural Stations and in 1939 an oil palm research station was started near Benin on an area of over 1,600 hectares.

In the mid-1930s Sumatran palm oil exports rapidly overtook those of Nigeria, but West Africa as a whole maintained her place as the predominant exporter of palm kernels. In 1936, Nigeria was providing nearly half and Sierra Leone nearly one-tenth of the world's exports. Exports from British territories in Africa in 1936 were over 410,000 tons. The territories under French rule had not developed plantations but, though only small exporters of oil, they provided substantial quantities of kernels for the European market, a peak export figure of over 170,000 tons being reached in 1936.

The Second World War and after

The war put the whole of the Far Eastern industry out of the export market from 1942 to 1946. Shipping difficulties prior to 1942 were already affecting trade in this region. Exports from French African territories fell to very low levels. Exports of palm oil from Zaire increased during the war period, but shipments of kernels were curtailed. There was also a slight curtailment of kernel exports from Nigeria and Sierra Leone, and palm oil shipments did not maintain the rate of expansion of 1939 to 1942.

The exports of palm oil and kernels from producing countries in the prewar years, war years and postwar years are given in Table 1.1.[32, 33] During the middle and late 1930s prices remained low and there was little encouragement to continue new plantings at the previous high rate. Soon after the war, however, prices of both oil and kernels reached four or five times their prewar level and they remained comparatively steady for 20 years, the lowest United Kingdom price for palm oil (in 1962) being only 65 per cent of the early postwar peak (Table 1.2).[34, 32] Prices again rose steeply at the end of 1973 and remained very high throughout 1974, falling back again only partially in 1975.

These circumstances made palm oil and kernel production an attractive proposition and from 1962 there was a rapid expansion of planting which was further encouraged by the fact that palm oil prices, though fluctuating, showed an overall upward tendency. The most spectacular increase in production was seen in Malaysia, but expansion was not confined to the traditional countries or regions. In 1972 there were at least twenty-four countries exploiting the palm commercially apart from those having a small production from grove palms. The main reasons for the expansion were the increasing world demand for oils and fats, special local demands, the relative stability of world oils and fats prices, the potential profitability of the crop, and the need and wish in many countries to diversify their agricultural economy.[35]

The expansion of exports was to some degree held back by internal disorders (as in Nigeria and Zaire) and, with rising populations, by increased local consumption. In some countries where almost all the production was previously exported an increasing proportion became absorbed in local manufacture, and a much higher proportion of palm kernel oil was extracted locally (Table 1.1 and Figs 1.2 and 1.3).

Table 1.1 *Exports of oil palm products from producing countries* (thousand tons* per annum)

Countries	1909–13	1924–7	1928–31	1932–5	1936–9	1940–1	1942–5	1946–9	1950–3	1954–7	1958–61	1962–5	1966–9	1970–3	1974
PALM OIL															
Africa															
Nigeria	82	120	129	125	136	130	131	134	173	186	176	132	47	8	—
Sierra Leone	9	4	3	3	1	1	—	2	2	†	—	—	—	—	—
Ivory Coast	6	7	7	17	19	9	8	5	13	17	1	1	1	36	89
Dahomey	13	17	13								10	11	10	12	11
Zaire	2	17	32	47	67	62	84	100	134	146	165	123	121	94	33
Angola	2	4	3	4	3	4	6	12	11	10	11	16	13	8	4
Other African	8	12	12	11	21	9	9	11	10	8	14	15	12	10	14
Asia															
Indonesia	—	12	43	115	202	161	—	35	115	126	113	116	158	214	277
Malaysia	—	1	3	15	46	51	—	39	50	56	87	120	250	608	886
Total	122	194	245	337	495	427	238	338	508	549	577	534	612	990	1,314
PALM KERNELS															
Africa															
Nigeria	172	258	253	292	334	307	321	324	384	439	425	394	223	191	182
Sierra Leone	47	71	60	72	74	45	38	63	73	59	56	54	47	49	29
Ivory Coast	6	13	11	69	76	48	47	55	76	70	15	12	10	20	37
Dahomey	34	47	31								52	41	6	6	—
Liberia	—	—	—	—	—	†	9	18	10	15	8	8	16	9	1

Table 1.1 – *continued*

Countries	1909–13	1924–7	1928–31	1932–5	1936–9	1940–1	1942–5	1946–9	1950–3	1954–7	1958–61	1962–5	1966–9	1970–3	1974
Africa – continued															
Zaire	7	72	70	55	89	37	53	64	87	49	28	6	2	—	—
Angola	6	7	6	6	6	6	8	12	11	10	9	15	14	8	8
Other African	45	79	92	75	82	54	61	59	76	72	97	83	73	69	50
Asia															
Indonesia	—	2	8	24	42	15	—	10	33	40	33	32	37	45	28
Malaysia	—	†	†	3	8	6	—	5	11	14	22	19	31	20	20
Total	317	549	531	596	711	518	537	610	761	768	745	664	459	417	355
PALM KERNEL OIL															
Africa															
Nigeria	—	—	—	—	—	—	—	—	—	—	—	1	32	33	38
Dahomey	—	—	—	—	—	—	—	—	—	—	—	4	19	23	16
Liberia	—	—	—	—	—	—	—	—	—	—	—	—	—	2	8
Angola	—	—	—	—	—	†	—	—	—	—	—	2	1	1	—
Zaire	—	—	—	—	—	†	3	13	15	39	54	37	43	43	33
Other African	—	—	—	—	—	—	—	—	—	—	1	1	4	1	—
Asia															
Malaysia	—	—	—	—	—	—	—	—	—	—	—	—	—	30	91
Total	—	—	—	—	—	*	3	13	15	39	55	45	99	135	203

* 1 long ton (2,240 lb) = 1.016 mt
† = very small.
— = nil.

Source: Commodities Division, Commonwealth Secretariat, and Oil World Publications.

Table 1.2 *Prices of palm oil, palm kernels and palm kernel oil* (Liverpool or London, per ton* to nearest £)

	(1) Palm oil (£)	(2) Palm kernels (£)	(3) Palm kernel oil (£)	Notes
1911 (Dec.)	29	18	38	(1), (2) Liverpool landed. (3) Liverpool ex mill.
1919 (Dec.)	86	39	92	,, ,,
1923 (Dec.)	38	20	41	,, ,,
1931 (Dec.)	20	12	23	,, ,,
1939 (Av.)	13	9	18	,, ,,
1945 (Dec.)	42		49	(1), (3) Govt. selling prices
1955 (Av.)	87	52	94	(1) 5% f.f.a.,† (3) crude, naked, ex mill.
1959 (Av.)	90	70	135	,, ,,
1963 (Av.)	81	55	101	,, ,,
1964 (Av.)	87	55	107	,, ,,
1965 (Av.)	99	65	129	,, ,,
1966 (Av.)	86	57	97	,, ,,
1967 (Av.)	82	59	109	,, ,,
1968 (Av.)	72	75	155	,, ,,
1969 (Av.)	78	65	130	,, ,,
1970 (Av.)	110	71	156	,, ,,
1971 (Av.)	110	61	141	,, ,,
1972 (Av.)	88	47	104	,, ,,
1973 (Av.)	157	109	190	,, ,,
1974 (Av.)	288	204	426	,, ,,
1975 (Av.)	200[1]	94	200[2]	,, ,,

* Long ton.   [1] 11 months av.   [2] 7 months av.

† Malaysian palm oil has normally commanded a premium of about £2 per ton over West African palm oil.

Source of recent data: Commodities Division, Commonwealth Secretariat.

A short account is given below of postwar development in the more important producing countries.

Indonesia (Sumatra)

A peak of 109,755 hectares was reached in 1940 but following the neglect of the war years the Sumatran plantations were only slowly brought back into cultivation and in the late 1940s and early 1950s planting was only on a very moderate scale. With the economic difficulties which followed independence, the Sumatran industry was not able fully to recover her place in world trade. The area under nominal cultivation had by 1956 increased by about 15 per cent over its 1938 level, but this was largely on account of plantings undertaken in the early war years; thereafter there

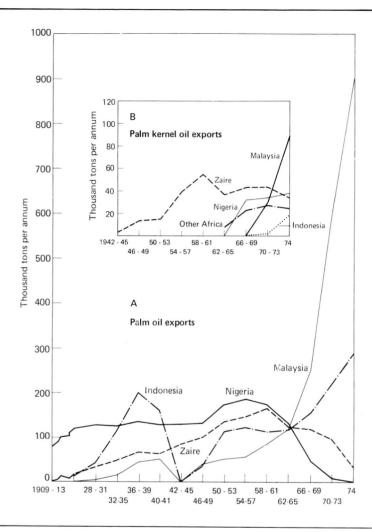

Fig. 1.2 Exports of palm oil and palm kernel oil from the principal exporting countries.

was almost no expansion. Moreover the area recorded as under production was not, even by 1962, as large as the prewar cultivated area and it was yielding at a low rate.

In the late 1960s the treecrop plantations in Indonesia were largely re-organized into estate groups (the Perusahan Negara Perkebunan or P.N.P.s); seven of these, situated in Sumatra, contained oil palm plantings covering a total area, in 1971, of about 90,000 hectares. The remaining

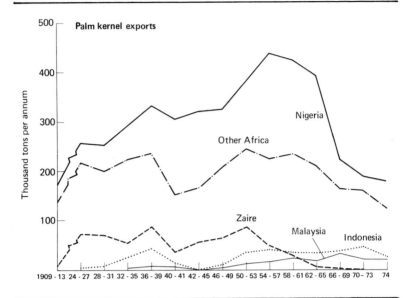

Fig. 1.3   Exports of palm kernels from the principal exporting countries.

area of about 36,000 hectares was still held by private companies. During the early 1970s there was considerable injection of capital by the World Bank and the Asian Development Bank for the rehabilitation of the P.N.P.s and their expansion. By 1973 this had assisted the rehabilitation of some 85,000 hectares and the new planting of 15,000 hectares. There is expected to be around 200,000 hectares by 1980 with a production of over 500,000 tons of palm oil. Rehabilitation and expansion of the private estates has been slower. Smallholders' schemes have been started and further large areas have been surveyed for expanding the industry.

An important recent development in Indonesia has been the conversion of large areas of *Imperator cylindrica* (lalang) to oil palms both by P.N.P.s and private companies. These areas had established themselves following indiscriminate forest felling for food cropping, but by ploughing or herbicide treatment, and with suitable manuring, it has proved possible to establish productive fields of oil palms.

Malaysia

Rehabilitation was faster in Malaysia after the war than in Sumatra and production was again in full swing by 1947. Acreages were slowly but steadily increased throughout the 1950s and thereafter planting proceeded at a fast rate. This was due not only to confidence in the future of the market for palm oil, but also to the need felt by plantation companies to diversify their interests and, in particular, to reduce their dependence on rubber. Much of the new planting in West Malaysia was therefore on land hitherto carrying stands of rubber or coconuts. The rate of planting in West Malaysia accelerated during the 1960s. In 1960 the planted area covered 55,000 hectares, by 1965 it was over 100,000[32] and by 1973 over 400,000 hectares. It was predicted that over 600,000 hectares would be under the crop by 1980 when production could be 1¾ million tons of palm oil.[36] A major part of this area was and is to be developed by the Federal Land Development Authority in whose settlement schemes the oil palm is playing an increasingly important role. In the well-known Jengka Triangle and other schemes, land is allotted at the rate of 4 hectares per settler but is worked by cooperative effort in a 'block-group' system. Plantation companies have been responsible for most of the other plantings, both from forest and rubber land, but in recent years there has also been much planting by smallholders intending to send their bunches to nearby estate mills. A high proportion of the available coastal alluvial soils has been taken up and much more planting has been taking place on inland soils, in some cases on quite steep land. In the East Malaysian state of Sabah the oil palm has become an important crop within 10 years; the planted area rose from around 400 hectares in 1960 to 68,000 hectares in 1975, and the Sabah Land Development Board has become responsible for about a third of the planted area.

Nigeria

Nigeria lost her foremost place in palm oil exports to Zaire in 1962 and regained it only temporarily in 1964–5. Total production, largely from the groves, was estimated by F.A.O. as 500,000 tons in 1971, but there are reasons for thinking this may be well below the actual production. Calculations cannot be based on kernel production since an unknown proportion of nuts are picked up and cracked without extraction of palm oil from the fruit. The decline of the oil palm industry in Nigeria through the early 1960s in terms of palm oil and kernel exports was the subject of much discussion, but it is probable that it was caused by increased domestic consumption following population growth and by the low producer prices allowed by the Marketing Boards which had a monopoly of purchase for export.[37] The producer prices given were little more than half the f.o.b. prices.

The outstanding achievement in Nigeria in early postwar years was the improvement of the quality of palm oil as judged by its free fatty acid (f.f.a.) content. This was achieved by the imposition of a large price differential between oil with less than 4.5 per cent, and later with less than 3.5 per cent, f.f.a. (Special or Edible grade) and oils of higher f.f.a. contents (Technical oil grades). The change-over was assisted by the establishment of Pioneer Oil Mills in the grove areas of eastern Nigeria. These found little difficulty in producing Special grade oil and small producers using the curb press soon found it necessary and profitable to fall into line. Exports from Nigeria in 1950 contained 0.2 per cent of the Special grade oil but by 1963 93.7 per cent of all purchases for export were of this grade.[38]

Conflicting policies were largely responsible for the failure of Nigeria to take advantage of the good market for palm oil. Large plantations were discouraged even in under-populated areas except for the limited number that could be financed by the Eastern and Western Nigeria Development Corporations. Some of the latter, particularly in eastern Nigeria, were well-sited and made a small contribution to exports, but others were poorly sited and went out of production.

By the end of the 1950s the work of the West African Institute for Oil Palm Research (set up with its Main Station near Benin City in 1952) was becoming better known, and regional governments began to see the need for, and the possibilities of, developing the industry on sounder lines. Development plans in the early 1960s included expansion of palm oil production by further Development Corporation plantations, by settlement schemes and by rehabilitation of palm grove areas.

The civil war which broke out in 1967 reduced the palm oil exports to a few thousand tons and palm kernel and kernel oil exports to half their former level. Following the war kernel and kernel oil exports recovered to some degree but palm oil exports, which came mainly from the areas of conflict, showed little sign of substantial recovery. With her large and increasing population it is unlikely that Nigeria will again play a large part in palm oil exporting until the large plantation and grove rehabilitation schemes being planned (1974) have been developed. It may be noted that in 1950 nearly half the palm oil entering world trade was extracted by village methods from the African palm groves. The Nigerian civil war removed most of this oil from the market and it seems unlikely that oil of grove origin will ever again form any substantial part of the trade.[35]

Zaire

The oil palm economy of Zaire was little affected by the war in comparison with other producing countries and exports of oil and kernels continued to expand. In 1949 a 10-year plan for the development of the country was published.[21] This recommended increased production from plantations and from the groves for export, local consumption and, in the case of kernels, local crushing and export of palm kernel oil. A production of 300,000 metric tons of oil and 130,000 metric tons of kernels was envisaged by 1959. It was intended that a considerable portion of this production should also be used for (i) local manufacture of margarine, (ii) production of soap, and (iii) production of palm kernel cake for local use and for export. The plan allowed for a considerable increase in small-acreage planting in addition to expansion of plantations financed from overseas.

This plan was in large measure fulfilled. In 1949 large plantations covered 103,000 hectares; by 1958 plantations covered 147,000 hectares of which 107,000 were in bearing and producing 77,000 metric tons of palm oil and 31,000 metric tons of kernels.[39] Yields from the best plantations were far above this low level brought about no doubt by poor production in the diseased areas of the south. Production of oil on certain estates reached 3 metric tons per hectare. By 1958 *small* plantations owned by Zairians covered 98,000 hectares of which 57,000 hectares were in bearing.

By 1959 Zaire was exporting over 180,000 tons of palm oil, which was 75 per cent of her recorded commercial production. In the years of unrest which followed, commercial production sank to 160,000 tons and exports to 80,000 tons (1965).[32] During the late 1960s and early 1970s there was a fair recovery, but it became apparent that Zaire would consume an increasing proportion of her production. By 1974 commercial production was estimated at 190,000 tons with an export of only 63,000 tons. Zaire did not take part in the expansion of planting that has been a feature of other producing countries, and established plantation companies were reluctant to increase their acreages. The great research station of the Institut National pour l'Etude Agronomique du Congo Belge, which had done so much for oil palm cultivation, ceased to operate on its previous scale. Exports have continued to decline.

The Ivory Coast

The Ivory Coast is climatically the most favoured of the smaller African countries, but no considerable plantation industry, as in Zaire, developed. Important development plans were put into operation in the early 1960s following successful plantings of Deli *dura* x *tenera* and *pisifera* material by the Institut de Recherches pour les Huiles et Oléagineux (I.R.H.O.) in areas of forest and of derived savannah. 'La Plan Palmier à Huile' allowed as a first stage for the planting in five selected areas of a total of 4,700 hectares between 1962 and 1965.[40] The scheme provided, in each area, for a central plantation block, a mill with a capacity of 12 tons of bunches per hour and the organization of 'satellite' holdings, owned individually or cooperatively, on the periphery of the block. The second stage, entrusted to the Société d'État pour le Développement du Palmier à Huile (S.O.D.E.P.A.L.M.), followed a search for suitable larger areas and allowed for the planting of 33,000 hectares between 1965 and 1968 with possible extensions to 75,000 or even 150,000 hectares by 1975. It was intended, through the system of plantations and satellite holdings, to immerse the local population in these ventures both as workers and owners. Considerable areas of selected progenies were established by the I.R.H.O. at La Mé to provide seed for the areas to be planted.

The scheme was remarkably successful. By 1969 out of 54,000 hectares planned 52,633 had been planted,[41] and by 1973 there were nine industrial plantations totalling 41,000 hectares and peripheral village plantings of 26,000 hectares. Many of the village plantings are of a high standard but the main problem facing the scheme in 1973 was the invasion of *Eupatorium odoratum* which became prevalent in the village plantings and from there invaded the plantation blocks.

The Ivory Coast was not a significant exporter of palm oil until 1970 but from that year exports steadily increased, reaching 100,000 tons by 1974, though much of the increasing production was being employed in local manufacture. Palm kernel exports have also been increasing in recent years.

Dahomey

The climate of Dahomey is too severe for high yields but the south of the country has supported little else but sparse areas of palm grove. Palm kernels constituted nearly half the total exports of the country in the early 1960s, but in later years an increasing quantity was crushed locally and by 1971 exports of palm kernel oil reached 24,000 tons with kernels only 12,000 tons.

In spite of the poor yield expectations a great effort has been made by the Société Nationale pour le Développement Rural du Dahomey (S.O.N.A.D.E.R.) to increase production and exports of palm oil products. This organization had by 1970 planted around 22,000 hectares in small farmers' cooperative schemes and was aiming to extend its plantations to produce about 40,000 tons palm oil by 1978.[41] Published figures of production show, however, that bunch yields have been lower than expected and that only around 1 ton of palm oil per hectare is being obtained from adult areas.[42]

Cameroon

Cameroon was only a small exporter of kernels in the early postwar period. However, with the very considerable expansion of planting by the Cameroons Development Corporation and by Unilever in West Cameroon and, more recently, with the inauguration of the Société Camerounaise de Palmeraies (S.O.C.A.P.A.L.M.) in East Cameroon, the country is expected to become a considerable exporter of oil palm products. By 1973 the C.D.C. had over 10,000 hectares and S.O.C.A.P.A.L.M. over 7,000 hectares under oil palms, further extensions were planned and exports of palm oil, kernels and kernel oil were beginning to increase.

Other African countries

Most other countries in Africa enter world trade through exports of kernels, or, more recently, kernel oil, although Angola had, until recently, a steady export of 10,000—15,000 tons of palm oil. These countries are extracting locally an increasing quantity of kernel oil. Sierra Leone has continued to provide about 60,000 tons of kernels.

South and Central America

The growing of the oil palm constitutes a new industry in Latin America. By the end of the 1950s adult producing plantations existed only in Venezuela, Colombia, Ecuador, Costa Rica, Nicaragua and Honduras. The areas in each case were small, and the palm oil and kernels were disposed of locally. About 1959 an increased interest began to be taken in the crop as it came to be realised that suitable growing conditions existed, and that both oils could replace the coconut and other oils which were being imported in increasing quantities.

Development has not been uniformly successful in Latin America and many difficulties have been encountered. Some unsuitable soils were initially used in Brazil while in some areas of Colombia and in Panama and Peru development has been checked by disease. Elsewhere lack of capital and the remoteness of otherwise suitable areas have hindered expansion.

Colombia

Plantations have been established in Colombia in widely distributed areas but principally in the Magdalena valley and in areas of the Pacific coastal plain. By 1966 15,835 hectares had been planted but in the 7 years to 1973 this total had increased only to 20,500 hectares. The failure of the industry to increase as rapidly as expected was due to serious outbreaks of Bud Rot and the Sudden Wither disease, Marchitez sorpresiva, which devastated the greater parts of two of the principal plantations. In spite of this, existing plantations have been expanded and the plantation devasted by Bud Rot has replanted many fields with the hybrid *E. guineensis* x *E. oleifera* which was found to be largely unaffected by the disease.

Production of palm oil was estimated to be 44,000 tons in 1973. The Government has drawn up plans for the financing of a further 20,000 hectares to be planted between 1974 and 1978 with the intention of raising production to around 150,000 tons palm oil by 1984.

Plantations in Colombia vary greatly in size. The largest covers 5,000 hectares,[43] but medium-sized estates of 100 to 400 hectares have been successful and small mills, suitable for these acreages, have been manufactured locally.

Ecuador

Development in Ecuador has been confined almost entirely to a small region north-west and south of Santo Domingo de Los Colorados on the Pacific plain.[44] The first plantation was established in 1953 but further planting did not get under way until the beginning of the 1960s when an Experiment Station was set up in the region by the Instituto National de

Investigaciones Agropecuarias. By 1975 planting was proceeding at the rate of about 3,000 hectares per year and nearly 15,000 hectares had been established. As in Colombia the size of holdings varies considerably. Many medium-sized estates of around 100 hectares have been established without mills in the hope that the bunches produced will be sold to nearby larger estates with mills. The provision of sufficient milling capacity for the industry as a whole has therefore presented a considerable problem. Production of palm oil was estimated to be around 10,000 tons in 1973,[45] and provided sufficient mills can be erected it is hoped to raise this production to about 50,000 tons by 1980.

Brazil

Development has proceeded at a slow rate in the areas south of Salvador in Bahia state and in the area around Belém in the state of Pará. It is in the former areas that the semi-wild palm groves are to be found, and in recent years modern mills have been set up to process bunches from the groves and from new plantings established within the same area. Development has not been uniformly successful in Bahia as some unsuitable soils were initially chosen for planting. Near Belém 1,500 hectares were established on a single plantation between 1968 and 1972 and yield prospects in this region are fair.

Costa Rica

On the Pacific coast large plantations have been developed by the United Fruit Company, the area planted reaching over 13,000 hectares by 1974. On this coast there is a dry season lasting from January to March. On the Caribbean coastal plain, where the rainfall is heavier and more evenly distributed, there have been some small plantings and high yields have been reported.[46] Further developments in this region are expected.

Honduras

The earliest of the American plantings was established in Honduras in 1943 by the United Fruit Company and the areas planted on the northern coastal plain now exceed 5,000 hectares. A unique scheme of communal plantings (cooperativas) was started in 1971 in the Aguán valley, and by 1975 the planted area had reached nearly 1,000 hectares. The palms in Honduras were very little affected by the hurricane and floods which devastated habitations and other crops on the northern coast in 1974.

Other American countries

Plantations also exist in south-western Mexico, Panama, Venezuela, Surinam and Peru. In the latter country a plantation has been established in the Huallaga valley as a nucleus for peripheral smallholdings.[47] The area has well-distributed rainfall but mean minimum temperatures are rather low. A small plantation established in Nicaragua between 1946 and 1950 was devastated by Bud Rot, but further developments in that country and in Guyana are expected.

Development methods

Of particular interest during the last decade has been the diversity of methods by which oil palm industries have been developed and the controversy that this has engendered.[48] Everywhere there has been a desire for smallholders to participate to a greater extent, but in the more traditional plantation countries such as Malaysia a doubt has persisted of the ability of the smallholder regularly to provide bunches of the correct ripeness to fulfil the more stringent quality demands of palm oil purchasers. For this reason the Malaysian Federal Land Development Authority in following the 'block-group method' of their rubber schemes did not go so far as eventually to allocate individual oil palm lots to settlers,[49] though recently there have been allocations of 40 hectares to be worked together by ten settlers' families. The Honduras Cooperativas will have 200–350 hectares each, to be cultivated by about 80–120 heads of families.

The idea of nucleus estates with peripheral smallholdings from which bunches are brought to a central mill has gained popularity in other regions and has been most successful in the Ivory Coast.[50, 41] It must be admitted, however, that in the latter country it is the estates which dominate whereas in a predominately smallholders scheme the nucleus estate would be expected to be relatively small and to provide processing facilities and an example of first-rate cultivation methods. The question of the proportion of the land to be occupied by the nucleus estate and the smallholdings is not an easy one however, since those financing a scheme are unwilling to erect a large mill unless they feel assured of receiving a sufficient tonnage of bunches for it. In addition to providing a mill the nucleus estate is able to organize transport of bunches, and perhaps also to undertake the clearing and planting of the smallholders areas and to provide planting material and advice. In a recent scheme in New Britain (Papua New Guinea), where the planting and harvesting of oil palms by smallholders has been conspicuously successful, the proportion of estate to smallholdings is 3 : 4.[51]

Development in Dahomey and in Cameroon stand in marked contrast: in the latter country the schemes so far undertaken entail estate planting only, whereas in Dahomey cooperative smallholder schemes have been the rule. In East Cameroon it has been argued that as there has been no tradition for oil palm planting by local farmers the initial developments should be on an estate scale.

A feature of development in Colombia and Ecuador has been the planting of medium-sized holdings financed by local entrepreneurs and having, or sharing, small locally-designed mills. This type of enterprise is of particular interest in view of the widely-held opinion in the plantation industry that only very large enterprises of 2,000 to 4,000 hectares are likely to be viable.

The importation and usage of oil palm products

Imports

The palm oil trade from West Africa has been traditionally, and remains, with Europe. The trade in kernels has followed a similar tradition. Only in three periods have countries outside Europe become large importers of oil palm products.

The United States entered the market for palm oil prior to the First World War, and between 1909 and 1913 was estimated to have taken about 20 per cent of the world's supply. Demand built up after the war and the United States was purchasing half the world's exports by 1930. Purchases continued at a high level until 1937, reaching 183,000 tons, but thereafter a decline set in and by the end of the Second World War imports had levelled off at less than 30,000 tons. A further fall from this level took place in the 1950s. The withdrawal of the United States as a large purchaser was due firstly to increased domestic production of vegetable oils, principally soya bean oil, for cooking fat and soap manufacture, and, later, to technological advances which reduced palm oil requirements in the steel industry.

The entry of the United States into the palm kernel trade was small and short-lived, starting in 1926 and ending with the Second World War. However, both immediately before and after the war the United States was a substantial importer of palm kernel oil, taking its supplies firstly from Western Europe and later direct from Zaire. In 1934–8 she took 60 per cent of the total imports of palm kernel oil, and in 1937 she took 90 per cent. When purchases then ceased the international commerce in palm kernel oil became negligible until production in Zaire started after the war. The United States then entered the field once more and by 1953 was purchasing two-fifths (22,500 tons) of the world's small export supply. This rose to nearly 40,000 tons in 1960. Considering that she was by far the largest exporter of vegetable oils this import seemed a curious anomaly. By Public Law 480 and various aid programmes, however, the United States so maintained the quantity and price of her own oil supplies (principally soya bean) that it often paid American manufacturers to import foreign oils instead of purchasing domestic supplies.[52]

Within Europe, the United Kingdom was always the principal importer of palm oil, taking nearly half of Europe's supply before the Second World War and a like proportion in the late 1950s. During the early 1960s this proportion fell to about one-third and later to less than one-third of Europe's imports; this was largely due to the decline in Nigerian production, since the United Kingdom has traditionally taken a large share of her supply. Other substantial importers were, and continue to be, Germany, Holland, Belgium, France, Italy and Portugal. New postwar entrants into the purchasing of palm oil are Japan, India and Canada, the latter having purchased fair quantities for a few years immediately before the war.

A new pattern of palm oil purchases has been emerging. During the 1960s Asia began to take a larger proportion of total exports. Although Indian purchases fell away, Iraq and Japan more than doubled their imports during the second half of the decade. By 1977 the United States had become the largest single importer, doubling her intake in two years.

The principal importer of palm kernels before the Second World War was Germany; she constantly took nearly half the world's supplies. After the war the United Kingdom became the principal purchaser, taking her supply almost exclusively from Nigeria and Sierra Leone. In the early 1950s she was taking well over half the world's supply, but from 1954 her imports tended to fall. Imports into France, Belgium and Holland rose and Germany maintained a high level of purchases.

In the early 1960s there was a decline in the imports of the principal importing countries, including the United Kingdom. However, Eastern Europe began to develop an import trade in palm kernels. During most of the 1960s there was a declining trade in palm kernels with some later recovery, but during the same period and in the early 1970s there were increasing imports of palm kernel oil into Western Europe and the United States.

Usage of palm products

Both palm oil and palm kernel oil are used in the manufacture of margarine, compound cooking fat and soap. The proportion in which the oils are used for these purposes depends on their quality and the supply of other vegetable oils. In manufacture, oils and fats are very largely interchangeable and if an oil used for a certain purpose is in short supply, another oil may be substituted and used for that purpose and the proportions in which the latter oil is used for various industrial processes will change.

Palm oil is used much more for the manufacture of edible products than previously because of the great improvement in the quality of the oil coming from West Africa. Besides this quality factor there were technological advances which made palm oil more easily adapted to margarine and cooking fat manufacture. Another use of palm oil is in the tin-plate industry.

Before the Second World War palm kernel oil was used almost entirely for soap-making. After the war it became used to a much greater extent in a wide variety of edible products. The uses of palm oil and palm kernel oil in the United Kingdom in 1957–9 were classified as shown in Table 1.3.[53]

In recent years there has been a substantial change in the usage of palm

Table 1.3   *Principal uses of palm oil and palm kernel oil in the United Kingdom 1957−9*

Product	Palm oil (%)	Palm kernel oil (%)
Margarine	30	15
Compound cooking fats	21	5
Soap	15	34
Candles, tin-plate industry, bakery trade, glycerine production	34	−
Confectionery, bakery trade, ice cream, mayonnaise, glycerine, detergents, pomades	−	46
	100	100

oil in the United Kingdom.[54] Usage in the three 'traditional outlets' − soap, margarine and compound cooking fats − has declined. The main reason for this decline has been the substitution of animal fats, tallow in the case of soap and lard in the case of the edible products. The substitution of marine oils also contributed to the decline at the beginning of the period. The sharp decline in the use of palm oil for soap-making is, of course, partly attributable to the improved quality of West African oil which has made it suitable for edible products. In 1971, however, 36 per cent of the palm oil imported into the United Kingdom was still being used for margarine and compound cooking fats. Other and new outlets are absorbing the remaining supply without difficulty: among these are fatty acid manufacture, additives to animal feeding stuffs, the potato crisp industry and the baking, biscuit and ice-cream trades.[55,56,57]

Palm oil refining has now started in producing countries, e.g. Malaysia, and a variety of products, including vanaspathi (vegetable ghee), are being supplied both to the local and overseas market.

Palm kernel oil usage has also undergone some change in recent years. Both palm kernel and coconut oils are required for their physical and chemical properties; they melt rapidly at a temperature just below that of the human body and when solid are hard and brittle. Moreover they are predominantly lauric oils. The decline in usage for margarine and compound cooking fats has been similar to that of palm oil. Usage in soap manufacture has however been maintained. Nut oils are particularly suited to toilet soaps and soap powders and as there is a tendency for this side of the trade to expand, the usage of palm kernel oil in soap manufacture may continue to be stable. Other outlets for palm kernel oil, particularly in the manufacture of detergents, have compensated for its reduced usage in the main edible products.

The place of oil palm products in world trade

The interchangeability of oils and fats in manufacture has already been mentioned. The oil palm has to compete in a market containing some

Table 1.4   *The position of oil palm products in world exports of vegetable and animal oils: net exports of oils, oil seeds and animal fats* (thousand tons,* oil equivalent per annum)

Oil	1934–8	1949	1954	1958–61	1962–5	1966–9	1970–3	1974†
Edible								
Groundnut	805	530	656	808	953	1,003	785	675
Soya bean	380	264	281	1,165	1,531	1,886	3,014	3,790
Cottonseed	190	118	344	265	337	203	317	350
Rapeseed	52	27	33	100	156	367	670	660
Sunflower	32	30	28	203	345	983	736	785
Sesame	68	35	40	55	72	85	98	100
Olive	134	41	112	149	159	178	283	250
Total	1,661	1,045	1,494	2,745	3,553	4,705	5,903	6,610
Edible–industrial								
Coconut	1,040	1,005	1,092	1,136	1,270	1,194	1,254	910
Palm kernel	315	347	394	407	358	314	324	375
Palm	437	488	570	568	531	617	999	1,325
Total	1,792	1,840	2,056	2,111	2,159	2,125	2,577	2,610
Industrial								
Linseed	610	257	827	430	437	394	431	265
Castor	85	82	102	150	186	204	219	225
Tung	78	60	46	57	41	50	53	48
Total	773	399	775	637	664	646	703	538
Vegetable oil total	4,226	3,284	4,325	5,493	6,376	7,478	9,183	9,758
Animal and marine								
Whale and fish	543	409	474	550	611	678	610	497
Butter, lard, tallow	841	915	1,283	1,721	2,059	2,232	2,519	2,703
Animal and marine total	1,384	1,324	1,757	2,271	2,670	2,910	3,129	3,200
Grand total	5,610	4,608	6,082	7,764	9,046	10,388	12,312	12,958
Palm and palm kernel oils as per cent of vegetable oils	17.8	25.4	22.3	17.7	13.9	12.4	14.4	17.4
as per cent of all oils and fats	13.4	18.1	15.9	12.6	9.8	9.0	10.7	13.1

* 1 long ton (2,240 lb) = 0.9844 tonne.
† Provisional.

Source: Commodities Division, Commonwealth Secretariat, and Oil World Publications.

thirteen principal vegetable oils and oil seeds, two types of marine oil, and three categories of animal fats. However, the versatility of palm oil and the very high oil yield per hectare should make it possible for the oil palm to compete very easily with other oils and fats within the world export market. The only substantial disadvantage the oil palm suffers is the relatively high extraction costs of oil and kernels on the plantations; but this disadvantage should be more than balanced by the high yields and by the fact that the main product, palm oil, does not need extraction off the plantations as is the case with oil from oil seeds.

When export statistics are studied, however, it is seen that the proportion of the world market held by oil palm products has not substantially altered since prewar days; indeed for most of the period their relative position in a gradually expanding market declined (Table 1.4). This decline took place during a period when the economic prospects of oil palm planting were fully appreciated and when expansion was, in fact, taking place in many parts of the tropics. That oil palm products did not take a larger part in the world trade in oils and fats was because of the economic uncertainty, restrictions and unrest in two principal producing countries and the contraction of supplies from palm grove areas. At the same time there was a rise in shipments of competing oils and fats such as soya bean oil, rape seed and sunflower seed oils, and lard and tallow. However, the early 1970s saw a marked change, with oil palm oils again constituting nearly 12 per cent of total exports of oils and fats. It was estimated that, with the large expansion of planting, export availabilities might treble during the decade to 1980 and palm kernel oil availabilities increase by more than 50 per cent. Under these circumstances palm oil exports might constitute 20 per cent of world exports while kernel and kernel oil exports would constitute a very much lower proportion of the total exports from the crop than hitherto.[58]

References

1. Raymond, W. D. (1961) The oil palm industry. *Trop. Sci.*, 3, 69.
2. Seward, A. C. (1924) A collection of fossil plants from South-eastern Nigeria. *Bull. Geol. Surv. Nigeria*, 6, 66.
3. Zeven, A. C. (1964) On the origin of the oil palm. *Grana palynol.*, 5, 50.
4. Cook, O. F. (1942) A Brazilian origin for the commercial oil palm. *Sci. Monthly*, 54, 577.
5. Corner, E. J. H. (1966) *The natural history of palms.* Weidenfeld and Nicolson, London.
6. Mauny, R. (1953) Notes historiques autour des principales plantes cultivées d'Afrique occidentale. *Bull. Inst. fr. Afr. noire*, 15, 684, and (1961) Tableau géographique de l'Ouest Africain au moyen âge. *Mem. Inst. fr. Afr. noire*, No. 61, Dakar, 587.
7. Rees, A. R. (1965) Evidence for the African origin of the oil palm. *Principes*, 9, 30.

8. **Zeven, A. C.** (1965) The origin of the oil palm (*Elaeis guineensis* Jacq.). *J. W. Afr. Inst. Oil Palm Res.,* **4,** 218.

9. **Crone, G. R.** (1937) The voyages of Cadamosto and other documents on Western Africa in the second half of the fifteenth century. Hakluyt Society, Series II, **80.**

10. **Chevalier, A.** (1934) La patrie des divers Elaeis, les espèces et les variétés. *Revue Bot. appl. Agric. trop.,* **14,** 187.

11. **Briey, J.** (1922) Le palmier à huile au Mayumbe. *Mem. Rapp. Matièr grass.,* **2,** 112.

12. **Waterston, J. M.** (1953) Observations on the influence of some ecological factors on the incidence of oil palm diseases in Nigeria. *J. W. Afr. Inst. Oil Palm Res.,* **1,** (1), 24.

13. **Schad, H.** (1914) Die geographische verbreitung der ölpalme (*Elaeis guineensis*). *Tropenpflanzer,* **18,** 359–91, 447–62.

14. **Zeven, A. C.** (1967) The semi-wild oil palm and its industry in Africa. *Agr. Res. Rpts.,* No. 689.

15. **Jumelle, H. and Perrier de la Bathie, H.** (1911) Le palmier à huile à Madagascar. *Matières grasses,* **4,** 2065.

16. **Portères, R.** (1947) Aires altitudinales des Raphias, du Dattier sauvage et du Palmier à huiles au Cameroun français. *Revue Bot. appl. Agric. trop.,* **27,** 203.

17. **Stilliard, N. H.** (1938) *The rise and development of legitimate trade in palm oil with West Africa.* Unpublished thesis, University of Birmingham.

18. **Dike, K. O.** (1956) *Trade and politics in the Niger Delta,* 1830–85, Oxford University Press. 250 pp.

19. **Waddell, H. M.** (1863) *Twenty-nine years in the West Indies and Central Africa.*

20. **Billows, H. C. and Beckwith, H.** (1913) *Palm oil and kernels, the consols of the west coast.* Charles Birchell, Liverpool.

21. *Plan Décennal pour le Développement Economique et Social du Congo Belge,* 2 Volumes (1949) Brussels.

22. **Hunger, F. W. T.** (1924) 2nd edn, De Oliepalm (*Elaeis guineensis*). Historisch onderzoek over den oliepalm in Nederlandsch-Indië. Brill, Leiden, 383 pp.

23. **Terra, G. J. A.** (1953) *Private communication.* W.A.I.F.O.R. Ref. 420.

24. **Rutgers, A. A. L.** (1922) *Investigation on oil palm at the General Experimental Station of A. V. R. O. S.* Ruygrot & Co., Batavia.

25. **Jagoe, R. B.** (1952) 'Deli' Oil palms and early introduction of *Elaeis guineensis* to Malaya. *Malay. agric. J.,* **35,** 3.

26. **Heurn, F. C. van.** (1948) De Oliepalm, in *De Landbouw in den Indischen Archipel,* eds. Hall, C. J. J. van, and Koppel, C., II A, pp. 526–98, Van Hoeve, The Hague.

27. **Leplae, E.** (1939) Le Palmier à huile en Afrique, son exploitation en Congo Belge et en Extrême-Orient. *Mem. Inst. r. colon. belge Sect. Sci. nat méd.,* **7,** (3), 108 pp.

28. **Godding, R.** (1930) Observation de la production de palmiers sélectionés à Mongana (Equateur). *Bull. agric. Congo belge.,* **21,** 1263.

29. **Ellis, W. D.** *et al.* (1925) Palm oil and palm kernels. Report of a committee appointed by the Secretary of State for the Colonies, September 1923, to consider the best means of securing improved and increased production. Colonial No. 10, H.M.S.O.

30. **Buckley, F. E.** (1938) The native oil palm industry and oil palm extension work in Owerri and Calabar provinces. *Third W. Afr. agric. Conf.,* **1,** Papers, 207.

31. **Mackie, J. R.** (1939) Annual Report of the Agricultural Department for the year 1937, Nigeria.

32. *Vegetable oils and oilseeds. A review.* Compiled annually by the Commonwealth Economic Committee's Intelligence Branch and published by H.M.S.O. 1948 onwards.

33. *Survey of vegetable oils and oilseeds.* Volume 1. 'Oil palm products.' Compiled by the Statistics and Intelligence Branch, Empire Marketing Board and published

by H.M.S.O., June 1932; and Vegetable oils and oilseeds. *Bull. Imperial Economic Committee.* H.M.S.O. 1938.

34. Review of the oilseed, oil and oilcake markets for 1945, etc. Publ. annually. Frank-Fehr & Co. London and Liverpool.
35. Hartley, C. W. S. (1972) *The expansion of oil palm planting.* In Advances in oil palm cultivation, eds., Wastie, R. L. and Earp, D. A. Incorp. Soc. of Planters, Kuala Lumpur.
36. Ho Sim Guan (1972) Malaysian palm oil. Paper presented at the Int. Ass. of Seed Crushers' Conference, Kyoto, 1972.
37. Hartley, C. W. S. (1963) The decline of the oil palm industry in Nigeria. N.I.F.O.R., Benin City, Nigeria. Mimeograph.
38. *Nigerian Trade Summary,* Dec. 1963, Chief Statistician, Lagos, Federal Republic of Nigeria.
39. Volume Jubilaire 1910—60. (1960) *Bull. agric. Congo belge.* 226. pp.
40. Boye, P. (1964) Le Plan 'Palmier à Huile' de Côte d'Ivoire. *Oléagineux,* 19, 1.
41. Carrière de Belgarric, R. (1970) Deux exemples de développement du palmier à huile: La Côte d'Ivoire et Le Dahomey. *Coopér. Développement,* 30, 15—28.
42. Dissou, Machioudi (1972) Développement et mise en valeur des plantations de palmier à huile au Dahomey. *Cahiers d'Etudes Africaines,* 12, (47), 485.
43. IRHO (1971) Rendement du palmier à huile sélectionné en Colombie et potentiel de production. *Oléagineux,* 26, 244.
44. Hartley, C. W. S. (1968) The oil palm in Ecuador, *Oil Palm News,* 5, 9.
45. Dow, K. (1972) Estudio de la situacion de los derivados de aceites de oleaginosas y el incremento nacesario en el cultivo con miras al autoabastecimento. *I.N.I.A.P. Boletin Tecnico,* No. 4.
46. Matamoros Ramfrez, F. and Gonzalez Soto, G. (1971) Elsayo comparativo de hibridos y variedados de palma Africana. *Bol. Tecn. Minist. Agr. Ganderia,* Costa Rica, 59.
47. Martin, G. (1969) Le programme de développement du palmier à huile du Huallaga au Pérou. *Oléagineux,* 24, 259.
48. Hartley, C. W. S. (1970) Oil palm research in Africa and Malaysia. In *Change in Agriculture,* ed., Bunting, A. H., Duckworth, London.
49. Taib Bin Haji Andak (1966) Land development in Malaysia under the Federal Land Development Authority — description of programme and techniques of development implementation. *Proc. Seminar Malaysian Centre for Dev. Studies,* Kuala Lumpur.
50. Phillips, T. A. (1965) Nucleus plantations and processing factories: their place in the development of organised smallholder production. *Trop. Science,* 12, 3.
51. Manderson, A.(1971) The Mosa experiment. Oil palm in Papua New Guinea. *Aust. External Territ.,* 11, 4, 22.
52. Faure, J. C. A. Address given at the International Association of Seed Crushers' Congress, Stockholm, 1961, 20 pp.
53. *African produce in the world market.* United Africa Co. Ltd reprinted from the *Stat. & Econ. Rev.* No. 25, March 1960.
54. Cheshire, P. C. *The market for oil palm products with particular reference to the United Kingdom market.* Oil Palm Conference, London, 1965.
55. Pritchard, J . L. R. (1975) Refining palm oil. *Oil Palm News,* 20, 5.
56. Jacobsberg, B. (1976) Malaysian International Symposium on palm oil processing and marketing, 1976, Review paper. *Oil Palm News,* No. 21.
57. Moolayil, J. (1976) Uses of palm oil. *Malaysian Int. Symposium on palm oil processing and marketing 1976.* Preprint. 23pp.
58. F.A.O. (1972) Prospects for supplies of palm oil and palm kernels in 1980. *F.A.O. Bull. Agric. Econ. Statist.,* 21 (4).

Chapter 2

The botany of the oil palm

Classification

The family of palms, the Palmae, has always formed a distinct group of plants among the monocotyledons. Although Bentham and Hooker's *Genera Plantarum* placed the palms with the Flagellariaceae and Juncaceae under the series Calycinae, Engler and Prantl's system allowed them a place by themselves under the order Principes. In the comparatively recent classification of Hutchinson, the Palmae remain alone, though in the order Palmales. Here the oil palm, *Elaeis guineensis* Jacq., is grouped with *Cocos* and other genera under the tribe Cocoineae. The anatomical studies of Tomlinson support this grouping.[1]

The genus *Elaeis* was founded on palms introduced into Martinique, the oil palm receiving its botanical name from Jacquin in an account of American plants.[2,3] *Elaeis* is derived from the Greek word 'elaion', oil, while the specific name *guineensis* shows that Jacquin attributed its origin to the Guinea Coast. From time to time other specific names have been attached to supposed species of *Elaeis*, but none has shown any signs of permanency other than *E. melanococca*, now named *E. oleifera*, and *E. madagascariensis*, the legitimacy of which is doubtful. The Index Kewensis lists fourteen names, the majority of which have disappeared from the literature. Many of them either refer to quite different palms or are synonymous with *E. guineensis*. Of passing interest is the American palm *E. odora*, Traill, also named *Barcella odora* and classified by some botanists with *Elaeis* in a subtribe Elaeideae.

Elaeis madagascariensis, Becc. was described by Odoardo Beccari[4] as a separate species on the basis of material sent to him by Professor Jumelle. This material was distinguished from *E. guineensis* because in the male flower the fused filaments of the staminal tube were shorter and the anthers were erect, instead of spreading, at anthesis, while the fruits were smaller and rounded and surrounded by larger bracts. In view of the wide variation in many minor characters in the oil palm it is doubtful if these differences justify the naming of a separate species; Jumelle himself only considered this palm might have the status of a variety distinguished mainly by its red fruit.[5]

Pl. 1 A dried spikelet and some fruit with decomposing mesocarp, a nut and a kernel of the oil palm – R. Dodonaeus, 1608.

There remain therefore three species, *Elaeis guineensis*, *E. odora* and *E. oleifera*; the latter will be described and discussed on a later page.

The early descriptions of the oil palm have been reviewed by Opsomer[6] who claims for Mathias de Lobel (Lobelius, 1538–1616) the earliest botanical description and illustration of the fruit which he named *Nucula Indica*. Lobelius reported that the palm was found in Guinea. The brief descriptions were published in his Plantarium seu Stirpium Historia of 1570 (London) and 1576 (Antwerp) and his Kruydtboeck of 1581.

Mention of the oil palm is scattered throughout the works of de l'Escluse (Clusius, 1526–1609)*; the descriptions in the revised edition of the herbal (*Cruydt-boeck*, Leiden) of 1608 of de Dodoens (Dodonaeus, 1516–85) and in Bauhin's *Historie des Plantes* of 1650 are also attributed to him. There is some confusion over description and nomenclature in the earlier works but in the later ones the palm is called *Nucula Indica*

---

* *Clusius, C.* Aromatum et simplicium aliquot medicamentorum apud Indos nascentium Historia . . . Antwerp, 1567, 1574, 1579, 1593 and 1605. Exoticorum libri decem . . . Antwerp, 1605. Curae posteriores . . . Antwerp, 1611.

Pl. 2   Jacquin's drawings of an oil palm, a fruit and a nut — N. J. Jacquin, 1763.

Secunda and *Palma Guineenis*. The descriptions taken together are remarkably complete and it is recorded that thin-shelled nuts appear among the
more numerous thick-shelled ones. A dried spikelet was illustrated in
Clusius's last work and in Dodonaeus's *Cruydt-boeck* and this probably
accounts for Clusius's erroneous belief that palm oil was extracted from
the kernel, turning red on the journey from Africa (Plate 1).

Opsomer states that nothing of consequence was written on the oil palm
for over 150 years between the time of Clusius and the first, and lasting,
modern botanical description of Jacquin (1763) (Plate 2). However
Sloane's *Catalogus Plantarum* (London, 1696), besides making first

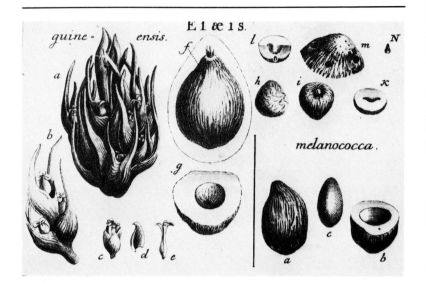

Pl. 3   Gaertner's drawing of a spikelet, flowers and fruit of *E. guineensis* together with the nut of his supposed *E. melanococca*, considered by Bailey, however, to be another *E. guineensis* nut – J. Gaertner, 1788.

mention of the spiny petioles, records the importation of the palm from the Guinea coast into Jamaica. It was from material from another West Indian island, Martinique, that Jacquin was to write his description and to name the plant. Jacquin's description is detailed, but he describes the flowers as either female or 'hermaphroditi steriles' and seemed unaware that flowers of the two sexes were in separate inflorescences. The production of male and female bunches was first recorded by Miller in his *Gardener's Dictionary* (London, 1768). Before the end of the century Gaertner, in his *De Fructibus et Seminibus Plantarum* (Stuttgart, 1788), gave a more detailed description of the flower parts, recording that the male and female flowers are on separate inflorescences (Plate 3).

Elaeis guineensis is a large feather-palm having a solitary columnar stem with short internodes. It is unarmed except for short spines on the leaf base and within the fruit bunch. The irregular set of the leaflets on the leaf gives the palm its characteristic appearance. The palm is normally monoecious with male or female, but sometimes hermaphrodite, inflorescences developing in the axils of the leaves. The fruit is a drupe which is borne on a large compact bunch. The fruit pulp which provides palm oil surrounds a nut the shell of which encloses the palm kernel.

The distinguishing of varieties of the oil palm has been attempted by many workers. These attempts have in most cases been unsatisfactory since in the wild state each palm is a hybrid in respect of certain of its characters. Most of the early attempts at classification are unworthy of mention since they were based on a very small acquaintance with the palm, and no knowledge of the inheritance of the characters described. Of interest, however, is the first description by Preuss[7] in 1902 of the *Lisombe* palm, a name used in the Congo, Cameroons and Nigeria for the thin-shelled *tenera* fruit form and employed to denote parental stock in quite recent times. Both Chevalier[8] and Jumelle[9] (1910 and 1918) divided the species into subspecies according to the outer appearance of the fruit, while Becarri extended Chevalier's classification. These classifications were unsatisfactory owing to their failure to attribute all the possible fruit variations to each subspecies, and it was left to Janssens[10] (1927) and Smith[11] (1935) to provide the first simple classifications which, in their essentials, have stood the test of time. Although nothing was known of the inheritance of the characters described, Janssens recognized that the fruit forms *dura* and *tenera*, distinguished by the thickness of shell, could be found among fruit types of different external appearance. Thus both the common fruit type *nigrescens* and the green-fruited *virescens* were divided by Janssens into three forms *dura*, *tenera* and *pisifera* (the latter called *gracilinux* − following Chevalier − when *virescens*). The white-fruited *albescens* was also recognized but only a *dura albescens* had been found. Similarly, although *dura* and *tenera* forms of the mantled-fruited *Poissoni* were found, no green-fruited mantled specimens were discovered. Smith, however, recognized both mantled and unmantled *nigrescens* and *virescens* fruit, called them 'types', and divided all four into thick-shelled and thin-shelled 'forms'. This simple procedure, described by Vanderweyen[12] as the most complete and logical of the empirical classifications, established the use, in English publications, of the fruit-type and fruit-form classification, thus eliminating the need of the term variety for material which might be heterozygous in many of its characters. That 'variety' was inappropriately used for the *tenera* form was recognized by Biernaert;[13] in the Far East Schmöle used the term fruit form as early as 1929.

Morphology and growth

The seed

The embryo of palm seeds is always small and the cotyledon is never erected as a green photosynthetic organ. Instead, the cotyledon apex becomes enlarged and, as the haustorium, absorbs the food reserves of the endosperm.[1] Thus in the oil palm the seed is adapted to support a developing seedling for many weeks after germination.

The oil palm seed[14, 15] is the nut which remains after the soft oily mesocarp has been removed, usually by retting, from the fruit. It consists of a shell, or endocarp, and one, two or three kernels. In the great majority of cases, however, the seed contains only one kernel since two of the three ovules in the tricarpelate ovary usually abort. Abnormal ovaries sometimes occur and four- or five-seeded nuts may, very rarely, arise from these. Nut size varies very greatly and depends both on the thickness of the shell and the size of the kernel. Typical African *dura* nuts may be 2 to 3 cm in length and average 4 g in weight (100 to the lb). Deli *dura* and large African nuts are larger, weighing up to 13 g. African *tenera* nuts are usually 2 cm or less in length and average 2 g (200 to the lb). Very small nuts weighing 1 g are not uncommon.

The shell has fibres passing longitudinally through it and adhering to it. The latter are drawn into a tuft at the base; this factor has been made use of in the construction of modern nut-cracking machinery. Each shell has three germ pores corresponding to three parts of the tricarpellate ovary, though the number of functional pores will of course correspond with the number of kernels developed. A plug of fibre is formed in each germ pore and these fibres are cemented together at the base to form a plate-like structure continuous with the inner surface of the shell.

Inside the shell lies the kernel. This consists of layers of hard oily endosperm, greyish-white in colour, surrounded by a dark brown testa covered with a network of fibres. Embedded in the endosperm and opposite one of the germ pores lies the embryo.

The embryo is straight and about 3 mm in length. Its distal end lies opposite the germ pore but is separated from it by a thin layer of endosperm cells, the testa and the plate-like structure referred to above. These three structures have been together called the operculum, but they are separate. In the quiescent state the bud is already well-developed laterally within the distal end of the embryo. In longitudinal section the apex with two differentiated leaves and the rudiments of a third can be distinguished, though the radicle is only poorly differentiated.[16] Opposite the bud there is a longitudinal split in the wall of the embryo (the 'fente cotylédonaire'). This part of the embryo is separated by a small constriction from the cotyledon which will develop into the haustorium. Within the cotyledon a system of procambial strands has developed.

The endosperm above the embryo is demarcated by a ring of cells of small size. When germination takes place the endosperm ruptures in this region and a disc consisting of endosperm, testa and the germ-pore plate is extruded from the germ pore together with the fibre plug (Plate 4).

The process of germination is illustrated in Fig. 2.1. The emerging embryo forms a 'button' (commonly called the hypocotyl but considered by Henry and other botanists to represent the petiole of the cotyledon[17]) which rapidly gains a plumular projection, while from the end of the embryo itself the persistent radicle emerges. The plumule and radicle both emerge through a cylindrical, persistent ligule close to the seed. It is of interest here to note[18, 19] that palm seedlings have been classified by Gatin and by Tomlinson according to seed and seedling structure and mode of germination. In certain palms the embryo is exserted from the seed by the growth of a cotyledonary extension organ termed the apocole. The

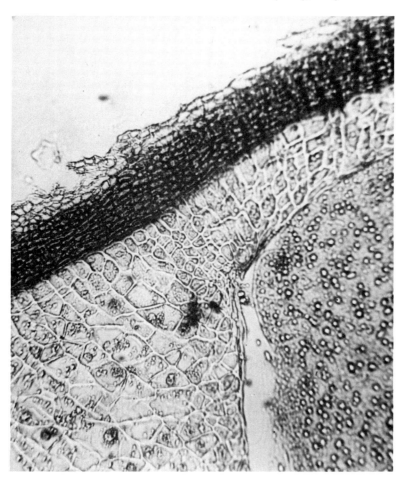

Pl. 4 Longitudinal section of a *tenera* kernel showing the distal end of the embryo enclosed by the endosperm and dark testa. Note that the endosperm is continuous above the embryo; rupture will take place through the smaller cells at the corner.

embryo may be carried some distance from the seed and buried as much as 2 feet below the soil surface. This appears to be an ecological adaptation to dry habitats and is exhibited by the *Borassus* palm whose range overlaps that of the oil palm. Such palms may have ligules or may not. Palms without a long apocole, however, usually have curved embryos and nonpersistent radicles. The oil palm does not therefore fit conveniently into any of the groupings, although in structural respects it most resembles those palms having ligules and elongated apocoles. The palm is adapted to a seasonal climate, but it appears that its ecological adaptation is a physiological one (see p. 136) and that a strongly developed apocole would be of no value to it.

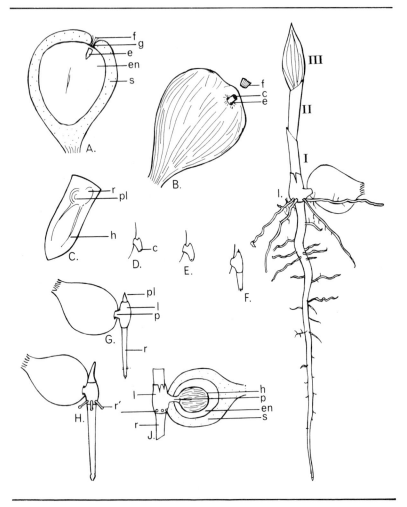

Fig. 2.1 Oil palm seed and early growth of seedling (Rees, 1960). **A.** Longitudinal section of seed through embryo. **B.** Just germinated seed. **C.** Median longitudinal section of embryo. **D, E, F, G.** Successive stages in the early growth of the embryo. **H.** Production of adventitious roots. **I.** Four-week-old seedling. **J.** Section of seed at stage I to show haustorium. c, cap of testa; e, embryo; en, endosperm; f, fibre plug; g, germ pore; h, haustorium; l, ligule; p, petiole; pl, plumule; r, radicle; r′, adventitious root; s, shell; I–III, plumular leaves.

Within the seed the haustorium develops steadily. This organ has a yellow pigment and is convoluted along the long axis of the nut thus providing a greater surface area for absorption. After about 3 months the spongy haustorium has absorbed the endosperm and completely fills the nut cavity.

The seedling

The seedling[20] has 3 months to establish itself as an organism capable of photosynthesis and absorption of nutrients from the soil.

The plumule does not emerge from the plumular projection until the radicle has reached 1 cm in length. The first adventitious roots are produced in a ring just above the radicle–hypocotyl junction and they give rise to secondary roots before the first foliage leaf has emerged. The radicle continues to grow for about 6 months by which time it has reached about 15 cm in length. Thereafter the numerous primary roots develop in its place.

Two bladeless plumular sheaths are produced before a green leaf emerges. The latter is recognized by the presence of a lamina, and it emerges about 1 month after germination. Thereafter, one leaf per month is produced until the seedling is 6 months old. The 'four-leaf stage', usually regarded as suitable for transplanting a prenursery seedling to the nursery, is reached 4 months after germination. A 2-month-old seedling is shown in Fig. 2.2.

After 3 to 4 months the base of the stem becomes a swollen 'bulb' and the first true primary roots emerge from it. These are thicker than the radicle and grow at an angle of 45° from the vertical. Secondary roots grow out in all directions. During this second period in the seedling's life the leaves become successively larger and change in shape. The leaves of the adult palm are pinnate, but this form is only reached in stages. The first few leaves are lanceolate with a midrib to half their length; two veins proceed from the end of this midrib to the tip of the leaf. In later leaves a split appears between these veins and the leaf becomes bifurcate. This type of leaf is quickly followed by leaves in which splits divide the laminae between the veins into leaflets or pinnae, although the latter are still joined to one another at the apex. Later still the leaflets become entirely free, though when the leaf opens the tip of the leaflet is always the last part to become entirely unattached.

Young pinnate leaves differ from maturer leaves − to be described later − in the following respects: the leaflets are inserted directly on to the midrib, without pulvini; the lower leaflets do not degenerate into spines; they are less xeromorphic than mature leaves.

The development of the stem and stem apex

In common with other palms, early growth of the oil palm after the seedling stage involves the formation of a wide stem-base without internodal elongation. A broad base is formed on which the stem column can rest firmly. Thus, although young palms in which stem-base formation is not yet complete may be blown over by high winds, the mature palm's stem is rarely disturbed. Where palms are found leaning over, or are lying on the ground, this is due to soil shrinkage following drainage or loss of topsoil, or both.

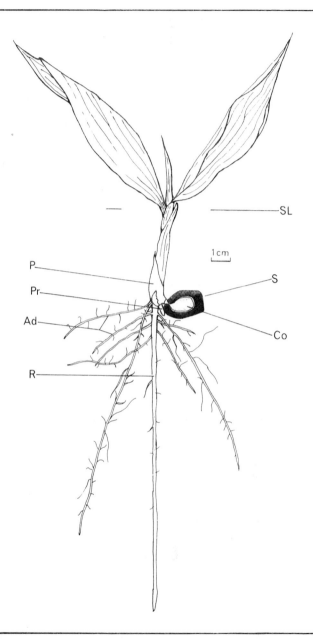

Fig. 2.2 Young seedling 2 months after planting. SL. Soil level. S. Shell. Co. Cotyledon (haustorium). P. Plumule. R. Radicle. Pr. Primary root. Ad. Adventitious roots. The nut is shown in section (*J. of W.A.I.F.O.R.*, 1956).

The palm has one terminal growing point. Very occasionally branched palms appear which develop two or more growing points or stems. It is believed that this branching is the result of damage to the apical cells resulting in the formation of two or more primordia with the independent power of growth and differentiation. The separate stems grow vertically, close together.

The apical meristem lies in a basin-like depression at the apex of the stem. In mature palms this depression is 10—12 cm in diameter and 2.5—4 cm deep. The apex itself is conical and is buried in the crown of the palm within a soft mass of young leaves and leaf bases commonly known as the 'cabbage', which is edible. The young leaves, which are yet to elongate, are largely composed of leaf bases with lateral extensions. The remainder of the leaf is reduced to small apical corrugations. The depression in which the apex lies is the result of the peculiar method of primary growth of palms which has been described by Tomlinson[1] as follows:

> . . . the stem virtually completes its thickening growth before inter-nodal elongation occurs. The apical meristem proper contributes little to the stem tissues but is largely a leaf-producing meristem. The tubular bases of the leaf primordia increase in diameter to keep pace with the increase in diameter of the nodes on which they are inserted. This thickening growth is brought about by the activity of a meristem which is continuous beneath successive leaf bases and in which cell-division is largely in a tangential plane. Since it brings about increase only in diameter it is known as a primary thickening meristem and internodal elongation only begins where its activity has ended, that is below leaves the bases of which have widened to reach its outer margin and where the stem has almost achieved its maximum diameter.

There are as many as fifty leaves from the centre of the depression to the highest point of the rim (Plate 5).

During the early years, while the wide stem base is being formed, the base of the stem assumes the shape of the inverted cone. It is from this cone that the adventitious primary roots are continually being formed both below ground and slightly above it. As soon as the internodes begin to elongate a columnar stem with adhering leaf bases is formed. Although each stem segment may be described as an internode plus leaf, it should be mentioned that the node is only indicated externally on old palms by the leaf scar; internally there is no boundary between adjacent internodes. The leaf bases adhere to the stem for at least 12 years, sometimes much longer. They fall away gradually starting to fall from the base, the crown or the middle of the stem. When all leaf bases except a few near the crown have been lost, the palm is said to be 'smooth-stemmed' instead of 'rough-stemmed'. In a palm grove, a palm rarely becomes smooth-stemmed until it has grown, at least partially, above the surrounding vegetation and is in

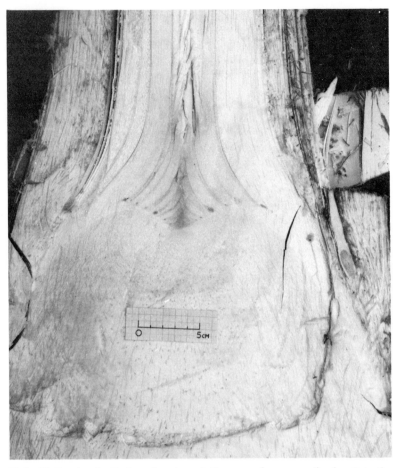

Pl. 5 Median longitudinal section through the apex of a young oil palm. Note the basin-like depression containing the apex and young unelongated leaves.

bearing. On the smooth-stemmed palms the scars of the leaf bases and those of the leaf sheaths (which encircle the stem) are clearly seen.

The manner in which the leaves are arranged with regard to the axis of the palm is known as its phyllotaxis. The leaves are produced at the apex in an orderly arrangement which, seen from above, is very roughly triangular. A fourth leaf in order of production does not, however, fall into place exactly above the first since the angle two successive leaves make with the axis (the divergence angle) varies about a mean of 137.5°. The arrangement therefore gives rise to sets of spirals or parastichies. This is illustrated in Fig. 2.3.[21, 22]

In well-grown plants two sets of spirals can be seen, eight running one way and thirteen the other.[22] Such an arrangement is described as (8 + 13).

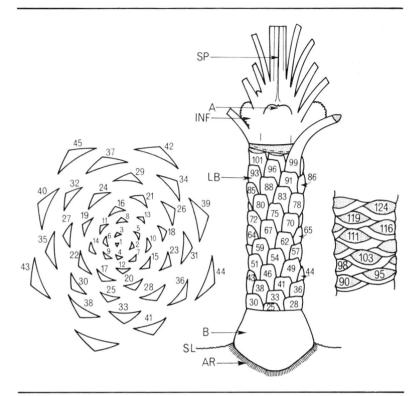

Fig. 2.3 *Left*, diagram of the phyllotaxy of the oil palm, adapted from Henry. *Centre*, diagrammatic representation of the stem; the sectioned upper portion shows the apex (A) surrounded by leaves, the spear (SP) above and the mature leaves with inflorescences (INF) laterally. The leaf bases are numbered in chronological order of formation from the base upwards and the 5, 8, 13 and other contact parastichies can be seen. Leaf bases have been omitted from the bulbous base of the palm (B). SL, Soil level. AR, Adventitious roots. *Right*, diagram of a portion of a stem from which the leaf bases have dropped. In contrast to the stem of the central diagram the 5 contact parastichy ascends from left to right, the 8 contact parastichy from right to left. The leaf sheath scars have been omitted for simplicity (*J. of W.A.I.F.O.R.*, 1961).

If the leaf bases are numbered in the order of leaf formation (the 'genetic spiral') this becomes clear, since, one way, every eighth leaf is seen to be in the same spiral while, the other way, every thirteenth leaf appears in the same (more nearly vertical) spiral. The distance between two leaves in a thirteen spiral or parastichy is a measure of the time of production of thirteen leaves and is termed thirteen plastochrones. Other parastichies can be seen on the palm (see Fig. 2.3) — three for instance, but the larger the parastichy number the more nearly it approaches the vertical. For instance a near vertical twenty-one parastichy can be distinguished on the diagram.

The foliar spirals are in either direction, left-handed or right-handed; in two surveys in Malaysia nearly 53 per cent of the palms were left-handed, but there was evidence that this character was not genetically determined.[23]

More modern studies of the phyllotaxis of the oil palm have been based on Richards's *Phyllotaxis Index* which in turn is calculated from the Plastochrone Ratio, the latter being the ratio between the transverse distance of a primordium from the centre and that of the immediately preceding primordium. More simply, an equivalent phyllotaxis index can be calculated from the radius of the palm cylinder and the longitudinal distance separating two consecutive leaf insertions on the genetic spiral. Modifications of this index result from differing rates of longitudinal and radial growth during development[24] and may be attributable to physiological factors.[25]

The rate of extension of the stem is very variable and depends on both environmental and hereditary factors. Under extreme shade, growth of both leaves and stem is very slow indeed; in dense plantations or secondary bush, stem growth may be very rapid and the palm will assume an elongated appearance. Under normal plantation conditions, and particularly with heterogeneous planting material, there are often marked palm-to-palm differences, but the average increase in height will be from 0.3 to 0.6 m per year.

In high forest, palms may reach a height of 30 m but elsewhere they reach no more than 15 or 18 m. It is not possible to tell the age of individual grove palms since under heavy shade seedlings and young palms grow very slowly indeed. It is believed that many palms may be 200 years old or more. Of planted palms, the two surviving original Deli palms at Bogor, Indonesia, are more than 120 years old.

The width of the stem, unclothed by leaf bases, varies from 20 to 75 cm. In the Deli palm the diameter is said to vary from 45 to 60 cm[26] but the stems of 'Dumpy' palm progeny are much wider. In plantations the stem, after the initial bulge, is often remarkably uniform in width, but uneven stems are commonly seen in palm groves. This uneveness is due to alterations in the usage of the surrounding land, and probably also to the scorching of the crown during firing of the cut surrounding bush or to excessive wine tapping.

The stem functions as a supporting, vascular and storage organ. A wide central cylinder is separated from a very narrow cortex through which the leaf traces pass. The cylinder has a wide peripheral zone of congested vascular bundles with fibrous phloem sheaths, and the intervening parenchyma cells are sclerotic; thus this zone provides the main mechanical support of the stem. The vascular bundles are much less congested in the central zone. In common with other palms there is no cambium or callus formation and, although some palms show slight secondary thickening by cell division and expansion, this activity is not exhibited by the oil palm. Starch grains and silica-containing cells (stegmata) are abundant.

The courses taken by the vascular bundles within the stem are naturally of importance in the supply of nutrients to the crown by long-distance translocation. Early nineteenth century workers made considerable progress in unravelling the vascular system, but a fuller understanding had to await the recent imaginative technique of Zimmerman and Tomlinson working with the small palm *Rhapis excelsa* while conjointly carrying out examinations of stems of larger palms.[27] The general pattern is believed to be essentially similar in the oil palm and other large palms and may be briefly described from these authors' work. Firstly the continuity of the vascular bundles through the stem has been demonstrated and secondly the relationships of the vertical bundles with the leaf traces and with each other has been clarified. All bundles maintain their individuality and proceed indefinitely up the stem, giving off leaf traces at intervals. In proceeding up the stem the bundle slants towards the centre from the periphery and then bends sharply back towards the periphery and divides into several branches. One branch is the leaf trace which proceeds into the leaf. Others go into the inflorescence, others 'bridge' into neighbouring bundles while another bends vertically again as the vertical bundle and the process is repeated. It is this course, being followed by many thousands of bundles, which accounts for the crowding at the periphery and the even but sparser distribution in the centre. Finally, it has been shown that in the central uncrowded part of the stem the bundles do not remain on one side of the stem but take a spiral or helical course.

The leaf

The growth of the seedling leaf and the arrangement of the leaves around the stem have already been described. In the crown of an adult palm a continuous succession of leaf buds or primordia are being separated laterally from the apical cone and the rate of development of these leaves has been studied.[28-30] Development is initially very slow. Some forty-five to fifty leaves are to be found in the crown; each leaf remains enclosed for about 2 years and then rapidly develops into a central spear and finally opens. The base of the leaf completely encircles the stem apex and in the adult leaf the base is persistent as a strong fibrous sheet.

The mature leaf is simply-pinnate, bearing linear leaflets or pinnae on each side of the leaf stalk.[31] The latter may be divided into two zones, the rachis bearing the leaflets, and the petiole which is much shorter than the rachis and bears only short lateral spines (Fig. 2.4). At the junction of petiole and rachis small leaflets with vestigial laminae (leaf blades) are found. Petioles vary greatly in length and in the Deli palm may be as long as 4 feet. Some petioles remain green for a considerable period.

The spines have been shown to be of two kinds which are named fibre spines and midrib spines (Fig. 2.5). The former are those on the petiole; they are very regular and are formed from the base of the fibres of the leaf sheath. The point at which these fibres break off is very regular, so the

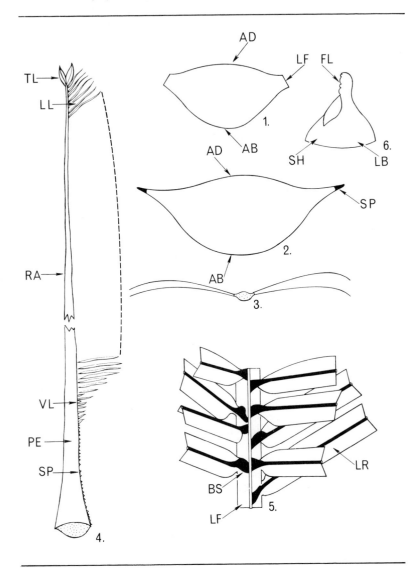

Fig. 2.4   The oil palm leaf. **1.** Cross-section of rachis. **2.** Cross-section of petiole. **3.** Cross-section of leaf viewed end-on, showing two-ranked insertion of leaflets. **4.** Diagram of oil palm leaf. **5.** Central portion of rachis from above, showing irregular leaflet insertion. **6.** Leaf apex of palm (*J. of W.A.I.F.O.R.*, 1962).

AD, Adaxial face; AB, Abaxial face; LF, Lateral face; SP, Spine; RA, Rachis; PE, Petiole; TL, Terminal pair of ovate leaflets; VL, Leaflets with vestigial laminae; LB, Leaf base; FL, Future green leaf; SH, Leaf sheath completely encircling apex and through which younger leaves and stem grow; BS, Basal swelling; LR, Lower rank leaflet.

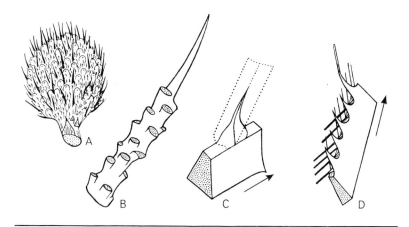

Fig. 2.5   Spines in the oil palm. Diagrams of **A**, fruit bunch with spiny branch tips; **B**, single branch of inflorescence with fruit removed; **C**, midrib spine, leaflet shown by dotted outline; **D**, fibre spines viewed from abaxial side. The arrows in C and D point to the leaf tip (Rees, 1963).

spines are nearly all the same length. Where the leaflets begin to occur they are poorly developed although they have the basal swellings of fully developed leaflets. The lamina of these poorly-developed leaflets frequently becomes torn away, leaving a spine which was originally the leaflet midrib. These spines have the same irregularity of set as have the fully developed leaflets on the leaf.[32]

The leaf stalk is a hard fibrous body which may be as long as 8 m. At the tip it is almost circular in cross section, but in the centre of the rachis it is asymmetrical with lateral faces where the leaflets are inserted. In the petiole the lateral faces disappear. The lower or abaxial face is much more strongly curved than the upper or adaxial face.

The leaflets are produced by the splitting of an entire leaf during the elongation of the leaf axis. Within the spear the leaflets are still attached to one another but are folded upwards and show clearly where the splitting is to occur. In an actively growing plant spears are produced one at a time and point vertically upwards. As the spear opens, another elongates rapidly to take its place. In severe dry seasons, however, and this is very noticeable in West Africa from January to March, many spears may elongate before the first of their number opens. In these circumstances it is not unusual to see half a dozen or more spears, many of them fully or nearly fully elongated, protruding from the centre of the crown. As soon as wet weather ensues the majority of these spears will open and the upper part of the crown takes on a light green appearance for a short period.

After the leaf has opened it is progressively displaced centrifugally as younger leaves emerge. Middle-aged leaves lie parallel to the ground with the tip bearing slightly downward. Usually the adaxial face of the rachis faces upwards, but sometimes the leaf twists into a vertical plane or intermediate position.

Typically, the leaflets inserted on the lateral faces alternate in upper and lower ranks. There is no exact regularity however, and two or more consecutive leaflets may appear in the same rank. Similarly within each rank the angle of insertion is often irregular, and very occasionally there is almost no 'ranking effect'. Generally, however, it is the provision of two ranks and the irregularity of leaflet insertion which gives the palm its shaggy appearance and distinguishes it, from a distance, from the coconut palm or *E. oleifera*. Individual leaflets are linear in shape and each leaf has a terminal pair. Leaflets number some 250–300 per mature leaf and are up to 1.3 m long and 6 cm broad. The leaflet midrib is often very rigid and the laminae sometimes tear backwards from the tip. This increases the 'untidy' appearance of the leaf. There is a small basal swelling, resembling a pulvinus but with no motor function, at the insertion of the leaflet on the rachis.

Apart from the leaflet variation noted above, more striking leaf variations are to be found. In the *idolatrica* palm the leaflets do not separate normally and an entire or semi-entire leaf is formed. The midribs of all the unseparated 'leaflets' are in one plane. There is still some doubt concerning the inheritance of the *idolatrica* character, but self-pollinated *idolatrica* palms have bred true and the character is thought to be recessive.[12, 33] Observations have suggested that the centre of distribution of the *idolatrica* palm lies between Ghana and the lower Niger. Many of these palms are found in Dahomey and western Nigeria. Westwards and eastwards specimens are rarer and are often found only in botanic gardens or agricultural stations (Plate 6).

Other leaf peculiarities occur but have been insufficiently studied. They may be due to genetic, nutritional or pathogenic factors. Some will be described in other sections of this book.

The number of leaves produced annually by a plantation palm increases to between thirty and forty at 5 or 6 years of age. Thereafter the production declines to a level of twenty to twenty-five per annum. Leaf production of grove palms is much lower (see Ch. 3).

In the axil of each leaf there is a bud which may develop into a male, female or, occasionally, a hermaphrodite inflorescence. Very rare cases have been known, however, in which a vegetative shoot is produced instead of an inflorescence. This has been termed 'vivipary' by Henry who has described[34, 35] an original palm at Okeita, Dahomey, the shoots taken and developed from it, and similar palms in the Ivory Coast. While in some cases the shoots from 'viviparous' palms can be rooted and will produce similar vegetative-shoot-producing palms, in other cases no roots are formed and sexual buds are later produced. In the latter cases there are also considerable malformations of both the vegetative and sexual parts of the shoots.

Pl. 6 The *idolatrica* palm, with fused leaflets.

The root system

The seedling radicle is soon replaced by adventitious primary roots emanating from the radicle—hypocotyl junction and then from the lower internodes of the stem which are formed into a massive basal cone or bole. The latter retains the capacity for producing roots well above ground level. Roots sometimes develop on the stem up to 1 m above ground but these normally dry out before reaching the soil.

In the mature palm thousands of primary roots spread rapidly from the bole. New primaries are continually replacing dead ones. A number of studies of the root system have been made[36-45] and it has been shown that its vertical extent depends very largely on the presence or absence of a water table. Two extremes may be cited. In Malaysia, Lambourne[38] studied the roots of 11-year-old palms growing in soil where the water table was as high as 1 m from the surface in dry weather. In these circumstances no primaries penetrated below this depth and the majority of roots were in the surface 45 cm. Individual primaries were found to a distance of 19 m from the stem and absorbing portions of roots were found at all intermediate distances. In contrast, Vine[40] and Purvis[39] examined root systems in free-draining sandy soils and found that primary roots may descend to great depths. It is this unimpeded root system which will now be described.

Primary roots extend either downwards from the base of the palm or radially in a more or less horizontal direction (Fig. 2.6). The descending primaries, which proceed directly from under the base of the palm are fewer in number than the radiating primaries and carry much fewer secondaries. Ruer[41] has shown that these descending roots are for anchorage and play little or no part in the absorption of water.

The remaining primary roots appear from the base of the stem at all angles to the soil surface, but they tend to bend to the horizontal and few are found below 1 m. From these primary roots, 5—10 mm in diameter, secondary roots ascend and descend in approximately equal numbers, though with a slight preponderance of ascending roots. These secondaries are 1—4 mm in diameter and give rise in turn to horizontally growing tertiaries of 0.5—1.5 mm diameter and up to 15 cm in length. From these are developed the mass of quaternaries of up to 3 cm in length and only 0.2—0.5 mm diameter.

The ascending secondaries generally reach the surface of the soil while the descending ones may penetrate to a considerable depth.

The density of all classes of roots in the top 40 cm of soil usually decreases with distance from the palm, but with adult palms the total quantity of absorbing roots in successive surrounding circles increases at least to a radius of 3.50—4.50 m.[42] The greatest quantity of roots is in the top 15—30 cm of soil and most of the absorption of nutrients has been shown to be through the quaternaries and absorbing tips of primaries, secondaries and tertiaries to this depth.[43]

Roots of all classes show a positive tropism towards superior conditions of water and nutrient supply and, with rotting felled vegetation or heaps of palm leaves, or under a good *Pueraria* cover, this may lead to a high density of quaternaries in the centre of the interline.[44] For instance, with a *Pueraria* cover and on good alluvial soil in Colombia tertiaries and quaternaries increased with distance from the palm, but where there was a grass cover the quantity of these roots declined with distance. Similarly, the quantity of roots is much reduced under the paths along the lines. Where the rooting volume is reduced by quantities of concretionary gravel primary roots tend to become twisted and constricted, the root system lies nearer the surface and the tertiaries and quaternaries are coarser and more lignified.[43] Sub-aerial roots which grow up into loose decaying leaves are readily produced.

The anatomy of palm roots is described by Tomlinson and that of the oil palm in particular has been studied by Purvis. The primary root consists of an outer epidermis and lignified hypodermis surrounding a cortex in

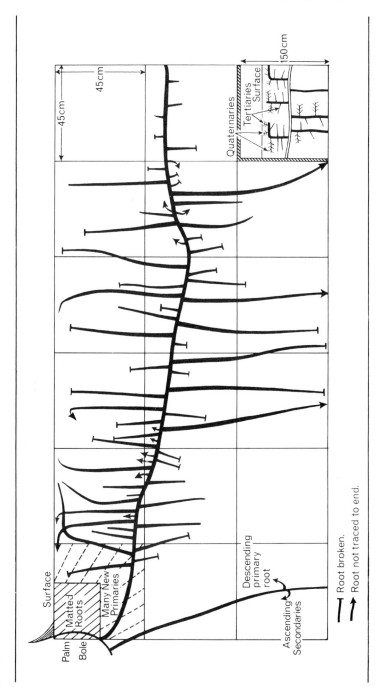

Fig. 2.6 Distribution of secondary roots from one primary root in a 3-year-old palm (Purvis). The drawing also shows the usual position of a descending primary root. *Inset* is a diagram of the development of tertiary and quaternary roots.

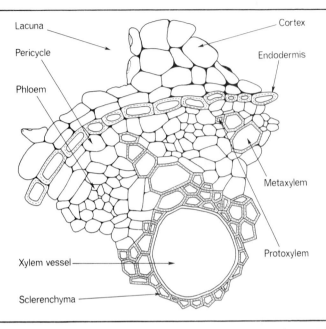

Fig. 2.7   Transverse section of middle-aged seedling root to show stele (Purvis).

which well-developed air lacunae are to be found. Within the cortex lies the central stele or vascular cylinder consisting of the surrounding lignified endodermis, the inner vascular strands of xylem and phloem and the pith or medulla which rapidly lignifies in old roots (Figs 2.7 and 2.8). The stele also contains lacunae. The secondary and tertiary roots have essentially the same structure as the primary roots. The unlignified tips of the growing primary, secondary and tertiary roots measure 3–4, 5–6 and 2–3 cm respectively. The quaternary roots are only 1–3 cm long, are produced in large numbers and are almost wholly unlignified. There are no root hairs and it is therefore reasonable to suppose that quaternary roots play the main part in the absorption of nutrients. Moreau has studied in some detail the lignification of oil palm roots and the anatomy of production of the substituting roots which are formed when a seedling root is damaged or affected by disease.[46] Such roots appear from just behind the point of damage or zone of disease and proceed in the same direction as the original root.

The roots of *E. guineensis* (and other palms) are characterized by the presence of pneumathodes. These, although appearing on both underground and aerial roots, have been supposed to ventilate the underground roots: direct physiological evidence for this is lacking. Yampolsky[37] found more pneumathodes on aerial than on underground roots in Sumatra, but the reverse is the case in West Africa. Moreover, they are commonest on

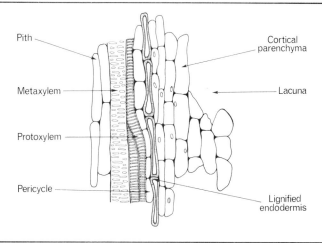

Fig. 2.8 Radial longitudinal section of old seedling primary root to show outer stele and inner cortex (Purvis).

seedlings grown in glasshouses or wherever the root system has been kept under water or in very moist conditions.

In pneumathode-forming root shoots the epidermis and hypodermis rupture and the stele and cortex extrude. The latter then proliferates and its parenchymatous cells become suberised or, if the pneumathode is aerial or subjected to dry conditions, lignified (Fig. 2.9). If the growing point is unharmed after the rupture of the epidermis it remains attached as a cap and sometimes a normal root may develop again.

The firm anchorage of the adult palm is not only due to the descending primary roots. The old roots are strong and elastic and persist in the soil long after they have died. When death of a root occurs the cortex degenerates leaving a tubular hypodermis with cortical fibres and the woody stele loose within.

The flowers and fruit

As the commercial products of the oil palm are obtained from the fruit, the flowering and fruiting habits of the plant are of prime interest and importance. The oil palm is said to be *monoecious*, that is to say male and female flowers occur separately — and in this case in distinct male and female inflorescences — on the same plant. Detailed investigation of the flowers has shown, however, that each flower primordium is a potential producer of both male and female organs though one or the other almost always remains rudimentary.[47] In very rare cases both the androecium and the gynoecium develop fully to give a hermaphrodite flower. Secondly, the formation of hermaphrodite inflorescences sometimes occurs, particularly

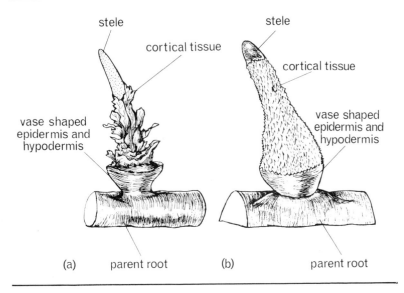

Fig. 2.9   Young, underground (a) and old, aerial (b) pneumathodes (Purvis).

in the early life of a palm. Thirdly, each female flower is flanked by two accompanying male flowers which normally do not develop but which very occasionally reach a stage of development at which pollen may be produced. These aberrations may be of importance to the plant breeder, and it should be borne in mind that, strictly, there is no female inflorescence; the term is in reality applied to the large number of 'normal' inflorescences which are found to be composed, when they reach sexual maturity at anthesis, of female flowers alone.

An inflorescence is initiated in the axil of every leaf but some inflorescences abort before emergence. Twin inflorescences in the leaf axil have been known. Each inflorescence is a compound spike or spadix carried on a stout peduncle 30—45 cm in length. Spikelets are arranged spirally around a central rachis in a manner which varies both with age and position on the rachis; however, equivalent phyllotaxis index measurements have shown little difference between male and female.[43] An inner and outer spathe tightly enclose the inflorescence until about 6 weeks before anthesis when the inner spathe begins to open. After a further 2 or 3 weeks the inner spathe splits; later both spathes fray and disintegrate and the inflorescence pushes its way through. Tattered remains of the spathes may, however, remain around the inflorescence too long and, particularly in young palms, this may hinder fruit setting. Six to ten long bracts are found below the lowest spikelet; two of them extend to the top of the inflorescence.

An inflorescence can be male, female or hermaphrodite and the order and proportions in which these are produced show little or no regularity. A widely varying number of female inflorescences is followed by an indefinite series of male inflorescences. Occasionally there is a hermaphrodite inflorescence between the two series and these inflorescences are commoner in young palms. The factors influencing the development of male and female inflorescences will be discussed in Chapter 4.

The number of spikelets per inflorescence varies greatly from palm to palm, but Beirnaert[49] has shown that the variation in this respect between inflorescences of a given palm is very small and is independent of the sex of the inflorescence. In thirty-seven adult palms in Zaire the average number of spikelets per inflorescence was found to range from 100 to 283 and in almost all cases the coefficient of within-palm variation was very small indeed. In hermaphrodite inflorescences the sum of the male, female and mixed spikelets is always close to the average number of spikelets for male or female inflorescences of the palm concerned. An examination of a few inflorescences from a very much larger number (1,476) of Zaire palms showed that, though the range was great (85 to 285 spikelets), inflorescences with 125 to 165 spikelets were most frequent. The number of spikelets per bunch is smaller in the case of Nigerian palms, while with Deli palms the variation is said to be between 100 and 200 per inflorescence.

The female inflorescence and flower

The female inflorescence reaches a length of 30 cm or more before opening. The female spikelets are thick and fleshy and develop in the axil of a spinous bract. The flowers are arranged spirally around the rachis of the spikelet; each is housed in a shallow cavity and subtended by a bract which is drawn up into a spine. At the end of the spikelet there is a spine of very variable length. The number of flowers in an inflorescence varies from palm to palm but in all cases there is a much larger number (twelve to thirty) on the central spikelets than on the lower or upper spikelets (twelve or less). The inflorescences will thus contain several thousand flowers.

Each cavity, or alveola, contains the female flower bud and the two accompanying small male flowers which normally abort. All three are enclosed by the floral bract (see Fig. 2.10). The tricarpellate ovary and rudimentary androecium of the female flower are enclosed by a double perianth of six sepaloid segments in two whorls; these in turn lie within two bracteoles. The sessile stigma has three lobes; these are hairy, and they exude moisture at the receptive stage. The sepals are about 2 cm long at anthesis. The rudimentary androecium has six to ten short projections; Beirnaert has described in detail the range of development of the rare hermaphrodite flowers which may occur both on the male and female inflorescences.

Occasionally two female flowers may develop, within a single pair of bracteoles, between the two accompanying abortive male flowers Another

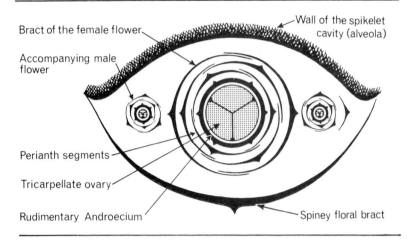

Fig. 2.10 Vanderweyen's diagrammatic representation of the female flower with accompanying rudimentary male flowers, in Beirnaert's *Introduction à la biologie florale du palmier à huile*, 1935.

unusual occurrence is the production of female flowers on the end of long peduncle-like stalks which are inserted directly on the rachis of the inflorescence. Inflorescence abnormalities are by no means uncommon in the oil palm and the tendency to abnormality must be taken into account by plant breeders (Plate 7).

The male inflorescence and flower

The male inflorescence is borne on a longer peduncle than that of the female inflorescence and contains long finger-like cylindrical spikelets. It is not spiney. The spikelet has bracts and terminal projections but these are much reduced in size. Spikelets measure between 10 and 20 cm in length.

A spikelet or finger of average size will have between 700 and 1,200 male flowers which are much shorter than the female flowers. Before opening, the sessile flower is completely enclosed in a triangular bract; it consists of a perianth of six minute segments, a tubular androecium with six, or rarely seven, anthers, and a rudimentary gynoecium with three projections corresponding to the trilobed stigma. Flowers begin to open from the base of the spikelet.

It has been shown in Malaysia that all flowers usually open within 2 days though during rainy weather opening may be prolonged to 4 days; most pollen is shed during 2 or 3 days following the start of anthesis and production ceases within 5 days. Viability of late-produced pollen is low.[50] Inflorescences produce 25–50 g of pollen.

Hermaphrodite or mixed inflorescences

There is a great variety of hermaphrodite or mixed inflorescences formed.

Pl. 7 Male and female inflorescences at two stages of development: **A.** Before anthesis within (1) and freed from (2) the spathes. **B.** At anthesis.

Below. Some abnormal pollen-producing flowers at the end of female spikelets.

Usually male, female and mixed spikelets appear on the same bunch but in widely differing proportions and positions. Some palms are more disposed to the production of these inflorescences than others.

The mixed spikelets have, in varying proportions, female flowers at the base and male flowers at the summit. In between, 'geminiflorous' male flowers, corresponding to the accompanying male flowers, lie close together with no female flower between them. Further up the spikelet these give place to the normal single male flowers. Spikelets can contain either all three types of flower group, or male flowers and geminiflorous flowers alone, or they may have the superficial appearance of a female spikelet but contain geminiflorous flowers.

A transition from femaleness to maleness may thus be (i) through the hermaphrodite flower, (ii) through the replacement of the female plus accompanying abortive male flowers by a pair of male flowers or (iii) through the development of part of the female spikelet as a male finger. It has been suggested that these transitions may be both genetic and physiological in origin.

Young palms occasionally produce a peculiar type of inflorescence which has been called andromorphic. This has all the appearance and structure of a male inflorescence before it opens. Examination shows, however, that the male flowers have been replaced by small solitary female flowers arranged in the manner of flowers in a male inflorescence. Small fruit develop from the flowers, but the carpels are not firmly joined together and the resulting fruit have three lobes corresponding to the three partially separated carpels. Male flowers are also to be found in andromorphic inflorescences though some are deformed.

The fruit and fruit bunch

A short description of the morphology of the fruit is given here. Fruit variation and the inheritance of fruit characters are discussed more fully in Chapter 5. It will by now be evident that the oil palm is a very variable plant and it is therefore not surprising that the fruit varies as much as, or more than, any other organ.

The fruit is a sessile drupe varying in shape from nearly spherical to ovoid or elongated and bulging somewhat at the top. In length it varies from about 2 to more than 5 cm, in weight from 3 g to over 30 g. The Deli fruit of the Far East are usually considerably larger than the fruit of Africa, though contrary to general belief fruit as large as Deli fruit are not uncommonly encountered in Africa.

The *pericarp* of the fruit consists of the outer *exocarp* or skin, the *mesocarp* or pulp (often incorrectly termed the pericarp — hence 'pericarp oil'), and the *endocarp* or shell. When measuring the pulp, the exocarp is included with the mesocarp. The endocarp together with the kernel forms the seed which has already been described.

There is one abnormal fruit type, variously known as *Poissoni*, mantled, and Diwakkawakka, in which fleshy outgrowths or supplementary carpels

Fig. 2.11 Longitudinal sections of mantled fruit; *left*, after Beirnaert (1941); *right*, after Janssens (1927), in which one of the supplementary carpels shows complete development.

surround the main part of the fruit. These have developed from the rudimentary lobes of the androecium of the female flower, but they are considered to be carpellary in character since they are often found to contain shell and kernel in the centre (Fig. 2.11). Mantled fruit are rare; in one area of Nigeria only 33 mantled bunches were found from among 20,291 bunches harvested from grove plots over a 4-year period. In Angola the frequency was found to be 9 palms in 10,000. Similar figures have been quoted elsewhere.[51]

In external appearance the fruit varies considerably, particularly when ripening (Plates I and IV, between pp. 78 and 79). Moreover, the exocarp of the external fruit tends to be more pigmented than that of the internal fruit. By far the commonest type of fruit is deep violet to black at the apex and colourless at the base before ripening. Such fruit has been described as 'ordinary' or *nigrescens*. A relatively uncommon type is green before ripening, and this is called green-fruited or *virescens*. The latter changes at maturity to a light reddish-orange though the apex of the external fruit remains greenish. The frequency of the *virescens* type was found to be 50 in 10,000 bunches in a grove area in Nigeria and 72 in 10,000 in Angola.

The colour of the ordinary or *nigrescens* fruit varies to an appreciable extent on ripening and there is evidence that this is connected with carotene content. This colour difference in ripening was recognized at an early date by Chevalier[8] who gave the names *communis* to 'fruit entirely red when ripe or with a small black or brown halo at the tip' and *semper-nigra* to 'fruit when ripe, black over the upper half but red at the base'. These differences are also recognized by some of the peoples in West Africa where different vernacular names are allotted to them (e.g. Abepa and Abetuntum in the Fanti and Twi languages of Ghana). The names *rubro-nigrescens* and *rutilo-nigrescens* have been proposed by Purvis.[52] The former fruit, when ripe, are defined as 'Cap − 00918, Garnet brown, sometimes tending to be darker, rarely extending over half the fruit; Base − 713, Indian orange, colour uniform to the base'. *Rutilo-nigrescens* is defined as 'Cap − Black, though it may show a brownish tinge at the edges, usually covering more than half the fruit; Base — colour not constant, tending to lighten towards base, the deepest being 13, Saturn Red'. The colours and colour numbers refer to the Wilson Horticultural Colour Chart.* Nevertheless it is not a simple matter to allot fruit with any certainty to one or other of these subtypes and fruit of intermediate appearance can invariably be found. Moreover, the cap colour is deceptive; there are no real black caps since even the darkest when seen through transmitted light are reddish in colour.

The above description refers to the colour appearance of fruit commonly or occasionally found in palm groves and plantations. There is, however, a much more fundamental colour variation due to presence or absence of carotenoids. The *albescens* fruit, characterized by 'absence' of carotene in the mesocarp, is extremely rare. Actually, this fruit does contain a very small quantity of carotene. It was first noted in Ghana, under the name Abefita, but was later named *albescens* by Beccari.[53] It has been subsequently found in Zaire, Angola, Nigeria, the Ivory Coast and other parts of Africa. In Angola the frequency was found to be only 3 in 10,000 and it may be rarer in other parts of Africa.

Albescens fruit may be of *nigrescens* or *virescens* type; in Zaire[54] the fruits are referred to as *albo-nigrescens* and *albo-virescens*. The difference is only in the cap of the fruit, the former's cap being dark brown to black in appearance, the latter's green. The rest of the fruit is ivory coloured, ripening pale yellow. Only a very few *albo-virescens* palms have been found. *Nigrescens* and *virescens* fruit contain varying quantities of carotenoids in the mesocarp and this will be referred to again in Chapter 14. Exterior fruit may have as much as twice the carotene content of interior fruit.

In internal structure the most important differences are to be found in thickness of shell (Plate I). As shells of all thicknesses from less than 1mm to 8mm can be found it might be thought that a division of fruit into thick-shelled and thin-shelled forms would be somewhat arbitrary. However, a rare shell-less form was early noted in Africa and named *pisifera* owing to the pea-like shell-less kernels found in fertile fruit. *Pisifera* palms always bear large quantities of female bunches. In many cases the majority of the bunches rot; these are known as infertile *pisifera* though the setting of a few fruit is of course necessary to identify them, as other abortive forms are encountered. Infertile *pisifera* tend to show strong vegetative growth. Fertile *pisifera* palms are less common.

Apart from the discovery of the *pisifera* it was also noted that in the majority of the thinner-shelled fruit there was a distinct ring of fibres embedded in the mesocarp but near to and encircling the nut. This can be clearly seen when the fruit is cross-sectioned. In 1935 Smith[11] recorded

---

* The Wilson nomenclature may cause some confusion since the full hue 13, an orange colour, is called 'Saturn red' while its deeper shade 713, which appears much redder, is named 'Indian orange'.

that 'the present dividing line between thick- and thin-shelled forms is that the fruit of the latter contain the mesocarp fibre ring and the nuts can be readily cracked'. Subsequent genetical studies have shown that the thin-shelled form with a fibre ring — the *tenera* form — is a hybrid of the shell-less *pisifera* and the common thick-shelled *dura* form which has no fibre ring. Internal fruit form may therefore be described as being either:

(*a*) *Dura*: shell usually 2—8 mm thick though occasionally less, low to medium mesocarp content (35—55 per cent but sometimes, in the Deli *dura*, up to 65 per cent); no fibre ring;
(*b*) *Tenera*: shell 0.5—4 mm thick, medium to high mesocarp content (60—96 per cent, but occasionally as low as 55 per cent); fibre ring; or
(*c*) *Pisifera*: shell-less.

The term *macrocarya* has been used for *dura* palms with shell thickness of 6—8 mm, but the term has largely gone out of use as it has no genetical significance. It must be said, however, that in many parts of West Africa (e.g. Sierra Leone and western Nigeria) fruit which could be described as *macrocarya* forms a large proportion of the crop. In a grove survey in eastern Nigeria 27 per cent of the *dura* palms were classed as *macrocarya* and there are undoubtedly much higher proportions elsewhere.

The mesocarp of all fruit contains fibres which run longitudinally through the oil-bearing tissue. This fibrous material usually constitutes about 16 per cent of the mesocarp but may vary from 11 to 21 per cent. The relative constancy of the fibre percentage has been used to devise a method for indirect estimation of oil percentage. The latter is found to vary from about 35 to 60 per cent. The fibres of the fibre ring in *tenera* fruit are dark in colour; dark fibres may also be distributed in other parts of the pulp though they are usually in the central section. Light-coloured fibres are distributed regularly throughout the mesocarp.

The fruit bunch (Plate II, between pp. 78 and 79, and Plates 8 and 9) is ovoid and may reach 50 cm in length and 35 cm in breadth. The bunch consists of outer and inner fruit, the latter somewhat flattened and less pigmented; a few so-called parthenocarpic fruit which have developed even though fertilization has not taken place (or possibly following partial abortion); some small undeveloped non-oil-bearing 'infertile fruit'; and the bunch and spikelet stalks and spines. In the parthenocarpic fruit endo-sperm and embryo are absent and the centre is usually solid. Bunch weight varies from a few kilograms to about 100 kg according to age and situation but in adult plantations mean weights are 10—30 kg. Well set bunches carry from 500 to 4,000 fruit. a mean of about 1,500 being usual, with a fruit-to-bunch ratio of 60 to 70 per cent. Ripening is usually from the apex downwards, the fruit becoming gradually detached.

The female inflorescence at anthesis is in the axil of the seventeenth to twentieth leaf from the central spear. By the time the bunch is ripe it is subtended by about the thirtieth to thirty-second leaf, but the bunch leans out from its subtending leaf on to a leaf in a lower whorl; it is not the leaf

Pl. 8   A bunch from a young Deli palm.

Pl. 9   A bunch from a young Nigerian *dura* palm.

subtending the bunch which supports it (Plate 10). The fruit develop steadily in size and weight from about the fifteenth to ninetieth day after anthesis. Oil formation in the kernel and mesocarp takes place towards the end of a period of maturation during which the shell hardens and the embryo becomes viable (see Ch. 14, p. 696).

Pl. 10   Peduncle of a ripe bunch cut to show that the bunch was subtended by a younger leaf than the one on which the bunch lay.

The American oil palm *Elaeis oleifera* (HBK) Cortés

This plant is of interest and importance both on account of its place in the present economy of South and Central America and of its possible value for hybridization with *Elaeis guineensis*. The palm has a procumbent habit and slow stem growth and these characters have encouraged hybridization trials. The names Corozo and *melanococca* have been popularly used for the palm.

Corozo is a name used in South American to denote the nuts of all Cocoid palms. In Central America and the Caribbean the spelling Corojo is used. Many species of palms are called Corozo with a qualifying adjective, and in Colombia *E. oleifera* is commonly known as Corozo noli or sometimes noli alone. An alternative name is Corozo colorado and other alternatives exist in Spanish-speaking America.[55, 64] In Brazil it is called Caiaué or Dende do Para.

Under the name Corozo the palm was first described by Jacquin (1763) as growing in Colombia, and he illustrated the fruit with calyx. On the basis of this description and figure, Giseke, in editing Linnaeus's *Praelectiones in Ordines Naturales* (1792), adopted Corozo as the palm's generic name. However, the name *Elaeis melanococca* soon became widely used for the palm owing to its employment in a description of fruit and seeds of both *Elaeis guineensis* and a supposed other species by Gaertner in his *De*

Fructibus et Semenibus Plantarum (1788). *Melanococca* signifies 'black-berried' and Bailey believed that the seed figured by Gaertner was in fact a seed of *E. guineensis*[3] (Plate 3). This has not been disputed, but Cook[56] claimed that Giseke did not intend to adopt the name Corozo. For many years Bailey's nomenclature, *Corozo oleifera*, was generally accepted,[57, 58] but it is now agreed[59, 60] that both the ease of hybridization with *E. guineensis* and the degree of divergence between the two species justify separation on a specific rather than a generic level. The specific name *oleifera* was transferred by Bailey from the earlier *Alfonsia oleifera* of Humboldt, Bonpland and Kunth (1816).

Elaeis oleifera is still by no means well known, though there are a number of recent descriptions. The first of them, by Vanderweyen and Roels,[61] is based mainly on four plants growing at Yangambi. The other descriptions,[62, 63, 64] the result of observations in America, suggest that some of the Corozo specimens in the Congo may be atypical or *E. guineensis* x *E. oleifera* hybrids.

Elaeis oleifera is found in the tropical countries of South and Central America and has been described or collected from Brazil, Colombia, Venezuela, Panama, Costa Rica, Nicaragua, Honduras, French Guiana and Surinam. In Colombia it is found in depressions between rolling areas of pasture land (Plate 11); elsewhere it is common in damp or even swampy situations or on the banks of rivers. In all these situations it can be found in pure and dense stand, but in pasture land or in some river-bank habitats it is also found dispersed or in small groups. In Nicaragua it is reported to be growing in the Escondido valley,[64] but the palm covers large areas on

Pl. 11   A natural grove of *Elaeis oleifera* in the Sinú valley area of Cordoba Province, Colombia.

Pl. 12   Two *Elaeis oleifera* palms in the Sinú valley, Colombia. Note the coiling, recumbent trunks.

the banks of the San Juan river between Lake Nicaragua and El Castillo. In Surinam, on poor, white, sandy soil there are dense stands, but the palms are of small size.

The main feature of the palm, and one which distinguishes it from *E. guineensis*, is its procumbent trunk. An erect habit may be maintained for as long as 15 years, but thereafter a procumbent habit is generally assumed though the crown is in an erect position and the erect portion is usually 1.5–2.7 m high (Plate 12). Trunks lying on the soil for a distance of over 7.6 m have been measured. In certain areas there are types which appear to remain erect until the palm reaches a height of over 3 m. The leaf bases persist for only a short period.

Apart from its procumbency there is evidence that *E. oleifera* grows much more slowly in length of trunk than does *E. guineensis*.[62] Procumbency is not due, as is sometimes asserted, to swampy conditions; other species of palm grow erect where *E. oleifera* is to be found.

The root development of the palm is similar to that of *E. guineensis*, but roots which may grow to 1 m in length are formed along the whole length of the procumbent trunk. It has been claimed that certain anatomical differences, namely greater lignification of the hypodermis and cortical parenchyma, less lacunae and the presence of tannins in the cells of the endoderm and phloem, account for resistance to certain diseases (see p. 605).[65]

The leaf of Corozo also readily distinguishes it from *E. guineensis*. All the leaflets lie in one plane and have no basal swellings, and the spines on the petiole are short and thick. The number of leaves is sometimes larger, and may reach as many as forty-four and average thirty. In good specimens the leaflets are larger than those of *E. guineensis*, being up to 1.9 m long and 12 cm wide.[66] There are usually more than 100 pairs.

The male inflorescence differs little from that of *E. guineensis*. The spikelets, of which there are between 100 and 200 varying in length from 5 to 15 cm, are pressed together until they burst through the spathe just before anthesis. Specimens in Zaire were longer and more slender than typical spikelets of *E. guineensis*. The male flower is somewhat smaller with shorter anthers; the rudimentary gynoecium is more developed and has three marked stigmatic ridges.

The female inflorescence is distinguished by a spathe which persists after it has been ruptured by the developing bunch. The spikelets end in a short prong instead of a long spine. The flowers are numerous and are sunk in the body of the spikelet; they are not subtended by a long bract as in the case of *E. guineensis*.

As a result of these characteristics the bunch of *E. oleifera* is surrounded by the fibres of the spathe, and contains no long spines (Plate III, between pp. 78 and 79). The persistence of the spathe has given rise to the view that the flowers cannot be entirely wind-pollinated. Bees are common around the male inflorescences and these and other insects may act as pollinating agents. Hermaphrodite inflorescences are found both in America and in planted specimens in Zaire; in the latter case inflorescences of the andromorphic type have been common. The bunches, being round and wide at their centre with a tendency to be pointed at the top, have a distinctly conical appearance. They usually weigh between 8 and 12 kg but occasionally reach 30 kg. The large number of small fruit, of which the normal ones alone may number more than 5,000, have been recorded in Colombia as weighing between 1.7 and 5.0 g. Parthenocarpic fruit, which are often present in even larger numbers and may constitute 90 per cent of all fruit, average 0.8 g. Higher fruit weights have been recorded though it is not certain that these were from pure *E. oleifera*. The mesocarp layer is thin and usually constitutes 29 to 42 per cent of normal fruit, though over 80 per cent in parthenocarpic fruit. Oil to mesocarp has been recorded as 23 to 38 per cent in Colombia, but in a more recent survey in four regions of that country the mean oil to fresh mesocarp was found to vary with locality from 16.7 to 22.6 per cent in normal fruit with lower oil percentages in parthenocarpic fruit.[66] Shell thickness varies from 1—3 mm and the shell forms between 43 and 53 per cent of the fruit. There is between 13 and 22 per cent of kernel, and nuts with two kernels are fairly frequent and those with three kernels occasional.

Fruit-to-bunch ratios are often low; the mean percentage normal fruit to bunch in a Colombian survey was found to be thirty-seven,[67] but in a more recent survey in four regions the mean normal fruit to bunch varied from 28.1 to 46.3 per cent with mean parthenocarpic fruit to bunch varying from 9.5 to 23 per cent. Within region variations were very great and some bunches were found with normal fruit to bunch as low as 8.9 per cent and

as high as 63.6 per cent.[66] Normal plus parthenocarpic fruit usually constitute less than 60 per cent of the bunch, and as parthenocarpic fruit form
such a high percentage of all fruit, kernel production is considerably lower
than indicated by normal fruit analysis and total fruit-to-bunch
percentages.

Variation

Surveys of the American oil palm in its natural habitats have shown that
the palm exhibits greater uniformity than does the African oil palm in
Africa. Although there are variations in habit of growth, leaf formation
and inflorescence and fruit characters seem to vary comparatively little.
There is some evidence that palms in Central America have higher mesocarp than those in Colombia.[70] As far as fruit colour is concerned about 90
per cent have orange fruit at maturity, these having developed from
immature fruit which were at first yellowish-green, then ivory coloured at
the base and orange above. Another and less common type has yellow fruit
at maturity which have developed from immature fruit at first bright
green, then turning olive-green and pale yellow.[67] The small palms in
Surinam are reported to have green immature fruit turning orange to red.[64]
There is no evidence that fruit forms comparable to *dura, tenera* and
pisifera exist in the populations of Corozo. Compared with *E. guineensis*
the oil has a higher oleic content and iodine value (see p. 701). The
carotene content is higher than that of the Deli palm but may be no higher
than that of many *E. guineensis* palms in Africa.

The above description indicates that the palm has value in hybridization
with *E. guineensis* on account of its slow growth in height and possibly
through the characteristics of its mesocarp oil.

Both mesocarp and kernel oils are extracted in America by crude
methods. In Colombia the oil is mainly sold for soap-making though in the
past it was employed domestically both for cooking and as a lamp oil.

The *E. guineensis* × *E. oleifera* hybrid

The two species have in the last few decades been widely hybridized on an
experimental scale and some information is now becoming available on the
hybrid's performance. The results available and their significance in selection and breeding will be discussed in Chapter 5 (p. 299). Botanically the
hybrid is characterized by leaves which are considerably larger than those
of either parent but retain the leaflet arrangement of *E. oleifera*. The
characteristics of the latter palm as regards height increment, parthenocarpy and fruit shape and colour are also retained in the hybrid. As both
the flower-subtending bracts on the spikelets and the end-prong are only
slightly longer than in *E. oleifera* the hybrid's bunches closely resemble
those of that species.[67] The number of the leaflets is intermediate.

The internal fruit characters of the hybrid naturally depend on the fruit form (*dura, tenera* or *pisifera*) of the *E. guineensis* parent, but they are also influenced by the tendency of *E. oleifera* to produce both normal fruit and large quantities of parthenocarpic fruit of two types: those with a small nut with a liquid-filled cavity and those which are smaller and have only a lignified central core including a rudimentary cavity. With normal fruit, mesocarp to fruit in the *dura* cross varies from under 40 to over 50 per cent, but fruit from *tenera* and *pisifera* crosses has given mesocarp percentages of 58 to 74 per cent.[66, 69] With parthenocarpic fruit mesocarp percentage depends on the degree of parthenocarpy. In a Malaysian trial large parthenocarpic fruit had a mean of 75 per cent mesocarp with 25 per cent shell, while the small type had 89 per cent mesocarp, the lignified core only accounting for 11 per cent of the fruit. Oil to mesocarp is intermediate between that of the parent species, and the distribution of fatty acids also appears to be intermediate[68] (see Ch. 14, pp. 698−700).

References

1. Tomlinson, P. B. (1961) *Anatomy of the monocotyledons. II. Palmae.* Oxford.
2. Jacquin, N. J. (1763, 1780) *Selectarum stirpium Americanarum historia.*
3. Bailey, L. H. (1933) Certain palms of Panama. *Gentes Herb.,* 3, Fasc II, 52.
4. Beccari, O. (1914) Palme del Madagascar. Florence, p. 55.
5. Jumelle, H. and Perrier de la Bathie, H. (1911) Le Palmier à huile à Madagascar. *Matières grasses,* 4, 6065.
6. Opsomer, J. E. (1956) Les premières descriptions du palmier à huile (*Elaeis guineensis*, Jacq.). *Bull. des Seanc. Acad. r. Sci. Colon. (outre Mer),* 2, 253.
7. Preuss, L. (1902) Die Wirtschaftliche bedeutung der Ölpalme. *Tropenpflanzer,* 6, 450−76.
8. Chevalier, A. (1910) *Les végéteux utile de l'Afrique tropicale française.* 7. *Documents sur le palmier à huile.* Paris. 127 pp.
9. Jumelle, H. (1918) Les variétés de palmiers à huile. *Matières grasses,* 11, pp. 4923, 4883 and 5005.
10. Janssens, P. (1927) Le palmier à huile au Congo Portugais et dans l'enclave de Cabinda. Descriptions des principales variétés de palmier (*Elaeis guineensis*). *Bull. agric. Congo belge,* 18, 29−58, 59−92.
11. Smith, E. H. G. (1935) A note on recent research on empire products. (Extract from Botanical section Rep., S. Provinces, Nigeria, Jan.−June 1935.) *Bull. imp. Inst., Lond.,* 33, 3, 371.
12. Beirnaert, A. and Vanderweyen, R. (1941) Contribution à l'étude génétique et biométrique des variétés d'*Elaeis guineensis* Jacq. *Publs. I.N.E.A.C.,* Série Sci., No. 27. Brussels.
13. Beirnaert, A. and Vanderweyen, R. (1941) Influence de l'origine variétale sur les rendements. *Publs. I.N.E.A.C.,* Com. No. 3 sur le Palmier à huile.
14. Notes on the botany of the oil palm. 1. The seed and its germination. *J. W. Afr. Inst. Oil Palm Res.,* 1, (3), 73 (1955).
15. Hussey, G. (1958) An analysis of the factors controlling the germination of the seed of the oil palm, *Elaeis guineensis* (Jacq.). *Ann. Bot.,* N.S., 22, 259.
16. Vallade, J. and Lucien, P. (1966) Aspect morphologique et cytologique de l'embryon quiescent d'*Elaeis guineensis* Jacq., *C. R. Acad. Sci.,* 262, 856.

17. **Henry, P.** (1951) La germination des graines d'Elaeis. *Revue int. Bot. appl. Agric. trop.,* 31, 349.
18. **Tomlinson, P. B.** (1960) Essays on the morphology of palms. 1. Germination and the seedling. *Principes,* 4, 56.
19. **Rees, A. R.** (1960) Early development of the oil palm seedling. *Principes,* 4, 148.
20. Notes on the botany of the Oil Palm. 2. The Seedling. *J. W. Afr. Inst. Oil Palm Res.,* 2, 92 (1956).
21. **Henry, P.** (1955) Note préliminaire sur l'organisation foliare chez le palmier à huile. *Revue gén. Bot.,* 62, 127.
22. Notes on the botany of the oil palm. 3. The stem and the stem apex. *J. W. Afr. Inst. Oil Palm Res.,* 3, 277 (1961).
23. **Arasu, N. T.** (1970) Foliar spiral and yield in oil palms (*Elaeis guineensis* Jacq.). *Malay. agric. J.,* 47, 409.
24. **Rees, A. R.** (1964) The apical organisation and phyllotaxis of the oil palm. *Ann. Bot. N.S.* 28, 57.
25. **Thomas, R. L., Chan, K. W. and Easu, P. T.** (1969) Phyllotaxis in the oil palm: arrangement of fronds on the trunk of mature palms. *Ann. Bot. N.S.,* 33, 1001.
26. **Jagoe, R. B.** (1934) Notes on the oil palm in Malaya with special reference to floral morphology. *Malay. agric. J.,* 22, 541.
27. **Zimmerman, M. H. and Tomlinson, P. B.** (1965) Anatomy of the palm *Rhapis excelsa.* I. Mature vegetative axis. *J. Arnold Arbor.,* 46, 160.
28. **Henry, P.** (1955) Sur le développement des feuilles chez le palmier à huile. *Revue gén. Bot.,* 62, 231.
29. **Henry, P.** (1955) Morphologie de la feuille d'Elaeis au cours de sa croissance. *Revue gén. Bot.,* 62, 319.
30. **Broekmans, A. F. M.** (1957) Growth, flowering and yield of the oil palm in Nigeria. *J. W. Afr. Inst. Oil Palm Res.,* 2, 187.
31. Notes on the Botany of the Oil Palm. 4. The Leaf. *J. W. Afr. Inst. Oil Palm Res.,* 3, 350 (1962).
32. **Rees, A. R.** (1963) A note on the spines of the oil palm. *Principes,* 7, 30.
33. **Zeven, A. C.** (1964) The *Idolatrica* palm. *Baileya,* 12, 11.
34. **Henry, P.** (1948) Un Elaeis remarquable: Le palmier à huile vivipare. *Revue int. Bot. appl. Agric. Trop.,* 28, 422.
35. **Henry, P. and Scheidecker, D.** (1953) Nouvelle contribution à l'étude des *Elaeis vivipares. Oléagineux,* 8, 681.
36. **Yampolsky, C.** (1922) A contribution to the study of the oil palm (*Elaeis guineensis,* Jacq.) *Bull. Jard. bot., Buitenz.,* 3, 107.
37. **Yampolsky, C.** (1924) The pneumathodes on the roots of the oil palm. (*Elaeis guineensis,* Jacq.) *Am. J. Bot.,* 11, 502.
38. **Lambourne, J.** (1935) Note on the root habit of oil palms. *Malay. agric. J.,* 23, 582.
39. **Purvis, C.** (1956) The root system of the oil palm: Its distribution, morphology and anatomy. *J. W. Afr. Inst. Oil Palm Res.,* 1, (4), 61.
40. **Vine, H.** (1945) Report of Chemistry Section, Agricultural Dept., Nigeria. Typescript.
41. **Ruer, P.** (1969) Système racinaire du palmier à huile et alimentation hydrique. *Oléagineux,* 24, 327.
42. **Ruer, P.** (1967) Repartition en surface du système radiculaire du palmier à huile. *Oléagineux,* 22, 535.
43. **Taillez, B.** (1971) La système racinaire du palmier à huile sur la plantation de San Alberto (Colombie). *Oléagineux,* 26, 435.
44. **Bachy, A.** (1964) Tropisme racinaire du palmier à huile. *Oléagineux,* 19, 684.
45. **Ruer, P.** (1967) Morphologie et anatomie du système radiculaire du palmier à huile. *Oléagineux,* 22, 595.
46. **Moreau, C. and Moreau, M.** (1958) Lignification et réactions aux traumatismes de la racine du palmier à huile en pépinières. *Oléagineux,* 13, 735.

47. **Beirnaert, A.** (1935) Introduction à la biologie florale du palmier à huile, *Elaeis guineensis* (Jacq.). *Publs. I.N.E.A.C.*, Série Sci. No. 5.
48. **Thomas, R. L., Chan, K. W.** and **Ng, S. C.** (1970) Phyllotaxis in the oil palm: arrangement of the male/female spikelets on the inflorescence stalk. *Ann. Bot. N.S.*, 34, 93.
49. **Beirnaert, A.** (1935) Introduction à la biologie florale du palmier Elaeis. Organisation de l'inflorescence chez le palmier à huile. *Revue int. Bot. appl. Agric. Trop.*, 15, 1091.
50. **Hardon, J. J.** and **Turner, P. D.** (1967) Observations on natural pollination in commercial plantings of oil palm (*Elaeis guineensis*). *Expt. Agric.*, 3, 105.
51. **Zeven, A. C.** (1973) The 'Mantled' oil palm (*Elaeis guineensis*, Jacq.). *J. W. Afr. Inst. Oil Palm Res.*, 5, 31.
52. **Purvis, C.** (1957) The colour of oil palm fruits. *J. W. Afr. Inst. Oil Palm Res.*, 2, 142.
53. **Beccari, O.** (1914) Contribute alle conoscenza della palma a olio. *Agricoltura Colon.*, 8, 255.
54. **Vanderweyen, R.** and **Roels, O.** (1949) Les variétés d'*Elaeis guineensis* Jacquin du type 'Albescens'. *Publs. I.N.E.A.C.*, Série Sci. 42, 6.
55. **Patino, V. M.** (1965) Private communication.
56. **Cook, O. F.** (1940) Oil palms in Florida, Haiti and Panama. *Natn. hort. Mag.*, 1940, 10.
57. **Bailey, L. H.** (1940) The generic name *Coroso*. *Gentes Herb.*, 4, Fasc X, 373.
58. **Salisbury, Sir E.** (August 1950) Private Communication. WAIFOR 424/89.
59. **Wessels Boer, J. G.** (1965) *The indigenous palms of Surinam*. E. J. Brill, Leiden.
60. **Corner, E. J. H.** (1966) *The natural history of palms*. Weidenfeld and Nicolson, London.
61. **Vanderweyen, R.** and **Roels, O.** (1949) L'*Elaeis melanococca*, Gaertner (em. Bailey). *Publs. I.N.E.A.C.*, Série Sci. 42.
62. **Blank, S. De** (1952) A reconnaissance of the American oil palm. *Trop. Agric. Trin.*, 29, 90–101.
63. **Ferrand, M.** (1960) Le Noli. *Oléagineux*, 15, 823.
64. **Meunier, J.** (1975) Le 'palmier à huile' américain *Elaeis melanococca*. *Oléagineux*, 30, 51.
65. **Arnaud, F.** and **Rabechault, H.** (1972) Premier observations sur les caractères cystohistochimiques de la résistance du palmier à huile au 'dépérissement brutal'. *Oléagineux*, 27, 525.
66. **Vallejo Rosero, J.** and **Cassalett, C.** (1974) Perspectivas del cultivo de los hibridos interspecificos de Noli (*Elaeis oleifera* (HBK Cortez) x Palma Africana de Aceite (*Elaeis guineensis*, Jacq.)) en Colombia. Instituto Colombiano Agropecuario, Colombia. Mimeograph.
67. **Hurtado, J. R.** and **Nuñez, G. R.** (1970) Estudio de la palmera Noli (*Elaeis melanococca*, Gaert.) y preliminares de su fitomejoramiento en Colombia. *Acta Agronica*, 20, 9.
68. **Hardon, J. J.** (1969) Interspecific hybrids in the genus *Elaeis*. II. Vegetative growth and yield of F_1 hybrids *E. guineensis* x *E. oleifera*. *Euphytica*, 18, 380.
69. **Obasola, C. O.** (1973) Breeding for short-stemmed oil palm in Nigeria. *J. Nig. Inst. Oil Palm Res.*, 5, (18), 43.
70. **Richardson, D. L.** (1976) Private communication.

Chapter 3

The oil palm and its environment

The environmental circumstances leading to the development of palm groves are in many respects different from those leading to the development of plantations. In the 1950s nearly half the palm oil entering world trade was extracted by village methods from the African palm groves; the 1960s saw a complete change however, and it is unlikely that oil of grove origin will ever again form a substantial part of the trade. Though large quantities of kernels may be expected from the groves for a long period to come, grove production of palm oil will be almost entirely for local domestic consumption.[1] In this chapter the environmental factors leading to the establishment and retrocession of groves are briefly described and this is followed by a fuller discussion of the environment suitable for plantations.

The development of groves

The ability of the oil palm to become semi-domesticated is said to have delayed its development as a planted crop.[2] This assertion is almost certainly true for Africa and for Brazil where the vast majority of palms have still only reached the stage of partial domestication. In the remote past the forests of Africa were largely uninhabited, being visited only infrequently by hunters and later by cultivators who made, but soon abandoned, small camps and clearings. The first stage of domestication took place in such clearings where discarded fruit were able to germinate and a few palms developed. As populations increased, the palm became a common relic of abandoned villages, compounds and farmland. Later, the practice of returning to former clearings to collect fruit became common and some natural selection is likely to have taken place during this process.

The vast majority of palms grew up without any conscious planting or selection on the part of man; there is evidence, however, that in certain localities the very limited natural selection in partially domesticated situations was augmented by the deliberate sowing of unselected seed and the transplanting of self-sown seedlings. The former practice is reported from the Ivory Coast and Togoland, while deliberate planting to extend the area

77

under palms (mainly for palm wine) seems to have been characteristic of Dahomey.

Some conscious planting of selected seeds and transplanting of the resultant seedlings has undoubtedly been practised in certain areas, usually those of high population. Use of *tenera* seed for such planting has resulted in certain areas, e.g. parts of eastern Nigeria and the Sales region of Angola, having a much higher proportion of *tenera* palms than is usual.

The groves of southern Zaire were first studied by Father Hyancinthe Vanderyst, an agriculturist and a missionary.[3] He postulated that in this region groves would develop near to the forest verges and banks of streams and rivers (where, he believed, palms had grown naturally) wherever the temperature and rainfall were within certain broad limits and, most important of all, provided a relatively dense human population had been or was becoming established. In an area such as southern Zaire where the palm was already present as part of the natural vegetation, the factor of greatest importance in the development of a grove was the existence of a dense population. The cutting down of forest areas for annual crop cultivation removed dense shade and created conditions suitable for the rapid establishment of the palm. Moreover, the sooner the farmer returned to the area to cultivate it again, the better the conditions for palm establishment, since long periods of secondary forest growth would reimpose shade and restrict seedling development. With an increasing population, therefore, the secondary forest gradually gave place to palm grove, while where areas were abandoned, owing in some parts of Zaire to sleeping sickness, the groves decayed and slowly gave place to forest once more.

Father Vanderyst also examined the occurrence of groves in the less-favourable plateau country of the Kwilu district and showed how high grassland, under-populated and subject to grass fires, was unsuitable for grove establishment, whereas in lower areas, where villages had been established in the past and fields had been cultivated, groves of varying density had grown up.

The greatest areas of palm groves, supporting large populations and providing the highest quantities of oil and kernels for export, are to be found in West Africa. All along the coast from Sierra Leone to West Cameroon, the oil palm has multiplied wherever population has increased, except in those areas, e.g. Ghana and western Nigeria, where cocoa has become the dominant economic crop or, as on the Accra plain, where rainfall has been well below 1,000 mm per annum. Population has, in fact, proved of more importance in the spread of the oil palm than has rainfall; in Dahomey, where the rainfall is often less than 1,200 mm per annum, the oil palm dominates the scene because there is no other suitable cash crop.

South-eastern Nigeria has been the most important area for grove exploitation although mid-west Nigeria has always produced as high an export of palm kernels. Studies in these regions provided information on the history and composition of groves and have made it possible to piece together the factors giving rise to groves of different types, to determine

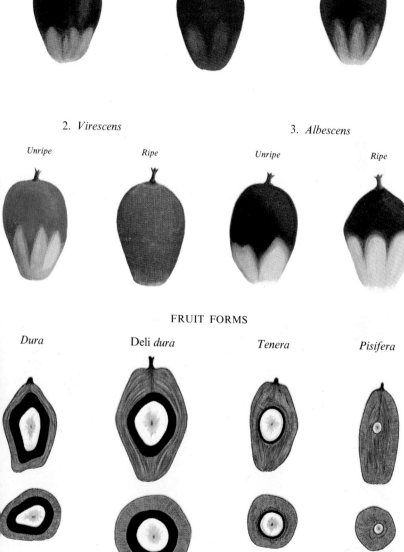

FRUIT TYPES

1. *Nigrescens*
Unripe

Rubro-nigrescens
Ripe

Rutilo-nigrescens
Ripe

2. *Virescens*
Unripe   Ripe

3. *Albescens*
Unripe   Ripe

FRUIT FORMS

Dura   Deli *dura*   *Tenera*   *Pisifera*

Pl. I The fruit types (external appearance) and forms (internal structure) of the oil palm.

Pl. II *Elaeis guineensis*: palm with a heavy crop at only 4½ years old in Guadalcanal, Solomon Islands.

Pl. III *Elaeis oleifera*: palm with many ripe and unripe bunches in the Sinú valley, Colombia.

Pl. IV The fruit of three *tenera*. A. Long-shaped Nigerian *tenera* 24.2864. B. A large-kernel *tenera*, 1.2229. C. *Tenera* from the cross *Elaeis oleifera* x *E. guineensis*, *pisifera* (note no fibre ring).

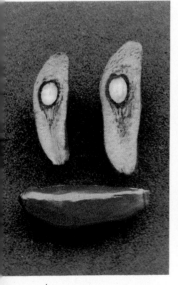

A

B

C

V Seedling showing acute N-
ciency symptoms on waterlogged soil
Brazil.

VI Young nitrogen-fertilized plants
wing on previously cultivated land in
a, Colombia.

VII Magnesium deficiency symp-
s:

Nursery seedling in Nigeria (below).
Young field plants in Colombia
right).
Orange frond of the adult palm leaf,
Sierra Leone.

Pl. VIII Symptoms which have been attributed to potassium deficiency:

A. Confluent Orange Spotting

B   C

Leaves and a leaflet from palms showing Mid-Crown Yellowing. Some of the lesions in C are similar to the Mbawsi symptom.

D

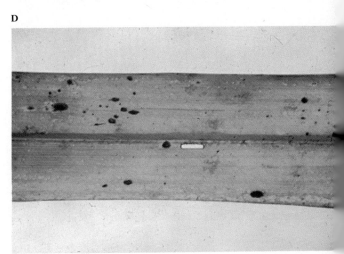

Pl. IX Two types of Anthracnose of prenursery seedlings:

A. *Botryodiplodia palmarum*.

B. *Melanconium* sp.

Pl. X A third type of Anthracnose of prenursery seedlings: **C.** *Glomerella cingulata.*

Pl. XI *Cercospora elaeidis* infection of the terminal leaflet of a heavily infected 6-month-old seedling.

Pl. XVI Leaves infected with *Pestalotiopsis* sp. in Colombia.

Pl. XVII The 'Peat Yellows' condition on palms growing on peat soil in Malaysia.

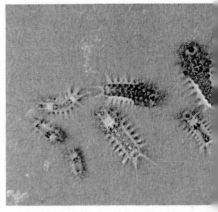

Pl. XVIII Nettle caterpillars in Malaysia:

A. *Setora nitens.*

B. *Darna trima.*

Pl. XIX Nettle caterpillars in Nigeria: *Parasa* sp.

their production and to consider their improvement and likely future.[4, 5, 6] A comprehensive review of early and recent studies has been provided by Zeven.[2]

Secondary rain forest with oil palms

As population increases, so more and more land is occasionally used for farming and for dwellings, but movement is so frequent and farming cycles so long that the predominant vegetation remains secondary forest. *Secondary rain forest with oil palms* is common throughout West and parts of Central Africa. Where it has been studied, near Benin in western Nigeria, the density of the human population is less than 50 per square mile and the number of palms with stems does not usually exceed 69 per hectare. The bearing palms are tall and have dropped their leaf bases − they are commonly called smooth-stemmed palms − and they often overtop the surrounding forest; smaller palms which have not yet dropped their leaf bases (rough-stemmed palms) are to be found in varying numbers but, being shaded by the forest, they contribute little to the yield. An area of 40.5 hectares, known as Block 18, which had been farmed on very long-term forest rotations up to 1940, showed a mean yield per hectare per annum over 12 years of only 104 bunches weighing 1,466 kg with a mean bunch weight of 14.1 kg. In 1952 there were 32 smooth-stemmed, 37 rough-stemmed and 44 seedling palms per hectare, but rough-stemmed palms contributed only 6 per cent of the yield. By 1963 the mean number of smooth-stemmed palms per hectare had increased to 41, but rough-stemmed palms had decreased to 7 per hectare. Thus competition from the forest following cessation of farming is followed by the death of many of the younger palms while only a few, usually the larger ones, drop their leaf bases and become smooth-stemmed. Eventually the smooth-stemmed palms will also decrease in number through infection by *Ganoderma* Trunk Rot and lack of replenishment from a dwindling stock of rough-stemmed palms. Other 40-hectare blocks adjacent to the one described above had by 1962−3 been reduced to total palm stands of between 12 and 30 per hectare.

Areas of occasionally farmed secondary forest as described above can be expected to give bunch yields of around 1−1.5 tons per hectare with relatively large bunches harvested by climbers from tall palms. This satisfies the immediate culinary needs of a small population and gives a small quantity of kernels for sale. Palm populations need to rise considerably before oil for export becomes available. In these areas, too, the proportion of poor *dura* palms is high, indicating that little or no selection, whether conscious or unconscious, has occurred. In the 40-hectare block already referred to, 96 per cent of the bearing palms were *dura*.

Palm bush

When an area of secondary forest with palms becomes more heavily

Pl. 13   Dense palm grove: looking from a felled area into a grove in South-east State, Nigeria.

populated and food-crop bush rotations become shorter, the palm population increases and the secondary forest becomes more open. Such areas are known as *palm bush* and are to be found in their typical form in the Warri and Asaba districts of mid-western Nigeria, where rainfall varies from 1,800–2,200 mm and the population density lies between 150 and 300 per square mile. This palm bush may be regarded as a transitional stage between secondary forest with palms and the more complicated grove structure of the really heavily populated regions of West Africa. It finds its counterpart, as far as population and palm density are concerned, in the drier fringes of the palm belt where the vegetation is often 'derived savannah' rather than the early stage of secondary rain forest. Such areas, with a rainfall of little more than 1,250 mm, are to be found in the 'middle-belt' of Nigeria and in the south of Dahomey. Population densities in these areas may reach or even exceed 300 per square mile; but yields are severely limited by the drought conditions which obtain for many months of the year.

The density of palms in the palm bush areas varies very greatly but is typically between 75 and 150 stemmed palms per hectare. Yields of around 2,200 kg bunches per hectare per annum may be reached and fruit-form distribution is similar to that in the secondary forest areas.

Dense groves and farmland palms

In the south-eastern corner of Nigeria, population pressure is high and the palm groves have reached their highest development. The palms are now in almost pure stand with small shrubs or food crops growing between them. The densest groves cannot be farmed, and the shrub growth is held in check only by lack of light. Farming cycles are short and in any area used for food-cropping the undergrowth is meagre owing to the frequency of cultivation. A situation has therefore arisen where groves of varying density cover large areas, or groves become separated from distinct areas of farmland which contain only a few palms. The whole association may be described broadly as *dense grove and farmland palms*. These areas develop in different proportions and with different densities of palms according to the original or acquired customs of the people. In Nigeria, the formation of the groves and the presence or absence of farmland palms has been primarily influenced by the practice of moving the homestead and by an aversion to felling palms. In certain areas this aversion has been partially overcome and a different dispersion of palms has resulted. Land in the dense-grove districts may be classified, following Waterston,[4] according to the history of the palms as follows:

Grove sub-type	Origin and characteristics
A. Primary compound palms	Palms growing around homesteads set up in areas of original forest.
B. Dense grove palms	Dense groves have developed around the old homestead following the movement of its inhabitants into a new forest clearing or into farmland.
C. Farmland palms in degraded groves	These areas have become largely farmland because of the reduction of the stand in the dense groves through disease or cutting out.
D. Open farmland palms	Palms often grow up in small numbers in areas which have been selected as farmland.
E. Compound palms near homesteads set up in farmland	When population pressure has become very high new homesteads must, perforce, be constructed in farmland.

It will be realized that this classification may proceed further, the land with E-type palms becoming a dense grove, similar to sub-type B, when the homestead is again moved. Moreover, in the oldest palm grove areas it is unlikely that A-type groves any longer exist, all compound palms being of E-type.

In south-eastern Nigeria compounds have in the past been abandoned 15 to 20 years after construction. Two reasons have been given for this: firstly, the trees around the compound have become so dense that light-requiring compound crops cannot be grown satisfactorily; secondly, by this time it is simpler to construct a new house than to repair a dilapidated old one. It is difficult to imagine the development and production of any other tropical crop being so inextricably interwoven with what might be considered the incidentals of village establishment and organization. But there seems little doubt that, while secondary forest and palm bush palms owe their origin to farming, the densest groves of the palm belt have originated in the dwelling habits and preferences of the people of Africa (Plate 13).

The number of palms on land which is farmed varies very greatly owing to two opposing tendencies. On the one hand increased frequency of farming in areas definitely allocated to food-cropping leads to the elimination of seedlings and hence the gradual reduction of palms to very small numbers. On the other hand, in some densely populated districts groves have been thinned out to provide areas which, though not devoid of palms, are open enough to support food crops. In the latter case, typified by an area surveyed at Asuten Ekpe[4, 5] abrupt changes from grove to farmland are seldom seen and it is impossible for the newcomer to distinguish one grove type from another. Statistics of palm number and yield tell their tale however.[5] While the dense groves of Type B have on average more palms than either the farmland areas (C and D) or the groves developing from farmland (E), when densities are similar the yields of the dense grove palms are higher than the yields of palms on land degraded by intensive farming (Table 3.1). The E plots tend to be of intermediate fertility, having been partially restored through proximity to human habitation. Thus, for all the unkempt and haphazard appearance of a typical grove area, such as Asuten Ekpe, each small area is showing, through its palm composition and yield, the history of its usage. Groves whose composition and yield were recorded at Abak, also in south-eastern Nigeria, were divided more abruptly from the bulk of the farmland, though the areas

Table 3.1 *The mean composition and yield of plots of different grove-types at Asuten Ekpe, south-eastern Nigeria*

Grove type	Number of palms per hectare			Yield of plots with 52–100 smooth-stemmed palms per hectare kg per hectare per annum 1949–51
	Smooth-stemmed	Rough-stemmed	Total	
B. Dense groves	94	106	200	2,914 (15.7)
C. Degraded groves with farmland palms	86	62	148	2,443 (14.8)
D. Farmland	57	35	92	1,784 (13.6)
E. Groves around compounds in farmland	52	62	114	2,403 (14.3)

The figures in brackets are the weights per bunch for the *dura* bunches in kg.

studied contained some farmland palms. Of particular interest was the comparatively recent desertion of villages in these areas.

In Sierra Leone, groves of a single type appear to cover wider areas, possibly because the people have lived in more settled homesteads. Inland, e.g. between Bo and Kenema, where areas of poor lateritic soil support a relatively high population, virtually the whole area, except the distinctive village sites with their large forest trees, is used for 'bush-rotation' farming, the length of the rotation being comparatively short. The whole area may therefore be described as *farmland with palms*, and the appearance of any particular plot of land and its palms depends on the stage in the rotation. When farming is started, the palms are scorched by the burning of the bush and have hardly recovered by the time farming ceases and reversion to bush begins. Thereafter, the palms are surrounded by higher and higher bush until the plot's turn for farming comes round again. The palms, and bunch yield, suffer considerably from the burning. Although palms appear fairly evenly scattered over wide areas, the stand per hectare is comparatively low[7] (Plate 14).

Pl. 14   A large area of 'farmland with palms' near Makeni, Sierra Leone.

Towards the coast, on better soils, thicker stands of palms are to be found in groves where the population is lower and bush rotations longer. Clearing for farming in these areas, e.g. Bissao, allows the smaller, rough-stemmed palms to start bearing and groups of seedling palms to develop. In contrast with the areas of farmland with palms, these *palm groves with infrequent farming* seem still to be developing and the yields from them will be similar to those of developing dense groves in Nigeria.

The characteristics of productive groves

Age and height of palms

There is no means of telling the age of individual palms since conditions of heavy shade will bring growth near to a standstill while elongation may be very rapid when this shade is reduced. The age of a group of palms forming a grove can only be determined by investigating the history of the particular area. For instance, in one large grove near Abak in Nigeria, some areas are known to have been occupied by villages up to the decade 1930—40. By 1960 most of these areas were covered by young dense grove which had not yet reached its full development. In a nearby area a village site had been abandoned in 1870; the resulting dense grove had long since died out giving place to farmland which, by 1950, was already sufficiently open for resettlement.[6]

From these considerations and from the general classification of grove types already given, Zeven[6] suggested that the age of palms in southern Nigeria would be as follows:

Grove type		Number of years since homestead was established	Age of palms in years
E1	{ Compound palms	0—11	Some old palms 70—90
E2	{ in farmland	12—20	A few 0—10 and 80—90
B1	{ Dense grove	21—30	0—10, thick young grove
B2	{ palms	31—60	11—40, dense older grove
C	Farmland palms in degraded groves	61—80	41—60, degraded grove
D	Open farmland palms	81—100	61—80, reverting to farmland

(E1 and E2, and B1 and B2, are early and late stages of E and B respectively)

The average height of palms in a grove is influenced by the age of the grove, density, climate and the amount of felling which has been practised. There is a decrease in annual height increment with age; rough-stemmed palms under 5 m may grow 120 cm per annum while very tall palms of over 20 m will grow only 15—20 cm. Mean heights in a large grove at Abak were 4.5 m for rough-stemmed and 15.5 m for smooth-stemmed palms. Palms of 18 m were common and two palms were over 24 m.[6] (Plate 15.)

Leaf production

The leaf production of grove palms is lower than that of planted palms and this is the initial reason for lower yields. Leaf production of shaded rough-stemmed palms is inhibited by low light intensity (with seedlings, leaf production is even lower), while advanced age and height accounts for low leaf production of smooth-stemmed palms. In a large grove at Abak mean

A

B

Pl. 15 Methods of climbing tall palms in West Africa: **A.** The two-rope method; note that one noose is around the right thigh, the other round the left foot. **B.** Climbing with a single rope.

annual leaf production varied in number from ten to nineteen with smooth-stemmed and ten to sixteen with rough-stemmed palms according to the amount of exposure to direct sunlight. In some situations seedlings had a mean annual leaf production of less than seven. The average life of a green leaf is 18 months with rough-stemmed and 21 months with smooth-stemmed palms.[6]

Inflorescence production

A high rate of floral abortion is a very important characteristic of grove palms. Abortion in mature plantation palms in Africa lies between 5 and 10 per cent, but even with smooth-stemmed grove palms in full sunlight the abortion rate will average between 10 and 20 per cent, while with reduced exposure to sunlight abortion will reach 80 to 100 per cent. However, the sex-ratio tends to be higher than with plantation palms and this partially counteracts the effect of abortion on total bunch production.

Bunch yield

For reasons already given grove yields are found to be low except where a high and even stand of relatively young smooth-stemmed palms exists. Individual palm yields may be very high, however. The mean bunch yield of smooth-stemmed palms growing in a compound was 184 kg per annum over a 4-year period. One palm yielded 411 kg in a year. Even in a dense B-type grove at Abak one palm yielded 272 kg bunches. All these palms are, of course, exceptionally favourably placed; once there is competition for light and nutrients, yields fall away rapidly. The relationship between grove-density and yield per hectare is therefore of considerable interest.

In a comparatively high-yielding grove near Abak in Nigeria the individual yields per annum of sixty-five 1-acre (0.405 hectare) plots were recorded over a 4-year period. On grouping the plots according to the density of their smooth-stemmed palms, average yields were found to increase with density even when stands were above the usual plantation optima (Table 3.2). Later work showed that in other groves actual densities were also well below optimum densities for yield. Optimum density varies from grove to grove for reasons which are not clear, though differences of age and soil fertility are likely causes. The main factors which distinguish a dense grove from a plantation of similar density are: (i) uneven height of producing palms and (ii) competition from an unproductive collection of rough-stemmed and seedling palms. It is possible that (i) leads to a higher optimum density (in some groves well over 250 smooth-stemmed palms per hectare) for groves than for plantations, and that (ii) is the most serious obstacle to higher yields. It is unlikely that any groves ever reach an optimum density since *Ganoderma* Trunk Rot seems to start affecting the older palms before this density is reached. Zeven considers that 'the non-yielding smooth-stemmed and the ill-developed rough-stemmed palms are a permanent obstacle to the development and produc

Table 3.2 *The relation between density and bunch yield in a palm grove near Abak, Nigeria*

Group	Number of smooth-stemmed palms		Mean annual bunch production			
			Per hectare		Per smooth-stemmed palm	
	per hectare	per acre	Weight	Weight per bunch	Weight	Weight per bunch
			(kg)	(kg)	(kg)	(kg)
1	175	71 and above	6,209	13.7	26.0	13.8
2	151–174	61–70	4,432	13.5	27.7	13.8
3	126–150	51–60	3,759	13.2	26.8	13.3
4	101–125	41–50	3,482	12.6	30.4	13.2
5	77–100	31–40	2,928	12.4	30.9	12.8
6	52–76	21–30	2,425	13.1	35.9	13.3
7	25–51	10–20	1,439	11.7	37.7	12.2
Mean		All plots	3,374			

tion of the neighbouring palms'.[6] The variation in yield due to previous land usage has already been shown (Table 3.1).

In general, annual yields of groves of the B-type in Nigeria may be expected to average about 2,800 kg bunches per hectare, but in small areas and under exceptional circumstances yields of 4,500 or even 6,500 kg may be obtained. A similar range of yields has been found in the Ivory Coast and Zaire.

In practice the actual out-turn of oil and kernels from a grove depends on the harvesting customs of the people. These may lead to harvesting under-ripe or, alternatively, to the lapse of many weeks between harvests, resulting in loss of loose fruit. The potential output of a grove yielding 2,800 kg bunches per hectare will be between 400 and 450 kg oil and between 220 and 280 kg kernels. Extraction methods may result in half the oil being lost. Even with the use of a hydraulic hand-press the eventual out-turn of a hectare of grove is unlikely to be more than 400 kg oil and 280 kg kernels.

Fruit form and type

Fruit are predominantly poor quality *dura*, and the quantity of *tenera* bunches is thought to be a measure of the amount of selection which has proceeded in the area. *Virescens* and mantled (*Poissoni*) fruit are always very few. Variation in recorded groves and in groups of similar grove plots in south-eastern Nigeria, is shown in Table 3.3. The Ufuma grove (731-1) has a particularly high percentage of *tenera* palms and a few *pisifera*. This is thought to indicate deliberate *tenera* planting in the past though the grove is now in Hardy–Weinberg equilibrium (see p. 265). When the green-fruited virescens palms appear they are often grouped.

Table 3.3 *Distribution of fruit forms and types in Nigerian groves* (percentage of those palms identified)

Locality	Grove or group of grove plots	Dura (%)	Tenera (%)	Pisifera (%)	Virescens (%)	Poissoni (%)
Abak	502−2	93.6	6.4	−	1.0	−
Abak	503−1	92.2	7.8	−	0.5	−
Abak	502−1	91.7	8.3	−	0.9	−
Abak	502−4	89.7	10.3	−	−	−
Abak	501−1	88.9	11.1	−	0.1	−
Abak	504−1	87.5	12.5	−	−	−
Asutan Ekpe	560−1−B	87.3	12.7	−	0.2	0.01
Abak	505−1	85.3	14.7	−	−	−
Asutan Ekpe	560−1−A	83.7	16.3	−	−	−
Asutan Ekpe	560−1−C	82.8	17.2	−	−	−
Asutan Ekpe	560−1−D	81.2	18.8	−	0.4	−
Asutan Ekpe	560−1−E	80.2	19.8	−	1.1	−
Ufuma	731−1	53.0	43.0	4.0	−	−

A study of ten Ivory Coast groves carried out by random sampling gave *tenera* percentages varying from 0 to 41, with 0 to 5 per cent *pisifera* and up to 3 per cent *virescens*. Mantled fruit was not encountered and there was only one specimen of *albescens*.[8]

Mean fruit to bunch ratio in groves may vary from forty-five to seventy though it is usually high in the tall smooth-stemmed palms. Mean fruit composition in predominantly *dura* groves may be very similar in widely separated groves. Fruit in groves near Abak and Benin in Nigeria for instance had mean mesocarp contents of 45 and 42 per cent respectively with 14 per cent kernel in both cases. In the Ivory Coast survey mean mesocarp to fruit percentages varied between groves from thirty-five to forty-nine with *dura* palms and forty-four to seventy with *tenera*; mean kernel percentages varied from thirteen to twenty-three with *dura* and eleven to thirty-two with *tenera*. These differences of fruit form and composition are thought to demonstrate human selection, whether intentional or not, in favour of the *tenera* and good fruit composition. Other factors encouraging an increase in the proportion of *tenera* are the sale of predominately *dura* fruit in the market, the use of *dura* palms for wine tapping, and the larger number of fruit per palm in the *tenera*. Acting against *tenera* increase are the low male inflorescence production of *pisifera* palms and, possibly, the inferior protection of the *tenera* seed in germination, particularly in dry climates.[2, 8]

Retrocession and decline of the groves

Whereas it is increasing pressure of population which brings the groves to their land dominance and optimum yield, so also it is a further increase in population which leads to their decline. The increase in population brings a further demand for farmland and leads to a decrease in the length of food-crop rotations, i.e. fewer years fallow, wherever farming is undertaken. This demand may have several different effects on the palm population.

Thinned grove

It has already been mentioned that in certain areas where fairly uniform groves tended to cover the ground, thinning had to be resorted to in order to provide sufficient light for food crops. Such areas tend to have a higher proportion of rough-stemmed palms than have untouched dense groves, and all grove sub-types tend to become sparse enough for farming.

Contraction of grove areas

In other areas of increasing population farmland tends to become separated from the dense groves and regeneration in farmland does not take place. This process of separation is assisted both by shorter bush rotations with more frequent burning, and by the modern tendency to build houses in villages near roads in preference to the old habit of moving to a new compound in farmland and thus restarting a palm 'cycle'.

As the population becomes progressively larger the farmland becomes exhausted and the areas of grove recede. Derived savannah with a few islands of grove palms then develops and expands until finally the grove islands are merely village groves surrounding permanently sited compounds. In these groves some planting is carried out so that a certain amount of regeneration may take place. Areas of derived savannah are found typically in the northern parts of eastern Nigeria and in the Ivory Coast. Their development is hastened by the practice of burning the grassland; this destroys the young seedling palms and prevents the older palms from developing. These areas naturally become depopulated and their re-development is a considerable problem. In the Ivory Coast the soil has proved intrinsically fertile enough for plantations of palms to be established. Such rehabilitation is being tried on the much poorer soils of eastern Nigeria, but success is by no means assured.

Farmland with palms

The *farmland with palms* association of Sierra Leone was described on p. 83. It may be presumed that the oil palm gained access to farmed land in the ordinary way, but that as population pressure increased the land surrounding a village site would suffer a reduction in the number of years fallowing. With long bush rotations the palms would be suppressed by shade and forest competition, but as rotations shortened so palm numbers and production would come to a maximum though always, under this system, suffering from the burning of the felled bush before food-cropping. Already, in Sierra Leone, this optimum stage has been passed as the rotations are now short, covering some 5 to 6 years, and probably some thinning has taken place. The effect of frequent burning is serious; besides the prevention of seedling establishment, in the first year of food cropping only a few green leaves remain on the old palms and it is nearly 2 years before a full crown has been re-established. The effect of this on yield will be obvious and one must expect a continuing reduction both of palms and of yield per palm.[7]

It may thus be expected that where large areas of land are given over to bush rotation farming, the oil palm, though previously covering the ground in quantity, will eventually die out. There is some evidence, however, that Sierra Leone bush rotations have naturally selected a palm which is partially fire-resistant and grows slowly. Palms from Nigerian mixed selected seed have been compared in trials both with unselected material from Nigeria and with unselected material from Sierra Leone. The Sierra Leone unselected palms, besides yielding less, were found to be slower growing than the mixed selected material, and they showed a characteristic premature withering of the leaves.

Sparse grove

An extension of the thinned grove is the sparse grove to be found around Porto Novo in Dahomey in an area of low rainfall.[2] Frequent arable-cropping is essential to support a high population; rotations are so short and the climate so severe that the bush regrowth is small and the palms are therefore not affected by burning as in Sierra Leone. Densities are maintained at under 100 per hectare by felling on the one hand and natural regeneration and some planting on the other. A grove with only forty palms per hectare was estimated to yield just over 1,550 kg bunches per hectare.

The presence of thinned grove in a traditionally high oil-producing region and of sparse grove in a dry region where the oil palm is the only possible export crop, suggest that the people of West Africa can find ways of at least arresting the decline of the groves, though only a very low level of production will be maintained. Wherever firing of grassland or the severe burning of felled bush have become ingrained in the customs of the people the oil palm cannot, however, survive even at a low level of production.

Groves in Zaire

Owing to the vast land area of Zaire and the relatively small population, groves are for the most part at the developing rather than the declining stages.

Although groves and small village plantings exist in large numbers in the climatically suitable areas of the north, the really large areas of grove, said to cover nearly 500,000 hectares, are to be found in the south, in the

Kwango and Kwilu districts, where there are many drawbacks to oil palm cultivation. The development of groves in this region has already been discussed at the beginning of this chapter; the region consists of relatively fertile valleys alternating with plateau areas covered by blown Kalahari sand and supporting only a meagre grass cover. The groves in the valley areas are thus in long narrow belts, and although in some places they are being converted to savannah by too frequent cropping, they are mostly more uneven and less dense than the groves of eastern Nigeria, and they alternate with forest.

Exploitation of the groves was organized by large companies rather than, as in Nigeria, by the people themselves. Villages were moved to the grass plateaux in the 1920s as a measure against sleeping sickness and from then onwards there was a chronic shortage of harvesters. Groves thus tended to be only partially harvested and were often choked with secondary undergrowth. Before independence there were groves in the possession of companies and harvested by their employees, and groves owned by the people but from which the bunches were, by agreement, sold to the companies.

The Brazilian groves

Very little is known of the early development of the Brazilian groves. The main palm belt is some 3 to 5 miles wide and runs from the Ilha de Itaparica in the Bay of Salvador to the south of Maraú in the State of Bahia. *Elaeis guineensis* is also to be found in scattered groups in many parts of the State as far as 18°S, and in small areas further north in the States of Sergipe, Algoas, Pernambuco and Para (though not in the basin of the Amazon).

It can be seen from the air that the groves in the main palm belt are sporadic and vary greatly in density. Along the road running north and south of Taperoa, however, the groves are often continuous and are tall and dense as in eastern Nigeria. This coastal region was one of the first to be settled by the Portuguese and was worked for sugar cane by slaves from Africa. The groves are believed to have become naturally established when the estates became infertile through continuous cropping and the sugar planters moved further inland. The groves have probably been maintained by occasional cropping of the land with manioc (cassava) and other food crops and by light grazing with cattle. The undergrowth is high in many areas and is cleared from time to time. In a simple trial, cleared and properly maintained plots gave a mean yield over a 2-year period which was more than double that of control plots. There was evidence of wide seasonal fluctuations of yield in this region, but in a favourable year the mean yield of the cleared plots was 4,765 kg per hectare.[9]

Dura, tenera and *pisifera* palms exist in the Brazilian groves. Both *virescens* and two types of *nigrescens* fruit, corresponding to Purvis's *rubro-* and *rutilo-nigrescens*, are also distinguished. Analysis of a sample of

440 bunches suggested that the range of bunch and fruit characters is similar to that in African groves. Fruit weight varied from 4 to 25 g, *dura* fruit had 33 to 65 per cent mesocarp, while *tenera* fruit with mesocarp percentages up to 86 were encountered with shell as low as 7 per cent. Some *dura* had analyses similar to that of a good Deli palm.[10]

Grove rehabilitation and improvement

This account of the palm groves demonstrates that only low and costly production can be expected from them, and it is being increasingly asked whether they should not be replaced either *in situ* or in adjoining forest areas by plantations. The problems of improvement and replacement have been comprehensively discussed by Zeven[2] and will only be briefly mentioned here.

Rehabilitation without replanting

Methods of improving the yield of groves without replanting have frequently been suggested and tried out, but the complexities of grove organization have usually precluded their widespread adoption.[11,12] The early trials, which have been reviewed by Zeven,[2] usually entailed clearing of the undergrowth where thought necessary, removing epiphytes, pruning necrotic leaves and thinning the grove to varying stands per unit area, usually between 135 and 200 per hectare or 55 and 80 per acre. Proper controls were often lacking, results were variable and the economy of the procedure doubtful. It is necessary to show that an increase of produce will cover the cost of clearing, cleaning, thinning and harvesting; the latter cost may actually be increased owing to a higher proportionate production being obtained from the taller palms which are the chief producers and therefore the ones to be left in.

Where 'improved' grove plots have formed part of a general rehabilitation experiment, yield increases have often been attributable to edge effects, i.e. adjacent replanted or thinned and cleared plots have reduced competition for light, and the edge palms of the adjoining grove have shown startling benefit. For instance, in a thinning experiment in a grove at Grand-Drewin in the Ivory Coast a control plot initially yielding 1,655 kg per hectare was producing a mean of 6,845 kg per hectare per annum 2 to 3 years later. Plots thinned down from over 200 palms per hectare (over 80 per acre) to 135 or 150 per hectare (55 or 61 per acre) gave yields 40 per cent higher than the control.

There is no doubt that many areas of grove can have their yields increased by keeping the undergrowth down and thinning out so that there is less competition for nutrients and light. In most respects this is simply a spacing effect of the kind shown in plantation spacing experiments and described in Chapter 9, but in addition there is in the grove the effect of palms 'overtopping' their neighbours. Large established grove palms often respond to a marked degree to a combination of thinning, clearing and cultivation. For instance, in an experiment at Obio-Akpa in Nigeria, cultivation and the reduction of the number of smooth-stemmed palms in one of the treatments by 67 per cent was followed by only a small reduction in plot yields in the following 3 years; this was then followed by a 34 per cent increase in yield in the next 3 years and an 86 per cent increase over pretreatment yield in the year after that.

It is difficult to draw overall conclusions from these scanty results. Some dense groves are already suffering from too high a population of palms and thinning will improve their performance; other groves will not be in this condition and marked results cannot be expected. Where replanting cannot be undertaken for the time being it may be worth while to seek out for thinning those groves choked with an excess of palms. In areas where replanting can be undertaken, however, it is very doubtful if thinning will ever prove a popular practice except where it is combined with the replanting of selected material in a general replanting scheme.

Grove improvement by manuring

The application of fertilizers to grove palms is almost never practised and has rarely been tried. In an unusually productive improved grove at Grand-Drewin in the Ivory Coast, potassium chloride applied at the rate of 1 kg per palm gave a 24 per cent increase in yield. No responses were obtained to nitrogen or phosphrous.

At Abak in Nigeria various quantities of K, Mg and bunch refuse were applied to the smooth-stemmed palms in a fertilizer experiment; responses were only obtained to K although the grove lay in an area of both K and Mg deficiency. A single application of 3.4 kg (7½ lb) potassium sulphate per palm gave a mean response over 6 years of 20 per cent; a higher application had no further effect. Although the 3.4 kg application was economic, it is obviously more advantageous to apply the fertilizer to palms capable of a higher production of oil in the bunch. However, it is of value to know that in any rehabilitation scheme in which some old palms are retained for a number of years, the old palms are likely to benefit from any general potassium dressing.

The improvement of grove areas by planting

Very early attempts to replace the grove palms by planted palms were largely unsuccessful owing to poor planting techniques, unimproved material, lack of fertilizers, and sometimes disease. Since the war, however, the establishment in such countries as Nigeria, the Ivory Coast and Sierra Leone of plantations in grove areas showed that there was no agricultural impediment to replanting, and the increase in oil production to be obtained merely by the substitution of good *tenera* for poor *dura* material has provided a further incentive. At Abak in the heart of the potassium and magnesium deficient Acid Sands areas of eastern Nigeria, plantings on old grove land were by the mid-1960s giving bunch yields of around 12 tons per hectare in the sixth year after planting. This was three to five times the yield of groves in the area and in terms of palm oil would be five to eight times grove production.

The main practical consideration in the grove areas was how to retain a reasonable income while the planted palms were coming into bearing; several methods were suggested for partially or gradually removing the stand and some of these were embodied in an experiment carried out in the grove at Obio-Akpa in eastern Nigeria which has already been referred to. This experiment had five treatments as shown below and its results are given in Table 3.4.

Treatment A. Undisturbed palm grove. As already stated, these plots showed considerable fringe improvement, but the elimination of a guard strip in all treatments reduced this effect.

Treatment B. Complete clearing of grove and replanting at 29 feet (8.83 m) triangular.

Treatment C. Thinning by eye judgement to about forty smooth-stemmed palms per acre (ninety-nine per hectare). Planting in the spaces left.

Treatment D. Thinning on the basis of pretreatment yields to about twenty-eight smooth-stemmed palms per acre (sixty-nine per hectare). Planting in the spaces left.

Treatment E. Thinning to half plantation density. Planting fully at plantation density 29 feet (8.83 m) triangular. Removing old palms when young ones come into bearing.

The thinning and planting were carried out in 1953 and the young palms came into bearing in 1957.

The period 1961–4 may be regarded as the first 4 years of mature yields for the *planted palms*. During that period those palms in the E and B plots gave nearly double the bunch yield of the untouched grove (A), a 60 per cent higher yield than the thinned grove palms of the D plots, but only an insignificantly higher yield than the grove palms of the C plots thinned out by eye judgement. It is thus evident that if thinning out is done to obtain an even stand, a considerable increase in bunch yield may sometimes be obtained; but the differences between the grove palm yields of the C and D plots indicate how variable results may be.

By 1965–6 the total yields of the completely replanted plots B and E had overtaken the total yield of the C plots but it would be many years

Table 3.4 *Palm grove replanting experiment. Obio-Akpa, Nigeria* (yield of bunches per hectare)

Treatment	1953–6	1953–60			1961–4			1965–6			1953–66
	Grove palms (kg)	Grove palms (kg)	Planted palms* (kg)	Total (kg)	Grove palms (kg)	Planted palms (kg)	Total (kg)	Grove palms (kg)	Planted palms (kg)	Total (kg)	Total (kg)
A. Control grove	12,129	28,104	–	28,104	16,585	–	16,585	11,114	–	11,114	55,804
B. Complete replanting	–	–	23,441	23,441	–	27,855	27,855	–	18,210	18,210	69,507
C. Thinning and planting. Eye judgement	10,545	43,785	2,210	45,995	26,109	4,161	30,270	14,934	3,018	17,952	94,218
D. Thinning and planting from pretreatment yields	8,285	29,425	7,326	36,751	17,976	10,802	28,778	10,954	6,289	17,243	82,771
E. Replanting under thirty grove palms/acre cut out in 1958	7,882	14,819	16,819	31,638	–	31,704	31,704	–	19,375	19,375	82,717

* Planted palm yields are from 1957, when they came into bearing.

before B and E could catch up in cumulative bunch yield. When oil and kernel contents of grove and planted palms are taken into account the superiority of B and E become more apparent and Zeven[2] concludes that complete replanting with partial felling of the old stand is to be recommended provided only thirty-seven grove palms per hectare (fifteen per acre) are retained and these are felled 4 years after planting.

The climate and soils of the oil palm regions

The climatic features of areas of highest production may be summarized as follows:

1. A rainfall of 2,000 mm (80 in) or more distributed evenly through the year, i.e. no very marked dry seasons.
2. A mean maximum temperature of about 29°−33°C (85°−90°F) and a mean minimum temperature of about 22°−24°C (72°−75°F).
3. Constant sunshine amounting to at least 5 hours per day in all months of the year and rising to 7 hours per day in some months.

It is in such a closely defined climate that the oil palm is cultivated in Indonesia and Malaysia. But in both these territories soil differences, and of course fertilizer practice, are responsible for quite substantial differences of yield. Moreover, in certain parts of these oil palm regions night temperatures may be a little lower than the above means and in some months of heavy rainfall sunshine hours may occasionally be below 4 per day; years of comparatively low rainfall are also not unknown.

While it may well be true that the highest yield will always be obtained on good soils under the climatic conditions stated above, the oil palm has been profitably cultivated in regions where rainfall is poorly distributed or very high, where temperatures are low in certain months or where sunshine hours are well below those of the Far East.

There are both botanical and economic reasons for the cultivation of the oil palm in areas where both climatic and soil conditions are much less favourable than those in the Far East. In the first place the plant is well adapted to areas of summer rainfall and winter drought. Though bunch production is reduced by drought conditions, 3 months without rain does not markedly reduce the health of the plant; growth of the bud continues, but spear leaves tend to remain unopened until the onset of wet weather and midday stomatal closure prevents excessive moisture loss.

Secondly, the oil palm is such a high producer of oil that even under the poorest plantation conditions, which may be exemplified by those parts of West Africa having 3 months without rain, few hours of sunshine during the wet season and poor, rapidly-drying soils, production compares favourably with that of other oil-bearing crops. Thus plantation yields under very favourable Malaysian conditions have been 25 to 30 tons of bunches per hectare per annum while Nigerian estates have been producing only 8 to 11 tons. But the latter yield supplies, from *tenera* material, over 2 tons of palm oil with ½ ton of kernels, a production exceeding in terms of oil that of the best coconut estates. Such production is rendered the more economic by the fact that labour in the low-production regions has been less expensive. It will be seen therefore that any definition of the limits of what might be termed favourable climatic conditions for the oil palm is extremely difficult.

Climate

In view of the difficulties of defining the limits of a 'suitable climate' for the oil palm, as explained above, it is proposed to consider each factor in turn and to discuss its magnitude in relation to other factors. Meteorological data relevant to this discussion are given in Tables 3.5 to 3.7 Centres have been chosen to show the contrasts which exist both within and between regions.

Rainfall and the water balance

The effect of prolonged drought on reducing the yield of the oil palm is well known and the physiological effects will be described in the next chapter. In areas where 2 to 4 months drought are the rule, e.g. Dahomey, southern Zaire and parts of Nigeria, there is a tendency for there to be big yield fluctuations from year to year with a year of very low yield occurring at intervals of 4 to 6 years.

Dahomey presents an extreme example of a country with a large production of oil and kernels, but with a low and poorly distributed rainfall. Four months (November to February) are almost devoid of rain and the average annual rainfall is 1,232 mm at Pobé Experiment Station. However, on certain soils of a high water-holding capacity and which overlie underground water, production is at least double that of palms on soils not so favoured and, with good planting material, a yield of over 12 tons of bunches per hectare has been achieved.[13]

In Nigeria, water supply has been shown to be deficient through uneven rainfall distribution even in areas where the total annual rainfall exceeds 2,000 mm. Yields are correlated with dry-season rainfall and with measures of 'effective sunshine' which take into account the distribution of such rain as falls in the dry season (see p. 187). In areas adjoining Dahomey where the rainfall is similar to that at Pobé, yields are very low, agreeing with those on the poorer Dahomey sites.

In southern Zaire (6–7°S) a dry season of 2 to 4 months obtains, with total precipitations of 1,400–1,900 mm. Yields have been very variable and disease has been the dominant feature of these plantings.

Dry seasons may be severe in several parts of tropical America where the oil palm is grown. Some instances may be given. Firstly in the north of

Table 3.5 Rainfall at centres of oil palm cultivation (millimetres)

Centre		Lat. and Long.	No. of years	Jan.	Feb.	March	April	May	June	July	Aug.	Sept.	Oct.	Nov.	Dec.	Annual	
Asia																	
Malaysia (West)	Bagan Datoh	4°N 100°45'E	10	136	135	139	108	100	98	102	118	178	228	276	219	1,837	W Coast
	Telok Anson	4°2'N 101°1'E	70	238	189	240	262	172	111	107	132	174	278	289	268	2,513	Coastal plain
	Paya Lang	2°35'N 102°40'E	15	224	140	160	176	194	153	109	118	133	193	223	210	2,033	Inland S
	Ulu Remis	1°15'N 103°30'E	17	282	160	262	246	205	138	159	166	180	239	222	247	2,507	Inland S
	Jerangau	4°59'N 103°9'E	10	363	182	152	159	204	170	196	255	316	314	519	804	3,634	E Coast State
Malaysia (East—Sabah)	Mostyn	5°N 118°5'E	18	207	118	145	206	244	212	156	188	226	213	202	205	2,322	E Coast
	Beluran	4°3'N 117°30'E	27	520	354	265	125	190	249	206	234	255	241	230	408	3,278	NE Coast
Indonesia (Sumatra)	Medan	3°35'N 98°41'E	58	114	91	104	132	175	132	135	178	211	259	246	229	2,487	E Coast 20 km
	Tindjowan	3°6'N 99°29'E	26	156	109	141	154	140	111	109	170	202	244	205	171	1,912	E Coast 22 km
	Marihat Baris	2°58'N 99°6'E	21	311	223	287	305	296	214	201	277	358	452	411	292	3,627	60 km Inland
Africa																	
Sierra Leone	Njala	8°6'N 12°6'W	39	12	22	79	127	251	364	418	517	437	338	180	38	2,822	Inland
Ivory Coast	La Mé	5°3'N 3°5'W	45	37	63	128	147	270	454	221	100	100	182	176	89	1,907	
Ghana	Aiyinasi	5°N 2°20'W	10	37	71	140	191	378	751	293	64	116	246	129	98	2,511	S West
Dahomey	Pobé	6°6'N 2°4'E	45	16	42	111	149	185	216	123	61	118	155	47	12	1,231	
Nigeria	N.I.F.O.R. Benin	6°30'N 5°40'E	33	14	29	98	161	192	254	350	221	306	223	58	10	1,916	
	Umudike	5°29'N 7°33'E	36	22	51	113	204	267	273	312	253	310	262	84	18	2,168	Eastern region
	Abak	5°5'N 7°40'E	18	29	47	131	196	237	310	357	317	384	300	134	30	2,472	
Cameroon	Lobé	4°30'N 9°10'E	8	31	80	174	196	190	377	610	750	649	374	115	31	3,577	
	Idenau	4°5'N 9°E	12	90	150	290	330	530	1,150	1,340	1,400	1,560	1,030	390	170	8,430	Coast
Zaire	Yangambi	0°49'N 24°29'E	30	85	99	148	150	176	127	146	169	181	235	183	123	1,822	N Congo basin
	Kiyaka	5°S 19°E	10	175	114	220	225	97	9	22	45	123	220	243	185	1,668	Kwilu
America																	
Brazil	Belem, Para	1°28'S 48°27'W	21	317	405	453	425	253	166	162	105	123	126	101	235	2,871	
	Taparoa, Bahia	13°32'S 39°6'W	5	118	135	305	215	183	193	131	152	86	116	123	142	1,899	Coast
Colombia	Aracataca	10°35'N 74°9'W	13	2	8	9	52	216	178	125	182	275	344	239	31	1,661	North
	San Alberto	7°40'N 73°30'W	10	59	37	113	220	305	242	209	187	240	357	343	92	2,412	
	El Mira, Tumaco	1°33'N 78°41'W	4	277	296	313	390	421	412	182	169	169	176	125	210	3,120	
	Bajo Calima	3°59'N 76°52'W	7	289	224	281	405	549	467	458	434	569	359	496	373	5,093	Pacific coast
Ecuador	La Concordia	0°05'N 79°20'W	8	506	468	614	545	340	220	97	79	75	65	48	156	3,213	
Costa Rica	Quepos	9°26'N 84°9'W	19	46	19	27	103	338	394	446	395	424	653	323	132	3,300	Pacific coast
Honduras	Tela	15°43'N 87°29'W	27	255	146	77	90	103	122	181	240	207	339	410	356	2,526	North coast

Table 3.6 *Sunshine at centres of oil palm cultivation (hours per day)*

Centre	Lat. and Long.	No. of years	Jan.	Feb.	March	April	May	June	July	Aug.	Sept.	Oct.	Nov.	Dec.	Annual Hrs/day	Annual Total
Asia																
Malaysia																
Kuala Lumpur	3°7'N 101°42'E	17	6.2	7.4	6.5	6.3	6.3	6.6	6.5	6.3	5.6	5.3	4.9	5.4	6.1	2,230
Chemara, Johore	1°15'N 103°30'E	6	3.5	5.1	5.0	5.7	6.1	5.1	5.4	5.0	4.1	4.4	3.9	3.7	4.8	1,729
Sumatra																
Medan	3°35'N 98°41'E	21	5.4	7.1	7.0	7.2	7.7	8.1	8.1	7.5	7.0	6.2	5.9	5.4	6.9	2,508
Sabah																
Mostyn	5°N 118°5'E	12	6.0	6.7	7.0	7.3	6.3	6.4	6.8	6.6	6.5	6.1	6.5	6.2	6.5	2,384
Africa																
Sierra Leone																
Njala	8°6'N 12°6'W	32	7.2	7.3	6.7	6.1	6.0	5.1	2.9	2.0	3.5	5.6	6.1	6.4	5.4	1,971
Ivory Coast																
La Mé	5°3'N 3°5'W	8	5.2	6.1	6.5	6.4	5.2	2.9	3.3	3.2	3.1	5.0	5.7	5.9	4.9	1,781
Dahomey																
Pobé	6°6'N 2°4'E	8	6.0	7.0	6.4	6.1	5.9	4.7	3.3	3.3	3.7	5.2	6.4	6.6	5.4	1,963
Nigeria																
N.I.F.O.R., Benin	6°30'N 5°40'E	15	5.6	6.0	4.9	5.3	5.4	4.2	2.6	2.4	2.6	4.2	6.0	6.4	4.6	1,692
W. Cameroon																
Idenau	4°5'N 9°10'E	9	5.2	6.7	4.7	5.0	4.7	2.6	1.6	1.0	1.3	2.4	3.5	4.5	3.6	1,306
Zaire																
Yangambi	0°49'N 24°29'E	10	6.6	6.8	6.0	6.1	6.0	5.5	5.0	4.4	5.2	5.1	5.5	5.7	5.6	2,054
Kiyaka	5°S 19°E	6	5.1	5.2	4.7	5.5	6.8	8.9	8.3	7.8	6.4	6.1	5.5	4.8	6.3	2,287
America																
Brazil																
Belem, Para	1°28'S 48°27'W	21	5.2	3.9	3.2	4.0	5.3	7.5	8.0	8.0	6.7	7.2	6.1	6.2	6.0	2,195
Iguape, Bahia	12°30'S 39°W	4	7.1	8.2	6.9	6.3	4.6	4.5	4.6	6.4	7.1	7.6	7.7	5.6	6.4	2,323
Colombia																
Aracataca	10°35'N 74°9'W	13	8.6	8.6	8.8	7.7	7.2	6.8	7.8	6.9	7.1	7.2	6.9	8.4	7.7	2,792
San Alberto	7°40'N 73°30'W	4	7.3	5.4	3.8	4.6	5.3	5.3	6.4	6.0	6.0	6.0	5.1	7.0	5.7	2,070
Bajo Calima	3°59'N 76°52'W	7	3.9	4.1	3.3	3.0	3.4	3.7	3.1	3.4	3.9	3.3	3.3	3.6	3.4	1,243

Table 3.7 Temperature at centres of oil palm cultivation (mean, mean maximum and mean minimum °C)

Centre		Lat. and Long.	No. of years		Jan.	Feb.	March	April	May	June	July	Aug.	Sept.	Oct.	Nov.	Dec.	Av.
Asia																	
Malaysia	Telok Anson	4°2'N 101°1'E	24	Mean	27.1	27.7	27.9	28.0	28.2	28.0	27.7	27.7	27.5	27.3	27.1	27.1	27.6
				M. max	31.7	32.3	32.8	32.8	32.8	32.8	32.5	32.4	32.0	31.7	31.4	31.4	32.2
				M. min.	22.6	22.8	23.0	23.3	23.5	23.2	22.9	23.0	22.9	23.0	22.9	22.8	23.0
	Kluang	2°1'N 103°19'E	13	Mean	24.8	25.5	25.7	25.8	25.9	25.8	25.6	25.3	25.2	25.2	25.1	24.9	25.3
				M. max	29.5	31.1	31.8	32.1	31.7	31.2	31.0	31.0	31.0	31.1	30.6	29.8	31.0
				M. min.	21.7	21.8	22.1	22.2	22.4	22.0	21.8	21.6	21.6	21.8	22.0	21.9	21.9
Sumatra	Medan	3°35'N 98°41'E	10	Mean	25.4	26.0	26.4	26.5	26.8	26.5	26.4	26.1	25.9	25.7	25.4	25.3	26.0
				M. max.	29.9	31.3	31.5	31.6	31.7	31.4	31.8	31.2	30.9	30.1	29.7	29.6	30.9
				M. min.	22.2	22.1	22.5	22.8	23.2	22.7	22.4	22.3	22.4	22.6	22.5	22.3	22.5
Africa																	
Sierra Leone	Njala	8°6'N 12°6'W	32	Mean	26.1	27.4	27.9	27.2	27.3	26.1	25.2	24.6	25.6	26.2	26.3	26.2	26.4
				M. max	32.3	33.0	33.3	32.8	32.6	30.7	28.9	28.2	29.5	31.0	31.3	31.4	31.5
				M. min.	19.8	20.7	21.3	21.8	21.8	21.5	21.4	21.5	21.6	21.3	21.3	20.5	21.2
Ivory Coast	La Mé	5°3'N 3°5'W	15	Mean	27.2	27.8	28.2	28.0	27.5	25.8	25.2	25.0	25.6	26.6	26.9	26.9	26.7
				M. max.	32.1	32.7	32.7	32.4	31.2	28.8	28.1	28.2	28.4	30.1	31.3	31.5	30.6
				M. min.	22.7	22.4	22.8	22.8	22.9	22.3	21.3	21.4	22.2	22.5	21.9	21.9	22.3
Nigeria	N.I.F.O.R., Benin	6°30'N 5°40'E	12	Mean	26.3	27.5	27.4	26.9	26.5	25.6	24.6	24.3	25.0	25.6	26.2	25.8	26.0
				M. max	30.9	32.7	32.4	31.5	30.9	29.4	27.6	27.4	28.3	29.6	30.7	31.2	30.2
				M. min.	21.6	22.3	22.4	22.2	22.0	21.7	21.5	21.3	21.8	21.6	21.6	21.4	21.8
Cameroon	Idenau	4°5'N 9°10'E	6	Mean	25.9	26.6	26.8	26.7	27.5	26.0	24.8	24.4	24.6	25.2	25.7	25.5	25.8
				M. max.	29.7	30.3	30.3	30.7	31.2	28.9	27.8	26.8	27.3	27.9	28.3	28.0	29.0
				M. min.	22.2	22.9	23.3	22.7	23.7	23.1	21.9	21.9	22.0	22.4	23.1	23.0	22.7
Zaire	Yangambi	0°49'N 24°29'E	10	Mean	24.1	24.1	24.3	24.4	24.1	23.6	23.1	23.0	23.2	23.3	23.5	23.4	23.7
				M. max.	30.2	30.8	30.6	30.3	30.1	29.5	28.5	28.4	29.2	29.1	29.3	29.0	29.6
				M. min.	19.6	19.4	19.9	20.3	20.0	19.8	19.3	19.5	19.4	19.5	19.7	19.5	19.7

Table 3.7 – *continued*

Centre	Lat. and Long.	No. of years		Jan.	Feb.	March	April	May	June	July	Aug.	Sept.	Oct.	Nov.	Dec.	Av.
America																
Brazil																
Iguape, Bahia	12°31′S 39°W	8	Mean	25.5	26.8	26.6	26.4	25.2	23.6	22.5	21.9	22.1	23.7	24.3	25.5	24.5
			M. max.	30.6	32.3	31.9	31.1	29.2	27.1	26.2	25.8	26.4	28.6	29.1	31.0	29.1
			M. min.	20.3	21.4	21.2	21.6	21.2	20.1	18.7	17.9	17.7	18.8	19.5	20.4	19.9
Belem, Para	1°28′S 48°27′W	28	Mean	25.7	25.4	25.5	25.9	26.0	26.0	25.8	26.1	26.0	26.2	26.4	26.1	25.9
			M. max.	30.9	30.3	30.1	30.4	31.3	31.8	31.7	32.1	31.7	32.0	32.0	31.7	31.3
			M. min.	22.6	22.6	24.0	23.0	27.7	22.6	22.2	22.3	21.9	22.1	22.2	22.4	22.5
Colombia																
Aracataca	10°35′N 74°9′W	13	Mean	27.3	27.4	27.5	28.9	28.4	27.8	28.0	27.9	28.1	27.5	27.5	27.8	27.8
			M. max	33.3	33.2	32.8	34.6	33.8	33.0	33.7	33.5	33.9	33.7	33.0	34.0	33.5
			M. min.	21.4	21.7	22.3	23.2	22.9	22.7	22.4	22.3	22.4	22.2	22.1	21.7	22.3
Barrancabermeja	7°4′N 73°52′W	23	Mean	29.3	29.6	29.7	29.4	28.9	28.9	29.2	28.8	28.7	28.7	28.6	29.1	29.0
			M. max	33.1	33.7	33.6	32.9	32.5	32.7	33.3	33.1	32.5	31.9	32.1	32.9	32.9
			M. min.	25.5	25.6	25.8	25.8	25.3	25.2	25.0	24.6	24.8	24.5	25.1	25.3	25.2
Bajo Calima	3°59′N 76°52′W	7	Mean	26.5	26.8	27.1	26.9	26.6	26.6	26.7	26.7	25.8	26.0	26.1	25.7	26.5
			M. max	30.5	30.6	31.2	30.5	29.7	29.9	30.5	30.4	30.6	29.8	29.8	29.5	30.2
			M. min.	22.5	23.0	23.0	23.3	23.6	23.3	22.9	23.0	22.9	22.3	22.4	21.9	22.8
Ecuador																
La Concordia	0°05′N 79°20′W	8	Mean	25.2	25.2	25.5	25.8	25.1	24.3	23.6	23.5	23.8	23.6	23.3	24.1	24.4
			M. max.	29.0	29.5	30.0	30.2	29.0	27.8	27.5	27.5	27.6	27.5	27.2	27.9	28.4
			M. min.	21.3	20.9	20.9	21.3	21.2	20.5	19.6	19.4	20.0	19.7	19.3	20.3	20.4
Honduras																
Tela	15°43′N 87°29′W	8	Mean	23.2	23.7	25.1	26.2	26.8	27.0	26.7	26.8	27.2	25.9	24.5	23.5	25.6
			M. max.	27.4	28.2	29.6	30.6	31.2	31.5	30.9	31.1	31.6	30.1	28.5	27.4	29.9
			M. min.	19.0	19.3	20.4	21.6	22.5	22.9	22.5	22.5	22.6	21.7	20.8	19.6	21.2

Colombia at Aracataca (11°N) there are 5 months, December to April, which provide a mean of only 100 mm of rain; the remaining mean of 1,524 mm falls in the period May to November. In some years the total precipitation is below 1,000 mm. In many parts of the area, however, water tables are high and irrigation water can be provided.

In the Palma-Sola area in Venezuela (10½°N) somewhat similar conditions obtain. Rainfall averages 1,450 mm, 4 months, January to April, usually having less than 150 mm of rain. Although the soil is water-retentive and subsoil water is to be found at 180−360 cm below the surface in all seasons, the dry weather is sufficiently severe to inhibit the opening of spear leaves as in West Africa. Irrigation is being tried.

On the Pacific coast of Costa Rica (9½°N) the rainfall pattern is similar to that of northern Colombia, but the total annual rainfall is double, i.e. 3,300 mm. There are usually 3 months, January to March, with less than 100 mm, but occasionally the dry season can extend over the 5-month period December to April. Here clay soils overlie a water table 120−180 cm below the surface, but the yield pattern is similar to that of West Africa, i.e. low yields coincide with the dry season and high yields with the beginning and middle of the wet season.

In recent years the concept of the water balance has played a much greater part both in assessing the suitability of areas for planting and in seeking the causes of yield fluctuations in already established regions.[1] Direct relationships have been found between the magnitude of water deficit estimates and bunch yields.[14, 15]

In calculating annual water deficits the year is divided into ten-daily or monthly periods and the following formula is applied:

B = Res + R − Etp

where B is the balance at the end of the period, Res is the soil water reserve at the beginning of the period and R and Etp are the rainfall and potential evapotranspiration during the period. The balance, B, is carried forward as the soil reserve at the beginning of the following period but this reserve has of course a maximum equal to the available water or field capacity.

The main difficulty with this method is determining the best estimates of available water stored in the soil and potential evapotranspiration. Available soil water will vary both with the physical properties of the soil and the rooting depth of the plant. Thornthwaite has estimated for perennial plants that except in areas of shallow soil the water storage capacity available to mature plants with fully developed root systems varies around a mean that is equivalent to 10 cm of rainfall.[1, 16] Studies on Acid Sands soils in Nigeria indicated that 100 mm are available at field capacity in the top 122 cm of soil,[17] while in Ghana it is also considered that available water in the top 100−120 cm of the majority of forest soils is not more than 100 mm.[18] However, 150 mm has been suggested in Malaysia[19] while 200 mm in the top 200 cm of soil has been used by the

Institut de Recherches pour les Huiles et Oléagineux (I.R.H.O.) for water-deficit calculations in West Africa and elsewhere.[14]

Estimates of potential evapotranspiration by Thornthwaite's method[20, 21] have been widely used. His estimates depend on mean monthly temperatures and day length; Fig. 3.1 shows Thornthwaite's mean monthly Etp with monthly rainfall at five centres of oil palm development. Other estimates have been employed. In Nigeria the use of Penman's method and lysimeter determinations suggested that Thornthwaite's figures might be some 25 cm per annum too high,[17] while Malaysian work has shown that evaporation is well correlated with saturation deficit and with sunshine hours and that computations based on these measures more nearly follow the variations in observed evaporation from pans.[19] Finally, I.R.H.O., after comparing Thornthwaite's figures with direct determinations and taking into account irrigation experiment results, have used the value of 150 mm for months with less than 10 rainy days and 120 mm for months with 10 rainy days or more.[14]

Van der Vossen,[18] who considered both available water and Etp to be overestimated in the I.R.H.O. system, compared water deficit estimates obtained by that system with those based on the lower figures of other authors and found that although mean annual water deficits were much higher when calculated by I.R.H.O. methods there was a very exact linear regression between the two sets of values. For comparative purposes therefore I.R.H.O. values are useful and in combination with a soil suitability classification have been used for forecasts of the adult bunch yields to be obtained in thirty-five combinations of soil and climate (Table 3.9).[22] It should be observed that the use of the soil classification, to be described later in this chapter, is to a large extent a method of correcting the constant 200 mm soil reserve maximum adopted. Table 3.8 shows the mean water deficit calculated by I.R.H.O. methods for a number of centres of oil palm cultivation. Figure 3.2 shows the relation between water deficit and bunch yield for I.R.H.O. Class I soils. In constructing curves the I.R.H.O. has used the deficit 28 months previous to the year of yield[15]

Table 3.8 *Annual water deficit at a number of centres of oil palm cultivation, using I.R.H.O. constants for ETP and available soil water* (millimetres)

Centre	Mean annual rainfall	No. of years	Water deficit		
			Mean	Highest	Lowest
Pobé, Dahomey	1,201	32	520	1,041	269
N.I.F.O.R., Nigeria	1,916	33	355	465	164
La Mé, Ivory Coast	1,993	22	254	703	28
Yangambi, Zaire	1,835	20	24	165	0
Bagan Datoh, W. Malaysia	1,837	10	169	375	0
Ulu Remis, W. Malaysia	2,300	14	5	67	0
Mostyn, E. Malaysia	2,322	18	12	158	0
San Alberto, Santander, Colombia	2,453	11	129	281	0

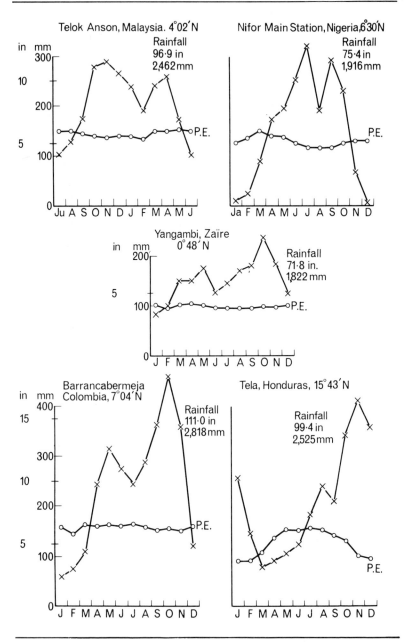

Fig. 3.1   Mean monthly rainfall and potential evapotranspiration in five oil palm districts.

Table 3.9 *Bunch production estimates for various soil classes and levels of water deficit (Olvin, I.R.H.O.)* (metric tons bunches per hectare)

Soil class	Water deficit (mm)						
	400	*350*	*300*	*250*	*200*	*100*	*Nil*
I	12	13	14	16	18	24	27
IIa	10	11	12.5	14	16	20	25
IIb	8	9.5	11	13	16	20	25
III	6	7.5	9	11	13	16	22
IV	4	5	6	8	9	12.5	16

following the work of Sparnaaij *et al.*[17] The latter authors evolved a parameter, 'effective sunshine', which was in effect a drought indicator. Effective sunshine measurement and its employment in predicting yield variations is described in Chapter 4 (p. 187) and its relation to nutrition is discussed in Chapter 11 (p. 532).

Oil palms are successfully cultivated in several areas of very heavy rainfall where the latter is always in excess of evapotranspiration. In the coastal areas of the West Cameroon over 9,000 mm per annum are recorded and estates inland receive over 5,000 mm. Along the Pacific plain of tropical South America as far as Quevedo in Ecuador, and on the Caribbean side of Central America, areas of very high rainfall are also encountered. In the former region this high rainfall is associated with constant cloudiness and the areas around Buenaventura in Colombia and Santo Domingo de los Colorados in Ecuador are among the dampest in the world.

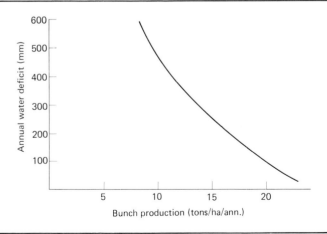

Fig. 3.2 Relation between the annual water deficit and bunch production on Class I soils (I.R.H.O.).

The effect of very high rainfall on the yield of the oil palm has not been investigated. Such rainfall occurs on areas of widely differing soil types, but is necessarily accompanied by constant leaching of the soil, by erosion and sometimes by very low solar radiation. Thus even the young volcanic soils of Ecuador are found, under these circumstances, to be deficient in magnesium, and nitrogen deficiency in young palms is common. Whether the rather disappointing yields often obtained under high rainfall conditions are due to such secondary effects only, or whether the constantly wet conditions are themselves harmful has not been determined. In Colombia there is some evidence on the Pacific coast of excessive bunch rot and poor fruit to bunch ratios. Soils of a very water-retentive composition may become waterlogged after months of very heavy rain even when the terrain is undulating. Under these circumstances young palms may be severely checked, the leaves becoming yellow and flaccid, but they recover after a period of dry weather. Such conditions exist in certain years on the 'Massape' soil of Bahia, Brazil, and on poorer inland soils in Malaysia.

Sunshine and solar radiation

The importance of a high level of solar radiation for the growth and bunch production of the oil palm has been inferred from several separate observations, but the exact requirements, either in terms of radiation or in hours of sunshine, for optimum yield are unknown. The importance of sunshine has been inferred from the following facts:

(a) Shading palms of all ages reduces growth and net assimilation rate.
(b) Shading adult palms reduces the production of female inflorescences.
(c) Pruning the leaves of adjacent palms increases the production of female inflorescences.[23]
(d) A positive, though not high, correlation has been found between annual sunshine data and yield in the 12-month period 28 months later.
(e) In countries such as Nigeria with marked seasonal differences in hours of sunshine, the differentiation of female inflorescences, as shown by the sex ratio at flowering 2 years later, is much higher during the months with many hours of sunshine than during the months with few.[24]

It is on such evidence as the above that the oil palm has been called a light-loving plant or heliophile. When all this has been said, and the serious effect of shade on production noted, there are certain anomalies which remain difficult to explain. In the northern hemisphere the seasonal variations in yield appear to remain unaffected by marked differences of sunshine variation. Seasonal yield variation at Benin in Nigeria (6°N) is similar to that at Yangambi in Zaire (1°N), but in the latter place sunshine hours only fall below 5 hours per day in 1 month of the year, while at Benin there are only 2½ hours per day over a period of 3 months. At Aracataca in Colombia sunshine is only below 7 hours per day in 3 months of the year, but seasonal distribution of yield is the same as at the other centres. Thus even where, as at Yangambi and Aracataca, high sunshine levels are reached during the wet season, a comparatively low proportion of female inflorescences is still differentiated during the period.

Again, sunshine hours at Yangambi approximate to those of plantations in Malaysia and rainfall distribution is even. Yet bunch yields at Yangambi have been nearer to those of low-sunshine Nigeria than to those of the Far East.[25] The only climatic difference between Yangambi and the Malaysian plantations is one of temperature and this difference is quite marked (Table 3.7).

As might be expected, areas of very high rainfall have few hours of sunshine. At Calima in Colombia where rainfall averages over 6,000 mm per year, daily sunshine is recorded as varying from a mean of 2.1 hours in July to 4.1 hours in February with a total of only 1,243 hours in the year. Figures as low as 750 hours have been suggested for parts of the Pacific plain in Ecuador. Records of adult palms in these areas show that yields higher than those of the seasonal parts of West Africa can be obtained. From this it can be inferred that the effect of 3 months' drought is more serious than a reduction by about 50 per cent of the hours of sunshine and that, provided water is available to the plant throughout the year, the palm will tolerate a considerably lower level of radiation than obtains in the Far East.

It is unlikely that more will be learnt about this subject until more satisfactory methods than the recording of sunshine hours have been more widely adopted for the measurement of radiation, and the effect of various light intensities on assimilation have been studied with the oil palm. It has been suggested, for instance, that in Ecuador, where sunshine hours are very low, the total sun and sky radiation may not be correspondingly low owing to the comparatively thin layer of cloud which covers the sky for long periods. At Benin in Nigeria where radiation records have been available for several years it has been noticed that there is an inverse relationship between hours of sunshine and total sun and sky radiation over the period October to April. Whereas in the middle of this period the sunshine is sufficiently strong to burn the Campbell-Stokes recording card for 6 to 7 hours per day, the sky is hazy with dust blown from the Sahara, and although there is little or no rain the total radiation is not very high. However, at the beginning and end of the wet season, in April to May and October to November, there are less hours of sunshine recorded but, with clearer skies, the total radiation is higher. In general, at Benin, it has been found that sunshine hours must drop below five per day before there is an appreciable fall in total radiation, but that radiation may also be low when the atmosphere is dust-filled in the dry season. These findings are shown graphically in Fig. 3.3 and it will be seen that while sunshine hours and radiation follow the same trends from April to November, this is not so from December to March.

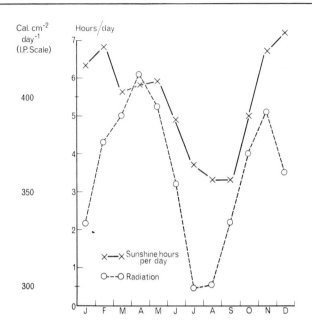

Fig. 3.3 Sunshine hours and total sun and sky radiation at N.I.F.O.R., Benin, Nigeria; mean monthly figures for the year 1958—63.

Temperature

Temperature is probably a much more important factor in determining growth and yield than has been generally realized. Henry[26] has shown that growth of young seedlings is totally inhibited at 15°C, and that growth at 25°C is seven times, and at 20°C three times, as rapid as at 17.5°C. He estimates optimum temperature for growth to be 28°C.

It is usual in countries of the humid tropics for mean minimum temperatures to be lower inland than near the coast or on a coastal plain. Low minimum temperatures are also found in areas of high latitude and, of course, where the elevation exceeds about 200 m. Thus, while mean minimum temperatures in most oil palm-growing regions are over 21°C (70°F) in all months of the year, a high latitude area such as Tela in Honduras (15°43'N) has 5 months with temperatures below 21°C and 2 months around 19°C (66.5°F). This is accompanied by a shortened daylength and a mean maximum of only 27°C (81°F). In such a climate, which includes periods of even colder weather, the crop is reduced to very small proportions in the early months of the year and nearly 90 per cent of it is harvested in the 7 months from June to December. Investigations of the effect of temperature on the duration of bunch development and analogy with other crops suggest that this uneven yield distribution will be due partly to delayed development of bunches during the colder months. In addition, there may be an effect on abortion towards the end of the cold period leading to a minimum of flowering from June to July and a corresponding minimum of fruiting at the beginning of the following year. It is also possible that there are sex ratio effects.

South of Salvador in Brazil ($13°-14°$S) even colder mean minimum temperatures are recorded. At Iguape, in the years 1961–3, there were 8 months with a mean minimum of below $21°$C and 3 months (August to October) with a mean minimum below $18°$C ($56°$F). Mean maxima may be as low as $22°$C in some months. Little is known yet of the effect of these low temperatures on production, but in the groves of Bahia the strongest period of fruiting is said to be from November to March.

The range of latitudes at which the oil palm can be grown differs from continent to continent owing to the effects on both temperature and rainfall of air and oceanic currents and of the great land masses.[27] The range is wider in America than in West Africa where distance from the equator tends to be directly correlated with length of dry season. In Asia the oil palm areas are largely outside the influence of continental land masses, being separated from them by expanses of ocean. Thus the Asian oil palm belt is relatively narrow and climatic variations within it not great.

Comparatively low night temperatures are characteristic of the areas far inland in Zaire. At Yangambi all months of the year have a mean minimum between 19.3 and $20.3°$C ($66.7°-68.5°$F). This feature of northern Zaire climate is the only one which distinguishes it markedly from that of the Far East. Mean maximum temperatures are also somewhat low, averaging $29.6°$C ($85°$F).

Mention was made in Chapter 1 of the presence of the oil palm at relatively high elevation in Africa, where low night temperatures may also be expected to show their effect. It has been reported that in Sumatra palms at above 500 m come into bearing at least a year later than palms in the lowlands, and that their early yields are correspondingly reduced.

Soils

It was early realized that the oil palm can be grown on a wide range of tropical soils and that the natural regulation of a water supply to the roots was a more important factor than the intrinsic worth of the soil. Bunting, Georgi and Milsum,[28] while suggesting that the most suitable soil was a 'loose alluvial loam overlying a friable clay subsoil', rightly confined their attention to four characteristics which in Malaysia would render the soil unsuitable. Later authors have drawn attention to the fact that in climatically marginal areas the characters of the soil become of greater relative importance. Naturally, in these areas it is a physical property of the soil, the available water within the rooting depth, which is of greatest importance, but the previous usage of the land and its nutrient reserves must also be taken into account.

Whenever possible, flat or gently undulating land should be chosen for the oil palm. Satisfactory yields can be obtained from hilly land in many countries, but costs of establishment and production will be increased. As will be seen in later chapters, plantations must be designed for easy transport to the oil mill of large quantities of bunches and loose fruit. The operations which will be much more expensive in hilly country will be road construction and maintenance, harvesting and field maintenance including maintenance of paths; additionally, inspection and supervision will be more difficult, transport vehicles will be subject to greater wear and tear and mechanization may prove impossible. Apart from these matters of expense there are the difficulties of preparing the land for planting without encouraging erosion, and the maintenance, in the later life of the planting, of a cover sufficient to prevent further erosion. A great deal can of course be done to prevent erosion and encourage covers, and on really steep land the palms themselves are normally planted on platforms, but all this work adds further to the expense.

While these factors favour the use of flat land it must not be thought that, where hilly land is available and climatic and soil conditions are otherwise suitable, such land should not be employed. Profitable plantations have often been established in hilly terrain. In some countries, e.g. Malaysia, a slope limit for cultivation is imposed by law, and oil palms can normally be planted on slopes below this limit. It is where large areas for development exist and there is scope for choice that flat land should be preferred.

It is sometimes claimed that the existence of palm groves or groups of semi-wild healthy palms will be an indication of suitable soil. This is a dangerous assumption for a number of reasons. Firstly, palms growing alone, or in small groups have the advantage of maximum soil volume and light, and are often in favoured positions. Secondly, adult palms are much more tolerant of waterlogged conditions than young palms, and may have grown up at a time when waterlogging was less severe. The introduction of young transplants, without attention to drainage, into grove areas carrying apparently healthy plants may therefore prove initially disastrous. Thirdly, a grove may develop in a small area where river or underground water supplements a meagre rainfall. This grove will not be representative of a wider area. Lastly, the vegetative appearance of a group of palms, i.e. their trunks, leaf-spread and the health of their leaves, is not an indication of their bunch-producing capacity.

On the other hand, it may be possible by control of the vegetation and the use of correct fertilizers to renovate land by planting oil palms; and in certain savannah areas of West Africa, notably around Dabou in the Ivory Coast, the land, though degraded by food-cropping and now actually devoid of grove palms, has sufficiently good basic characteristics for such renovation. Lalang (*Imperata cylindrica*) areas have been similarly renovated in Indonesia and Malaysia. In cases of this kind both soil examination and pilot schemes are desirable before large-scale planting is begun.

Unfavourable soil features

Soils which are unfavourable to the oil palm and which must be avoided are as follows:

1. *Poorly drained soils.*   Soils may be poorly drained because of the high water level of adjoining streams, rivers, etc., or they may be poorly drained because of their own structure. In the former case amelioration may be purely a matter of engineering while in the latter case, of which there are many examples, it may be impossible in some areas, or at times of very high rainfall, to drain the soil even when frequent and deep drains are provided. Soils of this kind are encountered in inland areas of undulating land both in Malaysia and Brazil. The effect of the lack of movement of water through the soil is most marked in young palms. If the soil can be sufficiently drained to bring the young palms through their early years, later growth and production may prove satisfactory. This is because the root system of the oil palm itself has a 'drying-out' effect through transpiration and through its effect on structure, though the former effect will only be noticeable under conditions of temporary poor local drainage. In areas where a high water table cannot be lowered, however, successful plantations cannot be established.

2. *Lateritic soils.*   The term is here used to mean soils containing concretionary ironstone or plinthite[66] usually in the form of gravel, but sometimes in thick bands in the subsoil or, following erosion, becoming exposed at the surface. Soils with a small proportion of laterite may be quite satisfactory, but larger quantities of gravel or sheets of laterite near the surface lead to reduced rooting volume and rapid drying out of the soil in dry weather. Palms may, therefore, suffer drought conditions even in a climate where there is usually a satisfactory water balance.

3. *Very sandy coastal soils.*   The oil palm does not grow satisfactorily on very sandy sea coasts where quite reasonable growth and yields are often obtained from the coconut. Occasionally very sandy soils are encountered inland and these are equally unsuitable.

4. *Deep peat.*   Though palms can be successfully established where 90−120 cm (3−4 ft) of peat overlie a good clay subsoil, a satisfactory stand cannot be obtained on deep peat of 250 cm (10 ft) or more. Contraction of the peat takes place on draining and the palms are unable to develop a root system sufficient for anchoring them in the soil. Palms lean

in all directions, some fall over, and production is low and access often difficult.

5. *Other unsuitable soils.*   There are a few other unsuitable soils, usually of limited area, which will be mentioned under individual regions.

The oil palm on the principal groups of tropical soils

The above description is of a negative nature. But it must be realized that for almost any crop in the tropics soil surveys are scanty and their assessment incomplete. Great tracts of the wet tropics are totally unsurveyed. It is not many years since Charter, in considering 'areas of more or less undamaged forest' where the oil palm is most likely to be grown, stated that

> although it is possible to map soils of any area . . . it is not possible to make accurate assessments of the value of the soil for any particular crop since we have not the necessary knowledge of the agronomy of tropical crops.[29]

He therefore stressed the importance of recognizing, describing and mapping soils and assessing them by reference to the experience of practical agriculturalists and by formal agricultural experiments. Coulter has also drawn attention to the difficulties of extending soil surveys into land use surveys owing to the scanty information available, and has stressed the need for studies of fertility, through crop field experiments, on soils which have been surveyed, and for studies of the chemical and physical factors controlling fertility.[30] Most provisional classifications of the soils of oil palm countries, though often following in broad outline the United States Department of Agriculture system, are adapted to their own local circumstances. However, under the early U.S.D.A. system the vast majority of areas bearing oil palms will be either latosols, alluvial soils or laterite soils (ground-water laterite).[31, 32] The undesirable features of these latter soils have already been mentioned, but when laterization is not extensive they may be considered akin to latosols. The majority of the latter would be Oxisols or Ultisols under the U.S.D.A. 7th Approximation and are referred to in much recent literature as ferralitic soils.

Latosols.   The latosols, which may be various shades of red, brown and yellow in colour, comprise all those soils in climates suited to the oil palm which are characterized by (*a*) low silica sesquioxide ratios of the kaolinitic clay fraction, (*b*) medium to low cation exchange capacities, (*c*) a relatively high degree of aggregate stability, (*d*) a usually low content of primary minerals except quartz. In other respects the latosols vary widely according to parent material and the amount of leaching; they are in fact the soils which develop over the great variety of parent material found in the wet tropics, and they cover vast areas. Their profiles are poorly differentiated though stone lines or some laterite concretions may sometimes appear. Colour may be indicative of parent material — soils from volcanic rocks are often reddish-brown — or may indicate the degree of leaching. Hence latosols used for oil palm cultivation are often classified by colour and derivation, e.g. 'Reddish-brown latosols derived from basic igneous rocks' or 'yellow latosols, strongly weathered and derived from sandstones', etc.*

Alluvial soils. Alluvial soils form a very important group for the oil palm in spite of their great variations in fertility. Coastal and riverine alluvial soils are commonly planted with the oil palm in Asia and America. The marine clays when predominantly montmorillonitic, not kaolinitic, are particularly productive. These soils may present considerable drainage problems, and where there has been an accumulation of organic materials they are often overlain by muck soils or by pure peat.

Alluvial soils in broad river systems, though in many ways similar to coastal alluvium, tend to have discernible sand factions and to be kaolinitic and micaceous; in narrow inland valleys river alluvium tends to have the characteristics of the sedentary soils round about and may be very poor.

Classification of soils in production forecasting

With the opening up of large areas for oil palm cultivation in various parts of the world a useful method of classing soils has been developed by Olivin.[22] This is based to a large extent on an assessment of the unfavourable factors already mentioned. Soils encountered in prospection are classed according to four of their properties, viz., texture, quantity of gravel or stones, water permeability or lack of drainage, and chemical composition, the latter being considered of much the least importance. The textural classes are seven in number from sand (clay, 5 per cent) to heavy clay or excessively silty soils; four levels in the horizon are taken into account. The gravel classes are six in number and follow the percentage gravel and stones in the four soil levels. For chemical composition classification pH, organic matter and exchangeable cations are taken into account. These four methods of classing the soil are drawn together into an 'agronomic classification' of which the main features are shown in broad outline overleaf. For the detailed classification of each character Olivin's paper[22] should be consulted. It will be seen that while texture and chemical status play some part in this classification it is mainly the quantity of gravel or the hydromorphic condition of the soil which will

---

* These distinctions of colour, particularly when associated with differences of rainfall, leaching and physical and chemical composition, find their place in certain nomenclatures in Africa, e.g. Charter's ochrosols and oxysols in Ghana,[33] the ferrisols and ferralsols of Sys[34, 35] in Zaire, and the classification of ferrallitic soils by French workers.[36, 37]

Soil Class	Characteristics			
	Texture	Gravel and stones	Drainage	Chemical status
I	Sands to clays	None	Good	Organic — good Exch. cations — good
IIa	Sands to clayey sands	None or very little	Good to 90 cm	Organic — medium Exch. cations — medium
IIb	Sands to clays	Some gravel	Good to 60 cm	Exch. cations — medium
III	Sands to clays	Gravelly	Poor	Organic — medium Exch. cations — poor
IV	Leached sand or very heavy clay	Very gravelly	Deep peat or very bad	Poor

determine its agronomic class. The latter classes are used in combination with water deficit determinations to estimate adult yields (Table 3.9).

Soils of the oil palm regions

Asia

Malaysia

The latosols of the Malay peninsula on which oil palms are grown are sedentary soils derived from igneous and sedimentary rocks; their manurial needs seem to depend largely on their derivation.[38] (Table 3.10, A, B and C.) the majority are Oxisols or Ultisols under the U.S.D.A. 7th Approximation.

Soils derived from volcanic types of basic and intermediate igneous rocks of andesitic or basaltic composition are comparatively rare. These are red-brown or red in colour with deep, uniform profiles and are of two distinguishable series, the *Kuantan* series being a reddish-brown sandy clay and the *Segamat* series being a red clay. No large areas of these soils have yet been used for the oil palm, but they are likely to be suitable in most respects. In spite of their high clay content they have an excellent crumb structure, and are very free draining.

Soils derived from acid igneous rocks, granite or granodiorite, vary with the structure of the parent rock and with topography. These are often considered the best of the commoner inland soils, and, although strongly leached and having a low quantity of exchangeable cations, are deep and of good structure. They are suitable for oil palms because of their fair waterholding capacity, depth and ready response to fertilizers, though they may dry out rather too readily in very dry weather. Perennial crops are assisted by the gradual release of nutrients in the lower horizons of the profile. For oil palms there is a primary deficiency of potassium on these soils, which can easily be rectified. Responses to other fertilizers may then follow and very satisfactory yields be obtained. Most common are the *Rengam* and *Jerangau* series which are reddish-yellow sandy clays; the less common *Tampin* series is a coarser-grained yellow sandy soil which drains rather too easily. The remaining yellow latosols are mostly derived from sedimentary and metamorphic rock and cover the largest area under cultivation. The parent material may be sandstone, shales, phylites, schists and quartzite, the latter name having been wrongly applied in the past to cover all the soils derived from sedimentary rocks. As a result of the great variety of parent material these soils vary from free-draining sandy loams (e.g. *Kedah* series) to very heavy silty clays (*Batu Anam* series) in which drainage is more or less impeded and the periodic presence of water tables quite high in the profile is shown by characteristic mottling. Laterite bands are not uncommon. The value of these soils for oil palm cultivation is equally variable and deficiencies of nitrogen, phosphorus and other nutrients are common. Many soil series are coming to be recognized, of which the *Serdang* series and the shale-derived *Munchong* series are the most suitable for oil palms. The latter, a clay soil is of good structure and occurs fairly widely[39] (Table 3.10, C).

A smaller group of yellow latosols developed over older alluvial material on low-lying dissected terraces and platforms have been shown to exist. It is important that these should be recognized since they are of lower fertility than the soils derived from sedimentary rocks.

The laterite soils of Malaysia, usually allotted to the *Malacca* series, need no special description. Their use for oil palm cultivation depends on the quantity of concretionary material in the soil. The largest areas of these soils lie in the north in Kedah, and in the south-west in Malacca, Negri Sembilan and North Johore.

The most important oil palm areas lie on the coastal marine alluvial clays, described by Panton as 'low humic gley' soils owing to their gley horizons developed close to the surface (Table 3.10, D). The clay fraction is predominantly montmorillonitic. Typically these soils have a friable brown organic clay topsoil overlaying a brownish-grey subsoil with, below, a horizon of grey silty clay with yellow and red mottling. Below this is a permanently wet blue-grey horizon with a pH as low as three. Initially, drainage is the most important factor in the usage of these soils for oil palms and in fact, it alters these soils considerably. Not only does any superficial peat layer disappear, but the gleyed and waterlogged horizons are altered. It has been recorded that if, on inspection, the gleyed horizon remains closer to the surface than 90 cm (3 ft), drainage needs to be improved. Although there is a wide variation in fertility there are large areas of coastal clay which possess high reserves of nutrients, and, in spite of high production, they do not respond to fertilizers so readily as inland soils. Areas are also found which provide excellent topsoil for prenursery beds or as a filling for polythene bags.

Table 3.10 *Some Malaysian profiles. Soil series of diverse origin*

A. *Basic Igneous Rock*

1. Segamat (*Basalt derived*)

Depth (inches) (cm)	2–9 5–23	9–16 23–40	16–26 40–63	26–37 65–93	37–48 93–122
Clay (%)	90	93	95	95	95
Silt (%)	7	4	2	2	2
Sand (%)	3	3	3	3	3
C (%)	1.76	0.88	0.66	–	–
N (%)	0.22	0.10	0.09	–	–
pH	5.1	4.9	5.0	5.0	5.0
Exch. K m eq/100 g	0.17	0.08	0.05		
Exch. Ca m eq/100 g	0.25	0.08	0.08	<0.05	<0.05
Exch. Mg m eq/100 g	0.93	0.59	0.50	0.42	0.40
Cation exch. capacity (C.E.C.) m eq/100 g	14.9	11.1	10.6	8.6	6.8
P, easily sol. in NaOH, ppm	89	82	78	55	35

2. Kuantan (*Basalt derived*)

Depth (inches) (cm)	0–3 0–8	3–12 8–30	12–24 30–61	24–36 61–91	36–48 91–122
Clay (%)	63	70	71	69	67
Silt (%)	10	9	10	10	10
Sand (%)	27	21	19	21	23
C (%)	3.14	1.21	0.77	0.77	0.49
N (%)	0.15	0.04	0.04	0.03	0.03
pH	4.2	4.1	4.7	4.7	4.6
Exch. K m eq/100 g	0.19	0.07	0.06	0.06	0.06
Exch. Ca m eq/100 g	0.42	0.03	<0.05	<0.05	<0.05
Exch. Mg m eq/100 g	0.74	0.20	0.16	<0.05	<0.05
C.E.C. m eq/100 g	20.2	11.5	10.1	9.2	8.2
P, easily sol. in NaOH, ppm	177	203	242	223	287
P, conc. HCl sol., ppm	885	750	825	845	1,260

The broad areas of river alluvium are not topographically separated from the coastal clays, but the soils are kaolinitic sandy clays whose properties are in most respects nearer to those of the sedentary rather than to the marine clay soils. The distinction between these contiguous alluvial soils is not often made, but Ng Siew Kee (see below) has shown how necessary it is to distinguish them.

Adjacent to the coastal clay soils, small areas of 'acid sulphate' soil formed under brackish water conditions are sometimes to be found. In this soil sulphates have been reduced to sulphides under anaerobic conditions, and the subsequent production of sulphuric acid produces pH values of below 3. Palms planted on this soil after drainage have shown serious deterioration, bunch production either ceasing or continuing at a very low level. Aluminium ions are responsible for the toxicity, and drainage

Table 3.10 — *continued*

B. *Acid Igneous Rock*

3. **Rengam** (*Granite derived*)

Depth (inches) (cm)	0−3 0−8	3−12 8−30	12−24 30−61	24−36 61−91	36−48 91−122
Clay (%)	43	46	58	65	66
Silt (%)	6	6	2	4	4
Sand (%)	51	48	40	31	30
C (%)	1.49	0.71	0.49	0.40	0.46
N (%)	0.16	0.11	0.07	0.06	0.06
pH	4.6	4.2	4.2	4.5	4.5
Exch. K m eq/100 g	0.41	0.17	0.12	0.09	0.09
Exch. Ca m eq/100 g	0.08	<0.05			
Exch. Mg m eq/100 g	0.33	<0.05			
Cation exch. cap. (C.E.C.) m eq/100 g	7.3	5.0	5.3	6.0	6.3
P, easily sol. in NaOH, ppm	39	22	19	19	18
P, conc. HCl sol., ppm	161	113	121	114	124

4. **Jerangau** (*Granodiorite*)

Depth (inches) (cm)	0−3 0−8	3−12 8−30	12−24 30−61	24−36 61−91	36−48 91−122
Clay (%)	36	50	57	59	57
Silt (%)	4	3	4	2	4
Sand (%)	60	47	41	39	39
C (%)	4.23	1.33	0.83	0.64	0.52
N (%)	0.21	0.09	0.06	0.05	0.05
pH	4.0	4.2	4.2	4.4	4.5
Exch. K m eq/100 g	0.20	0.08	0.07	0.06	0.06
Exch. Ca m eq/100 g	<0.05				
Exch. Mg m eq/100 g	0.33	<0.05			
Cation exch. cap. (C.E.C.) m eq/100 g	13.7	8.7	7.5	5.8	5.6
P, easily sol. in NaOH ppm	72	51	48	60	29
P, conc. HCl sol. ppm	211	221	224	239	232

increases the adverse effects on the palms. Growth and yield are improved by raising the water table.[40]

The effect of parent material on fertility in Malaysian soils has always been a subject of controversy. Very recently, on the basis of potassium and other studies, Ng Siew Kee[41] has deprecated the division of inland sedentary soils into fertility groups based on geology, and has claimed that the examination of the cation exchange capacity, exchangeable cations and clay mineralogy enables a distinction to be made only between recent marine clays and the rest of the soils. Examination of the analyses of typical profiles given in Table 3.10 lends support to his contention that a classification based solely on parent material *without plant response data* will be misleading (cf. Charter). However, with the oil palm, plant response data on the recognized soil series are accumulating, and certain broad

Table 3.10 — *continued*

C. *Sedimentary Rock*

5. Serdang *(Sandstone)*

Depth (inches)	0–3	3–12	12–24	24–36	36–48
(cm)	0–8	8–30	30–61	61–91	91–122
Clay (%)	22	34	36	42	42
Silt (%)	2	2	2	2	2
Sand (%)	76	64	62	56	55
C (%)	1.27	0.52	0.40	0.27	0.30
N (%)	0.10	0.06	0.04	0.03	0.03
pH	4.7	4.5	4.5	4.6	4.7
Exch. K m eq/100 g	0.14	0.10	0.08	0.08	0.10
Exch. Ca m eq/100 g	0.08	0.05	<0.05		
Exch. Mg m eq/100 g	0.42	0.42	<0.05		
Cation exch. cap. (C.E.C.) m eq/100 g	7.3	6.8	6.3	6.8	6.7
P, easily sol. in NaOH ppm	37	38	39	39	54
P, conc. HCl sol. ppm	62	75	88	97	105

6. Munchong *(Shale)*

Depth (inches)	0–3	3–12	12–24	24–36	36–48
(cm)	0–8	8–30	30–61	61–91	91–122
Clay (%)	63	69	71	70	73
Silt (%)	8	4	6	6	6
Sand (%)	29	27	23	24	21
C (%)	2.75	0.41	0.17	0.07	0.04
N (%)	0.26	0.11	0.09	0.07	0.04
pH	4.3	4.3	4.7	5.1	5.5
Exch. K m eq/100 g	0.30	0.12	0.09	0.09	0.09
Exch. Ca m eq/100 g	0.04	<0.05			
Exch. Mg m eq/100 g	0.42	0.30	0.33	0.33	0.33
Cation exch. cap. (C.E.C.) m eq/100 g	11.0	10.0	9.8	8.0	6.6
P, easily sol. in NaOH ppm	32	18	15	14	11
P, conc. HCl sol. ppm	92	92	76	81	—

distinctions can be made. For instance, it is well known that much better yields have been obtained on granodiorite-derived *Jerangau* series soils than on some shale-derived *Batu Anam* soils. How far this is due to differences in physical properties and hence to water relations, and how far to the soils' nutrient-supplying ability it is difficult to say (and the two are obviously interrelated), but the fact that fertilizer responses can be very different on sedentary soils of different derivations has been apparent in fertilizer experiments (see p. 566). What has probably received insufficient attention so far for oil palm cultivation are the differences between the sedentary soils in their drainage characteristics and their water-holding capacity.

Accounts of Malaysian soil series and a good appreciation of Malaysian

Table 3.10 — *continued*

D. *Alluvium*

7. Briah (*River flood plain alluvium*)

Depth (inches)	*0–4*	*4–14*	*14–43*
(cm)	*0–10*	*10–36*	*36–109*
Clay (%)	62	65	62
Silt (%)	35	33	31
Sand (%)	3	3	6
C (%)	3.34	0.32	0.23
N (%)	0.28	0.09	0.06
pH	4.5	4.5	4.4
Exch. K m eq/100 g	0.52	0.14	0.23
Exch. Ca m eq/100 g	1.43	0.50	1.18
Exch. Mg m eq/100 g	2.20	1.94	5.56
Cation exch. cap. (C.E.C.) m eq/100 g	25.8	21.0	20.0
P, easily sol. in NaOH ppm	79	73	77
P, conc. HCl sol. ppm	225	115	100

8. Selangor (*Marine clay*)

Depth (inches)	*0–6*	*6–12*	*12–24*	*24–41*	*41–53*
(cm)	*0–15*	*15–30*	*30–61*	*61–104*	*104–135*
Clay (%)	80	79	81	70	68
Silt (%)	18	17	17	18	20
Sand (%)	2	4	2	12	12
C (%)	1.31	1.04	0.82	1.01	1.44
N (%)	0.20	0.18	0.11	0.11	0.09
pH	4.7	4.3	4.3	5.4	7.7
Exch. K m eq/100 g	1.57	0.93	0.78	0.80	2.34
Exch. Ca m eq/100 g	4.8	3.9	5.1	8.1	10.7
Exch. Mg m eq/100 g	14.2	10.0	10.7	13.8	17.2
Cation exch. cap. (C.E.C.) m eq/100 g	32.5	32.5	30.0	30.9	52.0
P, easily sol. in NaOH ppm	65	84	125	128	42
P, conc. HCl sol. ppm	226	207	244	322	335

soils for oil palm planting have been provided by Leamy and Panton[42] and Ng Siew Kee.[41]

In *Sabah* oil palm development is taking place on riverine and coastal alluvium, on reddish-yellow latosols derived from sedimentary sandstones and shales and, to the north-east, on brown, reddish-brown and red latosols derived from a variety of rocks formed from lavas of successive volcanic eruptions. Yellow-brown soils derived from volcanic ashes are also found in this area, the soils of which have been surveyed and described by Paton.[43] Soils derived from basalt show some of the characteristics of the allied Malaysian soils; they are friable, uniform, reddish-brown clays which drain freely. Some of these soils have low contents of exchangeable cations but can be very productive; the high clay content, good structure and

Table 3.11   *Profile analysis: Brown Latosol developed over Quaternary lava*
Sabah, Malaysia (Paton). Table subfamily. Forest, Rainfall >2,000 mm.

Depth (inches)	0–4	4–15	15–33	33–72
(cm)	0–10	10–38	38–84	84–183
Clay (%)	36	76	83	78
Silt (%)	22	21	14	20
Sand (%)	25	0	0	0
pH	5.2	5.0	5.0	5.2
Exch. K m eq/100 g	–	0.12	0.20	0.46
Exch. Na m eq/100 g	–	0.09	0.08	0.08
Exch. Ca m eq/100 g	–	0.57	0.19	0.25
Exch. Mg m eq/100 g	–	0.51	0.09	0.34
Total exch. cations m eq/100 g	–	1.29	0.56	1.13
Exch. cap. (C.E.C.) m eq/100 g	–	11.7	10.8	10.8

possibilities of replenishment from parent sources appear to compensate for their relative paucity of nutrients (Table 3.11). Marked year-to-year yield variations have been experienced and are thought to follow water balance variations; Paton[43] has described these soils as being extremely porous and almost 'fluffy' and drought effects might therefore be expected early in any period of low rainfall, especially where the soil is shallow and overlies boulders.

Indonesia: Sumatra

No recent systematic account of the soils of Sumatra have been published. These soils are (*a*) 'liparitic' latosols (also described as Red-Yellow Podzolic soils) derived from recent rhyolitic tuffs, (*b*) latosols derived from older volcanic deposits (basaltic, andesitic, etc.), (*c*) latosols derived from tertiary sediments of both marine and riverine origin, (*d*) black soils developed over tertiary coral and shells. (*e*) alluvial soils, sometimes covered with peat, of the coastal plain and river estuaries.[44–46]

In the East Coast oil palm region soils derived from ryolitic volcanic acid tuffs have a relatively thick humic layer, are red to yellow in colour, and loose and crumbly in structure. They are Ultisols and in Indonesia are commonly known as Red-Yellow Podzolic soils,[66] and they are the most important and widespread soils supporting oil palms in the region. Their mineral content was high and the supply of potassium good, but they have been depleted of nutrients by long cultivation. They are characterised by often containing large quantities of minute quartz grains. On the less well-drained sites the soils tend to be more yellow.

South of the volcanic areas, and in Atjeh, there are large areas of latosols developed over tertiary sediments of sandstone and shale. These are sandy clays or sandy clay loams, yellow in colour and commonly referred to in Indonesia as Yellow Podzolic soils; they are poor in nutrients, particularly phosphorus,[46] and oil palms frequently show signs of magnesium deficiency. They are however suitable for oil palms, when well fertilized, except where impervious clay layers are present or where there is an excessive quantity of laterite gravel. These soils sometimes grade into yellow-grey soils of very varying texture and with some plinthite. Mixed soils of volcanic and sedimentary origin occur between the separate areas of the two derivations. Soils developed over older volcanic rocks are also encountered; they are more variable and are usually well supplied with nutrients.

In the lower-lying areas of the Sumatran East Coast plain the successful growing of the oil palm depends, as on the Malaysian coastal clays, on good drainage. A wide range of soils are encountered, including Humic Gley and Grey Hydromorphic soils, sandy Regosols and peat. Some areas of 'acid sulphate' soils, similar to those on the west coast of Malaysia, also occur.[44]

The soils developed over coral and black earth are mixed with volcanic ash or other volcanic material. Suitability depends largely on the depth of the soil.

Africa

The general statement has been made that freely drained tropical soils have a uniformity of profile, are deep and of a clayey texture.[47] This is because in the process of soil formation from rock in a climate suited to the oil palm the silica and cations tend to be leached out leaving high proportions of kaolinitic clay and hydroxides of iron and aluminium. Weathering is so active that most mineral particles either remain coarse or form clay; low silt to clay ratios are characteristic.

As will be noted in Table 3.10, most soils used for oil palms in Asia are clays. This is not so in the principal areas in Africa though the same soil-forming processes have been at work. The basic reasons for the formation of large areas of sandy clays or clayey sands, often with less than 20 per cent clay in the upper layers of the soil, are not entirely clear but are related to parent material. In Zaire for instance large expanses of country both in the north and south have parent material consisting of coverings of wind-borne sand with comparatively low proportions of clay (*nappes de recouvrement éolien*).[34] In West Africa, parent material of coarse unconsolidated sandstone is common and the amount of clay in the soil may sometimes be correlated with the extent to which the sandstones are interbedded with clay.[48]

There is one other important contrast between the oil palm regions of Asia and Africa. While in the former regions the climate tends to be uniform and changes of parent material are frequent and abrupt, in Africa very extensive areas are covered with the same parent material and there are substantial climatic differences within these areas. Thus, while differences due to parent materials are evident over the continent as a whole, differences due to climatic variations within the oil palm regions assume as great or greater importance. Furthermore, much of Africa has been land for a very long time and soil profiles may result from several cycles of weathering and erosion of parent materials overlying one another.[49]

Zaire

The oil palm is cultivated mainly in the north of the country between 1°S and 3°N in the provinces of Orientale and Equateur, and in the south in Kinshasa and Kasai provinces, at latitudes 4° to 7°S. The northern region has an even climate, but in the south there is a dry season lasting 3 to 4 months. The soils of these areas have been described by Kellogg and Davol[31] and more recently by Sys.[35]

The important soils of the *northern region* are latosols, designated by Sys as hygro-kaolinitic ferralsols, which are free draining, uniform and low in mineral reserves.

Table 3.12   *Profile analysis: Yangambi Latosol (Ferralsol)*
Zaire (Sys). Forest. An. rainfall 72 inches (1,822 mm)

Horizon and depth (cm)	A1—A3	B1	B2	C
	20	45	65	120
Clay (0—0.002 mm) (%)	26.3	30.0	38.7	35.6
Silt (0.002—0.05) (%)	2.1	2.0	2.1	1.7
Sand (0.05—2.0) (%)	71.6	68.0	59.2	62.7
C (%)	1.2	0.5	0.4	0.3
N (%)	0.10	0.6	0.04	0.03
pH	4.6	4.5	4.4	4.5
Exch. Ca, m eq/100 g	1.2	0.6	0.4	0.4
Exch. capacity m eq/100 g	3.5	4.5	4.6	4.3

Note: No data for exchangeable K or Mg are given for this profile. In a similar profile analysis by Kellogg and Davol Exch. K was 0.2—0.3 and Exch. Mg 0.1 m eq/100 g at all positions in the profile down to 72 inches (183 cm).

The most widely used of these latosols is of the *Yangambi* type (Table 3.12) which is derived from a 100—125 cm thick wind-borne covering which has a higher clay fraction than the other parent-material sands of Zaire. The soil is developed on gently undulating low plateau country, and, as is common in such terrain, the soils have a higher clay content towards the top of the slopes, becoming more and more sandy as the valleys are approached. Clay tends to be washed down the profile so that the lower horizons are appreciably less sandy. The higher parts of the plateau are considered very suitable for the oil palm. There is some colour variation from reddish-yellow to yellow, but typically a *Yangambi* forest soil is distinctly brown in the top 30 cm shading downwards to brownish-yellow at greater depths.

Latosols of the *Low Plateau* are found in the south-west of the region interspersed with swamp country along the Congo tributaries. These soils resemble the yellower of the Yangambi type soils in both their physical and chemical properties; the rainfall is somewhat higher in the region of these soils than in that of the Yangambi soils.

North of the Yangambi soils there is some oil palm cultivation on latosols derived from a diversity of rocks both igneous (granite) and sedimentary. Hence the colour, texture and level of exchangeable cations depend mainly on the parent material and may vary considerably. Laterite layers are common in some of these soils.

Recent alluvial soils carrying oil palms are found in this region near to the Congo and its tributaries and they cover a very wide area. The use of these soils for the oil palm depends on whether they can be drained. The clay fraction varies considerably and may be very high (70 to 90 per cent). Gley horizons are common. The cation exchange capacity is relatively high.

Also to be found in this northern region are some areas of soils derived from Salonga sand and Karroo rocks and which are described below.

The oil palm districts of *southern Zaire* between the Kwango and Kasai rivers are dominated by very poor sandy latosols, designated hygro- and hygro-xerokaolinitic arenoferrals by Sys, derived from the blown Kalahari and Salonga sands. The *Kalahari sand* ridges are found throughout the main oil palm area, are grass-covered and are unsuitable for the oil palm mainly on account of their very low water-holding capacity; the clay fraction is about 6 to 8 per cent. The *Salonga sands* lie within this area and to the north; though not quite so lacking in clay they are also unsuitable in this seasonal climate for oil palms.

Lying between the rounded Kalahari sand ridges and interspersed with the Salonga soils are the *Karroo latosols* (Sys's ferrisols). These soils, derived from sedimentary rock, are of a red colour and intrinsically fertile, but because of the topography and mixing with the sandy soils, satisfactory areas are limited. While very large areas of productive palm groves have developed in this region, plantations have been badly affected by disease and latterly very little planting has been done. When this soil appears in the northern oil palm region better use can be made of it.

West Africa

There being considerable climatic variations in the oil palm belt along the coast of West Africa, soils in this region are found to vary both with parent material and with climate. The majority of soils contain a high proportion of sand, and the effect of the ease of drainage and the drying out of these soils in the dry season is a main feature of oil palm cultivation in the region. Lateritic soils are found either very locally or in certain large expanses, e.g. in Sierra Leone. Almost all soils, except the best in Ghana, i.e. those usually employed for cocoa, are deficient in exchangeable cations, most conspicuously in potassium and magnesium. The amount of information available on the soils of the territories of West Africa varies considerably and published analytical data relate mainly to Nigeria and Ghana.

Nigeria and West Cameroon

The oil palm areas of *West Cameroon* contain soils derived from ancient basaltic flows. Rainfall being very heavy, some of these soils are not as rich as might otherwise be expected, and their clay content is surprisingly low. Marine alluvial soils exist seaward of the Cameroon mountain in areas of moderate and very heavy (8,000 mm) rainfall. These soils[50] are not as heavy as the Malaysian marine clays and they appear to be much affected by small differences of topography. While some are relatively free-draining, others are waterlogged gley soils requiring deep drains to bring them into a condition suitable for the oil palm.

On the west side of *West Cameroon* and across the border in *Nigeria*, estates have been established on latosols over basement complex rocks which consist mainly of granite and mica and quartzose schists, but which have also been influenced by gravelly drift. The profiles are immature and parent material in various stages of breakdown is much in evidence. The soils typically contain quantities of transported gravel and may in places be shallow. Some of the gravel is concretionary ironstone, but does not appear to be developing *in situ*. Hill creep is in evidence. In parts of some profiles the stones and gravel may exceed 50 per cent. Of the remaining soil, the clay fraction is usually higher than in other West African soils.[51]

When these soils are of a fair depth they are suitable for the oil palm, but they are very low in calcium, and are deficient in magnesium and probably in phosphorus; in contrast to other West African soils, potassium is normally in good supply in the higher parts of the profile. Unfortunately the areas covered by these soils tend to be rather hilly (Table 3.13).

The great areas of palm groves and of oil palm planting in Nigeria lie on Cretaceous and Eocene unconsolidated false-bedded coarse sandstones interbedded in varying degree with layers of clay. This great region, sur-

Table 3.13 *Profile analysis: Latosol derived from Basement Complex Granite, Hilltop*

Nigeria (Tinker). Recently felled forest. An. rainfall 115 inches (2,920 mm)

Depth (inches) (cm)	*0–2* *0–5*	*2–12* *5–30*	*12–23* *30–58*	*23–39* *58–99*	*39–60* *99–152*	*60–80* *152–203*
Stones and gravel (%)	17	23	50	44	28	21
Clay (%)	7	15	27	23	43	42
Silt (%)	3	5	3	7	6	7
Fine and coarse sand (%)	90	80	70	70	51	51
C (%)	1.7	0.9	0.7	0.4	0.4	0.4
N (%)	0.08	0.05	0.04	0.03	0.03	0.03
pH	3.6	5.2	5.4	5.5	5.3	5.5
Exch. K m eq/100 g	0.20	0.10	0.18	0.06	0.10	0.09
Exch. Na m eq/100 g	0.40	0.30	0.35	0.28	0.52	0.45
Exch. Ca m eq/100 g	0.96	0.20	0.16	0.06	0.24	0.22
Exch. Mg m eq/100 g	0.48	0.16	0.24	0.12	0.24	0.18
Total exch. cations m eq/100 g	2.04	0.76	0.93	0.52	0.10	0.94
Exch. capacity m eq/100 g	4.9	4.6	5.5	4.3	7.0	6.9

rounding the Niger delta and running some 300 miles from west of Benin to Calabar, still provides around half of the palm kernels of commerce though a much reduced proportion of the palm oil; it is as well, therefore, that the terms often used loosely in describing it should be correctly defined.

Benin sands.   This is a geological term that has been used for the parent material — false-bedded sandstones with subordinate beds of clay, of Eocene or Eocene and Cretaceous age.

Acid sands.   The latosols developed over the Benin sands. These have been divided into two main groupings,[48] the *Benin fasc* and the *Calabar fasc* which develop respectively in areas of high rainfall (*c.* 1,750 mm) with a severe dry season, and in areas of higher rainfall (2,000 mm) with a less severe dry season. These soils vary from slightly clayey sands to sandy clays, but in the top 30 cm the clay content rarely rises to 20 per cent and in the next 60 cm is rarely above 30 per cent. Attempts have been made further to classify these latosols into soil series by their clay content. In other respects the two groupings have been described as follows:

> The most obvious distinguishing characteristic is colour, Benin fasc soils being predominantly reddish, Calabar fasc predominantly yellow to yellowish-brown. Both fascs have a poorly differentiated profile, with reddish or yellowish-brown topsoil layers which shade off more or less sharply into reddish or yellowish subsoil respectively. The dominant colour becomes clearer and brighter down the profile. From about 6 ft (1.80 m) downwards the subsoil is of practically uniform appearance to a considerable depth, usually at least 20 ft (6 m).[52]

The differences between these two fascs have been compared with the differences between Charter's ochrosols and oxysols and in general the comparison holds good, particularly in respect of the ratio of calcium plus magnesium to potassium and in percentage saturation.

Thus the Benin fasc soils (Table 3.14) tend to be deficient primarily in potassium and the Calabar soils (Table 3.15) in potassium and magnesium. The satisfying of these requirements in the case of the oil palm leads to a demand for other nutrients (see Chapter 11). In other respects these soils have both advantages and disadvantages in oil palm cultivation. Firstly, they give a deep unrestricted rooting medium and thus allow the oil palm's anchoring roots to penetrate into the more clayey portion of the profiles; secondly, applied nutrients become easily available to the palm. On the other hand nutrients are very easily leached, particularly in the Calabar fasc, and it is difficult to rebuild the organic matter content of the topsoil if this has once been lost. Much of the region has been subject to cropping which rapidly reduced both the cation content and the organic content of the soil.

Table 3.14 *Profile analysis: Latosol, Acid Sands, Benin fasc*
Nigeria (Tinker and Ziboh). Oil palms. An. rainfall 72 inches (1,857 mm)

Depth (inches)	0–3	3–6	6–9	9–12	12–18	18–24	24–36	36–48
(cm)	0–8	8–15	15–23	23–30	30–46	46–61	61–91	91–122
Clay (%)	15	13	16	17	10	15	18	23
Silt (%)	2	1	1	1	1	2	1	1
Sand (%)	83	86	83	82	89	83	81	76
C (%)	1.3	0.7	1.0	0.3	0.4	0.3	0.3	0.2
N (%)	0.10	0.06	0.06	0.06	0.04	0.04	0.04	0.04
pH	5.7	6.5	6.4	6.0	6.4	6.1	6.1	5.7
Exch. K m eq/100 g	0.19	0.07	0.08	0.05	0.03	0.04	0.03	0.02
Exch. Na m eq/100 g	0.04	0.01	0.04	0.02	0.01	0.02	0.01	0.02
Exch. Ca m eq/100 g	1.56	0.72	0.66	0.41	0.16	0.16	0.16	0.16
Exch. Mg m eq/100 g	0.34	0.34	0.36	0.28	0.04	0.04	0.12	0.08
Total exch. cations m eq/100 g	2.13	1.14	1.14	0.72	0.24	0.26	0.32	0.28
Exch. capacity m eq/100 g	5.3	3.8	4.4	4.5	3.3	3.6	3.4	2.9

A further defect which is important in the circumstances of West Africa is the rapidity with which the upper, most sandy layer of the soil can dry out in the dry season. As such a large part of the palm's root system is in the superficial layers, water and nutrient uptake becomes severely restricted after some weeks of unbroken dry weather. These edaphic factors have had considerable influence on agronomic practices in the region of the Acid Sands soils.

The areas of these soils have been discussed in considerable detail by Vine[48] and by Tinker and Ziboh.[52] In some areas quite a sharp distinction can be found between the Benin and Calabar fasc soils but elsewhere

Table 3.15 *Profile analysis: Latosol, Acid Sands, Calabar fasc*
Nigeria (Tinker and Ziboh). Oil palms. An. rainfall 97 inches (2,452 mm)

Depth (inches)	0–3	3–6	6–9	9–12	12–18	18–24	24–36	36–48
(cm)	0–8	8–15	15–23	23–30	30–46	46–61	61–91	91–122
Clay (%)	7	8	7	7	10	18	24	20
Silt (%)	2	3	4	1	1	1	2	1
Sand (%)	91	89	89	92	89	81	74	79
C (%)	1.5	0.9	0.6	0.5	0.4	0.4	0.3	0.3
N (%)	0.09	0.06	0.05	0.04	0.03	0.04	–	–
pH	4.1	4.3	4.4	4.1	4.6	4.5	4.3	4.5
Exch. K m eq/100 g	0.08	0.04	0.03	0.03	0.02	0.02	0.02	0.02
Exch. Na m eq/100 g	0.04	0.04	0.02	0.03	0.02	0.01	0.02	0.04
Exch. Ca m eq/100 g	0.13	0.06	0.03	0.19	0.06	0.06	0.06	0.13
Exch. Mg m eq/100 g	0.12	0.07	0.05	0.07	0.00	0.00	0.02	0.02
Total exch. cations m eq/100 g	0.37	0.21	0.13	0.32	0.10	0.09	0.12	0.21
Exch. capacity m eq/100 g	5.8	5.0	3.9	3.5	3.4	4.8	4.9	4.0

'intergrades' occur. As with the Zaire soils, which are somewhat heavier than the Acid Sands, soils of the plateaux tend to have a higher clay content than soils on the long slopes towards rivers. It is a common experience that oil palm yields are higher on the plateaux.

Within the region of the Acid Sands soils, lateritic soils occasionally occur and one area is known where a layer of compacted ironstone is so thick that it acts as a barrier to root penetration. Such areas must of course be avoided. In general, the best areas of Acid Sands soils will be those which have not been subjected to severe cropping with annuals; are flat and therefore do not suffer the degradation of hill slope soils; are free from laterite, but are not subject to the very intense dry season characteristic of the north of the region. Such ideal conditions are now, with increasing populations, difficult to find, but it is probable that some areas in the south of Owerri Province and north of Port Harcourt most nearly fulfil the requirements.

To the west of the great areas of Acid Sands soils lie areas of basement complex soils with a rainfall which is marginal for the oil palm. Much of the area is suitable for cocoa and attempts to establish oil palm plantations have met with many difficulties.

Dahomey

As one moves through the coastal regions from the Niger delta towards eastern Ghana, the rainfall decreases and the climate soon becomes marginal for the oil palm. Dahomey is peculiar in that, in spite of unfavourable climatic conditions, it is still an important producer of oil palm products. Yields are naturally low and, in general, the soils which exist do not alleviate the adverse conditions. The most frequently encountered soil, derived from sedimentary deposits, is the '*terre de barre*', characterized by 2.5—4.0 cm of grey-brown sandy topsoil overlying a red deep clay subsoil with clay percentages varying from 30 to 65 per cent.[53, 54] This soil bakes very hard in the long dry season and the effective rooting volume is therefore low.

In some areas, on the edge of plateau country, underground water flows in from the plateau above an impermeable horizon. This natural irrigation helps to prolong the normal growth of the palms well into the dry season and results in much higher yields being obtained. Other soils in Dahomey are lateritic with compacted ironstone concretionary layers about 2 feet deep.

Exchangeable cations in the topsoil in Dahomey soils are somewhat higher than in the Acid Sands of Nigeria, but exchangeable potassium is low. Average figures have been given as: K 0.1, Mg 0.5—1.0, Ca 4.0—6.0, Total 5—7 m eq/100 g.

Ghana

Van der Vossen[18] has divided the rain forest zone of Ghana into unfavourable or marginal areas, suitable areas and favourable areas, with annual water deficits of above 400 mm, 250–400 mm and below 250 mm respectively. Over the greater part of the two latter areas the soils are derived from pre-Cambrian rocks, mainly phyllites and granites, but in the extreme south-west there is an area of Tertiary sandstones contiguous with that of the Ivory Coast. Charter's Ochrosols[33] are developed in the areas of lower rainfall (900–1,800 mm) and form the main cocoa soils of the country. Though some Ochrosols are to be found in Van der Vossen's 'favourable' regions most of their soils are the more strongly leached Oxysols with pH below 5 or Ochrosol–Oxysol intergrades.[55] Thus the areas most likely to be used for oil palms lie outside the regions most favoured for cocoa. Satisfactory and sustained yields have been obtained on Ochrosol–Oxysol intergrades over phyllites but on oxysols derived from Tertiary sands the maintenance of satisfactory yields has presented problems.[56] There is a marked phosphorus deficiency in these soils and in some areas consolidated concretionary ironstone layers are found at varying depths.

Ivory Coast

Although nine-tenths of the Ivory Coast is covered by soils derived from pre-Cambrian rocks, the bulk of the oil palm development is on soils derived from Tertiary sandstones in the south-east of the country. However, in other parts of the coastal region soils derived from the older rocks are found in a climate suited to the oil palm. Very few of the soils in the high rainfall area are red; the majority, whether overlying the sandstone or derived from older rocks, are typically yellow, or reddish-yellow latosols.[57] The sandstone soils, which are among the most productive in West Africa, show a remarkable uniformity. The topsoil is a brown clayey sand and the largely featureless profile becomes more clayey with depth to 1.5 m, below which the clay content usually becomes more or less constant at around 20 per cent. Exchangeable cation content is low, particularly that of potassium. Table 3.16 shows a typical profile analysis.

Table 3.16   *Profile analysis: Yellow Latosol over Tertiary Sandstone* Ivory Coast (Leneuf and Riou). Secondary forest

Depth (cm)	0–3	3–10	10–20	30–60	100–130	150–160	200–230
Clay (%)	10.5	11.5	13.2	14.7	19.0	20.2	19.0
Silt (%)	2.7	2.7	3.5	2.2	2.2	1.7	2.0
Sand (%)	83.0	83.8	81.9	81.5	75.6	74.5	75.8
pH	5.3	4.6	4.5	5.1	4.9	4.9	5.1
C (%)	2.8	1.6	1.1	—	—	—	—
N (%)	0.6	0.1	0.1	—	—	—	—
Exch. K m eq/100 g	0.07	0.04	0.03	0.02	0.02	0.02	0.03
Exch. Na m eq/100 g	0.01	0.01	0.01	0.04	0.01	0.01	0.02
Exch. Ca m eq/100 g	1.07	0.54	0.54	0.24	0.53	0.60	0.40
Exch. Mg m eq/100 g	0.60	0.30	0.30	0.18	0.22	0.12	0.12
Total m eq/100 g	2.59	1.31	1.30	0.72	1.13	1.07	0.83

Soils over granites and schists, often in regions of secondary forest and derived savannah, tend to contain quantities of ironstone concretions and quartz gravel.

Sierra Leone

For oil palm cultivation Sierra Leone is handicapped by having too seasonal a climate. A severe dry season lasting for 3 to 4 months is followed by many months of excessive rainfall. The soils may be divided into (*a*) latosols derived from the unconsolidated recent sandy sediments (Bullom sands) of the coastal areas, rather similar to the Acid Sands of Nigeria, and (*b*) ground-water laterite soils inland, mostly overlying basement complex non-sedimentary rocks. The far interior of the country is too dry as is the northernmost coastal area. Inland, riverine alluvial soils exist, but do not cover large areas in the climatic zone suited to the oil palm.

Little attention has been paid to the latosols, but it is believed that their problems will be similar to those of the Acid Sands in Nigeria. In certain areas of these soils productive palm groves are to be found.

A larger area is covered by lateritic soil and some profiles have been studied by Tinker.[58] These soils are developed over a variety of rocks, among which are the granites and gneiss; the heavy wet season followed by an intense dry season has no doubt been responsible for irreversible laterite formation. The quantities of stone and gravel found in these soils make them unpromising, and the intensity of cultivation for annual crops has reduced their fertility. In some areas a type of derived savannah, known locally as *Lophira* bush from the presence of a fire-resistant tree of that name, has developed. Most of the lateritic soils are drift soils on flat areas (Table 3.17) or hill creep soils; hard pans and large pieces of vesicular iron-stone reaching boulder size are not uncommon. In colour these soils tend

Table 3.17   *Profile analysis: Lateritic soil*
Sierra Leone (Tinker). Oil palms. An. rainfall over 100 inches (2,500 mm)

| Depth (inches) | 0–5 | 5–15 | 15–37 | 37–78 |
(cm)	0–13	13–38	38–94	94–198
Gravel (%)	1.3	7.0	65.6	73.0
Clay (%)	8.4	13.6	18.6	34.6
Silt (%)	3.9	2.4	3.4	2.7
Sand (%)	87.7	84.0	78.0	62.7
C (%)	1.4	0.8	0.7	0.5
N (%)	0.09	0.05	0.05	0.05
pH	5.7	5.4	5.6	5.6
Exch. K m eq/100 g	0.11	0.05	0.05	0.05
Exch. Ca m eq/100 g	0.94	0.14	0.20	0.14
Exch. Mg m eq/100 g	0.46	0.09	0.10	0.10
Total exch. cations m eq/100 g	1.69	0.48	0.69	0.82
Exch. capacity m eq/100 g	5.00	4.00	3.88	5.00

to be greyish-brown on top shading downwards to orange or reddish-brown or sometimes to yellow-brown. The top 15 cm (6 in) is usually a clayey sand, sometimes stone-free; clay usually increases with depth particularly in hill creep soils where it may reach 40 to 50 per cent. The stony material may be dispersed or in hard layers and, besides the irregular and pisolitic concretionary ironstone gravel, pieces of decomposing parent material are also found.

The nutrient status of these lateritic soils is very low as the result of leaching by the heavy rains. Potassium is low to very low and magnesium and calcium are also low. There is also strong evidence of phosphorus and nitrogen deficiency with the oil palm.

A proper assessment of the Sierra Leone soils for oil palm planting cannot yet be made. Early plantations had a history of poor techniques and lack of fertilizers, but there was evidence at Masanki that palms planted on latosols grew and produced more satisfactorily than on lateritic soils. Nevertheless, there have been suggestions that on certain lateritic soils young palms can grow reasonably well if supplied with the fertilizers needed.

The poor appearance of wild palms on the lateritic soils has long been a subject of comment. These have a characteristic premature withering of the old leaves and are obviously poor producers. However, a progeny trial on river alluvium at Njala indicated that this characteristic was at least partly genetic. It is not unreasonable to suppose that in this somewhat marginal part of Africa there has been a natural selection of palms able to maintain themselves on stony soils under conditions of severe seasonal drought, frequent fire damage and poor nutrient supply, and that when planted under more favourable soil conditions they cannot grow as vigorously as palms selected in a more favourable environment.

America

D'Hoore has drawn attention to the pedological differences between tropical Africa and America and his comments have considerable relevance to oil palm cultivation.[59] Broadly, tropical Africa contains wide depressions filled with sediments of continental origin and fringed with crystalline basement complex. In America, the important 'depressed regions' are ancient marine basins filled in by sediments of which only the upper layers are of comparatively recent continental origin. Examples of interest for oil palm planting are the Urabá Gulf region and the Magdalena valley of Colombia, the Maracaibo region, and the upper parts of the Orinoco and Amazon basins. Continental basins comparable with those of Africa are, however, to be found in the regions of the Brazilian and Guiana Shields, but are likely to be of less interest.

Perhaps the outstanding feature of the lowland soils of tropical America is the constant provision from the Andes chain and the ranges of Central America of nutrient elements being fed to the depressions by erosion. Thus the majority of areas where oil palm planting is being expanded are either flat and covered with alluvial or colluvial soils or are provided with young soils of recent volcanic origin.[60]

The several areas of oil palm development on the narrow plain to the west of the Andes are subjected to very varying rainfalls. In *Ecuador*[61] the soils have developed over successive deposits of volcanic ash under conditions conducive to erosion and leaching. The young soils (regolatosols) so developed are highly porous. Organic layers are found buried at depth by later deposits, and it is characteristic that horizon differentiation of the profile has seldom proceeded very far before the soil is buried under a fresh accumulation of detritus. Some soils have an exceptionally high organic content. Nutrient status varies considerably, but is high in most localities though the magnesium supply is often inadequate.

The soils of the heavy rainfall Pacific areas of *Colombia* are quite different. River alluvium of fair fertility is encountered in river valleys, but elsewhere the soils are leached latosols derived from Tertiary sedimentary rocks and their nutrient status may be expected to be only fair. They are usually reddish-yellow sandy clays or silty clays and drainage is often a problem. Soils derived from similar sedimentary deposits are also found in parts of the Magdalena and Zulia valleys, in the Urabá Gulf area and on the eastern side of the Andes; these are less fertile than the broad expanses of alluvial soils of these areas which are well suited to oil palm cultivation. Colombia appears to contain wide flat areas of suitable soils, mostly alluvial, interspersed with areas, sometimes hummocky or peaty, which are less suitable. With the high rainfall and the erosion from the mountains which is still occurring on a large scale, drainage of the alluvial soils is likely to remain a problem. Oil palms also grow well in northern Colombia on the alluvial soils when irrigation water is available in the dry season.

In general, in the whole of the Andean region soil fertility tends to be relatively high.[62] Nitrogen is often needed, as is magnesium, but the need for other nutrients has not been fully ascertained.

The landscape of tropical *Brazil* is dominated firstly by ancient erosion surfaces bounded by scarps and giving place to younger erosion surfaces at a lower level, and secondly by the Amazon basin. The soils of the areas where oil palms are found in groves or are being planted have not been investigated in sufficient detail for a proper account to be given here. The majority may be expected however, in contrast to soils near the Andes, to have clay minerals of low activity and to be low in cation exchange capacity. Local differences are, however, very great:[63] in the palm grove areas near Taperoa in Bahia State a reddish-yellow sandy clay latosol predominates, but further inland, where planting has been undertaken, yellowish-brown sandy soils are found at plateau level, but very heavy red clay ('massape') soil with considerable mottling is common at lower levels on rolling country. There are no doubt many variants of these soils which differ so markedly in their physical features.

In the lower part of the Amazon basin plantations have been established near Belem and there are experimental plantings elsewhere. There is an even and sufficient rainfall in the area around Belem but the yellow latosols developed over Tertiary sediments have low clay contents and sometimes contain considerable quantities of concretionary ironstone gravel. These soils are markedly deficient in phosphorus. On the northern bank of the lower Amazon and its tributaries a variety of soils are encountered. In the Amapa territory, in the direction of the Guiana shield, clayey sands are found with laterite layers over tertiary sediments. These give place to soils with a higher clay content, though still with concretionary layers, overlying the basement complex of granite and gneiss. There is some parallel here with soils of African oil palm areas, but the presence of laterite is more frequent and may be due to the greater seasonal drying out of the soil; dry seasons lasting 4 to 6 months are common.

In *Central America* the oil palm is being grown on coastal and valley alluvium in Honduras (North coast) and Costa Rica (Pacific coast), on coastal and inland soils overlying recent sandstone and mudstone in Panama, and on inland soils derived from volcanic material in Costa Rica. Little is known about the nutrient status of these soils for oil palms, except that there is a general deficiency of nitrogen, while the soils derived from sedimentary rocks under the high rainfall conditions of Panama may be expected to be generally low in exchangeable cations. Further areas of good alluvium and of soils derived from volcanic material exist in areas of high rainfall in Central America.

Surveys for oil palm development

With the considerable expansion of oil palm planting systematic methods of prospecting for suitable territory and for the siting of plantations have been developed. Ochs and Olivin[64] have described the methods and organization of semi-detailed surveys for determining the feasibility of an area for oil palm planting and detailed surveys for determining land usage and layout. In most cases such surveys have been preceded by a reconnaissance prospection. These authors point out that their methods differ considerably from soil surveys in the strict sense since they are directed to the eventual provision of detailed plans of the allotment of land for fields, roads, mills, etc., as well as the delimitation of areas unsuitable for planting.

In the reconnaissance prospection existing soil, meteorological and topographical documents are studied, available paths or paths to be cut for the prospection are chosen, and reconnaissance maps are constructed. Estimates are made of a 'utilization coefficient' based on the proportion of Class I and II soils (see p. 113) and a 'coefficient of subdivision by classes' which is a measure of the homogeneity of the area. The utilization coefficient is a comparative measure only at this stage and does not represent the proportion of the land which, after detailed survey, will actually be used for planting.

Broadly similar methods have been advocated[42] and employed in Malaysia, é.g. for the large Jenka Triangle Development scheme,[65] though the recognition of soil series takes the place of the agronomic classification of Olivin[22] and plantation siting is determined by reference to topography and an experienced assessment of the value of the series. Layout planning is described in Chapter 8 (p. 366).

References

1. **Hartley, C. W. S.** (1973) The expansion of oil palm planting. In *Advances in oil palm cultivation*, ed, R. L. Wasteie and D. A. Earp, Incorp. Soc. of Planters, Kuala Lumpur.
2. **Zeven, A. C.** (1967) *The semi-wild oil palm and its industry in Africa. Agr. Res. Repts.* No. 689.
3. **Vanderyst, Father Hyac.** (1919) Contributions à l'étude du palmier à huile au Congo Belge. 4°. Origine des palmeraies du Moyen-Kwilu. *Bull. agric. Congo belge*, 10, 70.
4. **Waterston, J. M.** (1953) Observations on the influence of some ecological factors on the incidence of oil palm diseases in Nigeria. *J. W. Afr. Inst. Oil Palm Res.*, 1, (1), 24.
5. **Hartley, C. W. S.** (1954) The improvement of natural palm groves. *J. W. Afr. Inst. Oil Palm Res.*, 1, (2), 8.
6. **Zeven, A. C.** (1965) Oil palm groves in Southern Nigeria. Part I. Types of groves in existence. *J. Nigerian Inst. Oil Palm Res.*, 4, 226, and Part II. Development, deterioration and rehabilitation of groves. *Ibid.* 5, 21.
7. **Hartley, C. W. S.** (1961) *The oil palm in its Sierra Leone setting.* Mimeographed paper, W.A.I.F.O.R.
8. **Meunier, J.** (1969) Étude des populations naturelles d'*Elaeis guineensis* en Côte d'Ivoire. *Oléagineux*, 24, 195.
9. **Savin, G.** (1966) Restauration de la palmeraie naturelle de Bahia. Résultats préliminaires. *Oléagineux*, 21, 431.
10. **Hartley, C. W. S.** (1968) Report on oil palm research and development in Brazil. *Commun. Tecn. CEPLAC*, 17, 1–32.
11. **Sparnaaij, L. D.** The Palmeries of the Kwango district, in *Notes on a visit to Research Stations, Oil palm plantations and palmeries in the Belgian Congo during April, 1958.* Hartley, C. W. S. Mimeographed W.A.I.F.O.R. paper.
12. **Desneux, R. and Rots, O.** (1959) Vers une exploitation plus intensive et plus rationelle des palmeraies subspontanées du Kwango. *Bull. agric. Congo belge*, 50, 295.
13. **Ochs, R.** (1963) Recherches de pédologie et de physiologie pour l'étude du problème de l'eau dans la culture du palmier à huile. *Oléagineux*, 18, 231.
14. **Surre, Ch.** (1968) Les besoins en eau du palmier à huile. Calcul du bilan de l'eau et ses application pratiques. *Oléagineux*, 23, 165.
15. **Anon.** (1969) Recherches sur l'économie de l'eau à l'IRHO. L'eau et la production du palmier à huile. *Oléagineux*, 24, 389.
16. **Thornthwaite, C. W.** (1948) An aproach towards rational classification of climate. *Geogr. Rev.*, 38, 55.
17. **Sparnaaij, L. D.** *et al.* (1963) Annual yield variation in the oil palm. *J. W. Afr. Inst. Oil Palm Res.*, 4, 111.
18. **Van der Vossen, H. A. M.** (1969) Areas climatically suitable for optimal oil palm production in the forest zone of Ghana. *Ghana J. agric. Sci.*, 2, 113.

19. **Nienwolt, S.** (1965) Evaporation and water balances in Malaya. *J. trop. Geogr.,* 19, 34.
20. **Thornthwaite, C. W. and Mather, J. R.** (1955) The water balance. *Publs Clim. Drexel Inst. Technol.,* 8, 1, 104 pp.
21. **Thornthwaite, C. W. and Mather, J. R.** (1957) Instructions and tables for computing potential evapotranspiration and the water balance. *Publs Clim. Drexel Inst. Technol.,* 10, 3, 185–311.
22. **Olivin, J.** (1968) Étude pour la localisation d'un bloc industriel de palmiers à huile. I. Les critères de jugement. *Oléagineux,* 23, 499.
23. **Sparnaaij, L. D.** (1960) The analysis of bunch production in the oil palm. *J. W. Afr. Inst. Oil Palm Res.,* 3, 109.
24. **Broekmans, A. F. M.** (1957) Growth, flowering and yield of the oil palm in Nigeria. *J. W. Afr. Inst. Oil Palm Res.,* 2, 187.
25. **Vanderveyen, R.** (1952) *Notions de culture d'Elaeis au Congo Belge.* Brussels.
26. **Henry, P.** (1957) Quoted by Micheau, P. *Oléagineux,* 16, 523.
27. **Ollagnier, M.** *et al.* (1970) The manuring of oil palms in the world. *Fertilité,* 36, 3.
28. **Bunting, B., Georgi, C. D. V. and Milsum, J. N.** (1934) *The oil palm in Malaya,* Mal. Planting Manual No. 1, Kuala Lumpur.
29. **Charter, C. F.** (1957) The aims and objects of tropical soil surveys. *Soils Fertil.,* 20.
30. **Coulter, J. K.** (1964) Soil Surveys and their application in tropical agriculture. *Trop. Agric. Trin.,* 41, 185.
31. **Kellogg, C. E. and Davol, F. D.** (1949) An exploratory study of soil groups in the Belgian Congo. *Publs. I.N.E.A.C.* Série Sci. No. 46.
32. **Owen, G.** (1951) A provisional classification of Malayan soils. *J. Soil Sci.,* 2, 20, and *J. Rubb. Res. Inst. Malaya,* 13, com. 274.
33. **Charter, C. F.** (1956) The nutrient status of Gold Coast Forest Soils. *Report of Cocoa conference,* London, Sept. 1955, 40.
34. **Sys, C.** (1961) La cartographie des sols au Congo. Ses principes et ses méthodes. *Publs. I.N.E.A.C.* Série Tech. No. 66.
35. **Sys, C.** (1960) Carte des Sols du Congo Belge et du Ruanda-Urundi. Congo Belge et Ruanda-Urundi. A. Sols. *Publs. I.N.E.A.C.* Brussels.
36. **Aubert, G.** (1964) The classification of soils as used by French pedologists in tropical or arid areas. *Sols afr. (African Soils),* 9, 1, 107.
37. **D'Hoore, J. L.** (1965) *Soil classification in Africa.* Symposium on Soil Resources of tropical Africa, London. African Studies Ass. of the U.K. Mimeograph.
38. **Panton, W. P.** (1963) *1962 soil map of Malaya.* Dept. of Agriculture Federation of Malaysia. Mimeograph.
39. **Ng Siew Kee.** *Oil palm soils.* Private communication.
40. **Bloomfield, C., Coulter, J. K. and Kanaris-Sotiriou, R.** (1968) Oil palms on acid sulphate soils in Malaya. *Trop. Agr. Trin.,* 45, 289.
41. **Ng Siew Kee.** (1965) The potassium status of some Malayan soils. *Malaysian agric. J.,* 45, 143.
42. **Leamy, M. L. and Panton, W. P.** (1966) *Soil survey manual for Malayan conditions.* Bull. 119. Div. of Agriculture. Min. of Agric. & Coop., Malaysia.
43. **Paton, T. R.** (1963) *A reconnaissance soil survey of soils of the Semporna Peninsula, N. Borneo.* H.M.S.O.
44. **Junus Dai and Soepraptohardjo, M.** (1973) Soil survey for agricultural development in particular reference to the rationalized land use program in North Sumatra. Soil Research Institute, Bogor. Mimeograph.
45. **Dell, W. and Arens, P. L.** (1957) Inefficacité du phosphate natural pour le palmier à huile sur certains sols de Sumatra. *Oléagineux,* 12, 675.
46. **Venema, K. C. W.** (1952) Oil palm production and oil palm soils in Indonesia. Part 11. The Atjeh soils. *Potash Rev.,* subject 27, Jan. 1952.

47. **Watson, J. P.** (1962) Leached pallid soils of the African plateau. *Soils Fertil.,* 25, 1.
48. **Vine, H.** (1956) Studies of soil profiles at the WAIFOR Main Station and at some other sites of oil palm experiments. *J. W. Afr. Inst. Oil Palm Res.,* 1, (4), 8.
49. **Paton, T. R.** (1961) Soil genesis and classification in Central Africa. *Soils Fertil.,* 24, 249.
50. **Brzesowsky, W. J.** (1962) Podsolic and hydromorphic soils on a coastal plain in the Cameroon Republic, *Neth. J. agric. Sci.,* 10, 145.
51. **Tinker, P. B. H.** (1962) Some basement complex soils of Calabar Province, Eastern Nigeria. *J. W. Afr. Inst. Oil Palm Res.,* 3, 308.
52. **Tinker, P. B. H. and Ziboh, C. O.** (1959) A study of some typical soils supporting oil palms in Southern Nigeria. *J. W. Afr. Inst. Oil Palm Res.,* 3, 16.
53. **Furon, R.** (1950) *Géologie de l'Afrique.* Payot, Paris.
54. **Surre, Ch. and Ziller, R.** (1963) *Le palmier à huile.* Maissonneuve et Larose, Paris.
55. **Brammer, H.** (1962) Soils. In *Agriculture and Land Use in Ghana,* p. 88. Oxford University Press.
56. **Van der Vossen, H. A. M.** (1970) Nutrient status and fertility responses of oil palms on different soils in the forest zone of Ghana. *Ghana J. agric. Sci.,* 3, 109.
57. **Leneuf, N. and Rion, G.** (1963) Red and yellow soils of the Ivory Coast. *Sols afr. (African Soils),* 8, 451. (C.C.T.A. Publication.)
58. **Tinker, P. B. H.** Tour of Sierra Leone by WAIFOR Soil Chemist. Mimeographed paper, W.A.I.F.O.R. 1962.
59. **D'Hoore, J.** (1956) Pedological comparisons between tropical South America and Tropical Africa. *Sols afr. (African Soils),* 4, 4.
60. **Hartley, C. W. S.** (1965) Some notes on the oil palm in Latin America, *Oléagineux,* 20, 359.
61. **Frei, E.** (1957) Informe al Gabierno del Ecuador sobre Reconocimientos edafologicos exploratorios. F.A.O. Information Paper No. 585. Rome.
62. **Bennema, J., Camargo, M. and Wright, A. C. S.** (1962) Regional Contrast in South American soil formation in relation to soil classification and soil fertility. *Int. Soc. Soil Sci., Trans. Joint Meeting Commissions IV & V,* Nov. 1962, p. 493.
63. **Savin, G.** (1966) Le palmier à huile dans l'État de Bahia (Brézil). *Oléagineux,* 21, 271.
64. **Ochs, R. and Olivin, J.** (1969) Étude pour la localisation d'un bloc industriel de palmiers à huile. II. Les techniques de prospection. *Oléagineux,* 24, 125.
65. **The Jengka Triangle Report.** Tippetts—Abbott—McCarthy—Stratton and Hunting Technical Services Ltd, 1967.
66. **Mohr, E. C. J., van Baren, F. A. and van Schuylenborgh** (1972) Tropical Soils; a comprehensive study of their genesis. 3rd edn, Mouton, The Hague, 481pp.

Chapter 4

Factors affecting growth, flowering and yield

Bunch production in the oil palm is ultimately limited by the initiation of only one inflorescence in the axil of each leaf. This inflorescence may be male, female or hermaphrodite, and it may abort. The number of leaves which are produced and the proportion of them which provide mature female inflorescences which subsequently set their fruit are factors of the greatest importance. The causes of variation in these factors may be genetic or environmental, and for a full understanding of them a study of the physiology of the oil palm is required. Much of the earlier physiological work was undertaken in Africa[1] and was directed to the solution of specific problems. More recently, however, much more attention has been given to physiological processes both in Africa and Asia and, in particular, the use of growth analysis has given a greater understanding of the palm's potential. It is proposed in this chapter to describe and discuss, in the light of these studies, the factors affecting growth and development at each stage of the plant's life.

The physiology of germination

Germination under natural conditions

Under natural conditions seeds become distributed through the forest and grove areas by a variety of means. Unharvested bunches or individual fruits will fall to the ground. Fruit will drop from bunches being carried through the groves or forest, or fruit may be removed and haphazardly distributed by forest animals, often squirrels or monkeys. Such seed does not germinate readily; on the contrary it has been shown that it requires certain seasonal conditions and, while awaiting these, it may be subjected to considerable destruction.[2, 3] Examination of seed in various sites in Nigeria from high forest to plantations, swampland, palm grove and open country shows that vast quantities of seed are eaten by rodents, bored by beetles or suffer delignification of the shell to become 'white' seed (Table 4.1). Of the two boring beetles, the Scolytid, *Coccotrypes congonus*, is of

Table 4.1   *Condition of ungerminated seed at end of a period of 2–2½ years under different habitats (Rees)*

Percentage of seed remaining in each treatment

Condition of seed	Type of habitat			
	Bare soil	Grassland	Plantation	Forest
Eaten by rodents	7.3	6.4	7.7	8.4
'White seed'	4.3	6.5	16.2	14.6
Perforated by beetles	3.0	1.3	74.7	74.9
Unperforated	85.4	85.8	1.5	2.2
Viable	3.0	10.0	3.5	1.0

greater importance than the Bruchid, *Pachymerus cardo*, and its activity is much more pronounced in habitats such as forest, thick groves or plantations where the amount of light penetration is low.

While rodent and Scolytid activities are major factors in reducing viability under natural conditions, the germination of the remaining seed depends markedly on season and probably to a lesser extent on site and treatment of the vegetation. In general, seed remains dormant throughout the wet and dry seasons in West Africa, but germination takes place during a short period of about 6 to 10 weeks at the start of the rainy season; there is a lag period of around 60 days after the first heavy rainfall before germination starts.

Differences between sites, after the differential effect of Scolytid activity has been taken into account, are more difficult to detect; the differences between the curves in Fig. 4.1 are almost entirely due to perforations of the seed by *Coccotrypes*. However, it is constantly observed that large numbers of seeds germinate when forest is cut down for farming, and seeds have been known to germinate on rotting bunches still attached to the palm. It is likely that the heat effect to be mentioned later is playing its part in these instances, and it is often suggested that the high temperature requirement for rapid germination has a survival value in a region with a severe dry season. Nevertheless, 50 to 70 per cent of seed stored at optimum moisture content for germination, but at temperatures similar to those prevailing in forest or plantation, will germinate though germination is very slow.[2]

In Malaysia, where the palm has hitherto been confined to plantations, it is beginning to 'escape' into surrounding secondary forest. Recently a palm which had been growing for 20 years or more was found in a forest area subject to periodic clearing by the indigenous people. In the palm grove area of Brazil (13°S) the period of minimum rainfall is the winter period of minimum temperature. Serious drought periods seldom occur, however, and the months of high temperature have a plentiful rainfall. Under these circumstances there has appeared to be little hindrance to spread of the palm by unaided germination.

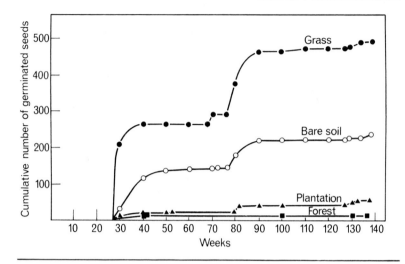

Fig. 4.1 Germination of oil palm seed planted in the open in Nigeria at the end of the wet season (November 1959) until June 1962. Black circles − grass planting; open circles − bare soil planting; triangles − palm plantation; squares − high forest (Rees, 1963).

Factors inducing germination

That a high temperature was needed for the satisfactory germination of oil palm seed was realised early in the history of the plantation industry.[4] Sand beds, well watered and exposed to the sun, proved sufficiently satisfactory in the Far East until valuable selected seed began to be produced. In West Africa, with its relatively sunless climate, germination was poor more often than not, and in spite of various systems of applying heat (which will be described in Ch. 6), a germination of between 25 and 60 per cent was as much as could usually be expected until a full physiological study had pointed the way to more refined methods.

The physiology of germination of oil palm seed has been extensively investigated by Hussey[5, 6] and Rees.[7-10] The seed is comprised of the shell and the kernel in which is embedded the embryo (see p. 41). The excised embryo is non-dormant and when placed on moist filter paper begins to elongate within 24 hours. Embryos still in contact with portions of the kernel will also begin to germinate provided they are free to elongate. Elongation is possible if the operculum is removed or the portion of the kernel near the cotyledon is cut away. Under these circumstances elongation is hastened in oxygen.

In early experiments, excised embryos cultured on nutrient media did not develop sufficiently to produce viable seedlings. Though a shoot or roots developed, the haustorium failed to enlarge and it was concluded that the embryo was dependent on as yet unknown growth substances supplied by the endosperm.[5] Later workers[11, 12] however, were able to produce fully developed seedlings from excised embryos. Rabéchault, using a comprehensive nutrient medium and maintaining a temperature of 30°C, distinguished two periods of development, 'germination' and 'seedling development'. The stages of germination were distinctive and the haustorium developed into a sickle-shaped body. By the end of this period, which occupies 4 to 5 weeks, both a plumule and a radicle have appeared. Subsequently normal leaves and roots develop. Seedlings can be developed without the addition of β-indolacetic acid to the medium, but most satisfactory growth is obtained with concentrations of 10^{-7} to 10^{-6}. Light accelerates differentiation, but retards growth. The reaction of embryos to gibberillic acid has also been investigated,[13] and has been found to be irregular. However, in a liquid medium the response is more regular and a concentration of 10^{-5} increases the speed of differentiation and the growth of the plumule and first leaves, although there appears to be no effect on root growth.

Although naked kernels cannot be conveniently germinated on a large scale, it was experiments with kernels which first opened the way to an understanding of the germination process.[5] Early work in various parts of the world showed that high temperature was as essential for germinating kernels as for germinating nuts, and that both germinated at approximately the same speed. Hussey showed that, at high temperatures (38°–40°C), increasing the oxygen tension markedly accelerated the process. At ambient temperatures, however, increasing the oxygen tension was ineffective except when high temperature had previously been applied. The results of two of Hussey's classic experiments are given in Tables 4.2 and 4.3.

These experiments showed the importance of oxygen supply and demonstrated that germination was inhibited until some process induced by high temperature had taken place. Subsequent physiological work

Table 4.2 *Effects of oxygen on the germination of* tenera *kernels at various temperatures (Hussey)*

Cumulative germination percentage

Temperature		Weeks	
		4	8
25°C	Air	0	0
	Oxygen	0	0
33°C	Air	0	0
	Oxygen	0	0
40°C	Air	0	5
	Oxygen	34	81
45°C	Air	0	0
	Oxygen	0	0

Table 4.3 *Effect of high temperature and oxygen on the germination of* tenera *kernels transferred to differing treatments after a period of 4 weeks (Hussey)*
Cumulative germination percentage

First 4 weeks			Second 4 weeks		
Initial treatment		Germination	Transferred to		Germination
40°C	Oxygen	32	40°C	Oxygen	78
			28°C	Air	67
			28°C	Oxygen	69
40°C	Air	0	40°C	Air	5
			28°C	Air	9
			28°C	Oxygen	58
28°C	Oxygen	0	28°C	Oxygen	1
			40°C	Air	3
			40°C	Oxygen	66
28°C	Air	0	28°C	Air	0
			40°C	Air	0
			40°C	Oxygen	45

showed that the optimum moisture and temperature conditions for germination are as follows:[10]

Moisture content	— dura seed	21—22 per cent
(dry weight basis)	tenera seed	28—30 per cent

These optima correspond to saturation moisture contents; Rabéchaut *et al.*[14] showed this to be 21.5 per cent overall for Deli *dura*, the approximate moisture contents of components being shell 21 per cent, kernel 23.5 per cent and embryo 135 per cent.

Temperature	— kernels, wet	40°C
	— seed, dry	42°C

The first temperature figure was deduced from the fact that germination at 36°C was very slow, at 39.5°C or 40°C it was comparatively rapid while at a constant temperature of 42°C no germination was observed with *tenera* kernels.[5] Under certain circumstances, and for short periods, seed can endure high temperatures, however. Seed exposed to a temperature of 50°C for 48 hours shows no ill effect on subsequent germination, though this temperature applied to wet seed for more than 3 days results in loss of viability. When *dry* seed is heat-treated at 42°C, subsequent germination at optimum moisture content in unimpaired.

The relationship of temperature, moisture and oxygen supply will now be considered. There was early evidence that cooling during or after heat treatment produced a flush of germination and hence that the application of heat was not necessary throughout the period of germination.[15] This was consistent with the belief that lowering the temperature would increase the oxygen supply to the embryo tissues because of the increased solubility of oxygen at lower temperatures and the high temperature coefficient of respiration. Seeds heat-treated at 39.5°C for 80 days at optimum moisture content show a rapid flush of germination when transferred to ambient temperatures; 50 per cent of final germination is reached after 5 to 6 days and germination is complete within 3 weeks. If heat is applied at moisture contents too low for germination and the seed is then simultaneously cooled and brought to the optimum moisture content, a similar flush is obtained, though this is delayed for about 3 weeks. Results of an experiment with *dura* seed demonstrate these facts (Table 4.4).

Table 4.4   *Germination of* dura *seed given 80 days' heat treatment at 39.5°C at a range of moisture contents (Rees)*

| | Moisture content | Germination after cooling at | | Days to 50% of final germination at optimum moisture content after 80 days |
| | | Original moisture content | Optimum moisture content | |
	(%)	(%)	(%)	
1.	24.2	12	71	23.5
2.	21.3	95	92	6.5
3.	18.6	0	97	17.5
4.	14.2	0	97	25.0

The dry heat-treated seed may be stored for up to 6 weeks before being brought to the optimum moisture content and a similar germination curve will then be obtained (Fig. 4.2). It is thus clear that the high-temperature reaction is irreversible. While it is not necessary for heat to be applied at or even near to the optimum moisture content, there is evidence that the heat treatment is not satisfactorily completed if the moisture content of the seed is either very high or very low. The figures presented in Table 4.4 show that with over-watering (Treatment 1) subsequent germination is both impaired and delayed. Seed set in a germinator at 39.5°C dry down to about 10 per cent moisture content at constant rate. If the seed is exposed, this occurs in about 5 days and an equilibrium moisture content of *dura* seed of about 7.5 per cent is soon attained. When kept in polythene bags, the rate of moisture loss is of course much slower. It has been shown that seed drying down to 7.5 per cent moisture content in the germinator subsequently gives poor germination at optimum moisture content. The seed is not killed, however, and if the ungerminated nuts are given further heat treatment at a higher moisture content, normal germination will be obtained. If *dura* seed initially at 17.5 to 18 per cent moisture is set in polythene bags at a germinator temperature of 39.5°C, the final moisture content will be about 8.5 to 9.0 per cent and the high temperature reaction will proceed normally.

During the course of work on dry-heat treatment of seed the optimum temperature for germination was reexamined. It was found that at a temperature of 42°C the length of the heat period could be reduced to 60 days. In practice, however, this is so near the lethal temperature that

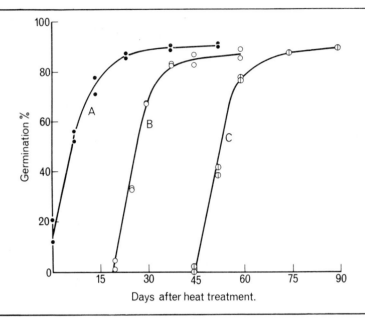

Fig. 4.2 The effect on germination of wet and dry heat treatment and storage of heat-treated seed (Rees, 1962). **A.** Seed heat-treated at optimum moisture content. **B.** Heat-treated dry. **C.** Heat-treated dry but stored dry for 25 days before bringing to the optimum moisture content.

$39°-40°C$ is better employed. Stored seed germinates more readily than fresh seed and this indicates that the high temperature reaction is proceeding at ambient temperatures though at a much slower rate. It has been estimated[10] that the temperature coefficient (Q_{10}) of the high-temperature reaction lies between 3.5 and 5, and that at ambient temperatures (*c.* 27°C) it would take about 300 days for completion. This is borne out by the results of seed storage experiments which have shown that up to 70 per cent germination may occur in seed stored at ambient temperatures and a moisture content around the optimum for germination.

The examination of the effect of heat treatment in relation to time was taken a stage further by French workers[16] who subjected seed stored for 6 months to high temperatures for relatively short periods of time as follows:

Temperature	Period of heat pretreatment and condition of seed
40°C	80 days, wet and dry
45°C	35, 40, 45, 50 days, wet and dry
50°C	6, 8, 10, 12 days, wet and dry
55°C	2, 4 days wet, 2, 4, 6, 8 days dry
60°C	3, 6, 12 hours, wet and dry

In each case the shortest pretreatment was the least deleterious. Germination percentages of between 43 and 53 per cent were obtained after 80 days following the shortest pretreatment at 45°, 50° and 55°C; 73 per cent germination was obtained after 3 hours' wet heat pretreatment at 60°C. A curious feature of these trials was the fact that further heat treatment of 20 days at 40°C produced a secondary flush of germination, this being particularly marked when the first pretreatment had been carried out with *dry* seed. The germination percentage after dry-heat pretreatment at 40°C was abnormally low (49 per cent) and the further 20 days heat treatment raised the germination percentage to 79; this suggested that some other factor, e.g. origin of seed, storage or moisture levels, may have influenced the results of all the trials. Since the temperature effect is proceeding, though slowly, at ambient temperatures, the period of high temperature required at 40°C for this stored seed would be expected to be well below 80 days. Nevertheless, this work clearly showed that 'shock' heat treatment at temperatures lethal when applied over long periods can induce the temperature effect in oil palm seed in periods very much shorter than those needed when non-lethal temperatures are used.

On the basis of these results and his own experiments, Rees pointed out that the relation between the pretreatment temperature and log hours of high-temperature pretreatment required for satisfactory germination is linear.[17]

It will thus be seen that oil palm seed has certain moisture, oxygen and temperature requirements for germination and that the latter requirement may be considered separately from germination itself and can be satisfied as a pretreatment. The moisture requirements have been defined, while the oxygen requirement is best satisfied by supplying the optimum moisture for germination after the seed has been brought to ambient temperature. The three paths to germination are illustrated below (after Rees). The

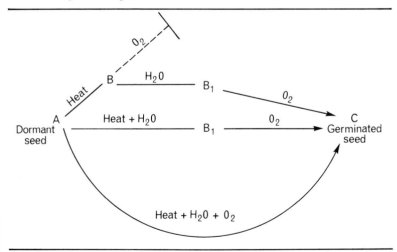

upper path — including dry-heat treatment (A to B) — has considerable practical advantages; the middle path is most rapid; the lower path is now seldom used.

In many cases of poor germination a proportion of the embryos are abnormal through incomplete development, malformation, necrosis or doubling or tripling of the embryo. It has been found that there is a relationship not only between the percentage of abnormality and germination but also between abnormality and seedling development. Seed lots with more than 15 per cent abnormal embryos tend to give germination percentages of around 60 per cent while the percentage of abnormal and non-developing seedlings will be as high as 50 per cent. There is a direct relationship between the amount of abnormality and the length and method of pollen storage[18] (see. p. 227).

The rate of growth of the oil palm

The yield of crops depends largely on the rate of development and the maintenance of the leaf surface.[19] This is very clearly shown in the oil palm where production will depend on the number and size of leaves produced and the proportion of them providing female inflorescences; and, as will be seen later, the maintenance of the highest possible leaf surface is an important factor in determining the sex ratio, which is the proportion of female to total inflorescences.

Physiologists have considered the rate of expansion of the total leaf surface from germination in two phases: the time from emergence to the unfolding of the first leaf, and the rate of growth of the leaf surface relative to the area already present. In the oil palm the first phase may occupy 4 to 6 weeks and measurement of growth in the second phase is difficult until at least the two-leaf stage has been reached some 2 to 3 months after germination.

The relative growth rate of leaves depends both on the rate of differentiation of new primordia, which in the oil palm are contained in the single apex, and on the rate of expansion of these primordia. These rates may be affected by many external influences, but there is evidence that the final leaf area is determined very largely by occurrences at the primordial stage. It has been concluded therefore that adverse conditions, particularly at germination or at the time of early growth, will be reflected in subsequent poor development. On physiological grounds, therefore, optimum conditions during germination and early seedling growth are of great significance.

For crop studies, one is not concerned only with the growth of leaf area, for it is assimilation and the weight of produce which is of ultimate importance. The net assimilation rate over a given period is the rate of increase of dry weight per unit leaf area and it is usually expressed in grams per square decimetre per week.[20, 21] Though net assimated rate was at first thought to be relatively constant,[22] later studies have shown that both this factor and the development of the leaf surface may vary considerably.[23, 24] Such studies were started with the oil palm in West Africa in the 1950s but in recent years use has also been made of growth analysis methods in Malaysia, and these are helping in the understanding of the palm's development and reaction to environmental changes.

The growth of young seedlings

The young seedling is at first dependent on the kernel, and removal of this organ before 5 weeks after germination causes a severe reduction in relative leaf area growth rate* which lasts for about 3 weeks. Removal of the kernel at 7 weeks has only a small effect and it is believed that kernel reserves are completely exhausted by about 14 weeks after germination.[25] Rees[26] determined in Nigeria the net assimilation rates, relative growth rates and the leaf area ratios for prenursery seedlings grown in full daylight and under three levels of shade. It should be observed that the leaf area ratio (F) is the relative growth rate (R_W) divided by the net assimilation rate (E_A). The results showed that with plants not already adapted to shade, i.e. not maintained under the various levels of shade from germination, there was a reduction of net assimilation rate and relative growth rate and an increase in leaf area ratio with increasing shade. When adapted to shade these relationships were no longer apparent although in the highest degree of shade (11 per cent of full daylight) the net assimilation rate was

---

* The mean values of the growth analysis functions employed are calculated as follows:

1. *Relative growth rate*: rate of increase in dry weight per unit of time.
$$\frac{\log_e W_2 - \log_e W_1}{t_2 - t_1}$$ per cent per day. Symbol — R_W.

2. *Net assimilation rate*: rate of increase of dry weight per unit leaf area.
$$\frac{(W_2 - W_1)(\log_e A_2 - \log_e A_1)}{(t_2 - t_1)(A_2 - A_1)}$$. g/dm²/week. Symbol E_A.

3. *Leaf area ratio*: leaf area per unit weight.
$$\frac{(A_2 - A_1)(\log_e W_2 - \log_e W_1)}{(\log_e A_2 - \log_e A_1)(W_2 - W_1)}$$. cm²/g. Symbol F.

4. *Relative leaf area growth rate*: rate of increase in leaf area per unit of time.
$$\frac{(\log_e A_2 - \log_e A_1)}{t_2 - t_1}$$ per cent per day. Symbol — R_A.

5. *Leaf area index*: area of leaf laminae per unit area of land
$$\frac{\text{leaf area}}{\text{ground area}}$$ (ratio). Symbol — L.

6. *Crop growth rate*: rate of increase of dry matter per unit area of land = EL. Symbol — C.

W_1 and W_2 and A_1 and A_2 are total dry weights and leaf areas, respectively, at times t_1 and t_2.

always reduced. The plants were transplanted at the two- to three-leaf stage and this had an effect on the net assimilation rate, reducing the latter to a very low figure. This result, repeated with nursery seedlings (see later), suggests an effect of death of part of the root system and shows why the correct handling of young seedlings is of prime importance for the continuance of normal assimilation and growth.

This study also suggested that, even at an early stage, the small seedling palm will develop more efficiently in full daylight (for net assimilation rate is a measure of photosynthetic efficiency), but that it can adapt itself to partial shade. This fact may be important for the survival of the palm in palm groves and farming associations, since temporary conditions of partial shade may give the seedling the opportunity to grow and establish itself above the surrounding vegetation. Under conditions of deep shade, however, the seedling will be unable to compete.

No seasonal effects on growth were noted with young seedlings.

Nursery seedling growth

The first estimations of net assimilation rates of nursery seedlings were made in the course of a nursery growth study by Wormer[27] using Deli crosses planted in the relatively dry conditions of Dahomey. The values obtained ranged from 0.04 to 0.30 $g/dm^2/week$. Plants about 21 months of age from germination had developed an average of sixty-eight leaves of which eighteen had died, fifteen were fully grown, eight were in course of rapid growth and the remaining twenty-seven were less than 2 cm long. The first visible inflorescence bud was to be found in the axil of leaves ranging from the twenty-fifth to the forty-first in order of growth. There was a suggestion that the relative growth rate was, for a period of 10 months, correlated with light intensity. A leaf area index (leaf area divided by ground area) of 4.45 was reached after 13 months (56 weeks) in the nursery (17 months from germination) at a spacing of 80 cm (31½ in) triangular. Higher values, up to 7.5, were recorded for older plants. It should be observed that these estimations were carried out with material of Far Eastern origin in a watered nursery in a very dry and seasonal part of the West African oil palm belt (average annual rainfall 1,231 mm).

The growth of nursery seedlings of African origin was studied by growth analysis methods in a field nursery in Nigeria[28] where the annual rainfall is over 1,900 mm and falls almost entirely in one prolonged wet season of 7 to 8 months' duration. The values are not easily comparable with those of Dahomey since, apart from the climatic and genetic differences already referred to, the Dahomey results were based on the fresh weight of the aerial parts of the plants only. In general, however, early growth rates seem to be higher in Nigeria since whole plant dry weight increased from under 1 g to nearly 42 g, and leaf area from 82–2,035 cm^2 in 28 weeks. Leaf area index had reached 0.55 by the end of the wet season (31 weeks) with a spacing of 61 cm (2 ft) square.

The results of growth analyses of young nursery plants in Nigeria again suggest that the imbalance of the shoot to root ratio following transplanting leads to a lowering of the net assimilation rate. The least possible disruption of plant and soil during transplanting will shorten the period of low net assimilation rate and low leaf-area ratio.[28] Values of the former function (E) and of relative growth rate are again low, but there is a time trend of increasing E associated with increasing solar radiation towards the end of the wet season. Values then reach about 0.3 g/dm²/week.

With watered older nursery plants in the dry season net assimilation rates are very similar, indicating that at this stage E is largely unaffected by age or season except for the association with solar radiation already mentioned.[29] Values of growth functions of a nursery planted in early May and sampled between October and the following March are given in Table 4.5.

Table 4.5   *Growth from October to March in a nursery planted in May*

Dry weight, leaf area and leaf-area ratio (F) for the sampled plants; mean leaf-area index (L) at each sampling; net assimilation rate (E_A), relative growth rate (R_W) and crop growth rate (C) for each period (Rees, 1963).

Periods of 4 weeks	Sample	Mean dry weight per plant g	Mean leaf area per plant dm²	F cm²/g	L	E_A g/dm²/week	R_W %/day	C g/dm²/week
	1	55.2	24.0	43.4				
					0.6			
1	1	(53.7)	23.0			0.271	1.70	0.22
	2	84.9	37.4	44.1				
					1.0			
2	2	(89.7)	40.7			0.238	1.51	0.29
	3	135.3	60.2	44.5				
					1.4			
3	3	(111.8)	45.8			*	*	*
	4	139.6	69.2	49.6				
					1.8			
4	4	(132.0)	63.8			0.289	1.74	0.55
	5	211.3	76.5	36.2				
					2.0			
5	5	(188.4)	69.7			0.220	1.15	0.55
	6	258.0	92.2	35.7				
					3.0			
6	6	(369.8)	131.2			0.175	0.84	0.62
	7	456.0	152.9	33.5	4.1			

* Anomalous results for this sampling not included.

Note: Dry weight figures in brackets relate to plants whose leaf area was measured at one sampling time, but which were harvested at the next sampling time. The weights in these cases were estimated from the leaf area.

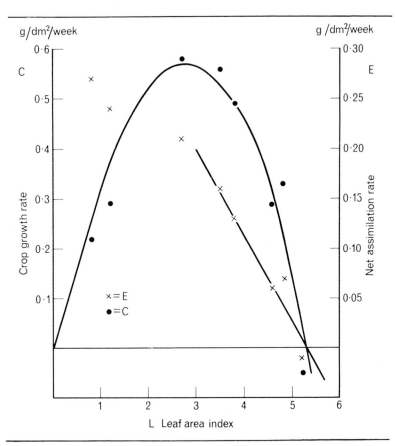

Fig. 4.3   Relationship between leaf area index (L) and crop growth rate (C) and net assimilation rate (E) in a nursery spacing experiment (Rees, 1963).

The plants were spaced at 61 cm (2 ft) square and it will be seen that a leaf area index of 4.1 was reached in 10 months.

The importance of correct spacing is well illustrated by the relationship between crop growth rate and leaf area index.[30] With closely spaced plants the net assimilation rate soon starts to fall and reaches zero at a leaf area index of about 5.4. With crop growth rate, which is the product of the net assimilation rate and the leaf area index and thus gives the rate of increase of dry matter per unit area of land, the relationship is curvilinear, so that an optimum leaf area index for crop growth appears. These relationships are shown in Fig. 4.3. Beyond the optimum a point will be reached where the net rate of growth of leaves is zero, new leaf production being balanced by the death of older leaves and the rate of production of non-photosynthetic parts falling to zero.

The practical factors determining the best time for transplanting to the field will be considered in another chapter, but it is clearly advisable that the plants should be growing at their maximum rate. L for optimum growth per plant in the Nigerian trial was under 3.0. If spacing is too close, maximum growth per palm will occur when the palms are either too small for transplanting or, as would be the case in West Africa, when the season is unsuitable for transplanting. It has been calculated that, in normally planted West African field nurseries, to attain optimum plant growth at a leaf area index of about 2.0 in mid-April, the maximum spacing should be about 71 cm (2.8 ft) square. Field experiments have confirmed this deduction (see p. 343).

It is important to note a distinction between crop growth rate as measured here and as measured in adult palm studies to be discussed later. In seedling investigations C has been measured as the change in the biomass (the standing crop) during a period of time, in the same manner as is usual in studies with annual crops. This is not the same as the total organic matter incorporated in the period since some of the leaves produced will have died off and not been measured. With adult palms, however, the crop growth rate is measured as the total dry matter production over a period of time. It is proposed here to distinguish this from C by using the symbol CGR.

From the growth data obtained in the Nigerian nursery and figures of the total hemispherical radiation, an estimate of carbon fixation was made.[29] It was found that, compared with temperate climate plants, the oil palm is comparatively inefficient, utilizing only 0.8 per cent of total radiation and 1.8 per cent of light energy. In terms of carbon fixation (assuming carbon to be 40 per cent of dry weight) the annual value is 3.3 g carbon/m^2/day.

Some studies of seedling growth in 'polybag' nurseries have been carried out in Malaysia.[25] Net assimilation rates in the first year of growth were about the same as in Nigeria (0.23−0.25 g/dm^2/week) and crop growth rates (C − biomass increase) have been similar. However, with older nursery polybag seedlings, at 12 to 18 months, the leaf area index for maximum C was found to be around 8, which is very considerably higher than shown in Fig. 4.3. The explanation may lie in the difference of age of the nursery palms since those in the Nigerian experiments were 5 to 10 months old with leaves at first at the bifurcate or partially fused stages whereas by 12 months leaves have become fully pinate, there is greater light penetration and the canopy is gradually approaching that of the adult palm. One might expect therefore that L for maximum growth would also be nearer that of the adult palm. The trend of C with age remains to be studied in detail, however.

Water supply and the growth of seedlings

Of the greatest importance in the growth of seedlings, whether in polybags or field nurseries, is the supply of water. Water shortage in field nurseries is most serious on the light soils in the seasonal climate of West Africa; but watering is usually required at some period in nurseries in all oil palm-growing regions. With the advent of the polybag nursery watering has become of even greater importance. In Malaysia, for instance, field nurseries could usually be left for long periods without watering, but with the restricted rooting volume of the polybag this is no longer possible.

Controlled investigations of water requirements have been mainly with field nurseries. Investigations were carried out at Pobé in Dahomey[31] where a very severe dry season is associated with a mean annual rainfall of 1,231 mm and at Benin in Nigeria[32] where the wet season rainfall is abundant (1,900 mm) but the dry season covers 3 to 4 months. In each case measurements of field capacity, wilting point and hence available soil water were made, though methods of estimation were different. Both soils were of low water-holding capacity.

Evapotranspiration in the nursery during the dry season was estimated approximately at Pobé from data of initial and final soil water. The estimations were for plots either unwatered or watered with very small quantities (corresponding to 7−8 mm) applied to the base of each palm once a week. Evapotranspiration was thereby doubled and it was clear from the relative growth of the seedlings and from soil studies that very little of the water was utilized by the plants. Other plots were mulched with straw or bunch refuse and similarly watered or left unwatered. Estimated evapotranspiration was reduced by the mulches and seedling growth was improved more by the mulches than by the small quantity of water applied to the bare soil. This work showed the value of mulches on Dahomey soils, but also showed that water supplies were very limited in the dry season and that watering on a much larger scale than that tried would be needed.

Later work on stomatal opening suggested that the Molisch infiltration method, modified for the use of twelve isopropanol and water mixtures, might be utilized for determining whether watering was needed or not. After an initial watering the stomata will remain fully open in the daytime for a period. After a certain number of rainless days, if the degree of stomatal closure indicated by the eighth mixture (65 per cent isopropanol) is reached, watering is held to be required. Fairly large applications are made since frequent small applications are wasteful of water through evaporation and are expensive. After an application of about 80 mm of water the frequency of watering needed is determined by tests of stomatal opening. The practical use of this method in areas with marked dry seasons has been recommended.[33]

At Benin, evapotranspiration was obtained for the whole nursery season from an evapotranspirimeter installed in a nursery and planted with nine uniform seedlings at the same time as the surrounding nursery area was planted. Evaporation figures were corrected for rainfall, ground cover percentages were estimated from photographs taken vertically and, in one season, estimates of leaf area index were also made.

Evapotranspiration over two seasons is shown in Fig. 4.4, in which log.

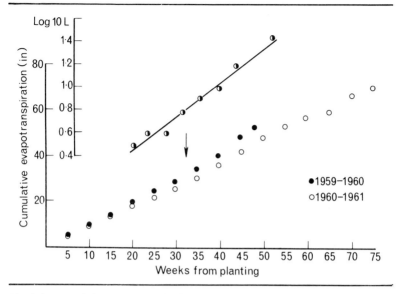

Fig. 4.4   Cumulative evapotranspiration in a nursery in Nigeria during the 1959–60 and 1960–1 seasons, together with the leaf area index (as log. 10L) for the 1960–1 season. The arrow marks the point of 1 inch (25.4 mm) water deficit as determined gravimetrically in the 1959–60 season (Rees and Chapas, 1963).

leaf area index x 10 for the later season is also shown. The nursery period was from May to March, but was extended in the second season. Ground cover progressed as follows:

Cover	1959–60 weeks	1960–1 weeks
20%	22	20
50%	33	34
90%	42	53

The data indicate an annual consumption of water in an irrigated nursery of around 1,300 mm. Clearly a great deal of the water loss must occur from the soil in the early stages of growth. Examination of seasonal changes showed that the evapotranspiration rate increased through the very dry months of February and March to a peak of 152 mm per month in early April, falling again to a minimum of 61 mm in August.

The soil data obtained in this study showed that the water-holding capacity of this nursery, representative of wide areas in the African palm belt, is low:

Depth (cm)	30	61	91	122	152	183
Water-holding capacity (mm)	56	130	201	269	350	419
Available water (mm)	20	53	81	104	135	160

As a result, a soil-moisture deficit is rapidly established, even at quite low depths, in the dry season, and root penetration is halted. The evapotranspiration figures, indicating an annual consumption of about 1,300 mm, agree well with the empirically determined irrigation guide of 25 mm per week; and the seasonal differences would support the recommendation of rather higher rates of watering during the very dry periods. It can be seen that the water available in the soil is far too little to keep the plants transpiring at the potential rate for a dry season which may last for as long as 16 weeks. However, the fact that actual evapotranspiration in the early part of the dry season was less than half the potential rate indicates the efficiency of stomatal restriction of transpiration.

Measurements of the consumptive use of water by polybag seedlings have not been made though the practical aspects of watering a polybag nursery have been studied and are described in Chapter 7.

Growth of the transplanted palm

Even under the conditions of drought described above the palm does not wilt. The leaf has a high proportion of lignified tissue; the cells of the epidermis have a thick cuticle and overlie a hyperdermis which is more highly developed on the upper or adaxial surface. The stomata are on the lower or abaxial surface at a mean density of 146 per mm.[34] They are semi-xeromorphic, that is to say they have a structure which is adapted for the prevention of desiccation over long periods. This, incidentally, supports the view that the palm originated in a region with a strongly seasonal climate. Moreover the stomata of the Deli palm differ in no respect from those of palms in Africa. The guard cells of the stomata are thick-walled with external thickened ridges which lie pressed together for their whole length when the stomata close; at the same time subsidiary cells meet between the guard cells and the stomatal cavity.

Unwatered palms in drought conditions in a nursery show pronounced midday stomatal closure which prevents death. On transplanting to the field, however, root damage may be such that even the semi-xeromorphic structure of the leaves is insufficient to prevent drying out and subsequent death of the plant. Young plants transferred to the field in very dry weather often show a characteristic blackening of the youngest fully-opened leaf and it may be presumed that this is because this leaf has not yet fully developed its protective structure. It has been noted that, in young palms, stomata on old leaves open more widely than those on younger leaves.

The main objective in transplanting, therefore, must be to transfer an intact, undisturbed and developing root system so that the transpiration stream may be returned to normality as soon as possible. The means of achieving this will be described in Chapter 9.

A well-transplanted palm will start to develop new leaves very soon after transplanting. With very large seedlings, these may initially be smaller

than those last developed in the nursery; this is due to a temporary imbalance between the leaf and the root systems, but does not in any way
invalidate the planting of large seedlings, as will be shown in Chapter 9.

As the root system of a recently transplanted palm is necessarily small,
periods without rain may in any part of the world delay its development.
In soils retentive of moisture and where competition from surrounding
covers is kept at a minimum, rainless periods of several weeks may have
little effect on the palm. On some sandy or stony soils even in the non-
seasonal climate of the Far East, growth may be checked and premature
withering of the older leaves may be caused by quite short periods without
rain. Of more serious consequence are the first few dry seasons in the
seasonal climate of West Africa. Here again the defence of stomatal closure
comes into play and some studies have been made of the process and
causes of stomatal movement in young palms.

Normally, a sharp, light-induced opening of the stomata at dawn is
followed by maximum opening from 09.00 hours to a little after 17.00
hours. The stomata then close equally rapidly at nightfall. Opening is most
easily estimated by a modification of the Molisch infiltration method using
mixtures of isopropanol and water as follows:

Solution number	1	2	3	4	5	6	7	8	9	10	11
% isopropanol	0	10	20	30	40	50	60	70	80	90	100
% water	100	90	80	70	60	50	40	30	20	10	0

When closed, solution 11 fails to penetrate the stomata and the degree of
closure is said to be 12. When fully open to 4 μ diameter, solution 4 penetrates, penetration being detected with the naked eye by the appearance
of scattered dark spots of intercellular flooding. The degree of opening
may be recorded on the 4 to 12 scale or this may be transformed to
percentages.

Towards the end of a dry season, the typical curve of opening (see Fig.
4.5) is altered by pronounced though partial closure during the midday
period. This is followed by a short period of reopening before final
closure. During the transition from dry to wet season, slight midday
closure occurs but this is not evident at the transition from wet to dry
season as the effect of the drought does not show itself until well into the
dry season. The percentage stomatal opening at various dates in 1956 and
1957 at Benin in Nigeria is shown in Table 4.6.

Studies of stomatal movement and of aperture have been made in
relation to evaporation, transpiration, soil humidity, light intensity and
temperature.[35–41] During the opening period early in the day and the
closure period in the evening, aperture is positively correlated with log.
light intensity. The extent of midday closure is correlated with total daily
evaporation and this relationship continues after rain even though the level
of closure is reduced. The relationship of aperture with soil humidity and
transpiration follow the expected course: soil desiccation is accompanied
by reduced transpiration and reduced aperture. In pot studies in Dahomey

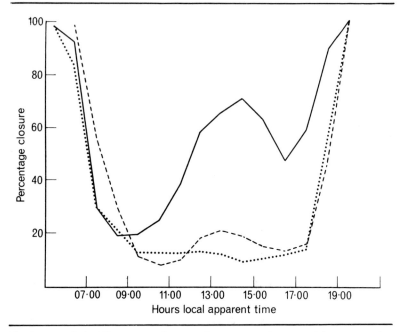

Fig. 4.5 Typical daily curves of stomatal movement (Rees, 1961) ———— dry season, ... wet season, - - - - - - - intermediate.

transpiration has been shown to fall to a very small quantity as the permanent wilting point is reached, but to rise to its maximum, depending on the hour of the day, at a point somewhat nearer to the wilting point than to field capacity.[37]

A relationship between midday stomatal closure and temperature has been found in 4-year-old palms in the West African dry season.[38] Watered and unwatered palms were used, and the aperture–temperature curves were not the same in each case, showing that two factors, temperature and

Table 4.6   *Stomatal opening at various dates, Benin, Nigeria (Rees)*

Date	Opening over 24 hr as per cent of max. possible opening	Season
17.1.56	25.8	Dry
17.2.56	26.1	Dry
19.3.56	38.9	Wet
18.4.56	40.5	Wet
15.5.56	44.3	Wet
19.6.56	42.0	Wet
6.11.56	44.9	Wet
14.12.56	45.4	Wet
6.2.57	33.1	Wet-Dry

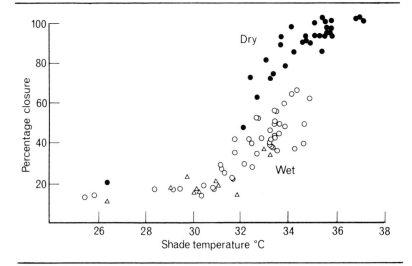

Fig. 4.6 Midday stomatal closure in watered and unwatered palms, and air temperature under the palms. Black circles − a palm under dry conditions; open circles − a palm under wet (watered) conditions during the same dry-season period; triangles − the first palm, previously under dry conditions, under wet conditions later in the season (Rees, 1961).

water strain, influence midday closure (Fig. 4.6). Closure is marked between 30° and 35°C shade temperature under the palms. It appears that when water is not limited there is some closure at high temperatures, but when soil moisture falls below a critical level there is a more sudden increase of closure which disappears following watering. Two special circumstances influence these relationships. Firstly, watering at the end of the dry season does not reduce the extent of the closure from the 'dry' to the 'wet' values; this is believed, from observations, to be due to the death during a long dry season of much of the root system in the upper layer of the soil. Secondly, low temperatures with very low humidity, as experienced in the 'Harmattan' period in West Africa, cause closure.

It is clear from the above that transpiration in the young palm can be much reduced in areas subject to a severe dry season. This inevitably delays growth and development, and it is perhaps surprising that there is such comparatively little difference between seasonal and non-seasonal regions of similar total rainfall in the development of the young palm and in the length of time it takes to come into bearing. Planting early in the season by sound techniques helps a plant to establish a deep root system by the time the dry season arrives, and this will clearly reduce the period during which the plant will suffer midday water stress. It is probably this fact more than any other that has enabled plantations in West Africa to be brought into bearing some 1½ years earlier than previously, though a largely unconscious selection of planting material for precocity must not be overlooked.

While much can be done to combat the effects of successive dry seasons, a more uniform high rainfall will always be more conducive to a sustained high growth rate, and soils retentive of the rainfall they receive are to be recommended. Ochs[39] has suggested that such soils can be detected by tests of stomatal closure and has shown that stomatal movement is affected by the type of cultivation employed. At Pobé a relationship was established between production from a number of soil types and a 'drought index' based on the number of days in the dry season on which the stomata reached a certain degree of closure. It was also shown that a *Pueraria* cover caused earlier and intenser stomatal closure than was found in plots cultivated with maize or left bare of cover during the dry season. In a root cutting experiment in the Ivory Coast it was shown that stomatal closure is related to the quantity of lateral primary roots severed.[40] This emphasizes the importance of water supplies to the oil palm and suggests that the leaf area of a cover should be reduced before the dry season and that inter-row implements must not be allowed to damage the roots.

It has been shown that stomatal closure can occur in the more even climate of Malaysia and, as in Africa, the degree of closure depends both on water stress and temperature. Corley,[41] who measured stomatal resistance to diffusion of water vapour, estimates that the light-saturated rate of photosynthesis is reduced by over 75 per cent when stomatal resistance exceeds 20 sec/cm, and he considers that yields might be reduced by 10 per cent if stomatal closure occurred on 40 days in the year. He also showed that irrigation significantly reduced stomatal resistance.

The rapid growth that can be attained by the transplanted palm under the best conditions of soil and climate is exemplified by data from the coastal alluvium of Malaysia.[42] Table 4.7. shows the total dry weights, dry matter production and growth parameters of transplanted palms in the period between transplanting and coming into bearing. Crop growth rate (CGR) here is the total organic matter incorporated and not, as with nursery seedlings, a biomass change.

It will be noted that these palms already had a small production of bunches 30 months after transplanting to the field. Although the leaves did not yet overlap at 30 months leaf area index was already 3.5.

The growth of the adult palm

A study of the growth of the adult palm is of particular importance in view of the large differences in bunch yields found between regions where the palm is planted as an economic crop. Superficial observation does not suggest that there is any obvious association, from region to region, of leaf production with yield. Well-grown fields of palms in Africa yielding less than half the bunch weight of fields in the Far East often give the appearance of having a similar or even denser canopy at the same spacing. Studies

Table 4.7 *Dry weights and growth of transplanted palms on coastal alluvium in Malaysia (Corley et al.)*

Months from trans- planting	Leaves	Trunk	Roots	Bud	Male flowers	Bunches	Total
	Mean total dry weights: kg per palm						
18	54.0	4.7	8.8	2.8	–	–	70.3
30	132.7	13.8	15.9	5.9	–	–	168.3
	Dry matter incorporated: kg/palm/an.						
18	40.4	4.0	8.0	2.5	–	–	54.9
30	96.1	9.1	7.1	3.1	2.3	43.4	161.1

	Growth parameters (density 148/hectare)			
	Leaf area (m^2)	E_A $(g/dm^2/wk)$	L	CGR (tons/hectare/an.)
18	77.4	0.137	1.14	8.10
30	235.0	0.132	3.47	23.8

of growth and dry matter production in the adult palm were first carried out in Nigeria and recent work in Malaysia has shed light on the differences of production between the two regions.

For the purpose of growth studies the palm may conveniently be divided into:

1. Leaflets
2. Leaf rachises including petioles
3. Unopened spear leaves
4. Growing point or 'cabbage'
5. Trunk
6. Roots
7. Inflorescences and fruit bunches

Accurate measurements of these organs have been made for palms of different ages in both West Africa and Malaysia and the rate of dry matter production, crop growth rate and other growth parameters have been estimated from the data.[42, 43] Some of the results obtained are compared in Tables 4.8 and 4.9. For obvious reasons destructive methods cannot be widely employed, but Corley *et al.*[44] have now evolved non-destructive methods of estimating annual dry matter production which can be used in comparisons of growth between progenies, environments or agronomic treatment, and which make possible a considerable extension of growth analysis methods in oil palm investigations. These methods cover only leaves, trunk and bunches, but these organs were found to constitute over 96 per cent of total annual dry matter production.

Table 4.8 *Dry weight of vegetative parts and leaf area of producing palms of different ages in Nigeria and Malaysia (Rees and Tinker, and Corley et al.) (kg per palm)*

Age	Country	Leaves	Trunk	Spear and cabbage	Roots	Total	Leaf area per palm (m²)
7	Nigeria	67.4	82.3	4.5	—	(154.2)	202.8
6.5	Malaysia	118.4	78.9	8.8	40.6	246.7	238.9
10	Nigeria	90.7	171.0	5.9	—	(267.6)	254.6
10.5	Malaysia	161.1	145.3	10.8	49.0	366.2	330.3
14	Nigeria	86.0	238.7	4.5	—	(329.2)	252.0
14.5	Malaysia	168.2	232.6	11.3	68.9	481.0	309.5
17	Nigeria	95.0	280.0	10.7	128.0	402.4	297.0
17.5	Malaysia	140.2	290.7	11.9	61.6	504.4	347.9
20	Nigeria	150.0	439.0	9.0	—	—	426.8
22	Nigeria	111.4	389.0	10.8	—	—	448.0
27.5	Malaysia	115.4	300.5	8.8	130.8	555.5	251.9

As a result of high correlations found between certain measurements and the actual dry matter content of the various organs the following formulae have been adopted in Malaysia:[44]

Leaf area, $L = b (n \times lw)$, where n = number of leaflets, lw = mean of length x mid-width for a sample of the largest leaflets, and b, the correction factor = 0.55. This factor was found to range from 0.51 to 0.57 with age groups from 1 to 2 to 8 to 11 years,[45] but for most comparisons the figure of 0.55 is used.

Leaf dry weight, kg, $W = 0.1023P + 0.2062$, where P = petiole width x depth in cm². The width and depth of the petiole are measured at the junction of the rachis and petiole, i.e. the point of insertion of the lowest leaflet.

Trunk dry weight, $Tr = VS$, where V is the volume of the trunk and S is the density of dry trunk in g/l. The annual growth increment of a trunk is represented by a cylinder the volume of which is estimated from its diameter, after removal of the leaf bases, and its height. Thus $V = \pi (d/2)^2 h$, while the density $S = 7.62T + 83$, where T is the age of the palm in years from transplanting. While this method is accurate for measuring dry weight increases it is not directly applicable for measuring the total dry weight of a palm since the base is conical and the density of dry trunk has been found to increase from apex to base.

Bunch dry weight, $D = 0.5275 F$, where F is fresh bunch weight in kg. There may be some difficulty in applying this formula owing to differences of fruit to bunch ratio and the effect of the latter on percentage dry

Table 4.9 *Annual dry matter production, net assimilation rate (E), leaf area index (L) and crop growth rate (CGR) of palms of different ages in Nigeria and Malaysia (Rees and Tinker, and Corley et al.)*

Age	Country	Annual DM production, per palm							E	L	CGR
		Leaves	Trunk	Roots	Bud	Male flowers	Bunches	Total			
		(kg)	(kg)	(kg)	(kg)	(kg)	(kg)	(kg)	(g/dm²/wk)		(t/ha/an.)
7	Nigeria	55	–	–	–	–	25	–	–	3.0	–
6.5	Malaysia	109	18	2	1	6	100	236	0.189	2.9	28.6
10	Nigeria	62	31	–	–	–	35	128	0.097	3.7	18.8
10.5	Malaysia	136	21	2	1	6	108	274	0.156	4.0	33.4
14	Nigeria	54	19	–	–	–	29	102	0.077	3.8	15.0
14.5	Malaysia	116	22	5	0	6	90	239	0.149	3.8	29.2
17	Nigeria	66	22	–	–	–	48	136	0.088	4.4	20.1
17.5	Malaysia	119	19	5	0	6	108	257	0.142	4.2	31.3
20	Nigeria	86	18	–	–	–	30	134	0.068	5.6	19.4
22	Nigeria	84	17	–	–	–	28	129	0.077	4.8	19.0
27.5	Malaysia	92	1	5	0	6	48	152	114	3.1	18.2

matter. Where large differences in fruit-to-bunch ratios exist the formula $D/F = 0.3702\ X + 0.2865$, where X is the fruit-to-bunch ratio, may be used.

For use in treatment or progeny comparisons there is no reason to believe that the above formulae will not have general application. Nevertheless for greater accuracy the correction factors should be checked and adjusted before adopting them in other parts of the world.

From the gross weight of plant parts, either directly measured or estimated through the above formulae and from records of bunch and leaf production and trunk height increment, it is possible to estimate the annual production of leaf, trunk and bunch dry matter. Measurement of root production presents some difficulty. Increases in root dry weight have been estimated in Malaysia from estimates of the total weight of roots of palms of different ages, but this takes no account of the production of roots which have 'replaced' old roots which have died. Male inflorescence production can be estimated from inflorescence counts and sample dry weights. There will also be a small increase in spear and cabbage ('bud') weight until the palm reaches 15 to 20 years of age. The sum of all these yield components in terms of dry matter can then be related to leaf area to give the net assimilation rate, E, or to the land area to give the crop growth rate, CGR. Table 4.9 shows the annual dry matter production of the Nigerian and Malaysian palms analysed in Table 4.8.

Figures from Malaysia and Nigeria have been presented together in Tables 4.8 and 4.9 to show the differences of dry matter content and incorporation and of growth parameter values that may be found between palms on areas with marked water deficits and those on first class soils with little or no water deficit. The figures must however be viewed with caution and some further differences between the two areas chosen for sampling should be emphasized. Firstly, in the Nigerian area the density was 147.8 per hectare (29 ft triangular) while in the Malaysian area it was 121.3 per hectare (32 ft triangular); secondly the area containing the older Nigerian palms was shown to be potassium deficient.

From Table 4.8 it will be noted that trunk dry weights were similar throughout. The dry weight of leaves was markedly higher in Malaysia, but two factors must be taken into account. Though certain other work does not suggest this, it is possible that wider spacing may have favoured greater leaf dry matter production. Secondly, differences of pruning policy may have affected the comparison. The large difference of root dry weight of 17-year-old palms may partly be attributable to differences of extraction techniques used or to soil differences; with the closer spacing and lower overall productivity the Nigerian root system might have been expected to be of lower total weight per palm.

Turning to dry matter production per annum (Table 4.9), account must again be taken of the differences of density. Incorporation of dry matter in the trunk was higher with the Nigerian palms both on a per palm and a per hectare basis. Leaf production was considerably lower on a per palm basis but differences were much less on a per hectare basis particularly at the higher ages. Bunch dry matter production was two to four times higher in the Malaysian palms on a per palm basis and up to three times higher on a per hectare basis.

With the higher density, leaf area indices (L) tended to be higher with the Nigerian palms, but the lower dry matter production, particularly in bunches, gives rise to considerably lower net assimilation rates (E) and crop growth rates (CGR). Among the Nigerian palms the highest L value recorded was 6.2 in a 20-year-old palm.

In Malaysia, correlations were found between L and both CGR and bunch yield in a study with twelve progenies,[45] and it was suggested that one method of increasing yields would be to promote a higher L by higher density planting.[42] Experience with early density experiments shows however that there are obvious limitations to this method. Rees[30] pointed out that bunch or oil yield could be increased either by increasing CGR without change in the proportions of bunch in the total crop increase, or by increasing the proportion of bunch without change in CGR, and in recent growth analysis in Malaysia interest has centred largely on the apportionment of total dry matter production between bunches and the vegetative organs (leaves and trunk). Comparisons between Malaysian data and those from Nigeria have shown that the large differences in CGR are mainly attributable to incorporation of larger quantities of dry matter in bunches and that the vegetative dry matter (VDM) production is broadly similar.[46, 42] In Table 4.10 the VDM, CGR, E, L and Bunch Index (BI) of various groups of palms are compared. The table includes data from a density experiment in Malaysia which has helped to throw further light on the growth pattern of the oil palm.[47]

Under the good growing conditions of Malaysia CGR increases rapidly to about 30 tons/hectare/an. at around 7 years at usual plantation spacings, but at high density a CGR of up to 40 tons/hectare/an. is attained. However, it has been shown that this increase of CGR with density is achieved through the constancy of VDM per palm, and the higher VDM per hectare more than compensates for decreasing bunch dry matter (Y) per palm until a ceiling is reached at densities of over 300 palms/hectare with L reaching 10–12 (Fig. 4.7). While CGR increases curvilinearly with density, VDM per hectare increases linearly over the same range.

Two other important facts about VDM production have emerged. Firstly, as will be seen from Table 4.10, there is little evidence of lower VDM in areas of lower Y. Secondly, pruning experiments in Malaysia have shown[25] that VDM is not reduced even by severe pruning. It has therefore been postulated that the palm first satisfies its requirements for vegetative growth and only when conditions make additional assimilation possible can 'excess' dry matter be diverted to bunches. Adverse conditions for bunch production, whether these be water deficits as in Nigeria or high densities in which the palms are in severe competition with each other, result in a lower net assimilation rate (E), and this is reflected in a lower

Table 4.10 *Bunch and vegetative dry matter (Y and VDM) production, Bunch Index (BI) and growth parameters of groups of palms in Nigeria and Malaysia*[*]

Country	Age	Density (palm/ha)	Y (kg/palm/an.)	VDM[†] (kg/palm/an.)	Y (t/ha/an.)	CGR (t/ha/an.)	BI	L	E (g/dm²/wk)
Nigeria	10–17	147.8	37	85	12.6	18.0	0.30	3.9	0.087
Nigeria	20–22	147.8	29	103	15.2	19.5	0.22	5.2	0.073
Malaysia (coastal)	6.5–17.5	121.3	103	132	16.0	28.5	0.44	3.6	0.160
Malaysia (coastal)	27.5	121.3	47	93	11.2	17.0	0.34	3.1	0.114
Malaysian progenies on inland soil:									
Low BI	10	138.1	44	124	17.2	23.2	0.26	4.1	0.112
High BI	10	138.1	122	101	13.9	30.8	0.55	5.1	0.117
Mean of 10	10	138.1	93	107	14.8	27.6	0.46	4.7	0.115
Malaysian density experiment on inland soil:									
Low density	7	112	105	88.4	9.9	21.7	0.54	3.9	0.107
Normal density	7	145	103	93.1	13.5	28.4	0.53	5.1	0.106
High density	7	184	82	93.5	17.2	32.3	0.47	6.5	0.096
High density	7	227	59	89.0	20.2	33.6	0.40	7.2	0.090

[*] Data from Rees and Tinker (1963), Corley, Gray and Ng (1971), Corley, Hardon and Tan (1971) and Corley (1973).
[†] Leaf and stem DM only.

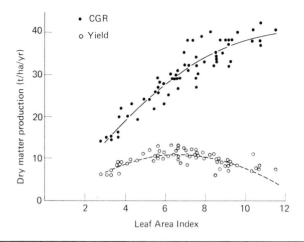

Fig. 4.7 Crop growth rate, yield and leaf area index of 6-year-old palms in Malaysia, the points being plot means and lines fitted quadratic curves (Corley, 1973).

bunch index (*BI*), which is the ratio of *Y* to *Y* + *VDM*. Thus in any particular situation the optimum density for CGR will be higher than for *Y* (Fig. 4.7).

When considering *CGR* and photosynthetic efficiency the high energy content of oil synthesized in the bunches must be taken into account. While the highest density will produce the highest *CGR*, this density will not provide the highest photosynthetic efficiency since the dry matter produced contains a relatively low proportion of oil-bearing bunches. Thus for 7-year-old palms in Malaysia when *L* at optimum *CGR* was 9.6, *L* at optimum photosynthetic efficiency was only 8.3.[47]

At high densities leaf area per palm is reduced. This is due to a reduction in the number of leaves per palm, probably because leaves at the base of the canopy die when shaded below their photosynthetic compensation point.[47] Leaf area per leaf and leaf area ratio (*F*), however, are relatively constant with density though varying with age. A relationship has been found between optimal density for current yield and leaf area per leaf and this is discussed in Chapter 9 (p. 413) when the practical aspects of determining density are considered. Corley[47] has presented diagrammatically the relationship between *L* and *CGR*, *Y* and *VDM* per hectare and concludes that *Y* could be increased by increasing *E*, through changes in photosynthetic rate or canopy structure, or by changing the *VDM/L* linear relationship as shown in Fig. 4.8. It will be seen that if the *VDM/L* relationship is depressed from line 1 to line 3, both *Y* and the *L* for optimum *Y* are increased. *VDM* per unit *L* is related to leaf area ratio (*F* — taken as new leaf area produced (m²) per kg *VDM*), so if *VDM* is reduced both *F* and *BI* will increase with the increase in yield which is shown in

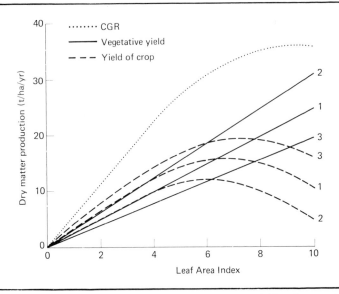

Fig. 4.8 Diagram showing relation of crop growth rate, yield and vegetative dry matter production to leaf area index, and the effect of changing *VDM* with constant *CGR* (7-year-old palms − Corley, 1973).

Fig. 4.8. Corley concludes that *VDM*, *F* and *E* are therefore the important parameters through which yield is determined. The exploitation of these findings in the improving of yields by selection and breeding is discussed in Chapter 5 (p. 262), but it will be plain from the above that *BI* will be an important parameter for bunch yield selection.

Growth analysis is now being employed in the interpretation of fertilizer responses. The first results from Malaysia[48] suggest that, in general, fertilizers increase *CGR* through increases in *L*. This results in an increase of both *VDM* and *Y* with no alteration in *BI*. However, nitrogen also increases net assimilation rates and the number of new leaves per palm but may reduce *F*. The lack of fertilizer effect on *BI* is of particular interest in view of the fact that this factor is strongly influenced both genetically and by water deficit.

The responses of *CGR* to fertilizers without change of *BI* has stimulated interest in the relationship of *VDM* to *Y* (perhaps as an alternative to leaf analysis). It was noticed that *VDM* reaches around 130 kg/palm/an. in the highest yielding plots of fertilizer experiments and on the best coastal soils in Malaysia (see Table 4.10). It was also shown that in fertilizer experiments correlations between *VDM* and *Y* can be as high or higher than correlations between leaf nutrient levels and yield.[49] However, the relation between *VDM* and *Y* at the higher levels of *Y* is not convincing and much more work will be required before use can confidently be made of this relationship. Growth analysis in its practical relation to nutrition and density will be discussed further in Chapter 11 (p. 558).

Further work on the growth of the oil palm is likely to involve inter-actions between genotypes, densities and fertilizers and different climatic conditions.

Perhaps the most remarkable feature of the growth of the oil palm is its leaf area duration (D). This is the integral of the leaf area index over the growth period, is used as a measure of photosynthetic potential, and is measured in units of time. It thus takes into account both the magnitude of the leaf area and its persistence in time. Watson[24] describes it as a measure of the ability of the plant to produce and maintain leaf area and hence of its whole opportunity for assimilation, and he points out that in conditions of constant net assimilation rate dry matter accumulation would be proportional to leaf area duration.

With perennials with a fairly constant leaf area index, as in the oil palm, D is calculated by multiplying L by the 52 weeks of the year. Thus when the L of normally spaced plantation palms is about 5.0, D will be 260. This is much higher than any value obtained from other crops, though values for other evergreen perennials come closest to it.

Mention has been made of photosynthetic efficiency in a Malaysian density experiment. The data was based on estimates, from hours of bright sunshine, of total radiation of 376, 383 and 377 cal/cm^2/day in the 3 years studied.[47] On this basis the highest photosynthetic efficiency achieved by any treatment was 3.2 per cent of photosynthetically active radiation (taken as 40 per cent of total). However, the efficiency at optimum oil yield per hectare was around 3 per cent for 7-year-old palms, still considerably higher than an estimate of 1.2 per cent for adult palms in Nigeria.[29] Comparisons between West Africa and Malaysia remain some-what unsatisfactory however owing to the variation of radiation estimates. An estimate of 470 cal/cm^2/day has been used in Malaysia for coastal areas,[42] giving an efficiency of only 2.2 per cent, while in Nigeria the mean of 6 years of Kipp Solarimeter measurements was 360 cal/cm^2/day with very considerable variations between months (see p. 108). It may reason-ably be concluded, however, that where growing conditions are good the efficiency of conversion of light energy absorbed is as high as in most agricultural crops.

Leaf production in the adult palm in relation to climate

Leaf production in numerical terms in relation to bunch production will be discussed in the next section. Very little is known of the numerical production of leaves in relation to climate. In adult palms in West Africa annual leaf production varies from eighteen to twenty-seven. Early records from Malaysia gave twenty as a normal leaf production for 10-year-old palms.[50] Annual production was recorded as sixteen to twenty for the Cameroons, twenty to twenty-four for Sumatra[51] and eighteen to twenty-six for Zaire.[52]

In West Africa and in southern Zaire the dry season is sufficiently intense to restrict the opening of leaves. Whereas the normal rate of opening may be two per month, opening may be halted for 2 to 3 months and six or more leaves may elongate in the crown but fail to open. This is a common sight in West Africa at the end of the dry season, but is only rarely seen in the Far East or America. While there is no delay in the laying down of leaf primordia to correspond with the dry-season delay in leaf opening, there is some evidence that, within West Africa as a region, annual leaf production is low in areas with an annual rainfall as small as 1,250 mm. In Nigeria, palms of similar genetic origin gave the following averages of leaf production for the thirteenth to nineteenth year after planting:

	Place		
	Umuahia	*Ogba*	*Ibadan*
Average annual rainfall (mm)	2,108	2,032	1,219
Average annual leaf production	23.1	23.1	20.5

Substantial differences of height increment of palms of the same genetic origin were noted between the Ivory Coast (annual rainfall 1,907 mm) and Dahomey (1,231 mm) indicating differences of leaf production.

Flowering and yield in the oil palm

Flowering

Although many early observations were made concerning the seasons of highest or lowest production, the first study of the flowering and production of the oil palm in relation to climatic factors and leaf production was that of Mason and Lewin[53] in Nigeria in 1925. Unfortunately their study covered periods of only 1½ and 3 years at a centre (Moor Plantation, Ibadan) of low rainfall. However, these workers were able to recognize the importance of sex ratio (ratio of female to total inflorescences, though Mason and Lewin used male to total inflorescences in their paper). They also noted that inflorescence abortion took place between the central spear stage and anthesis and was most pronounced in the dry season. A number of much more comprehensive studies have now been carried out, first in Africa, later in Malaysia. There is a great need for flowering studies in tropical America since the palm is grown there under greater extremes of rainfall, temperature and daylength than elsewhere.

Leaf development.　The progress of leaf and inflorescence development is shown by dissection of the adult palm. Forty to fifty leaves are found between the stage of initiation in the bud and the central spear stage (the stage at which the leaflets are starting to open at the top of the spear leaf in the centre of the crown). Palms in Nigeria with forty-five to fifty leaves before the spear had an annual production rate of twenty-two to twenty-four leaves; 2 years therefore elapsed between a leaf's initiation and its becoming a spear leaf.[54] As the palm may carry up to forty fully grown leaves, 3½ to 4 years may elapse between initiation and death. In non-seasonal climates leaves will open regularly at the rate of about two a month. There is evidence that development is more rapid with younger bearing palms in favourable climates. Four-year-old bearing palms in Malaysia had only thirty-six to forty-four leaves younger than the spear.[25]

In seasonal climates, as mentioned earlier, opening is delayed by the dry season and leaf production is then severely reduced. What occurs within the crown can be seen from Fig. 4.9 (after Broekmans). Normally, very

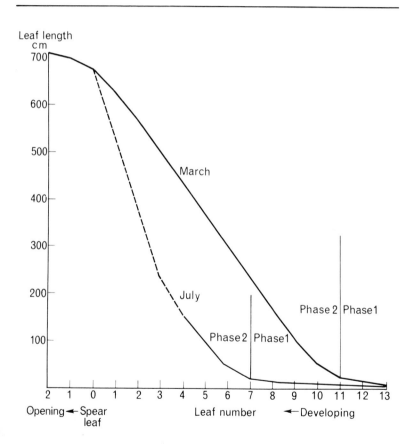

Fig. 4.9 Curves of leaf development in the crown in the dry and wet seasons in Nigeria (Broekmans).

little growth is made by the leaf until it reaches the stage of being sixth or seventh in order from the spear leaf when it may be said to be still in phase 1 of its development. Phase 2 is a stage of very rapid development from a length of a few centimetres to a full-grown spear leaf of, perhaps, 7 metres in length. There will be six or seven leaves in this stage of active growth with a spear leaf opening every fortnight. This is illustrated by the July curve. After several months of drought the central spears will cease to open normally and leaves which would have passed over into the fully opened state if conditions had been favourable will remain in phase 2. Those leaves in phase 1 are small and drought conditions do not therefore affect their development. They pass over from phase 1 to phase 2 at the usual rate, but their growth rate becomes progressively retarded. Though the ones next to the spear may elongate to nearly its length, they, like the spear, remain 'waiting' and unopened until rain falls, when several leaves will open in quick succession. As will be seen from the March curve in Fig. 4.9, up to ten leaves may accumulate in phase 2; while the usual number per month have entered phase 2, none have left it, to become opened leaves, for several months.

Leaf production determines potential bunch production and factors affecting leaf production will affect actual bunch production. Dissection of adult palms has shown that an inflorescence primordium is produced in the axil of every leaf, and that normal though slow development takes place at least up to a time between the spear-leaf stage and the stage of anthesis.

With regard to seedlings, Beirnaert claimed that leaves produced in the first two years were an exception to the rule that all leaves have an inflorescence bud. Dissection has shown, however, that inflorescence buds are present in every leaf axil.[25] Though the first primordia must normally abort, the production of normal inflorescences after a palm has been only one year in the field indicates that inflorescence buds are being extensively produced in the axils of leaves in the nursery and that the whole process of development is much faster than with the adult palm.

Leaf production in a young palm increases rapidly to a maximum before falling off to the levels already mentioned.[55] Examples of leaf production in fields planted in Nigeria in 1940 and 1941 are given below:

Age in years	2	3	4	5	6	7	8	9	10	11	12	13
1940 planting	18	27	27	29	29	28	25	23	23	24	21	–
1941 planting	–	–	27	30	31	28	25	24	25	23	23	23

In Malaysia, leaf production of Deli palms increases to thirty-two to thirty-three per annum at 3 to 4 years, levelling off at between twenty-two and twenty-six per annum at 7 to 8 years of age in the field. In Nigeria, *tenera* palms within a progeny open up to 3 per cent more leaves per annum than the *dura* palms while *pisifera* palms show a similar increase over *tenera*. In Malaysia the tendency of *tenera* palms to open more leaves has also been noted.

Inflorescence initiation.   The inflorescence bud is microscopically visible in the axil of the fourth leaf from the apex.[56] It appears, therefore, that the leaf and the inflorescence are initiated at the same time and the rate of initial inflorescence production is thus the same as leaf production and is subject to all the same influences. As with the leaf, the inflorescence bud will reach the central spear stage after 18 to 24 months. A further 9 to 10 months will elapse before the flowers open (anthesis). Initiation to anthesis therefore takes about 27 to 35 months and to fruit ripening 33 to 40 months.

Sex differentiation, sex ratios and inflorescence abortion.   Sex differentiation is the most important process in the development of the oil palm and it is interesting to note that botanists were already speculating on the secrets of this process many years ago. O. F. Cook wrote in *Oil Palms of Florida, Haiti and Panama*:[57]

> In the life of a dioecious plant a single determination of sex may be supposed to take place, and that at the beginning of embryonic development, as with the higher groups of animals, while in a monoecious plant like *Elaeis* or *Alfonsia* [*E. oleifera*] a series of sex-determinations is called for, in advance of the development of the successive inflorescences. . . . The production of two kinds of inflorescences in the monoecious palms might also be considered as an alteration of sex and from that point of view may be worthy of special study. . . . It seems not impossible that such alterations might be influenced by external conditions or by fertilizers applied to the soil. . . .
>
> As in other groups of plants, the changes from bisexual to unisexual flowers presumably are accomplished by suppression or reduction, the pistils being suppressed in forming male flowers, and the stamens suppressed in forming female flowers. . . .
>
> Sexual specialization of the palms are accompanied by simplification and reduction of inflorescences forming greater protection of the reproductive system. . . . The palms that have the inflorescence reduced to a cluster of short simple branches, as in *Elaeis* and *Alfonsia* [*E. oleifera*], would represent rather advanced stages of specialization.

High yield demands the differentiation of a high proportion of female inflorescences. Determination of the time of such differentiation and the factors influencing it have presented the most difficult and intricate problems. If the stage at which differentiation takes place is known, the factors influencing it can the more easily be sought. The time of sex differentiation is however, only gauged by inference. Eye examination of buds dissected out shows only that sex is determined at least 12 to 18 months before anthesis (see p. 176); determining a time of sex differentiation earlier than this is more difficult, and it may of course vary.

Broekmans argued that[54]

> as inflorescences reach the anthesis stage in approximately the same order in which they were initiated, they have also passed through the stage of sex-differentiation in the same sequence. Therefore the seasonal changes in the sex-ratio of inflorescences at anthesis reflect the conditions which have prevailed during the stage of sex-differentiation.

Now it so happens that in the seasonal climate of Nigeria changes in sex ratio are well marked (see Fig. 4.10), a peak of sex ratio occurring during the dry and relatively sunny months of January to March and a fall occurring during the second half of the year when there is constant rain and overcast conditions. Thus high sex ratio and conditions of high light intensity correspond.

Several attempts have been made to interpret these changes. Beirhaert[52] suggested that the ratio of photosynthesis to mineral uptake was a determinant, high levels of photosynthesis being expected during periods with much sunshine which, in Africa, would be during the dry season and at the beginning and end of the wet season. This hypothesis was based largely on the observation that shading or pruning reduced sex ratio.

Broekmans[54] developed this hypothesis a stage further by pointing out that, as his dissections indicated that sex differentiation must occur some time between initiation (at around 33 months prior to anthesis) and 18 months before anthesis, the most likely time would be 24 months, since only under these circumstances would the supposed effect of high sunshine in inducing a high sex ratio be seen in flowering records 2 years later; and, as mentioned above, he demonstrated (Fig. 4.10) that periods of high sex ratio at anthesis did in fact consistently fall in the dry season when sunshine was at its maximum.

Broekmans further considered that the severity of the dry season would be in opposition to the beneficial influence of high sunshine and that therefore if rainfall in the dry season was very low this would be reflected in low maxima of sex ratio 2 years later. He was able (Table 4.11) to demonstrate in two fields in Nigeria that dry-season rainfall was positively correlated with both maximum and average sex ratio in the sex ratio cycle 2 years later. A correlation was also found with number of female inflorescences. The sex ratio cycle used in these determinations was from one minimum to the next (see Fig. 4.10) and the period varied from 10 to 14 months.

In this Nigerian work it was assumed that abortion of inflorescences would not affect the apparent sex ratio as determined at anthesis; it will be shown below, however, that this assumption was not justified.

Investigations of floral abortion have been carried out in Zaire, Nigeria, and Malaysia. In the axil of leaves which have not produced an inflorescence at anthesis the remains of an inflorescence can be found. The stage of abortion is nearly constant; in an investigation in Nigeria 80 per cent of

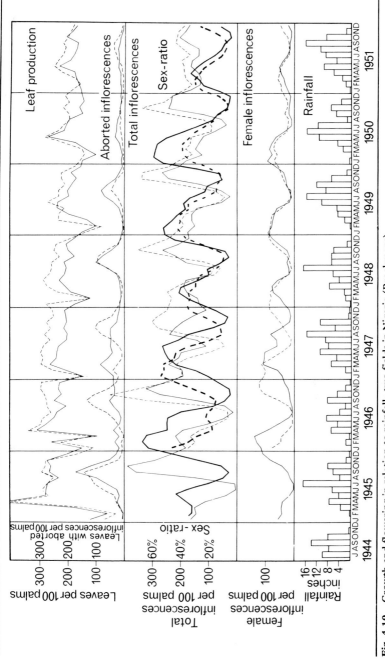

Fig. 4.10 Growth and flowering in relation to rainfall: two fields in Nigeria (Broekmans).

Table 4.11   *The relationship between dry-season rainfall (November to April) and maximum and average sex ratios and number of female inflorescences per sex ratio cycle 2 years later (Broekmans)*

Rainfall, Nov.—April		Sex ratio cycle	Experiment 2—1			Experiment 6—2		
			Sex ratio		♀ Inflor-escences	Sex ratio		♀ Inflor-escences
			Max.	Av.		Max.	Av.	
Period	(in)	Period	(%)	(%)	No.	(%)	(%)	No.
1941—2	15.4	1943—4	75.2	60.6				
1942—3	12.1	1944—5	43.3	22.5	3.8			
1943—4	15.3	1945—6	65.0	38.1	8.4			
1944—5	12.0	1946—7	55.6	28.0	7.6			
1945—6	12.0	1947—8	42.4	25.8	4.1	37.3	27.1	6.5
1946—7	16.6	1948—9	53.8	32.8	6.4	44.4	28.1	7.2
1947—8	14.0	1949—50	60.7	29.4	6.1	39.6	23.7	5.5
1948—9	9.1	1950—1	33.7	18.2	3.0	31.3	18.6	4.0
1949—50	9.3	1951—2	33.3	17.9	3.2	29.7	20.9	4.2
1950—1	14.4	1952—3				48.0	27.9	5.9
1951—2	13.5	1953—4				32.4	21.6	4.7
1952—3	16.2	1954—5				36.6	24.1	4.8
Corr. coefficient, rainfall:			0.83	0.74	0.75	0.73	0.70	0.63
sex ratio or ♀ inflores-			$P =$	$P =$	$P =$	$P =$	$P =$	$P =$
cences			0.01	0.05	0.05	0.05	0.1	0.1

abortive inflorescences in African palms measured 6.0—12.9 cm, in Deli palms 5.0—11.9 cm. Deli palms in West Africa show a much higher abortion rate than African palms, but there are large progeny differences. Figure 4.11 shows the progression in length of inflorescences subtended by the leaves of two Deli palms. The aborted inflorescences, it will be seen, are all of similar size. It is clear that the time of abortion coincides with the time at which the inflorescence is just starting to develop at a rapid rate. This is the critical stage in inflorescence development, and its determination in time makes it possible to seek the causes of abortion. The curve in Fig. 4.11 shows that the critical stage is entered at the eighth or ninth leaf from the central spear, anthesis being reached at about the seventeenth to twentieth leaf. As the central spear stage to anthesis occupies 8 to 10 months, and leaves are being produced at the rate of about two a month, the critical stage is 4½ to 5½ months before anthesis. This period has been confirmed in Malaysia.[58]

In Nigeria, Broekmans found that abortion was highest in very young palms (about 25 per cent), falling to rates between 5 and 11 per cent with age, and that abortion was at a peak with leaves which had been at the spear stage around September (Fig. 4.10). Clearly, therefore, the conditions causing the maximum abortion were those associated with the dry season, and limited water supply is likely to be the main cause. In the dry season there is also a delay in inflorescence production corresponding to the delay in phase 2 of leaf production (see Fig. 4.9). This delay, together

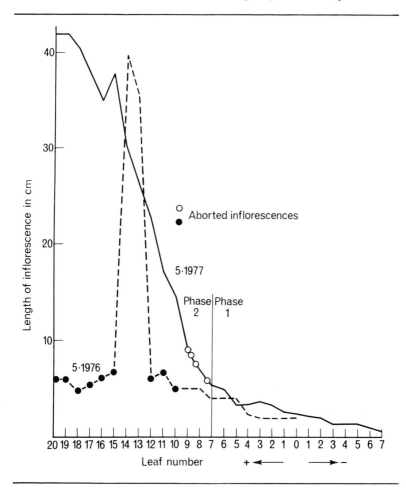

Fig. 4.11 The development of inflorescences in two Deli palms in Nigeria. Leaf O is the central spear (Broekmans).

with the relatively high abortion rate, leads to a low production of inflorescences at anthesis 8 to 10 months after the spear stage (September) and 4½ to 5½ months after the dry season (January to February), i.e. in June and July. This is followed by a maximum of flowering in October and November, corresponding to the maximum of leaf production which accompanies the beginning of the rains.

Typical abortion rates in West Africa are illustrated by the following data:

Age (years)	4	5	6	7	8	9	10	11	12
Progeny A, abortion %	24	24	19	18	11	17	11	5	7
Progeny B, abortion %	38	35	28	21	15	20	17	6	11

In Malaysia abortion rates are often low but vary considerably both between areas and individual palms. Corley *et al.*[58] quote four palms where abortion over a 2-year period averaged 12 per cent, but in one palm, which produced forty-six female inflorescences, was only 4 per cent. In experiments on coastal clays a mean annual abortion rate of 10 per cent was found but the range was from 2 to 28 per cent. In some Deli *dura* the rate was as high as 50 per cent.[59]

Broekmans considered the possibility that sex ratio of aborted inflorescences might differ from the ratio in which inflorescences are differentiated. However, he assumed that, since aborted inflorescences usually occur in an unbroken series and the sex ratio at time of highest abortion appeared to be about average, the sex ratio of aborted inflorescences would also be about average. In Malaysia[58] it has now been shown that there may be a preferential abortion of female inflorescences. Palms which were pruned to sixteen leaves showed a marked fall in female inflorescence production and apparent sex ratio compared with palms on which forty leaves were maintained:

Leaves	Period after pruning (months)	No. inflorescences per palm		Sex ratio (%)
		Male	Female	
16	1–3	2.39	3.21	57.3
40	1–3	1.95	3.30	62.9
16	4–6	1.01	1.60	61.3
40	4–6	0.99	3.35	77.2

Since sex is determined much earlier than 4 to 6 months before anthesis but abortion occurs around that time it must be concluded that female inflorescences were preferentially aborted.

In an area planted at double normal density in Nigeria the degree of shading was altered for certain palms by heavy pruning of 0, 1, 3 or 4 adjacent palms, and the effect of this treatment on apparent sex ratio and abortion was observed. Pruning was done every year either in December, April or August. Reducing the shade by pruning surrounding palms reduced abortion and increased apparent sex ratio. However, Corley *et al.*[58] showed that, for palms with surrounding palms pruned in December, if it is assumed that all the aborted inflorescences were female, reducing shade had no effect on sex ratio. If this assumption is applied to palms with surrounding palms pruned in April or August it is found that there is actually a rise in sex ratio with increasing shade. The data, set out in Table 4.12, must be viewed with some reserve since there were only twenty-four palms under observation per treatment, but the results, taken together with the effects found in pruning experiments, strongly suggest that both environmental and cultivation changes will have their primary effect on abortion and that effects on sex differentiation are complex and difficult to trace.

Table 4.12 *Effect of degree of surrounding shade on inflorescence abortion, apparent sex ratio and sex ratio adjusted on assumption that all aborted inflorescences were female (from data of Sparnaaij)* (percentage)

Month of pruning	No. of surrounding palms pruned	2nd year after initial pruning			3rd year after initial pruning		
		Abortion	Sex ratio		Abortion	Sex ratio	
			Apparent	Adjusted*		Apparent	Adjusted*
Dec.	4	12.7	25.5	35.0	12.1	26.3	35.2
	3	12.5	18.5	28.7	23.4	14.1	34.2
	1	18.3	19.4	34.1	21.2	15.3	33.2
	0	17.8	14.6	31.3	25.7	13.7	35.9
April	4	21.6	23.1	39.7	27.1	28.5	47.9
	3	31.4	22.5	46.8	27.0	24.1	44.6
	1	34.0	22.7	49.0	37.0	19.5	49.3
	0	40.5	22.4	53.8	42.3	21.1	54.8
August	4	6.8	14.6	20.4	11.6	20.0	29.3
	3	11.0	12.3	21.9	19.6	14.6	31.3
	1	17.8	15.8	30.8	25.6	16.9	38.2
	0	35.8	19.1	48.1	34.1	15.0	44.0

* Assuming all aborted inflorescences were female.

The main causes of high abortion rate have been shown to be those factors which affect dry matter production in the palm, viz. over pruning or defoliation by pests, mutual shading through high density planting or moisture stress.[60] There is no direct evidence that abortion rate is directly controlled by carbohydrate supply though it seems reasonable to suppose that Beirnaert's suggestion, already mentioned (p. 170), is more applicable to abortion rates than to true sex ratio. Auxin applications have been followed by decreased abortion but this effect may be secondary since the initial response has been reduced yield through the production of small parthenocarpic bunches.[25]

It will be clear from the above discussion that all previous data and deductions on sex ratio must be viewed with the probability in mind that aborted inflorescences were either all female or predominantly female. Nevertheless it has been possible to draw some conclusions about the time of sex differentiation and the factors influencing it from the data available in Nigeria and Malaysia. If Broekmans's data in Fig. 4.10 are examined it will be seen that the peak of inflorescence production in Nigeria falls in October/November at a time when female inflorescence production has not yet risen and when abortion (as indicated at time of anthesis) is low. Thus this peak is largely of male inflorescences and it is this fact which results in low sex ratios at that time. Corley[60] has suggested that this strongly male flowering owes its origin to the dry season 20 months earlier and states that in Malaysia abnormal peaks of male inflorescence production have been encountered 19 to 21 months after severe drought. Support has been given to this thesis by two experiments in Malaysia in which high density plantings were thinned. It was found that, even when it was assumed that all aborted inflorescences were female, there was a significant effect on sex ratio 17 and 20 months respectively after thinning. Dissections within the areas of these experiments showed that these times corresponded with the time of initiation of the first bracts subtending the spikelets. These dissections also showed that the time scale of inflorescence development may vary appreciably (Table 4.13) and, from dissections in other areas in Malaysia, Corley concluded that the interval between sex differentiation and anthesis, as indicated by the time of bract initiation, varies from 17 to 25 months with palms of 4 to 11 years of age, though within this age span no trend with age can be seen.[25]

Table 4.13   *Stages and time-scale of inflorescence development* (adapted from data of Corley)

Stages of development	Range of months before anthesis
0 Inflorescence initiation	27−35
1 Outer spathe initiation	20−28
2 Inner spathe initiation	18−26
3 Initiation of first bracts subtending spikelets; sex differentiation	17−25
4 Initiation of fourth bract	15−20
5 Spikelet initiation	11−15
6 Spikelet differentiation distinct	8−11
7 Abortion	3−6
8 Anthesis	0

It should be mentioned that though Stage 6 in Table 4.13 is given as 8 to 11 months, Broekmans stated that sex differentiation could be determined up to 18 months before anthesis while Beirnaert claimed to have determined differentiation by microscopical observation of about two-thirds of all inflorescences present, i.e. up to 40, or about 20 months before anthesis.

It will be seen that Broekmans's finding of a relation between dry season rainfall and maximum and average apparent sex ratio 2 years later (Table 4.11) does in fact agree with the development schedule of Table 4.13. Sparnaaij *et al.*[61] in their yield prediction work (see p. 187) found even closer agreement between their 'effective sunshine' measured from 1 September to 31 August and apparent sex ratio from 1 July to 30 June 22 months later. This was an index of drought and lends further support to the contention that sex is determined and influenced some 17 to 25 months before anthesis. Sparnaaij *et al.* attempted to take into account also the effect of drought on abortion by including in their weather parameter the dry season rainfall of the same and the previous year to the year of sex ratio (since abortion effects are seen in 4½ to 5 months and any one sex ratio year will be effected by two dry seasons). This only improved their correlations (p. 189) marginally. It may perhaps be concluded that seasonal drought regularly has an abortive effect on inflorescences at the critical stage and that the degree of abortion, which in any case appears to decrease with age, is not substantially altered within the scale of droughts annually encountered in Nigeria, but that the prolongation of the drought period does have an appreciable positive effect on the differentiation of male inflorescences and so alters the true sex ratio.

Little is known about development of the inflorescence in very young palms, most of the work having been done with palms nearing or having reached their adult yield. If leaf production rises to thirty-six per annum in young palms sex differentiation may be expected to take place only some 11 to 19 months before anthesis, or even sooner if there are fewer leaves between the spear leaf and the leaf with an inflorescence at anthesis than is usual with the adult palm. Broekmans recognized that in young palms the period from sex differentiation to anthesis might be shorter than in adult palms and he drew attention to the fact that the peak in sex ratio may coincide with the peak in total inflorescences (due to minimal abortion) in October to November, and that this in turn gives rise to the large peaks of bunch production often found in very young fields early in the year. This is accentuated by the fact that the period from anthesis to ripening tends to be shortened in the dry season.

Recent dissections in Malaysia of palms 15 months old in the field have shown that the bract initiation stage was 9 to 12 months before anthesis; the results of castration experiments also suggest that for these very young palms the time between sex differentiation and anthesis may be as short as 9 months.[25]

The development of leaves and inflorescences in the adult palm producing two leaves per month and a young palm producing three leaves per month is illustrated diagrammatically in Fig. 4.12.

Curiously, the peak of yield and hence of sex ratio seems to be earlier in tall grove palms in West Africa than in adult planted palms. The explanation for this is almost certainly that these tall old palms have a slower rate of leaf production than adult plantation palms. Zeven[62] showed that in typical areas of dense and degraded groves the production of palms receiving direct sunlight was between fifteen and nineteen leaves per annum. Under these circumstances the rate of leaf production would be around 1½ per month and the period from differentiation to anthesis and to bunch production would be prolonged by 9 to 10 months. This would bring production in West Africa into the early months of the *following* year.

Flowering cycles.   Leaf and flowering data so far considered in relation to age or season are based on populations of palms; but any influence of climatic factors impinges on individual palms which are subject to successive periods, often termed cycles, of male and female flower production.

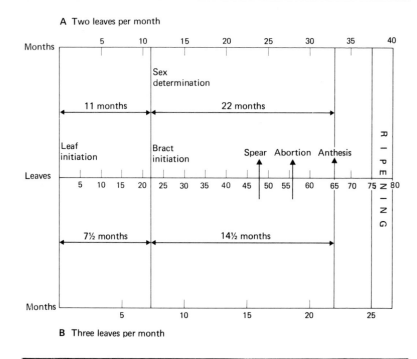

Fig. 4.12   Development of leaves and inflorescences on the assumption that: **A.** the adult palm is producing two leaves per month and sex-differentiation is 22 months before anthesis; **B.** the young palm is producing three leaves per month.

These periods alternate, sometimes with hermaphrodite inflorescences appearing between a run of male and a run of female inflorescences. The lengths of the male or female periods vary widely, and in the higher-yielding regions and progenies the female periods tend to be long drawn out. From the examination of certain data from Zaire and Nigeria, Haines and Benzian[63] concluded that periods of 4 or 5 months predominate. This is equivalent to runs of eight to ten inflorescences. Other workers have found that female phases follow a Poisson distribution with short phases predominating and single-inflorescence phases common, and they conclude that inflorescence sex is semi-randomly determined.[64, 65]

The flowering characteristics of the individual will be determined by its genetic constitution, its position in the field, particularly as regards the growth of its neighbours, and by other factors. Haines and Benzian detected cycles of high yield in individual palms and populations at Ndian Estate in West Cameroon (high rainfall, but distinct dry season); for the individual palms 3-year cycles were detected and it was postulated that this phenomenon could be accounted for by the 'fundamental' 9-month male—female cycle producing a 'beat' of 3-year periodicity when the seasonal cycle of 12 months was superimposed. For the whole population, however, a 5-year cycle was discerned. In this case it was supposed that a high nutritional demand during the ripening of bunches in a female period might lead to the differentiation of large numbers of male inflorescences which in their turn would reach anthesis in 24 months leading to low demand during the following months, i.e. 'a reversed yield movement would occur 30 months later'. This in its turn would lead to female differentiation and the high yield to high yield period would then be 30 + 30 months, or 5 years.

Corley[65] found 40 per cent of palms showed an approximate 3-year yield cycle but this could not be generally attributed to a cycle of sex ratio. He concluded that the only sex ratio cycle undoubtedly occurring was an annual one caused by exogenous factors and that evidence for endogenous cycles was lacking.

There is also however the effect of possible abortion cycles on cycles of flowering, and it has been postulated that large numbers of developing bunches from 1 to 3 months after anthesis can induce high rates of abortion which would be reflected in a low number of bunches at this same stage of development some 7 to 8 months later, thus giving rise to a 15-month abortion cycle. Some evidence of this has been found in a pruning trial in Malaysia[65] but the unpruned control palms followed the normal 12-month abortion cycle.

Nutritional and hormonal aspects of flowering.   It has already been mentioned that Beirnaert[52] suggested that an increased female inflorescence production or high sex ratio was determined by high photosynthetic activity or high carbon/nitrogen ratios. Sparnaaij[55] supposed, from this, that nitrogen applications would reduce sex ratio. However, he demonstrated only that two progenies in Nigeria reacted differently to different levels of nitrogen, and in Malaysia it has often been shown that nitrogen applications increase bunch production. The effect of mineral nutrition on yield is discussed in Chapter 11.

Very little work has been done on the hormonal control of leaf and flower production. In Malaysia growth regulators (2,4,5-TP, GA_3 and NAA) were found to reduce yield through increasing parthenocarpy. This was followed by decreased abortion and, after 3 to 4 years, increased sex ratio not accounted for solely by the decreased abortion.[25] Various interpretations can be put on these effects.

Applications to nursery seedlings which were continued on the plants after transplanting to the field showed that while Ethrel delayed flowering, GA and CCC caused earlier flowering. CCC increased the number of male inflorescences appearing before the first female; this would be expected with earlier flowering because young palms normally have an initial run of males. GA also produced earlier female flowering.

It has often been pointed out that the oil palm has many means —

abortion, sex change, decreased or increased bunch size, leaf production changes, stem girthing — whereby adjustments are made to changes in photosynthetic supply. The successive production of large numbers of bunches, i.e. a long female cycle, has effects both on the abortion rate (as mentioned above) and perhaps also on the sex ratio of succeeding inflorescences, and this is an important factor responsible for the wide palm to palm variation in length and timing of male and female cycles. It might thus be expected that the palm would compensate for any growth regulator treatment effects by various means, but particularly by abortion and sex ratio changes, unless the rate of photosynthesis is partly dependent on the demand for dry matter by developing bunches. Corley[60] has pointed out that if photosynthesis is increased by demand then increasing the latter through treatments to reduce abortion and increase sex ratio and bunch size will itself be effective. If not, however, present understanding of the physiology of the oil palm suggests that, apart from the selection of genetically productive palms, the supply and efficient use of dry matter can best be assured by maintaining a suitable leaf area index through planting at optimal density, controlling leaf pests and diseases, manuring sufficiently, maintaining soil moisture supplies and avoiding over pruning.

Bunch production, pollination and bunch failure

The factors influencing female inflorescence production and sex ratio have been discussed above and clearly these will be of the greatest importance in determining bunch yield. However, there are a number of factors, physiological and pathological, which effect the proportion of inflorescences reaching anthesis which will set and ripen their fruit. The rotting of bunches between anthesis and ripening is termed *bunch failure.* This term covers failure from whatever cause and should not be confined, as suggested by some authors, to failure of 'adequately pollinated bunches' since paucity of pollen is a major cause of bunch failure and the distinguishing of this from other causes is always difficult and often impossible. Nevertheless there is evidence that bunch failure may occur although sufficient pollination for normal fruit set has taken place.

Bunch failure is much more common in young than in mature palms, but with the majority of *pisifera* palms it tends to be extensive throughout the palm's life. It is associated with high sex ratios whether in the *pisifera* or in other fruit forms.

In the Far East it has been assumed that bunch failure in the early years is mainly caused by lack of pollen. In areas where natural groves do not exist and where extensive areas of young palms have been planted, the production of male inflorescences may be very low indeed; it has been noted that, at certain times of year, male inflorescence production in young fields may fall to nil. Male inflorescence counts in a recent planting gave the following numbers per hectare:

Age (years)	Male inflorescences	Age (years)	Male inflorescences
6—7	7.2	10—11	19.8
7—8	9.9	11—12	20.7
8—9	21.0	12—13	19.5
9—10	17.8		

Pollination

The oil palm is almost exclusively wind-pollinated. In Malaysia the abundance of pollen on the male inflorescences attracts a number of insects particularly the three bees *Apis indica, A. dorsata* and *Melipona laeviceps.* They do not, however, visit female flowers, and the mild smell of aniseed emitted from these flowers is thought to owe its origin to a primitive ancestor.[66] Jagoe made counts of pollen grains falling on slides set in mature plantations. In one area where male inflorescences were noted at 8.5 m (28 ft) and 15.2 m (50 ft) from the slides, an average of only just over 1 grain per square inch per hour was recorded, this being the equivalent to 94 grains per square inch (6.45 cm^2) over the 3-day period when a female inflorescence is receptive. In other areas deposition rates of 167 and 109 grains per square inch (6.45 cm^2) per 3 days were recorded and were considered sufficient for pollination. In these cases male inflorescences were noted in fair numbers (seven and thirteen) from 9.1 m (30 ft) to 68.6 m (225 ft) from the slides. It was concluded that under these conditions and where palms were exposed to prevailing winds, windborne pollen was likely to be sufficient for optimum pollination. It was recognized, however, that the typical dense crown of leaves of a young palm would form an effective screen against windborne pollen as a result of which its female inflorescences might only be lightly pollinated. Jagoe suggested that the screen of leaves might act as a natural protection against over-pollination.

Recent work in Malaysia has shown that pollen falling on inflorescences as much as 6 days before they become receptive is capable of effecting a good set, but that pollen remaining on an inflorescence for more than 6 days rapidly loses its viability.[67] Spore-trap counts showed that pollen is dispersed principally in the afternoon and hardly at all during the night and early morning when relative humidity is at its maximum. Rainfall was also shown to reduce pollen density appreciably and wet inflorescences ceased to disperse pollen. The density of pollen in the atmosphere is greatest at bunch level and dispersal by strong prevailing winds during the afternoon is of importance. The spore-trap investigations showed that dispersal under these conditions was significant for at least 35 metres.

Long-range dispersal of pollen above the canopy is of little importance and the density of atmospheric pollen is dependent both on the number of male inflorescences present in an area and on the rainfall. Peaks of pollen density coincide with the presence of male inflorescences but these peaks will be depressed if there are a large number of rainy days. Other factors affecting the dispersal of pollen are the dense low canopy of very young palms, the height of inter-row covers and planting density.

The probability of bunch failure or poor bunch set through lack of pollen was recognized from the earliest days of the industry in the Far East and the practice of 'assisted pollination' was in vogue for a period between the wars and has, since the war, become standard practice in young areas on the majority of plantations. The practice of assisted pollination is discussed in Chapter 10 (p. 466); its widespread use had followed the low male inflorescence production of young plantings, particularly of modern *dura* x *pisifera* material, and the fear that there would otherwise be catastrophic losses of crop through bunch failure. Such catastrophes have occurred though how far they have been caused by lack of pollen has not always been easy to establish. In this chapter the effects of the lack or the supply of pollen on growth and flowering are considered.

An indication of the effect of pollination failure and subsequent bunch rot on the behaviour of the palm may be deduced from the behaviour of the infertile *pisifera* and of palms which have been subjected to ablation of inflorescences. In the former case the palm tends firstly to divert its assimilates to the production of more and larger leaves and a more massive trunk and secondly to continue to produce predominantly female inflorescences. Similarly, the ablation of female inflorescences initially increases vegetative growth and the number of female inflorescences produced. From what has been said earlier about the distribution of assimilates within the palm these results are not unexpected. Wherever there is widespread bunch failure through a poor supply of pollen this tendency to failure will prolong itself by the consequently reduced male inflorescence production.

When additional pollen is supplied there is an increase in fruit set, fruit-to-bunch ratio, bunch weight and oil-to-bunch ratio with a decrease in mean fruit weight and oil to mesocarp.[68] The magnitude of these effects has been difficult to gauge since in standard field experiments there is bound to be pollen drift from treated to control plots; moreover the production of more male inflorescences in the treated plots provides further pollen to drift into the controls. However, some spectacular results of assisted pollination were recorded in early trials. Yield increases of 100 to 150 per cent were obtained in the first 4 years of the palm's bearing life and increases of the order of 20 per cent were still obtained when the palms were 11 years old. These results were obtained by an increase in the number of bunches set and by an increased fruit-to-bunch ratio. As might be expected, there was no effect on fruit composition, but the size of the fruit was slightly diminished by assisted pollination.[69] A slight decrease in percentage oil to fruit was recorded in Sumatra.[70]

Many recent experiments have shown that the supply of pollen by regular rounds of assisted pollination initially increases bunch yield by a decrease in the number of rotted bunches and by increased fruit set and bunch weight; subsequently, however, male inflorescences increase and the

Table 4.14 *Effect of assisted pollination on yield components of palms on Coastal Clay in Malaysia (Gray)*

Treat-ment	1962*	1963	1964	1965	1966	1967
	Number of bunches per hectare					
0	860	3,737	3,480	2,282	2,472	1,778
p	886	3,339	3,162	2,087	1,838	1,489
	Mean bunch weight, kg					
0	4.17	6.31	7.35	13.53	10.90	13.89
p	4.31	7.31	8.63	13.94	13.80	15.80
	Weight of bunches per hectare, tons					
0	3.59	23.58	25.58	30.88	26.94	24.70
p	3.82	24.41	27.29	29.01	25.36	23.53
	No. of male inflorescences per hectare					
0	615	299	983	336	756	682
p	560	284	1,005	563	983	1,060
	Mean leaf weight (leaf no. 1), g					
0				5,866	7,087	
p				5,505	6,632	

* 6 months.
0 = No assisted pollination, p = assisted pollination, twelve rounds per month.

number of bunches decreases and this often leads to an overall reduction in bunch yield. An example from the work of Gray[59, 71] is given in Table 4.14, but in the examination of any such data the influence of pollen drift on the controls should be borne in mind. It will be noted that by the second year the number of inflorescences was sufficiently reduced by assisted pollination to offset the positive effect on bunch weight. The effect on bunch number was so early that it can be assumed that pollination increased the female inflorescence abortion rate. In the later years pollination actually reduced bunch yield and Gray has shown that over a 7-year period total yields may be 5 per cent lower with pollination.[71] Corley, however, has suggested that the higher oil-to-bunch ratio obtained from better set bunches will offset the reduction of bunch yield.[60] Using data on dry matter contents of bunch components and comparing treatments giving respectively 47 and 61 per cent fruit to bunch, he shows (Table 4.15) that 100 kg carbohydrate used for bunch production would produce more mesocarp oil with the high fruit to bunch percentage even though the weight of bunches produced would be less. The dry bunch weight decreases with the increased F/B because carbohydrate is being used in respiration for synthesis of the additional oil. From this Corley suggests that more complete pollination is always likely to be beneficial.

This thesis, of course, supposes that the carbohydrate is going to be used for bunch production. However, experiments suggest that a part at least is, following assisted pollination, diverted to male flower production.

Table 4.15 *Estimated weight of bunches and oil obtainable from 100 kg carbohydrate with bunches of 47 and 61 per cent fruit to bunch (Corley)*

Fruit to bunch (%)	Weight of bunches			Weight of oil		Oil to bunch	
	dry (kg)	fresh (kg)	(%)	(kg)	(%)	calculated (%)	actual* (%)
47	72.4	152.9	100	27.6	100	18.1	18.4
61	69.8	136.2	89	30.2	109	22.2	22.3

* Oil percentages obtained in an experiment of Wong and Hardon in which treatments gave 47 and 61 per cent F/B respectively.

It is clear that much work remains to be done on pollination and embryo development before sound bases can be found for the practice and timing of assisted pollination; techniques and the practical aspects of timing are discussed in Chapter 10.

The problems of fruit set are complicated by the incidence of the fungus *Marasmius palmivorus* which is frequently found in conjunction with bunch rot and was sometimes thought to be its cause. While *Marasmius* may become pathogenic after living saprophytically on rotting bunches in an area of widespread bunch failure, its ability to initiate rotting has not yet been demonstrated.

In the areas of high sex ratio and serious bunch rotting attention has naturally been concentrated on pollination problems. Bunch failure also occurs, however, in areas in Africa where an abundant natural supply of pollen exists. Cases of serious bunch failure have occurred, e.g. in Ghana, in young areas surrounded by forest and in these areas scarcity of pollen may be a contributory factor, but in the majority of situations such vast numbers of pollen-producing palms exist that pollen supply cannot be held responsible. In these situations bunch failure rarely exceeds 10 to 15 per cent with adult palms though it is usually higher in young palms. The percentage survival of female and hermaphrodite inflorescences in two progenies in a field in Nigeria are given below:

Year of harvesting	1	2	3	4	5	6
Progeny A, survival %	54	66	78	73	81	85
Progeny B, survival %	45	64	90	87	88	93

Though variations such as those shown above do occur, significant differences in bunch failure between progenies have not been found.[55]

The distinguishing of bunches which fail through inadequate pollination from those which fail from other causes is difficult. Turner and Bull[72] maintain that although failure of pollinated bunches may occur at any stage it is most frequent between 2 and 4 months after anthesis, and they suggest that the causes are physiological. It has been postulated that the palm may set and partially develop more bunches than it can subsequently support through its supplies of carbohydrate, mineral nutrients or water. Sparnaaij suggested that weather, i.e. water supplies, was likely to be the most important factor, while Malaysian workers have suggested 'over-bearing', i.e. that heavily bearing palms which have been in a female cycle for a long period may have over-taxed the palm's resources before the mechanisms of abortion and sex-determination have had time to come into play. Evidence for this general thesis is small, though an increase in bunch failure following very heavy pruning has been reported.[60] Turner and Bull believe that Bunch End Rot may be a form of bunch failure.

Fruit form and bunch production

There are differences in bunch production between the *dura, tenera* and *pisifera* forms which owe their origin to differences in sex ratio. A typical example from Nigeria is as follows:

	Dura	*Tenera*	*Pisifera*
Sex ratio over a 7-year period	27.9	31.1	45.7

As a result of these sex ratio differences *dura* palms produce fewer but heavier bunches than *tenera* palms. The production of shell-less *pisifera* fruit depends on the ability of the bunches to survive to maturity. The so-called infertile *pisifera* produces rotten bunches in which only a very few fruit set, and it has a very high sex ratio.

Under West African conditions the Deli palm produces a small number of very large bunches, having a very low sex ratio. Progeny differences are, of course, large. Most progenies show a tendency to reduced sex ratio on moving into drier districts though some, by maintaining a high sex ratio, seem to be adapted to drier conditions.

Yield variation within and between years: yield prediction

Much has already been said about annual variation in sex ratio and abortion, and these factors are largely responsible for annual bunch yield variations. In seasonal climates the annual yield usually has only one peak, the time of the peak depending, as has already been shown, on the age and rate of leaf production of the palms and, in the adult plantation palm, on climatic conditions about 28 months before fruit ripening. In non-seasonal climates, there are occasionally two peaks of production in the year, though one tends to be much the more prominent; there is considerable variation in the magnitude of the peaks. In Fig. 4.13 the monthly production of adult fields in different territories is illustrated.

Yield prediction can only be attempted if a good deal is known of the factors controlling yield. The general relationship between climate, and in particular the water deficit, and yield levels has been discussed in Chapter 3 (p. 102). Annual variations have tended to be greater in Africa than in the Far East, climatic variations have been more clear cut, and meteorological data more complete. For these reasons a much greater scientific interest has been taken in annual yield variation in Africa. Apart from the

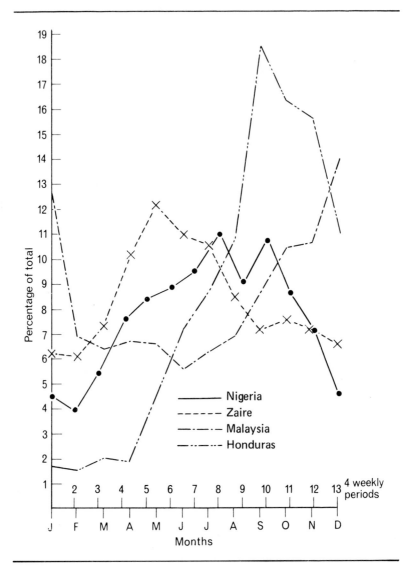

Fig. 4.13 Monthly or 4-weekly distribution of bunch yield in adult fields in Nigeria, Zaire, Malaysia and Honduras. Means of 5 years, except for Honduras where the figures are means of 12 years and relate to 4- or 5-week periods.

work of Broekmans, five studies have been made in Africa on the relation of climatic factors to annual yield.[64, 73–77] An examination of these studies, together with a further investigation in Nigeria, showed that though the conclusions drawn from the data presented were not always in line one with another, there was general agreement that climatic conditions were optimal during the drier parts of the year. Sparnaaij *et al.*[61] were able to reconcile the deductions drawn by previous authors and to provide parameters which were much more closely correlated with future yield; they postulated that the carbohydrate status of the plant largely depends on the sunshine which has been available, provided that the plant's metabolism has not been restricted by some other factor. They therefore directed their attention to obtaining reliable estimates of 'effective sunshine' which they defined as total sunshine during the period of water sufficiency plus a fraction of the sunshine in the drought periods. It will be realised that such measures of effective sunshine must, in seasonal climates, be directly related to dry season rainfall; but they proved much more sensitive as prediction factors than did the rainfall data of previous authors.

It has been seen that, in adult plantations, the stage of anthesis is some 17 to 25 months after sex differentiation. Ripening takes place 4½ to 6 months after anthesis, i.e. some 21 to 31 months after sex differentiation. Broekmans's relationship of rainfall and bunch number was for the period November to April (roughly covering the dry season) and the yield during the second calendar year following, for example:

November—April rainfall	Number of bunches in calendar year
1960—1	1963

The exact weather period taken for correlating with yield is not of great significance, but as the period of high sex ratio lies in the dry season in the seasonal climate of West Africa, the significant period must 'straddle' the dry season 2 years earlier. Sparnaaij *et al.* chose the period 1 September to 31 August to compare with the calendar year 28 months later. In the first place a correlation was found between annual sunshine data and yields 28 months later, though the correlation was not high. A more definite relationship was found when the sunshine figures for the drought periods were excluded, and it was apparent that the useful sunshine contribution of the drought periods could only result in minor differences. It was realised that the physiological effect of drought on the carbohydrate status of mature oil palms not having been investigated, any estimate of the contribution of drought-period sunshine must be tentative. Nevertheless the methods developed were highly successful. After a consideration of potential evapotranspiration calculated by the Thornthwaite formula, and estimates by Penman's method and lysimeter determinations, a figure for P.E. of 50 inches per annum or 1 inch (25 mm) per week was used in drawing up an estimate of the length of the drought period. Two estimates of effective sunshine (E.S.) in drought periods were then made as follows:

E.S. (*a*): Dividing the total duration of sunshine in hours during a drought period by the average water deficit factor, assuming that the deficit increases by unity each week. A cumulative deficit factor for 5 weeks is:

$$\frac{1 + 2 + 3 + 4 + 5}{5}$$, and this is divided into the sunshine in hours.

E.S. (*b*): Estimation of effective sunshine by the same method but on a weekly basis, viz:

$$\frac{S_1}{1} + \frac{S_2}{2} + \frac{S_3}{3} + \frac{S_4}{4}$$, etc. where S_1, S_2, etc. denote the sunshine in successive weeks of drought.

In drawing up the total effective sunshine it was assumed that in the drought period:

1. Maximum water reserve in the soil was 4 inches (100 mm), this being derived from an estimate of the root constant.
2. Depletion of soil water was 1 inch (25 mm) per week irrespective of reserves.
3. Drought started when reserves fell below 1 inch (25 mm).
4. A weekly rainfall of below ½ inch (13 mm) had no effect on drought.
5. A weekly rainfall of ½−1½ inches (13−38 mm) did not break the drought but did not increase the deficit factor.
6. A weekly rainfall of over 1½ inches (38 mm) broke the drought.

A simplification of the effective sunshine calculations and one which is probably only suited to Nigeria where it may be assumed that the November to April weekly dry-season sunshine is constant, was also successfully employed. This is termed 'active weeks' and is calculated by deducting the dry weeks from the 26 weeks (weekly units) of the November to April season and then adding for each drought period (as defined at 3 above) 1½, 2, 2½, or 3 units according to whether the length of the drought period was 2, 3−5, 6−8 or 9+ weeks. Highly significant correlations between all these factors and annual yield 28 months later were found at Benin in Nigeria for a 12-year period and the highest correlations were found with active weeks.

As mentioned earlier (p. 176), short-term climatic influences were also included, i.e. the influence on abortion of the dry season immediately preceding the harvest which is gathered at the end of one calendar year and the beginning of the next. This is the time, in West Africa, of minimum yield and Broekmans found a significant correlation between dry-season rainfall and bunch yield in the time of minimum harvest about 1 year later. As, however, this is the time of minimum yield, it is unlikely to have a large effect on the total calendar-year yield and the adding to the effective sunshine data of rainfall data for the yield year and the preceding year (both of which may have an effect on abortion in the calendar year of yield) did not appreciably improve the correlations. The correlations obtained by the various methods at Benin are given below (with significance denoted by asterisks):

Correlations with yield 28 months later and sex ratio 24 months later	Effective sunshine hours (a)	Effective sunshine hours (b)	Effective sunshine (b) + 2 years dry season rain	Active weeks
Number of bunches per palm	0.70*	0.70*	0.71**	0.79**
Weight of bunches per palm (lb)	0.84***	0.84***	0.88***	0.95***
Sex ratio	0.77**	0.78**	0.79**	0.79**

The relationships are also shown in graphical form in Fig. 4.14.

As mentioned in Chapter 3 (p. 103), the time interval of 28 months used by Sparnaaij *et al.* has been employed in determining general relationships between water deficits and yield while the parameter effective sunshine itself has been found useful in investigations of the relation between environmental and nutritional factors[77] (p. 532). Both effective sunshine and active weeks are in fact good drought indicators. A negative correlation has been found between the annual yield data employed in Fig. 4.14

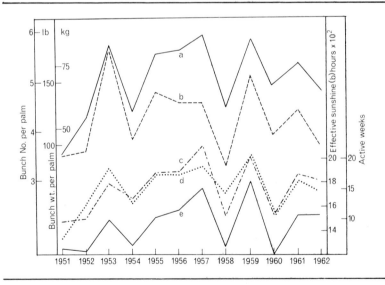

Fig. 4.14 Relation between number and weight of bunches, effective sunshine (b) and active weeks during the period 1951–62.

a, mean weight of bunches per palm; b, mean number of bunches per palm; c, effective sunshine (b) plus 2 years dry season rain; d, number of active weeks; e, effective sunshine (b) (Nigeria — Sparnaaij, Rees and Chapas).

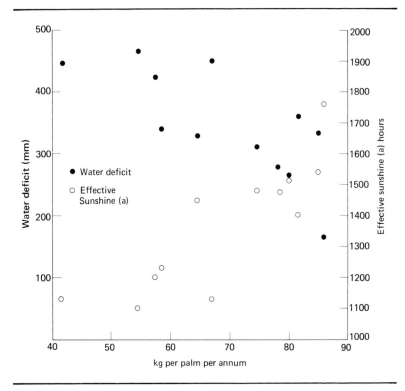

Fig. 4.15   Relationships between the water deficit during the dry season and the effective sunshine measured from 1 September to 31 August and bunch yield during the calendar years 28 months later (Nigeria — yield data from Sparnaaij *et al.*; water deficit computation by I.R.H.O. methods).

and the water deficit during the dry season 2 years earlier, but this correlation ($r = -0.74$), though significant, is less close. Figure 4.15 shows the relation between yield per palm, water deficit and effective sunshine (*a*) and the water deficit/yield relationship may be compared with that given in Fig. 3.2.

In Asia, studies of the effect of climatic factors on yield have been much more recent. The approach has necessarily been different; the complete absence of a calculated water deficit being usual, interest has largely centred on the effect of such drought periods as may from time to time occur. However, the frequency of these periods varies widely both with geographical situation and time. For instance, seven estates in Sumatra between latitudes 2°30′ and 3°35′N had measurable water deficits in a mean of 5.2 out of the 8 years 1968 to 1975, whereas three estates south of this region showed deficits in the same period in a mean of only 2 years. In the same region there is some evidence also that periods of more severe drought have been more frequent in recent years than towards the beginning of the century. The Asian data so far examined suggests that a deficit has a greater effect, through increases in the abortion rate, in the year after the deficit than it has, through alterations to the sex ratio, in the following year. Nevertheless the general pattern of low yields 8 to 12 and 21 to 26 months after a drought period, with an increase in male flowering in the latter period, are found.[78] There is also a suggestion that sudden drought in a normally water-sufficient area may cause abortion of inflorescence buds at the time of sex differentiation, i.e. 17 to 26 months before anthesis according to palm age.

Soil type and drainage intensity are of importance in periods of sudden drought in Asia. It has been shown on the coastal clays in Malaysia for instance that plantations which are able to maintain high water tables can be largely unaffected by periods of drought even where calculated water deficits are as high as 300–600 mm.[78]

An ambitious scheme for yield prediction by means of a mathematical model taking account of rainfall, calculated available soil water, potential evapotranspiration, temperature and global energy (calculated from sunshine hours and rainfall intensity) has been developed for Malaysia.[79] A measure of soil fertility was also included through estimation of 'plantation vigour' from previous yield performance in the areas concerned. It is not clear from the published account how far this model has been successful in predicting actual monthly yields, though significant correlations were found. The study confirmed that soil water changes had the greatest effect on yield variability; temperature also had some effect, but variations in global energy made virtually no contribution to yield differences.

In general the pattern of responses to climatic factors is not so clear in Asia as in Africa; and the effect of additional factors such as successive periods of drought and 'compensation' for losses through unusually high yields following the drought effects need to be taken into account and therefore require study.

References

1. **Prévot, P.** (1963) Données récentes sur la physiologie du palmier à huile. *Oléagineux*, **18**, 79.
2. **Rees, A. R.** (1963) Some factors affecting the germination of oil palm seeds under natural conditions, *J. W. Afr. Inst. Oil Palm Res.*, **4**, 201.
3. **Rees, A. R.** (1963) A note on the fate of oil palm seed in a number of habitats. *J. W. Afr. Inst. Oil Palm Res.*, **4**, 208.
4. **Bücher, H. and Fickendey, E.** (1919) *Die Ölpalme*. Berlin.
5. **Hussey, G.** (1958) An analysis of the factors controlling the germination of the seed of the oil palm, *Elaeis guineensis* (Jacq.), *Ann. Bot.*, N.S., **22**, 260.
6. **Hussey, G.** (1959) The germination of oil palm seed: Experiments with *tenera* nuts and kernels. *J. W. Afr. Inst. Oil Palm Res.*, **2**, 331.

7. **Rees, A. R.** (1959) The germination of oil palm seed: The cooling effect. *J. W. Afr. Inst. Oil Palm Res.,* 3, 76.
8. **Rees, A. R.** (1959) The germination of oil palm seed: Large scale germination. *J. W. Afr. Inst. Oil Palm Res.,* 3, 83.
9. **Rees, A. R.** (1960) The germination of oil palm seeds – A review. *J. W. Afr. Sci. Ass.,* 6, 55.
10. **Rees, A. R.** (1962) High temperature pretreatment and the germination of seed of the oil palm, *Elaeis guineensis* (Jacq.). *Ann. Bot.,* N.S., 26, 569.
11. **Bouharmont, P.** (1959) La culture d'embryons d'*Elaeis guineensis* Jacq., *Agricultura, Louvain,* 7, 297.
12. **Rabéchault, H.** (1962) Recherches sur la culture 'in vitro' des embryons de palmier à huile (*Elaeis guineensis* Jacq.) I. Effets de l'acide β indolylacétique. *Oléagineux,* 17, 757.
13. **Bouvinet, J. and Rabéchault, H.** (1965) Recherches sur la culture 'in vitro' des embryons de palmier à huile. II. Effects de l'acide gibberillique. *Oléagineux,* 20, 79.
14. **Rabéchaut, H, Guénin, G. and Ahée, J.** (1967) Absorption de l'eau par les noix de palme (*Elaeis guineensis* Jacq. var. *Dura* Becc.) 1. Hydration des différentes parties de graines amenées à teneurs globales en eau determinées. *Cah. Orstom, sér. Biol.,* No. 4, 31.
15. **Henry, P.** (1951) La germination des graines d'Elaeis. *Rev. int. bot. appl. Agric. trop.,* 31, 349.
16. **Labro, M. F., Guénin, G. and Rabéchault, H.** (1964) Essais de levée de dormance des graines de palmier à huile (*Elaeis guineensis* Jacq.) par des températures élevées. *Oléagineux,* 19, 757.
17. **Rees, A. R.** (1965) Private communication.
18. **Noiret, J. M. and Ahizi Adiapa, P.** (1970) Anomalies de l'embryon chez le palmier à huile. Application à la production de semences. *Oléagineux,* 25, 511.
19. **Milthorpe, F. L.** (1956) The relative importance of the different stages of leaf growth in determining the resultant area, in *The growth of leaves,* pp. 141–8. Butterworth, London.
20. **Gregory, F. G.** Physiological conditions in cucumber houses. *Exp. and Res. Station, Chestnut, 3rd Ann. Rep. 1917.*
21. **Briggs, G. E., Kidd, F., and West, C.** (1920) A quantitative analysis of plant growth. *Ann. appl. Biol.,* 7, 103.
22. **Heath, O. V. S. and Gregory, F. G.** (1938) The constancy of the mean net assimilation rate and its ecological importance. *Ann. Bot.,* N.S., 2, 811.
23. **Watson, D. J.** (1956) Leaf growth in relation to crop yields, in *The growth of leaves,* pp. 178–90. Butterworth, London.
24. **Watson, D. J.** (1947) Comparative physiological studies on the growth of field crops. I. *Ann. Bot.,* N.S., 11, 41.
25. Oil Palm Genetics Laboratory. Progress Reports, 1968–72, Layang-Layang, Malaysia.
26. **Rees, A. R.** (1963) An analysis of growth of oil palm seedlings in full daylight and in shade. *Ann. Bot.,* N.S., 27, 325.
27. **Wormer, Th.** (1958) Croissance et développement du palmier à huile. *Oléagineux,* 13, 385.
28. **Rees, A. R. and Chapas, L. C.** (1963) An analysis of growth of oil palms under nursery conditions, I. Establishment and growth in the wet season. *Ann. Bot.,* N.S., 27, 607.
29. **Rees, A. R.** (1963) An analysis of growth of oil palms under nursery conditions. II. The effect of spacing and season on growth. *Ann. Bot.,* N.S., 27, 615.
30. **Rees, A. R.** (1963) Relationship between crop growth rate and leaf area index in the oil palm. *Nature, Lond.* 197, 63.
31. **Wormer, Th. M. and Ochs, R.** (1957) Humidité du sol et comportement du palmier à huile. *Oléagineux,* 12, 81.

32. **Rees, A. R. and Chapas, L. C.** (1963) Water availability and consumptive use in oil palm nurseries, *J. W. Afr. Inst. Oil Palm Res.,* **4,** 52.
33. **Ochs, R.** (1963) Utilisation du test stomatique pour le contrôle de l'arrosage du palmier à huile en pépinière. *Oléagineux,* **18,** 387.
34. **W.A.I.F.O.R.** (1956) Fourth Annual Report, 1955—6, p. 102.
35. **Rees, A. R.** (1957) W.A.I.F.O.R. Fifth Annual Report, 1956—7, p. 114.
36. **Rees, A. R.** (1958) Field observations of midday closure of stomata in the oil palm, *Elaeis guineensis,* Jacq. *Nature, Lond.* **182,** 735.
37. **Wormer, Th. M. and Ochs, R.** (1959) Humidité du sol, ouverture des stomates et transpiration du palmier à huile et de l'arachide. *Oléagineux,* **14,** 571.
38. **Rees, A. R.** (1961) Midday closure of stomata in the oil palm, *Elaeis guineensis,* Jacq. *J. exp. Bot.,* **12,** 129.
39. **Ochs, R.** (1963) Recherches de pédologie et de physiologie pour l'étude du problème de l'eau dans la culture du palmier à huile. *Oléagineux,* **18,** 231.
40. **Ruer, P.** (1969) Système racinaire du palmier à huile et alimentation hydrique. *Oléagineux,* **24,** 327.
41. **Corley, R. H. V.** (1973) Midday closure of stomata in the oil palm in Malaysia. *M.A.R.D.I. Res. Bull.* **1,** 2: 1—4.
42. **Corley, R. H. V., Gray, B. S. and Ng Siew Kee** (1971) Productivity of the oil palm (*Elaeis guineensis* Jacq.) in Malaysia. *Expl. Agric.* **7,** 129.
43. **Rees, A. R. and Tinker, P. B. H.** (1963) Dry matter production and nutrient content of plantation oil palms in Nigeria. I. Growth and dry matter production. *Pl. Soil,* **19,** 19.
44. **Corley, R. H. V., Hardon, J. J. and Tang, Y.** (1971) Analysis of growth of the oil palm (*Elaeis guineensis* Jacq.) I. Estimation of growth parameters and application in breeding. *Euphytica,* **20,** 307.
45. **Hardon, J. J., Williams, C. N. and Watson, I.** (1969) Leaf area and yield in the oil palm in Malaysia. *Expl. Agric.,* **5,** 25.
46. **Ng Siew Kee, Thambo, S. and De Souza, P.** (1968) Nutrient content of oil palms in Malaya. II. Nutrients in vegetative tissues. *Malay. agric. J.,* **46,** 332.
47. **Corley, R. H. V.** (1973) Effects of plant density on growth and yield of the oil palm. *Expl. Agric.,* **9,** 169.
48. **Corley R. H. V., and Mok, C. K.** (1972) Effects of nitrogen, phosporus, potash and magnesium on growth of the oil palm. *Expl. Agric.,* **8,** 347.
49. **Lo, K. K., Chan, K. W., Goh, K. H. and Hardon, J. J.** (1972) Oil palm — the effect of manuring on yield, vegetative growth and leaf nutrient levels. In *Advances in oil palm Cultivation,* p. 324. Incorp. Soc. of Planters, Kuala Lumpur.
50. **Bunting, B., Georgi, C. D. V. and Milsum, J. N.** (1934) *The oil palm in Malaya.* Kuala Lumpur.
51. **Fickendey, E. and Blommendaal, H. N.** (1929) *Ölpalme.* Deutscher Ausland-verlag, Hamburg and Leipzig.
52. **Beirnaert, A.** (1935) Introduction à la biologie florale du palmier à huile. Publs. I.N.E.A.C., *Sér. Sci.,* No. 5.
53. **Mason, T. G. and Lewin, C. J.** (1925) Growth and correlation in the oil palm (*Elaeis guineensis*). *Ann. appl. Biol.,* **12,** 410.
54. **Broekmans, A. F. M.** (1957) Growth, flowering and yield of the oil palm in Nigeria. *J. W. Afr. Inst. Oil Palm Res.,* **2,** 187.
55. **Sparnaaij, L. D.** (1960) The analysis of bunch production in the oil palm. *J. W. Afr. Inst. Oil Palm Res.,* **3,** 109.
56. **Henry, P.** (1955) Morphologie de la feuille d'Elaeis au cours de sa croissance. *Revue gén. Bot.,* **62,** 319.
57. **Cook, O. F.** (1940) Oil palms of Florida, Haiti and Panama. *Natn. hort. Mag.,* **19,** 10.
58. **Corley, R. H. V., Williams, C. N. and Rajaratnam, J. A.** (1971) Developmental and physiological aspects of sex-expression and yield in the oil palm. University of Malaysia Seminar, 1971, Mimeograph.

59. **Gray, B. S.** (1969) A study of the influence of genetic, agronomic and environmental factors on the growth, flowering and bunch production of the oil palm on the west coast of Malaysia. Thesis, University of Aberdeen.

60. **Corley, R. H. V.** (1973) Oil palm physiology – a review. In *Advances in oil palm Cultivation*, p. 37. Incorp. Soc. of Planters, Kuala Lumpur.

61. **Sparnaaij, L. D., Rees, A. R. and Chapas, L. C.** (1963) Annual yield variation in the oil palm. *J. W. Afr. Inst. Oil Palm Res.*, 4, 111.

62. **Zeven, A. C.** (1965) Oil palm groves in Southern Nigeria: Part 1. Types of grove in existence, *J. Nigerian Inst. Oil Palm Res.*, 4, 226.

63. **Haines, W. B. and Benzian, B.** (1956) Some manuring experiments on oil palms in Africa. *Emp. J. exp. Agric.*, 24, 137.

64. **Hemptinne, J. and Ferwerda, J. D.** (1961) Influence des précipitations sur les productions du palmier à huile. *Oléagineux*, 16, 431.

65. **Corley, R. H. V.** (1976) Oil palm yield components and yield cycles. Int. Agric. Oil Palm Conference, Kuala Lumpur, 1976.

66. **Jagoe, R. B.** (1934) Observations and experiments in connection with pollination of oil palms. *Malay. agric. J.*, 22, 598.

67. **Hardon, J. J. and Turner, P. D.** (1967) Observations on natural pollination in commercial plantings of oil palm (*Elaeis guineensis*). *Expt. Agric.*, 3, 105.

68. **Wong, Y. K. and Hardon, J. J.** (1971) A comparison of different methods of assisted pollination in the oil palm. Chemara communication No. 9, quoted by Corley, see ref. 60.

69. **Jack, H. W. and Jagoe, R. B.** (1929) Preliminary note on variation of individual fruits of the oil palm grown under avenue conditions. *Malay. agric. J.*, 17, 35.

70. **Opsomer, J-E.** (1932) Notes sur l'Elaeis à la Côte Est de Sumatra. *Bull. agric. Congo belge*, 23, 422.

71. **Gray, B. S.** (1966) The necessity for assisted pollination in areas of low male inflorescence production and its effect on the components of yield of the oil palm. *Planter, Kuala Lumpur*, 42, 16.

72. **Turner, P. D. and Bull R. A.** (1967) *Diseases and disorders of the oil palm.* Incorp. Soc. of Planters, Kuala Lumpur.

73. **Bredas, J. and Scuvie, L.** (1960) Aperçu des influences climatiques sur les cycles de production du palmier à huile. *Oléagineux*, 15, 211.

74. **Devuyst, A.** (1948) Influence des pluies sur les rendements du palmier à huile enregistrés à la station de La Mé de 1938 à 1946. *Oléagineux*, 3, 137.

75. **I.R.H.O.** Rapport Annuel, 1957.

76. **Michaux, P.** (1961) Les composants climatiques du cycle annuel de productivité du palmier à huile. *Oléagineux*, 16, 523.

77. **Ruer, P.** (1966) Relations entre facteurs climatiques et nutrition minérale chez le palmier à huile. *Oléagineux*, 21, 143.

78. **Turner, P. D.** (1976) The effects of drought on oil palm yields in South-east Asia and the South Pacific region. Int. Agric. Oil Palm Conference, Kuala Lumpur, 1976.

79. **Robertson, G. W. and Foong, S. F.** (1976) Weather-based yield forecasts for oil palm fresh fruit bunches. Int. Agric. Oil Palm Conference, Kuala Lumpur, 1976.

Chapter 5

Oil palm selection and breeding

Oil palm breeding has as its aim the production of plants giving the maximum quantity of palm oil and kernels per hectare. The oil palm breeder must therefore concern himself with all the circumstances of bunch production, with the quantity of oil and kernels per bunch, and with resistance to disease.

Several factors concerning growth and bunch production make the work of the breeder difficult. Firstly, though the propagation of the palm by tissue culture has been achieved, practical methods of large-scale vegetative propagation have yet to be devised. Secondly, the monoecious character of the palm makes cross-pollination general and, as with other cross-fertilized plants, the value of parents will therefore ultimately be determined by the performance of their progeny in crosses. Thirdly, two products, oil and kernels, must be taken into account, and in countries where palm kernel oil is extracted the separate values of three products may have to be considered. Nearly all breeding has been directed towards palm oil production since in high quality *tenera* fruit the quantity of oil is four to eight times that of the kernels. However, this is not in itself sufficient reason for breeding entirely for oil, as will be shown later.

The early history of selection

The establishment of plantations on a large scale was an achievement of the 1920s, both in Zaire and in the Far East, and workers in these territories were not slow to start investigations on the improvement of the crop by selection and breeding. In Zaire, selection was in the hands of the Institut National pour l'Etude Agronomique du Congo Belge (I.N.E.A.C.), while in the Far East the work was undertaken by the large plantation companies of Indonesia and Malaysia and by the Algemene Vereniging van Rubberplanters ter Oostkust van Sumatra (A.V.R.O.S.) and the Department of Agriculture, Malaya.

Owing to the great differences in the material being used, the approach to improvement in the two regions was also different. In Africa the poor quality of the *dura* fruit was apparent, and although there was some hesitance in breeding exclusively for *tenera* palms, the presence of fine specimens of the latter led to an early concentration on the discovery and reproduction of high quality *tenera* material. This policy reaped its reward with remarkable rapidity.

In the Far East, the relatively high quality of the illegitimate *dura* progeny emanating from the Deli ornamental avenues tended largely to confine the very early work to mass selection* for the provision of seed for further planting. In 1922 A.V.R.O.S. stated emphatically that the Deli type must remain the standard oil palm for Sumatra until breeding had done its work with the offspring of newly imported varieties. Under these circumstances the general procedure was for mass selection to be carried out for extending plantings, and for separate breeding programmes to be undertaken both with outstanding Deli individuals and with imported material.

Indonesia (Island of Sumatra)

Little information exists regarding the yield and quality of the original unselected plantation population of Deli palms and the few publications from Indonesia were concerned largely with crosses between Deli and imported material. Schmöle[1] accepted Blommendaal's figures of 62 to 63 per cent mesocarp, 30 per cent shell and 7 to 8 per cent kernel to fruit as being 'average components' in 1929, but committed himself to the somewhat contradictory statement that the Deli was of more or less constant standard with a percentage shell varying from 20 to 40. He also suggested 100 kg per palm as an average annual yield of the first generation on good soil. It may be remarked that 40 per cent shell is quite a usual figure for African *dura* fruit, and the composition of the early Deli material may not have been so 'standard' as is often claimed.

The mystery of the origin of the four Buitenzorg (Bogor) palms has been referred to in Chapter 1; but the descendance and selection of their progeny in the 70 years between their coming into bearing and the establishment of plantations is also somewhat mysterious. It has been stated that the similarity of the four palms is 'generally acknowledged', but only two palms remain and there are no published figures of their fruit and bunch analysis. Recent observation suggests that the fruit composition was not outstanding. Illegitimate progeny of these old palms, planted in West Africa in 1958, indeed showed a remarkable similarity in appearance and in the form and set of the bunches, but the fruit analyses were not

---

* The term is used according to Hayes, Immer and Smith's definition: selection of some desired character or characters, where progeny of the plants selected are grown in bulk. With the majority of mass selection in the oil palm, even in the early days, controlled pollination was used.

similar to the modern Deli, being more akin to some of the *dura* as found in the Africa groves. The progeny of later generations of Delis planted in West Africa have always shown typical Deli composition.

The standards of selection adopted when planting the first estates from illegitimate seed taken from the ornamental avenues which had been established in the last two decades of the nineteenth century are not known. Stoffels[2] states that the best mother-palms used for the original Pulu Radja Estate and Tamiang plantings were those in the Saint Cyr and Tandjong Morawa avenues, but that, owing to an insufficiency of planting material, seed had also to be taken from poorer specimens in the gardens of Medan.

In spite of the good quality of the Deli material there was interest in importations from Africa right from the start. There are records of importations to Bogor in 1914—15 and to several estate groups and A.V.R.O.S. a few years later. Some of this material proved an embarrassment as it was of inferior quality; and much of it, being poor African *dura*, was eliminated at an early stage not only from government selection stations, but also from the estates of plantation companies. As mixed plantations existed in the early days it seems most likely that some subsequent plantations were 'contaminated' with characters inherited from this material.

One of the first attempts to select within Deli material through the evaluation of proper records was made at Marihat Baris estate where 2,000 palms, planted in 1915, were recorded from the time they came into bearing. In late 1922 self-pollination of fifteen palms was carried out in cooperation with A.P.A. (A.V.R.O.S.) and progenies were planted in the selection areas at Sungei Pantjur and later at Polonia during the period 1924 to 1931. The factors sought were high bunch yield, high bunch number, mesocarp thickness of 4 mm or more, and freedom from Crown disease. The selected palms were selfed and crossed for seed production as well as for the laying down of the F_1 generations at the selection stations. It has been reported that, in the event, this selection and breeding gave rise to high-yielding palms with comparatively few, though heavy, bunches. The selections were made at a very early age and the progeny were characterized by early bearing, suggesting that precocity was a factor unconsciously selected and inherited. It is possible that the exceptionally early bearing of some later fields in Sumatra and Malaysia may be partly due to this factor and not entirely to improved agronomy as is usually assumed.

Selection undertaken by estate groups led to progenies with larger numbers of relatively small bunches, but the total oil yield per palm proved to be of about the same order. Examples of this work may be given from two estates. Fifty parents were selected from 24,500 palms which had been individually yield recorded. They were originally selected for high yield, bunch weight, fruit-to-bunch ratio and mesocarp percentage. Oil determinations were introduced later. These good characters were in

Table 5.1   *Selection of Deli* dura *in Sumatra* (means per palm per year)

Es-tate	Genera-tion	Years from plant-ing	Weight of bunches (kg)	No. of bunches	Wt. per bunch (kg)	Fruit to bunch (%)	Meso-carp to fruit (%)	Oil to meso-carp (%)	Oil to bunch (%)
A	F_0 selections	6–14	224	10.5	21.3	66.1	67.2	48.8	21.6
A	F_1 progenies	11–17	213	9.2	23.2	67.4	65.7	49.8	22.0
B	F_1 progenies	11–17	236	10.0	23.6	64.9	67.5	51.1	22.3

large measure reproduced in their progenies (F_1) from which up to 4 per cent of the palms were selected for legitimate seed production for further estate plantings. The averaged data are shown in Table 5.1. It will be seen that both in yield and fruit characters the standard of the selected palms, which were the elite of their generation, was reproduced or even bettered in the progeny; in the case of Estate A the F_1 generation was a replanting whereas the F_0 was on virgin soil.

These figures illustrate the standard of Deli material being obtained by selection in Sumatra in the 1930s, and they show how the Deli palm obtained its reputation both as a high yielder and as a palm of good and uniform composition.

Another plantation group followed initial Deli selection with two generations of selfing or with one generation of selfing followed by an F_2 of crosses. For this S.O.C.F.I.N. (Société Financière de Caoutchoucs) selection and breeding work at Mopoli and Bangun Bandar few details are available, but in the F_1 selfs progeny yields 50 to 60 per cent above those of the initial populations, and in the F_2 selfs and crosses 20 per cent, 12 per cent and 40 per cent above the F_1 populations, were claimed.[3] These figures are impressive, particularly those of the F_1 selfs, but it is of course impossible to tell what proportion of the yield increases must be attributed to improved planting methods or the use of more fertile areas. The yields and bunch analysis of the selected F_1 palms at the two centres were strikingly similar. Bunch yield was about 200 kg per palm per annum, fruit to bunch ratios were 60 to 61 per cent and mesocarp to fruit 68 to 69 per cent.

Serious breeding work with material imported from Africa was started by A.P.A. soon after the First World War, imports being received from many parts of the continent; trials of Deli crosses with Deli x import *dura* and *tenera* and with crosses within African material were put in hand at once and the majority of progenies were planted at Sungei Pantjur and Polonia.[4] The high sex ratio and low bunch weight of the imported material was soon apparent.

The pedigree of the important Sumatran imported material is confusing, so it is set out schematically below:

The Path of 'Import' Tenera in Sumatra

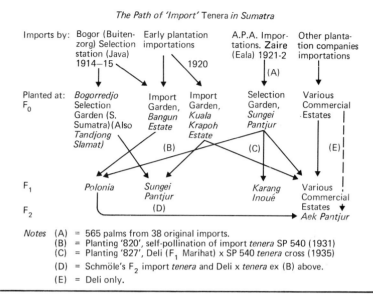

Notes (A) = 565 palms from 38 original imports.
(B) = Planting '820', self-pollination of import *tenera* SP 540 (1931)
(C) = Planting '827', Deli (F_1 Marihat) x SP 540 *tenera* cross (1935)
(D) = Schmöle's F_2 import *tenera* and Deli x *tenera* ex (B) above.
(E) = Deli only.

An existing record shows that the well known SP540 *tenera* at Sungei Pantjur was part of a consignment of seed sent by the Director of the Eala Botanic Gardens in Zaire. The seed was recorded as Var. 'Djongo', indicating that it came from the famous Djongo *tenera* palm (see p. 211) which was to give rise to a high proportion of the best *tenera* lines in its own country. It is of interest to note here that seed from Eala was imported into Colombia some 15 years later and provided one parent of unquestionably 'Yangambi-type' *tenera* to be found at Calima. Thus it is almost certain that a great quantity of good *tenera* in all three continents is descended from the Djongo palm.

Unfortunately no satisfactory yield comparisons of these *tenera* progeny with selected Deli *dura* could be made in Sumatra owing to the different dates of planting and age of nursery material. At that time certain 'standard' bunch yield criteria were adopted following discussions between plantation companies, and a 'standard' of 147 kg per palm (324 lb) was considered normally attainable by the eighth or ninth year from planting. In general, the yields of pure *tenera* were below the standard at this age but some of the mantled *tenera* and selected Delis were above it.

The material planted at what may be termed the F_1 and F_2 stations may be briefly described as follows:

Polonia

Deli material. Two lots of selections were made for an F_2 of the Marihat material. The second of these provided adult oil to bunch percentages of 18 to 20 and yields some 40 per cent above the F_1 mean.

Tenera. Selection was largely concentrated on the progeny of palm SP540 from Sungei Pantjur. Seven parents were selected mainly on their bunch and fruit analyses which gave oil to bunch percentages as follows:

	Year	Number of analyses	Oil to bunch (%)
7 'import' *tenera* from selfing of SP540, planted 1931	1937–8	12	23.0
	1938–9	48	29.0
	1939–40	11	31.2
	1940–1	5	29.6
Parent palm SP540	1935–9	7	23.9

It may be noted that the 1931 planting of the SP540 *tenera* selfs at Polonia consisted of 123 plants, but the segregation of *dura*, *tenera* and *pisifera* forms was not recorded. However, these seven F_1 probably represented a selection of about 9 per cent of all the *tenera*. The parent SP540 analysis suggests a fruit-to-bunch ratio of about 60 per cent with mesocarp to fruit of 80 per cent. Interest was also taken in mantled *tenera* originating from Tandjong Slamat. These gave high bunch yields but their oil-to-bunch ratio was only 21 to 24 per cent. One, Pol.3244, had remarkably heavy bunches, though this was a general feature of Polonia selections.

One *pisifera* palm, Pol.3184, is of interest as it is to be found in the pedigree of many Sumatran and Malaysian progeny. This came from the Bangun *tenera* stock and was largely fertile, but with the usual low fruit-to-bunch ratio.

Sungei Pantjur

The selection areas were cut out in 1938–9 and only two palms, SP540 *tenera* already referred to and SP2041 Deli *dura* from Marihat, were used as parents. Bunch production of SP2041 was 557 kg during the fifth to eighth years inclusive (from planting) compared with 314 kg from twenty-one other Marihat Deli *dura* Its oil-to-bunch ratio was 19.3 per cent in 1938–9.

Karang Inoué

Schmöle's Deli *dura* Pol.3003 crossing with SP540 *tenera* is of outstanding interest at this station. It may be assumed that Pol.3003, which was of the Marihat F_1 progeny, was a good Deli with a fruit-to-bunch ratio of over 60 per cent, shell to fruit of 30 per cent or less and kernel below 10 per cent. Something has already been said of the analysis of SP540. The equal numbers of *dura* and *tenera* progeny in this planting (numbered 827) analysed as follows:

	Mesocarp to fruit (%)	Shell to fruit (%)	Kernel to fruit (%)
Dura	53.8	35.5	10.7
Tenera	83.5	11.1	5.4

It must be remembered that the *dura* progeny represent, genetically, a Deli x African *dura* cross. Many a *tenera* with over 80 per cent mesocarp to fruit has been produced from an African *dura* parent with around 50 per cent mesocarp. A Deli *dura* x African *tenera* cross would normally be expected to result in a down-grading in fruit composition of the *dura* half of the progeny as seems to be the case in the above results. It is true that the gain from the *tenera* composition would more than balance this loss on the *dura* side, but in the issue of Deli x *tenera* seed the probable inferiority in composition of the *dura* half of the progeny is apt to be forgotten. More is said of this later in this chapter (p. 255).

Aek Pantjur

Immediately after the Karang Inoué *dura* x *tenera* planting, other important Deli and African crossings and selfings were laid down in fields A.P.7/8 and 9 at Aek Pantjur. These were largely of Polonia parentage, and the seven *tenera* parents descended from SP540 were used. There are two unusual features of the F_1 SP540 selfing and the F_2 selfings and crossings within the F_1 of SP540. In the first place it was reported in 1953 that only two *pisifera* had been identified in the F_1.[4] Secondly, the Import *dura**derived from the SP540 F_1 selfings and crossings had an exceptionally good analysis for an African *dura*: the fruit weight averaged 21 g and the mesocarp to fruit averaged 64 per cent. These figures are higher than those of *dura* in Africa which produce the very best *tenera* progeny. Thus SP540, which had an exceptional origin, was an exceptional plant giving larger, not smaller, fruit in its offspring than a typical Deli and as high (or higher for that time) a proportion of pulp as is found in a good Deli.[5] The Deli x Import *tenera* were intermediate in fruit weight and mesocarp percentage, but it was their African *dura* parentage which was superior in these respects, not the Deli *dura* parentage as would normally be expected. From this it must be assumed that the mesocarp percentage of the *dura* in the Karang Inoué Deli x Import *tenera* cross was lowered, not raised, by its Deli parentage. Yet, owing to the lack of *pisifera* in the SP540 progeny there remains the possibility that there was some out-pollination with Deli pollen in the original selfing. However that may be, it is clear that the seven *tenera* in the F_1 were selected for a fruit size above that of the average Deli and for factors giving rise to unusually high mesocarp percentages in their *dura* progeny. This may be seen in Table 5.2.

---

* The term is used here and in Sumatran parlance to indicate *dura* descended only from importations after 1900 and which have not resulted from any crossing with the Deli.

Table 5.2   *Fruit and bunch analysis of Deli, Import dura and tenera, and of crosses planted at Aek Pantjur 1939—40. Analyses 1950—2 (Pronk)*

Category of palm	Fruit to bunch ratio* (%)	Weight per normal fruit (g)	Mesocarp to fruit (%)	Nut to fruit (%)	Number of palms
Deli *dura*	58.8	15.4	56.8	43.2	102
Deli Import *dura*[b]	57.4	20.1	59.3	40.7	109
Import *dura*[a]	55.3	20.9	64.2	35.8	49
Deli *tenera*[b]	53.2	15.6	80.7	19.3	101
Import *tenera*[a]	56.2	18.5	87.9	12.1	86

* normal + parthenocarpic fruit. [a]derived from *tenera* selfs and crosses, F_2 from SP540. [b]derived from Deli x *tenera* crosses.

Interesting differences of bunch production were also found, but as all the 'Import' palms were descended from a single parent no general conclusion could be drawn concerning the yield potential of Deli, Import and Deli x Import material. Table 5.3 shows the yields of the two plantings at Aek Pantjur and the results of similar crossings carried out with other material on an estate.[4]

Table 5.3   *Bunch yield of Deli, 'Imports' and crosses at Aek Pantjur and an estate in Sumatra (per palm per annum)*

Category of palm	Number of bunches			Weight per bunch			Weight of bunches					
							kilogrammes			as per cent of Deli		
	A.P. 1939	A.P. 1940	Es-tate	A.P. 1939 (kg)	A.P. 1940 (kg)	Es-tate (kg)	A.P. 1939 (kg)	A.P. 1940 (kg)	Es-tate (kg)	A.P. 1939 (%)	A.P. 1940 (%)	Es-tate (%)
Deli *dura*	5.0	6.8	4.4	19.0	18.6	35.8	94	126	158	100	100	100
Deli Import *dura*	6.6	9.9	5.7	19.0	19.8	31.6	125	195	180	133	155	114
Import *dura*	7.3	11.8	7.8	10.6	12.1	24.9	78	143	194	82	113	123
Deli *tenera*	7.3	10.3	6.1	17.5	18.5	27.3	128	190	167	135	151	106
Import *tenera*	8.9	11.7	8.6	11.1	10.4	21.6	98	122	187	104	96	118

A.P. yields: 5 years. Estate yields: 3 years. Ages: A.P. 10 to 15, Estate 17 to 20 years old.

The weight per bunch was highest in the Deli, but in the cross it was usually nearer to the Deli figure than to that of the import material. The number of bunches was consistently higher in the import material, but at Aek Pantjur the cross was near enough to the import to make the overall yield much higher in the crosses than in either the Deli or the import. This was not so on the estate, where the imports were slightly ahead on overall yield.

With regard to the *pisifera* from the A.V.R.O.S. programmes, those descending from SP540 were largely infertile. A number of fertile *pisifera* arose from the Bangun imports and from the mantled *tenera* programme. The Bangun progeny, *pisifera* Pol.3184, has already been referred to. Fruit-to-bunch ratios of the fertile specimens were usually around 35 per cent.

Interesting *tenera* programmes were also carried out by a number of estate groups. In one case about 240 palms in a supposedly Deli planting

Table 5.4 *Selection from illegitimate Deli* tenera *on an estate in Sumatra* (means per palm per annum)

Generation on Estate B	Years from planting	Weight of bunches (kg)	No. of bunches	Weight per bunch (kg)	Fruit to bunch (%)	Mesocarp to fruit (%)	Oil to mesocarp (%)	Oil to bunch (%)	Oil per palm (kg)
F_0 Illegitimate Deli *tenera*	8–11	224	12.9	17.4	59.5	83.5	51.4	25.6	57
F_1 *tenera* from D × T Cross	9–14	231	12.8	18.0	59.4	86.4	50.1	25.8	60

were found to be *tenera*. These were illegitimate progenies from Deli palms which had been growing near to African material. In this case, therefore, *tenera* selection was within Deli *tenera* material right from the start. The production and analysis of the whole population is not known, but from those of the original selection of forty-seven palms, i.e. nearly 20 per cent, it appears that the standards must have been very fair. These selections were crossed with selected *dura* and the progeny, half *dura*, half *tenera*, were planted in the same estate. Table 5.4 shows the production and analysis of the original illegitimate Deli *tenera* and of the progenies on the same estate.

The improvement between the two generations is in mesocarp percentage. At the same time it should be noted that fruit-to-bunch ratios are low in comparison with the *dura* in Table 5.1. This has reduced considerably the advantage to be gained from moving from the Deli to Deli *tenera*.

The *tenera* of another estate group, S.O.C.F.I.N., owed their origin to the Kuala Krapoh importation from which an illegitimate generation was established at Padang Pulau in 1924. An F_1 generation of selfs was planted at Bangun Bandar Estate in 1934. There were very few *pisifera*,[3] suggesting some out-pollination, and bunch production reached 17 tons per hectare at 6 to 7 years from planting.

Crossings of Deli and *tenera* selections were laid down at Mopoli in 1927 and Bangun Bandar in 1936. The Mopoli crosses were from the original Kuala Krapoh *tenera* and some Delis from an early planting at Tandjong Genteng Estate; the Bangun Bandar crosses were from Padang Pulau illegitimate *tenera* crossed with Deli from another early estate, Senadam. An F_2 generation of *dura* × *tenera*, was obtained from the Mopoli material and planted in 1942–3.

The only *tenera* bunch analysis data available from these estates are for the best palms in the F_0 and F_1 fields at Padang Pulau and Bangun Bandar respectively. These are as follows:

	No. of palms	Age	An. bunch yield (kg)	Fruit to bunch (%)	Mesocarp to fruit (%)	Oil to mesocarp (%)	Oil to bunch (%)	An. oil per palm (kg)
F_0 Pg. Pulau	5	11–14	128	55.6	81.1	47.5	21.4	27
F_1 Bangun Bandar	10	6–9	196	62.6	78.9	56.2	27.8	54

The striking differences between these chosen palms, which must not be confused with full generations, is in the fruit-to-bunch ratios and oil-to-mesocarp percentages. This is only one of a number of indications that to obtain the full benefit of a change in fruit form when passing from the Deli *dura* to the *tenera* it is necessary to minimize the *tenera* tendency for a low fruit-to-bunch ratio and to raise the oil to mesocarp content above that of the parent *dura*.

Malaysia

The early estates in Malaysia obtained open-pollinated seed from the ornamental avenues and other plantings in Sumatra and so the Deli palm became the established plantation oil palm. The history of the first plantings is a little confused but the first estate, Tennamaram, seems to have obtained its seedlings from Sumatra via a nursery at Rantau Panjang estate, while the second, Elmina, obtained seed from an avenue planted at Rantau Panjang in 1911–12.[6] (Plate 16)

In 1922 an avenue of eighty-six palms was established at the Central Experiment Station at Serdang. The seed came from palms in Kuala Lumpur of unknown origin, but of Deli type. These palms showed considerable variation in yield and quality, but the ten best palms were widely used for seed production and their descendants have found their way into later commercial breeding programmes in Malaysia and into programmes in Africa. It is of interest, therefore, to note the data on which the selection of these ten palms was based (Table 5.5).[7] The bunch yields are of little value for comparison with the yields of palms in fields, since the former were from palms in single, widely-spaced rows and had received assisted pollination for several years. Fruit analysis data are of interest, however, since they approximate to Schmöle's 'average components' and show a higher kernel percentage than has in later years been extracted from commercial Deli fruit (Plate 17).

Table 5.5 *The ten selected Serdang Avenue Deli* dura *palms (2½ years data, Jack and Jagoe)*

Palm no.	Bunch yield per annum (kg)	Fruit to bunch ratio (%)	Mesocarp to fruit (%)	Kernel to fruit (%)
27	234	67.2	60	7.5
37	190	67.6	58	6.9
3	193	65.7	64	6.9
7	192	65.1	62	8.0
5	187	66.0	63	6.7
9	188	62.8	60	8.0
23	187	61.2	56	9.3
57	161	66.3	64	6.7
65	151	68.1	61	7.6
67	151	59.3	67	6.1
Mean	184	64.9	61	7.4

Pl. 16   A typical avenue of oil palms from which so much of the industry of the Far East developed. This avenue is in Perak, Malaysia.

Pl. 17   The Serdang avenue palms in 1962.

Though many estates made use of the Serdang material, others made special importations of Sumatran seed. An important group of estates in Lower Perak, for instance, was planted with the Marihat material.

The oil palm industry made a slower start in Malaysia than in Sumatra and yield studies and selection were not under way until the 1930s; serious trial of the value of importations did not start until after the war. Jack and Jagoe[7] studied a block of 589 10-year-old palms at Elmina estate and found that in 1930 bunch yields varied from 0 to 309 kg per palm with a mean of 117 kg. The standard deviation was 121 and the coefficient of variation was 47 per cent. There seemed, therefore, to be a considerable field for selection. Even the twenty best palms which formed the prelimi-

Table 5.6   *Twenty best Deli palms in the Elmina Estate block from which final selections were made*

Palm no.	Bunch yield per annum (kg)	Mesocarp to fruit (%)	Kernel to fruit (%)	Shell to fruit (%)
214	294	64	4.9	31.1
270	248	68	5.7	26.3
207	239	64	6.1	29.9
267	230	62	8.0	30.0
242	228	61	8.6	30.4
268	223	70	4.9	25.1
360	219	62	5.5	32.5
243	206	59	9.0	32.0
180	204	59	5.2	35.8
20	200	58	6.4	35.6
208	198	55	10.6	34.4
141	195	61	6.0	33.0
179	195	62	6.5	31.5
120	188	76	3.9	20.1
27	187	63	6.0	31.0
97	184	63	5.2	31.8
36	183	62	4.9	33.1
152	179	74	2.9	23.1
88	175	62	5.4	32.6
94	170	63	5.8	31.2
Mean	207	63	6.7	30.5

nary selection still showed much variation in yield and fruit composition as is shown in Table 5.6. No attention appears to have been given to fruit-to-bunch ratios, a figure of 60 per cent being taken as normal.

These figures illustrate the standard of the early Deli material available in Malaysia and its variability, and there is no reason to believe that it was any different from the Sumatran material from which the early selections, already described, were made. Conscious attempts had been made to select from the Sumatran avenues those palms appearing to give the largest bunches and fruit with the thickest mesocarp, but these parent palms were open-pollinated.

One feature of the Elmina data which was not noted at the time was the negative and highly significant correlation (-0.72) between mesocarp and kernel percentage. Whether this correlation existed in the population as a whole is not recorded but it would appear that when shell percentage to fruit was in the 20 to 25 per cent range mesocarp percentages were high and kernels very small, whereas with large or even medium-sized kernels, shell percentage was usually over 30 per cent. This suggests a fair uniformity of shell thickness in selected Delis.

It was also noted at Elmina that high-yielding palms might sometimes be in pairs or small groups and that therefore environmental factors as well as heredity were concerned in their performance. This 'position effect' was later to be taken into account in making the final Elmina selections and in several other selection programmes.

Jagoe[8] had described how in the course of his selection work at Elmina two of the palms under particular observation were seen to have an unusually large girth and a slow height increase in spite of a high rate of leaf production. One of these palms, the well-known Dumpy E.206, was included among the final selections. In yield and some other characters the palm was not quite up to the selection standards set for other palms, although the bunches were large and the fruit-to-bunch percentage above average. Its position in the field was unfavourable and its yield was not therefore thought to have reached its full potential. Selfed and crossed progeny of E.206 were planted at Serdang just before the war; this was fortunate as the original mother palms on Elmina Estate were all cut out during the Japanese occupation of Malaysia (Plate 18).

The selfed progeny of Dumpy E.206 was quite uniform and the larger girth and small annual height increment were transmitted to a remarkable degree as shown in Table 5.7.

Table 5.7   *Height and girth data of Dumpy E.206 and its progeny*

	Girth at 122 cm (4 ft) (cm)	Height to base of crown (cm)	Age Years
Original E.206	287	335	15
'Normal' palms at Elmina	226	518	15
F₁ Selfed E.206	272	292	12
Cross E.206 x E.268	249	406	12
Cross E.206 x E.152	221	411	12
Selfed E.152	206	417	13
Selfed E.268	249	475	13

The bunch yield of E.206 selfed progeny proved to be high where not overshaded by other palms, but the fruit-to-bunch ratio varied considerably and there were abnormalities of fruit set. The intermediate girth and height data of the Dumpy crosses suggested that dumpiness was an expression of quantitative genetic factors and that E.206 and its selfed progeny exhibited a high degree of homozygosity in this respect though not in fruit characters. Some of the latter were of a very high order in the E.206 x tall crosses, so a programme of 'back-crosses' between these and their half-sibs, the E.206 selfed progeny, were started. True back-crosses could not be made owing to the destruction of the parent palm. The so-called 'back-crosses', together with F₂ Dumpy selfs, were distributed to a number of estates and considerable populations of these palms exist.

Selection within Deli populations was only undertaken before the war by two plantation groups. In the larger, selection was undertaken firstly among palms owing their origin to Serdang Avenue palms. It has been

Pl. 18 E.206 Dumpy F_1 self at Serdang. 26 years old. Note the characteristic 'blunt' leaf tip of this progeny.

recorded[9] that palms Nos 5 and 7 were of most value and their progenies were presumably most extensively used. Table 5.5 shows that the fruit quality of these palms was in no way outstanding, shell percentage being 30 per cent in both cases. Selections were also made among large numbers of palms grown from Sumatran open-pollinated Deli seed. In the period

1935–40, 380 progenies were laid down by inter-crossing 175 palms of Serdang legitimate and Sumatran illegitimate origin and planting in 'genetic blocks', the term being used to signify blocks of palms of known percentage. Individual yield records of the parents was undertaken, but fruit and bunch analysis was lacking or incomplete.

Selection undertaken by another plantation group has not been described in such detail and the basic material was of different origin to that described above, being legitimate and illegitimate Deli material selected for production at a single source in Sumatra. This F_1 material was planted between 1928 and 1940 by progenies of 100 to 150 palms each. During the latter part of this period crosses were made between the best palms to form a further generation and certain of these progenies were kept after the war to form the basis of future Deli breeding programmes; the latter included several hundred acres of inbred material and crosses. The commercial Deli seed issued was then obtained from palms whose progenies had done well in selfings and crosses.[10] In general the standards adopted led to the breeding of progenies with comparatively few, but large, bunches. Differences in this respect are well illustrated by a trial of two Deli progenies from the Serdang Elmina collection with two progenies each from the two plantation groups. The data given are for 1962 when the palms were 10 years old:

Jerangau Station	Number of bunches	Weight per bunch (kg)	Weight of bunches (kg)
Two Serdang progenies	8.9	22.2	197
Two plantation 'A' progenies	12.5	17.8	210
Two plantation 'B' progenies	9.1	18.8	171

One of the two Serdang progenies (14/2 x 10/8) was outstanding in its consistent production of large bunches; in the whole bearing period from 1956 to 1962 the bunch weight was 30 per cent higher than the mean bunch weight of the plantation progenies.

It is disappointing that it is not possible to measure the degree of improvement in yield and quality of the Deli palm in the inter-war period. More exact analytical work was carried out in Sumatra than in Malaysia and larger numbers of palms were recorded. Claims were made that bunch production in the F_1 was 50 per cent above the production of the fields from which the parents were selected, but how far improved cultural practice was responsible for enhanced yield it is impossible to say. There are no reports, as in Africa, of trials of selected versus unselected material. The general standard of later production does not suggest that the recording of tens of thousands of palms was likely to produce much better progenies than the recording of several hundred, and in view of the restricted origin of the Deli this is perhaps not surprising.

Zaire

The history of selection and breeding in Zaire is the history of the emergence of the *dura* x *pisifera* cross. This discovery closed one epoch of selection and heralded another. The early epoch in Zaire was, however, quite different from that of the Far East, and it was the early realization of the value of the *tenera* by Ringoet and the meticulous and inspired work of Beirnaert, carried through under the difficult conditions of Central Africa, that made it possible for the *tenera* era to develop within 30 years of the establishment of the first plantations.

Beirnaert, whose death in 1941 deprived the oil palm world of its leading figure, described the early Zaire selection work in both its practical and theoretical aspects in considerable detail.[11-13]

Ten open-pollinated *tenera* bunches were used for the establishment of the *Palmeraie de la Rive* at Yangambi in 1922. One was from a palm — the famous Djongo (= the best) — at Eala which had 55 per cent mesocarp and 30 per cent oil to bunch, and the other nine were from groves at Yawenda. The planting was at close-spacing and *dura* and poor types of *tenera* were gradually eliminated. Further fields were planted in 1924 and 1927 with open-pollinated *tenera* seed from a plantation at Gazi where 16,000 palms had been under observation. Fields planted in 1929 and 1930 were from seed of *tenera* palms taken in the *Rive* after the elimination of *dura* palms; this was a second generation of illegitimate *tenera* material.

After the thinning out, the remaining palms in these fields were subjected to a system of 'control', i.e. they were yield-recorded, and those proving to be high yielders were evaluated by an elaborate process of bunch analysis, the sum of the exterior, interior, parthenocarpic and abortive fruits being taken as the measure of the numbers of flowers produced by the inflorescence. The examination of all these data resulted in the choice of possible mother palms for breeding; these were known as *Candidats Arbres-mères*, or C.A.M. The work was carried on in stages with the elimination of many palms at each stage. Finally, after 3 years of observation when the palms were mature, the following minimum standards were laid down and resulted in the choice of actual *Arbres-mères* (A.M.):

1. Average bunch production   140 kg/annum
2. Mesocarp percentage to fruit   75 %
3. Average weight of exterior fruit   12 g
4. Weight of kernel   1 g
5. Yield of palm oil   33 kg/annum
6. Normal growth

These standards were not applied rigidly and exceptional performance in one respect allowed a palm to be selected even though it was below standard in another. This method of selection was based on certain theoretical and practical concepts.[11] Beirnaert drew attention to the fact that, at that time, the mode of inheritance of characters combining to give a high oil yield was unknown. The parents would, however, undoubtedly be polyhybrids and selfing palms or crossing two palms with similar characters would lead to the production of individuals superior, inferior or equal to the parents. If, it was argued, parents showing high values in some respects and low values in others were crossed with parents showing the inverse of this, then a great many favourable genes and some homozygosity could be expected in the progeny. The crossing of palms with complementary characters was therefore advocated and selfing was only to be permitted with palms showing a good performance in all respects. Selfing was recommended for the F_2 generation using palms with the expected high performance in each desirable character.

The standards laid down are of much interest and they were to have considerable effect on future generations of *tenera* palms throughout the world. The low mesocarp percentage standard reflects the initial difficulties in finding high quality *tenera*; later breeding raised this figure considerably. Bunch production was set at an exactingly high figure for conditions in Zaire; among the original selections it was not often achieved.

Of particular interest was the insistence on large fruit and a kernel of at least 1 g in weight. It was postulated that 'the best type is that which gives the largest absolute quantity of oil; in other words, the fruit which has the greatest absolute quantity of mesocarp and kernel'. This can be disputed. Oil yield is the product of bunch weight and oil per bunch. The absolute quantity of mesocarp in the individual fruit (as distinct from its percentage) is, theoretically, of no consequence. Ten bunches weighing 10 kg each and giving 22 per cent oil to bunch can provide 22 kg of oil whether the individual fruit weigh 8 g or 16 g. This oil will be provided by larger numbers of smaller fruit or vice versa, and there appears, neither on theoretical nor experimental grounds, to be any virtue in large fruit *per se*. The only requirement in this particular example is that the fruit shall have at least 73.4 per cent mesocarp on the assumption that there will be 60 per cent fruit weight to bunch and 50 per cent oil to mesocarp.

However, Beirnaert also considered that a large number of flowers per bunch was an important factor and, naturally, if a narrow standard for this factor was to be laid down then the size of fruit would be important. In fact, however, the number of flowers per inflorescence in the C.A.M.s varied widely and the size of fruit and the number of flowers tended to be inversely related. There was also a considerable variation in the number of flowers setting fruit. It was admitted that larger fruit size may sometimes be indicative of unfavourable characters, e.g. small numbers of bunches or poor fruit set, but it was claimed that larger fruit would have advantages in milling.

All these considerations led to the selection of a definite type of fruit, the *Yangambi-type*, which was large, ovoid, had a large thin-shelled nut placed a little above centre and a wide basal portion of mesocarp (Fig.

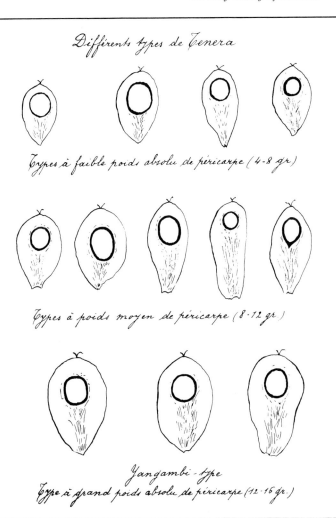

Fig. 5.1 Beirneart's drawings of different types of Congo (Zaire) *tenera* (1933).

5.1). These distinctive characters of the early Yangambi-type *tenera* fruit have made them relatively easy to identify and descendants of the first selections can be found in many parts of the world, e.g. West Africa, Malaysia, Colombia. Very small numbers of this type of fruit existed in the original material. For instance, of 200 palms with 70 to 80 per cent mesocarp the distribution of mesocarp weight per exterior fruit was as follows:

Weight of mesocarp in exterior fruit (g)	4–8	8–12	12–16	16–18
Frequency (%)	36	51	11	2

Table 5.8 *Some data of the best represented* tenera *parents in the* F_1 T x T *generation at Yangambi*

Palm no.	Weight of bunches per annum 1931—3* (kg)	Weight per fruit (ex-terior) (g)	Weight per kernel (g)	Num-ber of flowers per bunch	Meso-carp to fruit (%)	Kernel to fruit (%)	Fruit to bunch (%)	Oil per palm (kg)
5 R	120	10	0.5	—	89	5.5	55	28
7 R	161	18	1.4	700	85	7.5	67	42
16 R	136	10	1.2	1,180	79	11.0	—	33
19 R	136	20	1.4	540	87	7.0	—	34
25 R	136	9	0.6	900	87	6.4	66	31
37 R	117	11	0.7	670	88	7.0	62	27
68 R	152	19	1.1	980	82	6.2	65	39
79 R	122	7	0.9		75	13.7	54	25
85 R	114	12	1.1	1,180	79	8.8	63	29
121 R	137	16	1.6		77	10.2	60	34
130 R	149	16	1.4	1,100	81	11.2	66	34
176 R	?	13	1.1		78	12.0	62	24
229 R	153	16	1.3	560	83	8.8	66	42
255 R	101	11	0.9		88	7.2	56	25
261 R	133	8	0.8		76	11.0	51	25
267 R	117	9	0.7	2,400	73	8.7	57	23
397 R	164	11	1.2		81	9.5	70	45
302/1	158	10	1.1	1,000	80	10.5	62	38
53/3	218	10	1.5	2,300	75	15.0	63	47
244/4	163	10	1.2	900	72	12.7	60	33
171/6	218	15	1.0	1,600	85	6.5	44	37
Mean	145	12.4	1.1		81	9.4	60	33

* Partly estimated from 3 years' data.

The frequency distribution of mesocarp percentage of a group of 167 palms in the *Rive* illustrates the quality standards of *tenera* palms existing at that time and the potentiality of the best of them:

Percentage mesocarp to fruit	65—70	70—75	75—80	80—85	85—90	90—95
Frequency (%)	6	13	29	33	17	2

Table 5.8 gives data of certain of the Yangambi mother palms recorded between 1931 and 1933 and which were stated in 1941 to be among the best represented in the F_1 fields of the I.N.E.A.C. which were planted from 1934 onwards. Of the selected palms in the *Palmeraie de la Rive* (R) over 70 per cent were descended from the Djongo of Eala.

It is of interest to compare these selections of illegitimate material coming from original open-pollinated selections with the mean of the F_0 generation of Deli palms in Table 5.1 and with individual palm data in Table 5.5. All three lots of palms represent a starting point in their respective countries for selection and for crossing and selfing to breed an F_1 generation. The Deli *dura* fruit characters were much more uniform than the Zaire *tenera* characters. Yield, mainly affected by environment, was higher in the Deli as was fruit-to-bunch ratio and weight per bunch, but mesocarp percentage was naturally much lower and little notice was taken of the kernel. These two groups of palms were soon to converge.

West Africa

Very little need be said about prewar selection in West Africa since it was on a small scale and lacked the support of any large organizations. In the *Ivory Coast* a start on similar lines to that of Zaire was made: the progeny of sixty open-pollinated *tenera* palms were planted at La Mé between 1924 and 1930. The grove mother palms had very poor composition, but the best of the progeny were selfed between 1934 and 1938 to give an F_1 generation which proved to be of low production and was not further used, though some of the parents were later to be employed in postwar breeding schemes of the I.R.H.O.[14]

In *Nigeria*, the early work was concentrated on a population of some 800 palms of different forms and types planted at Calabar in 1912 to 1916 with later supplying. *Dura, tenera, virescens* and mantled fruits were used to provide seed and, except for the *tenera* planting, only one parent of each form or type was used.[15] Palms of all forms and types appeared in the progeny and green-fruited *dura* and *tenera* were common. Table 5.9 shows the number of palms recorded and selected and gives data on the yield and quality of the latter.

The much higher bunch yield of the *dura* palms led to the assumption that *dura* would in general yield more highly than *tenera* palms; however, this result appears to have been quite fortuitous and no doubt due to the

Table 5.9 *The Calabar parents. 1922–8 data*

Form and type	Number of palms recorded	Number selected	Bunch weight of selected palms		Fruit to bunch	Meso-carp to fruit	Shell to fruit	Kernel to fruit
			per palm/ annum (kg)	per bunch (kg)	(%)	(%)	(%)	(%)
Dura, nigrescens virescens	376 ⎱382 6 ⎰	9	96	11.8	65	49	32	12
Tenera, nigrescens virescens	36 ⎱43 7 ⎰	10	67	9.2	56	64	18	10
Mantled (*Poissoni*)	24	3	74	5.6	50	–	–	–
Total or Mean	449	22	80	–	–	–	–	–

Note: the fruit analysis relates only to eight *dura* and eight *tenera*.

qualities of the single original grove *dura*. Certainly the original selection from the groves was fortunate in its *dura* and unfortunate in its *tenera*; the latter gave rise to comparatively low-yielding palms with poor fruit-to-bunch ratios and low mesocarp percentages, but large kernels. A series of selfs and crosses of these palms was planted on four stations in Nigeria in 1930 and sterile *pisifera* palms soon appeared in the *tenera* x *tenera* material. The *dura* continued its superiority as a yielder into this F_1 generation and this result biased seed distribution in favour of the *dura* for quite a long period. Although there was later shown to be little justification for this bias, it did prevent a period of large scale issue of *tenera* x *tenera* seed leading to 25 per cent steriles, as had been the case in Zaire.

Another prewar selection of *dura* and *tenera* parents was carried out among palms of grove origin at Aba. Controlled pollination was not fully successful and populations of several thousand largely out-pollinated palms were established at the Oil Palm Research Station near Benin in 1939–41. This formed a useful source of further breeding material and, with the large numbers of *tenera* produced, gave no support to the theory that the latter were inherently lower yielding than *dura* palms. The *tenera* progeny were derived from four Aba *tenera* parents. The *dura* material proved to be of little interest.

Large quantities of seed issued within West Africa owe their origin to these two programmes and to later generations from them. Mention should be made of the outstanding Calabar *dura*, CA256 (or 551.256), which produced good selfed progeny and entered into many crosses. The palm is over 50 years old and still in existence; its bunch analysis, exceptional for African *dura*, was as follows:

Fruit to bunch	Mesocarp to fruit	Shell to fruit	Kernel to fruit	'Loss'
63.5%	52.2%	31.2%	10.3%	1.9%

Early analysis in Nigeria was done by a boiling and pounding method which led to a 'loss' figure of 2 to 4 per cent appearing in the analysis; mesocarp percentages were thus lower by a few per cent than figures obtained by present methods which are the same as those used in other territories. The frequency of green-fruited specimens in Nigerian seed issues is largely due to the inheritance of this character from two Calabar *dura*, CA(551).341 and 375, of good composition, and from some green *tenera* crosses and selfs.

The emergence of the *dura* x *pisifera* cross

The hybrid nature of the *tenera* was discovered by Beirnaert when he examined the fruit form of the progeny of his selfed and crossed *Arbremères*. The latter had been selfed, and also large numbers of complementary crosses, as already explained, had been made. The majority of the parents used are shown by their numbers in Table 5.8. For several years it had been apparent that a substantial proportion of the progeny of seed issued to estates was sterile. A count of 29,454 palms in fifteen blocks at Yangambi showed that 24.34 per cent were *pisifera* and this, overall, did not differ significantly from the 'expected' 25 per cent. In the examination of the descendants of twenty-five mother palms only seven had percentages of *pisifera* differing significantly from 25 per cent. The range was from 19.74 to 33.20 per cent. Examination of the productive palms showed that the *tenera* always constituted 50 per cent of all the palms, while the percentage of *dura*, though usually 25 per cent, varied in the same manner as the *pisifera*. This variation is discussed on p. 248.

Examination of the progeny of *dura* x *tenera* crosses showed that there were no *pisifera* and for the majority of these crosses the segregation was *dura* 50 per cent, *tenera* 50 per cent. It was also noted that *tenera* seed taken from grove palms surrounded, as to about 90 per cent, by *dura* palms gave progeny in the proportion:

357 *tenera* (50.50%): 336 *dura* (47.52%): 14 *pisifera* (1.98%),

the latter being accounted for by assuming that a small quantity of *tenera* pollen might naturally fertilize female *tenera* inflorescences.

Dura x *pisifera* adult fields did not exist at that time, but the hybrid nature of the *tenera* became clear from the above results. By 1938 it was known that as much as 25 per cent steriles could be expected in *tenera*-derived material and it was becoming plain, after the discarding of some far-fetched theories, that there was a 'correlation between the shell thickness of the parent and the percentage of *pisifera* palms in the progeny'.[16] The true position was realised in 1939, but a full explanation of the presence of steriles was not published until the issue of M. Beirnaert's 'La problème de la stérilité chez le palmier à huile' in the first wartime Leopoldville edition (1940) of the *Bulletin Agricole du Congo belge*.[17] In this little-read paper M. Beirnaert not only showed clearly the inheritance of the shell-thickness character, but he also brought forward evidence against the theory, current in French West Africa at that time, that the *tenera* was a degenerating form of the oil palm. At the same time he stated without hesitation the steps to be taken to prevent further sterility, namely that *dura* x *tenera* and *tenera* x *dura* seed should replace *tenera* x *tenera* seed for a short period, after which full *tenera* production should be assured by the issue of *dura* x *pisifera* seed. The full information obtained from the Yangambi F_1 plantings was published in 1941, after M. Beirnaert's death.[18]

Confirmation of the 'Congo theory', as it was at first called, was not long in coming from other territories, though publication was tardy. The Calabar *tenera* selfs and crosses in Nigeria in due course showed their general agreement,[19] and *dura* x *pisifera* crosses including Delis as parents were seen to provide *tenera* progeny.[20, 21] In Sumatra many *dura* x *tenera* and several *dura* x *pisifera* crosses had been carried out before the war and instances of 50 per cent *dura* and 50 per cent *tenera* in the progeny of the former were already known. Confirmation of *tenera* production from *dura* x *pisifera* crosses had, however, to await the postwar period. Errors of pollination technique were not uncommon in West Africa, and were occasional in Sumatra. The success of the huge programme in Zaire was undoubtedly due to the very high standard of skill and discipline of the special workers, recruited from outside Zaire, who undertook the pollinations, germination and planting.

The techniques of selection and breeding

To select individual palms for breeding it is necessary to measure their bunch yield and to analyse the bunches for their oil and kernel content. To breed from selected palms controlled pollination must be undertaken.

The recording of bunch yield presents no real problem provided supervision is sufficient to ensure the observance of a few simple rules. Harvesting must be regular, a fixed standard of ripeness must be employed (see Chapter 10), and bunches must be weighed at the foot of the palm and marked or labelled there with the palm number before they are removed for analysis and processing. Their loose fruit must be carried with them. The harvester records the date, the palm number, the number of bunches (occasionally more than one bunch is ripe on a single palm on the same day), and the weight of the bunch or bunches. Field books with duplicate sheets are suitable (Fig. 5.2). The information is then transferred to individual palm record books or cards.

FIELD 551			
Date	Palm No.	No.	Wt.
30.10.64	719	1	16.3 kg
	725	2	33.6
	731	1	9.1
	734	1	10.7
	737	2	25.9
	750	2	21.3
	752	1	11.8
	773	1	9.1
	783	1	13.6
	791	1	9.1
	795	1	17.1

Fig. 5.2 An example of a simple harvesting record sheet for individual palm recording.

Bunch analysis

A special building is required as an analysis or field laboratory where bunch composition is determined. Many old buildings have been adapted for the purpose by organizations undertaking selection and breeding, but the one illustrated in Fig. 5.3 was specially designed for the purpose. The procedures used have not differed widely one from another; the ones described here[22] are an adaptation of those used at Yangambi in Zaire[23] and have the advantage of being suited to the handling of up to 3,000 bunches per month at a labour expenditure of less than 0.2 man-days per bunch without mesocarp oil analysis, the procedures for which will be mentioned later. In the past much bunch analysis was carried out to the mesocarp stage only, the assumption being that oil constitutes about half the mesocarp. Considerable variation does exist, however, and oil analysis is now considered essential. Modern methods have made it practicable to carry out oil analysis on all bunches.

Analysis must be carried out so that the minimum moisture loss occurs during the process; this is particularly important where oil determinations are to be made. Bunches must be form-determined (*dura, tenera* or *pisifera*) on their arrival with their loose fruit at the laboratory, and they must be dealt with on the same day (if for oil analysis) or the following morning. After weighing on a Berkel scale the spikelets are removed and if their total weight is over 6 kg their weight is reduced by a random sampling method to around 5 kg (A). The weighed stalk and remainder of the spikelets (B) are discarded; these weights are all recorded on a card which then accompanies the 5 kg sample (A) (see Fig. 5.4).

If mesocarp oil is to be determined the retained spikelet sample A is immediately processed, but if no oil determination is required spikelets may be stored for 3 days to loosen the fruit; mesocarp content is found to be reduced by this delay by only 0.5 to 1.0 per cent.

On the spikelets there are to be found (i) fully fertile fruit, (ii) parthenocarpic fruit in which an oil-bearing mesocarp has developed although there has been no fertilization and hence no embryo, kernel or shell development, and (iii) infertile fruit which are small, colourless and non-oil bearing. The latter, which form a very small proportion of the total weight, are included in the weight of the empty spikelets. (i) and (ii) are sometimes estimated separately, and for exact estimations of mesocarp, mesocarp oil and kernels to bunch, separate estimations are necessary.

From sample A the *fruit-to-bunch ratio* can be calculated, i.e.

$$\text{Fruit-to-bunch ratio} = \frac{(\text{Normal} + \text{parthenocarpic fruit}) \times 100}{\text{Total bunch weight}}$$

For fruit analysis it has been shown that a 500 g sample is sufficient for bunches weighing less than 50 lb and even for bunches above that weight this size of sample has proved satisfactory. Normal fruits are taken by a sampling procedure and put on a scale until a 500 g sample has been

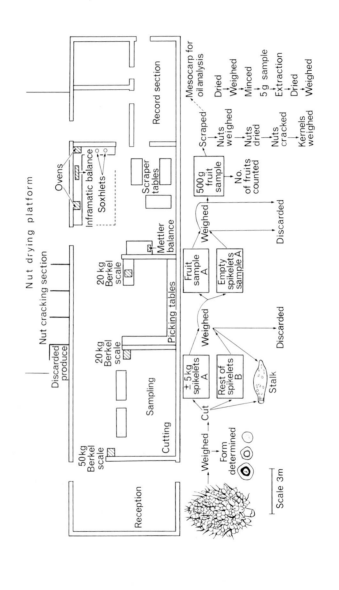

Fig. 5.3   A plan of a field laboratory for fruit and bunch analysis, with diagrammatic representations of the processes.

Bunch Analysis Card				
Palm No:	A 374	Fruit form: D T P		
Bunch Wt.	14.3 kg	Date harvested 21.3.65		

Bunch Wt. A	5.1 ''	Fruit Sample Wt.	500 g
'' '' B	5.3 ''	No. of fruits	42
Wt. Stalk	3.9 ''	Wt. per fruit	11.9 g
	14.3		

		Nut weight		248 g
Fruit Wt. A	4.5 ''		g	%
Spikelet Wt.	0.6 ''	Mesocarp	252	50.4
	5.1	Shell	185	37.0
% F/B	64.3	Kernels	63	12.6
			500	100.0

Oil Analysis: Palm No. A 374   Harvest 21.3.65
Drying: Wt. of container & wet mesocarp   g
  ''   ''   ''   & dry   ''   g

                          Wt. dry mesocarp   g

SAMPLES: A   B   *Mean*
Wt. of samples
Balance reading
% Oil/dry mesocarp
% Oil/wet mesocarp (Sox.)
% Oil/wet mesocarp (indirect)

Fig. 5.4   A bunch analysis card including oil analysis.

obtained. The number of fruits is recorded and so the *weight per fruit* can be estimated.

Mesocarp is cut and scraped from the fruit by a sharp knife. Removal of mesocarp can also be done by boiling and pounding, but this method has been found to give low mesocarp figures as all the fibres cannot easily be removed. If oil analysis is to be undertaken it is necessary for the fruits to be sampled and weighed soon after the bunch has arrived. A delay after weighing the 500 g sample is not important however. After the mesocarp has been removed the nuts are weighed following 1 day of surface drying, and the *mesocarp percentage to fruit* is then calculated by difference. Early weighing of fruit and of nuts is important for obtaining an accurate mesocarp measurement on which to base subsequent oil to mesocarp calculations.

The nuts are subsequently air dried for 3 or 4 days to facilitate cracking. After cracking, the kernels are weighed and the shell weight determined by difference. *Kernel to fruit* and *shell to fruit* percentages may thus be obtained. This is not an entirely accurate method as it has been shown that the kernels also lose weight during the drying process, and that the percentage of kernels to fruit may, therefore, be underestimated. In a study in Malaysia[24] the following initial distributions of moisture and dry matter in fresh nuts were found:

	Per cent of fresh nuts				Moisture	
	Moisture in kernels	Dry kernels	Moisture in shell	Dry shell	per cent of fresh kernels	per cent of fresh shell
Deli *dura*	6.7	19.0	11.8	62.5	26.1	15.9
Tenera	14.8	37.1	9.1	39.0	28.5	18.9

Oven drying at 80°C caused a more rapid moisture loss from shell than from kernels in both fruit forms, but particularly in the case of *dura*. After 4 hours drying, moisture in the kernels was reduced to 5 per cent with *dura* and 9 per cent with *tenera* fruit, and it was calculated that if the kernel weights after this amount of drying were to be accepted in the fruit analysis then the kernel percentage to fruit would be underestimated by about 1 in the case of *dura* and 1.5 in the case of *tenera*. This work suggests that with the usual 3 to 4 days air drying underestimation of kernel percentages (and overestimation of shell) will not be so large; but it should nevertheless be recognized that there will be some underestimation of kernel content in all methods relying on drying before cracking and on calculating shell content by difference.

Oil analysis of the mesocarp may be undertaken by:

1. Indirect estimation through a knowledge of the water content
2. Direct extraction
3. A specific gravity determination

Indirect estimation. Indirect methods of estimating mesocarp oil content have been based on two assumptions: 1. The existence of a relationship between dry non-oily matter in the mesocarp and the mesocarp percentage of the fruit, a knowledge of the latter enabling the dry matter to be estimated.[25] Simple determination of the moisture content then, it was claimed, enables the oil content to be calculated by difference. This method would be expected to show a distinct difference between fibre to mesocarp percentages for *dura* and *tenera* material. This is not general experience and the method has therefore not been widely used. 2. A linear relationship between moisture content and oil content with a fixed dry matter percentage. Various formulae have been put forward, the first, Y (oil/fresh mesocarp) $= 87.38 - 1.08\ X$ (water/fresh mesocarp, being that of Vanderweyen *et al.* in Zaire.[26] Later investigators in West Africa considered the method to be admittedly rough and obtained equations so near to

$$Y = 84 - X$$

that this approximation, which assumes a constant fibre percentage of 16, was considered satisfactory.[27, 28] No significant difference was found between the mean fibre content of *dura* and *tenera* pulp.

For this method the cut mesocarp obtained by the previous sampling methods may be reduced by quartering to duplicate samples of 100 to 150 g. These are accurately weighed and dried in an oven at 105°C to constant weight. This takes 18 to 24 hours. Duplicates should not differ by more than 2 per cent in their moisture content. Recent research (see p. 257) suggests that dry-fibre determinations may be of particular importance in selection for oil to mesocarp in the *tenera*.

Direct extraction.   Direct extraction by Soxhlet extractors has been widely practised both for bunch analysis and mill control. The objections to the method have, in the past, been that the whole process is slow, the size of the sample small and an adequate sampling system difficult to determine. For these reasons oil analysis was often neglected and selection sometimes based on mesocarp plus kernel content alone. Much larger Soxhlets are now used however, and adequate techniques and sampling procedures have been worked out. This, together with the discovery of wider variations in mesocarp fibre percentage in selected progenies, caused all workers to abandon the indirect method and revert to oil extraction. It was shown that there was a variation of as much as two to three in the mean percentage fibre to mesocarp from progeny to progeny, and that seasonal within-progeny differences may be much higher.[22]

In using large Soxhlet extractors (600 ml capacity), the mesocarp content of the 500 g fruit sample is calculated by subtracting the nut weight from the fruit weight. This avoids error due to moisture loss during depulping. Oven-drying of the mesocarp is then undertaken at 105°–110°C for 24 hours and duplicate 5 g samples are prepared on an inframatic moisture balance in filter paper containers plugged with cotton-wool. The infra-red lamp fitted above the balance, which is designed to measure percentage loss, prevents moisture uptake during weighing. The duplicates may then be put with other samples into separate Soxhlets for extraction. Twenty-four samples can be housed in each Soxhlet and extraction with isohexane proceeds for 18 hours. After air drying until all isohexane has evaporated the residual fibre is oven-dried for 24 hours, weighed and the *mesocarp oil content* is calculated from the loss in weight. Duplicates differing by more than 3.1 per cent are discarded.

Using two ovens, one Paladin mincer, one inframatic balance, two 600 ml Soxhlets with electric heating mantles, forty-eight 1-litre aluminium containers with tight-fitting lids for dry samples and one distillation apparatus for isohexane recovery, over 6,000 duplicate samples, representing bunches, can be oil-determined per annum.[22]

A cold solvent extraction method has recently been developed and it is claimed that where large numbers of analyses are required this method is much less expensive.[29] Large quantities of petroleum ether 60°–80° are used and the solvent is recovered by distillation. The method relies on eight changes of solvent over a period of 4 days, but to reduce distillation the solvent is transferred from one partially extracted lot of samples to another on a shelf system.

Oil determination by measurement of specific gravity.   A method of measuring mesocarp oil content by specific gravity determinations is in use in some laboratories. The method is not claimed to be as accurate as Soxhlet extraction but it has the advantage that each of a large number of determinations can be completed in a relatively short time. If a given weight of solvent of known specific gravity is added to a given weight of dried mesocarp and the mixture is ground in a high-speed grinder, the oil and solvent may be filtered out. The specific gravity of the filtered mixture will be found to depend on the amount of oil in the solvent and hence in the original sample. The solvent used must have high extraction properties, a high evaporation point and a specific gravity as different from that of the oil as possible.

This 'oléomètre' method is carried out as follows:

A sample of fruit, say 500 g, is taken from a bunch in the manner already described and after removal of the pulp the nuts are weighed and the pulp weight obtained by difference. A sample of 40 g mesocarp is partially dried in an oven at 105°C for 3 hours and allowed to remain in the oven overnight. The mesocarp, with 75 cc orthodichlorobenzene, is then placed in the cup of a high-speed grinder. Grinding is continued for 3 minutes and the mixture removed after a further minute. The oil plus the orthodichlorobenzene is filtered through a hand-pressure screw filter into a tall measuring tube and then allowed to cool to 25°–27°C. The specific gravity of the mixture is read off on a density metre and the oil percentage to mesocarp is obtained by reference to tables, corrections for temperature being made as necessary. Tables can be easily constructed for each lot of solvent; mixtures of 75 cc solvent and a number of different quantities of oil, say 14 g, 16 g and 18 g, are made up and their densities at a given temperature read off.

Results from the oléomètre tend to be slightly lower than from Soxhlet determinations and, by occasional Soxhlet checks, correction can be made. For comparisons, however, the oléomètre method seems quite satisfactory.[30]

Records.   As previously mentioned, all weighings should be recorded on a card which moves with the bunch and samples, and on which all the weighings are first recorded and percentage determinations subsequently made. After fruit scraping, the mesocarp and nut samples go different ways; a separate card must therefore be used for oil analysis, or a perforated card can be employed, the bottom portion being torn off to go forward with the wet pulp. This method may assist in reducing errors of identification.

The information from the bunch analysis card or cards is later transferred to the individual palm record card an example of which is shown in Fig. 5.5.

PALM NO. 364		PARENTAGE	Calabar tenera, virescens
EXPT. 32-2		TYPE & FORM	Tenera
PLANTED 1941		POSITION	

FRUIT AND BUNCH ANALYSIS

Date Harv.	No. of Anal	Total Bunch Wt. Kg	Wt. of Fruit Kg	% F/B	500g Sample No. of Fruit	% Mes	% Shell	% Kern	% Oil/m indirect	% Oil/m soxhlet
1955–56*	13	291.72	175.96	60.3		82.8	9.0	5.1		
22.5.59		45.000	27.9000	62.0	67	90.2	6.4	3.4		51.0
8.9.61		13.330	7.603	57.0	79	91.6	5.6	2.8	53.4	50.1
18.9.61		17.150	10.588	61.7	59	89.0	6.8	4.2	51.2	52.7
29.8.62		21.600	15.298	70.8	66	89.8	6.2	4.0		
5.10.62		16.625	9.965	59.9	81	87.4	7.4	5.2	50.1	52.7
23.4.63		9.800	4.786	48.8	63	89.4	6.6	4.0	51.6	54.4
Mean				60.1	69	89.6	6.5	3.9	51.6	52.2

*Old method of analysis (boiling and pounding)

BUNCH PRODUCTION

Year	No. Bunches	Wt. Bunches kg	Year	No. Bunches	Wt. Bunches kg	Year	No. Bunches	Wt. Bunches
1946	5	15.9	1953	9	172.3			
1947	15	79.8	1954	5	84.4			
1948	7	58.0	1955	10	218.1			
1949	10	122.0	1956	5	106.6			
1950	10	132.0	1957	5	96.1			
1951	18	59.0	1958	5	96.1			
1952	7	59.9						

REMARKS: Selected for the Main Breeding Programme in 1958

Fig. 5.5   Individual palm record card.

Controlled pollination

Controlled pollination for breeding must be distinguished from assisted pollination in which pollen is placed on unprotected bunches to ensure or enhance fruit set for production. The effects of the latter practice have been discussed in Chapter 4.

The production of plants of known parentage is not a simple procedure in the oil palm. A very large female inflorescence has to be isolated, pollen has to be collected from a large male inflorescence without contamination, and the number of stages of the work from isolation and collection to the final transference of the plant from the nursery to the field is so great that very special measures have to be taken to prevent errors both of technique and identification.

When crossing *dura* with *tenera* or *pisifera*, or crossing *tenera* with *tenera* or *pisifera*, the proportions in which the fruit forms segregate in the progeny are known and therefore provide some check on the work done. It is not, however, possible to detect errors of pollination in *dura* x *dura* crosses, particularly in the Deli palm (except when other forms appear) unless some other distinctive heritable factor, e.g. dumpiness, is involved. In the early days of oil palm breeding many bunches said to be control-pollinated were in fact out-pollinated due to the crude methods used or lack of constant supervision. The discovery of the facts of fruit form inheritance provided a further stimulus for the devising of sound methods of controlled pollination and, although the essential requirements have remained unaltered, many refinements of technique have been introduced. It is not proposed here to describe in detail all the methods used, but only to state the essential requirements and to describe up-to-date methods.

The collection and storage of pollen

Male inflorescences must be bagged 7 days before the flowers open. Pollen-providing palms must therefore be visited frequently and the progress of male inflorescence production noted. At anthesis the inflores-cence will be in the axil of the seventeenth to twentieth leaf from the central spear. The spathes of the inflorescence are cut away and the spike-lets are thoroughly sprayed with a 40 per cent formaldehyde solution diluted by reducing to 1 part in 10 parts of water. This solution kills any foreign pollen or insects adhering to the spikelets. The inside of the bag to be used in similarly treated, the mouth being held downwards during the operation. Before carrying out any work connected with pollen the hands should also be sterilized with methylated spirits.

Bags have been made of many kinds of material, strong brown paper or heavy canvas being common. The former is subject to damage in the field while the latter requires to be impregnated with some substance to ensure that pollen does not pass through the cloth. In Zaire, where results showed that the standard of isolation was very high, khaki drill was dipped in rubber latex and vulcanized by dusting with sulphur. One corner was open along the seams but was tied up tightly with cord when not opened for observation or pollination. In Nigeria canvas bags 24 inches (61 cm) long by 18 inches (46 cm) wide and soaked in old engine oil to make them impermeable were used, but they were superseded by Terylene bags. Both types of bag have celluloid windows for observation, one measuring 4 inches (10 cm) square and set in the middle of the front and one 2 inches by 1 inch (5 x 2.5 cm) at the top of the rear of the bag as placed on the bunch; but bags used for male inflorescences may measure 24 inches by 16 inches (61 x 41 cm) and be provided with a window in the front only.

In the bagging operation a collar of cotton lint, well sprayed with formalin and dusted with D.D.T. and pyrethrum powders, is placed around the peduncle, and the mouth of the bag is tied over this collar with strong twine. The bag may need supporting by tying its upper corners to nearby leaf petioles.

The pollen has to be removed under sterile conditions and dried before storing. By using a bag of Bondina Terylene material the male inflorescence may be dried and shaken out while still in the bag thus obviating the need for removing it from the bag and shaking out the pollen in boxes or chambers where, in spite of formalin spraying, contamination may take place. After withdrawing the empty inflorescence, the pollen is dried at $35°-40°C$ for 24 hours in the bag. The pollen may then be sieved through a special sieving chamber and passed straight into a storage tube without any exposure to the outside atmosphere.

In a breeding programme, pollen may need to be stored for several months. Testing pollen for viability is usually done by germination of the grains on a maltose agar medium, but it has been shown that actual germination on stigmas is higher than on the artificial medium, and that good fruit set may be obtained when viability, as shown by laboratory tests, is as low as 10 per cent.[31] Pollen may be stored satisfactorily to this level of viability over calcium chloride at tropical room temperatures for 6 to 8 weeks. Prolongation of this period may be obtained by refrigeration to $±5°C$ or by storage in a partial vacuum.[32, 33] More recently it was noted that certain bunches giving low germination contained seed with atrophied embryos, and a relationship was found between the number of the latter and the length of time pollen is stored. This relationship was particularly clear with pollen stored for 1, 2, 3 and 4 months by ordinary methods. Where a partial vacuum was used, the effect of time of storage on germination was less pronounced, and when a complete vacuum was used the deleterious effect of storage was eliminated. The cause of this effect is unknown, but there seems no doubt that the use of differently stored pollen, or even of genetically different pollen similarly stored, may have differential effects not only on germination but also on early seedling growth. Such effects have been particularly noted with seed produced by fertilizations with stored pollen sent from one part of the world to another.

Standard methods and apparatus have now been adopted at seed-producing centres for vacuum drying pollen after initial drying over calcium chloride or in an incubator.[34, 35] The pollen is sealed in ampoules and stored in deep-freezers at $-18°$ to $-20°C$. Rehydration for use takes about 24 hours; a sharp fall in viability is usual after 10 days, as with fresh pollen.[34]

Preparation of the female inflorescence

A female inflorescence should be bagged at least 1 week before the first flowers are expected to open. Gauging this time is a matter of experience but entails regular inspections of each bunch. If there is any premature opening of the flowers, the bunch must be discarded. A flower is receptive when the lobes of the stigma are well separated but are still pink. On changing to red the stigma is believed to be no longer receptive, but to be safe the bag should be left on the bunch for 3 to 4 weeks after pollination; the stigma will then be black. Observations have shown that the receptive period for the flowers on an inflorescence normally lasts for 36 to 48 hours. Flowering starts at the base of the inflorescence and continues to the top. Occasionally receptive flowers may be found on an inflorescence for a week. Usually, therefore, if pollination is carried out on 3 successive days, maximum fertilization will result.

Spraying the inflorescence and bagging is carried out in the same manner as for the male inflorescence, i.e. a cottonwool collar well sprayed with formalin and dusted with D.D.T. is placed round the stalk and the bag tied over it with strong twine and supported above by tying to adjoining petioles. Labelling is done at bagging, and the pollination particulars should be added to the label at the time of pollination. Canvas bags soaked in engine oil and fitted with celluloid windows as already described are suitable, but the lighter Bondina Terylene bags have also been used in West Africa. For female inflorescences they measure $48-51$ cm by $73-76$ cm and are provided with windows at the front and at the top of the rear side of the bag. Tests have shown that isolation with these bags is complete even though they are air porous. These tests take the form either of isolation without pollination or isolation followed by the inducing of parthenocarpic development of fruit by application of α-2,4,5-T at 100 ppm. The latter method is to be recommended as the inducing of parthenocarpy ensures the development of any flower which may have been fertilized but which might fail to develop on a mainly non-developing bunch. Bunches treated with 2,4,5-T have to be examined fruit by fruit to detect any developing kernel and embryo.

Some doubt has been thrown on the efficacy of Bondina Terylene under Malaysian conditions. It has been possible to blow pollen through the bags under pressure and small quantities of pollen have been detected inside the bags after exposure to heavy rain. Although double bagging may be resorted to, it is now more usual to treat the bags with a dilute (1:20) spray of acid-stable emulsified wax.[36]

Pollination

In bags with celluloid windows there are small round holes covered over with adhesive tape. When pollination is to be carried out the pollen may be placed in a test-tube, carrying a cork fitted with two L-shaped pieces of glass tubing stopped with cottonwool; or a bulb puffer with a projecting nozzle may be used. The puffer should be of single-spout type which does not inhale plantation air through the bulb and is only allowed to inhale through the spout when in the bag. Before pollination the pollinator sprays the bag with formalin and sterilizes his hands with methylated spirits. Each hole is then opened in turn and the pollen blown (in the case of the test-tube by blowing into one L-shaped piece while inserting the other in the hole) into the bag while the nozzle or glass tube is so manipulated as to ensure that pollen reaches every part of the inflorescence. The adhesive tape is replaced as the nozzle or tube is withdrawn. The bag is then shaken to help the pollen settle on all the flowers. Pollination is normally carried out in the morning (Plate 19).

Palms being used as seed-producers need a good deal of attention in order to facilitate pollination. Old leaf bases and debris have to be removed and the fibres and spines on the leaves subtending or near the bunch need to be cut away. Spines should be removed up to the lowest leaflet only and as little damage as possible must be done to the leaves since any pruning would be likely to reduce further female inflorescence production. However, light cuts may be made on each side and on the under surface of the leaf petiole of the subtending leaf so that it may be gently forced downwards to make access to the female inflorescence easier. If carefully done this does not seem to harm the leaf as a photosynthetic organ and premature withering will not take place.

Variation in the oil palm

In the 1920s it was considered that, in selection in the oil palm, it was necessary to differentiate between the Deli palm of the Far East and the palms of Africa on account of their difference of variation. Not only was the fruit form of the Deli constant, but variation of shell percentage did not seem to be very great. As late as 1930 it was stated that it was not known whether such variations as existed within the Deli were the result of genetic differences or of variations in environmental conditions.[1] Nowadays the term Deli must be taken simply to indicate that the material concerned emanates from the *dura* populations of the early plantations of Sumatra and Malaysia.

As to African material, both in Africa and within the small quantities of material exported from Africa to other countries, it was obvious that very great variation existed, particularly in fruit and bunch characters, and Belgian and other workers set about the examination of these variations with great thoroughness.[12, 37, 38]

Pl. 19   Controlled pollination: **A.** Pollinator with bag for isolating the inflorescence.
B. Spraying formaldehyde to kill extraneous pollen before pollination takes place.

This section describes variation as it exists within fields and progenies. The genetic element in this variation and its employment in selection and breeding is discussed later.

Bunch production

In Sumatra it was found that in reasonably uniform plantations the bunch yield of the highest yielding palms was twice or three times the average yield of all the palms. Very similar variation was found in Zaire where in a field of 2,000 illegitimate palms 0.8 per cent of these gave a yield a little less than three times the average of the field. The most outstanding palm in a field may sometimes have four to five times the average per palm yield of the field. On one of the earliest planted estates in Malaysia the coefficient of variation of individual palm yields for a 3-year period was found to be 41 per cent.[39]

Clearly it must be the plant breeder's aim to reduce this palm-to-palm variation in production and to eliminate those factors causing low yield. The presence of large numbers of low-yielding palms besides reducing average production per palm leads to higher expenditure per ton in maintenance and harvesting. It is important, therefore, to have some knowledge of yield distribution within fields of palms. The variation is not, of course, entirely due to genetic factors. Most fields contain several non-producers; these are often palms which have been damaged or supplied in the early years and the shade thrown by surrounding palms and competition for nutrients has retarded their growth and inhibited flowering. Even when these and other environmental factors which lessen uniformity are taken into account however, there is still a very large variation in palm yields.

A few examples will now be given of the degree of variability to be found in fields of palms in Africa and Malaysia which may be regarded as 'foundation stock' or F_1 from illegitimate material. These examples and the differences to be seen between them are purely illustrative; they suggest trends rather than establish firm disparities.

In Africa, coefficients of variation per palm for annual bunch yields have been computed for several fields. In Nigeria these coefficients varied from 40 to 70 per cent whether the material was illegitimate and unselected or from selfings of individual· *dura*.[40, 41] Coefficients obtained from three-palm plots in the Ivory Coast suggested similar per palm variations in that country.[42] There was some early evidence that yield variability might be reduced by selfing, and much selfed *dura* material was used for early agronomic experiments in West Africa, but later it was seen that this was not a general feature of selfing. With progeny of this type it was shown that the coefficients of variation tended to decrease with age and with number of years' records. Single-year yields were positively skew, this being most marked in the early years. When totalled over 4 years the yield distribution became normal and the coefficient of variation dropped to 28 to 36 per cent.

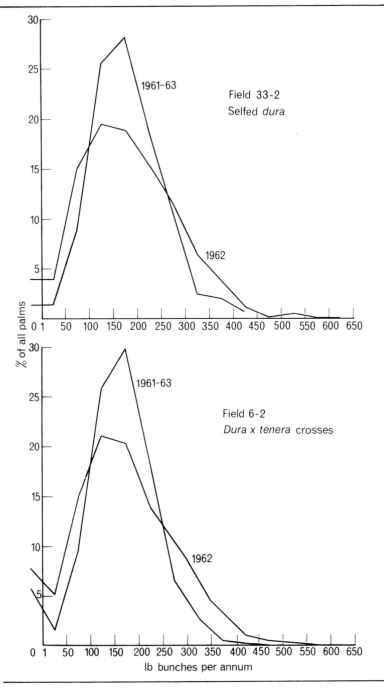

Fig. 5.6 Bunch yield frequency curves of two fields in Nigeria; single year (1962), and 3-year mean (1961–3).

Figure 5.6 shows the frequency distributions of mean annual yields in the 3-year period 1961–3 of two fields planted in 1941, one (33-2) with selfed *dura* and the other (6-2) with plots of eighteen *dura* x *tenera* progenies. The skewness of the single-year yields in 1962 can still be detected. It is clear that these two kinds of material do not differ appreciably in their variation.

Figure 5.7 shows the frequency distribution of mean annual yields per palm in 1961–3 of a field planted in 1952 with a large number of *dura* x *pisifera* crosses, together with the frequency distribution of two separate crosses within the field. The coefficients of variation of the latter are 31.5 per cent (Progeny 24) and 46.5 per cent (Progeny 40). In this case the per palm yield variation in Progeny 24 is clearly less than for the whole field while the variation in Progeny 40, though higher yielding than Progeny 24, appears to be considerably wider.

Data on the variability of number of bunches per palm is meagre; from that presented by Chapas[41] for two-palm plots, however, it seemed probable that the coefficients of variation would be somewhat lower than those of bunch yield but that there would be less tendency for a decrease in variability with age. The former deduction is not borne out by the two

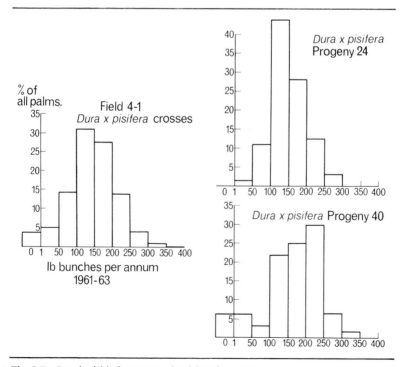

Fig. 5.7 Bunch yield: frequency of weight of bunches per palm per annum in a field in Nigeria and of two progenies within it.

progenies 24 and 40 previously referred to; the coefficients of variation for bunch number for the 3-year period are 36.0 and 48.1 per cent respectively, i.e. 4.5 and 1.6 per cent higher than for bunch yield by weight.

In examining variability data of weight per bunch the relationship of bunch number to bunch weight must be borne in mind. Total bunch yield is the product of number of bunches and weight per bunch, the former depending on leaf production, sex ratio and abortion rates. Bunch number and weight per bunch are negatively correlated; when a palm sets a small number of bunches these tend to develop to large size. Estimates of mean bunch weight per palm are thus not based on the same number of observations. But bunch weight does nevertheless seem to be strongly affected by heredity. In the case of the two Nigerian progenies already referred to the coefficients of variation per palm in weight per bunch over the period 1961—3 were 28.7 and 26.9 per cent against a coefficient of 42.6 per cent for a similar-sized random sample from the field as a whole. The histograms in Fig. 5.8 show the altering variation in 1962 and 1963; it will be seen that the variation is less in both years for these progenies than for the field as a whole. Thus in the case of these progenies the comparatively low variation in weight per bunch seems to be counteracting the high variation in bunch number to give the variations in bunch yield per palm seen in Fig. 5.7.

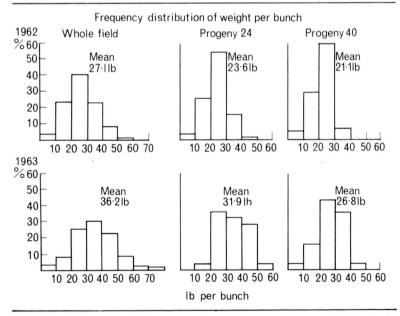

Fig. 5.8   Frequency distribution of weight per bunch of palms in a field in Nigeria and of two progenies in the field.

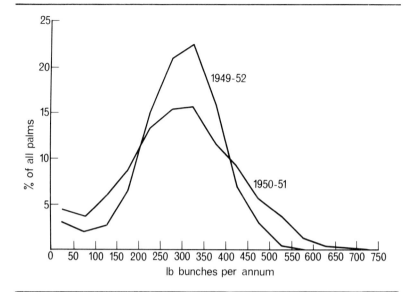

Fig. 5.9 Bunch yield frequency curves of a field of Deli progenies in Malaysia: single year (1950—1), and 3-year mean (1949—52).

Similar data are presented for a field of largely F_1 Deli progenies on an inland estate in Malaysia. The field contained thirty-six progenies derived from thirty-two parents; many progenies thus had at least one parent in common with other progenies. Moreover thirteen parents were themselves derived from ten Serdang avenue palms. Variation in the individual palm yields for single years was not as noticeably skew as in West Africa, and the variation in mean annual yields per palm over a 3-year period was rather less than in any of the West African fields cited.

Figure 5.9 shows the bunch yield frequency curve for 3 years and for a single year; the lower variability in the 3-year means is apparent. Bunch number variability is similar.

Variability of bunch yield within the Deli progenies was lower than in the Nigerian *dura* x *pisifera* progenies cited, even when there had been inbreeding. In one F_1 progeny (Fig. 5.10) the coefficient of variation for 3 years' per-palm yields was only 21.5 per cent while in an F_2 progeny derived by selfing Serdang palm S5 and crossing sibs in the F_1 generation the coefficient of variation was still only 26.0 per cent. The coefficient of variation in a 5 per cent sample of the field as a whole was 30.2 per cent. Though considerable yield variability still exists in this Deli material there is certainly a suggestion of less variability than in African material; but further studies are obviously required.

Variability in weight per bunch also appears to be less in the Deli palm in Malaysia than in African palms in West Africa. The two progenies cited

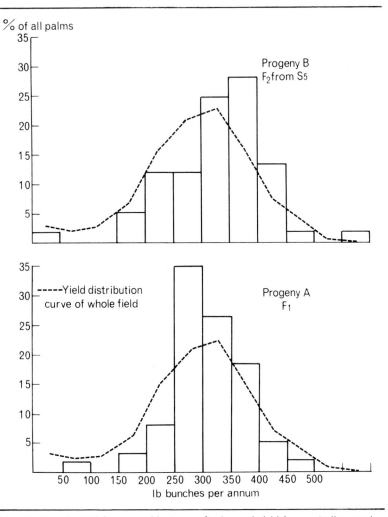

Fig. 5.10 Bunch yield frequency histograms for 3 years' yield for two Deli progenies in Malaysia. An F_2 progeny bred from Serdang S5, and an F_1 progeny from illegitimate parents.

above had mean bunch weights of 16.9 kg (37.3 lb) and 17.2 kg (37.9 lb) respectively over a 3-year period and gave coefficients of variation of 23.0 and 24.3 per cent. The field as a whole showed a sample coefficient of variation of 26.3 per cent (Fig. 5.11).

Coefficients of variation of bunch yield, bunch number and weight per bunch of Deli and African lines in the Ivory Coast were found to be no lower than those of their crosses (see p. 279). For 4 years' yields, coefficients of variation lay between 21 and 26 per cent for groups of more than

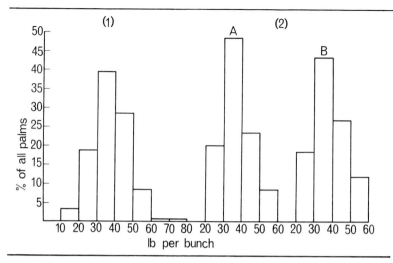

Fig. 5.11 Frequency distribution of weight per bunch: 1. 7 per cent sample of palms in a field of Deli progenies in Malaysia; 2. two progenies — A and B — in the field. Three-year means per palm.

200 palms of different origins and their crosses. Coefficients of variation for number of bunches and for weight per bunch averaged 32 and 29 per cent respectively.[43]

Figure 5.12 shows the contrast in bunch production per palm in the non-seasonal climate of Malaysia and in the seasonal climate of West Africa.

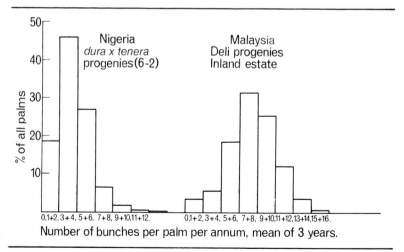

Fig. 5.12 Frequency distributions of number of bunches per palm, Nigeria and Malaysia.

Variation in production with age and fruit form

Bunch number decreases rapidly as the palm ages, both on account of a lower leaf production and a reduction of the sex ratio, while weight per bunch increases even more rapidly until the product of the two factors levels out at 'maturity'. This stage was reached in the earlier fields of palms when they had been bearing between 6 and 8 years, but with better planting methods and more precocious planting material a mature total bunch yield may now be reached much earlier as may be seen from the examples given in Table 5.10.

Table 5.10   *Yield increases with age* (tons bunches per hectare)

Climate*	Strong dry season			Weak dry season	No dry season	
Country	Nigeria			Ivory Coast	Malaysia	Ecuador
Planting year	1941	1952	1959	1961	1967	1964
Material	Dura	d × p	d × p	d × t	d × p	d × p
Year from planting†						
3	—	—	—	7.7	12.8	13.8
4	—	2.7	6.4	8.3	24.5	17.2
5	2.3	6.4	7.4	10.8	31.3	17.8
6	5.7	8.4	10.8	17.4	30.5	20.9
7	5.9	10.0	8.6	16.2	33.1	18.8
8	7.3	9.0	10.3	11.5	—	—
9	7.0	11.3	—	12.4	—	—
10	7.2	8.8	—	—	—	—
11	9.2	9.5	—	—	—	—
12	12.3	15.3	—	—	—	—

* See water deficits in Table 3.6.
† Year 0 = planting year.

Sources: N.I.F.O.R. data and Annual Reports, I.R.H.O. Annual Reports, Malaysian and Ecuadorian plantation data.

Yield progression with age has been different in areas with little or no water deficit and those with strong dry seasons. In the latter, modern material tends to progress steadily to a mature yield at the eighth or ninth year after planting; if conditions are favourable optimum yields may be obtained earlier, but any unusually severe dry season may delay full bearing. In areas of no water deficit modern *tenera* material may advance very rapidly to a mature yield and bunch production of around 20 tons per hectare in the second bearing year has not been uncommon in recent times.

While precocity of bunch yield has undoubtedly been increased by improved cultural practices, a strong genetic component has been established by Blaak[135].

Bunch number depends largely on sex ratio, a character which varies widely between individual palms, between progenies and between palms of different fruit form. In Africa *pisifera* palms have a much higher sex ratio than *tenera* palms while *tenera* palms have a higher sex ratio than *dura* palms. Two cases in Nigeria may be quoted.[44]

Field	Planted	Sex ratio in			Period of observation	Total number of palms
		dura (%)	*tenera* (%)	*pisifera* (%)		
1—1	1939	27.9	31.1	45.7	1943—9	814
6—2	1941	29.4	31.4		1945—55	1,636

Although there are wide variations, palms of African origin when planted in Africa or in the Far East usually give a higher sex ratio and number of bunches than Deli palms. Examples from fields in Nigeria[44] and Sumatra[4] are:

		Sex ratio	
Period	Planted	Deli palms	African palms
Africa (Nigeria)			
1946—9	1941	29.1	38.0
1950—4	1941	20.4	33.5

		Number of bunches per palm		
Period	Planted	Deli palms	African dura	African tenera
Far East (Sumatra)				
1949—54	1939—40	32.8	56.2	56.6

Individual palm bunch weights will vary from a few kilograms when first in bearing to 10 to over 100 kg in weight when mature, depending on locality, fertility and inherited characters. A typical well-grown mature field in Africa may have bunches averaging 20 kg. In a typical productive mature field in Sumatra or Malaysia, Deli x African bunches may average 20—30 kg with a range from 6—80 kg per bunch. In exceptionally productive land such as the coastal clays of Malaysia, average bunch weights as high as 30—40 kg may be attained. The variation will be from palm to palm, progeny to progeny and between fruit forms. Deli bunches are heavier than African bunches both in the Far East and in Africa; for example, in a Deli and African progeny trial in Nigeria and in a trial in Sumatra[4] the bunch weights at 10 to 12 years old were as follows:

Year of record	Planted	No. of palms	Deli (kg)	African (kg)	
Nigeria 1952—3	1941	1,182, 498	14.8	12.5	
			Deli (kg)	Dura (kg)	Tenera (kg)
Sumatra 1949—54	1939—40	487, 56, 104	18.6	12.0	10.4

It is clear that it is characteristic of the Deli palm to produce fewer but heavier bunches than the general run of Africa palms from whatever region they may come. Exhaustive studies in Zaire and elsewhere confirm that *dura* palms give a lower mean number of bunches but a higher mean weight per bunch than *tenera* palms.

Bunches vary considerably in the length of the spines at the end of the spikelets, African palms tending to have long spines and Deli palms short spikes. Many African palms can be found with short spikes however, for example, the Yangambi-type *tenera.*

Fruit production per bunch

The percentage of fruit per bunch is an important yield factor. The usual range is between 60 and 65 per cent but F/B ratios below 60 per cent are not uncommon while progenies with percentages of over 70 per cent are sometimes found. With some palms the fruit-to-bunch ratio is a consistent character. With other palms this is far from the case. Young palms often show very low fruit-to-bunch ratios sometimes amounting to no more than 40 per cent. Fruit-to-bunch ratios of *pisifera* palms depend on their fertility. Complete bunch rot, or a bunch with only a few fruit, is the rule in the so-called infertile *pisifera.* In fertile *pisifera,* i.e. those setting sufficient fruit to give a recognisable non-rotten bunch, ratios may be as low as 25 per cent and only very rarely reach 60 per cent. *Tenera* palms have, on average, a lower fruit-to-bunch ratio than *dura* palms. Shell weight is partly responsible for this form difference.

The percentage fruit to bunch depends partly on the number of spikelets which, as already mentioned (p. 61), varies widely. There is no difference between the mean number of spikelets of *dura* and *tenera* bunches, but the mean number of spikelets on *pisifera* bunches is significantly higher than on bunches of the other fruit forms. The number of flowers per spikelet differs to a very small degree between fruit forms, the ascending order of number being *dura, tenera, pisifera.* Fruit set, however, tends to be higher in the *dura* than in the *tenera.*[18]

Fruit type and form variations

The main fruit types (external appearance) and fruit forms (internal structure) have been described in Chapter 2. For the inheritance studies shortly to be described, the form and type classification is not entirely satisfactory since colour, though it may be noted externally, is an internal factor of chemical composition. The classification used by Belgian workers in Zaire has much to commend it and is set out below with an indication of when a 'type' or 'form' is concerned:

1. Presence or absence of supplementary carpels
   (*Poissoni* or 'mantled')   2 types
2. Presence or absence of shell and/or of fibre
   ring (*dura, tenera, pisifera*)   3 forms
3. Presence or absence of anthocyanin in the
   exocarp (*nigrescens, virescens*)   2 types
4. Presence or absence of carotenoids in the
   mesocarp at maturity (*albescens*)   2 types

These characters being independent, there can be eight types or type combinations of fruit, and twenty-four type-form combinations. The variety of specific fruit characters is therefore large before one begins to consider the variation within each character. Fruit form and type differences do not give rise to 'varieties' in the botanical or systematic sense; the differences are ones of horticultural form.

The *Poissoni* or mantled character has already been described (p. 64). It may be added here that from six to ten supplementary carpels are formed; often only shell is found within the flesh but sometimes endosperm is also present. There is a well-known illustration of Janssens[45] in which a complete shell and kernel, perhaps with embryo, is shown in one of the carpels and described as an 'exterior fruit' (see Fig. 2.11). From the point of view of the plant breeder the *Poissoni* must now be regarded as an aberrant type which he would wish to keep out of his collection.

Something has already been said about the variations within the *nigrescens* and *virescens* types (p. 65). There are variations in carotene content of the mesocarp and it is possible that this is connected with difference of fruit colour as already described. The examining of large quantities of *nigrescens* fruit for differences of carotenoid content and for quantitative differences of anthocyanins in the exocarp has not been undertaken. Ripe fruit of the Deli palm normally appear to be of a paler orange colour than that of most African fruit, and oil from the Deli is easier to bleach.

Carotene contents of plantation palms (estimated as α carotene) have been found to vary in Nigeria from 200 to 1,100 ppm of extracted oil. Oil from grove palms has varied from 600 to 2,875 ppm carotene. The reasons for these differences of carotene content are still obscure, but may be connected with degree of ripeness and over-ripeness, and with the incidence of light on the bunches. There is evidence of differences between progenies in carotene content,[46] and between oil from African palms and oil from Deli palms. The oil from palms considered to be *albescens* usually contains less than 60 ppm carotene but as much as 90 ppm is sometimes found.

Variation within fruit forms

In Chapter 2 the *dura* and *tenera* forms were described as follows: *dura* — shell usually 2—8 mm thick though occasionally less, low to medium mesocarp content (35 to 55 per cent but sometimes, as in the Deli palm, up to 65 per cent), *no fibre ring; tenera* — shell 0.5—4 mm thick, medium to high mesocarp content (60 to 96 per cent but occasionally as low as 55 per cent), *fibre ring*. The fibre ring only gradually came to be recognized as the ultimate criterion. Previously shell thickness was taken as the main classifying measurement, Beirnaert and Vanderweyen defining the *tenera* as fruit with a shell-thickness of 0.5—2 mm. The remarkably consistent segregations in these authors' classic work were obtained through using *tenera* of low shell thickness (i.e. below 2 mm), thus assuring that only genuine *tenera* were being used and that the progeny would not be hard to classify. Later, it was recognized in Zaire that *dura* of shell thickness less than 3 mm, and *tenera* of shell thickness up to 3 mm, could be found.[47] Studies in West Africa have confirmed that the overlap of the range of thickness of shell of the *dura* and the *tenera* may be as much as 2 mm. Shell measurements are made, preferably with sliding calipers, at two opposite points of the shell where it has its maximum diameter.

In grove populations a slightly skew distribution of shell thickness is found. Fruit with shells between 3.0 and 5.5 mm are most common; the very thick-shelled fruit is rather more common than the very thin-shelled fruit and at least 10 per cent of the shells are above 5.5 mm in thickness. This may be seen in the histogram in Fig. 5.13 (taken from Beirnaert and Vanderweyen's data). The distribution of shell thickness among Deli palms in Sumatra is remarkably similar, only the small percentage with less than 2.5 mm thickness and the very thick shells being almost absent. The histogram shows that Deli *dura* with less than 2.5 mm shell thickness exist, however. It is the thickness of mesocarp, not differences of shell thickness, that normally distinguishes the Deli *dura* from the African *dura*, though African *dura* with the composition of good Deli *dura* are to be found.

The third histogram in Fig. 5.13 shows the wide distribution of shell thickness to be obtained from the illegitimate progeny of grove *tenera*. A selection of such material closely conforms in practice to random *tenera* x *dura* crosses since the majority of the pollen in the groves is of *dura* origin. *Tenera* and *dura* peaks can be distinguished, but owing to the wide variation in shell thickness of the parents there is considerable overlapping.

By weight, shell in the fruit is found to vary in Africa as follows:[14]

dura	25—55 per cent shell
tenera	1—32 per cent shell

Mesocarp variation has been given at the beginning of this section. Contrary to general belief the mean kernel content of *tenera* palms is the same as that of *dura* palms. Indeed, with nearly 7,000 *tenera* and over

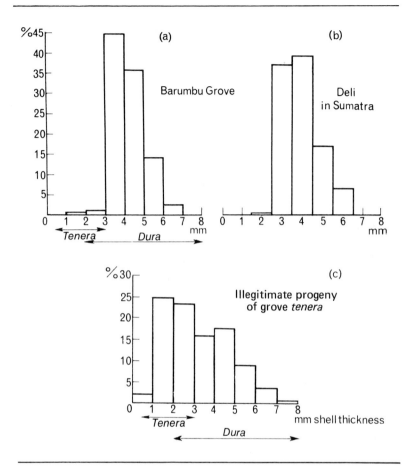

Fig. 5.13 Distribution of shell thickness in (a) a Congo grove (b) an early field of Deli palms in Sumatra, and (c) illegitimate progeny of Congo *tenera* palms. (Drawn from the data of Beirnaert and Vanderweyen, 1941.)

3,000 *dura* Beirnaert and Vanderweyen found the average kernel percentage to fruit actually higher in the *tenera*, though this was due entirely to the higher kernel content of *tenera* interior fruit. The usually lower kernel content of *tenera* fruit now encountered is due to selection and breeding for high mesocarp content; this has reduced both the shell and kernel contents. *Tenera* with large kernels still exist however. Sparnaaij *et al.*[14] give the range of kernel content for *good* African fruit as:

dura	7–20 per cent
tenera	3–15 per cent

Pl. 20 The wide variety of shape and size of African fruit.

With Deli *dura* the range is 4 to 11 per cent though percentages below 7 are uncommon.

Variation in weight and shape of fruit has already been described (p. 64 and Plate 20). Fruit composition is of course more important than size of fruit, and oil seed crushers are as satisfied with a given bulk of small kernels as with a similar weight of large ones. However, large fruit have certain advantages; they are easier to handle and analyse, loose fruit fall from the bunch more readily, but are not so easily overlooked by harvesters, and they are perhaps more adapted to machine processing. Beirnaert drew attention to the great differences of fruit weight between *tenera* palms and strongly favoured the large ovoid fruit with 12–18 g mesocarp and 1.2–1.8 g kernel. He suggested that such fruit were more uniform than small fruit, but gave no quantitative reason for their choice; no claim was made that large-fruited bunches would on average have a higher overall weight or a superior analysis in terms of oil and kernels than bunches with small fruit. As already described, this preference led to the production of the Yangambi-type *tenera* which is so distinctive wherever it is planted, which is attractive in appearance, has a good analysis and is not encumbered by long spines. In many respects the bunches are similar to Deli bunches.

Variation in the size and conformation of *pisifera* fruit is similar to that of *tenera* fruit, but in the palm as a producer the chief variable is fruit set.

Variation in sterility of fruit

Sterility occurs predominantly in the *pisifera*, the majority of palms of this fruit form bearing few fertile fruit and showing a strong vegetative development. *Pisifera* have been classified as (i) fertile — these palms are relatively few and their value in breeding is discussed on another page; (ii) showing partial sterility — small numbers of fertile fruit per bunch, vegetative development less than in (iii); and (iii) sterile — giving, occasionally, a few fruit, but the bunches rotten and vegetative development strong. Intermediates between these categories exist and palms tend to become less infertile as they age. In sterile fruit there is no development of the ovule, or ovular development is retarded. Abnormalities of the tissues surrounding the ovule also occur.[48] Sterility sometimes occurs in the *tenera* and has been reported in the *dura*.

Variation in vegetative characters in relation to breeding

Marked differences of vegetative development and leaf habit have always been apparent in the oil palm but until comparatively recently few systematic studies have been made.

De Berchoux and Gascon[49] studied the differences in vegetative characters between a Deli progeny from Johore Labis Estate, Malaysia (JL1133 x JL1113), a local Ivory Coast La Mé *tenera* x *tenera* cross (L2T x L11T), and a cross derived from Yangambi (Zaire) material (2469B x 1473B) all growing at La Mé. It is presumed that these progenies were chosen as being typical of the material they represented; forty-two to fifty palms per progeny were studied. While the expected difference between Deli and African material emerged, i.e. the Deli had fewer but longer, wider and heavier leaves, more leaflets and wider petioles, the differences between the Ivory Coast and Yangambi progenies are of particular interest. About the same number of leaves were produced by both progenies, but the Yangambi progeny's leaves were about 40 per cent heavier and the weight of the leaflets and their number approached those of the Deli (though the rachis was lighter). Leaf length was, however, no greater in the Yangambi than in the Ivory Coast progeny and the advantage of the former was found in its wide leaves and larger number of leaflets. Some of the data is shown in Table 5.11. The results suggest that the early selection in Zaire provided palms of vegetative development somewhat different to that of the Deli palm but of equal photosynthetic potential; but in view of the limited nature of this study general deductions cannot be drawn.

Through the growth analysis techniques developed by Corley in Malaysia it has been shown that important differences exist between individual palms and progenies in leaf area and vegetative dry matter production and in the growth parameters leaf area index and net assimilation rate. The physiological aspect of this work has been discussed in Chapter 4 and its value in breeding will be considered later in this chapter (see p. 262).

Table 5.11   *Extremes of vegetative development at La Mé, Ivory Coast. Some data of Deli, Ivory Coast and Zaire progeny (De Berchoux and Gascon)*

Derivation and progeny	Deli dura	African tenera crosses	
	SOC1386 (JL1133 x JL1113)	La Mé LM6 (L2T x L11T)	Yangambi YA3 (2469B x 1473B)
Annual leaf production, 1960–3	19.3	24.2	24.4
Number of leaves on the palms, per palm, mean of 4 years	34.6	40.6	41.6
Weight of 17th leaf, kg, 4-year average	8.1	4.8	6.6
Weight of leaflets, 17th leaf, 1963, kg	3.1	2.3	3.0
Length of 17th leaf, mean of 2 years, m	6.60	5.63	5.64
Width of 17th leaf at mid point, m	2.15	1.78	2.07
Number of leaflets on 17th leaf, 1963	350	309	344
Width of petiole, 17th leaf, 1963, cm	9.0	6.7	8.2

The inheritance of bunch and vegetative characters

To effect an improvement of the yield of products of the oil palm it is necessary to have a knowledge not only of the variability of production factors, but also of the relative extent to which these factors are inherited. Variance due to genetic factors may be exhibited through:

1. the simple Mendelian inheritance of specific qualitative characters,
2. additive genetic variance,
3. non-additive genetic variance.

Although important facts are known regarding the inheritance of specific fruit characters, studies of the inheritance of quantitative characters in the oil palm are recent, while the study of combining ability, which is often considered to be of great importance in a cross-pollination crop, has so far been of a general nature and detailed progress will not be made until fully scientific breeding programmes have proceeded much further.

Inheritance of specific fruit characters

1. Presence and absence of shell

An account has already been given of the emergence of the *dura* x *pisifera* cross in the inheritance of fruit forms in *tenera* x *tenera* selfs and crosses.

This cross may be illustrated as follows:

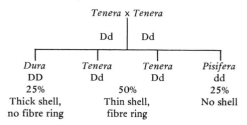

The *tenera* is thus a hybrid (Dd) between the thick-shelled *dura* and the shell-less *pisifera* forms and, apart from its generally thinner shell, it exhibits itself by the ring of fibres embedded in the mesocarp and running, at varying distances from the shell, longitudinally from the top of the fruit to its base.

It follows, therefore, that other fruit form crossings or selfing will result in the following segregations:

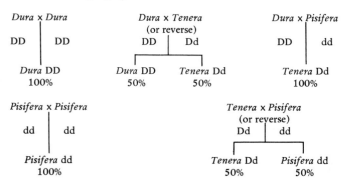

In the original Zaire *tenera* x *tenera* selfings and crossings certain divergencies from the $1:2:1$ D:T:P ratio were found.[18] While *tenera* progeny remained nearly constant at 50 per cent of the total, a few *tenera* mother palms gave as much as 35 per cent or as little as 15 per cent *dura* with a correspondingly low or high proportion of *pisifera*. Such cases can emerge through contamination of *tenera* pollen with small quantities of either *dura* or *pisifera* pollen, e.g.

		D	T	P
	x *Tenera* =	25	50	25
Tenera (Mother palm)				
	x *Dura* =	30	30	
	Total (%)	34.4	50.0	15.6

But detailed examination showed that the cases in Zaire not only occurred in certain *tenera* x *tenera* lines, but when *tenera* which produced about 15 per cent *pisifera* in selfings (Type II) were crossed with *tenera* giving the normal 1 : 2 : 1 segregation (Type I) the segregation was intermediate, i.e. 20 per cent *pisifera* and 30 per cent *dura* were produced. Similarly *tenera* producing 15 per cent *dura* and 35 per cent *pisifera* in selfings (Type III) when crossed with *tenera* giving normal segregation (Type I) gave 20 per cent *dura* and 30 per cent *pisifera*. Finally *tenera* giving 15 per cent *pisifera* (Type II) when crossed with *tenera* giving 35 per cent *pisifera* (Type III) gave the normal 1 : 2 : 1 segregation. A few examples from the work of Beirnaert and Vanderweyen may be given:

Tenera × *Tenera*	*Segregation of fruit forms (%)*			*Number of palms*
	Dura	*Tenera*	*Pisifera*	
Type II selfs and crosses	32.6	51.3	16.1	353
Type III selfs and crosses	14.8	49.2	36.0	390
Type II × normal (Type I)	28.8	50.9	20.3	645
Type III × normal (Type I)	19.5	51.4	29.1	492
Type II × Type III	74.2		25.8	275

These variations from the normal have not as yet been recorded on a large scale elsewhere and no fully satisfactory explanation has been provided although de Poerck suggested an hypothesis based on the possible presence of a series of alleles modifying the proportions of the three fruit forms in *tenera* × *tenera* progeny.[50]

2. Presence or absence of pigmentation of the exocarp

Inheritance of the presence of anthocyanins in the exocarp has been studied in Zaire and Nigeria. *Nigrescens* or ordinary fruit are black capped, but vary considerably in appearance (p. 65). In green fruit (*virescens*) absence of anthocyanin does not appear to be absolute; there is evidence of traces of an anthocyanin which may be distinct from the one normally encountered in the ordinary fruit.[47]

There is evidence that the green-fruited character is monofactorial and dominant, but the number of green-fruited palms found in natural populations is so small that it might normally be expected to be found in the heterozygous condition. In Zaire a green-fruited palm, assumed to be heterozygous, gave the following progeny:[18]

| (1) | (2) |
| *Virescens* × *Virescens* | *Virescens* × *Nigrescens* |

In Nigeria an open-pollinated green-fruited bunch gave rise to 46 per cent *virescens* and 54 per cent *nigrescens*. Later work was not quite so conclusive. One *virescens* × *virescens* self and two crosses gave approximately 3 : 1 ratios as follows:

Virescens   *Nigrescens*
  67   18

while nine *virescens* x *nigrescens* crosses gave near to the expected 1 : 1 ratio:

108   91,

assuming *virescens* heterozygosity. However, in two crosses, one *virescens* x *virescens* and one *nigrescens* x *virescens*, involving the use of a common *virescens* parent, No. 1.439, unexpectedly high proportions of *virescens* were found, viz:

1.439 *virescens* x *virescens*   18 : 2
nigrescens x 1.439 *virescens*   16 : 7

This parent had been used in other crosses giving the expected ratio. It is possible therefore that in some cases other factors are involved.

3. Presence or absence of carotenoids

The *albescens*, or 'white-fruited', type can be 'ordinary' or green-fruited as to its exocarp character. It was early noted in Zaire that illegitimate *albescens* gave either no *albescens* or 1 to 5 per cent *albescens* in the progeny. It was therefore supposed that absence of carotenoids was a recessive character.[51] As already noted, there is not a complete absence of carotene in the *albescens* palm.

4. Presence or absence of supplementary carpels

This character, *Poissoni*, 'mantled' or diwakkawakka, appears to be inherited mono-factorially and to be dominant. Certain mantled palms in Zaire gave 100 per cent mantled progeny when selfed; others gave 75 per cent. Crosses between mantled and ordinary palms gave 50 per cent mantled and 50 per cent ordinary.[47] The inheritance was therefore assumed to be as follows:

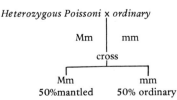

In Nigeria mantled *tenera* palms showing a high degree of out-pollination also showed approximately 50 per cent of each type in their progeny. Zeven quotes other illegitimate progeny in support of dominance.[52]

Experience in Sumatra has not given full support to the theory of dominance.[53] Two *Poissoni* selfs, one *tenera*, one *dura*, gave a mean segregation of 51.3 per cent mantled and 48.7 per cent ordinary. At the same time a *Poissoni tenera* (3240) crossed with four separate Deli *dura* gave a mean of 51.1 per cent mantled and 48.9 per cent ordinary. Some rather curious results were also reported. A certain non-*Poissoni tenera* (3317) when crossed with ordinary Deli parents gave varying proportions, up to one-third, *Poissoni* progeny. When selfed, this *tenera*, which belonged to a family in which mantled palms occurred, gave one-quarter mantled palms. Another ordinary *tenera* suddenly changed to the production of mantled bunches. It should be noted that the results of the *Poissoni* x ordinary Deli crosses agree with the Zaire results and, on the assumption of heterozygosity in the *Poissoni*, imply dominance. Other trials[4] in Sumatra in which mantled *tenera* were crossed with ordinary *tenera* and Deli *dura* in five separate plantings supported the theory of dominance if it was assumed that the mantled parents were heterozygotes. The 1 : 1 segregation was not far from achievement both as a whole and when the *dura* and *tenera* progeny were examined separately. The behaviour of the *Poissoni* selfs mentioned above and the non-*Poissoni tenera* aberrations cannot, however, be explained.

The *Poissoni* character was regarded with favour in early breeding work in Sumatra. There was evidence that *Poissoni* Deli *tenera* and even *Poissoni* Deli *dura* would produce a higher proportion of mesocarp than the ordinary Deli. Extensive crossings with *Poissoni tenera* and *dura* were therefore made. Later, when the inheritance of the *tenera* characters was understood, *Poissoni* palms dropped out of use and today they are little more than a curiosity.

The inheritance of specific vegetative characters

The only specific vegetative characters which have been given attention are the *idolatrica* leaf form and trunk 'dumpiness'. With regard to the former, evidence is conflicting and it has been thought to be both dominant and recessive. The former theory is suggested by the fact that an open-pollinated bunch from an ordinary palm in Zaire which had some *idolatrica* neighbours gave 16.7 per cent *idolatrica* progeny. Of thirty-four progeny of a selfed *idolatrica* in Zaire, twenty had been identified as *idolatrica* when they were quite young seedlings.[18] Intermediate forms were found in Nigeria, but controlled pollinations also suggested dominance. The breeding true of a selfed *idolatrica* palm in Sumatra has, however, suggested that the character might be recessive.[54]

Of much greater importance is the inheritance of dumpiness. This has already been described in the case of the Malaysian dumpy E.206; the growth of the F_1 selfs suggested a high degree of homozygosity for this character in the original parent. This supposition was supported by results from the F_2 generation in which F_1 sibs from the original E.206 selfing gave progeny with uniformly small height increment and large girth. Thus the F_2 on one estate had an annual height increment of 16.3 cm with girth of 358 cm at 7 years of age, while non-dumpy palms under equally fertile conditions had shown annual height increments of 22.9–35.6 cm at 5 to 8 years with girths of 256–267 cm[55] The dumpy character was not strongly transmitted in the F_1 dumpy x normal cross but in the so-called back-cross (F_1 dumpy x normal) x F_1 dumpy self, the characters were inherited as shown below:

Estate	Progeny	Age	An. height increment	Girth
H	Dumpy 'back-cross'	8	30 cm	302 cm
	Normal Deli	8	36.6 cm	257 cm
Jerangau	Dumpy 'back-cross'	5	25.4 cm	267 cm*
Exp. St.	Normal Deli	6	36.8 cm	267 cm

* At 3 feet; otherwise at 4 feet.

Studies in the Ivory Coast with groups of progenies of Malaysian, Zaire and Ivory Coast origin showed that measures of leaf production, internode length and stem growth tend, in crosses between these groups, to be intermediate to the same measures in crosses within the groups.[56]

With the movement of the oil palm into higher latitudes the resistance of the stem and stem base to strong winds has assumed some importance. In general the palm stands up well to the winds preceding tropical storms in West and Central Africa and South-east Asia. Oil palms of all ages survived the severe force of hurricane Fifi in Honduras in 1974, when adjoining banana plantations were completely flattened. In Colombia a sudden tornado in the Magdalena valley in 1968 demonstrated that there were distinct progeny differences in resistance to high winds.[57] In general Deli *dura* x La Mé *pisifera* progenies were less disturbed than Deli *dura* x Yangambi *pisifera* progenies. Most affected were palms of 2½ to 4½ years of age in the field, a period when there is likely to be the greatest imbalance between the trunk and canopy and the roots. At all ages there was a significant correlation between height of trunk and the amount of disturbance (measured by the angle of inclination). This provides an additional reason for breeding palms which grow slowly in height.

Heritability and genotypic values

Some of the specific characters of fruit form and type whose inheritance has been discussed above are governed by single gene pairs and they are referred to as qualitative characters. It has been seen in the previous section, however, that characters such as bunch yield, bunch number, weight per bunch, fruit-to-bunch ratio and many of the factors of fruit composition show continuous variation. These characters, which are known as quantitative or metrical characters, do not segregate in discrete classes and they are usually governed by a large number of genes.

The variability observed in quantitative characters is due partly to differences in genetic factors and partly to differences of environment. The concept of heritability was developed to enable the relative importance of genetic and environmental factors to be determined and to show whether selection progress is relatively easy or difficult to obtain. Heritability is the genetic variance in relation to the observed, or phenotypic, variance. The genetic variance has an additive portion, which is caused by genes acting independently in affecting the character concerned, and a non-additive portion which is concerned with general and specific combining ability. It is not possible here to give any detailed account of these two phases of genetic variance.

The heritability of any character may be measured by the regression of the progeny values on the values of the parents. Where the relationship is between progenies and, in each case, two known parents, the mid-parent values are used. When data of single parents only are available, heritability is taken as $2b$ where b is the regression coefficient.[58] It should be realised that heritability figures are likely to differ from environment to environment, from population to population and from generation to generation.

The first studies of heritability were those of Blaak[59] and Menendez and Blaak[60] working in Nigeria on material established by Toovey and others, but more recently several other studies have been made, both in Africa and Malaysia, and some of these have embraced vegetative characters as well as those of fruit and bunch. Before discussing these studies an example from Nigeria of heritability in the oil palm may first be given.

Twenty-five parent *tenera* palms were pollinated with a mixture of *tenera* pollen from those of their number which had the highest fruit quality. The regression of the mean mesocarp to fruit percentage of their *tenera* progeny (y) on the mesocarp percentage of the parents themselves (x) was calculated from the data in Table 5.12 and is:

$$b_{yx} = \frac{82.17}{263.64} = 0.3116 \pm 0.080. \text{ Heritability } 2b_{yx} = 0.623, \text{ or } 62.3\%.$$

This regression is highly significant and it shows that, for this population under the existing environment, a difference of 1.0 per cent in the mesocarp content of the female parents may be expected to give a difference of 0.312 between the mean mesocarp content of the progenies. This is illustrated by the regression line in Fig. 5.14. It may also be noted that, although the male parents are not known and therefore the heritability can only be calculated from the single parents, the use of mixed *tenera* pollen from high quality parents appears to have raised the level of mesocarp content from a mean of 71.4 to 77.7 per cent.

Table 5.12   *The mesocarp to fruit content of twenty-five* tenera *mother palms and their progenies (Nigeria)*

Progeny no.	Mother tree (x) (%)	Progeny (y) (%)
69	69.1	75.7
70	75.0	79.3
71	74.4	79.7
72	68.3	75.9
73	75.5	78.4
74	71.3	79.2
75	79.1	78.0
76	73.2	78.2
77	73.2	79.4
78	66.6	75.0
79	71.1	75.6
80	70.3	79.8
81	66.9	77.0
82	75.4	79.8
83	68.9	77.1
84	74.7	78.4
85	69.8	74.8
86	71.3	78.7
87	68.0	77.7
88	72.5	79.3
89	70.8	77.5
90	65.6	76.5
91	70.8	78.6
92	68.0	75.0
93	73.7	78.8
Mean	71.4	77.7

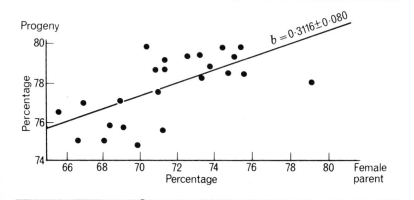

Fig. 5.14   Regression of the mean mesocarp to fruit percentage of progeny on the mesocarp to fruit percentage of female parents: Ufuma, Nigeria.

Bunch composition heritability

The work on heritability in Nigeria[60] has shown that most fruit and bunch characters are strongly inherited and that therefore selection for those characters leading to high oil production is justified. Supporting data have been provided from the Ivory Coast in crosses between Deli palms and palms of Zaire and Ivory Coast origin.[61]

Fruit-to-bunch ratio.   This is an extremely important character and very variable. It has assumed an added importance with the advent of the *dura* x *pisifera* cross since *tenera* bunches have on average a lower fruit-to-bunch ratio than *dura* bunches. This is not always due to a poorer set; it is largely due to the increase in mesocarp to bunch failing to compensate in weight for the lower shell to bunch. However, if full advantage is to be taken of the fruit composition of the *tenera*, then the fruit-to-bunch ratios must be maintained and a high account must be taken of this in the *dura* parent. Unfortunately this character is the least strongly inherited and environmental factors play a large part in fruit set. Within-field heritabilities varying from 0.19 to 0.34 were recorded in Nigeria.

With later material in Nigeria Van der Vossen[62] found significant fruit-to-bunch heritabilities for *tenera* of 0.55 with *tenera* x *tenera* progenies and of 0.18 within *dura* x *tenera* progenies. In the Ivory Coast overall heritabilities of fruit to bunch were low and parent/progeny correlations insignificant. However, within certain groups of parents, e.g. in the La Mé *tenera* x Deli lines, heritability was found between *tenera* parents and their *tenera* offspring, and Van der Vossen has concluded that the generally lower heritabilities found by Meunier *et al.*[63] are explained by the lack of distinction between fruit forms. Meunier *et al.* consider however that the discovery of heritability in certain groups of progenies where a character is only weakly heritable may indicate non-additive variance or specific combining ability. This view agrees with Ooi's finding with Deli material in Malaysia[134] where, however, other data gave heritabilities of 0.36 and 0.55.[91]

Gascon and de Berchoux[64] found a fairly high and significant correlation at Le Mé in the Ivory Coast between the fruit-to-bunch ratios of *dura* and *tenera* in the same progeny. Twenty-five progenies were Deli *dura* x *tenera* crosses, while the remaining sixty-one may have been African *dura* x *tenera* crosses or *tenera* crosses or selfs. The regression lines showed that although *dura* palms had a higher fruit-to-bunch ratio than the *tenera* in the same progeny, the difference between the two values diminished as the F/B ratio rose, e.g. a *dura* F/B ratio of 50 per cent corresponded to a *tenera* F/B ratio of 43 per cent, but at the higher end of the regression line the correspondence was *dura* 75 per cent: *tenera* 71 per cent. This finding increases further the importance of high F/B ratios in *dura* parents.

Mean fruit weight.   Meunier *et al.*[63] have shown that this is perhaps the most heritable of all the bunch composition characters, with a mean heritability of 1.01 on the mid-parent values.

Mesocarp to fruit. High values of heritability, amounting to 0.62 and above, were obtained for this character in Nigeria.

Examination of the data of more recent Nigerian material has given heritabilities for *tenera* of 0.96 and 0.80;[62] Ivory Coast figures are lower at 0.52 to 0.56,[61, 63] while a heritability of 0.59 has been recorded for Deli material in Malaysia.[91]

As with the fruit-to-bunch ratio, so also there was found to be a highly significant correlation, within progeny, between *dura* and *tenera* mesocarp percentages of the La Mé progenies mentioned above. The inclusion of about one-third Deli crosses may have affected the result, although all the *dura* in these progenies appear to have had relatively low proportions of mesocarp compared with Delis in the Far East; the *dura* mean was 46.9 per cent and the *tenera* mean 71.6 per cent.

Mesocarp percentage is not a simple factor and it is affected by fruit size, nut size and shell thickness all of which are themselves heritable characters. Large Deli fruit with relatively thick shells but mesocarp percentages of over 60 do not appear to give rise to *tenera* with a higher mesocarp percentage than that of *tenera* from many African *dura* with much lower mesocarp percentages. Although a more detailed study of this question is clearly required, some examples may be given:

1. Five Deli palms and five African *dura* palms were all crossed with the same *pisifera* in Nigeria. Although the Deli *dura* parents had a higher average mesocarp and lower shell percentage than the African *dura* parents, early bulk analysis of the *tenera* progeny showed similar mean mesocarp percentages; but the African *tenera* had lower shell and higher kernel content than the Deli *tenera* as can be seen below.

	F/B (%)	Meso-carp (%)	Shell (%)	Kernel (%)	Mesocarp + kernel (%)
Mean of 5 Deli *dura* parents	68.2	61.8	28.3	7.8	69.6
Mean of their *tenera* progeny	68.8	73.0	17.0	9.9	82.9
Mean of 5 African *dura* parents	66.2	56.4	32.2	11.3	67.7
Mean of their *tenera* progeny	67.8	73.6	15.2	11.2	84.8

The generally low mesocarp percentage in the *tenera* is unusual and much lower shell percentages might have been expected.

2. In a standard African *dura* x *tenera* cross (6.594 x 5.1450) used in comparative trials in seven fields in Nigeria[65] the mesocarp to fruit percentages of the *dura* and *tenera* progeny were as follows:

Dura progeny — Mean: 45.7%. Range between fields: 43.1—47.3%
Tenera progeny — Mean: 81.6%. Range between fields: 80.3—82.9%

The difference in mesocarp percentage between the *dura* and *tenera* sibs is very wide; from the regression lines obtained from the mesocarp values of the La Mé sibs a ratio of about 70 per cent mesocarp to fruit would be expected from *tenera* sibs of *dura* with 45 to 46 per cent mesocarp.

3. In a comparison in Nigeria of the standard cross and another African cross with four La Mé *tenera* x Dabou Deli crosses the following mean analysis data emerged (Expt. 14—4):

	Mesocarp to fruit percentage	
	Two Nigerian progenies	La Mé tenera x Dabou Deli
Dura palms	46.7	47.4
Tenera palms	87.2	75.1

It seems likely that the inheritance of mesocarp to fruit must be considered in conjunction with nut size and shell thickness. More will be said on this subject when additive genotypic or breeding values are considered (see p. 260).

Shell to fruit.   The heritability of shell percentage from *dura* mother palms to *dura* progeny was found to be very high in Nigeria: 0.83 to 1.32. From *tenera* to *tenera*, heritability was still 0.62 in the earlier and 1.06 in the later Nigerian determinations. Heritability from *dura* to *tenera* in *dura* x *pisifera* crosses was lower, at 0.53, but still highly significant. Shell percentage heritability is clearly of the greatest importance as within the single fruit it is high oil plus kernel that is required. Shell thickness has also been shown to have a high heritability.[65] This and inheritance from *dura* to *tenera* forms are considered further on p. 260.

Kernel to fruit.   Heritability in Nigerian material was found to be relatively high, varying from 0.41 to 0.74. In the Ivory Coast values of 0.29 to 0.65 were recorded,[63] and the correlation of within-progeny *dura* and *tenera* kernel to fruit percentage was high ($r = 0.81$) and significant.[64] Kernel percentages were usually lower in the *tenera*, but at the higher end of the regression line this difference tended to disappear.

There seems no reason to believe that kernels cannot be as large, or form as high a percentage of the fruit, in *tenera* as in *dura*. The early Zaire selections had large kernels, but since those days there has been a tendency to increase the mesocarp at the expense of both kernel and shell. This is discussed further on p. 273.

Oil to mesocarp.   In the early days of oil palm selection and breeding too little notice was taken of this factor in the mistaken belief that it approximated to 50 per cent and was not very variable. More efficient techniques for estimating oil-to-mesocarp percentages have shown that this is not the case however, and it must be regarded as a potentially important selection factor. There are of course practical difficulties in its assessment. (i) The bunch must be cut fully ripe and only ripe fruit must be used in the estimation. (ii) The oil content of the fruit of young palms is always low; the mature level is often not reached until the fourth or fifth year of bearing. (iii) There are seasonal variations in oil percentages, fruit in the peak production months tending to have a higher oil content than fruit harvested at other times. When these variables have been accounted for, oil percentage of mesocarp may still vary between palms from 40 to 62 per cent. In West Cameroon an estimate of 0.56 heritability has been made for *D* x *T* material and in Malaysia 0.48 has been recorded,[123] but in both the recent Nigerian material[62] and in the Ivory Coast[61] oil to mesocarp heritability did not reach a significant level though in the latter case a few individual parents seemed to transmit relatively high oil to mesocarp to their progenies. Van der Vossen[62] has studied the differences of oil and dry fibre content of *dura* and *tenera* full sibs. He found that, with inclusion of the fibre of the fibre mantles, the dry fibre content of the *tenera* was higher than that of *dura* fruit (Table 5.13). Nevertheless there was not a corresponding difference in the oil content of the whole mesocarp because the oil content of mesocarp outside the fibre mantle was found to be significantly higher in the *tenera* than in the *dura*. Highly significant correlations were found between the *dura* and *tenera* full sibs in both oil and fibre contents, and estimations of the proportion of genetic variance to total variance (wide sense heritability) suggested that both oil and dry fibre to mesocarp are heritable though dry fibre heritability is likely to be higher.

This work suggests that progress in oil to mesocarp selection is only likely to be made when determination methods are improved, and Van der Vossen recommended that dry-fibre contents should be determined by a

Table 5.13 *Oil and dry fibre content of fresh mesocarp in full-sib progenies* (Van der Vossen)* (per cent)

	Tenera				Dura	
	Oil[1]	Oil[2]	Fibre[1]	Fibre[2]	Oil	Fibre
Mean	52.6	48.7	15.0	16.8	49.8	15.3
Range	47.3–56.7	42.0–52.3	14.1–17.2	15.3–19.0	44.2–55.6	14.4–16.6
Coeff. of varia- tion	4	5	5	5	5	4

[1] Oil or fibre outside the fibre mantle.

[2] Oil or fibre in whole mesocarp.

* The data is for twenty-nine progenies. Computations for twenty-one progenies showed significant within-progeny correlations between *dura* and *tenera* for both oil and fibre, and significant differences between the fruit forms in fibre content of the whole mesocarp, in oil of the mesocarp outside the fibre mantle (in favour of the *tenera*), and oil in the whole mesocarp (in favour of the *dura*).

limited number of direct analyses. As this factor has been found to be relatively constant, within palms, irrespective of age and stage of ripeness, the data obtained can then be used in conjunction with the indirect method (i.e. water content determinations) applied to mature palms and fully-ripe fruit.[62]

Determination of the mean oil to mesocarp content of progenies of Deli, local and Yangambi (Zaire) origin available at La Mé have given some interesting data.[66] The Deli material was most variable both within and between progenies, but seven progenies obtained direct from S.O.C.F.I.N. in Malaysia all had mean oil to mesocarp percentages of over 50, one giving the very high mean of 56.6 per cent. Of the African material, that of Yangambi origin was markedly superior to, and more uniform than, that of La Mé. The means were 54.7 and 50.9 per cent respectively and the ranges 4.7 and 7.9 per cent. La Mé parentage did not appear to lead to any overall improvement in Deli crossings. The oil percentages in Deli x Yangambi *tenera* and *pisifera* crosses, however, assumed very nearly the same range as that of the Yangambi progeny themselves, suggesting that oil to mesocarp in this material, of Zaire origin, is heritable.

Bunch yield inheritance

Until recent studies were undertaken the inheritance of bunch yield was only inferred. Firstly, increases in yield from one generation to the next following the selection of high yielding parents were seen in many parts of the world to be marked but there was no means of telling how far this was due to better agronomic practices and how far to heredity. Secondly, comparisons in replicated trials showed that mixed progeny from selected parents gave a higher bunch yield than the mixed progeny of unselected material.[19, 67] In a series of trials of this nature in Nigeria significant increases in yields over 4-year periods of between 12 and 50 per cent were obtained, but there was a tendency for these differences to be less marked after about 8 years of bearing. This suggested that selection may have been partly, and unconsciously, for precocity. In five trials planted over successive years a mean annual yield increase of selected over unselected material of 22 per cent was obtained during the period from bearing to the fifth to eighth bearing year. That precocity is heritable was clearly shown by Blaak.[135]

Heritability of bunch yield was studied in Nigeria by means of a number of parent—progeny comparisons.[59] In all these comparisons the yield of only one parent was available; one study concerned *dura* x *pisifera* crosses, while the other two concerned *dura* parents from two separate groves pollinated with a mixture of *dura* pollen. The heritabilities of bunch yield were:

Field 4—1	*dura* x *pisifera* crosses.	
	dura parents — *tenera* progeny	0.29
Field 15—1	Ufuma *dura* parents — *dura* progeny	0.36
Field 15—1	Aba *dura* parents — *dura* progeny	0.23

In the more recent studies of the Ivory Coast and Nigerian material the heritability of bunch number (0.50–0.51) was found to be higher than for bunch weight (0.13–0.21), while values for total bunch yield were extremely low and insignificant.[62, 63] Within the Deli population in Malaysia low heritability for bunch yield was also suggested by the absence of additive genetic variation,[68, 69] and the few estimates that have been made confirm this. Ooi *et al.*[69], using data from four Deli *dura* breeding experiments, found that while there were significant differences between females within males in progeny bunch yield and its components, the genetic variability present was largely non-additive. These findings imply severe limitation to increases in bunch yield potential and this is discussed further in the next section of this chapter.

Although mean bunch weight is only weakly heritable, in the Ivory Coast it was shown that individual parents could transmit a high mean bunch weight to their progeny in a range of crosses.[61] In general, bunch number and mean bunch weight are negatively correlated, and this fact accounts for the low heritability of bunch yield.

Genotypic or breeding values

To make full use of the ability of the plant to transmit characters from one generation to the next it is necessary to disentangle the genetic component of variance from the environmental component. It has been shown that the genetic component in the oil palm is largely additive[43, 70] and that therefore the mean of a full-sib progeny for any character will give a reliable estimate of the genotypic value of the mid-parent, i.e. the mean of the values of the two parents. If crosses are made between three palms and if the mean values for any character for each of the three crosses is known then the genotypic value for each parent can be calculated. The example given by Van der Vossen[62] may be used, viz:

Kernel to fruit.   Progeny 4.3488 x G145 = 9.4 per cent
2,381D x G145 = 12.4 per cent
4.3488 x 2,381D = 10.0 per cent

Designating the parents T_1, T_2 and D_3, $\dfrac{A(T_1) + A(D_3)}{2} = 9.4$, $\dfrac{A(T_2) + A(D_3)}{2} = 12.4$ and $\dfrac{A(T_1) + A(T_2)}{2} = 10.0$, where A = the genotypic value. Thus $A(T_1) = 7.0$, $A(T_2) = 13.0$ and $A(D_3) = 11.8$, and the genotypic value for kernel to fruit for any other palm can be determined if the mean value in its cross with any of the above three palms is known, e.g. T_1 x T_4: progeny 4.3488 x 14.892 = 8.0, so $A(T_4)$, i.e. the genotypic value of palm 14.892 = 9.0.

Other methods have been used for assessing genotypic or breeding values. One statistical method treats a cross as being a combination of treatments, each parent comprising one treatment; this method has been preferred because additivity is not assumed.[71]

While bunch quality factors such as kernel to fruit will not be much affected by age or the different circumstances under which parents and their progeny may be grown, bunch yield components are much affected. Strictly therefore comparison between phenotypic and genotypic values for individual parents should be made for bunch yield components using data of similar ages and growth conditions for progenies and parents. This is not difficult as far as age is concerned if full records of parent palms are available, but correction factors may be needed for climatic variations and progeny trial situations.

Using Nigerian data Van der Vossen found that there was a strong or medium correlation between parental phenotypic and genotypic values, particularly *tenera* values, for bunch quality factors and for number of bunches. Correlation coefficients for mean bunch weight were lower and for bunch yield were barely or not significant.[62] These findings conform in general with those of the heritability studies and were confirmed in a more detailed assessment of material from the same programme; genotypic values for quality factors calculated from different sets of values showed close agreement with each other.

Genetic correlations between bunch yield and quality components

It has been generally held that all components are inherited independently. However, in the case of bunch quality components it is obvious that a change in the percentage of one directly affects the percentages of the others. It is not surprising therefore that negative correlations are found between the additive genotypic values of mesocarp and shell. However, the correlation between kernel and shell is positive.[62] Of greater interest is the negative genotypic correlation found between number of bunches and mean bunch weight. The significance of this finding for breeding policy is discussed later in this chapter.

Inheritance from the *dura* to the *tenera*

The correlations found in the Ivory Coast between bunch quality values in the *dura* and *tenera* have been mentioned in the previous pages.[64] With more recent material, however, the correlations have been somewhat lower.[62, 63] Van der Vossen, using Nigerian data, estimated the *tenera* additive genotypic values of *dura* parents and found that, except for kernel percentage, there were no significant correlations between the phenotypic value of a *dura* parent and its *tenera* genotypic value, and heritability estimates from *dura* to *tenera* based on the regression of genotypic on phenotypic values were equally insignificant. In contrast, the correlations and heritabilities from *dura* to *dura* and *tenera* to *tenera* were, as already described, marked and significant (Table 5.14).

With mesocarp percentage these findings confirmed the anomalies found between Deli *dura* and African *dura* in their *tenera* progeny of

Table 5.14   *Correlations between phenotypic and genotypic values for yield and bunch quality components (Van der Vossen)*

Character	Fruit form comparison	Correlation coefficient	Mean values			
			tenera	dura	*Deli* dura	d + t
Number of bunches	t + d	0.632***				P 9.8 A 9.2
Mean bunch weight (kg)	t + d	0.441***				P 6.6 A 5.1
Bunch yield (kg)	t + d	0.297*				P 59.2 A 46.0
Single fruit weight (g)	P_t–A_t	0.567**	P 8.0 A 6.5			
Fruit to bunch (%)	P_t–A_t P_d–A_d P_d–A_t	0.577*** 0.179 0.045	P 63.8 A_t 65.6	P 66.7 A_t 66.6 A_d 67.7	P 68.2 A_t 72.0 A_d 72.5	
Mesocarp to fruit (%)	P_t–A_t P_d–A_d P_d–A_t	0.912*** 0.872*** 0.249	P 83.1 A_t 79.0	P 55.9 A_t 77.2 A_d 50.7	P 59.7 A_t 77.8 A_d 55.9	
Shell to fruit (%)	P_t–A_t P_d–A_d P_d–A_t	0.924*** 0.744*** 0.176	P 9.0 A_t 12.5	P 33.0 A_t 12.9 A_d 37.5	P 30.9 A_t 13.8 A_d 34.7	
Kernel to fruit (%)	P_t–A_t P_d–A_d P_d–A_t	0.718*** 0.584*** 0.541***	P 7.9 A_t 8.5	P 11.1 A_t 9.9 A_d 11.8	P 9.4 A_t 8.4 A_d 9.5	

P = Phenotypic value, A = genotypic value. d and t indicate whether the values are *dura* or *tenera*. * P < 0.05, ** P = <0.01, *** P = <0.001.

which an example was given on p. 255. It will be seen in Table 5.14 that although the *dura* mesocarp values for the Deli are higher than for African *dura*, the genotypic (A) *tenera* values are similar for both groups. Thus the superior fruit composition of the Deli palm is of no advantage over African *dura* in breeding *tenera* material.

The explanation of this phenomenon has now been provided by Van der Vossen[62] working on an assumption of Sparnaaij[70] that both the shell and the fibre mantle of the *tenera* must be taken into account. Volumetric determinations showed that if only that part of the mesocarp lying outside the fibre was accounted in the *tenera*, then the correlation for mesocarp volume between *dura* and *tenera* full sibs was high (0.82***). Moreover the correlation was even higher (0.94***) between shell volume in the *dura* and shell + fibre mantle volume in the *tenera*, although the correlation between shell only in the two fruit forms was insignificant. Thus although the amount of lignification (*dura*, 100 per cent; *tenera*, partial, and *pisifera*, none) is inherited monofactorially, the shell thickness or percentage is determined by polygenes and is exhibited in the *dura* by the shell itself and in the *tenera* by the shell plus fibre mantle.

This work also showed that, volumetrically, the ratio of shell to shell + fibre mantle in the *tenera* varied from 0.29 to 0.50, the latter figure being usual where the kernel is relatively large and there is the same kernel-to-fruit ratio in both *dura* and *tenera* full sibs. When the kernel-to-fruit ratio is much lower in the *tenera* sibs, as appears to be the case where the kernel percentage of one parent differs markedly from that of the other parent, then the ratio shell/shell + fibre mantle is much lower.

Van der Vossen concludes that the *tenera* genetic value of a *dura* for shell to fruit (and mesocarp to fruit) can only be obtained from the mean of its *tenera* full-sibs or through *dura* x *tenera* crosses.

The inheritance of quantitative vegetative characters

The study of progeny characteristics by de Berchoux and Gascon[49] already referred to (p. 245) included two crosses between Deli *dura* and African *tenera* palms. One cross was a La Mé *tenera* x Deli from Dabou (L2T x D10D) the other a Pobé (Dahomey) *tenera* x Deli from Johore Labis, Malaysia (P7T xJL1273). The La Mé parent was one of the parents of the La Mé *tenera* cross whose characters have already been described. While only four parents were involved and only one parent was to be found in the within-derivation crosses used for comparison, the Deli and La Mé characteristics were considered sufficiently extreme and well known for broad deductions to be made. It was found that the Deli x La Mé cross was intermediate between the Deli x Deli and La Mé x La Mé crosses in the following respects: number of leaves on the palm; weight of leaf 17; weight and number of the leaflets; length of leaf 17; breadth of leaf 17; breadth of petiole. The Pobé x Deli cross did not show such clearly intermediate characters, the cross being in some respects nearer to the Deli.

In the course of this work it was noted that vegetative development, when expressed in terms of photosynthetic material, was not uniformly related to yield, some progenies having a much higher bunch yield per unit weight of leaflets than others. It was suggested that there were thus differences of photosynthetic efficiency and that the latter was heritable. In discussing crop growth rates and net assimilation rates in Chapter 4 (p. 164) it was suggested that increased yields must be sought in increases of net assimilation rates and the diversion of the increase in dry matter to bunch production, and it was suggested that these increases would be likely to be attained by selection and breeding.

Growth parameters and their relation to bunch yield were discussed in Chapter 4. In three progeny trials in Malaysia which involved the crossing of each of a number of *tenera* palms with a number of Deli *dura*, measurements were made of bunch yield (Y), vegetative dry matter production

(*VDM*), bunch index (*BI* = *Y/Y* + *VDM*), net assimilation rate, leaf area index (*L*), leaf area and leaf area ratio (*F*).[72] Estimates of genetic variability were computed for each parameter. The results suggested that heritability may be high for vegetative dry matter, bunch index and leaf area index Heritability of the other factors, though not negligible, was less consistent. Some evidence has also been provided for genetically controlled variation in photosynthetic rate.[73] It was suggested that bunch yield per hectare could be improved by selecting palms with high *BI* and low *VDM* and planting at higher densities than is at present usual.

Methods of selection and breeding

The advent of the *dura* x *pisifera* cross gave rise to a sudden and substantial means of improvement in the yield of palm oil per hectare. In Zaire, which was certainly ahead of other countries, it was not considered that satisfactory selection, following the concentration of *tenera* material in genealogical fields, could begin before 1939. Selection and breeding programmes are thus largely postwar developments and they have often been based on severely limited material.

Prospection

In order to obtain material of sufficient genetic variability to improve the crop through selection and breeding it is necessary to prospect on as large a scale as practicable. The earliest important prospection of oil palms carried out to obtain high quality *tenera* progeny was that carried out in Zaire in the 1920s and already described. Postwar prospection in Zaire included later generations of the same material, palms of local origin on estates and palms in grove areas, though in southern Zaire prospection of grove material was undertaken for a short period only and was disappointing. It was in Zaire that the most genuine and successful attempt was made to apply the general principle that selection should be over a wide area with the object of capturing plants with those genes which carry high yield and bunch quality characters.

Provisional 'Candidats Arbre-mère' (C.A.M.s), which were largely *tenera* on estates and *dura* in the groves, were selected after fruit and bunch analysis and yield recording.[74] In the area of northern Zaire only one *tenera* palm would be selected as a C.A.M. out of up to 35,000 palms examined.[75] The rigour of this selection may be seen by the following comparison between the average composition of *dura* and *tenera* material used for multiplication at Yangambi up to 1956 and the C.A.M. material selected from prospection in two districts.

Of particular note is the high mesocarp to fruit percentage of the *dura* C.A.M.s.

Material and prospection	Fruit to bunch (%)	Mesocarp to fruit (%)	Oil to mesocarp (%)	Oil to bunch (%)
Average Yangambi *dura*	70.0	50.0	45.0	16.0
Average Yangambi *tenera*	65.0	78.0	45.0	22.0
Prospection material:				
Dura mother palms: Elisabetha (Yangambi origin)	70.3	70.6	49.7	24.5
Tenera mother palms: Elisabetha (Yangambi origin)	62.5	92.5	56.0	32.3
Tenera mother palms: Likete (Local origin)	67.0	92.5	56.0	34.7

In southern Zaire (Kwango) postwar prospection was carried out mainly on plantations established in the 1930s from *tenera* seed obtained from the palm groves. In this region, owing to its poor soil and relatively unfavourable climate, high bunch yields were not expected, so every effort was made to seek both *tenera* and *dura* palms of very high quality adapted to the region. About 420 hectares were searched, but only seventeen provisional selections were made. The particulars of three *tenera*, four *dura* and one fertile *pisifera* are given in Table 5.15. Included in the table are data of the one *tenera* selected in the groves of the Kwango region. This work showed that good material was to be found, but only in small quantity and that the previous concentration of open-pollinated *tenera* progeny on estates had helped to provide palms with some valuable characters.[76]

The advantages to be gained by thorough systematic prospection is clearly shown by the results obtained both in the prewar and postwar

Table 5.15   Tenera, dura *and* pisifera *palms selected in southern Zaire (Kwango) after the 1954–6 prospection*

	Tenera				Dura				Pisfera
	T19	T138	T332	Grove selection	D62	D64	D65	D73	P21
Bunch Production per year (kg)	119	139	146	83	168	142	165	155	184
Mean bunch weight (kg)	17	20	17	—	23	35	27	51	37
Fruit to bunch (%)	70	74	68	68	71	72	76	70	67
Mesocarp to fruit (%)	87	87	89	88	58	63	56	62	89
Oil to mesocarp (%)	53	—	—	50	55	54	—	57	50

work in Zaire. That work of this fundamental kind was not seriously attempted on a large scale elsewhere in Africa was due to three causes: (i) the unrewarding composition of the predominantly thick-shelled *dura* groves to be found on the flanks of the palm belt, (ii) the lack of staff or enthusiasm for this arduous task, and (iii) the view that more immediately promising results could be obtained from the use, in crosses, of the restricted material already available.

In Nigeria up to 1961 only very small areas had been prospected. These consisted of the palms at Calabar derived from fruit of a range of forms and types, an old grove area at Aba covering 11 hectares and another grove area of 49 hectares at Ufuma. Both the Calabar and Aba plots were in areas of poor composition and very rigorous standards could not, because of the small number, be applied in prospection. The Ufuma grove was, however, chosen because of the very high proportion of *tenera* palms (43 per cent). This grove underwent various thinning processes, but finally about 2 per cent of the original stand was selected. Data of the selected *tenera* palms are given in Table 5.16.

Table 5.16 *Yield and analysis of grove* tenera *palms at Ufuma*

Palm	Adjusted bunch yield (% of mean)	Fruit to bunch (%)	Mesocarp to fruit* (%)	Shell to fruit (%)	Kernel to fruit (%)	Mesocarp to fruit of progeny (%)
203	246	56	68	14	15	76
1,187	222	52	75	13	10	80
141	220	61	75	15	8	79
965	218	53	67	18	13	75
918	217	62	73	12	12	79
1	210	53	69	17	12	76
1,912	207	56	66	11	21	77
1,658	206	53	70	15	11	75
262	189	61	76	12	11	78
1,744	184	53	68	16	14	78
977	181	58	71	15	11	76
295	181	64	71	15	11	79
992	168	56	70	14	14	80
1,605	150	52	75	13	10	78
143	153	56	74	13	9	80
912	150	56	73	13	11	78
1,513	152	55	69	16	12	77
1,776	152	58	73	13	11	79
1,682	156	54	71	17	10	79
1,928	156	53	71	17	10	78
1,959	142	55	74	13	11	79
1,059	141	57	67	19	12	77
778	138	60	79	11	7	78
1,859	135	58	71	14	13	76
1,952	136	57	68	16	14	75

* Analysis by boiling and pounding methods: see text, p. 267.

Table 5.17 Analyses of tenera fruit samples obtained in a prospection of eastern Nigerian groves

Area	Village	Number of tenera	Average fruit weight (g)	Range of fruit weight (g)	Mesocarp (%)	Shell (%)	Range of shell (%)	Kernel (%)
Awka	Egwuoba	102	9.5	14.7–5.5	78.8	10.6	4–20	10.7
Ufuma	Akpugo	77	8.9	12.1–3.6	80.3	7.2	3–18	12.4
	Ufuma	31	8.6	14.8–5.2	81.1	7.8	3–17	11.1
	Akpu	47	12.3	15.3–7.8	87.4	4.2	2–8	8.4
	Ekwulobia	221	9.0	17.7–4.4	80.0	7.2	2–17	12.2
Nnewi	Ozubulu	73	10.2	18.0–5.3	81.9	7.6	2–21	10.3
	Nnewi	16	8.2	10.6–6.5	83.3	7.4	2–21	8.5
	Azia	28	10.0	15.9–6.5	77.6	9.6	3–25	12.6
	Ezinifite	19	7.7	11.3–4.0	73.4	11.7	3–25	15.0
Owerri	Orodo	70	12.3	18.2–6.6	88.6	3.7	1–11	7.6
	Mbieri	34	10.4	14.6–7.0	82.8	7.1	3–24	9.9
	Umuokauna	49	11.3	16.9–7.7	78.8	8.6	3–21	12.6
	Ilile	13	7.7	9.0–5.2	74.4	14.4	8–22	11.4
Okigwe	Umukele	69	10.9	17.2–5.8	83.1	7.5	3–15	9.4
	Okigwe (East)	43	9.1	13.5–5.9	81.5	7.0	2–18	11.6
Aba (south)	Ogwe	17	13.0	17.5–8.4	74.1	9.6	6–12	16.2
	Ihie	27	11.7	20.0–8.4	74.0	12.1	7–25	13.9
	Azumini	33	9.8	12.0–7.4	68.7	10.8	4–25	20.6
Abak	Opoikot Anang	22	7.1	—	77.1	—	—	—
	Obio Akpa	241	8.3	—	75.2	9.4	—	15.4
	Obonung Ikot	59	9.8	—	75.3	6.7	—	18.0
	Ekeparakwa	319	8.9	—	84.7	6.1	—	9.2
Umuahia	Olo Umuahia	97	12.6	—	84.0	8.2	—	7.8
Ikot Ekpene	Ikot Ekpene	100	9.9	—	81.5	8.1	—	10.4
Arochuku	Arochuku	38	8.4	—	63.0	—	—	—
	Ikot Ekpika	106	6.2	—	83.3	6.8	—	9.9
	Ibiaku	119	7.1	—	86.2	11.5	—	2.3
Coast	Opobo	104	7.0	—	71.3	18.1	—	10.6
	Ndiya	135	8.8	—	73.4	12.5	—	14.1
	Eket	86	9.8	—	73.5	—	—	—
	Oron	133	5.3	—	70.5	10.0	—	19.5

Yields were adjusted for effects of density and fertility. Analysis was by the boiling and pounding method which gives a rather low estimate of mesocarp to fruit. The range in the parents may be taken to be about 70 to 85 per cent instead of 66 to 79 per cent. Even allowing for this, however, the fruit analyses were not impressive and selection was clearly not as rigorous either in percentage of palms selected or in characters as in Zaire; however, the use of mixed pollen from the best *tenera* palms provided progeny with higher mesocarp to fruit percentages than those of the parents. This type of grove prospection was regarded at the time as only a beginning of proper exploitation of those groves of eastern Nigeria which had a high proportion of *tenera*; detailed prospection of small isolated areas was unlikely to be as fruitful as prospection over a wide field.

In the early 1960s a preliminary prospection of *dura* and *tenera* fruit in many grove areas in eastern Nigeria was undertaken. Early *tenera* collections showed the fruit analyses given in Table 5.17. The mean fruit weights and shell percentages of *dura* and *tenera* material are summarized for the main areas in Table 5.18.

Table 5.18 *Mean fruit weight and shell percentage of samples obtained during prospection in eastern Nigeria*

Area	Dura		Tenera	
	Av. weight per fruit (g)	*Shell to fruit* (%)	*Av. weight per fruit* (g)	*Shell to fruit* (%)
Ufuma	11	36	10	7
Nnewi	12	35	9	9
Owerri	14	36	11	11
Okigwe	16	37	10	7
Aba (south)	13	39	11	10
Abak	9	36	8	8
Arochuku	–	–	7	9
Coast	9	37	8	13

These tables showed clearly that there was still to be found in eastern Nigeria *tenera* material of very high quality and, no doubt, of potentially high yield. The variations between areas, and even between villages, are interesting. Some areas do not appear to contain appreciable numbers of high-mesocarp fruit, but the kernel percentage may be high; in others, e.g. Aba and Coast, many samples have high kernel and rather high shell percentages. Fruit size varies greatly and although no very large fruit emerged in this survey, fruit well up to the size of Deli fruit were frequently encountered. Indeed some of the *dura* fruit seen were very similar to Deli fruit and there was a great variation in spine length in the spikelets.

In the Ivory Coast, production material was developed from limited local prospections in that country and Dahomey, and from progenies of Zaire and Malaysian origin. Material designated as 'La Mé origin' was developed from twenty-nine palms of a grove at Bingerville while 'Pobé origin' stemmed from thirty-eight *tenera* selected in groves near Porto-Novo.[77]

A systematic prospection was later carried out by Meunier[77] in groves in the Ivory Coast. These groves are in general less dense and more scattered than those in Nigeria. About 100 palms were taken at random in each of eleven areas. The proportion of *tenera* varied from 0 to 41 per cent and only seven *pisifera* were found. *Virescens* palms were rare, only one *albescens* was found and no mantled palms were encountered. Areas with the highest proportion of *tenera* tended to have the best mean fruit composition. Where *dura* of good composition were found, the *tenera* were also of good composition. There were wide differences between groves in bunch and fruit weight as well as composition and populations could be grouped according to their fruit and bunch characteristics although members of a group were not necessarily adjacent to each other. Some data from this prospection is given in Table 5.19. This is not directly comparable with the earlier Nigerian data in Tables 5.17 and 5.18 since in the latter prospection the palms were not taken at random.

It was concluded from the Ivory Coast prospection that the Yocoboué, Sassandra group was close to the imported Zaire material ('Sibiti') as far as mesocarp to fruit and bunch weight were concerned, whereas other groups were similar to the original F_0 La Mé material from Bingerville or to the Pobé material from Dahomey. Sufficient variability was shown to exist for improving the *tenera/pisifera* stocks at present being used in breeding and selection.

The most recent prospection has been carried out in Nigeria under the joint auspices of N.I.F.O.R. and M.A.R.D.I. About twenty palms, taken at random, were examined at forty-five sites and bunch, fruit and leaf parameters were measured.[78] Only palms with very poor fruit characters were rejected. There were wide differences between sites in all characters and in the proportions of *dura* and *tenera*. The latter varied from 0 to 62 per cent. High frequencies of *tenera* were common in sites in the southeast, thus confirming that this area is a most fruitful source of *tenera* material. *Virescens* palms were found at thirteen sites, mantled palms at one.

There were significant correlations between *dura* and *tenera* for all characters measured except fruit length, nut diameter and shell thickness. The lack of correlation in the case of shell thickness is of particular interest in the light of the work on mesocarp and shell by Van der Vossen[62] already described (p. 261).

There was a variation between sites in mean mesocarp percentage from 41.6 to 52.9 with the *dura*, and 61.2 to 77.9 with the *tenera* (omitting sites with less than three *tenera*). The overlap of shell thickness between the *dura* and *tenera* was quite marked: some *tenera* were recorded with shell thickness of 2–5 mm and *dura* with thickness of less than 1 mm. Although the site mean for *dura* shell thickness was usually 1.9 mm and

Table 5.19 *Some data from a random selection of grove palms in the Ivory Coast (Meunier)*
Mean bunch weight and composition of palms from groves grouped according to their fruit and bunch characteristics

Grove group	Fruit form	Mean bunch wt. (kg)	Fruit to bunch (%)	Mesocarp to fruit (%)	Shell to fruit (%)	Kernel to fruit (%)	Mean fruit wt. (g)
Yocoboué–Sassandra	*Dura*	14.7	58.4	47.7	39.5	12.8	7.0
Abobo–Dabou–Bingerville	*Dura*	11.0	55.3	43.3	42.9	13.8	7.5
Tabou–Danané	*Dura*	10.7	53.3	45.8	41.4	12.8	8.7
Soubré–Man	*Dura*	10.0	55.1	36.2	46.8	17.0	6.0
F_0 of La Mé*	*Dura*	10.1	60.9	40.2	45.2	14.6	7.4
Yocoboué–Sassandra	*Tenera*	12.8	54.1	68.4	20.5	11.1	6.2
Abobo–Dabou–Bingerville	*Tenera*	9.9	52.4	64.2	22.7	13.1	6.0
Tabou–Danané	*Tenera*	9.2	47.1	62.6	24.6	12.8	6.9
Soubré–Man	*Tenera*	8.4	52.3	50.0	33.6	16.4	4.6
F_0 of La Mé*	*Tenera*	9.2	56.6	61.6	25.3	13.1	6.0

* Mean composition of original 'La Mé' F_0 open-pollinated material from Bingerville for comparison.

above there were a substantial number of *dura* with shells of between 1 and 2 mm thickness. Seed from this prospection was planted in Nigeria and Malaysia.

It will be noticed that up to the time this prospection was started emphasis had been mainly on prospection for good quality *tenera* fruit. However, unlike the prospection in the Ivory Coast, where selections were subsequently made only from the Yocoboué grove which showed the highest fruit quality characters, the collectors in Nigeria did not use their data to 'select' palms. The objective was to obtain a more or less representative sample of the Nigerian palm population and thus to obtain as much variability as possible in all characters. The prospectors were concerned with capturing characters other than fruit quality (in which most progress has already been made) and with the threat of genetic erosion through the clearing of large areas of palm groves for development and food cropping; they therefore put genetic conservation before the immediate prospects of obtaining further fruit quality improvement. With changing selection criteria, e.g. the new emphasis on bunch index and on progeny and density relationships, there is little doubt that this is, in the long run, the more productive approach. It was pointed out that the performance of individual palms in the widely varying conditions of palm groves cannot, except in the case of fruit characters, reflect their genetic potential, and that a random or semi-random method should therefore be adopted.[124, 125]

The world-wide importance of this prospection work should be realized. With the universal adoption of the *dura* x *pisifera* cross, the countries hitherto cultivating the Deli palm have become as dependent on the procuring of high quality African *tenera* with good bunch and other qualities as any other countries in the world.

Prospection and bunch analysis started in the Brazilian groves south of Salvador show that high quality *tenera* and *dura* palms also exist there; these may well be adapted to the environment of coastal Bahia.[79]

Prospection for good Deli *dura* foundation stock was undertaken in Sumatra and Malaysia by rather different methods. For instance in Sumatra 24,500 palms were yield recorded on one group of estates and forty of them were crossed and selfed to provide the foundation stock of 2,760 palms for selection and breeding and seed production. In Malaysia, large numbers were not recorded in the first instance but apparently good producers were noted in the fields over thousands of hectares; these were yield-recorded and analysed prior to establishing progeny as foundation stock in genetic blocks.[9] No very recent large-scale prospections have been undertaken; the inbred nature of the present Deli population has been repeatedly stressed and preliminary evidence of its lack of genetic variance has been obtained in studies in Malaysia.[68, 69]

However, it should be noted that this evidence relates to bunch yield, number and weight, and further improvement of the Deli *dura* in respect of fruit composition cannot be ruled out. Some marked differences in composition of *tenera* progeny of Deli mother palms have been noted (see p. 292).

With the predominant planting of Deli *dura* x *pisifera* material the maintenance of fields of Deli mother palms of high potential becomes of paramount importance and all oil palm growing countries will need to give attention to the adequacy of their stocks. In most countries only publicly-owned research stations will be in a position to maintain and replant large fields of this material.

Prospection for *Elaeis oleifera* material is discussed on a later page (p. 301).

Selection standards

Selection standards adopted in the various centres of selection have been largely arbitrary and have been determined by the demand for seed. Selection standards for further breeding have often been determined by the amount of land available for progeny trials. This is perhaps inevitable in a long-term crop with a low stand per hectare. Another determinant is, of course, the standard of the material available for prospection and selection. For instance, the local material in the Ivory Coast was comparatively poor and selection in that country was to a great extent from material imported from the Far East and from Zaire via Sibiti in Congo (Brazzaville).

The demand for seed in most countries has been such that much of the seed supply has perforce been from palms of good yield and analysis and perhaps of good parentage or within good progenies, but not of proved progeny performance. An extra demand for seed, besides lowering standards, tends to raise the amount of seed provided from parents which have no recorded progeny. This practice need not be condemned since, as already shown, the majority of characters selected have a reasonable heritability. In Nigeria, a large additional demand for planting material led to the grading of seed on the basis of quality standards, progeny standards and 'progeny testing'. First quality (later called Special) N.I.F.O.R. *dura* x *pisifera* seed, for instance, was provided from *dura* parents belonging to legitimate progenies of known good performance crossed with *pisifera* pollen parents which had been progeny tested to the extent that they had *already* produced good *tenera* palms in crosses with *dura* parents. The *dura* standards were laid down as the following minima:

Fruit to bunch	65%
Shell to fruit	35%
Mesocarp + kernel to bunch	45%
Minimum bunch yield	150 lb (61 kg) per annum

Minimum quality standards for the *dura* when of Deli origin were 3 per cent more exacting, e.g. F/B 68 per cent. Other organizations have not published their standards, perhaps because of the difficulties already described. Bunch yield standards, or yield claims for progeny, are valueless except in the context of the selection field in the country of origin, since yield potential varies greatly both within and between countries. A given Deli x *pisifera* progeny might provide 200 kg bunches per palm per annum on alluvial clay in Malaysia, 120 kg in the Ivory Coast and 80 kg on the Acid Sands in Nigeria. Bunch yield standards can therefore only be adjusted to local conditions and will be subject both to the pressure of selection and the conscious weighing of the relative value of bunch composition and bunch yield. Composition, since the advent of the *tenera*, has tended to hold the field, but much more attention is now being paid, particularly in low sex ratio countries, to bunch yield. A method of evaluating the yield of a palm one or more of whose neighbours are missing is given in Chapter 9 (p. 417).

The term 'progeny-tested' is often used without definition or standardization (see p. 278). The standards of these tests must for some time to come be somewhat arbitrary and subject to the same kind of variations and pressures as described above for selection standards.

Bunch production or bunch quality?

It has been pointed out that the move from *dura* to *tenera* planting gave substantial increases in both oil production and economic returns, that quality improvement within the *tenera* is likely to be of a lesser order, and that therefore more benefit is to be gained from improved bunch yield. Van der Vossen showed that if each component of oil yield were increased separately by its standard error then the returns from increases in either number or weight of bunches would be five to thirteen times greater than the returns from increases in fruit to bunch, mesocarp, mesocarp oil or kernel oil.[62] His computations were based on calculated genotypic mean values of *tenera* selected in Nigeria and recorded over the first 3 years of bearing; moreover the material showed high heritabilities for fruit components but low heritability for bunch yield. Thus it was doubtful if representative economic conclusions could be drawn from the data, and it is probably more realistic to enquire what improvement in returns can be expected in adult fields from different increases or decreases in bunch yield, shell percentage to fruit and oil percentage to mesocarp. Progeny trials have indicated substantial bunch yield gains where individuals from separate populations have been crossed, but differences between progenies within such crosses are not of the same order. The uncertainties of bunch production breeding have been stressed by Hardon[80] but further progress can be expected through widening the field of selection, maintaining genetically diverse populations and making use of the heritability of vegetative characters.

Table 5.20 shows the returns to be obtained in an area with a low level of yield such as Nigeria by improving bunch yield by 33 per cent and other factors from relatively low levels to possible high levels. Increasing yields by similar percentages in high yield areas such as Malaysia would of course have the same proportionate effect on returns. It will be seen that

Table 5.20 *Returns to be obtained by improving bunch yield* and bunch quality*

Tons/hectare palm oil and kernels, and economic units, obtained by improving each of the factors bunch yield, F/B, shell to fruit, oil/mesocarp, while keeping the other factors constant.

	Bunch yield (t/ha)	Fruit to bunch (%)	Shell (%)	Mesocarp (%)	Kernel (%)		Oil to mesocarp (%)
	12 M 2.52 K 0.63 Eu 30.2	58 M 2.52 K 0.63 Eu 30.2	14.0	77.0	9	M 2.52 K 0.63 Eu 30.2	47 M 2.52 K 0.63 Eu 30.2
	13 M 2.72 K 0.68 Eu 32.6	61 M 2.65 K 0.66 Eu 31.8	11.5	80.5	8	M 2.63 K 0.56 Eu 30.8	50 M 2.68 K 0.63 Eu 31.8
	14 M 2.94 K 0.73 Eu 35.2	64 M 2.78 K 0.69 Eu 33.3	9.0	84.0	7	M 2.75 K 0.49 Eu 31.4	53 M 2.84 K 0.63 Eu 33.4
	15 M 3.15 K 0.78 Eu 37.7	67 M 2.91 K 0.72 Eu 34.9	6.5	87.5	6	M 2.86 K 0.42 Eu 32.0	56 M 3.00 K 0.63 Eu 35.0
	16 M 3.36 K 0.84 Eu 40.3	70 M 3.04 K 0.76 Eu 36.5	4.0	91.0	5	M 2.98 K 0.35 Eu 32.6	59 M 3.16 K 0.63 Eu 36.6
Eu per cent increase lowest to highest	33	21				8	21
Eu per cent increase by raising three or two bunch quality factors lowest to highest					61	33	

* Bunch yield levels used correspond to those in more seasonal parts of West Africa, e.g. Nigeria.

M = Mesocarp oil. K = Kernel. Eu = Economic units based on 10 Eu per ton mesocarp oil, 8 Eu per ton kernels.

increasing bunch production without improving quality is more rewarding than improving any single bunch quality factor. However, if several quality factors can be improved simultaneously then the effect of the latter will be as great or greater than that of improving bunch yield. Nevertheless, bunch quality improvement has severe limits and there are progenies in existence near those limits; but the consistent production of large quantities of only these high quality bunches has not yet been achieved in practice by seed producers. Clearly the breeder still has to give attention, in each situation, to both bunch yield and bunch quality breeding and to try to provide sufficient high bunch quality progenies to maximize oil plus kernel production to bunch.

Oil or kernels?

The kernel has been almost entirely neglected in the breeding of the oil palm since the days of Beirnaert, who had the wisdom to record it as one of his breeding criteria.[11] Broekmans made a positive plea for the kernel in the introduction of his 'C' factors: C1 was the content of kernel plus mesocarp to bunch, i.e. $(1 - \text{shell/fruit}) \times \text{F/B}$.[67] Though often appearing in tables of fruit composition, kernel production has seldom seemed to have been a positive aim. In fact, however, at prices which have been ruling since the war, kernels make a contribution to the economy of a plantation of no less importance than that of oil though this fact is masked by the higher price per ton of the latter product.

The ordinary producer will contend that he wants the maximum oil to bunch as palm oil has a high price, let us say $500 per ton. He does not want any of this replaced by kernels at, say, $400 per ton. Assuming F/B, shell to fruit and oil to mesocarp remain constant, an increase of 3.25 in the kernel to bunch percentage (5 per cent kernel to fruit) would lead to the following theoretical changes in gross return per 100 tons bunches:

	Tons per 100 tons bunches	
	A (High oil)	*B (High kernel)*
Fruit to bunch (F/B)	65	65
Mesocarp/bunch (mesocarp/fruit %)	52 (80%)	48.75 (75%)
Shell/bunch (shell/fruit %)	6.5 (10%)	6.5 (10%)
Kernel/bunch (kernel/fruit %)	6.5 (10%)	9.75 (15%)
Oil to bunch	26	24.375
Gross return:	($)	($)
Oil @ $500	13,000	12,188
Kernel @ $400	2,600	3,900
Total	15,600	16,088

In practice, kernel extraction efficiency being usually a little higher than oil extraction efficiency, the difference in gross return in favour of B will be rather greater than the $488 shown. The two analyses are quite possible, indeed usual, and at the 5 : 4 price ratio the high-kernel type B is more profitable. The oil palm breeder is then presented with a difficult problem — the parameters mesocarp to bunch, oil to bunch, mesocarp + kernel to bunch (C1), even oil plus kernel to bunch, are not fully adequate. An increase in oil percentage must mean a similar increase in the non-oily half of the mesocarp and, with shell percentage unchanged, must be accompanied by a reduction of kernel by about 2 x oil increase, thus obliterating the price advantage.

A number of general conclusions may be drawn.

1. If high-mesocarp fruit is being bred, then the factor oil to mesocarp becomes of greatest importance; an increase from 50 to 55 per cent will appreciably reduce the advantage kernel has over mesocarp. Conversely oil percentages of less than 50 per cent are disadvantageous and fruit with large kernels would in this case be preferable. This brings out the importance of oil to mesocarp determinations which should not be masked in oil to bunch data.

2. It is probably not possible to reduce shell percentage to as low a figure in high-kernel *tenera* as in high-mesocarp *tenera*; therefore the highest mesocarp plus kernel percentage will be obtained by high-mesocarp fruit. If this can be combined with oil to mesocarp well above 50 per cent this type of fruit is likely to give the highest gross return. This point has not yet been generally reached in breeding however. Two

high-quality palms in West Africa may be quoted in which the shell averaged 8 per cent of fruit:

Percentage:	Mesocarp	Shell	Kernel
Palm 14.437	85.8	8.0	6.2
Palm 38/0401	79.0	8.0	13.0

With similar F/B ratios and oil/mesocarp, 38/0401 would give the higher monetary return; in fact this palm had a much higher F/B ratio and a high oil content. To evaluate these two palms on mesocarp percentage or on palm oil to bunch would be gravely to underestimate the value of 38/0401. On the other hand a palm with a shell content as low as 3 to 4 per cent, e.g.

	Mesocarp	Shell	Kernel
Palm 4.493	90.6	3.8	5.6 per cent

will give a very high monetary return, and an alternative composition such as

	85.0	3.8	11.2

giving a higher gross return, would be impossible to find. Even the thinnest shells usually constitute 50 per cent of the nut weight and cases like palms 4.493 and 38/0401 where the shell is 40 per cent and 38 per cent of shell + kernel are rare. In the above hypothetical alternative to 4.493 the shell is 25 per cent of shell + kernel and this is unknown.

One must conclude therefore that a high kernel percentage gives the highest gross return unless shell percentages can be reduced to about 5 per cent of fruit or below; breeding should not be dominated by considerations of mesocarp or oil to bunch or oil yield per hectare as these measures do not indicate the full and accurate production yield of the oil palm.

Breeding — general considerations

The oil palm is a monoecious plant which is naturally cross-pollinated. It has alternating cycles of male and female inflorescences and it is rare for pollen to be produced on a palm when female flowers on the same palm are receptive. The most important other monoecious crop is maize, and probably more is known about the genetics of maize than of any other plant. It was early found in the breeding of maize that the most satisfactory production and quality could be attained by the concentration of desirable characters in inbred lines followed by the use of these lines to provide productive hybrids.[81]

Although oil palm breeders have been aware of the similarities between their plant and maize, breeding has not, for a variety of reasons, been dominated by the work on maize. It is known that Beirnaert in Zaire had for many years been considering the difficulties of following the classical methods of maize breeding.[16] He considered that high productivity being the consequence of the presence through Mendelian inheritance of a large number of characters, the concentration of all or some of these characters by the classical method of inbred lines would take a very long time and, though one might readily obtain homozygosity in one factor such as high number of bunches, it would be difficult to obtain by inbreeding many or any palms containing all the productivity factors. He proposed, therefore, to omit this stage and to cross within his basic material palms which had complementary characters, e.g. palms with productivity characters of the type AA Bb cc dd would be crossed with aa bb Cc DD. This and other complementary crosses would be carried out on a large scale to produce palms of the type Aa Bb Cc Dd. Later, selfing would be carried out to fix these characters in homozygous lines.[13] By 1938 Beirnaert, still unaware of the nature of the *tenera*, had at his disposal about 100,000 palms, largely *tenera*, and but for the problem of sterility and the fruit-form inheritance discovery shortly to be made, selection and breeding might have gone further forward on these lines than in fact it did.

In practice, the classical system of breeding naturally cross-pollinated crops has had to be modified in the case of the oil palm for reasons which will shortly be discussed; first, however, inbreeding itself must be considered.

Inbreeding

The comparisons between inbreeding and crossing have been few and one has to fall back on 'general experience' to provide some account. The main palm belt of Africa runs from the western end of the west coast through Nigeria, Cameroon, to southern Zaire and Angola. It is in the two extremities of this belt that a marked depressing effect of inbreeding has been noted. *Tenera* palms selected from local plantations and palm groves in the Kwango region of southern Zaire showed serious degeneration in their selfed progeny.[23] Germination was poor and seedlings both in the prenursery and nursery were irregular and deformed.

In the northern part of Zaire, at Yangambi, where a great concentration of *tenera* material was effected, selected *tenera* were both selfed and crossed among themselves. In general there was not much difference in the appearance of the young palms, but the yields of the selfed progenies were usually considerably lower than those of the crosses,[23] e.g.

Self or cross	Bunch yield per palm	
	5th year in field (kg)	6th year in field (kg)
1393 B selfed	24.5	31.4
2381 D selfed	39.1	63.6
1393 B x 2381 D	51.7	67.6
1393 D x 69 Mab	53.2	48.2
69 Mab x 2381 D	51.7	76.0

In recent years the depressing effect of inbreeding on yield has been shown in several breeding programmes. In Malaysia, Hardon[82] has estimated the inbreeding coefficient, Fx, for a range of Deli *dura* progenies on the assumption that breeding from the original Bogor palms to the establishment of basic material in Malaysia was equivalent to three half-sib matings, giving an inbreeding coefficient of 0.305. Significant negative correlations between the coefficient and bunch yield were obtained, though this was composed of a strong negative correlation for mean bunch weight and a weak positive correlation for number of bunches. However, other data provided by Hardon[83] suggested that yield depression would not be significant where inbreeding coefficients are low (below about 0.3), particularly in the Deli *dura*; and that these low levels could be tolerated where considerations of adaptability of the material to the environment and the characteristics of the individual parents become overriding.

In Nigeria early results of a breeding programme showed clearly the effect of inbreeding both in Deli *dura* and African material (Table 5.21). The narrower differences between the Deli *dura* selfs and crosses can be attributed to the fact that the Deli *dura* parents were already closely related.

In the Ivory Coast, experience has perhaps been intermediate between that of southern Zaire and that of Yangambi. While widespread deaths of selfed material have not occurred, the growth of selfs has been inferior to that of crosses. Deli selfs have also grown poorly and have suffered severely from Crown disease. With the Deli it is possible to postulate that many undesirable growth and yield characters have already been eliminated in the process of breeding from closely related material.

Data from the Ivory Coast also indicate a strong inbreeding yield depression with selfing within populations, though only inter-population crosses were available for comparison.[84] The differences between within-population selfs and between-population crosses were no greater than those shown for *dura* crosses and selfs in Table 5.21.

Table 5.21 *Cumulative bunch yields of selfs and crosses in Nigeria** (tons/hectare)

	Deli dura		Dura	
	No. of progenies	Cumulative yield	No. of progenies	Cumulative yield
1. Crosses	6	20.97	7	28.22
2. Selfs	4	18.08	3	14.40
3. Crosses	6	19.17	–	–
4. Selfs	5	16.70	–	–
5. Crosses	–	–	1	39.85
6. Selfs of parents of 5	–	–	2	23.13

* Data from N.I.F.O.R. Fifth Annual Report.

These facts lead to a conclusion which hardly needs stating: that inbred or selfed progenies should not be used for production. Their value in breeding programmes will be considered later.

Modern breeding systems

A number of factors has tended to divert oil palm breeding away from the maize analogy and to bring the ideas and exposition of the animal breeders, through D. S. Falconer,[58] to the fore. Oil palm breeding is breeding from a pair of individuals, each occupying many hundreds of square feet of space, for the production of large though manageable families of hybrid offspring. Though both parents have male and female flowers, the *dura* has taken on the role of·mother palm and the *pisifera* the role of father. It is through the distinctive qualities as well as the yield performance of the hybrid that parents may be chosen for continued breeding into further generations.

With outbreeding annual crops there is no possibility of using the parents after the performance of the progeny has been noted; with animals and with the oil palm the parents can be re-employed for many years and, indeed, an oil palm can be employed a great many more times than most animals. However, it must not be thought that seed production breeding can depend solely on 'proved' parents. The disadvantages of relying entirely on *progeny testing* have been outlined by Falconer as follows:

> In practice progeny testing suffers from the serious drawback of a much lengthened generation interval because the selection of the parents cannot be carried out until the off-spring have been measured. The evaluation of selection by progeny testing is apt to be rather confusing because of the inevitable overlapping of generations, and because of a possible ambiguity about which generation is being selected, the parents or the progeny. The progeny, whose mean is used to judge the parents, are ready to be used as parents just when the parents have been tested and await selection. Thus both the selected parents and their progeny are used concurrently as parents.[58]

This certainly occurs in oil palm breeding though there are ways of avoiding the difficulties and confusion.

Heterosis and the Deli palm

Considerable importance has been attached to the phenomenon of heterosis (hybrid vigour) displayed by crosses made from material whose origins are quite different. It has been shown that heterosis is the opposite of inbreeding depression and that if crosses are made at random between inbred lines developed from a certain population the mean value of any character in the progeny will be the same as in the original population.[58] Without selection, therefore, the crossing of lines in a large population cannot be expected to make any permanent change in the population mean. Heterosis can, it is believed, be explained in terms of quantitative inheritance. Where crossing of *selected* individuals from different populations has been successful this has been due to the selection itself, to the effects of the utilization of additive genetic variance of independent characters exhibited in general combining ability and to a lesser degree to specific combining ability. Falconer has pointed out that though the amount of heterosis shown in a particular cross depends partly on the differences of gene frequencies in the two populations concerned, wide crosses often fail to show a general heterosis though it may be exhibited in some characters. French workers have extolled the value of 'inter-origin' crosses but have confined their evidence to comparisons involving Deli *dura* x African *pisifera* or *tenera* crosses.[85] The performance of their 'inter-origin' crosses is attributed to the additive effect of a favourable combination of factors for weight per bunch and number of bunches from the parents.[43] Even if only mid-values are obtained in the progeny there may be an overall gain in yield, for example:

	Weight per bunch (kg)	Number of bunches per annum	Product (kg)
Deli	15.0	6.0	90.0
African	10.0	9.0	90.0
Cross	12.5	7.5	93.75

With young palms giving lighter bunches the increase may be greater, as exemplified by data from La Mé:[85]

(5th–8th year)	Wt. per bunch (kg)	No. of bunches per annum	Wt. of bunches per annum (kg)	Increase over product of mid-values (%)
Deli x Deli	14.2	6.0	84.3	
African x African (Ivory Coast)	7.3	14.4	100.7	
African x Deli	11.8 (10.8)	10.3 (10.2)	119.3 (110.2)	9.1

The figures in parentheses are the mid-values and their product and it will be noticed that in the case of weight per bunch the cross gives a value nearer to that of the Deli. This is a common experience and is exhibited to an even greater extent in the Deli cross with selected material of Zaire origin. Two years' data with young palms gave the following results:

	Deli	Zaire (via Sibiti)	Cross	Mid-value
Weight per bunch (kg)	14.5	7.4	12.4	11.0

Similar differences between imported Deli and local material have been found in other parts of Africa.[86] The low production of the Deli is due to a relatively low sex ratio not fully compensated for by its large bunches. There is little doubt that the introduction of African material into the Far East has been responsible, through its high sex ratio, for raising bunch production in the earlier years; but Hardon has pointed out that the original Deli palms provided a population peculiarly well suited to Malaysia and Sumatra and that outbreeding does not give the marked bunch yield increases in adult years which have been observed in Africa.[82] The introduction of the Deli into Africa on the other hand brings factors for low sex ratio to an area where environmental conditions already favour low sex ratios and where selection is largely geared to the discovery of high sex ratio palms. Fully satisfactory exploitation of the high bunch weight character of the Deli must therefore include a search for Delis giving a much higher than average sex ratio under African conditions. Alternatively African palms must be found giving weight per bunch similar to Delis, but with a higher sex ratio. The Deli palms in Africa have been somewhat restricted in number and parental origin and it is possible that more recent importations may be as important to Africa as the importation of further African material is to the Far East. A proportion of the Deli material imported from Malaysia is known to have come from selections involving high bunch weight but comparatively low sex ratio, and it is therefore not improbable that there are Delis selected on other lines which will be much more suitable for Africa than many of those already employed.

While French workers have confined their theory and practice to Deli x African population crosses, other breeders have adopted or advocated the use of crosses between populations or sub-populations which do not necessarily include the Deli. Van der Vossen considers that as members of sub-populations are likely to be genetically more diverse, intercrossing will raise the ceiling of physiological efficiency through the accumulation of genes of general physiological efficiency in the offspring. Using parents whose genotypic values for bunch number and weight have been calculated he showed that the actual and predicted bunch yields of their progenies were closely similar where no inbreeding had occurred, but that with inbreeding actual yields were well below predicted yields. Neither predictability nor high yield in the out crossings depended on a Deli palm being one of the parents. He concluded that it was of basic importance to maintain and increase genetically diverse sub-populations and to employ only interpopulation crosses in estimating genotypic values.[62]

Selection and breeding in practice

Africa

Practical breeding programmes in Africa have been much influenced by a recurrent selection programme started in Zaire and planted in 1956 and which was based on the material of the early selection programmes described at the beginning of this chapter. The basic material was six *tenera* selections which were selfed and crossed in all combinations, the intention being to breed from *dura* and *pisifera* in those selfs whose parents had shown good progeny in their crosses.[23, 75] The resulting *tenera* material, besides being used for production, would also be planted in *dura* x *pisifera* comparative trials. This is illustrated in Fig. 5.15A.

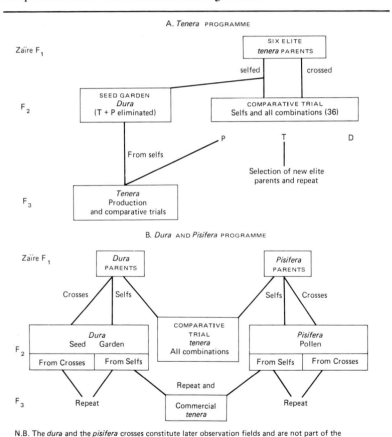

N.B. The *dura* and the *pisifera* crosses constitute later observation fields and are not part of the original programme

Fig. 5.15 A representation of modified recurrent and reciprocal selection in the I.N.E.A.C. (Zaire) breeding programmes.

It can at once be seen that, because the parents for production must be *dura* and *pisifera* to give 100 per cent *tenera*, the recurrent selection procedure is modified to allow for: (i) the recurrence to take place from the original comparative trial of selfs and crosses (because the oil palm is perennial there is no need to run a generation of inbreeding and then to inter-cross before choosing the next recurring generation); (ii) seed production for plantation use to take place in a special area; (iii) *pisifera* parents, which cannot exhibit their potential yield and quality characteristics themselves, to be chosen in the selfs in the comparative trial on the basis of the performance of their *tenera* parents as parents in the crosses; (iv) *dura* parents to be selected for production breeding mainly on the performance of their *tenera* sibs. For instance if high yielding and quality *tenera* are found in the progeny T4 x T6 then seed would be produced for distribution by crossing selected *dura* within the progeny of the T4 selfing in the special seed garden with *pisifera* pollen from the T6 selfing in the comparative trial. Two questions arise. Can the *dura* and the *pisifera* be judged as parents from the performance of their *tenera* parents and half-sibs? Is inbreeding necessary or desirable?

Another part of the I.N.E.A.C. programme involved 'purification' of the two lines by inbreeding to give homozygous *dura* and *pisifera* lines for crossing. This scheme has a number of advantages and drawbacks: (i) the comparative test was D x P and selected parents provided the actual *tenera* it was hoped to reproduce for as long a period as the parents survive; seed is limited by the number of parents successfully used in the comparative trials; (ii) seed provided from *dura* selfs and *pisifera* selfs in the next generation is not identical to that from the original tested parents; (iii) the *pisifera* line is of fertile *pisifera* unless the inbreeding is done in a *tenera* line; it is known that the thinner-shelled *tenera* palms come predominantly from infertile *pisifera*, and breeding within selfed *pisifera* lines is technically difficult; (iv) the original parents were chosen for their individual qualities as *dura* and *pisifera* palms and it is not known whether they will be the best producers of *tenera* progeny. The *dura* and *pisifera* programme is illustrated in Fig. 5.15B.

This programme is in reality a form of *reciprocal* (or *reciprocal recurrent*) selection used commonly in both animal and plant breeding. Two sources, in this case *dura* and *pisifera* palms, form the starting point and they are crossed in comparative trials. In animal selection the parents would await the results of these trials before being selected and the selected individuals would be crossed within their own line or source to produce another generation for testing. The cycle is then repeated. In annual crop breeding the selfed or 'within-source' seed can be held over and the selected portion planted in the third year after appraisal of the comparative trials; the progeny is then inter-crossed within the same source for further between-source crossings in the next generation.[81] In palm breeding all the parents used for crossing may be used in the inbreeding or 'source-breeding' programmes so that the progeny of selected palms as well as the selected palms themselves may be used in large programmes of seed production.

Other oil palm breeders in Africa have adopted forms of reciprocal selection though the source material has been predominantly *tenera* on the one side and *dura* on the other, and the methods of choice have varied. As already described, Gascon and de Berchoux[85] in the Ivory Coast adopted the Deli as almost the sole *dura* source though a few Deli x African *dura* were included in their programme. The other side of the programme consisted largely of *tenera*, but included some *pisifera*. These varied considerably in quality and were obtained from four separate geographical origins. Taking *tenera* (*pisifera*) as source A and Deli *dura* (pure or crossed) as source B, breeding and seed supply proceeded as follows. The planting consisted of eighty-five Deli x Deli and seventy-nine *tenera* x *tenera* selfings and crosses with 446 Deli x *tenera* or *pisifera* crosses made to compare the ability of the parents to provide productive progeny in crosses. Following appraisal of the results of this comparative trial the best parent *tenera* and, separately, the best parent Deli *dura* are crossed among themselves, thus providing a further selection generation. In commercial seed production the *pisifera* bred from source A are crossed with *dura* bred from source B and the progeny will thus come solely from palms whose parents have shown their performance in the comparative trials. Selection is repeated in the second selection generation through a further comparative trial and a repetition of the selection process. The system is illustrated in outline in Fig. 5.16.

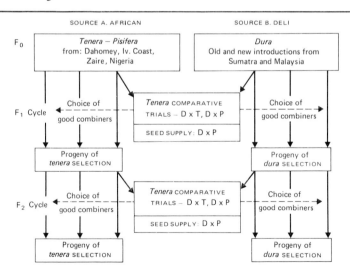

N.B. Although this programme is predominantly Deli x African *tenera* and *pisifera*, the A and B sources do contain a few progenies of mixed origin

Fig. 5.16 Outline of reciprocal selection system adopted at La Mé, Ivory Coast.

A second cycle of selection and breeding was later started which included not only selections from the first cycle but also the exploitation of introductions previously made and material of *Elaeis oleifera*. This cycle also includes crossings between the progenies of selfings made in the first cycle particularly those of Dabou Deli D10D and La Mé *tenera* L2T and top crosses of either of these palms to the progeny of the other.

Data from these programmes has been largely presented as means of hybrid populations. The Deli *dura* have been crossed with three groups of *tenera/pisifera* emanating respectively from La Mé, Sibiti/Yangambi and N.I.F.O.R., Nigeria. In general, bunch production has been higher in the La Mé and N.I.F.O.R. crosses than in the Sibiti/Yangambi crosses though the quality of some of the latter has been good and fruit size is large. Shell content in the N.I.F.O.R. crosses was low with percentages down to 5 to 8 in certain progenies. Oil to mesocarp was highest in the La Mé crosses but the N.I.F.O.R. crosses gave the highest overall palm oil per hectare.[84, 87]

A breeding programme adopted at the Nigerian Institute for Oil Palm Research, N.I.F.O.R., contained in its design the elements of both recurrent and reciprocal selection. It will be seen from Fig. 5.17 that *dura* seed palms can come from the *tenera* selfs as in the Zaire programme, but they can also come from the *dura* selfs and crosses as in the Ivory Coast programme. In the former case the testing of parents will have been through the *tenera* x *tenera* trials, in the latter through *dura* x *tenera* trials; but *tenera* for production can be provided through the combined information of these tests as suggested by Sparnaaij *et al.*,[14] e.g. if *tenera* palms T1, T4 and T6 gave outstanding progeny in

crosses T1 x T4
    and T1 x T6 in the *tenera* x *tenera* trials, and *dura* and *tenera* palms
        D2, D5 and T5 gave outstanding progeny in
crosses D2 x T5
        D5 x T5
        D2 x T6
        D5 x T6 in the *dura* x *tenera* trials,

then seed for distribution can be produced by *dura* x *pisifera* crosses from the following sources:

Dura sources	*Pisifera* sources
1. F_0 palms D2 and D5	F_1 palms in T5 and T6 selfed
2. F_1 palms in T1 selfed	F_1 palms in T4 and T6 selfed
3. F_1 palms in T4 and T6 selfed	F_1 palms in T1 selfed
4. F_1 palms in D2 selfed	F_1 palms in T5 and T6 selfed
5. F_1 palms in D5 selfed	F_1 palms in T5 and T6 selfed

In examples 2 and 3, the material comes solely from *tenera* breeding, but one F_0 parent concerned (T6) has also done well in the *dura* x *tenera*

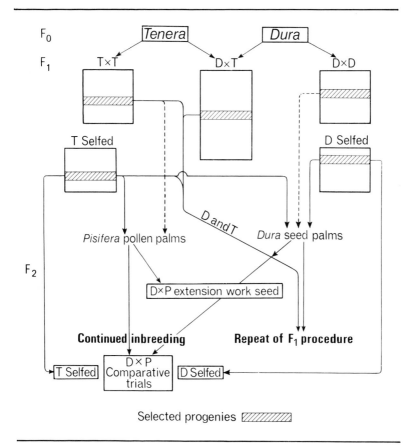

F_0 **Tenera** **Dura**

F_1 $T \times T$ $D \times T$ $D \times D$

T Selfed   D Selfed

F_2

Pisifera pollen palms   *D and T*   *Dura* seed palms

$D \times P$ extension work seed

Continued inbreeding   **Repeat of F_1 procedure**

$D \times P$
T Selfed | Comparative | D Selfed
trials

Selected progenies

Fig. 5.17 The N.I.F.O.R. breeding programme.

trials. In examples 1, 4 and 5 the parent material has been selected from the results of both the *tenera* and *dura* breeding.

Sparnaaij *et al.*[14] also suggested that if inbred depression prevents or restricts selection in the selfings (as early results, e.g. Table 5.21, suggest may be the case) the desired *tenera* production palms may additionally be produced from parents within both appropriate crossings and selfings, e.g.

Dura source	*Pisifera* source
F_1 palms from D2 or D5 selfs	T5 x T6
F_1 palms from D2 x D5	T5 x T6
F_1 palms from D2 x D5	T5 or T6 selfs

Breeding from crosses may prove at least equal to breeding from selfs; Falconer[58] states that 'since the variation in general combining ability is attributable to additive variance in the population from which the lines are derived, selection should be effective without inbreeding'.

The Nigerian programme allowed for an inbreeding section (not to be confused with selfings in the repeat of the main F_1 procedure — see Fig. 5.17) with *dura* x *pisifera* comparative trials. *Dura* and *tenera* within the selfings would be again selfed, the *pisifera* in the comparative trials being the sibs of these *tenera*. This programme has not yet been undertaken.

The N.I.F.O.R. programme was designed to continue into further generations as in the I.N.E.A.C. and I.R.H.O. programmes, but unlike the latter the *dura* source is not confined to Deli palms. However, the latter were not crossed with African *dura* in the *dura* programme since it was thought best to determine separately the value of the best Delis available and to compare them with good Africa *dura*. All crosses in the trials, whether comparative or within fruit form, were either compensatory, i.e. a defect in one parent was compensated by excellence in the same character in the other parent, or 'excellence' crosses, where palms with similar outstanding characters are crossed to improve them still further.

In spite of its design, it has been stated[62] that in this programme no methods of reciprocal recurrent selection are applied, since the programme is basically one of selection for good genotypic value starting from base populations of *dura* and *tenera* palms, and the special 'assortative mating' between groups of palms[70] is used. Early data from the programme were used by Van der Vossen[62] for computing genetic values, but the actual methods of further breeding to be employed will follow an appraisal recently undertaken.

Many very high quality *tenera* palms with shell percentages below 7 have emerged and the mean *tenera* analysis of five *tenera* crosses planted in 1959 and one *tenera* self may be given as an example:[88]

	Fruit to bunch (%)	Mesocarp	Shell to fruit (%)	Kernel	Single fruit weight (g)
Five *tenera* crosses	69.4	81.8	9.8	8.8	7.2
2.3495 self	72.2	86.7	6.1	7.1	7.5

Material from Nigeria bred in the Ivory Coast has given *tenera* progenies containing palms with 83 to 90 per cent mesocarp and only 4 to 9 per cent shell.

Asia

The course of oil palm breeding in the Far East was severely checked by the war. Breeding by A.V.R.O.S. was continued under the new name R.I.S.P.A. (Research Institute of the Sumatra Plantation Companies Association). The breeding methods adopted were akin to those of Africa in that further breeding from Deli *dura* lines was carried on simultaneously with breeding from *tenera* lines. Test crosses of *dura* x *tenera* and *dura* x *pisifera* were also carried out or planned.[4]

The early selfings and F_2 sib crossings of SP540 have already been described (p. 200). F_1 *tenera* parents POL3468, 3258, 3520 and 3409 were outstanding in various characters or in their progeny, and an inbreeding programme was continued into the F_3 with these and other palms at Aek Pantjur. *Tenera* of other origins were handled in the same manner. In the course of this breeding, pure African *dura* progeny superior in some respects to selected Delis arose and programmes for the selection of these for the breeding of new generations of pure African *tenera* were put in hand. Plans were also made to breed from a few semi-fertile *pisifera* which had arisen in the later progeny of the *tenera* programmes. Although these programmes entailed the separate maintenance of Deli *dura* and African *tenera* lines, a programme of selfing Deli *tenera* was included.

In recent years Marihat Research Station has been set up to serve two groups of nationalized estates and a considerable breeding and seed production programme has been put in hand. Special efforts have been made to trace out the pedigree of the Deli and imported material bred during previous decades by commercial estate groups and R.I.S.P.A. Much use is being made of *pisifera* descended from *pisifera* EX5 of Bah Jambi which had given good progeny in the past. At Aik Pantjur R.I.S.P.A. has planted areas of selfed dumpy E.206 (from Malaysia) x SP540 or *pisifera* descended from SP540. The latter palm is still (1976) growing at Sungai Pantjur. The progenies, besides showing the dumpy character, are high yielding and have large fruit. Unfortunately, however, no data have been published on the yield performance and analysis of either the parents or progenies in these Sumatran programmes, but improvement is being sought firstly through breeding programmes in which the purity of the Deli and *tenera/pisifera* lines are maintained and secondly through the introduction of considerable quantities of diverse material from overseas to combine with the local material.[126]

In Malaysia there was no institution to carry on the breeding work started by Jagoe before the war and the areas available to the Department of Agriculture were not large enough for the carrying out of programmes of the size of those in Africa. Quantities of material, both in the form of pollen and seed, were however imported from Africa and Sumatra, and a Cooperative Breeding Programme was started with a number of plantation companies in 1956. This has been described by Haddon and Tong,[55] and consists of a number of small Main and Minor Breeding Lines. These include further breeding with the dumpy and other Elmina Delis, and selection within Deli *tenera* progenies for both *tenera* and *pisifera*.

Commercial breeding programmes have been carried forward with two aims in view: (i) the breeding of the highest yielding and quality material for the companies themselves, and (ii) the sale of seed to an expanding industry. The early histories of two commercial programmes have already been described (p. 208). In the first of these programmes Deli breeding was continued by back-crosses of the Serdang x Sumatran material on to outstanding parent palms of Sumatran origin.[9] The Deli *dura* line is shown below.

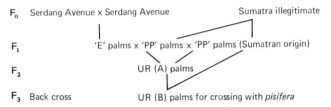

F_0 Serdang Avenue x Serdang Avenue   Sumatra illegitimate

F_1 'E' palms x 'PP' palms x 'PP' palms (Sumatran origin)

F_2 UR (A) palms

F_3 Back cross   UR (B) palms for crossing with *pisifera*

Choice of seed palms was soundly based through both progeny and individual selection, i.e. palms had to be of progenies giving a yield higher than the mean of progenies in any block, and they themselves had to yield more highly than the mean of the progeny to which they belonged. Particular attention was paid to fruit-to-bunch and mesocarp-to-fruit ratios, but shell percentage was not mentioned as being considered. The full productive value (oil + kernel) of the palms does not therefore appear to have been known, and this may account for some of the characteristics of the fruit eventually produced. At the UR(B) stage, selection was based on the same criteria but individuals had to exceed, in bunch yield, the mean of their plot, as well as the progeny to which they belonged, and the oil-to-bunch ratio had to exceed 20 per cent. Kernel and shell percentages were again not mentioned.

The *tenera* lines, which were in Yangambi, Zaire, were in a 1935 planting of some of Beirnaert's F_1 material (see p. 211); they were received in Malaysia as *pisifera* pollen. The *tenera* stock was never tested in Malaysia and the pollen was used to provide an F_1 Deli *tenera* generation by crossing in the F_2, UR(A), generation. These Deli *tenera* were subjected to selection on the basis of bunch yield, as before, and on fruit to bunch, mesocarp to fruit, oil to mesocarp and oil to bunch. At this stage a decision was taken, on grounds already described, to induce the production of fertile, quarter-Deli *pisifera* by crossing selected *tenera* with fertile *pisifera* from Serdang (Malaysia), Nigeria and Zaire. This was successful and palms in this F_2 which were not fertile *pisifera* were cut down. This *pisifera* material was therefore selected only on the fact of fertility and the performance of one parent. The F_2 fertile *pisifera* were for some time used for commercial seed production by crossing with the Deli F_3.

Breeding in Malaysia was much expanded in the 1960s by a large exchange programme organized by the Department of Agriculture in Sabah and by the setting up of a breeding and genetics centre, The Oil Palm Genetics Laboratory (O.P.G.L.), by four of the principal plantation companies.

The Sabah programme allowed for the collection and exchange of *dura* and *tenera* crosses and selfs from many sources in Malaysia and Africa and for linked Deli *dura* x *tenera* comparative trial crosses.[89, 90] Most of the latter were produced by importing *tenera* pollen from Africa. The programme included 192 progenies but owing to difficulties of transport only 152 were provided up to the end of 1969 when the programme was virtually completed. The material was distributed to all participants and as far as Malaysia is concerned led to an increase in the availability of good Deli progenies and to a much-needed injection of *tenera/pisifera* sources. Consolidated results of this programme have not been published, but some examples of yields and analysis from one of the centres of planting are shown in Table 5.22 and are probably typical and indicate that high yields and good bunch composition can be expected from selected Deli palms crossed with *pisifera* bred from the West African *tenera* introduced.

The Oil Palm Genetic Laboratory was established to give a firmer genetic basis for selection and breeding and to assist and interpret the selection and seed production programmes of the contributing companies. The O.P.G.L. was taken over by the Malaysian Agricultural Research and Development Institute (M.A.R.D.I.) in 1973. The work undertaken,[91, 92] apart from the improvement of techniques, has consisted of:

1. The study of existing breeding populations.
2. The introduction of new material.
3. The establishment of progeny trials and selection programmes for participating companies, including the exploration of experimental designs suited to oil palm breeding work.
4. A study of all factors, including growth parameters, which may be related to bunch yield or oil production.
5. Investigation of hybridization between *E. guineensis* and *E. oleifera*.

Some of the work under 1 and 2 above has already been mentioned in connection with heritability and prospection. With regard to selection and breeding methods, a system of family and individual mass selection in *dura* x *dura* and *tenera* x *tenera* crosses has been advocated[80] together with the enlargement of the genetic variability of both populations by introducing unrelated material and avoiding inbreeding. Details of the scheme of breeding and the method of parental selection for seed production have not been published. *Dura* x *tenera* test crosses are not employed and it is stated that *pisifera* selection is, in the first instance, based on the performance of the *dura* and *tenera* within the same progeny.[93] Progeny testing of individual *pisifera* is carried out in a number of polyallele experiments; however it was pointed out by Hardon *et al.*[80] that as this entails the recording of another generation the results should be compared with the progress obtained from a second round of family selection in *tenera* x *tenera* crosses.

An innovation in the Malaysian breeding work has been the introduction of the growth analysis parameters, particularly vegetative dry matter (VDM) and bunch index (BI), in an attempt to improve bunch yield per hectare. The heritability of these characters has been referred to on p. 263.

Table 5.22 *Some yield* and analysis data from the Sabab breeding programme material (West Malaysia, Briah series. O.P.R.S., Banting)*

Crosses	Mean of six Deli crosses	D × T	D × T	D × T	D × T	D × T	D × T	T × T	T × T
Female parent	Deli	D2–32/112	D2–1/2	D2–1/2	D4–93/76	D4–93/76	D3–57/77	LC(T)3	UR(T)3
Male parent†	Deli	LC(T)3	LN(T)2	W(T)4	W(T)4	W(T)1	UR(T)3	LC(T)2	UR(T)3 (Self)
Bunch yield‡ per palm/an. (kg)	156	176	164	200	189	212	185	182	140
No. of bunches per palm/an.	17.5	20.4	21.5	24.7	23.3	17.4	19.1	21.4	21.0
Mean bunch weight (kg)	9.2	8.6	7.6	8.1	8.1	12.2	9.7	8.5	6.7
Fruit to bunch (%)	63.7	59.6	59.1	60.2	61.0	66.3	58.4	59.5	56.1
Mesocarp to fruit (%)	63.9	82.9	78.3	82.6	85.6	83.6	83.8	84.1	76.5
Shell to fruit (%)	28.6	10.0	14.2	10.9	9.4	10.0	8.3	9.6	14.3
Kernel to fruit (%)	7.4	7.0	7.4	6.4	5.0	6.4	7.9	6.2	9.2
Oil to mesocarp (%)	48.7	47.6	49.4	49.3	48.5	48.0	48.7	48.9	48.2

* Yield data for 3 years of bearing (years 4 to 7, year 0 = planting year).

† Origin of *tenera*: LC – Cameroon; LN – Nigeria; W – Nigeria (N.I.F.O.R.); UR – Malaysia. D = *dura* (T) = *tenera*.

‡ Bunch yield of *tenera* palms only in both D × T and T × T crosses.

Hardon *et al.*[80] have presented some data which suggest that palms selected for high yield also have a higher VDM and leaf area and may therefore be more competitive and require a reduced density. Palms with a high BI had a leaf area equal to the mean and, of course, a low VDM. This suggests tolerance of competition and should allow planting at higher densities leading to a higher yield per unit area.

It appears not unlikely that much past breeding work has favoured palms with a high VDM and leaf area and that this may account for the declining yields which have sometimes been recorded, particularly on fertile volcanic soils, when the palms have closed in and are in competition with each other. Such competition results in a lowering of net assimilation rate and this is reflected in a lowering of the bunch index (see p. 162). Such a situation may be met by reducing density, but a method more likely to give optimum yields would be to breed palms which maintain a high BI at a high density.

Details of other commercial seed breeding in the Far East have not been published. In 1963 one company was providing seed from its own selected Deli *dura* and from fertile *pisifera* of Zaire and Ivory Coast origin. Tests of the *dura* x fertile *pisifera* crossings were being carried out simultaneously and these included tests of the *pisifera* as such owing to 'some uncertainty in botanical determination of the *pisifera*'.[10]

The results obtained in breeding in Malaysia have shown very clearly the importance of selecting good *pisifera* (Plate 21). *Pisifera* of Zaire origin and descending from the Sumatran *tenera* SP540 and *pisifera* POL3184

Pl. 21 *Tenera* fruit from Deli *dura* x *pisifera* crosses. The effect of different *pisifera* on shell thickness in the progeny can be clearly seen.

Table 5.23 *Some yield and analysis data from Malaysian breeding trials (O.P.R.S., Banting)* (yields — mean of years 4 to 8)

	1	2	3	4	5	6	7
Crosses	D x P	D x P	D x P	D x P	D x P	D x P	T x T
Female parent	*Deli*	*Deli*	*Deli*	*Deli*	*Deli*	*Deli*	*Sum/Z*
Male parent*	*Sum/Z*	*Sum/Z*	*Sum/Z*	*Sum/Z*	*Serd.*	*Serd.*	*Sum/Z*
No. of progenies	6	5	7	4	10	5	20
No. of palms	503	419	588	336	839	418	392
Bunch yield† per palm/an. (kg)	224	204	197	203	186	179	135
No. of bunches per palm/an.	23.6	20.6	21.3	22.4	18.3	17.6	20.0
Mean bunch weight (kg)	9.5	9.9	9.3	9.2	10.2	10.2	6.8
Fr. to bunch (%)‡	58.5	59.0	57.7	58.8	54.3	57.1	58.6
Mesocarp/fr. (%)	80.0	78.8	83.7	81.6	76.6	73.2	84.4
Shell/fr. (%)	10.0	10.7	8.2	8.1	12.7	15.5	7.3
Kernel/fr. (%)	10.0	10.5	8.1	10.2	10.7	11.3	8.2
Oil/mesocarp (%)	49.3	50.7	50.1	49.8	49.5	48.8	48.7

* Sum/Z — four different *pisifera* of Zaire origin through Sumatran palms SP540 and POL3184. Serd. = *pisifera* descended from Serdang palms. Twelve *tenera* were used in the T x T crosses.

† Bunch yield for *tenera* only in T x T crosses.

‡ Low fruit to bunch, possibly due to poor pollination.

have been particularly successful in giving high yield, quality and precocity of bearing. Some data from *tenera* derived from Deli x Sumatran Zaire *pisifera* are given in Table 5.23 together with data from other D x P and *tenera* crosses.[94] With regard to the latter, the effect of inbreeding on yield levels will be noted, but the possibilities of further improving fruit quality are also clear. It should also be noted that the bunch yield of the first group of progenies represents a per hectare yield of over 32 tons.

Latin America

The great majority of the early plantations in Latin America and many of the plantings undertaken from 1959 owe their origin to seed introduced in 1926 by the United Fruit Company and planted in the Experimental Garden at Lancetilla near Tela in Honduras.[95, 96]

The collection consisted of African material already in existence at Bobos in Guatemala and seed collected by Dr Fairchild, together with Deli material collected from various sources in Java, Sumatra and Malaysia. The original collection also included out-pollinated *tenera* imported from Africa. When it was seen that the Java and Sumatra material was more productive than that from Africa or Malaysia the latter material was cut out, though not before some Malaysian illegitimate progenies had been established at San Alejo and seed from Java and Sumatra palms had been issued to Quepos in Costa Rica and Patuca in Colombia. Thus the predominantly Deli material which was passed on to other plantations in America may have carried some African *dura* characters, though the younger plantings of the United Fruit Company, and the first plantation in Ecuador, were planted with Java and Sumatra material obtained long after the African palms had been cut down.

Until the late 1950s oil palm plantings in America were dominated by material from these 1926 introductions. No systematic selection and breeding was undertaken and such analytical data as exist suggest that this Deli material had a rather higher kernel-to-mesocarp ratio than is normally found.

Some other small introductions into America have occurred. Of particular importance are the progeny of seed brought from Zaire by Mr Florentin Claes in 1931 and imported by Dr M. J. Rivero from Eala in 1936 and planted at Palmira, near Cali in Colombia. From these introductions Dr Patiño made *tenera* selections, and a second generation which included some legitimate progenies was established at Calima.[97, 98] The *tenera* are of the Yangambi type. *Pisifera* palms also exist in this second generation of Zaire material and the whole population is now being used in selection and breeding programmes. Work being carried out by the Instituto Colombiano Agropecuario comprises selection of *dura, tenera* and *pisifera* at Calima, selection of Deli *dura*, from Honduras, at Aracataca, and includes *dura* x *tenera* and *dura* x *pisifera* crossings for production. A new breeding programme has been started at the El Mira station at Tumaco.

In Venezuela the La Esperanza plantation has an interesting collection of planting material owing its origin to Yangambi, Zaire, to Sumatra and to West Africa. Exceptionally fertile and high production *pisifera* palms exist at La Esperanza through breeding from a thick-shelled *tenera*. In Ecuador a breeding programme was started with imported material and seed is being produced. Nigerian *pisifera* pollen was first employed, but locally-bred *pisifera* have now been selected.

The recent spate of planting which started around 1960 has been based either on the Honduras material obtained direct or secondhand, or on importations from the Ivory Coast, Cameroon, Malaysia and Surinam. These importations have been largely *dura* x *pisifera* crosses, and they show considerable promise under the conditions obtaining.

The plantations in Costa Rica and in Honduras contain additional interesting planting material, namely African *dura* x *pisifera* material from Zaire and *tenera* progenies from Nigeria. Both these lots of material may be useful for future breeding work in America.

Very recently, owing to the incidence of Bud Rot disease, commercial planting of the *E. guineensis* x *E. oleifera* hybrid has started in Colombia, and a selection programme of *E. oleifera* with linked *E. guineensis* x *E. oleifera* crosses has been started in the areas around Monteria. In Costa

Rica a collection of *E. oleifera* progenies from more than thirty localities in South and Central America has been established.

Selection of *pisifera* parents

The selection of a *pisifera* parent is regulated by the performance of its parent or sib *tenera*, or by its own performance as a parent of *tenera* palms. But there are a number of special considerations in the selection and usage of a *pisifera*.

Firstly the *pisifera* is a pollen parent and provided it yields a sufficiency of male inflorescences it can be used for the pollination of many female parents over a period of time. The female *dura* parent is more limited in the production of progeny, producing, say, six bunches of 1,000 seeds per year in Africa and double that amount in the Far East, i.e. enough, with normal germination and nursery losses, for about 4 hectares per bunch or 25 to 50 hectares per year's production. By diluting the pollen with talc it is possible to obtain a very much larger number of progeny from the male parent. Fifty to one hundred grammes of pollen may be collected from a single inflorescence; but 0.5 g or less may be sufficient for pollinating one female inflorescence, so 100 bunches giving 100,000 seeds for the establishment of, say, 400 hectares can be produced from one male inflorescence. With the most recent techniques[99] it is believed that as little as 0.05 g pollen may be required for a single pollination and the area covered by the progeny of one inflorescence of one pollen parent could be increased to 4,000 hectares.

Some infertile *pisifera* palms are shy of producing male inflorescences and indeed it is characteristic of the *pisifera* to produce large numbers of abortive female bunches. There are ways of inducing these palms to produce male inflorescences however, and the relative paucity of the latter will not eliminate the potential productive predominance of the male parent. For reproductive potential one male inflorescence equals at least 100 female inflorescences.

These facts tend to give greater importance to the selection of the male than of the female parent. However, the potential of the *pisifera* is harder to gauge than that of the *dura*; whereas the potential of the latter may be estimated from its own bunch production and analysis as well as those of its *tenera* sibs, the yield of the *pisifera* is more or less strongly affected by abortion, and fruit analysis gives little or no assistance. On the other hand the very fact that a *pisifera* can provide its genes (through pollen) in so much greater quantity than can a *dura* means that much more use can be made of a *pisifera* already 'proved' by its progeny than can be made of a *dura* similarly proved. Thus in breeding-cum-production programmes, *pisifera* palms which have themselves produced outstanding *tenera* progenies will still be being used when *dura* seed trees of a later generation are in production. The former may not be such potentially good parents as some of the *pisifera* of the later generation, but their worth will be more fully known. Nevertheless it is questionable whether it will really be better to use these *pisifera* rather than those obtained in a later generation of outstanding *tenera* x *tenera* progenies.[80]

As partially fertile *pisifera* of high bunch production exist, there has naturally been a desire to measure their production and make use of them in breeding. Several reasons for employing such fertile *pisifera* have been put forward apart from the ability to measure their yield which will be referred to later. Firstly, with the usual almost-sterile *pisifera* it is necessary to obtain a few properly developed fruit in order to identify them with certainty as *pisifera*; infertile *tenera* and *dura* have been known and it is of course essential that these should be recognized and not confused with *pisifera*. Secondly, oil to mesocarp contents can be measured in fertile *pisifera* and this perhaps provides the strongest argument for their use; it has not however been shown that high oil to mesocarp cannot as easily be transmitted through infertile *pisifera* from *tenera* which are outstanding in this factor.

The main arguments against the use of fertile *pisifera* are that they are limited in number, that their own bunch yield is no real indication of their power to transmit factors for bunch yield, and that they are normally related to thick-shelled *tenera* and carry heritable factors for shell thickness. The latter postulate arose from the discovery of very fertile *pisifera* among the progeny of supposed thin-shelled *dura* which later were identified as thick-shelled *tenera*. Later, in Nigeria, a negative correlation was found between the occurrence of fertile *pisifera* and the thinness of shell of their *tenera* sibs.[14] Two other programmes confirmed this result. A set of thick-shelled *tenera* crosses, besides giving a good example of the high heritability of shell thickness, provided a large number of fertile *pisifera*. In the other programme different types of *tenera* crossed with fertile *pisifera* gave the following results:[65]

Cross	*Fertile* pisifera as a percentage of bearing palms
Deli *tenera* (Deli x fertile P) x fertile *pisifera*	52
Thick-shelled *tenera* x fertile *pisifera*	36
Thin-shelled *tenera* x fertile *pisifera*	23

Doubts as to the usefulness of yield data from fertile *pisifera* are engendered by the fact that fertility varies so greatly. *Dura* palms have on average a lower sex ratio than *tenera* palms but a higher fruit-to-bunch ratio. The sex ratio of *pisifera* palms is very high because they are either not setting their bunches or setting bunches with low fruit-to-bunch ratios. Sterility is, in fact, a relative term in the *pisifera*.[48] In the Zaire prospection of estates derived largely from *tenera* parents about two-thirds of nearly 900 *pisifera* identified aborted nearly all their fruit.[74] Of 147 bunches cut from palms setting some fruit the distribution of those with over 10 per cent F/B was as follows:

Number of bunches	Per cent fruit to bunch
14	10—20
43	21—30
30	31—40
15	41—50
7	51+

Further analysis of the 7 palms giving bunches with 51 or more per cent F/B showed that, though there were two palms with bunches of over 60 per cent F/B, bunches from the same palm did not give regular ratios, for example

	Palm	
	439	409
Fruit-to-bunch ratios:	50.4%	52.5%
	46.9%	47.9%
		40.0%
		36.5%

Prospection in estate fields in southern Zaire revealed a remarkably fertile *pisifera*,[100] numbered P.21. This was the illegitimate progeny of a grove *tenera*, probably of poor quality. Bunch yield was 127 kg per year, F/B 66.7 per cent and the fruit had a large kernel constituting 10.7 per cent of the fruit; 38.7 per cent germination was obtained from the kernels. Such *pisifera* are very rare and in general it cannot be expected that their bunch yield will reflect the yield of their *tenera* progeny.

Very recently a special plea has been made for *pisifera* x *pisifera* production breeding in Malaysia.[101] Two Serdang fertile *pisifera*, S112 and S29/36, were crossed. Germination was satisfactory and 377 seedlings were successfully brought to bearing. Unfortunately, although the bunch production was said to be prolific, no overall yield records were provided. Analysis of two palms showed F/B ratios of 68.7 and 71.8 per cent and kernel to fruit of 18.7 and 15.2 per cent. Oil to mesocarp was, at 36 to 39 per cent, lower than that of adjacent *tenera* palms of the same age. In view of the positive correlation between kernel and shell (see p. 260) it is not unexpected that fertile *pisifera*, derived as these are from thick-shelled *tenera*, will have large kernels; and one of the main problems might be the extraction of the mesocarp and kernel oil simultaneously and the disentanglement of the valuable palm kernel residue from the mesocarp fibre. It should also be noted that the mesocarp percentage of the two *pisifera* analysed was no greater than that of many good *tenera*.

The choice of breeding systems

The question of whether to employ for the oil palm a system based on modifications of reciprocal recurrent selection or a system of individual and family mass selection has been a matter of some controversy. Against the use of reciprocal recurrent selection it is claimed[80] that it is mainly additive genetic variance which is to be exploited and that (*a*) the limited number of parents in the two base populations which can be tested may result in random loss of genetic variability, (*b*) the alternate cycles of progeny testing and selection may cause gene frequencies to oscillate rather than show progress, (*c*) the high degree of inbreeding in existing base populations, especially the Deli *dura*, does not allow a wide enough choice of parents. Great stress is put on the need to create new and genetically variable populations.

In favour of the use of a form of reciprocal recurrent selection it is argued that,[102] though mass selection may give progress for the more heritable characters showing additive genetic variance,[61] for the important but less heritable characters such as bunch yield it is probable that marked non-additive genetic variance is involved. Some evidence for this has been provided[63] and has been referred to earlier (pp. 254 and 259). It is also argued that progenies can show combining ability which is not foreshadowed by their phenotypic values and that therefore the comparative trials have an important part to play.

Without delving too deeply into these opposing positions it may be observed that the form of mass selection proposed and employed, though it has not been set out in any full detail, does contain provision for the progeny testing of individual *pisifera*, though not for *dura* x *tenera* trials.[80] Secondly, the schemes which do allow for the latter trials also have programmes for introducing new *tenera* and *dura* material to the two sides of the programme in each generation. Thus views on the breeding of the oil palm may not in practice be as divergent as they seem; in all programmes efforts are being made to introduce new material and some form of progeny testing is carried out. Van der Vossen[62] has proposed that the general schemes adopted at several centres may be modified so that new populations, from prospections or from other centres, may be used in test crosses (D x T, T x T) with so-called standard populations, i.e. those already being employed (Deli, Yangambi, etc.). In this case the two sides of the scheme would not be *dura* and *tenera*, they would be populations of different origin whether *dura* or *tenera*. He also suggests, as is apparent from the growth analysis work, that density trials will need to be introduced into breeding programmes.

What has clearly emerged from the work already cited and from an assessment of the large breeding programme conducted in Nigeria[103] is that, whereas for fruit and bunch characters additivity predominates and both phenotypic selection and the estimation of genotypic values can be usefully employed in any breeding scheme, with bunch yield the deviations from additivity (although small for bunch number) are considerable.

West *et al.*[127] have presented data from Nigeria showing significant deviations from additivity with bunch yield and weight per bunch in seven groups of progenies and they consider that their results reinforce the argument of Meunier and Gascon[102] that any form of selection which takes account only of additive genetic variance will neglect advances which can be made by utilizing specific combining ability. Within their breeding material they found numbers of progenies which produced much higher or much lower bunch yields than would be predicted from additive inheritance alone. For bunch yield and weight per bunch therefore breeding programmes must be able to assess specific combining ability, and the schemes adopted in the Ivory Coast and Nigeria, with their *dura* x *tenera* and *dura* x *pisifera* comparative trials, are well adapted for this purpose. It must be realized however that the wide spacing at which the crop is planted puts a severe limitation on the number of parents that can be compared. Finally, and also for bunch yield, the assessment of bunch index should be included in future programmes with the object of maximizing bunch yield per hectare.

Vegetative propagation

In the last few years substantial progress has been made in the discovery and testing of methods of vegetative propagation through tissue culture. The successful completion of this work will make it possible to chose clones which would give high bunch yields with known bunch indices and these could be planted at suitable densities; in addition, bunch quality would be uniform with bunches giving high extraction rates of oil and kernels.

Work has been undertaken in several laboratories to develop tissue culture methods and attempts have been made to establish cell cultures from roots, the rachis, the inflorescence, the apical meristem and the base of the petiole.[104–7, 128] Most recently it has been reported that success has been attained with leaf material and this would clearly provide the largest quantity of material per adult donor palm.

Using seedlings, Jones established in England cultures from different regions of the young plants.[107] Those from regions close to the original embryonic axis gave white globular structures similar to the so-called 'embryoids' produced from apical meristem by Rabéchault *et al.*[105] These cultures, after passing through callus stages, developed both roots and shoots. Plantlets have been successfully developed from this material and transferred to Malaysia and raised in nursery polybags at Layang-Layang.[128] Plants of two different clones have been matched with seedlings prior to transfer to and comparison in the field.

In the growth cabinets the first leaves of clonal plants are narrow and grass-like and the production of a leaf similar to the first seedling leaf is delayed. Later however the leaves appear to be quite normal and growth analysis data of the growth cabinet plantlets showed all growth parameters to be similar to those of comparable seedlings. In the nursery comparison with seedlings the two groups of plants were very similar in shape and young leaf production, and it was only noticeable that the clonal plants had had many more very small leaves the stumps of which were visible. While this work represents very considerable progress towards clonal propagation, it should be emphasized that it remains (1) to establish that the growth, flowering and fruiting habits of the clones will be normal; (2) to find the region of the adult donor palm from which the largest quantities of material can easily be obtained and developed; this may be the leaves; (3) to develop methods of raising large quantities of plantlets for distribution. Work on these subjects is now (1976) in progress in Europe, Malaysia and West Africa.

Corley has considered the possible impact of this work on production.[129] He estimates that if the main objective was to increase bunch index and leaf area ratio with consequent or accompanying decreases in VDM production, bunch yields of 40 tons per hectare could be achieved in Malaysia. If these characters are combined in clones with the high fruit quality characteristics which are already quite common, then production of 9 to 11 tons of palm oil with 1.3 to 2.2 tons of kernels per hectare per annum should be commercially possible. However, plant breeders will also be required to combine within productive clones factors for qualities such as short trunk, disease resistance and improved oil composition; and the possible susceptibility of clones to pest and disease attack will need to be guarded against. Obviously, tissue culture has provided breeders with a challenging new tool and the idea that there will be less breeding work to do may be discounted.

The employment of *Elaeis oleifera* in breeding

Interest in breeding from *Elaeis oleifera* has increased in recent years owing principally to its apparent resistance to a serious form of Bud Rot in Colombia. However, the hybrid with *E. guineensis* is also of considerable promise because of its slow growth in height, its possible special adaptation to climates with strong dry seasons, its possible resistance to other diseases and its production of oil with a higher proportion of unsaturated fatty acids than is contained in palm oil from *E. guineensis*. This oil would probably command a higher price and in that case a rather lower yield of oil per hectare would be acceptable.

As there are no *tenera* in *E. oleifera*, breeding for production entails the crossing of selected *E. oleifera* with selected *E. guineensis pisifera*. There has been much haphazard planting of the hybrid in small plots in various parts of the world, mostly with *dura* or *tenera* as the male parent; but what is most needed is the establishment in all oil palm countries of base populations of *E. oleifera* for subsequent selection and crossing. Such trials as exist indicate that the pure *E. oleifera* yields poorly under plantation conditions, but that the hybrid gives bunch yields at least comparable with those of *E. guineensis*. In the Cooperative Breeding Programme organized by the Department of Agriculture in Malaysia the hybrids grew well and in one trial two progenies gave a mean annual bunch yield of 203 kg per palm per annum over a 4-year period against 190 kg per palm for four *E. guineensis tenera* (Deli x P) progenies. In Nigeria four hybrid progenies gave a mean of 55 kg/palm per annum over the first 5 bearing years against 56 kg/palm/an. for *E. guineensis* palms, and 77 kg/palm in the fifth year against 71 kg/palm for the *E. guineensis*.[108]

The botanical characteristics of *E. oleifera* and the hybrid have been described in Chapter 2 (p. 69). It is not easy to gauge, from the scanty information available, the potential of the hybrid in terms of oil and kernels to bunch or oil and kernel yield per hectare under different climatic conditions. In Malaysia, with a Deli *dura* x *E. oleifera* cross, it was shown that parthenocarpic fruit may constitute more than half the total fruit and that the oil to mesocarp contents of normal and of small and large parthenocarpic fruit are not identical.[109] Clearly, bunch analysis in the hybrid is complicated and kernel to bunch percentages will be low. Analysis of an *E. oleifera* x *pisifera* hybrid in Nigeria gave 66.8 per cent mesocarp and 7.3 per cent kernel to fruit, but it was not stated whether this was for normal fruit or for a random sample of all the fruit on the bunches. For the Malaysian *dura* hybrid mesocarp in normal fruit was 53 per cent and in large and small parthenocarpic fruit 76 and 89 per cent respectively, while kernel percentage for normal fruit was 8.8. With a *pisifera* hybrid therefore one might expect a high overall figure of mesocarp to fruit with a low overall kernel to fruit of not more than 3 per cent. Analyses from two fields in Colombia gave overall mesocarp to bunch percentages of 52.8 and 48.5, figures similar to those of good *E. guineensis tenera*.[110]

Oil to mesocarp percentages are low (about 30) in *E. oleifera* and in the hybrid are intermediate between those of the parent species, though there seems to be a high variability. It may well be possible to select *E. oleifera* palms showing relatively high oil to mesocarp and this will then be reflected in the hybrid. It is questionable whether breeders should attempt to reduce parthenocarpy; this is largely a matter of whether it is decided to maximize oil to bunch or to encourage a higher kernel content.

The chromosome number of both species of *Elaeis* is 32 and there is no difficulty in crossing them. However, Hardon and Tan[109] found that seed set, germination and seedling survival from germinated seed were all well below what is normally expected with *E. guineensis*. In examining the inheritance pattern in the hybrid it was found that there was dominance of *E. oleifera* in height increment, fruit shape and colour, parthenocarpy and arrangement of leaflets on the rachis, with over-dominance of leaf length and size of leaflets (and possibly bunch yield); intermediate values between parent species in mesocarp, shell and oil per fruit and chemical composition of the oil; and partial dominance of *E. oleifera* in spikelet morphology and size of subtending bracts. The over-dominance of leaf length and size of leaflets is of particular note, and it has been thought in the past that this would demand a reduction in planting density. However, it has been shown in many plantings that sex ratio has remained high at the normal density of 143 palms per hectare. The very large leaves tend to have long petioles so that the proximal leaflets are at a greater distance from the stem than in *E. guineensis*.

A programme of 'back-crossing' the *tenera* hybrid with the *tenera* form of *E. guineensis* has been described from Nigeria.[131] As might be expected, the progenies were heterogeneous though falling into three morphological categories: those appearing like *E. guineensis*; those looking like *E. oleifera*; and those combining characters of both species. The majority (85 per cent of those included in the study) were, as expected, of the first category, while those of the last category showed the most vigorous growth and had bunches more like *E. oleifera*. The fruit in general resembled that of *E. guineensis*. Not all palms could be classified but many aborting palms were thought to be *pisifera*; two fertile *pisifera* were identified. *Tenera* palms showed between 66 and 79 per cent mesocarp, with 13 to 26 per cent shell and 39 to 48 per cent oil to mesocarp, and there was much variation in fruit size and structure with differences in positioning and concentration of the fibre ring (which does not appear distinctively at all in the hybrid). The mean height of the *E. guineensis*-like palms at 15 years of age was only 87 cm, a very low figure for the palms' age, but no height data of adjoining *E. guineensis* palms was recorded for comparison.

While slow growth in height, resistance to Bud Rot and the production of a high proportion of unsaturated fatty acids are clear advantages, the hybrid suffers from certain defects. It has been estimated that its low fruit-to-bunch ratio and low percentage oil to mesocarp give an oil-to-bunch extraction of only 17 per cent against 22 to 23 per cent with the best *E. guineensis tenera* commercial material.[111] Secondly, many hybrids have, in Africa, been found to be heavily attacked by Freckle (*Cercospora elaeidis*). This suggests that systematic programmes of prospection for *E. oleifera* material are needed in which both its good and bad qualities are taken into account; and that such prospections should be associated with hybridization programmes for the combination of the desired characters. Some systematic prospection within *E. oleifera* populations has been undertaken in America.[110] In the Monteria region of Colombia prospection and breeding was undertaken by I.R.H.O. workers in collaboration with the Instituto Colombiano Agropecuario and quantities of both *E. oleifera* and hybrid seed were planted in that country and in the Ivory Coast. The results of this programme have yet to appear. An extensive survey was also carried out by workers from Costa Rica and Honduras throughout the scattered grove areas of Central and South America; the prospection sites have been mapped and classified collections have been established for future breeding in Costa Rica.[130] It is generally agreed that much further prospecting is needed and steps are now (1976) being taken to mount further expeditions to the most promising areas and to coordinate the efforts of a number of institutions.[133] With the increased planting of *E. guineensis* in America the danger of collecting *E. oleifera* seed out-pollinated by nearby *E. guineensis* plants must be kept in mind.

Breeding for oil composition

The work with *E. oleifera* stimulated interest in improving oil composition, i.e. producing a more fluid oil with a higher proportion of un-saturated fatty acids. Data from various sources of the fatty acid composition of both *Elaeis* species are given in Chapter 14 (pp. 698—703). Evidence has been presented of differences in unsaturated acid contents of various *E. guineensis* base and hybrid populations in the breeding programmes in the Ivory Coast. Marked differences between progenies within populations have also been found.[112] With eighteen lines from the La Mé population and nineteen lines from the Yangambi—Sibiti population crossed with twenty-nine Deli *dura*, the mean percentages of myristic, palmitic, stearic, oleic and linoleic acids in the crosses was intermediate between those of the parent populations and in all cases remarkably close to the mean of the populations concerned.[132] The La Mé population had, on average, a much higher oleic to palmitic ratio than either the Yangambi—Sibiti or the Deli populations. Heritability determinations in the Deli x *tenera* crosses showed significant values (varying from 0.43 to 0.87) for iodine value, total unsaturated acids, and oleic acid in both population crosses and significant values for other acids in the Deli x Yangambi—Sibiti crosses. It was estimated that 58 to 59 per cent unsaturated acids could be reached by choice of the most favourable parents in *dura* x *pisifera* crosses using the La Mé population and that this would represent an improvement of 6.5 per cent over the mean of the Deli x La Mé population.

In hybridizing the two *Elaeis* species the wide variations in oleic and linoleic contents of *E. oleifera* fruit can also be exploited to provide hybrids with an unsaturated fraction of over 70 per cent. Oleic acid contents of up to 59 per cent and linoleic contents of up to 15 per cent are reported.[111]

Results of breeding and selection

Information on the material developed in breeding programmes, in comparison with unselected or previously selected material, has been scanty, though data relating to various progenies or groups within the progenies has been available and some examples have been given in previous pages.

Information on the first I.N.E.A.C. *dura* x *pisifera* issues was published in 1956—7.[113] This gave bunch yields per annum in the second and third years of production as varying from 9 to 13 metric tons per hectare which showed, for Africa, a startling precocity. Full quality data were not presented, but the mean oil-to-bunch ratios were 23.3 and 20.9 per cent in the second and third years respectively.

A short account has been given of the quality, in terms of oil to fruit, of seed issued at various times by the I.N.E.A.C.[114] The progeny of seed issued contained both *dura* and *tenera* palms until 1951 as first *tenera* x *tenera* and then *dura* x *tenera* seed was issued. A gradual improvement was claimed in the standard of the *tenera* palms being grown, and at the end of the period 1951–9 the average standard lay between two sample types which were described as having the following analyses:

Sample	Mesocarp/fruit	Oil/fruit
1 *Tenera*	93%	43.7%
2 *Tenera*	73%	37.1%

It was stated that periodical analyses showed the mean oil-to-fruit content to lie between 40 and 42 per cent.

Data relating to I.R.H.O. Deli x *pisifera* material has also been published. Yields of 120–123 kg per palm per annum have been cited for palms 5 to 10 years of age in the Ivory Coast.[99] These are calculated from experiment station progenies and partly from *tenera* in *dura* x *tenera* crosses; they are therefore not comparable with the I.N.E.A.C. data which relate to over 800 hectares of commercial planting. Quality data relating to some 60 acres of Deli x *pisifera* ex-Yangambi–Sibiti progenies and 120 acres of Deli x *pisifera* ex-La Mé showed the clear differences between the Yangambi and La Mé *tenera* ancestry. The former cross is characterized by *tenera* with large bunches and fruit and with high mesocarp-to-fruit and oil-to-mesocarp ratios; the latter by more numerous but smaller, spiny bunches still carrying the comparatively low mesocarp-to-fruit character of the La Mé selections, though having satisfactory oil-to-mesocarp and kernel-to-fruit ratios.

Seed issued in Nigeria has from time to time been compared with unselected material and bunch yields have been shown to be some 20 to 25 per cent higher.[65]

Improvements due to breeding in Malaysia are hard to assess because the new generation is *tenera* and the generation it is replacing was Deli *dura*. The *tenera* material is undoubtedly more precocious, giving higher yields in the first few years, but Hardon and Ooi[83] have drawn attention to the good adaptation to the Malaysian climate of the Deli *dura* and to the high mature bunch yields obtained, and they suggest that in the adult years the bunch yield improvement will not be great.

Very recently (1973) a comparison of current planting material has been organized on an international scale to cover a range of environments. The results obtained should be of considerable interest and utility.

Breeding for disease resistance

De Berchoux and Gascon[115] have shown that in the Ivory Coast progenies of purely Deli parentage are highly susceptible to Crown disease. When Deli palms are crossed with local Ivory Coast material, however, the progeny assume the immunity of their local parent; but African material from Zaire does not always have this effect in crossings with the Deli palm (see p. 618).

Work in Nigeria on resistance or tolerance to Vascular Wilt disease showed that, although no progenies were wholly immune, they could be grouped according to their susceptability or tolerance.[116] Similar work has been carried out in the Ivory Coast[117] and, more recently, new techniques have been devised in England.[118] This work is discussed on p. 631.

An outbreak of Dry Basal Rot in a progeny trial at the N.I.F.O.R. main station was confined to the progenies of a certain parent. Subsequent inoculation trials at the seedling stage confirmed that there were definite lines of tolerance and susceptibility and suggested that resistance might be inherited monofactorially.[119]

In West Africa differences of susceptibility to Blast have long been apparent, Deli material and progenies from the Ivory Coast providing the largest number of casualties. Surveys in Nigeria showed two groups, the susceptible progenies and a range of tolerant ones.[120] The latter range was obtained from the selfings and crosses of two parents selected as resistant and by crosses between them and susceptible parents. Resistance was clearly inherited and dominant overall. Blaak has confirmed these findings in West Cameroon. He found that Deli progenies were highly susceptible, but that West African progenies showed a wide range of tolerance.[121] Crosses between Deli *dura* and the more tolerant West African material could reduce incidence to levels of 1 to 6 per cent from over 45 per cent for the Deli and a mean of 15 per cent for West African material.

Breeding for drought resistance

Morphologically the oil palm is well adapted to resist drought (see p. 152), but the effect of drought on bunch yield is severe. In Dahomey, where water deficits are often in excess of 600 mm, the normal effects of drought, such as the accumulation of unopened spear leaves and premature leaf withering, are often exaggerated. Marked differences between progenies in a drought 'sensibility index' have been recorded.[122] This index is simply a numerical assessment of the drought effects recorded in a population or progeny. These preliminary findings have encouraged both the selection of progenies showing low sensibility indices and further investigations to determine methods of measuring drought resistance through submitting seedlings to high osmotic pressures or high temperatures.[122]

The purchasing of seed

At the present stage of selection and breeding it will not always be possible for the buyer of seed to obtain all the information that it is desirable to have, but the following guide may be suggested.

1. Prospective yields in terms of palm oil per hectare are sometimes given. These are of limited value, since oil yield depends on bunch yield and the latter varies widely with environment both within and between countries.
2. The seed should, if possible, be obtained as progenies with information on parental achievement and analysis. Allowing for losses, seed from one progeny bunch will plant 4 to 6 hectares. A programme of several hundred hectares will not entail a very great number of parents.
3. A clear definition of the terms 'proved seed' or 'proved parentage' should be obtained if these terms are used by the seller.
4. The following data relating to parents and sibs should be obtained:
   (*a*) Bunch yield at maturity over a given number of years with statement of location, soil, rainfall distribution, water deficit and sunshine.
   (*b*) Bunch analysis: F/B; mesocarp, shell and kernel to fruit; oil to mesocarp. (This enables calculation of oil plus kernel to bunch.)
5. The female *dura* parent data are probably of least importance since their value is most difficult to assess, so *tenera* analyses of the following are most desirable:
   Pisifera: *tenera* mother palm and mean of *tenera* sibs.
   Tenera test progeny: Mean of progeny of *tenera* mother palm x *dura* mother palm.

References

1. Schmöle, J. F. (1930) The selection of Oil Palms (*Elaeis guineensis* Jacq.). *Proc. 4th Pacif. Sci. Congress*, Java, 1929, Proc., 4, 185.
2. Stoffels, E. (1934) Contribution à l'étude de la sélection d'*Elaeis guineensis* à Sumatra. *Revue Bot. appl. Agric. trop.*, 14, 93.
3. Belgarric, R. Carrière de (1951) Notes sur la sélection du palmier à huile à Sumatra: Résultats obtenus par la SocFin. *Oléagineux*, 6, 65.
4. Pronk, F. (1955) *De Veredeling van de oliepalm door het Algemeen Proefstation der AVROS*. Mimeograph, 29 pp.
5. Pronk, F. (1953) Vergelijkend tros-analytisch onderzoek van enkele typen van de oliepalm (*Elaeis guineensis* Jacq.). *Bergcultures*, 22, 527−33 and 573−81.
6. Jagoe, R. B. (1952) Deli oil palms and early introductions of *Elaeis guineensis* to Malaya. *Malay. agric. J.*, 35, 3.
7. Jack, H. W. and Jagoe, R. B. (1932) Variation in fruiting ability of oil palms. *Malay. agric. J.*, 20, 16.
8. Jagoe, R. B. (1952) The 'dumpy' oil palm. *Malay. agric. J.*, 35, 12.
9. Chemara Research Station, Layang Layang (1960) *Oil palm planting material, Ulu Remis* (D x P). Mimeograph.
10. Socfin Company Ltd (1963) *Oil palm planting material*. Mimeograph.
11. Beirnaert, A. (1933) La sélection du palmier à huile. *Bull. agric. Congo belge*, 24, 359−80, 418−58.

12. **Beirnaert, A.** (1933) Les bases de la sélection du palmier à huile. *Journée Agron. Colon.*, 124.
13. **Beirnaert, A.** (1933) Les méthodes de la sélection du palmier à huile. *Journée Agron. Colon.*, 135.
14. **Sparnaaij, L. D.** *et al.* (1963) Breeding and inheritance in the oil palm. Part I. The design of a breeding programme. *J. W. Afr. Inst. Oil Palm Res.*, 4, 126.
15. **Smith, E. H. G.** (1929) The oil palm (*Elaeis guineensis*) at Calabar. *8th An. Bull. Dep. Agr., Nigeria.*
16. **Toovey, F. W.** (1938) *Report on a visit to the Belgian Congo to study certain aspects of the oil palm industry.* Mimeographed report, Agric. Dept., Nigeria.
17. **Beirnaert, A.** (1940) Le problème de la stérilité chez le palmier à huile. *Bull. agric. Congo belge.* (Leopoldville Edn.), 31, 95.
18. **Beirnaert, A. and Vanderweyen, R.** (1941) Contribution à l'étude génétique et biométrique des variétés d'*Elaeis guineensis* Jacq. *Publs. I.N.E.A.C.* Série Sci., No. 27.
19. **Hartley, C. W. S.** (1957) Oil palm breeding and selection in Nigeria. *J. W. Afr. Inst. Oil Palm Res.*, 2, 108.
20. **Vanderweyen, R.** (1953) Le croisements '*Dura* x *Pisifera*' et ses premiers resultats. *Bull. Inf. I.N.E.A.C.*, 2, 123.
21. **Toovey, F. W. and Purvis, C.** (1956) Segregation from the *dura* x *pisifera* cross. *J. W. Afr. Inst. Oil Palm Res.*, 1, (4), 106.
22. **Blaak, G.** *et al.* (1963) Breeding and inheritance in the oil palm. Part II. Methods of bunch quality analysis. *J. W. Afr. Inst. Oil Palm Res.*, 4, 146.
23. **Sparnaaij, L. D.** (1958) *Oil palm breeding and selection in Belgian Congo.* W.A.I.F.O.R. mimeographed report.
24. **Mollegaard, M.** (1970) The effect of drying of oil palm nuts in respect of crackability of same and estimation of the percentage of shell and kernel to fruit. *Oléagineux*, 25, 139.
25. **Dyke, F. M. and James, F. O.** (1925) Contribution à l'étude chimique des fruits d'*Elaeis guineensis. Bull. agric. Congo belge*, 16, 516.
26. **Vanderweyen, R., Rossignol, J. and Miclotte, H.** (1947) Considérations sur les teneurs en eau et en huile de la pulpe des fruits d'Elaeis. *C. R. de la Semaine Agricole de Yangambi.* Common. No. 54, 730.
27. **Dessasis, A.** (1955) La détermination de la teneur en huile de la pulpe de fruit d'*Elaeis guineensis. Oléagineux*, 10, 823.
28. **Chapas, L. C.** *et al.* (1957) The determination of the oil content of oil palm fruit. *J. W. Afr. Inst. Oil Palm Res.*, 2, 230.
29. **Blaak, G.** (1970) L'extraction de l'huile, à froid, dans l'analyse des régimes des palmier à huile. *Oléagineux*, 25, 165.
30. **Servant, M. and Henry, J.** (1963) Détermination de la richesse en huile de la pulpe du fruit de palme. *Oléagineux*, 18, 339.
31. **Broekmans, A. F. M.** (1957) Studies of the factors influencing the success of controlled pollination of the oil palm. *J. W. Afr. Inst. Oil Palm Res.*, 2, 133.
32. **Devreux, M. and Malingraux, C.** (1960) Pollen d'*Elaeis guineensis* Jacq. Recherches sur les méthodes de conservation. *Bull. agric. Congo belge*, 5, 543.
33. **Henry, P.** (1959) Prolongation de la viabilité du pollen chez *Elaeis guineensis. C. r. hebd. Séanc. Acad. Sci., Paris*, 98, 722.
34. **Hardon, J. J. and Davies, M. D.** (1969) Effects of vacuum drying on the viability of oil palm pollen. *Expl. Agric.*, 5, 59.
35. **Bénard, G. and Noiret, J. M.** (1970) Le pollen de palmier à huile. Recolte, préparation, conditionnement et utilisation pour la fécundation artificielle. *Oléagineux*, 25, 67.
36. **Menendez, T. M.** (1969) N.I.F.O.R. Third Annual Report, 1966–7, p. 72.
37. **Smith, E. H. G.** (1930) The oil palm at Calabar. 2nd Conf. W. Afr. Agr. Officers, 1929, Paper 16, 181 (*Gold Coast Dept. Agr. Bull.* 20).

38. **Smith, E. H. G.** (1933) Further yields from the Calabar plantation oil palms. *10th An. Bull. Dept. Agr.*, Nigeria, 1—18.
39. **Bunting. B., Georgi, C. D. V. and Milsum, J. N.** (1934) *The oil palm in Malaya.* Planting Manual No. 1. Dept. of Agriculture, Kuala Lumpur.
40. **Webster, C. C.** (1939) A note on a uniformity trial with oil palms. *Trop. Agric. Trin.*, **16**, 15.
41. **Chapas, L. C.** (1961) Plot size and reduction of variability in oil palm experiments. *Emp. J. exp. Agric.*, **29**, 212.
42. **Ollagnier, M.** (1951) Forme, dimension des parcelles et nombre des répétitions dans les essais culturaux sur l'arachide et sur le palmier à huile. *Oléagineux*, **6**, 707.
43. **Gascon, J. P. et al.** (1966) Contribution à l'étude de l'hérédité de la production de régimes d'*Elaeis guineensis* Jacq.: Application à la sélection du palmier à huile. *Oléagineux*, **21**, 657.
44. **Broekmans, A. F. M.** (1957) Growth, flowering and yield of the oil palm in Nigeria. *J. W. Afr. Inst. Oil Palm Res.*, **2**, 187.
45. **Janssens, P.** (1927) Le palmier à huile au Congo Portuguaise et dans l'enclave de Cabinda. *Bull. agric. Congo belge*, **18**, 29.
46. **Nwanze, S. C.** (1961) W.A.I.F.O.R. Ninth Annual Report, 1960—1, p. 100.
47. Proceedings of the Conference on Oil Palm Research at the Oil Palm Research Station, near Benin, Nigeria, Dec. 1949. Mimeographed report.
48. **Henry, P. and Gascon, J. P.** (1950) Les palmiers à huile du type *pisifera* et la stérilité. *Oléagineux*, **5**, 29.
49. **De Berchoux, C. and Gascon, J. P.** (1965) Caractéristiques végétatives de cinq descendances d'*Elaeis guineensis* Jacq. *Oléagineux*, **20**, 1.
50. **De Poerck, R.** (1942) Comment expliquer les disjonctions anormales de certains *tenera? Bull. agric. Congo belge*, **33**, 206.
51. **Vanderweyen, R. and Roels, O.** (1949) Les variétés d'*Elaeis guineensis* Jacquin du type *albescens. Publs. I.N.E.A.C.*, Série Sci., No. 42.
52. **Zeven, A. C.** (1973) The 'mantled' oil palm (*Elaeis guineensis* Jacq.). *J. W. Afr. Inst. Oil Palm Res.*, **5**, 31.
53. **Janssen, A. W. B.** (1959) Miscellaneous notes on estate agriculture in Sumatra, No. 7. Oil palm productivity and genetics. Chemara Research Station.
54. **Fickendey, E.** (1944) Die Züchtung der Ölpalme (*Elaeis guineensis* Jacq.). *Z. PflZücht.*, **26**, 136.
55. **Haddon, A. V. and Tong, Y. L.** (1959) Oil palm selection and breeding: A progress report. *Malay. agric. J.*, **42**, 124.
56. **Noiret, J. M. and Gascon, J. P.** (1967) Contribution à l'étude de la hauteur et de la croissance du stipe d'*Elaeis guineensis* Jacq. Application à la sélection du palmier à huile. *Oléagineux*, **23**, 85.
57. **Taillez, B. and Valverde, G.** (1971) Sensibilité aux ouragans de différents types de croisements de palmier à huile. *Oléagineux*, **26**, 753.
58. **Falconer, D. S.** (1960) *Introduction to quantitative genetics.* Oliver & Boyd, London.
59. **Blaak, G.** (1965) Breeding and inheritance in the oil palm. Part III. Yield selection and inheritance. *J. Nig. Inst. Oil Palm Res.*, **4**, 262.
60. **Menendez, T. M. and Blaak, G.** (1964) W.A.I.F.O.R. Twelfth Annual Report, 1963—64, pp. 68—9 and unpublished data.
61. **Noiret, J. M. et al.** (1966) Contribution à l'étude de l'hérédité des caractéristiques de la qualité du régime et du fruit d'*Elaeis guineensis* Jacq. Application à la sélection du palmier à huile. *Oléagineux*, **21**, 343.
62. **Van der Vossen, H. A. M.** (1974) Towards more efficient selection for oil yield in the oil palm (*Elaeis guineensis* Jacq.). *Agr. Res. Rpts.*, **823**, Wageningen, Holland.
63. **Meunier, J., Gascon, J. P. and Noiret, J. M.** (1970) Hérédité des caractéristiques des régime d'*Elaeis guineensis* Jacq. en Côte d'Ivoire. *Oléagineux*, **25**, 377.

64. **Gascon, J. P. and de Berchoux, C.** (1963) Quelques relations entre les *dura* et *tenera* d'une même descendance et leur application à l'amélioration des semences. *Oléagineux*, **18**, 411.

65. **Menendez, T.** (1964−5) *W.A.I.F.O.R. Twelfth Annual Report*, 1963−4, pp. 49−75; *N.I.F.O.R. First Annual Report*, 1964−5, pp. 59−73.

66. **Bénard, G.** (1965) Caractéristiques qualitatives du régime d'*Elaeis guineensis* Jacq. Teneur en huile de la pulpe des diverses origines et des croisement interorigines. *Oléagineux*, **20**, 163.

67. **Broekmans, A. F. M.** (1957) The production of improved oil palm seed in Nigeria. *J. W. Afr. Inst. Oil Palm Res.*, **2**, 116.

68. **Thomas, R. L., Watson, I. and Hardon, J. J.** (1969) Inheritance of some components of yield in the Deli *dura* variety of oil palm. *Euphytica*, **18**, 92.

69. **Ooi, S. C., Hardon, J. J. and Phang, S.** (1973) Variability in the Deli *dura* breeding population of the oil palm (*Elaeis guineensis*, Jacq.). I. Components of bunch yield. *Malaysian Agric. J.*, **49**, 112.

70. **Sparnaaij, L. D.** (1969) Oil Palm. In *Outlines of perennial crop breeding in the tropics*. Landbouwhogeschool, Wageningen. Misc. papers 4.

71. **Ward, J. B.** (1973) Oil palm selection and breeding. In *Unilever Plantations Group Annual Review of Research*, 1971, Part II, pp. 4−42; and Green, A. H., private communication.

72. **Hardon, J. J., Corley, R. H. V. and Ooi, S. C.** (1972) Analysis of growth in oil palm. II. Estimates of genetic variances of growth parameters and yield of fruit bunches. *Euphytica*, **21**, 257.

73. **Corley, R. H. V., Hardon, J. J. and Ooi, S. C.** (1973) Some evidence of genetically controlled variation in photosynthetic rate of oil palm seedlings. *Euphytica*, **22**, 48.

74. **Vanderweyen, R.** (1952) La prospection des palmeraies congolaises et ses premiers résultats. *Bull. Inf. I.N.E.A.C.*, **1**, 357.

75. **Pichel, R.** (1957) L'Amélioration du palmier à huile au Congo Belge. Conf. Franco−Britannique sur le palmier à Huile. *Bull. agron. Minist. Fr. d'outre mer*, **14**, 59, and *Bull. agric. Congo belge*, **48**, 67.

76. **Desneux, R.** (1957) Prospection des palmeraies et sélection du palmier à huile au Kwango. *Bull. Inf. I.N.E.A.C.*, **6**, 351.

77. **Meunier, J.** (1969) Étude des populations naturelles d'*Elaeis guineensis* en Côte d'Ivoire. *Oléagineux*, **24**, 195.

78. **Rajanaidu, N.** (1974) Some aspects of the variation of the genus *Elaeis*. Typescript Thesis, Dept. of Botany, University of Birmingham.

79. **Hartley, C. W. S.** (1968) Report on oil palm research and development in Brazil. *Comun. Tecn. CEPLAC*, **17**, 1.

80. **Hardon, J. J., Mokhtar Hashim and Ooi, S. C.** (1973) Oil palm breeding: a review. In *Advances in oil palm cultivation*. Incorp. Soc. of Planters, Kuala Lumpur.

81. **Hayes, H. K., Immer, F. R. and Smith, D. C.** (1955) *Methods of plant breeding*. McGraw-Hill, New York and London.

82. **Hardon, J. J.** (1970) Inbreeding in populations of the oil palm. (*Elaeis guineensis* Jacq.). *Oléagineux*, **28**, 449.

83. **Hardon, J. J. and Ooi, S. C.** (1972) To what extent should inbreeding be avoided in oil palm seed production? *Com.* (*Agron.*) No. 9. Kumpulan Guthrie Sdn. Bnd.

84. **Gascon, J. P., Noiret, J. M. and Meunier, J.** (1969) Effets de la consanguité chez *Elaeis guineensis* Jacq. *Oléagineux*, **24**, 603.

85. **Gascon, J. P. and de Berchoux, C.** (1964) Caractéristiques de la production d'*Elaeis guineensis* (Jacq.) de diverses origines et de leurs croisements. *Oléagineux*, **19**, 75.

86. **Broekmans, A. F. M. and Toovey, F. W.** (1955) The Deli palm in West Africa. *J. W. Afr. Inst. Oil Palm Res.*, **1**, (3), 51.

87. **I.R.H.O.** *Rapports Annuels*, 1969—73.
88. **N.I.F.O.R.** *Second Annual Report*, 1965—6, p. 80.
89. **Hartley, C. W. S.** (1962) *Report on a visit to North Borneo.* Mimeograph.
90. **Dept. of Agriculture, State of Sabah** (1970) *Annual Report for the year 1969.* Kota Kinabalu.
91. **Oil Palm Genetics Laboratory,** *Annual Reports, 1966 to 1970.*
92. **Hardon, J. J. and Thomas, R. L.** (1968) Breeding and selection of the oil palm in Malaya. *Oléagineux,* 23, 85.
93. **Hardon, J. J.** (1969) Developments in oil palm breeding. In *Progress in oil palm.* Incorporated Society of Planters, Kuala Lumpur, p. 13.
94. **Banting Oil Palm Research Station,** *Annual Report,* 1972.
95. **Trafton, M.** (1951) *The African oil palm in Honduras.* Bull. 2, United Fruit Co's. Trop. Res. Dept., La Lima, Honduras.
96. **Hartley, C. W. S.** (1965) Some notes on the oil palm in Latin America. *Oléagineux,* 20, 359.
97. **Patiño, V. M.** (1948) Información preliminar sobre la palme de Aceite Africana (*Elaeis guineensis*) en Colombia. *Estación Agro-forestal del Pacifico de Calima-Buenaventura,* Serie Bot. Apl. Vol. 1 (2). Dec. 1948.
98 **Patiño, V. M.** (1958) La Palme de Aceite Africana (The African oil palm). *Economica Colombiana,* Oct. 1958.
99. **Bénard, G. and Malingraux, C.** (1965) La production de semences sélectionnées de palmier à huile à I.R.H.O. Principe et Réalisation. *Oléagineux,* 20, 297.
100. **Desneux, R.** (1958) Un palmier 'Pisifera' remarquable. *Bull. Inf. I.N.E.A.C.,* 7, 95.
101. **Tang Teng Lai** (1971) The possible use of *pisifera* in plantation. *Malaysian Agric. J.,* 48, 57.
102. **Meunier, J. and Gascon, J. P.** (1972) La Schéma général d'amélioration du palmier à huile à l'I.R.H.O. *Oléagineux,* 27, 1.
103. **West, M. J.** (1976) The analysis of yield data of the NIFOR oil palm main breeding programme and the choice of new parental material. Supplementary report to the Min. of Overseas Development. Mimeograph.
104. **Stavitsky, G.** (1970) Tissue culture of the oil palm as a tool for its vegetative propagation. *Euphytica,* 19, 288.
105. **Rabechault, H., Martin, J. P. and Cas, S.** (1972) Recherches sur la culture des tissus de palmier à huile. *Oléagineux,* 27, 531.
106. **Smith, W. K. and Thomas, J. A.** (1973) The isolation and *in vitro* cultivation of cells of *Elaeis guineensis. Oléagineux,* 28, 123.
107. **Jones, L. H.** (1974) Production of clonal oil palms by tissue culture. *Oil Palm News,* 17, 1; and *Planter, Kuala Lumpur,* 50, 374.
108. **Obasola, C. O.** (1973) Breeding for short-stemmed palms in Nigeria. *J. W. Afr. Inst. Oil Palm Res.,* 5, (18), 43.
109. **Hardon, J. J., Tan, G. Y. and Hardon, J. J.** (1969) Inter-specific hybrids in the genus *Elaeis,* I and II. *Euphytica,* 18, 372 and 380.
110. **Vallejo Rosero, R. and Cassalett, C.** (1974) Perspectivas del cultivo de los hibridos interspecificos de Noli (*Elaeis oleifera* (HKB Cortez) x palma Africana de aceite (*Elaeis guineensis,* Jacq.)) en Colombia. Instituto Colombiano Agropecuario, Colombia. Mimeograph.
111. **Mennier, J. and Boutin, D.** (1975) L'*Elaeis melanococca* et l'hybride *E. melanococca* x *E. geuineensis.* Premières données. *Oléagineux,* 30, 5.
112. **Gascon, J. P.** (1975) Amélioration de la production et de la qualité de l'huile d'*Elaeis guineensis* Jacq. *Oléagineux,* 30, 1.
113. **I.N.E.A.C.** (1956 and 1957) Rendements obtenue en plantation par l'utilisation de graines d'*Elaeis* sélectionées à Yangambi et issues du croisements *dura* x *pisifera. Bull. Inf. I.N.E.A.C.,* 5, 271 and 6, 329.
114. **Poels, G.** (1959) La qualitée des fruits de palme produits par les agriculteurs congolais. *Bull. Inf. I.N.E.A.C.,* 8, 309.

115. **De Berchoux, C. and Gascon, J. P.** (1963) L'arcure défoliée du palmier à huile. Éléments pour l'obtention de lignées résistantes. *Oléagineux*, **18**, 713.

116. **Prendergast, A. G.** (1963) A method of testing oil palm progenies at the nursery stage for resistance to Vascular Wilt disease caused by *Fusarium oxysporum* Schl. *J. W. Afr. Inst. Oil Palm Res.*, **4**, 156.

117. **Renard, J. L., Gascon, J. P. and Bachy, A.** (1972) Recherches sur la fusariose du palmier à huile. *Oléagineux*, **27**, 581.

118. **Locke, T. and Colhoun, J.** (1974) Contributions to a method of testing oil palm seedlings for resistance to *Fusarium oxysporum* Schl. fsp. *elaedis* Toovey. *Phytopath. Z.*, **79**, 77.

119. **Robertson, J. S.** (1962) *W.A.I.F.O.R. Tenth Annual Report*, 1961–2, p. 82.

120. **Robertson, J. S.** (1963) *W.A.I.F.O.R. Eleventh Annual Report*, 1962–3, p. 78.

121. **Blaak, G.** (1969) Prospects of breeding for Blast disease resistance in the oil palm (*Elaeis guineensis* Jacq.). *Euphytica*, **18**, 153.

122. **Maillard, G., Daniel, C. and Ochs, R.** (1974) Analyse des effets de la sécheresse sur le palmier à huile. *Oléagineux*, **29**, 397.

123. **Ooi, S. C. and Abdul Wahab bin Ngah** (1976) Oil palm breeding – some aspects of selection. Int. Agric. Oil Palm Conference, Kuala Lumpur, 1976.

124. **Arasu, N. T. and Rajanaidu, N.** (1976) Oil palm genetic resources. Int. Agric. Oil Palm Conference, Kuala Lumpur, 1976.

125. **Arasu, N. T. and Rajanaidu, N.** (1975) Conservation and utilization of genetic resources in the oil palm (*Elaeis guineensis*, Jacq.). In *South East Asian Plant Genetic Resources* (ed. J. T. Williams *et al.*, Bogor), pp. 182–6.

126. **Lubis, A. U. and Kiswito** (1975) New perspectives in oil palm breeding in Indonesia. In *South East Asian Plant Genetic Resources* (ed. J. T. Williams *et al.*, Bogor).

127. **West, M. J. *et al.*** (1976) The inheritance of yield and of fruit and bunch composition characters in the oil palm – an analysis of the NIFOR main breeding programme. Int. Agric. Oil Palm Conference, Kuala Lumpur, 1976.

128. **Ong Hean Tatt** (1976) Studies into tissue culture of oil palm (1976) Int. Agric. Oil Palm Conference, Kuala Lumpur, 1976.

129. **Corley, R. H. V., Barrett, J. N. and Jones, L. H.** (1976) Vegetative propagation of oil palm *via* tissue culture. Int. Agric. Oil Palm Conference, Kuala Lumpur, 1976.

130. **Richardson, D. L.** (1976) Private communication.

131. **Obasola, C. O., Obesesan, I. O. and Opute, F. I.** (1976) Breeding of short-stemmed oil palm in Nigeria. Int. Agric. Oil Palm Conference, Kuala Lumpur, 1976.

132. **Noiret, J. M. and Wuidart, W.** (1976) Possibilities for improving the fatty acid composition of palm oil – results and prospects. Int. Agric. Oil Palm Conference, Kuala Lumpur, 1976, and *Oléagineux*, **31**, 465.

133. **Meunier, J.** (1976) Les prospections de palmacées. Une nécessité pour l'amélioration des palmiers oléagineux. *Oléagineux*, **31**, 153.

134. **Ooi, S. C.** (1975) Variability in the Deli *dura* breeding population of the oil palm (*Elaeis guineensis* Jacq.) II. within bunch components of bunch yield. *Malaysian Agric. J.*, **50**, 20.

135. **Blaak, G.** (1972) Prediction of precocity in the oil palm (*Elaeis guineensis*, Jacq.). *Euphytica*, **21**, 22.

Chapter 6

The preparation, storage and germination of seed

An account of the physiology of germination has been given in Chapter 4 (p. 136) where it was seen that exact conditions of temperature, moisture content and, hence, oxygen supply were required for the satisfactory germination of seed. The erratic results that were obtained in the past, both in experiments and in germinating large lots of seed, were due mainly to a failure to control these factors with sufficient exactitude. Two other factors have, however, also been responsible for poor results. In the first place preparation and storage methods have often been far from ideal, and have resulted in reduced viability. Secondly, there is evidence of genetic differences in the germination of progenies.[1] Seed from some progenies has shown consistently poor germination, but this abnormality is probably far less common than was often supposed, many of the differences attributed to progeny being in reality attributable to small variations in preparation, storage and germination conditions. There is evidence that some selected material has partially lost the high-temperature requirement, possibly because of poor germination in earlier years and consequent high selection pressure for rapid germination. Some selected seed from West Africa, Central Africa and the Far East has given germination rates far higher than normal, and an instance has been quoted of one batch of West African seed which gave 61 per cent germination in 8 months when kept at a temperature of around 22°C.[2]

Preparation of seed

Fruit to be used for seed supplies should be harvested ripe. Baskets may be tied to the palms immediately under a ripening bunch or the bunch can be enclosed in a sack to catch the first-loosened fruit and prevent it being lost or damaged. The degree of ripeness of the fruit has no effect on germination percentage.[3] Seed from bunches under-ripe for harvesting has given as good germination as seed from very over-ripe bunches; nevertheless it is useful to set some standard to be adhered to in practice, and it is simplest to require the bunches to be at the normal stage of ripeness for harvesting.

In the preparation of seed for germination the fruit has first to be

311

removed from the bunch by hand picking. This is facilitated by first cutting up the bunch into its constituent spikelets and then allowing the fruit to loosen naturally for a day or two. The removal of the exocarp and mesocarp from the fruit is usually, though inaccurately, termed depericarping. It can be carried out by scraping off the mesocarp with a knife, by retting in water, or by rotary screened depericarpers. The first method results in less damage to the seed, but is impracticable for large quantities.

In the retting process the fruit may be immersed in either running or standing water. Early experiments on the effects of these methods were probably vitiated (as with other germination comparisons) by imperfect or varying germination techniques, but with the perfecting of these techniques useful information has now been obtained. There is no difference in final germination percentage or in the rate of germination* following depericarping by hand or by retting in standing or running water. The retting process is completed earlier in running water and this method avoids the housing of large numbers of evil-smelling retting tubs. On the other hand a convenient constant-flowing stream or special means of providing a flow of water may not be available for running-water retting. Retting for 10 days is usually sufficient, but there is no effect on subsequent germination of the length of the retting period up to 20 days. In Africa, one species of fungus and one bacterium predominate in the water during the retting process; towards the end of the retting period, the fungus, which is aerobic and grows on the surface of the water, excludes almost all other organisms. The faster retting in running water appears to be due to these conditions favouring the growth of the fungus against that of the bacterium.[3] If retting is carried out in standing water, the water should be changed once a day.

The retting process is completed by pounding the fruit with sand in a wooden mortar to remove adhering mesocarp. With *dura* fruit a clean product is obtained; with *tenera*, a quantity of the fibre remains adhering to the smaller seed.

In centres of large-scale seed production mechanical depulpers are used. These consist of hexagonal cages revolving at 30 rev/min. The walls are made of expanded wire metal mesh and water is fed to the cage to keep the fruit wet. The fibre removed from the seed is hosed off the mesh as necessary by pressure hoses.

Depulping is followed by the drying of the seed to a moisture content suitable for storage. Before drying the seeds may be treated by immersion and agitation in a solution of fungicide and bactericide such as Thiram and streptomycin for 3 minutes. There is no direct evidence that this treatment is necessary however, and as no adverse effects of methods of harvesting

---

* Various terms such as the rate or speed of germination have been used in the literature. The best measure is the number of days to 50 per cent of **final** germination percentage.

or depulping have been detected, it is thought that any loss of viability must be concerned with the drying process. Drying of seed in full sunlight may be very rapid and moisture contents below the normal air dry percentage may soon be reached. Although the surface temperatures reached are unlikely to be lethal, the rate of drying may be an important factor. Evidence for this is to be found in an experiment[4] on the storage of differently dried lots of *dura, tenera* and small *tenera* seed. Seed was dried, towards the end of a dry season, in the sun or in the shade for 1, 2 or 3 days. After 160 days' storage in 500-gauge polythene bags germination was still between 84 and 100 per cent for all lots of seed, but by the end of a year's storage viability had been almost totally lost in all sun-dried seed and in seed dried for 3 days in the shade. *Dura* seed dried for 1 or 2 days in the shade still gave 98 per cent germination; for *tenera* seed, however, more than 1 day's shade drying was deleterious. Seed should therefore be dried in the shade and care should be taken to see that even shade drying is not too prolonged. The usual practice is to spread the seed out on concrete or wooden floors or on wire racks for air-drying under shade for 2 days. With *tenera* seed a shorter period is advocated. Progenies are kept in separate trays. The seed is then ready for storage.

It may be mentioned here that no differences in viability or germination percentage have been consistently noted between seed from outer fruit and the rather smaller inner fruit, nor between seed of *dura* and *tenera* fruit form. *Tenera* seed may, however, be dried out too quickly if care is not taken and poor germination of some *tenera* samples is believed to be due to this. No seasonal effect has been generally found, nor does the age of the seed-bearing palm affect germination.

Outer seeds on a bunch tend to be larger and more are multi-kernelled. As already mentioned, there is no difference between the germination of large and small seed though very small seed often fails to germinate. A higher percentage of very small seed has no kernel or no embryo. Small numbers of 'white' seeds often appear which lack a normally lignified shell. Though these seeds germinate satisfactorily, the white-shell character, which is likely to be troublesome during shell and kernel separation in processing, may be inherited. Empty seeds and the majority of white seeds float in water and can thus be separated off and discarded.

Storage of seed

It was noticed by many early workers that viability deteriorated gradually, sometimes alarmingly, with length of storage time. The fact that stored seed germinates faster was not noticed so soon however, probably owing to the imperfections of germination techniques. Vanderweyen[5] stated that in Zaire there was a slight diminution of 'speed' of germination after 4 months' storage and that germination was markedly reduced after 10 months. Galt,[6] however, showed that stored seed germinates more readily than fresh seed and this has been interpreted as the effect of the high-temperature reaction proceeding at ambient temperatures. A later publication from Zaire by Marynen and Bredas[7] showed that the germination of *dura* seed had remained satisfactory for samples stored up to 12 months though with *tenera* seed germination was more erratic and seemed to diminish earlier. This was probably due to the more rapid drying out of *tenera* seed in store. Subsequent experiments in Zaire confirmed Galt's finding regarding the earlier germination of stored seed.[8]

A thorough investigation of storage had to await the devising of fully efficient germination techniques. Storage of seed for at least some months is often necessary because nursery seedlings of transplantable size may be required for planting out only at the beginning of a rainy season. An extreme case is to be found in parts of West Africa having a single wet season where planting in the field is best carried out between March and May and where nursery seedlings establish most easily when planted in April. Even in these areas, however, methods have been devised of establishing nurseries in October, so the storage period can be reduced from 12 to between 5 and 7 months. In other parts of the world nurseries may be established in many months of the year, and there is more latitude in the time of planting in the field. Under these circumstances, problems of seed storage are not of such importance.

Seed has been conveniently stored on a large scale in 44-gallon drums covered with a wooden lid. Such a drum will contain 25,000 to 30,000 African *dura* seed or rather smaller quantity of Deli *dura* seed. The effect on subsequent germination of storing large quantities of seed at both ambient (*c.* 27°C) and air-conditioned (22°C) temperatures (i) when stored after normal air drying for 2½ days in shade, and (ii) when stored at around the optimum moisture content (see p. 140) for germination has been investigated in an experiment.[9] Under all the treatments investigated, viability remained very high up to 12 months of storage. In the drums kept at ambient temperature with seed maintained or started at a moisture content near the optimum for germination, very considerable germination took place between 6 and 15 months of storage, amounting to between 56 and 72 per cent. Such treatment is therefore clearly impracticable.

Interest centres therefore on seed stored at ambient temperature after normal air drying to 14 to 17 per cent moisture or seed stored wet (21 to 22 per cent moisture) in an air-conditioned room at 22°C. In the experiment referred to, where the original moisture contents were estimated visually, there was a net loss of moisture with time in the seed at the top of the drums, though this was almost negligible with the seed stored wet. In the middle and bottom of the drums there was a marked rise in moisture content, the levels reached being largely maintained for 12 to 15 months. Some results from these two treatments in the experiment are shown in Table 6.1.

The results show clearly that viability can be maintained at ambient temperatures at moisture contents too low for germination or at low tem-

Table 6.1 *Viability (as indicated by germination percentage) and moisture content of seed stored under two sets of conditions (Rees)*

Original moisture contents*	Temperature	Position in drum	Months of storage				
			3 (%)	6 (%)	9 (%)	12 (%)	15 (%)
A. Germination percentage							
14–15%	27°C	top	100	96	88	89	75
	(ambient)	middle	100	95	96	95	95
		base	100	94	95	97	96
21–22%	22°C	top	100	99	97	98	100
	(air-	middle	93	94	91	92	98
	conditioned)	base	86	86	90	91	96
B. Seed moisture contents							
14–15%	27°C	top	15.5	14.4	13.2	14.2	13.6
		middle	21.4	18.2	18.0	18.2	19.1
		base	20.4	19.3	17.7	18.4	18.6
21–22%	22°C	top	23.0	21.8	20.4	21.8	21.0
		middle	24.5	22.1	23.0	24.3	24.6
		base	24.6	24.4	24.4	24.8	22.4

* Visual estimation.

peratures under higher moisture conditions. In another storage experiment[4] using these two temperatures, small lots of *dura, tenera* and small *tenera* seed were all stored at the optimum moisture contents for germination. Three storage treatments were used: (i) polythene bag under an inverted wooden box, (ii) polythene bag exposed, (iii) seed exposed to air. In all cases the seed dried out sufficiently to inhibit germination in storage. Between 92 and 100 per cent germination was obtained after 1 year's storage in all lots except two (one being destroyed by rats, the other unexplained). These experiments and the one referred to earlier (p. 313) clearly indicate that prestorage drying is likely to be more deleterious than any gradual drying after the start of storage; moreover storage temperature seems to be unimportant.

Where seeds are stored on a large scale without regard to careful drying, viability may be lost at over 1 per cent per week. It will be noted from Table 6.1 that this rate of loss was only approached in the top of the ambient-temperature drum when seed moisture content had fallen to below 14 per cent. Other work has shown that loss of viability during storage at about 1.2 per cent per week occurs at ambient temperatures and a moisture content at or below the air-dry level (about 14 per cent).[3, 10]

In conclusion, it can be said that seed may be satisfactorily stored at ambient temperatures without appreciable loss of viability if it is carefully dried in the shade for not more than 2 days. While drum storage has proved quite satisfactory most large seed producers now prefer to store seed in polythene bags in an air-conditioned atmosphere at 22°C. However, it must be emphasized that it has not been shown experimentally that air-conditioning has any advantage over ambient conditions. Seeds have also been satisfactorily stored on trays with air-conditioning to give a temperature of 20°—22°C and a humidity of 60 per cent.

Practical methods of germination

The requirements for a rapid and even germination of fresh seed are: (i) a temperature of 39°—40°C for 80 days at a moisture content too low for germination but not below about 14.5 per cent; (ii) transfer to ambient temperatures and the optimal moisture content for germination, i.e. 21 to 22 per cent for *dura* or 28 to 30 per cent for *tenera*. The optima for thinner shelled *dura* may be intermediate.[28] These moisture contents are achieved in practice by keeping the seed as wet as possible but with no superficial moisture; alternatively samples may be taken for moisture determination by oven-drying to constant weight. Satisfactory germination is obtained with Deli seed in Malaysia after only 40 days' heat treatment, and where seed has been stored for 3 months or more shorter periods will suffice.

These requirements can best be met by keeping the seed in polythene bags and placing them in a germinator in which temperature control to within 1°C can be assured. An electrically-controlled germinator has proved most satisfactory, but germinators heated by hot water pipes can be satisfactorily operated provided some means of temperature warning is installed. Many attempts have been made to accelerate germination by soaking in acids, alkalis or solutions of growth substances. Cracking the seed and prechilling has also been tried. No effects on germination of these treatments have been found.

Germinators

Germinators for oil palm seed are by no means new. The temperature requirement, though inexactly appreciated, has been known for a long time. Galt[11] gave a comprehensive account of the germination methods and germinators which were in use up to 1953; certain of these continued in use for many years.

In the Far East, where bare-ground temperatures in the daytime are usually consistently higher than in Africa owing to the much longer periods of sunshine, open sand beds were in general use for a long period. Pure sand or a sand layer over a layer of compost were employed, and the beds were often concrete-sided and raised off the ground.[12] Experiments showed that with Deli *dura* seed, germination of over 85 per cent could be expected in 6 to 8 months, though faster germination sometimes occurred, particularly when glass frames were placed over the beds.[13] Germination was often erratic as well as slow, however, and for this reason the method *is not recommended.*

Germinators in which heat is applied within the body of the germinator may be divided into (i) fermentation pits or boxes, (ii) oven-type or flue-type germinators heated by a wood fire, (iii) boiler and hot-water pipe germinators, (iv) electrically-controlled germinators. *The latter are recommended* wherever large-scale germination is to be carried out over many years.

Fermentation boxes

Fermentation of vegetable matter, as a method of heating the seed, appears to have originated in the Ivory Coast, but was developed in Zaire[14] and Dahomey. The use of the fermentation box also saw the introduction of wood charcoal as a medium for germinating the seeds. This was found to be much more satisfactory than sand, sawdust or mesocarp fibre,[15] but has now given way to the practice of maintaining naked seeds at the correct moisture content in polythene bags.

The well known Zaire fermentation box has been described by Vanderweyen[5] and by Marynen and Bredas.[7] About 20,000 African *dura* seed (30,000 *tenera*) can be germinated in a box 1.10 m wide and 3 m long. It can be constructed of concrete or have sides of wooden planks. The depth should be about 60 cm but one side is raised slightly above the other to accommodate a gently sloping detachable roof. About one-third of the box is sunk into the ground. The box is kept supplied with fermentable vegetable material to within a few inches of the top. Green material from various plants has been tried. A mixture of leaves and skins of bananas, grasses, legumes, etc., has been used in Dahomey. In Zaire *Pueraria phaseoloides* and *Galopogonium mucunoides*, coarsely shredded, were preferred, but when not available cut lalang (*Imperata cylindrica*), elephant or Napier grass (*Pennisetum purpurum*) and other grasses were used.

In this fermenting material are placed the wooden boxes containing the seed mixed with moist charcoal. Deep boxes measuring, internally, 60 x 20 cm and 35 cm deep are used. Six seed boxes are placed in each fermentation box. The thickness of the seed-plus charcoal mixture should be limited to about 12—13 cm (5 inches).

While good results could usually be obtained from the fermentation box when operated by practised hands, it was recognized[16] that its greatest inconveniences were (i) the need for constant knowledgeable supervision to maintain sufficiently high temperatures and to prevent over-heating (temperatures of around 50°C are sometimes reached), (ii) the difficulty of obtaining suitable green fermentable material. Both published results[7] and experience have indeed shown that the germination percentages obtained are very variable; the most successful fermentation boxes have been situated near to managers' houses.

To overcome this variation and to improve the handling of the seed, Desneux[16, 17] designed a fermentation pit 1.2 m wide and 75—90 cm deep which could be of any convenient length. A length of about 4 m accommodated 25,000 to 30,000 *dura* seed in five boat-shaped 'half-drums' (200-litre or 44-gallon drums cut longitudinally) lined with nylon netting. When the seed is examined for germination it is simply lifted out of the half-drums in the netting. In an attempt to maintain high but even temperatures in the pit three new features were introduced. Firstly, half drums with their contents are placed on top of, and partially in, the fermenting material and are exposed to the sun; at night or in rainy weather they may be covered with straw or some other protective material. Secondly, the fermenting vegetation is covered with 10 cm of lateritic soils; this slows down the fermentation process and makes it unnecessary to renew the vegetable material. Thirdly, the pit it orientated east—west to obtain maximum heating from the sun. The half-drums are first filled with charcoal alone and it is not until the temperature has been stabilized at about 38°C that the seed is introduced. The mixture is then covered with a thin layer of charcoal. Coarse charcoal of particle size 1—5 mm has been found most satisfactory.

The fermentation box and its later modifications (Fig. 6.1) are ingenious, but insufficiently exact temperatures and seed moisture contents are maintained for assured regularity of germination or for the use of the heat pretreatment. Results[17] show that germination, even when the final percentage is satisfactory, proceeds relatively slowly and often takes 5 to 6 months to complete. The system can therefore only be recommended for use by practised hands where the means cannot be made available for maintaining constant temperature for operating the dry heat-treatment system.

Oven-type and water-heated germinators

Various crude methods of applying the heat of a wood fire to seed embedded in sand or moist wood charcoal were used in the early days of oil palm culture in West Africa.[11] The first artificial germinator was adapted from a propagator used in East Africa for the vegetative propagation of coffee. This was later further modified to form the oven-type germinator which has been widely used on small stations. An oven-type germinator has been successfully employed for the dry heat treatment of seed in polythene bags, but constant reliable supervision of the temperature of the boxes in which the bags are kept has seldom been available. Oven-type germinators[11] are units consisting of a number of upper and lower chambers, the latter for wood fires and the former for holding charcoal-filled seed boxes or polythene bags which are placed on slatted shelves. The chambers are separated by sheet metal covered with a cement layer. A 'hot-room' germinator, developed in Zaire,[18] was an extension of the oven-type germinator. In this germinator a square flue passed through the hot room from the fire box to a chimney.

More satisfactory than the open-fire type of germinator are the various water-heated systems which have been developed in West Africa. The first

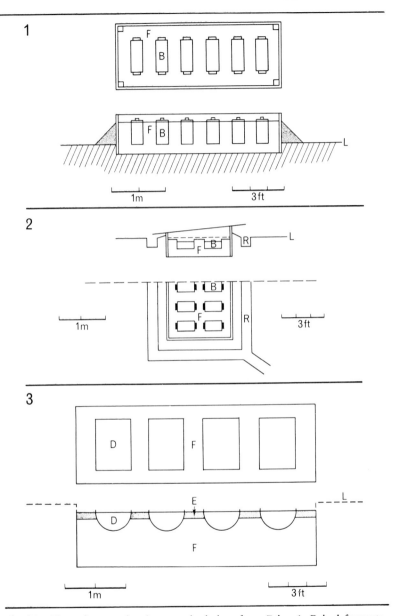

Fig. 6.1   Three fermentation box or pit designs from Zaire. **1.** Raised fermentation box with deep seed boxes (Vanderweyen). **2.** Roofed fermentation box with double row of shallower seed boxes (Marynen and Bredas). **3.** The Kiyaka half-drum fermentation pit (Desneux). E. Red earth layer. F. Fermenting vegetation. B. Seed and charcoal boxes. D. Seed and charcoal half-drums. R. Rainwater drain. L. Ground level.

of these was introduced in Dahomey[1] in 1931, and this type of germinator became general in Nigeria from 1949.[11] The germinator may be of any convenient size according to seedling requirements and is divided into the working shelter in which the boiler is situated, and the germination room. The latter is provided with large-diameter hot water pipes as in normal glasshouse practice. The boiler is of a type used for domestic hot water systems and it is fired by wood. The walls can be constructed of concrete blocks and a double ceiling must be provided. Any windows must be double-glazed. The hot water pipes run beneath the staging on which are placed the boxes containing the polythene bags of seed. Passage ways may run on both sides of central staging or down the middle of the room with staging on the sides.

Temperature control in water-heated germinators is easier than in open fire-heated germinators, but results still depend on the integrity of the operator and adequate supervision of his work. Improved control can be obtained by the use of two thermostats connected to warning lights which light up if the temperature rises above or falls below certain points. This does not entirely eliminate the 'human factor' however, and although water-heated germinators can be generally recommended, electrically-controlled germinators are the most satisfactory.

Electrically-heated and controlled germinators

A small electric germinator was first constructed at N.I.F.O.R. in Nigeria in 1958.[19] A description of the essential features of this germinator are given below. The building is constructed of concrete blocks to the dimensions in Fig. 6.2. The walls are double with a 13 cm (5 in) air gap and the ceiling is of double asbestos sheeting. The double doors are hollow with a 13 cm (5 in) air space, and they enclose an air lock where the fuse boxes, main and subsidiary switches, relays and time switch are housed. The internal working space is 3.9 x 2.2 x 2.0 m high and there are 61 cm (2 ft) wide slatted shelves on each side of a 1.22 m (4 ft) central gangway.

The placing of the heater, the air intake louvre, the oscillating fan and the air exit fan are of great importance and an even temperature on the shelves is not obtained unless the design is followed and the apparatus operated correctly. The design originally allowed for a humidifier which was tried out because it was thought that it might be possible to germinate naked seed on the shelves without the protection of polythene sheeting. It was found, however, that seed dried out in spite of use of the humidifier and full germination of viable seed was only obtained by use of polythene bags. Moreover, the constant high humidity in the germinator led to rapid depreciation of apparatus and materials.

The germinator is heated by six 500 W strip heaters mounted in front of a 1/6 hp fan, the whole unit being placed on the floor against the back wall. Aeration is achieved periodically by a 1/6 hp exhaust fan near the top of one wall and an automatic louvre near the heater at floor level; the louvre opens when the exhaust fan is in operation. The fan and louvre are

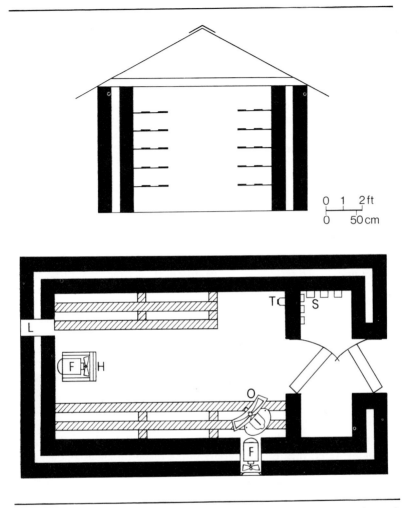

Fig. 6.2 An electrically-controlled germinator (Rees, 1959). Median section and plan. F. Fan. H. Heater. O. Oscillating fan. T. Thermostat. S. Switches and relays. L. Automatic louvre.

operated by a time switch. It was found that there was only a very slight drop in temperature when the extractor fan was in operation. The building is heated rapidly from cold by all six strip heaters; at 1°C below the setting of the thermostat three heaters are cut out by the operation of one-half of a two-stage thermostat mounted on the wall. Temperature distribution is achieved by the oscillating fan on the top platform. The right positioning of this fan will remove any 'islands' of too high or too low a temperature and reduce the temperature range to 1°C.

A small germinator of this type, holding up to 200,000 seeds in polythene bags is inexpensive to construct. Power consumption per week is less than 300 kWh.

It should be noted that in constructing electrically-controlled germinators of different sizes certain essential features must be retained, viz. double walls and ceiling, air lock with double doors and control panel, heater on floor and exhaust fan near top of shelves. The number of heater strips required, the stages of thermostatic control, and the obtaining of a uniform distribution of temperature will all require trial *before operation*, using oscillating fans in different positions until all 'pockets' have been eliminated. It is only when a uniform temperature of $39° - 40°C$ has been achieved that germination should be started.

Germination technique

Assuming that a germinator can be consistently operated to give the correct temperature for heat treatment, then germination may be carried out by one of two techniques which may be designated (i) wet heat treatment, and (ii) dry heat treatment. In the former case the whole process will be shorter (95 days against 120 days), but some germination will take place during the heating period and so a less uniform set of seedlings will be produced. Watering will be unnecessary throughout the heating period with the dry heat treatment technique so there will be considerable economies of watering time. Better results are almost invariably obtained with this technique as the need for gauging the optimum moisture content through the heat period is obviated.

Wet heat treatment

The seed should be soaked for 5 days in buckets or pans with a daily change of water, then dried carefully in the shade for about 2 hours until adhering water evaporates, but the seed is still almost black. Polythene bags to receive this seed are most conveniently made from 69 cm (27 in) wide 500 gauge lay flat tube. This is cut into 61 cm (24 in) lengths and one cut edge is welded by running a small flame along the edge which is held between two flat pieces of wood so that about 3 mm project. Five hundred to 700 *dura* seeds or about 1,000 *tenera* seeds can be put in these bags, the open end of which is then secured by a stout elastic band made from cut-up cycle inner tube. The bags are then placed in boxes of convenient size (30 x 20 x 10 cm) and put in the germinator where the temperature should be $39°C$. The seed is examined twice weekly and watered with a fine spray from a small hand mist sprayer to restore the correct moisture content (Plate 22). This may be defined as — *as wet as possible but with no superficial moisture*. Alternatively, samples may be taken for moisture determination by oven-drying to constant weight. Moisture contents should be 21 to 22 per cent for *dura* and 28 to 30 per cent for *tenera*. Any

Pl. 22 Small *tenera* seeds having their moisture content adjusted, on inspection, by a fine spray.

germinated seed is removed for sowing in the prenursery or nursery.

After 80 days the boxes and bags should be taken out of the germinator and put in a cool place and the correct moisture content maintained as before. A flush of germination will start a few days after cooling and, if watering in the germinator has been done with care, germination will be complete in about 15 to 20 days.

Stored seed requires less heat treatment than fresh seed, so if some of the seed being germinated has been stored for a few months, more will germinate before cooling and the flush of germination of the remainder will not be so pronounced.

Dry heat treatment

The technique of dry heat treatment entails the heating of seed at 39°C when it is in a condition too dry for germination, but not so dry that the heat reaction is impaired. The seed must be soaked for 5 days with a daily change of water, as for the wet heat treatment. It should then be dried for 24 hours in the shade, spread out in a single layer. A moisture content of 17 per cent has been found satisfactory in Malaysia with Deli *dura*,[29] but a rather higher content has been suggested in Nigeria.[28] *Excessive drying must be avoided.* The seed is then placed in polythene bags as previously described. These must be checked from time to time to see that no seam has become unwelded nor rubber band broken.

After 80 days the seed is removed, soaked again in water for 5 days with a daily change of water, then dried slightly in the shade for about 2 hours to evaporate surface water. The amount of drying needed before returning seed to the bags has been described as 'until their colour passes from shiny to dull black'.[30] At this stage the seed may be retreated with TMTD and streptomycin (see p. 312). It is then returned to the bags, kept in a cool place, and examined twice a week to maintain, by means of the fine spray, the correct moisture content for germination (Plate 22). There will be a delay of some 10 to 20 days after the 80 days; germination should then be rapid and be complete in 10 to 15 days (Plate 23).

Pl. 23   A flush of germination in a polythene bag, following the cessation of heat treatment.

For very small lots of seed 600 ml jars covered with 200 gauge polythene sheet may be used.

An additional advantage of the use of the dry heat treatment is the avoidance of Brown Germ disease of just germinated seedlings caused by *Aspergillus fumigatus*. This fungus shows optimum growth at a temperature near that of heat treatment, but when seed is germinating at ambient temperature following dry heat treatment, growth is usually too slow to cause serious infection before the non-susceptible stage is reached.

Length of period of heat-treatment

Though 80 days is normally required for fresh African *dura* seed, stored seed requires a shorter period and some lots of seed, whether fresh or stored, have been found to have lower heat requirements. Experience with selected Deli seed in Malaysia has shown that 40 days is normally sufficient to provide over 80 per cent germination and this period is now standard practice in some germinators.

To obtain the most complete and rapid germination possible with a given lot of seeds the wet-heat treatment may be combined with a sampling procedure devised by Tailliez.[20] This may be useful if seedlings are urgently required and there is reason to believe that the heat period requirement may be well below 80 days. Samples are taken from the germinator weekly and kept at ambient temperature. As soon as any such sample has attained 80 per cent germination (and this may be expected within 20 days if the heat requirement has been fulfilled) all the seed may be taken out of the germinator to ambient temperature. The method was devised for use with a wet charcoal substrate, but could also be employed with the polythene bag wet-heat method.

Germination and embryo abnormalities

Examination of the seeds of large numbers of Deli *dura* x *pisifera* crosses has shown that embryo abnormalities may consist of malformation of the whole or part of the embryo, incomplete development, double or triple embryos, a tendency to necrosis or the complete absence of an embryo. A direct relationship was found between both the percentage and speed of germination and the percentage of normal embryos.[21] Crosses with a low percentage of normal embryos show a high variability in embryo length, suggesting that a low embryo normality and hence low germination percentages indicate that a progeny is strongly heterogeneous. While this may be so, it should also be borne in mind that both the length of time of pollen storage and the methods of pollen preparation used have an affect on the percentage of normal embryos.

Noriet and Ahizi Adiapa[21] suggest that in seed production one cross in five should be examined. They test successive samples of fifty seeds and eliminate a cross unless 0, less than 4, less than 7, or less than 20 abnormal seeds are found in samples 1, 1 + 2, 1 + 2 + 3 or 1 + 2 + 3 + 4 respectively. This is equivalent to the imposition of a standard of normality of 90 per cent. If one or more crosses have to be eliminated all crosses made in the same month are checked.

Germination of *pisifera* seed

Plant breeders occasionally wish to propagate *pisifera* x *pisifera* progenies. *Pisifera* seeds are difficult to germinate because they tend to dry out rapidly and are susceptible to fungal and bacterial infection. A method developed in Malaysia has given germination percentages of fertile seeds of from 5 to 52.[22] The percentage of fertile seed varied between parents and between bunches but rarely exceeded 60 per cent.

The seeds are extracted from the fruit and cleaned to remove all mesocarp, great care being taken not to nick or bruise the testa. They are washed and rinsed in 1 per cent Clorox (0.05 per cent sodium hypochlorite) before being put into polythene bags. They are then treated in the germinator at 39°–40°C by the wet-heat method and individual seeds are removed as germinated or if showing any signs of fungal or bacterial infection. Superficial fungal growth can be removed by washing when the seeds are examined and the likelihood of penetration to the embryo is thus reduced.

In Ghana, as high a germination percentage as that quoted above was obtained simply by allowing bunches to lie under the palms during the wet season and protecting them from rodents with wire netting. In this case the mesocarp was clearly giving some protection to the kernel and the heat requirement was being fulfilled by the heat of fermentation.[23]

Germination of *Elaeis oleifera* seed

The germination percentages obtained with *E. oleifera* seed have usually been low, and in general the methods used for *E. guineensis* have been used. It has been reported that in Colombia very satisfactory germination has been obtained by adopting the dry heat treatment with 22 per cent moisture content for an initial 15 days, then soaking in warm water at $43°C$ for 15 minutes and following this with the wet heat treatment for 65 days.[31]

Despatch of germinated or preheated seed

Seed can be sent long distances after preheating or when germinated. In Malaysia, newly germinated seed is usually placed in 200 gauge polythene bags at the rate of 200 seeds per bag.[24] These are packed in boxes with vermiculite at the rate fifteen bags (3,000 seeds) per box of 18 x 12 x 11 inches (46 x 30 x 28 cm). The advantage of this method is that the seed can remain in the bags for up to 10 days before planting, though the plumule and roots should not be allowed to grow so long that they become twisted.

In the Ivory Coast a box has been designed for the rapid distribution of germinated seed by motor vehicle to distant plantations where prenurseries are being established.[25] The box consists of a framework 50 x 30 cm in which are inserted twelve shallow trays, 3.5 cm high, each containing 250 germinated seeds embedded in sawdust. Thus a consignment of 30,000 in ten boxes can readily be despatched in one vehicle.

Germinated seed should be despatched as soon as germination has started and if possible not later than the fourth day, when the root and plumule can just be distinguished (D of Fig. 2.1, p. 44). This will enable planting to be done up to 10 days later when the root and plumule are clearly differentiated (F–G of Fig. 2.1).

Germinated seed can be badly damaged by treatment with seed dressings containing fungicides or pesticides, and Brown Germ cannot be controlled by such methods (see p. 606). Gamma-BHC is particularly dangerous and should not be used either with seed or in prenurseries.

Seed can also be sent in polythene bags at the end of the dry-heat period and its moisture content can be adjusted to the optimum for germination on arrival at its destination.

If seed is sent from one country to another the sanitary regulations of the receiving country must be observed. Seed is normally treated with fungicides and pesticides but fumigation with methyl bromide has been found satisfactory provided the seed moisture content is not more than 10 per cent. Experiments have shown that such seed may be fumigated at 24°−30°C with 1 kg methyl bromide per 30 cubic metres for up to 18 hours without affecting viability, germination or growth.[26] Methyl bromide has also been employed for the fumigation of germinators between periods of usage. Mercurial dressings should not be used. Fungicide and pesticide treatment may consist of washing for 2−3 minutes in 0.2 per cent Thiram with 0.1 per cent Dipterex (trichlorphon) and including a spreader, e.g. Teepol, at 0.1 per cent. It is important to have the seed as clean as possible before treatment and it should therefore be prewashed in Teepol.

Locke and Colhoun[27] have shown that *Fusarium oxysporum* f.sp. *elaedis*, the pathogen of Vascular Wilt disease (see p. 627), can be seed borne. They isolated the fungus from seed despatched from West Africa to England and showed that even where routine seed disinfection had been carried out the seeds yielded large numbers of other fungi, though not *F. oxysporum*. This study showed that some current methods of disinfection are inadequate and that the most stringent procedures are required.

References

1. **Henry, P.** (1951 and 1952) La germination des graines d'Elaeis. *Revue Int. Bot. appl. Agric. trop.,* **31,** 565; **32,** 66.
2. **Rees, A. R.** (1962) High-temperature pre-treatment and the germination of seed of the oil palm, *Elaeis guineensis* (Jacq.). *Ann. Bot.,* N.S., **26,** 569.
3. **Rees, A. R.** (1962) Some observations on preparation and storage of oil palm seed. *J. W. Afr. Inst. Oil Palm Res.,* **3,** 329.
4. **Rees, A. R.** (1965) Some factors affecting the viability of oil palm seed in storage. *J. Nigerian Inst. Oil Palm Res.,* **4,** 317.
5. **Vanderweyen, R.** (1952) *Notions de culture de l'Elaeis au Congo Belge.* Brussels.
6. **Galt, R.,** W.A.I.F.O.R. Report (unpublished).
7. **Marynen, T. and Bredas, J.** (1955) La germination des graines d'Elaeis. *Bull. Inf. I.N.E.A.C.,* **4,** 156.
8. La conservation des graines d'Elaeis par la Division du Palmier à Huile. *Bull. Inf. I.N.E.A.C.,* **7,** 31 (1958).
9. **Rees, A. R.** (1963) A large scale test of storage methods for oil palm seed. *J. W. Afr. Inst. Oil Palm Res.,* **4,** 46.
10. **W.A.I.F.O.R.** (1961) Ninth Annual Report 1960−1, p. 92.
11. **Galt, R.** (1953) Methods of germinating oil palm seed. *J. W. Afr. Inst. Oil Palm Res.,* **1,** (1), 76.
12. **Lucy, A. B.** (1940) Experiments on the germination of oil palm seed. *Malay. agric. J.,* **28,** 151.
13. **Bunting, B., Eaton, B. J. and Georgi, C. D. V.** (1927) The oil palm in Malaya. *Malay. agric. J.,* **15,** 297.

14. **Beirnaert, A.** (1936) Germination des graines d'Elaeis. *Publs. I.N.E.A.C.* Série Tech. No. 4.
15. **Ferwerda, J. D.** (1956) Germination of oil palm seeds. *Trop. Agric. Trin.*, 33, 51.
16. **Desneux, R.** (1959) Une méthode simplifiée pour la germination des graines du palmier à huile. *Bull. Inf. I.N.E.A.C.*, **8**, 23.
17. **Desneux, R.** (1960) A propos d'une méthode simplifiée de germination des graines du palmier à huile. *Bull. Inf. I.N.E.A.C.*, **9**, 43.
18. **Desneux, R.** (1957) La germination des graines d'Elaeis en chambre chaude à la station de Kiyaka. *Bull. Inf. I.N.E.A.C.*, **6**, 11.
19. **Rees, A. R.** (1959) Germination of oil palm seed: Large scale germination. *J. W. Afr. Inst. Oil Palm Res.*, 3, 83.
20. **Tailliez, B.** (1970) Germination accélérée des graines de palmier à huile. Technique avec substrat. *Oléagineux*, **25**, 335.
21. **Noiret, J. M. and Ahizi Adiapa, P.** (1970) Anomalies de l'embryon chez le palmier à huile. Application à la production de semences. *Oléagineux*, **25**, 511.
22. **Arasu, N. T.** (1970) A note on the germination of *pisifera* (shell-less) oil palm seeds. *Malay. agric. J.*, **47**, 524.
23. **Wonkyi-Appiah, J. B.** (1973) Germination of *pisifera* oil palm seeds under plantation conditions. *Ghana J. agric. Sci.*, 6, 223.
24. **Bevan, J. W. L. and Gray, B. S.** (1966) Germination and nursery techniques for the oil palm in Malaysia. *Planter, Kuala Lumpur*, 42, 165.
25. **Sérier, J. B.** (1966) Transport à longues distances des graines germées de palmier à huile. *Oléagineux*, 21, 211.
26. **Mok Chak Kim** (1970) Effects of methyl bromide on oil palm seed. *Proc. Int. Seed. Test. Ass.*, 35, 243.
27. **Locke, T. and Colhoun, J.** (1973) *Fusarium oxysporum* F.sp. *elaeidis* as a seed-borne pathogen. *Trans. Br. mycol. Soc.*, **60**, (3), 594.
28. **Odetola, A.** (1974) A re-examination of large-scale germination of the oil palm seed at NIFOR. *J. Nig. Inst. Oil Palm Res.*, **5**, 79.
29. **Turner, P. D. and Gillbanks, R. A.** (1974) Oil palm cultivation and management. Incorp. Soc. of Planters, Kuala Lumpur.
30. **I.R.H.O.** (1970) Germination des graines de palmier à huile en sacs de polyéthylène. Méthode par 'chaleur sèche'. *Oléagineux*, **25**, 145.
31. **Chew Poh Soon** (1976) Private communication.

Chapter 7

The raising of nursery seedlings

Early bearing in the field is dependent on the transplanting of healthy seedlings from a nursery. A great deal of attention has therefore been paid to nursery techniques, and different systems have been perfected in different parts of the world in response to differences of climate, soil, disease incidence and management. Damage and exposure of the root system of the palm at any stage of the nursery process is followed by a slow recovery, and so all modern nursery systems allow for the roots to remain encased in soil from early growth until the seedling is planted in the field.

Seedlings are more difficult to raise in the seasonal climate of Africa than they are in the more uniform climate of the Far East and they are much more subject to nursery diseases. For this reason much more research has, until recently, been done on nursery methods in West Africa than elsewhere. For a long period the standard method of raising seedlings was to plant germinated seeds in prenursery beds or pots and when they reached the four- to five-leaf stage to transfer them to a specially prepared field nursery where they grew on for about a year. The raising of nursery seedlings in pots has many managerial advantages but was initially unsuccessful in Africa mainly owing to the high disease incidence which accompanied the heavy watering found necessary. In the Far East, however, satisfactory methods of raising seedlings in large polythene bags were developed in the mid-1960s and the field nursery has everywhere become the exception rather than the rule.

Nursery practice often depends, sometimes to the detriment of the plant's growth, on the form in which planting material is received. Usually, and understandably, it is the wish of the seed producer to get rid of his product as early as possible. However, the need for special germinators with skilled men to operate them often compels the producer to distribute germinated seed or small seedlings. In the Far East it was at one time common for the producer to raise seedlings to the two-leaf stage in a sand bed and then distribute them to estates where they would be planted into baskets or small polythene bags and kept for a few months before they were transferred to nurseries. This is not the ideal stage for moving

329

seedlings, and the distribution of germinated seed has been found more satisfactory (see Chapter 6, p. 326).

In Africa, the advent of the dry heat-treatment system led to the distribution of heat-treated but ungerminated seed to large subcentres where germination could easily be carried forward at ambient temperature near to prenursery and nursery sites. Distribution to the planter, who in this case has only a small acreage, is then accomplished when the nursery seedling is sufficiently developed for field planting. Where field nurseries were used the introduction of root-cutting (see Chapter 9) made this system much more successful than previously.

Prenurseries

Prenurseries may be constructed of bags, baskets or beds or the prenursery stage may be omitted altogether. Few comparative trials have been carried out and healthy plants can be raised in either bags or beds, so the decision is usually dictated by conversance or managerial convenience. The prenursery methods to be adopted depend in some measure on the stage at which the planting material is distributed. Seedlings at the two-leaf stage may be sent out from the distribution centre either packed in wet soil in boxes or as bare-root seedlings packed in polythene bags. In either case the seedling will have to be planted in a small basket or polythene bag, unless the prenursery stage is omitted altogether, and this extra transplanting causes a severe check to growth. Nevertheless, by careful handling and the use of a suitable potting mixture, an even set of seedlings can be raised for transference to the nursery at the four- to five-leaf stage some 2 or 3 months later.

The obtaining of a suitable potting or bedding soil may present some difficulty. A wide variety of soils and mixtures have been successfully used of which the following may be mentioned:

1. Sandy topsoil partially sterilized by heating over a fire (Nigeria).
2. Deep friable topsoil, overlying alluvial clay, undiluted or mixed with a small proportion of coarse river sand (Malaysia).
3. Stiff clay topsoil mixed with coarse river sand in proportions 3 : 2 (Malaysia).
4. Peat and sand mixed in equal proportions (Malaysia).
5. Sandy topsoil (Zaire).
6. Inland clay-loam topsoil mixed with river sand and well-rotted cattle manure (Malaysia).
7. Sifted forest topsoil (Ivory Coast).

In general a fertile topsoil sufficiently free-draining to prevent 'puddling' or sealing of the surface is required. In areas where soils tend to be too heavy the admixture of a proportion of coarse sand is always desirable. On the other hand many soils of good structure have been successfully used without mixing with sand and if too much sand is used the soil may break up on transplanting to the nursery.

Baskets and polythene bags

Baskets should be about 25 cm deep and 10 cm broad, but black polythene bags of 250 gauge and of similar size (25 x 10 cm lay-flat) are now preferred. If bare-root seedlings are being planted some of the soil medium to be used is placed at the bottom of the bag and the seedling is then held in position while further soil is packed round it, layer by layer, great care being taken to see that none of the primary roots are fractured in the process. Root fracture is the main cause of blackening off of seedlings transferred to bags. Some damage to the root system is inevitable in transplanting bare-root seedlings and, for this reason only, it may be necessary to provide some light shade for about 2 weeks after planting. Where germinated seed is planted no shade is required provided the supply of water is sufficient; however, if overhead irrigation is not installed or water supplies are likely to be restricted, then a light shade of palm leaves or shade cloth erected on frames may be required for a few weeks and should be removed in stages. It should be emphasized that entirely satisfactory prenursery plants have been raised without shade and that over-shading causes etiolation and depression of growth (Plate 25).

Baskets rot very rapidly under constantly moist conditions and they may sometimes be in an advanced condition of decay by the time the seedlings are ready for transference to the nursery. Great care has therefore to be taken in transporting them. Black polythene bags do not have this drawback. If the bags are properly filled in the first place, the growth of the root system tends to assist the formation of a compact ball of soil within the polythene film. When the polythene is torn away just before planting, this ball retains its shape and can be inserted in the nursery planting hole without fracturing. Basket plants are planted in the nursery in their baskets.

Bags should be filled to within half an inch from the top. The germinated seed is placed in a hole made with the finger; this hole is about 2.5 cm deep and in the centre of the bag. The seed is then covered in with soil. The radicle and plumule of the germinated seed should be clearly differentiated and care taken to see that the plumule is pointing upwards and the radicle downwards; twisted seedlings are thus avoided (Plate 24). A just-germinated seed, where only a 'button' is showing, may be returned to the polythene bag for a few days until it has developed further; if it is planted immediately it should be laid on its side, i.e. with its long axis horizontal. The radicle and plumule will then grow respectively downwards and upwards with the minimum of bending. One advantage of planting as soon as germination has taken place is that damage in handling is less likely to occur. A disadvantage, apart from the possibility of bent growth, is that infection by Brown Germ (*Aspergillus* sp.) may not be noticed. By the

Pl. 24   Planting a germinated seed in a prenursery.

time the plumule has developed, however, any infection will be obvious and the infected seed may be discarded. Each seed should be carefully examined for abnormalities of growth and development. If such abnormalities are rigorously excluded the amount of roguing needed in the prenursery will be reduced.

The use of polythene bags at the prenursery stage has been found particularly useful when germination is carried out centrally and nurseries are to be established far from the germinator and near to large plantings. Bags may be arranged in plank-walled 'beds'; in the Ivory Coast beds of twenty lines of 175 bags (3,500 bags) have been found convenient and the bags are transported to the nursery sites when the seedlings have reached the five-leaf stage.[1]

Prenursery beds

Germinated seed is usually planted in prenursery beds at or near ground level. The beds, which are 20—23 cm deep and for convenience should not be more than 1.2 m wide, require low retaining walls of brick or wood. The ends are removable to facilitate transplanting. Such beds will be satisfactory for an estate where a finite programme of planting is going to be carried out over a period of a few years; where, however, the raising of prenursery seedlings is going to be a continuous process it is more convenient to set aside an enclosed area for raising seedlings in raised trays (Plate 25). A permanent system of mist irrigation can then be constructed and the seedlings can be planted and later dug up with great ease, while the

Pl. 25 Overhead spray-lines in operation over a raised-tray prenursery in Nigeria.

marking of progenies is greatly facilitated. Experiments and experience have shown that a more uniform set of seedlings is obtained in trays than in beds. The trays should be raised 60 cm (2 ft) off the ground and be 15−20 cm deep; a convenient size is 120 x 60 cm.[2]

The best medium for prenursery beds or trays is a sandy topsoil, but any of the mixtures described at the beginning of this section may be used. The soil should be partially sterilized before being put in the beds or trays. This is carried out quite simply by baking on a steel sheet; weed seeds and insects are killed by this treatment.[3] In Zaire parathion at a concentration of 8 g to 10 l water has been used[4] and this method of insect control is also advocated in the Ivory Coast; in Malaysia however it is believed that the young growing point may be endangered and the application of insecticides as granules scratched into the surface soil is recommended.

The spacing of germinated seed in the bed or tray depends on the stage at which the seedling is to be transferred to the nursery. Transplanting to the nursery at the two-leaf stage has often been recommended in the past and in that case planting of germinated seed 5 cm apart has been thought suitable. However, experience and experiments have shown the advantages of transplanting at the four- or five-leaf stage and, for this, the plants must be spaced at 7.5 or 10 cm. Apart from the question of stage of growth, however, seedlings so spaced can be lifted and transferred to the nursery in

Pl. 26   Removing seedlings from a raised-tray prenursery in soil blocks.

soil blocks and their root systems are thereby protected (Plate 26). Wide-spaced seedlings do not, unless grown beyond the correct stage, become etiolated, and they respond well to fertilizers. At 7.5 cm, 128 seedlings may be raised in a tray of internal measurement 120 x 60 cm; this is equivalent to 177 seedlings per square metre of bed. At 10 cm apart 81 seedlings are accommodated per square metre. Planting of germinated seed is carried out in the same manner as when using bags.

Maintenance of the prenursery

Shading of beds or bags is not required when overhead irrigation is supplied (see Plates 25, 27) or when hand watering is adequate.[5] The deleterious effect of shade on growth at the prenursery stage has been demonstrated by workers in Zaire.[4] Mulching, on the other hand, has been shown to be beneficial and mulched seedlings subsequently grow better in the nursery.[5] The most satisfactory mulching material is finely divided bunch refuse. This may be applied soon after planting and the seedlings grow up easily through it. If bunch refuse is not available, sawdust, palm shell, groundnut husk or other fibrous material may be used.

The primary requirements for maintaining a steady growth after

Pl. 27 Large polythene bag prenurseries under irrigation in the Jengka Triangle area, Pahang, Malaysia.

emergence of the first leaf is adequate watering and a balanced supply of fertilizer. Mulching is a supplementary factor which conserves moisture, prevents soil compaction and, to a lesser degree, provides nutrients.

The frequency of watering depends, of course, on the amount and frequency of rain falling during the prenursery period. Adequate watering arrangements must be available, however, and provided the soil used is not too heavy and drainage is unimpeded, there is little danger of over-watering. In severe dry weather twice-daily watering will be needed at the rate of one 4 gallon watering can per two trays or 2 square metres of bed or, when bags are used, to 'run off'. Can-watering is inevitably uneven and in any permanent prenursery site an irrigation system should be installed. In Malaysia it is considered that irrigation equipment is justified for plantings of 400 hectares or more.[6] The simplest system is one employing polythene tubes and spray nozzles designed to give a fine mist spray over the whole prenursery area.[7] Tubes of 1½ inches (3.8 cm) bore are cut into 9–12 m lengths and are jointed together with fittings provided by the suppliers of the tube; they are connected directly on to the water supply which should have a pressure of 20 psi (1.4 kg per sq. cm) or more. The tubes are supported 1 m above the seedlings; frequent supports are required. The sprayline can be moved as needed or lines can be installed to cover the whole area. Pairs of two types of nozzles,* inserted alternately

---

* Nozzle types E and F (hole size No. 7) manufactured by British Overhead Irrigation Ltd have been used.

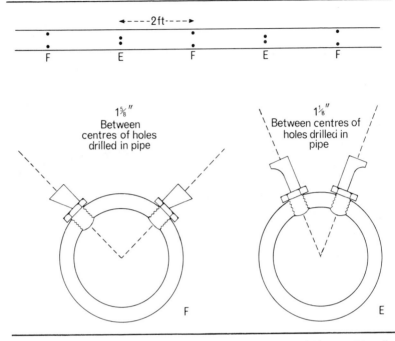

Fig. 7.1 The set of B.O.I.L. nozzle types E and F in polythene tubing for prenursery overhead irrigation.

and spaced 60 cm apart along the line, have been found to cover the area evenly. One pair is set in holes drilled into the tubing $1\frac{5}{8}$ inches (41 mm) apart, the other in holes $1\frac{1}{8}$ inches (29 mm) apart, as shown in Fig. 7.1. The nozzles are turned to spray outwards and they are usually situated at the top of the pipe. The equipment can be operated upside down, however.

If very fertile forest topsoil is used for filling prenursery beds or bags, fertilizer application may not be needed, but such soil is rarely obtainable and, in most media used, prenursery seedlings have shown marked responses to fertilizer, particularly to nitrogen. The amounts required are very small. Experiments in Nigeria and Malaysia have shown responses to nitrogen and phosphorus, particularly as ammonium phosphate, but not to potassium. A solution of 2 oz of ammonium phosphate in 4 gallons of water (57 g in 18 l) should be applied to four standard trays or per 4 square metres of bed every week. If sulphate of ammonia is used 2 oz in 4 gallons (57 g in 18 l) every 2 weeks may be applied. The beds or trays should be lightly watered afterwards to remove the fertilizer from the leaves. With bag seedlings in Malaysia urea has been used at ½ oz in 1 gallon (14 g in 4.5 l) of water per 100 seedlings per week.

Since seedlings are in the prenursery for such a short time very little weeding is usually required. Hand weeding is therefore resorted to. However, in the Ivory Coast it is claimed that application of ametryne (Gesapax) will reduce costs to 10 per cent of those of hand weeding.[8] This herbicide can be applied before the emergence of the plumule at a rate equivalent to 2–3 kg of the commercial product per hectare or after the emergence of the plumule but before the appearance of the first leaf. In the latter case the concentration is of importance and the herbicide is watered on at a strength of 0.1 to 0.3 per cent; for 10,000 seedlings an average of 18 g of the commercial product in 9 l water is sufficient.

In Africa, anthracnose fungi (see p. 607) may be troublesome in the prenursery and regular weekly spraying with Captan or Cuman organic fungicides starting about 6 weeks after planting has been found effective. In the Far East *Curvularia* Leaf Spot is occasionally encountered. Pests such as snails, grasshoppers and night-flying beetles are usually controlled by hand collection.

Nurseries

It is necessary to raise seedlings in nurseries firstly because germinated seed or very small seedlings cannot easily be protected in the field from insects and rodents, secondly because such very young material will produce an uneven stand, and thirdly because the planting of large seedlings shortens the time between the completion of field preparation and first cropping. When an area is being opened from forest and it has not been possible to prepare a nursery ahead of this operation, then the planting out of small nursery seedlings may be justified, but if nursery work is planned ahead, as it should and usually can be, large seedlings 12 to 14 months old in the nursery should usually be the aim, for these, in spite of an initial check in the field, will produce the earliest crop. Age of seedling at the time of transplanting to the field is discussed in more detail in Chapter 9.

The raising of nursery plants in baskets or pots instead of in a field nursery has been attempted, and for periods practised, in many oil palm countries since plantations started. Early experience in the Far East was not very promising and field nurseries were the rule until the mid-1960s. In Africa, experience varied. Vanderweyen pointed out that basket nurseries can be sited where convenient, the soil being transported if necessary and that the method facilitated transport and handling.[9] Marynen and Poels[10] claimed that basket seedlings grew better after planting in the field than field nursery seedlings, but their comparison was not with root-pruned nursery plants and their method entailed the planting of two baskets per point. In West Africa, Gunn et al.[1] found that the growth of seedlings in pots and baskets was greatly inferior to that of field nursery seedlings. The heavy watering required by basket plants was conducive to severe incidence of Freckle. In West Cameroon, basket nurseries were used for a period of many years but, in spite of heavy watering, Blast incidence was high and Freckle difficult to control. Field nurseries were subsequently reverted to.

The successful use of pots in the Far East has been due to the work of Gray, who, following experience in Sabah, carried out trials with large black polythene bags filled with friable topsoil from areas of coastal alluvium. He showed that, with heavy watering, large healthy seedlings could be produced in bags weighing only about 35 lb (16 kg); and he later demonstrated that it was possible to dispense with the prenursery stage by planting germinated seed straight into the large 'polybag'.[6] Subsequent trials showed that other soils could also be employed for polybag nurseries.

The superiority of the polythene bag over the basket is probably due to the fact that the drying out between waterings which occurs with the openwork basket and which is likely to hinder root development is much reduced with the use of polythene. In Malaysia, root development is good in polybags and although cases of a disease similar to Blast of Africa have occurred in polybag nurseries there is evidence that this was due to underwatering.

The use of polythene bags for nursery seedlings received some stimulus from their prior and successful use in the rubber industry; but the reasons for the adoption of polythene bags for established rubber budgrafts have no parallel in oil palm planting. In the former case the object has been to reduce the time from planting to first tapping by carrying out in the nursery work normally done in the field; and to ensure full success at transplanting so that the number of rubber trees planted will be only slightly greater than the number required when tapping begins.[11] With the oil palm, by contrast, there is no operation to be carried out either in the nursery or field other than root-pruning, and the latter cannot be done with bagged plants; a 'take' of near 100 per cent in the field is demanded of any transplanting system and is normally achieved; and the shortest period from planting to bearing is obtained by planting large seedlings however these are raised.

Experiments comparing polybag and field nurseries have been carried out in both Malaysia and Nigeria. In the Malaysian experiments, there were no significant differences between palms raised in polybags and those raised in field nurseries either in growth or early bunch yield, but the survival rate, which was generally very high, was slightly higher with polybag palms. Palms previously raised in prenursery beds were less advanced than those raised in prenursery bags, but this was due largely to the erosive effect of heavy rain on the plank-sided beds used.[12]

In the Nigerian experiment, polybag plants showed slower growth in height in the nursery than field nursery plants, but they had a lower Blast incidence, produced more leaves and more plants were adjudged transplantable. On planting into the field the polybag plants showed more rapid initial growth and after 18 months were significantly taller and had produced more leaves.[13, 14] Plants originally raised in prenursery bags grew better and had less Blast in the nursery than those grown in prenursery raised trays, but the growth difference did not persist into the field. The cultural practices used for field nurseries, i.e. irrigation, no shade and standard fertilizer treatment, were found applicable to polybag nurseries.[14]

In the Malaysian experiments, omitting the prenursery stage and planting germinated seed direct into large polybags gave larger seedlings on transplanting to the field at 13 months from germination than were obtained from using prenurseries, but this effect did not persist for more than a year in the field.[12]

Since polybags are now widely used for nurseries not only in Asia but also in Africa and America, it is now possible to assess the relative value of the two systems in practice. Bevan and Gray[6] have set out the advantages and disadvantages of polybags and field nurseries in some detail, and their main characteristics and requirements may be contrasted here as follows:

Field nursery	Polybag nursery
Site and soil	
Flat or gently sloping land. Soil must hold together at transplanting, but must be well drained. Site preparation includes cultivation. Favourable weather conditions at planting may be important unless irrigation system installed.	Site may be chosen for managerial advantages rather than agricultural ones. But large and regular water supply must be available. 'Cultivation' entails only clearing and levelling. Suitable soil must be transported to site. Relative independence of weather conditions.
Maintenance	
Little watering necessary under non-seasonal climatic conditions. Regular weeding necessary for whole area.	Regular watering essential. Weeding of polybags only.
Transplanting	
Root-cutting improves establishment. Digging out with ball of earth. Seedling must be moved and transported in wrapping or head pan. Great care required to avoid breakage of ball of earth with consequent delay in establishment.	Root-cutting not possible. Detachment of bag from ground (some roots penetrate from the bag). Seedling transported in bag without further wrapping. No shattering of ball of earth likely and hence transplanting less dependent on weather conditions and little chance of delay in establishment.

From the above it may be concluded that polythene bag nurseries should certainly be employed where the soil will not readily hold together in a ball of earth at planting time; many poor plantings have been due to this cause. On the other hand where a large and constant water supply cannot be assured, field nurseries should be used. Beyond these two cultural factors, the differences are largely ones of organization and cost, the balance of opinion favouring the polybag nursery on these grounds though only marginally in the case of costs.

The polythene bag nursery

Black polythene of 500 gauge is used for these bags. Different sizes are still being employed, but bags too wide in proportion to their height tend to become lopsided while those too high in proportion to their width may lean over. Bags 20 inches (50 cm) deep by 15 inches (38 cm) wide, lay-flat, have been found serviceable and are in general use; they are perforated with holes spaced 3 x 3 inches (7.5 x 7.5 cm) on the lower half of the bag for drainage.[6] When filled to the top they contain about 16 kg of nursery soil, the exact amount depending on the type of soil or mixture used (Plate 29).

Normally, the polybag nursery will be sited on a convenient level piece of land near to the essential water supply and, if possible, not far from the source of soil. Soil—sand mixtures should be similar to those used for pre-nurseries (p. 330). The bags are best filled at the source of soil and then transported to the nursery as this entails less handling of soil than filling at

Pl. 28   A large polythene bag nursery in Malaysia.

Pl. 29 A polybag plant about 75 cm (2.5 ft) high, 4 to 6 months before field planting. Bag 38 x 50 cm (15 x 20 in lay-flat).

the nursery site. The addition of cattle manure has been shown to improve growth in the bags.

Where an irrigation system is in use the bags may be spaced 3 x 3 ft (90 x 90 cm) triangular from the start (Plate 28). Closer spacing will cause etiolation of larger plants. With hand watering the bags can most economically be placed close together in rows three-bags wide until the seedlings are about 6 months old from the germinated seed stage. They are then separated to 3 x 3 ft (90 x 90 cm).

Polybag nurseries may be planted with either (i) prenursery seedlings,

(ii) bare-root seedlings from seed suppliers; or (iii) germinated seed.

Prenursery seedlings at the four- or five-leaf stage will be transplanted from bags or beds. In the former case the small bag is simply torn off and the seedling with its ball of soil is inserted in a hole dug in the soil in the large bag and the soil consolidated around it. In the latter case the method used will depend on the prenursery bed soil mixture. This is described in detail on p. 346.

Bare-root seedlings as supplied by seed suppliers are usually planted in a prenursery, but they may be planted direct into large polybags of the usual size. Shading will be required during the first few weeks and can be gradually removed. Planting is carried out in the same manner as in the prenursery (p. 331).

In the direct planting of germinated seed, methods are also the same as for the prenursery, but unless irrigation water can be supplied at the heavy prenursery rate then some shade is required until the one- or two-leaf stage is reached. The success of direct planting in Malaysia has seemed to depend on this shade which can be supplied most economically by staking palm leaflets in each bag to form an umbrella over the developing seedling. The system has been reported as being more expensive in labour and water in the first 4 months than the prenursery and nursery polybag system, but experiments have shown that this disadvantage is outweighed by the reduction of overall time in the nursery by about 2 months.[6] It is probably inadvisable to introduce the system until a considerable skill in nursery management has been attained; nevertheless, it is now the most commonly practised nursery method in all parts of the world.

Maintenance of a polybag nursery entails the weeding of the bags and the supply of sufficient water. With hand watering in Malaysia the minimum watering is considered to be four times weekly to 'run off' using, in the absence of rain, 2–4 gallons (9–18 l) per plant per week according to stage of growth, soil type and other factors.[6] With this heavy watering the soil surface needs to be disturbed from time to time to prevent compaction. Irrigation systems are discussed on p. 350. Water usage will of course be much greater with an irrigation system than with hand watering since in the former case the water will be applied to the whole area and not only to the soil in the bags. Water losses from bags will be higher than from an equivalent area of field nursery. However, provided there is sufficient water available, an irrigation system is to be preferred because it is cheaper to operate and usually gives better results. With hand watering the requirements per week are not as great in the early stages as when the plants are well grown and presenting a large leaf surface for transpiration.

Field nurseries

Field nursery sites should be as flat as possible, adjacent to a good water supply and not subject to prolonged waterlogging or impeded drainage; the soil should be fertile, friable and capable of forming a good ball of earth at transplanting time. Proximity to the areas of field planting and to access roads is desirable. If the nursery area is being opened from forest, felling, thorough rooting, stacking and burning must be carried out fairly rapidly before planting so that the area does not stand bare for any length of time. Concentrations of ash from the burn must be dispersed.

Following this standard treatment, the area must be cultivated and manured. The type of cultivation to be given and the manurial requirements depend very largely on the soil. In the deep sandy soils of West Africa it was found that deep cultivation (25 cm) with a disc plough increased growth and that growth was further assisted by the ploughing in of organic manure, usually in the form of bunch refuse, at the rate of 50 tons per acre.[5] The effect of organic manure is greater on areas which have already been used for cultivation. In the Ivory Coast subsoiling is recommended where digging or ploughing cannot be carried out deeply or the subsoil tends to be impermeable.[15] On clay soils, whether inland or coastal, in the Far East, the depth of ploughing is largely determined by the depth of topsoil, it being undesirable to bring to the surface large clods of subsoil. The value of organic manuring has also been demonstrated on Malaysian inland soils and the ploughing in of well-rotted cattle manure or composted bunch refuse is to be recommended when either is available.

Cultivation can, of course, be carried out manually with a spade or hoe, but this is generally less effective and the incorporation of organic manure more difficult. Ploughing is followed by levelling with a disc harrow or by hand labour.

The incorporation of inorganic manure before planting a nursery is not usually practised. Both in the Ivory Coast and in Malaysia the incorporation of about 3—4 cwt of rock phosphate has been recommended; in the latter country this practice has been passed over from the rubber industry where the value of rock phosphate in rubber nurseries has been established. For oil palm nurseries, however, more soluble fertilizers may be advantageous and mixtures of urea, triple superphosphate and sulphate of potash/magnesia have been used.

Plants spaced too closely in the nursery tend to become drawn up and weakly and when transplanted to the field are more likely to suffer wind damage than plants spaced wider apart. Experiments in Nigeria[16, 5] have shown that 2 feet (61 cm) is too close and that a density of more than about 17,000 plants per hectare (7,000 per acre) is disadvantageous. No advantage is to be gained by planting at triangular spacing at the nursery stage; at the same density per hectare square planting gives as satisfactory seedlings as triangular planting. Square planting has certain practical advantages especially where it is desired to plant separate progenies in lines or rectangular blocks.

Close spacing also makes disease control more difficult and this is of particular importance in Africa where *Cercospora* Leaf Spot is troublesome. The results of a density experiment, after the seedlings had been 11 months in the nursery, are given below:

Density	Height per palm (cm)	Weight per palm (kg)	Weight of base of palm (kg)	Score for severity of Cercospora Leaf Spot
17,290 palms/hectare	150	1.89	0.69	25.3
27,170 palms/hectare	161	1.36	0.48	51.1

Square planting at 75–90 cm (2½–3 ft) is therefore to be recommended, with wider spacing (90 cm, square or triangular) being advisable if the plants are to be left for 1 year or more in a soil of average to high fertility.

The layout of a field nursery will depend on whether or not irrigation equipment is to be used. With such equipment paths must be left along the lines of piping with occasional cross paths for inspection.

It is usual not to re-employ nursery sites as nurseries, but at experiment stations semi-permanent sites have been established, parts of the site being fallowed for several years between usage. This is quite satisfactory at least for three or four rounds over a 20-year period in spite of the removal of soil and provided plenty of bunch refuse is available for mulching or ploughing in. Special continuous nursery experiments were carried out in Nigeria[17] with the object of determining whether nurseries could be satisfactorily maintained on permanent sites in crowded grove areas where seedlings might be planted by the improved naked root method (see p. 404) and where large balls of soil would therefore not be removed. In each case the area was divided into two halves which were used in alternate years. Over a period of 10 years it was shown that satisfactory nurseries could be grown under this system where bunch refuse at 50 tons per acre was ploughed in and a mulch of this material supplied. Fertilizer requirements varied with locality.

Roguing in the prenursery

The idea that a large amount of roguing in oil palm prenurseries and nurseries is required has been widespread but is not borne out by the experience of those who have maintained a high standard of cultural practice. While there may be a very small percentage of stunted or abnormal seedlings in the prenursery which can only be accounted as genetical in origin, the majority of poor doers or abnormal plants may well be the result of cultural deficiencies, usually bad planting or insufficient water. The plants which must be removed are those suffering from a twisted growth or showing symptoms of the freak conditions Leaf Roll, Leaf Crinkle or Collante which are described in Chapter 13 (p. 611). Apart from these easily distinguishable conditions (Plate 30), plants which should be rogued include those which show poor development, abnormally

Pl. 30   Certain abnormalities of nursery seedlings: **A.** Leaf Crinkle. **B.** Leaf Roll. **C.** Collante.

erect or dumpy habit, or have been affected either by poor planting of the germinated seed or by being left too long in the prenursery and hence become etiolated. In raised trays, a larger number of healthy seedlings are often, owing to double seeds, obtained than the number of seeds planted.

There are fairly wide differences in the expectation of percentage transplantable prenursery seedlings. In Nigeria general experience has been that 95 per cent are transplantable but in the Ivory Coast the average losses from deaths and abnormalities have been given as 15 to 20 per cent. In Malaysia the expected loss is 5 to 7 per cent. It is possible to raise both seedlings from double seeds, but it is sounder practice to discard one, retaining the other with the seed attached.

Transplanting from the prenursery to the nursery

Various methods of transferring prenursery seedlings to a field nursery have been in use. If the seedling is in a basket, the latter can be planted into a field nursery since it soon rots away when planted in the ground. The other methods of transplanting are generally applicable also to a polybag nursery. Small polybag prenursery plants are carried to the planting hole when the polythene film may be ripped off and the cylinder of soil containing the plant inserted directly into the hole. With plants raised in a prenursery bed or raised tray, the method to be employed depends upon the consistency of the soil in the prenursery. If the soil holds together well, longitudinal and transverse cuts may be made in the soil between the plants 4 weeks before transplanting or at the time of transplanting. In the former case some proliferation of roots within the block of soil takes place before planting in the nursery and this may assist initial establishment and growth. At planting, the soil blocks with their seedlings are carefully lifted out of the prenursery and planted in the prepared planting holes (Plate 26).

When the consistency of the soil prevents the lifting of an intact block, a special tool may be used. Several of these have been devised of which the following may be mentioned:

1. *Plantoir Richard.*[18]   This is a small cylinder 14 cm high and 10 cm in diameter which is pressed down over the plant into the soil of the prenursery. The plant, which is at the four- to five-leaf stage but has been lightly leaf-pruned, must be exactly in the centre of the cylinder which has two small flaps serving as handles. After rotating the cylinder with the aid of the flaps, the cylinder and the ball of soil with the seedling may be lifted from the prenursery, the vertical roots being cut with scissors. Seedlings thus encased may be sent long distances and on arrival the cylinder is inserted in the nursery planting hole. A flat plate shaped in the form of a U and of slightly smaller diameter than the cylinder is then placed on the soil surrounding the seedling and held in place by a handle

attached vertically to it. The cylinder is then withdrawn by its handles and the soil round the seedling is firmed down.

2. *Square-sided scoops.*   Where a prenursery bed has a detachable plank at one end, this may be removed, and scoops which exactly fit the block of soil in which each seedling is growing may then be used to remove them. A simple three-sided instrument with a handle has been found quite satisfactory, but a more refined design includes a movable plate which lies flat against the middle side and can be pushed forward by a handle connected to it through a hole in this side. Thus, when the seedling reaches the planting hole in the scoop it is pushed out by the movable plate and drops into the hole with the soil intact.

Seedlings taken from a bed prenursery should be planted at once. If they have to be transported some distance they should be dug early in the day and use should be made of some transplanting tool or cylinder as already described. If it is impossible to carry them embedded in a block of prenursery soil then they may be very carefully lifted and dipped in a clay slurry and tied together in bundles before transport, but this method is not to be generally recommended.

In planting, holes a little larger than the block of soil to be inserted should be dug with a trowel. The soil block is then carefully inserted so that it does not shatter and the surface of the soil block is made level with the surrounding soil. Deep planting must be avoided. Seedlings not surrounded by a block of soil will require even more careful handling. The roots must be inserted in the hole dug by the trowel and the soil must be packed round them so as to ensure that they are not bent or broken.

If there is no rain after planting, the seedlings must be watered daily until rain falls. In Africa, it is necessary to spray the plants at this stage against Anthracnose and later against *Cercospora* Leaf Spot.

Time of planting

In seasonal climates and in areas subject to Blast disease, the time of planting a nursery is of considerable importance. Elsewhere nurseries may be laid down at any time of year though it is preferable to choose a period when rain is most likely to fall and which allows a lapse of 12 to 16 months before the prearranged field planting programme. Both in Zaire and in the Far East nurseries are commonly planted all the year round and nursery plants are taken for planting in the field in the wetter periods beginning in April or September.

In West Africa the field planting season runs from the very beginning of the rains, in March, to the time when the rains have fully set in in May or early June. In the Ivory Coast, where the rainfall is more even than in other parts of West Africa and completely dry months are rare, April is also preferred as the planting month even though the short dry season of

August and September is more pronounced than elsewhere and October and November rainfall is high.

Apart from the consideration of field-planting time, however, early planting of nurseries is desirable in Africa for the avoidance of Blast disease. Experiments in Nigeria have shown that the heaviest Blast attacks occur with seedlings which are planted, or are still small, in August or September. Planting as early as possible in the rainy season is therefore desirable and, in practice, April is found to be the best month. By the time the dry season is reached, in November, the plants have had 7 months in the nursery and are well forward. With irrigation throughout the months of December to March they continue to grow and by March they have reached a height of around 5 feet (1.5 m).

Blast disease in Nigeria tends to attack seedlings of a susceptible, i.e. small, size and usually only between the months of October and January. *Very* small seedlings planted in October and shaded are not susceptible and for this reason a 'dry-season nursery' planted in that month was used in Nigeria.[5] This nursery can be forced forward by irrigation, mulching and more frequent fertilizer dressings and seedlings become transplantable in late May when 70−100 cm (2−3 ft) high. This is not, however, the optimum size for transplanting, and dry-season nurseries are no longer commonly used. In the Ivory Coast, where the dry season is less severe, transplanting from prenurseries into polybag nurseries is commonly done as late as September, but heavy shade is used as an anti-Blast measure from the time of planting until February. Such shading causes etiolation and the system cannot be generally recommended for West Africa; it should only be used where Blast is a peculiarly intractable problem (see p. 615).

To summarize, in planning germination and nursery work, timing should be calculated back from the start of the planned time of field planting, as follows:

Germination (dry heat treatment)		3−4 months
Prenursery	4−5 months	
Nursery	12 months	
Total Nursery Period		16−17 months
Total		19−21 months

Thus if planting is planned to run from September to November, germinated seed should start to be sown in the prenursery in April or May of the previous year. This should ensure that the earliest seedlings have been at least 1 year in the nursery before field planting starts. In order to start prenursery sowing in April seed must be set in the germinator in late December or early January. However, when only 40 days' heat treatment are given[6] this period is shortened by over a month and seed could be set in February.

For the seasonal programmes of West Africa the following sowing and planting schedules have been used:

Sowing or planting	Main season	Dry season
Seed set in germinator	September	March
Prenursery	December—January	June—July
Nursery	April	October
Field	March—April	Late May, early June

If direct planting of germinated seed into the nursery bags is practised, the overall period is reduced by 2 or 3 months, and seeds will be set to germinate 2 to 3 months later than shown in the above schedules. With the single-stage nursery the seedlings are normally ready for transplanting in 11 to 12 months and germination should therefore be taking place exactly a year before the time it is intended to plant in the field.

Water supply

In countries of high and even rainfall very little watering will be required after the plants have settled in and the need for watering can be reduced by adequate mulching. Wherever there is a distinct dry season, as in West Africa, or when there is a period of several weeks of drought, watering is necessary, particularly if this drought period should fall in the first 6 months of the plants' growth.

Watering may be done by hand or by means of an irrigation system. The latter is much more efficient and is to be recommended wherever large nurseries are to be planted over a number of years.

Field nurseries

Mention has already been made in Chapter 4 of the water requirements of young seedlings. In West Africa the equivalent of 1 inch (25 mm) of rain per week has proved adequate for well-grown seedlings in field nurseries though newly-planted seedlings require more, and in the 'Harmattan' conditions of the dry season the requirement is even higher. In other oil palm growing regions the soils, with some exceptions, tend to be more retentive of moisture but, on the other hand, the overcast conditions of West Africa are not usually encountered and day temperatures and hours of sunshine are higher. It is therefore not safe to assume that the water requirement in these regions is any less than in West Africa.

With hand watering, a minimum of 1 gallon (4.5 l) per seedling per week should be given and in very dry weather this should be increased to 2 gallons (9 l). In the Ivory Coast, irrigation in the dry season at the rate of 20 litres (about 4½ gal) per week per square metre, equivalent to 2 cm of rain, was recommended with a planting distance of 80 cm.[15] Later, however, recommendations based on estimations of stomatal aperture were adopted[19] (see p. 150). Hand watering is costly, about 1,000 man-days per hectare per year being required under Nigerian conditions.

Irrigation systems for oil palm nurseries may be of the large rain gun type, needing pressures of around 100 psi (7 kg/cm²) and covering about 1 hectare at the rate of 25 mm per hour, or they may be of the smaller rotary type requiring lower pressures with a minimum pressure of 20 psi (1.4 kg/cm²). The latter types are usually more suitable unless very large nurseries are to be maintained over a long period of time. Trickle type irrigation has not been found satisfactory. Water requirements are large and a careful check must be made of the source of supply.

The following system operated satisfactorily in a field nursery in Nigeria over a period of more than 10 years.[7] Water is supplied by gravity feed from a high level tank giving a pressure of 20–30 psi (1.4–2.1 kg/cm²) with a static head of water of 50 feet (15 m). This pressure can of course equally well be provided by a suitable pump. A 3 inch (7.6 cm) main aluminium line supplies 2 inch (5 cm) lateral lines spaced 40 feet (12 m) apart. Low-angle sprinklers* on stand pipes 18 inches (45 cm) high are placed every 6 m along the lateral lines. With two main lines and four lateral lines of 120 feet (36 m) in length, 2 inches (5 cm) of water is supplied to 0.44 acres (0.18 hectare) in 4 hours. The apparatus is easily moved and a little less than 1 acre (0.4 hectare) can therefore be supplied with 2 inches (5 cm) per day, and about 3½ acres (1.4 hectare) covered twice a week, or 6 to 7 acres (2.6 hectare) twice every 10 days. Watering at this rate is only necessary when the seedlings are small and the weather dry; to seedlings well grown by the beginning of the dry season one application of 2 inches (5 cm) per week is given.

Not more than two men per day are required to operate a complete irrigation system for 3 to 7 acres (1.4–2.8 hectares) of nursery (Plate 31).

Polythene bag nurseries

The construction, layout and operation of irrigation systems for poly-bag nurseries have been described in great detail for Malaysia by Bevan and Gray[20] and for West Africa by Coomans.[21] The former authors base their designs on the assumed need for 0.33 inch (8.4 mm) rain or irrigation per day after allowing for losses due to evaporation, the spacing of bags 3 feet (91 cm) triangular from the start, and the provision of water by spray lines which supply the 8.4 mm in 1½ hours.

In their design for a 25-acre (10 hectare) nursery (see Fig. 7.2) they allow for spray lines shifted seven times during the day giving 1½ hours spraying in each position. Each spray line is of 3 inches (7.6 cm) diameter served by a 6 inch (15.2 cm) main pipe from the pump house. The long beds lying between the sprayline positions are 49.4 feet (15 m) wide and divided into five sub-beds to allow for paths along which the workers shifting the lines can move. The spray lines are 705 feet (215 m) long with twenty-four sprinklers; they have a throw of about 43 feet (13 m) radius and supply 3.6 gallons (16.4 l) per minute at 40 psi (2.8 kg/cm²). Single-nozzle sprinklers ($\frac{5}{32}$ in − 4 mm) are suitable. A nursery of this size will

---

* Junior low-angle sprinklers supplied by Messrs Wright Rain Ltd.

Pl. 31 Nursery irrigation: irrigation equipment in a field nursery, mulched with bunch refuse in Nigeria.

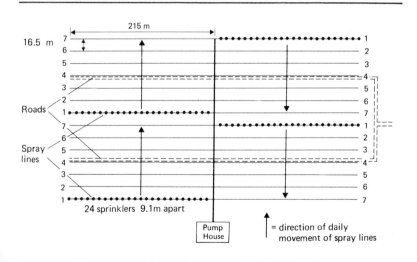

Fig. 7.2 Plan of nursery irrigation system covering 10 hectares and supplying about 100,000 seedling (Bevan and Gray).

provide about 130,000 seedlings which, after culling, would be sufficient for about 700 hectares (1,750 acres).

Under the system recommended by Coomans[21] the nursery is organized in two stages, first with the bags at 40 x 40 cm and later, at the end of February, at 70 x 70 cm. Under this system less water will be required but labour will be needed for reorganizing the nursery in February, and the eventual spacing of 70 cm, in contrast with 91 cm under the Malaysian system, may lead to etiolation. Water is provided on the assumption that evapotranspiration is 5 mm per day, that the well-watered bags (40 x 40 cm lay flat) have a capacity equivalent to 60 mm of water and that the lower limit of the usable water reserve is 15 mm. Thus in 9 dry days all available water would be used, so a daily log is kept both of evapo-transpiration and of the water supplied by rain, and the need for irrigation is gauged from the state of the water reserve in the bags and the capacity of the irrigation system to supply. If, for instance, only 30 mm can be supplied in a day the calculated reserve must not fall below 35 mm on the day previous to irrigation.

The designs recommended for this system allow for wider-spaced sprinklers (36 m) with a further throw (26 m radius) than the Malaysian system. In the absence of rain, bags at the original close spacing are supplied twice a week firstly at 20 mm with 3 hours' watering and secondly at 15 mm with 2 hours' watering. When set out at the eventual 70 cm spacing the irrigation system is extended and each plant is irrigated twice for 3 hours (20 mm per period) on one of the first 3 days of the week and then twice for 2 hours (15 mm per period) on one of the next 3 days. It will be noted that whereas in the Malaysian system the plants will receive 58.8 mm (8.4 x 7) per week, in the West African system the younger plants will receive less than this (35 mm) but the older plants more (70 mm). More recently Quencez[22] has estimated the water require-ment per plant at 4 mm per day in the first 2 months, rising, with increasing evapotranspiration, to 10 mm after 6 months in the nursery, and he recommends three periods of irrigation per week, each irrigation round providing 9 mm per plant up to 2 months of age rising to 23 mm at 6 months. With shaded nurseries the recommended quantities are halved. This author has provided detailed examples of irrigation requirements assuming intermittent rainfall and plans for an irrigation system suitable for the needs of a 4.5 hectare nursery for a 500 hectare field planting with bags placed at 70 cm triangular.[22] With the wider spacing used in the Malaysian nursery 7 hectares are needed per 500 hectares of field planting.

Post-planting cultural treatments

Shading of a nursery is never required except as an anti-Blast measure. In West Africa it has been noticed that plants shaded through the dry season increase more rapidly in height under the shade, but later, when the shade is removed and the rains have set in, they grow more slowly than unshaded plants.[23] Unless, therefore, the area concerned has peculiar problems of Blast control the plants should not be shaded.

Field nurseries

The mulching of field nurseries, whether with organic material or artificial mulches, improves seedling growth. The most effective mulch, and often the most easily obtainable, is bunch refuse. On a new estate far from other oil palm plantings bunch refuse may not be available. In this case any cut vegetation may be used, large-leaved grasses being particularly suitable, or an artificial mulch of black polythene sheeting may be laid down. It is not sufficient to maintain a mulch merely in a dry season. Mulching is most effective when applied 2 to 4 weeks after planting and maintained or replenished at least until the seedlings are fully covering the ground. There is some evidence that if the supply of bunch refuse is limited it is better used as a mulch than ploughed in before planting. Where soil potassium and magnesium levels are low, bunch refuse mulches may induce symptoms of magnesium deficiency (Orange Frond) and fertilizer dressings must then contain a sufficiency of magnesium.

Except where unusually fertile jungle soil is being employed for the first time, fertilizers, particularly nitrogen, are always required for maximum growth of plants in a field nursery. On the poorer sandy soils of West Africa applications of organic manure before planting, and of mulch after planting, seem to have a greater effect than fertilizers on growth. The need for nitrogen is perhaps most pronounced in the Far East. However, in almost all oil palm growing areas growth is improved by applications of N, P, K and Mg. Applications of N and K without Mg tend to induce magnesium deficiency symptoms which are very common in nurseries. Lack of nitrogen is clearly distinguished by the characteristic chlorotic appearance of the leaves. Applications of phosphorus as superphosphate have given increases of growth and enhanced the responses to nitrogen. Potassium applications improve growth as well as reducing the incidence of *Cercospora* Leaf Spot. Magnesium applications, although they have not been shown to improve growth, remove the symptoms of Orange Frond. A suitable fertilizer mixture contains:

1 part sulphate of ammonia
1 part superphosphate (18 per cent P_2O_5)
1 part sulphate or muriate of potash
2 parts magnesium sulphate (or 1 part anhydrous magnesium sulphate)

If ammonium phosphate is used as the source of P instead of superphosphate then the proportions may be $3:1:2:4$, though in this case the mixture will contain a higher proportion of N and P and a lower proportion of K and Mg.

In the Far East, fertilizer mixtures are often applied monthly, but experiments in Nigeria have shown that with nurseries planted in April it is sufficient to apply a total of 4 oz (100–120 g) per plant according to the following schedule:

May	½ oz (14 g) per plant	Age – 1 month
July	1½ oz (43 g) per plant	3 months
October	2 oz (57 g) per plant	6 months

In West Africa the dry season sets in in November and no further dressings are usually needed. In non-seasonal climates nitrogen in particular may become insufficient and dressings of the mixture may continue as follows:

2 oz (57 g) per plant	Age – 7 months
2 oz (57 g) per plant	9 months

For the special dry-season nursery under irrigation in West Africa more frequent applications of fertilizers are required: 1 oz per palm per month starting 1 month after planting.

Fertilizer mixtures applied to seedlings should be spread in a ring 2–3 inches (5–8 cm) away from the seedling for the first application, gradually widening to 6–8 inches (15–20 cm) from the seedling for the last application.

Polythene bag nurseries

With polybag nurseries in the Far East monthly application of a compound fertilizer of formula N12:P12:K17:Mg2 is recommended[6] starting about 1 month after transplanting from a prenursery and increasing the rate per plant from 7 g at 1 month to 14 g at 2 to 4 months, 21 g at 5 to 6 months, 28 g at 7 to 9 months and 42 g thereafter.

Now that the single-stage polybag nursery is more usual the manuring methods and schedules have to be properly adapted to the stage of growth. The first 3 to 4 months correspond with the prenursery stage. No fertilizer is given during the first month but from the beginning of the second month to the end of the third month, i.e. from the one- to three-leaf stages, urea can be watered on weekly at the rate of about 7 g in 5 l of water per 100 seedlings; this rate may be doubled at the four- to five-leaf stages and application of solid fertilizers, either compound or a mixture such as that given on p. 353, will then be applied from about the end of the fourth month. Monthly applications are then usual, starting with about 10 g per plant and gradually increasing the rate to about 35 g at 10 months. At all stages care has to be taken to see that the fertilizers do not cause leaf scorch and washing off with water after the manuring round is therefore recommended.

For West African soils Surre[24] has recommended, in addition to mixing the soil with some form of organic material, the watering in before planting and when the bag is half full of 75 cl of a solution of 4 kg urea, 3 kg KCl, 4 kg superphosphate and 1 kg kieserite in 200 l of water. This is followed by monthly applications of 5–10 g urea and 5 g kieserite per plant. In Nigeria satisfactory growth in polybags has been obtained by mixing soil with cattle manure and supplying fertilizers at the same rates and frequencies as in field nurseries (see p. 353).[14]

Weeding

Nursery weeding is usually carried out by hand. In field nurseries the need for weeding is considerably reduced by the methods of mulching already described. With polybags monthly weeding rounds are recommended in Malaysia[20] both for the bags and the intervening ground. Mulching is not normally practised with polybags; the surface soil is lightly scratched at the time of fertilizer application so the amount of weeding required should never be very great. In the Ivory Coast the treatment of the area, before the bags are positioned, with a total weed killer (sodium chlorate or dalapon), is advocated; and 1 month after final positioning of the bags the area is treated with paraquat or MSMA. For the soil in the bags three sprayings below the plant with Ametryne in 0.2, 0.3 and 0.4 per cent solutions respectively are recommended with application rates of 3 kg, 4 kg and 5 kg per hectare of nursery.[8]

The prevention of nursery diseases

Diseases of oil palm seedlings are described in Chapter 13, but a brief account will be given here of the disease-control measures taken in nurseries since these form an integral part of sound agronomic practice.

1. Leaf diseases

The Anthracnose fungi can be largely eliminated by wide spacing in the prenursery, careful transplanting and the use of organic fungicide sprays. Spraying should be carried out weekly in the prenursery after the two-leaf stage has been reached. The following sprays have been found suitable in West Africa:

Captan powder (Orthocide M 50)	2 kg to 1,000 l water
Cuman (Ziram)	1 kg to 1,000 l water

The addition of a sticker such as Tenec or Albolineum is recommended. Spraying is continued for 6 weeks after transplanting the seedlings into the nursery. Both sides of the leaves must be covered.

Cercospora Leaf Spot appears in African nurseries after the danger of Anthracnose is over. Its incidence is reduced by potassium manuring. Spraying with Dithane M45 or benomyl (see p. 609) should be carried out fortnightly from 6 weeks after the nursery has been planted. It is particularly important that the undersides of the leaves should be sprayed as well as the upper sides.

Curvularia Leaf Spot, though occasionally troublesome in Malaysia, is not so serious a disease as *Cercospora* and it may be treated in the same manner.

Freak leaf symptoms are often noticed in the nursery, particularly in the dry-season nursery in West Africa (Plate 31). The causes of these conditions — Leaf Crinkle, Leaf Roll and Collante — are unknown (see p. 611).

2. Blast disease

This disease, described on p. 613, has been killing seedlings in African nurseries for a long time. More recently, symptoms of Blast have been seen in the Far East. The disease may appear at any time, but in West Africa it normally attacks young seedlings between October and January, i.e. from the end of the rains until well into the dry season. It has been shown that (i) seedlings of a year or more old are rarely attacked,[25] (ii) seedlings planted in October and shaded escape attack,[5] (iii) the severity of the 'short-dry' season of August affects the incidence of Blast, and hence the provision of irrigation water during this season reduces Blast incidence[26] and (iv) shading a main-season nursery from October to February reduces Blast incidence.[25, 30]

Blast incidence has varied considerably from place to place and from year to year. Basket nurseries have often been among the worst hit. The disease has been particularly troublesome in the Ivory Coast where conditions seem to favour it. Progenies received in Nigeria from the Ivory Coast and Dahomey have proved particularly susceptible. In the Ivory Coast the disease is such a menace that shade is provided from October to February even with polybag nurseries.[24] In Nigeria the shading of nurseries was stopped in 1954 as it was believed that any possible reduction of Blast was counterbalanced by the more drawn-up and weakly seedlings produced. Later, improved cultural practices and the provision of irrigation during the short-dry season so improved the standard of growth that shading at such a late stage was not reintroduced. For the dry-season nursery it was conclusively shown, however, that shading the October-planted seedlings for 3 months almost eliminated Blast during that period. Shade can, at this stage, be provided inexpensively by single bent-over palm leaves. These gradually wither, letting more and more light through and when removed in December or January the seedlings make very rapid progress.

The above findings seem to apply as much to polybag as to field nurseries; in the Nigerian experiment referred to on p. 338 Blast incidence in unshaded nurseries was actually less with polybags than in the field nursery.[13] For West Africa as a whole the decision on whether to use shade must depend on experience of the locality concerned, the time of planting, the irrigation facilities available and the origin of the planting material. This is further discussed on pp. 614—16.

3. Pests

Pests are more troublesome in nurseries in Asia than in Africa and although regular prophylatic spraying is not usual a close watch has to be kept for nightflying beetles (*Apogonia* and *Adoretus* species) and crickets and grasshoppers and spraying must be started as soon as necessary (see pp. 655 and 679).

Costs

In this and the following chapters figures of costs of the operations described will be given in terms of man-days of labour. These are averaged figures obtained from various sources and are given as a guide to what has been accomplished in practice. It must be appreciated firstly that there is a considerable variation from country to country, according to terrain, soil, labour availability, capability and organization, and secondly that new practices are all the time tending to reduce labour requirements. Detailed estimates of labour usage in very large irrigated prenurseries and nurseries have been given by Bevan and Gray.[20]

Prenurseries (per 1,000 plantable seedlings) *Man-days*

	Africa			Malaysia
	Ivory Coast[2] *Bag prenursery*	*Zaire Prenursery beds*		*Prenursery beds with irrigation*
		(1)	(2)[19]	
Preparation of site	4.5			Clearing, site preparation, construction, irrigation installation 15.0
Bag or bed preparation and planting	5.4	6.0		Bed preparation and planting 1.9
Maintenance	1.9	5.0		Maintenance 9.3
				11.2
Total	11.8	11.0	6.9	26.2

Notes: Part of the preparation labour cost in the Ivory Coast and Malaysia is for erection of shade. Maintenance includes watering, application of fertilizer and pest control; the last two items entail much more labour in Malaysia.

Assuming that 80 per cent of these prenursery plants eventually reach the field, the cost in man-days per planted acre or hectare of the prenursery can be obtained by multiplying the above figures by 0.075 and 0.179 respectively, e.g. at 11.0 man-days per 1,000 seedlings costs per planted acre will be 0.825 md and per hectare 1.97 md.

Nurseries (per 1,000 transplantable seedlings) *Man-days*

	Africa				Nigeria[28] (Field)
	Ivory Coast		Zaire (Field)		
	Field[15]	Polybag[24]	(1)	(2)[27]	
1. Felling, clearing, burning, tillage and drainage	105		100	257	
2. Preparation and planting, including lining and shade	57	53	38	10	
3. Maintenance, including mulching, watering, application of fertilizers, pest control	114	60	47	45	
	276	113	185	312	150

	Malaysia	
	Polybag nurseries	Field nurseries
1. Clearing, levelling, drainage	8	38
2. Preparation of soil, fencing, etc.	16	15
3. Planting	38 (with filling)	20
4. Watering	60	35
5. Other maintenance including weeding, manuring, pest control	46	100
	168	208

Notes: Although in all the African areas the nurseries were opened from forest it is apparent that there were wide differences of labour requirements for clearing and preparation. In using the Ivory Coast field nursery data an assumption of 80 per cent transplantable seedlings has been made; the high figures for preparation and maintenance are due to shade erection (30 md) and frequent hoeing (67 md). It is apparent that with mechanical cultivation and without the use of shade, the labour cost of preparation, planting and maintenance can be reduced to below 50 man-days per 1,000 transplantable seedlings.

In examining the Malaysian figures it is important to remember that they are man-day figures only; in polybag nurseries the cost of the polybags themselves has been estimated as about one-quarter of the total cost and polybag nurseries may therefore cost more than field nurseries.[29]

The Malaysian figures for polybag nurseries are the mean of eight estates using hand watering, but there were very large differences between estates in both watering and planting labour. The areas were not opened from forest. The installation of irrigation should reduce labour considerably as two men per day can operate a movable irrigation system on 2 hectares of nursery.[7] Field nurseries appear to require more labour for preparation, less for planting and watering and more for maintenance.

Costs of nursery work per acre and per hectare of field planting can be obtained by multiplying by 0.06 and 0.143 respectively. At 150 man-days per 1,000 transplantable seedlings costs would thus be 9.0 man-days per acre and 21.5 man-days per hectare. Taken together, prenursery and nursery *work* form a very small proportion indeed of the capital cost of bringing a plantation into production. Economies which reduce the standard of the transplanted seedling are therefore misplaced as they are likely to be followed by a substantial reduction in early production.

References

1. **Ruer, P.** (1963) Les prépépinières de palmiers à huile en sachets de polyéthylène. *Oléagineux*, 18, 693.
2. Notes on the establishment of an oil palm nursery. *J. W. Afr. Inst. Oil Palm Res.*, 1, (1), 88, (1953).
3. **Toovey, F. W.** (1947) Oil Palm Research Station, Nigeria, Seventh Annual Report 1946−7, pp. 17−18.
4. **Dupriez, G.** (1956) Prépépinières d'Elaeis. *Bull. Inf. I.N.E.A.C.*, 5, 141.
5. **Gunn, J. S., Sly, J. M. A. and Chapas, L. C.** (1961) The Development of improved nursery practices for the oil palm in West Africa. *J. W. Afr. Inst. Oil Palm Res.*, 3, 198.
6. **Bevan, J. W. L. and Gray, B. S.** (1966) Germination and nursery techniques for the Oil Palm in Malaysia. *Planter, Kuala Lumpur*, 42, 165.
7. **Sly, J. M. A. and Sheldrick, R. D.** (1961) The practical aspects of irrigation of an oil palm nursery. *J. W. Afr. Inst. Oil Palm Res.*, 3, 273.
8. **Tailliez, B.** (1969) Le désherbage sélectif des prépépinières de palmier à huile. *Oléagineux*, 24, 541.
9. **Vanderweyen, R.** (1950) Pépinières en paniers ou en pleine terre. *I.N.E.A.C.-Réunion Congopalm*, Comm. No. 10, 58.
10. **Marynen, T. and Poels, G.** (1960) Considérations sur les méthodes de mise en place du palmier à huile. *Bull. Inf. I.N.E.A.C.*, 9, 81.
11. **Mainstone, B. J.** (1962) Dunlop polythene-bag planting technique. *Pltrs. Bull. Rubb. Res. Inst. Malaya*, 63, 154.
12. **Hew Choy Kean and Tam Tai Kin** (1969) A comparison of oil palm nursery techniques. In *Progress in oil palm*. Incorp. Soc. of Planters, Kuala Lumpur.
13. **Aya, F. O.** (1974) The use of polyethylene bags for raising oil palm seedlings in Nigeria. *J. Nig. Inst. Oil Palm Res.*, 5, (19), 7.
14. **Aya, F. O.** (1969 and 1971) N.I.F.O.R. Fourth Annual Report, 1967−8, pp. 25−7. Sixth Annual Report, 1969−70, pp. 31−2.
15. **Fraisse, A.** (1962) Les pépinières de palmiers à huile. *Oléagineux*, 17, 173.
16. **Gunn, J. S.** (1960) W.A.I.F.O.R. Eighth Annual Report, 1959−60, p. 41.
17. **Gunn, J. S.** *et al.* (1962) W.A.I.F.O.R. Tenth Annual Report, 1961−2, p. 32.
18. **Klaver, H.** (1961) Le transport des plantules de palmier à huile. *Oléagineux*, 16, 601.
19. **Ochs, R.** (1963) Utilisation du test stomatique pour le contrôle de l'arrosage du palmier à huile en pépinière. *Oléagineux*, 18, 387.
20. **Bevan, J. W. L. and Gray, B. S.** (1969) *The organisation and control of field practice for large-scale oil palm plantings in Malaysia*. Incorp. Soc. of Planters, Kuala Lumpur.
21. **Coomans, P.** (1971) L'arrosage des pépinières de palmiers à huile en sacs de plastique. *Oléagineux*, 26, 295.

22. **Quencez, P.** (1974, 1975) Arrosage par aspersion des pépinières de palmiers à huiles en sacs de plastique. *Oléagineux*, **29**, 405; **30**, 355, 409.
23. **Gunn, J. S. and Sly, J. M. A.** (1961) W.A.I.F.O.R. Ninth Annual Report, 1960—1, p. 34.
24. **Surre, Ch.** (1968) Les pépinières de palmiers à huile en sacs de plastique. *Oléagineux*, **23**, 573.
25. **Bachy, A.** (1958) Le 'Blast' des pépinières de palmier à huile. *Oléagineux*, **13**, 653.
26. **Robertson, J. S.** (1959) Blast disease of the oil palm: its cause, incidence and control in Nigeria. *J. W. Afr. Inst. Oil Palm Res.*, **2**, 310.
27. **Vanderweyen, R.** (1952) *Notions de culture d'Elaeis au Congo Belge*. Brussels.
28. **W.A.I.F.O.R.** (1953) The cost of establishing an oil palm plantation. *J. W. Afr. Inst. Oil Palm Res.*, **1**, (1), 92.
29. **Bull, R. A.** (1966) The production of oil palm seedlings for field planting. Sabah Planters' Association. Oil Palm Seminar. Mimeograph (1965); and, *Planter, Kuala Lumpur*, **42**, 248.
30. **Rajagopalan, K.** (1974) Influence of irrigation and shading on the occurrence of Blast disease of oil palm seedlings. *J. Nig. Inst. Oil Palm Res.*, **5**, (19), 23.

Chapter 8

The preparation of land for oil palm plantations

New oil palm plantations are usually established in areas of primary or high secondary forest and there has been much study of different establishment methods and their costs. Occasionally, however, oil palms may be planted on land under light secondary growths or where strong growing weeds such as lalang (*Imperata cylindrica*) or Siam weed (*Eupatorium odoratum*) have become dominant following food-cropping; under these circumstances rather different problems are encountered and these will be discussed later in this chapter.

The soil and terrain on which oil palms are likely to be successful have already been described. Usually, oil palm plantings are on flat or gently undulating land, but on inland soils quite steep slopes may sometimes be encountered. In these cases the normal triangular planting distances are maintained horizontally, but the palms may be planted on platforms. Planters accustomed to rubber cultivation have been anxious in some areas to adopt or maintain rubber terracing methods,[1] but the fact that the oil palm obtains water and nutrients predominantly from the topsoil and that interference with the rooting area leads to reduced growth suggests that terracing should not be adopted unless experimentally proved for the region and soil concerned (see pp. 156 and 370).

On flat areas of heavy clay soil and on undulating areas where the permeability of the soil is low and the natural drainage courses are sluggish or subject to flooding, preparations for planting include the design and construction of a drainage system which will prevent the stagnation of water in the upper layers of soil and allow for the rapid removal of floodwater after heavy rain.

By far the most expensive work in the establishment of a plantation is the actual felling of the forest and the clearing of it so that the seedlings may be planted out. The cost of this work may be as much as two-thirds the total cost of establishment. It is not surprising therefore that attempts have been made to reduce these costs.

Early consideration was given in Zaire[2] to reducing clearing costs by a combination of twin-row planting and directional felling. Under this system the rows were 'closed in' to form pairs of rows at the close spacing of 6 m — about 20 feet — apart, leaving a wider avenue of 10 m (33 ft) between each pair of rows; into these wide avenues the forest trees were directionally felled and moved in two stages, the smaller trees being dealt with first. The work both of felling and of clearing was much assisted by the use of hand winches. These methods somewhat reduced the labour required and provided fully cleared alternate avenues for ease of planting, maintenance and inspection; but the twin-row method did not find lasting favour owing to the unevenness of stand which tends to reduce yield per palm. Later work[3] showed the value of mechanical saws both in the felling and cutting up of large forest trees.

Elsewhere attempts have been made to reduce costs both by the practice of burning off the felled forest and by mechanization. In all this work the value of directional felling has been realized.

The practice of burning

The question of whether the felled forest should be burned or not has only recently been resolved, though it has been debated since the plantation industry was established. Early opinion in Africa was that burning would lower fertility. Falling yields in Zaire and Nigeria after 8 or 9 years of bearing were attributed to depletion of nutrients through the practice of burning; to this was added the loss of soil organic matter and the provision of conditions leading to erosion.[4, 5] Early workers in Zaire[6, 7] and Malaysia[8] supported these ideas, though in Malaysia it was later realized that, with the increasing employment of tractors on estates for all manner of work, the improved access obtained through burning followed by mechanized or hand clearing might outweigh any advantages of leaving the forest unburnt.[9]

The advantages of burning were seen by its advocates to be managerial and economic: a reduction in the costs of establishment and the creation of conditions where early work in the plantation was easier and supervision and maintenance therefore less time-consuming and costly (Plate 32).

The theory that burning 'lowers fertility' was further considered in soil studies, and, in Nigeria, field trials comparing burning and non-burning were laid down.[10] These trials were unique, and though all three of them were on Acid Sands soils (though one in an area of heavier rainfall some 250 miles from the others) their results are likely to be generally applicable. The yields and soil analysis results obtained are given in Table 8.1. No yield differences were found between burning and non-burning except in the first 4 years in one of the experiments at Benin where increased yields were obtained from the burnt plots. Increased early yields might well be expected owing to the larger release of some nutrients, particularly nitrogen, and the lack of competition from natural vegetation following burning; the latter effect was probably exaggerated in this experiment as maintenance during the year after planting was inadequate and this, of course, favoured the burnt plots where natural vegetation takes longer to re-establish itself.

The effects of burning on soil nutrient levels 20, 10 and 9 years after planting in the three experiments were described by Sly and Tinker.[10] In

A

B

Pl. 32   Before and after the burn at the end of the dry season in Nigeria.

Table 8.1 *Effects of burning on yield and soil analysis. Nigeria (Sly and Tinker)*
Fruit bunches per hectare per annum; soil data — depth in cm.

| Expt. | Age at soil sampling | Treatment | Yield | | Exchangeable cations | | | | | | | | | | | C (%) | | N (%) | |
			First 4 yrs (kg)	Adult years† (kg)	K m eq/100 g 0–15 (cm)	15–30 (cm)	Na m eq/100 g 0–15 (cm)	15–30 (cm)	Mg m eq/100 g 0–15 (cm)	15–30 (cm)	Ca m eq/100 g 0–15 (cm)	15–30 (cm)	Excb. cap. m eq/100 g 0–15 (cm)	15–30 (cm)	0–15 (cm)	15–30 (cm)	0–15 (cm)	15–30 (cm)
33–2 Benin	20	Burnt	4,860	9,947	0.045*	0.034	0.070*	0.064*	0.68	0.24	3.15	1.89	6.04	3.55	1.10	0.47	0.095	0.041
		Unburnt	5,242	9,730	0.038	0.021	0.039	0.033	0.63	0.27	3.82	1.89	6.27	3.43	1.20	0.49	0.100	0.041
3–4 Benin	10	Burnt	4,958*	6,549	0.070	0.047	0.048	0.038	0.69	0.43	2.18	1.07	5.87	4.93	1.18	0.73	0.098	0.061
		Unburnt	4,363	6,499	0.073	0.045	0.056	0.041	0.83*	0.47	2.75*	1.39	6.15	4.70	1.24*	0.68	0.106*	0.059
506–2 Abak	9	Burnt	2,617	6,038	0.073	0.049	0.044	0.042	0.15	0.08	0.14	0.10	6.01	5.16	1.14	0.79	—	—
		Unburnt	3,017	6,168	0.073	0.040	0.046	0.028	0.16	0.09	0.13	0.07	6.01	5.10	1.19	0.76	—	—

* Significant at P = 0.05.
† Means of 11 years (33–2); 6 years (3–4); 7 years (506–2).

general the differences between burnt and unburnt areas are small and un-important. Burning tends to conserve potassium, this effect showing up particularly in the later years. Where magnesium and calcium are in good supply burning lowers their levels to a small extent, but this does not occur when the levels are already low. Burning causes a slight depression of organic matter and nitrogen in the top 15 cm, a rapid nitrate flush occurs immediately after burning and this leads to an earlier loss of nitrogen. Under some conditions, therefore, an increased need for nitrogen ferti-lizers might be expected in the early years. It should be noted that in the experiments quoted above no leguminous covers were planted after burn-ing. With the rapid establishment of leguminous covers the loss of organic matter following burning might be reduced.

On theoretical grounds, Bocquet and Michaux[11] postulated that, in soils with a clay content of 20 per cent, if the amount of carbon in the top 20 cm is less than 2.2 per cent, the forest should not be burnt. It would not be possible to use this figure as a standard since, in general, C per cent is positively correlated with clay content;[12] moreover the loss (or relative constancy) of carbon in the topsoil following oil palm planting appears to vary widely. These authors quote cases of losses of organic carbon at a steady rate for 8 to 12 years after planting from primary or secondary forest with 2.5 to 3.5 per cent C in the top 10 cm of soil. In Nigeria, how-ever, a special study of soil changes under plantation conditions[13] showed no fall of organic carbon in a burnt area during a period of 15 years from 1 year after planting; indeed, under certain cultural conditions there was a tendency for organic carbon to rise. It must be borne in mind that high carbon content is not always a sign of fertility; that poor degraded soils may have a high organic matter content has been shown both in West Africa and Malaysia.[12, 14] It is not possible, therefore, to fix any soil criterion for the practice, of burning and the evidence available shows that, for all *usual* situations, the practice will not lead to any diminution of crop and will prove less costly than the opening of forest areas without burning.

Nevertheless, there is some evidence[15] that areas which have for many years had only a light covering are better left unburnt. One example is an area of palm grove and farmland with a poor structureless sandy soil in which farming had been practised in the past as part of a 'bush-rotation'. Palms were planted in this area under various cultural treatments. Burnt plots normally maintained with a natural cover showed a lower response to an NKMg fertilizer mixture first applied in 1957 than was shown by unburnt plots:

Mean annual yield of fruit bunches per hectare (Abak, Nigeria, Expt. 506—2)

	1955—7	1958—65	
	(kg)	No fertilizer (kg)	Fertilizer (kg)
Burnt	1,556	4,876	7,142
Not burnt	2,167	4,516	7,673

One other method of reducing establishment costs should be mentioned here. In many parts of the world oil palms are planted in districts containing primary or very old secondary forest. Quantities of very valuable timber may exist in such forests and proper exploitation may be profitable and provide a substantial contribution to the costs of establishment. Quite apart from this possibility however, the forest will supply a great deal of useful small timber for fence posts, bridges, etc., and this must be extracted to meet future requirements while the areas are being opened up.

Layout

The general layout of a plantation is determined by the terrain, the position of the mill and the length of 'carry' of bunches by harvesters to the road or rail side. If the area is flat or gently undulating, the collection roads or railway tracks should be laid straight and in an east–west direction and should connect with one or more sub-main north–south roads in turn leading to the main road to the mill. If the country is hilly the collection roads will, at least on parts of the estate, have to follow a winding course to avoid steep gradients. In very steep country the roads must be arranged so that the carry is a good deal less than on flat land since the harvesting work is more arduous. A railway system will not be suitable on a hilly estate. A typical estate layout for hilly land is shown in Fig. 8.1. The interlocking of drainage and transport systems on a flat plantation is shown in Fig. 8.3.

The maximum practicable length of carry is generally considered to be 200 m so the roads or railway tracks are 400 m apart or less. On some estates, however, rail tracks have been as much as 660 m apart. With the opening of new estates and the increase in operating costs much thought has been given to the type and arrangement of the transport system to be used. Between the wars many railways were laid down and these are still being operated economically. The capital cost of these systems is now considered prohibitive, however, and newly-opened estates are developing road systems.

It has been shown that the determination of the optimum frequency of collection roads is by no means a simple matter since it depends on labour costs and construction and maintenance costs, and consideration must also be given to the loss of crop through the excision of road areas from the total possible productive area; moreover the optimum frequency will vary according to whether land or capital is limited.[16, 17] As labour rates rise, the optimum road frequency for maximizing return on capital will tend to rise also, i.e. the distance between roads should fall. Calculations might show that a 200 m carry was optimal but, looking ahead, it might be wiser to accept a 160 m carry with a larger road area, higher road construction and maintenance costs, but easier harvesting. However, this general thesis might need modification if more extensive use could be made of inter-row

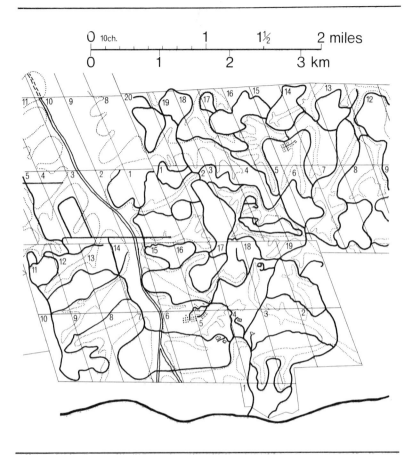

Fig. 8.1 Example of an estate layout: hilly estate. ········ Rough boundary of low-lying land. ———— Roads. ═══════ Main drainage course.

transport and mechanical harvesting.

Many new estates are being laid down with collection roads 320 m apart,[16] with intervals as low as 100 m on very hilly land. With 320 m intervals between collection roads the length of road per hectare will be 36 m, assuming sub-main roads every 2 kilometres and not including peripheral roads. With sub-main roads every kilometre there will be 41 m of road per hectare. A design which has been used in Africa and America consists of a square kilometre block (100 hectares) with collection roads at 250 m intervals giving four fields of approximately 25 hectares each; the maximum carry of 125 m is rather less than is usual in Asia and there are 50 m of road per hectare not including peripheral roads. With these systems the roads will occupy between 2 and 4 per cent of the total area.

Methods of land preparation

Various sequences of operations for land preparation have been followed in different parts of the world. The methods chosen often depend on local circumstances and experience. In the days of non-burning it was usual to underbrush (clear the ground vegetation and small saplings) and line out before felling. With the advent of burning, underbrushing was still the first operation though lining was often put off until after felling and burning. It was found in many areas, however, that even a severe burn did not consume the hardwood lining pegs and that there were several advantages in carrying out underbrushing and lining together with holing, where holing was necessary, in the shade of the tall trees.

Various sequences which have been employed after marking out the plantation are given in Table 8.2. It will be seen that the usual sequence is underbrushing, felling, burning and lining, but where attention is given to directional felling as in E, lining and the preparation of the planting rows precede felling and burning. It is important that, from planting, the young palms shall be easily accessible and this is assured only if large tree trunks and branches are not allowed to cross the row paths. This can be arranged either by felling the trees along the rows or by moving them into line when once felled. Clearly the latter method is more expensive and skill in felling is therefore important. Method B is designed to concentrate felled material which is not consumed by a light burn into narrow windrows in alternate interlines thus leaving the other interlines and 2 m on each side of each row absolutely clear except for the stumps of the very largest forest trees. This method increases further the ease of inspection and makes some mechanical maintenance possible; it is, however, expensive and if carried out entirely by hand labour would, in terms of man-days, be at least twice as costly as the more usual methods. Even with mechanical saws and mechanical stacking the method may be more expensive in Africa than ordinary hand methods.[18]

The usual operations carried out in preparing land for planting will now be considered separately. These do not include road building which has already been mentioned, nor drainage, which will be discussed later.

Underbrushing

This is done to make access to the area easy and, in the cases mentioned above, to allow lining (and holing) to be carried out. The work is done with cutlasses and axes and will require between 10 and 20 man-days per hectare according to the thickness of the undergrowth; all herbaceous undergrowth, lianas and young saplings 7.5 cm or less in diameter should be cut, the latter as low as possible.

Lining

Many methods of lining have been used and suggested. In the first place it

Table 8.2 *Methods of field preparation*

A. *Zaire*[19]	B. *Ivory Coast*[18]	C. *Ivory Coast*[18]	D. *Nigeria*
1. Underbrushing	1. Underbrushing	1. Underbrushing	1. Under-brushing
2. Felling forest trees	2. Marking the position of the lines on the base line and the first stacking line (windrow)	2. Felling	2. Felling
3. Heaping		3. Beating down and cutting up	3. Beating down
4. Burning	3. Felling directionally along the interlines and marking windrows	4. Heaping ⎱	4. Burning
5. Lining		5. Burning ⎰	5. Lining
6. Clearing paths	4. Stumping the interlines which are to be cleared and stumping along the lines	6. Clearing the lines	6. Opening paths along rows
7. Opening rides along palm lines		7. Sow cover crop	
8. Terracing	5. Beating down and cutting the large branches		7. Sow cover crop
9. Sow cover crop	6. Cutting up trunks		
	7. Carrying out a light burn of branches and debris		
	8. Verification and marking of the stacking lines		
	9. Heaping the logs into windrows in alternate avenues, by tractor		
	10. Levelling and final clearing		
	11. Sow cover crop		

E. *Cameroons*	F. *Malaysia*[20]	G. *Malaysia*[20]	H. *Malaysia*[21]
1. Underbrushing	1. Underbrushing	1. Rough lining	1. Under-brushing
2. Lining and clearing paths	2. Felling, cutting larger branches (dry out for 3 months)	2. Underbrushing	2. Felling
3. Holing		3. Felling, cutting larger branches	3. Burning
4. Felling small trees into interlines*	3. Burning	4. Pile branches on top of large timber (dry out 2–4 weeks)	4. Cutting up, stacking
5. Felling large trees into interlines*	4. Lining		5. Re-burning
	5. Clear paths along rows	5. Scorch burn	
6. Sow cover crop along lines	6. Holing	6. Lining	6. Lining
	7. Sow covers	7. Clear paths along rows	7. Clearing paths
7. Burning		8. Holing	8. Weeding
8. Realign and peg		9. Sow covers	9. Sow covers
			10. Holing

* Paths kept clear of debris.

must be realized that the rows are going to run north and south and that with triangular planting the palms in one row will not be opposite the palms in the adjacent row; the line from each palm to the nearest palm in the adjacent row will make an angle of 60° with the row and the distance between rows will be less than the planting distance. This orientation and planting arrangement ensures that the maximum sunlight falls on the individual palms.

Some lining methods allow for the making of a base line with palms marked off at the planting distance, say 9 m. From this base line, which should of course be N—S, equilateral triangles with sides of 9 m are built up by running lines at 60° to the base. On flat land, however, the road will be running E—W and it will be easier to have the base line parallel with the road or E—W boundary. The rows can then be marked off along the base line according to the interline distance which is the perpendicular distance from any of the three points of the 9 m-sided triangle to the opposite side. This is easily calculated by multiplying the planting distance by 0.866, e.g. for 9 m spacing the rows will be 7.8 m apart. Alternatively 9 m triangles can be made from the first two palms of the first row and the first palm of the second row (which will be 4.5 m from the base line) and the rows can then be extended from the base line and the palms marked off at 9 m intervals. The other rows can then be 'placed' by means of similar triangles touching the base line, and every point throughout the field is brought into alignment in three directions. Guide lines can be put in at an angle of 60° from the first line. Lining pegs should be made of some hard wood if they are to survive the burn.

It must be remembered, when lining on hilly land, that the planting distance is a horizontal distance and the lining ropes must therefore be kept horizontal when measuring from one stand to another. With a gradient of 1 in 3 the distance along the ground will be 5.3 per cent greater than the horizontal distance. Contour lining is not recommended since the unevenness of stand produced will lead to some reduction in yield. However, on very hilly land maintenance of harvesting paths and the sowing of covers along the contour may assist soil conservation and on erodible soils small contour bunds or silt pits may be constructed.

Platforms and terraces

Except in certain inland areas of Malaysia (usually following rubber) the oil palm has been very little planted on steep land. Although terracing has been advocated for Malaysia, there is strong evidence that palms on platforms establish and grow better than those on terraces (Plate 33). Apart from the very high cost of terrace construction, terrace soil becomes trampled by harvesters and forms a poor rooting medium for the palm, while the backward slope of the terrace often leads to pools of water collecting round the young palms, retarding their growth. In a hilly area of Malaysia where one part was terraced while on the adjoining area

Pl. 33   Palm planted on a platform on hilly land in Malaysia.

platforms of 2.5 x 3.0 m were constructed the terraced area came into bearing nearly a year later than the area with platforms although planted at the same time with the same material. On very steep terraced land it is difficult for harvesters to move between the palm and the bank behind. Advocates of terracing have put forward arguments which are mostly of a managerial and non-agronomic nature.[22] Nevertheless on some very steep land with slopes over 20° terracing may be unavoidable, though it is questionable whether such land should not have been left in forest (Plate 34). It is certainly unwise to enter into a large and expensive terracing project until the production from such areas *vis-à-vis* similar but non-terraced areas has been ascertained and the returns evaluated.

Pl. 34   Oil palms terraced on steep land in Malaysia.

Clearing paths

Where lining follows underbrushing (see E, Table 8.2) it may be possible to carry out the clearing of paths 1.5 m wide at the same time, especially if these paths are to follow the rows. Clearing paths is usually carried out later however, but also after lining (e.g. D and H), and their construction in the centre of every other avenue rather than along every line is advocated; these paths will act as essential inspection paths in the first few years and then become harvesters' paths.

Holing

It will be noted that holing is not included in some of the schemes set out in Table 8.2. This is because on the free draining sandy soils of West Africa it has been shown that holing prior to planting has no effect on establishment or subsequent growth and yield.[23] On heavy or stony land, however, holing is usually undertaken several weeks before planting, the holes being 60 or 90 cm cubes according to the soil type. Before planting, the holes should be refilled with surrounding topsoil and allowed to settle. The smaller hole in which the seedling's ball of earth is placed is not dug until the time of planting. Where holing is not carried out in advance of planting, the planting site must be well weeded so as to prevent any competition between the newly-planted seedling and surrounding weeds.

In Malaysia, experiments on the need for holing have not been carried out and several different practices exist. Holing prior to planting as described above is carried out on most soil types but it has been suggested that on heavy coastal clays it is disadvantageous to hole before planting as the holes may fill with rain-water and planting, filling and consolidation become difficult. Under these circumstances a hole sufficiently large to accommodate the ball of earth and to allow for proper firming of the soil around it is all that is needed.[24] Holes with dimensions 45 x 45 x 40 cm are recommended.

Felling

Costs of clearing planting lines will be considerably reduced if directional felling is practised. On flat or undulating land this felling should be N–S to avoid heavy trunks falling across the planting lines. On steep areas the forest is best felled along the contour. It is sometimes easier to fell the small trees first and later to fell the large trees in a separate operation.

A varying number per hectare of large buttressed trees will have to be felled by axing from a platform 2 m or more above ground level. The stumps of these trees, as already mentioned, will have to remain in the ground. Felling must be followed by the loping of any branches or the cutting up and moving of any trunks which fall across the cleared lines. These operations may, however, be left until after burning. There may be some trunks which are too large for cutting up and removing, but with good directional felling these should be very few.

With hand axing, felling may take up to 50 man-days per hectare, but if chainsaws are used as little as 5 man-days may be required. In Malaysia two chainsaw men with four axemen can fell more than a hectare a day[44] (see pp. 424–5).

Felling must be correctly timed. In regions with a distinct dry season the work will be done as early in the dry season as possible, so that burning may be carried out easily as soon as the material has dried out. Beating down and some cutting up of branches is often done before the burn (see A, C and D, Table 8.2), and the burn is so complete that little more than the clearing of planting rows is then needed. In non-seasonal countries such as Malaysia felling is done 8 to 12 weeks before burning, the latter being timed for one of the two drier periods of the year (e.g. February). If the felled forest is left longer there is danger of regrowth. The burn is often incomplete, so cutting up and stacking for reburning is undertaken after the burn (see H, Table 8.2) and this work may span the next dry period (e.g. June/July) so that planting is unlikely to begin less than 6 to 7 months after the first burn. In regions of very even rainfall however, where the seedlings may be planted almost throughout the whole year, felling can proceed at a steady rate regardless of season, burning follows some 3 months later and planting will be done as soon after stacking and reburning as possible.

Burning

In regions with a distinct dry season a good burn is usually achieved without difficulty unless clearing operations have been started too late and the dry weather is brought to an abrupt end by heavy rain. Indeed, the greater danger is usually the fire hazard to neighbouring planted areas where dry debris may be lying about or combustible weeds such as *Imperata* or *Eupatorium* have gained access. For this reason wide cleared fire traces are essential; these may vary from 20–40 m in width to prevent scorching, but however wide a trace may be there is always a chance of sparks setting alight adjoining areas and for this reason water carts or drums of water at the roadside should be ready so that any smouldering vegetation may rapidly be quenched. *Fire is a perpetual hazard to planted areas throughout the dry season in West Africa whether large burns are in progress or not; several large planted areas have been totally lost through failure to take elementary precautions.* Some damage to small areas has also resulted during a burn owing to a change of wind and inadequate fire traces. A fire may start in a felled area before it was intended and before adequate fire traces have been cut. For this reason, on any developing plantation, some form of patrol and emergency fire service is needed.

Firing in these areas is simple: a line of men sets fire to the felled forest at the windward side at a time of day when the prevailing wind is blowing moderately. The fire burns rapidly and fiercely through to the other side with the men following it; any portion which may, by chance, be missed as the fire proceeds on its course may then be kindled on the windward side.

Firing in non-seasonal regions such as Malaysia is carried out rather differently.[20] The men are drawn up at 20 m intervals along the side *away* from the prevailing wind. They then advance through the area into the wind setting fire, with firebrands and kerosene, to heaps of dried material at about 20 m intervals, sprinkling kerosene if needed. This method is necessary in a region where the felled forest is not so easily fired. Firing should be carried out at about 2 p.m. on a day when there is a breeze. On a still day the burn may be a failure. Firebelts of standing forest are favoured; these are felled, cleared and burnt separately after the main burn.

Post-burning operations

From earlier descriptions it will have been apparent that the amount of work remaining after the burn will depend on whether such operations as lining and path clearing have already been done and on the success of the burn or thoroughness of reburning. If lining and clearing the rows has been well done before the burn, the work of realigning and final pegging will be very light and the field will be ready for planting almost immediately. If lining follows burning, the methods already described must now be followed.

In Malaysia the work of cutting up, stacking and reburning is usually very considerable and the stacking is often done round the base of large buttressed tree stumps. On steep or undulating land this work is followed by lining and the construction of platforms or terraces. Platforms are usually about 4 m in diameter and have a slope back into the hillside of about 7°−8°. With slopes of 12°−20° such platforms are suitable and the normal plantation layout and planting distances can be maintained.[25] Above 20°, however, terracing will probably be necessary although, as already mentioned, it is doubtful if planting should be undertaken at all on such steep land. Leamy and Panton[26] define the steep land boundary in Malaysia as the line separating land with slopes greater and less than 20° and rightly assert that though land below the boundary is suited to tree-crop agriculture, land with slopes in excess of 20° is better suited to permanent forest.

Field preparation by mechanical methods

The great majority of oil palm plantations have been opened by hand labour, but in recent years consideration and trial have been given to mechanical means. Two main considerations must be borne in mind: (i) Will mechanical preparation be less expensive than hand methods? (ii) Is it intended to prepare the land so that subsequent maintenance can be by mechanical means? These two questions are inseparable since, *if mechanical maintenance is to be practised from the start*, a complete clearing, which can best be carried out with mechanical aids, will be essential. Therefore the cost of mechanical clearing must be considered in the light of subsequent intentions. In some countries where mechanical maintenance is not contemplated the operations needed can be done by hand more cheaply than through mechanical aids since very low sawing and the cutting up and movement of large timbers will not be required on an extensive scale. If, however, access to the fields by wheeled tractors and tractor-drawn implements is needed from the start, then the use of mechanical aids will hasten and cheapen the work.

No preparation method is entirely mechanical, but the work must be organized so that all medium and small trees are uprooted or cut at ground level and so do not form snags impeding the passage of wheeled and rubber-tyred tractors, and so that the unburnt trunks may be conveniently stacked to admit tractors to the area. Early work in Zaire was directed mainly to the movement of timber so that tractors could subsequently enter the area for inter-row cultivation.[27] Felling was therefore carried out mainly by axing, while the movement of timbers was undertaken by using winches operated behind tractors. Large trees were felled, and large fallen trunks were broken up by the introduction of explosive charges into holes bored with pneumatic drills which in turn were operated by tractor-mounted compressors. These methods were comparatively slow and the amount of labour employed was still considerable.

A number of more complete methods of mechanical clearing have recently been used, the first closely following Method B of the Ivory Coast already referred to in Table 8.2. Heavy track-tractors of at least 235 hp are fitted with special bulldozer blades which include an upward extension for the protection of the tractor and a projection at the lefthand end of the blade.[28] The blade measures 4.38 x 1.60 m (14.4 x 5.25 ft). Dealing with the smaller and medium-sized trees presents no difficulty, but buttressed trees of over 75 cm diameter have to be prepared for felling in two stages, firstly by breaking up the sides of the base of the trees with the bulldozer blade and extension, and secondly by breaking up the buttresses opposite the felling side. This is dangerous work in the hands of an unskilled operator. Under the conditions in the Ivory Coast two or three trees per hectare will still have to be felled by hand.[29] Experience has shown that with tree densities varying from 300 to 500 per hectare the time taken to fell by these methods will vary from 3 to 10 hours per hectare.[30] After felling, the larger trunks are cut up by mechanical saws to ease the subsequent work of windrowing. Methods of mechanical forest felling have recently been further advanced by the introduction of the 'tree crusher', a 50-ton tractor and blade which is capable of clearing a hectare in less than 1½ hours.[30]

After the forest is felled the area is marked out with lines whose distance apart is twice the distance between rows. For instance, with 9 m triangular planting the distance between rows being 7.8 m the lines would be 15.6 m apart. As can be seen in Fig. 8.2, the cut-up forest trees and stumps are then worked into the centre of the blocks formed by these lines and windrows of felled material fall into place between each pair of rows, the space within the pairs and along the rows being left absolutely clear. Felling is usually followed by a light burn and the sawing up with mechanical saws of the felled vegetation in order to make the subsequent windrowing easier. Stumping, which is a costly operation, may be done with the aid of projections to bulldozer blades, but the shearing of the stumps with a Rome KG spur blade is preferred.[45] Windrowing is estimated to take 3 tractor hours per hectare for forest of 300 to 400 trees per hectare of which only two to five are over 150 cm in diameter. Martin has given estimates of the additional time needed both for felling and windrowing denser forests.[45]

The system outlined above has the advantage of keeping every row clear for ease of inspection and it is also possible, from the start, to maintain every other avenue by mechanical means. While a good burn could greatly reduce the amount of material in the avenues containing the heaped debris, these avenues will inevitably become covered with creepers and all manner of plants will grow up within them. Maintenance will therefore be difficult until considerable rotting has taken place. The use of weed killers may, however, appreciably reduce this disadvantage. The method relies on the heaviest of tractors and very expensive equipment, and unless the size of the programme justifies high capital outlay or contractors with such

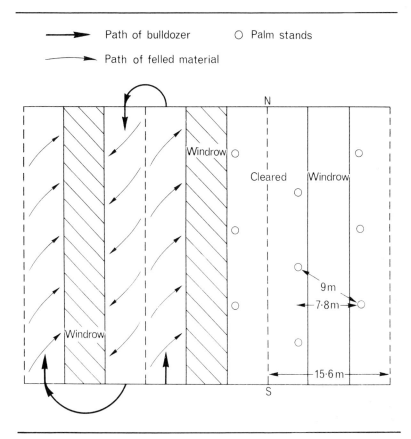

Fig. 8.2 Alternate avenue clearing and windrowing by bulldozer, Ivory Coast (after Huguenot).

equipment are available in the district, methods in which lighter and cheaper tackle are employed are to be preferred. The method also entails burning debris after the cutting up of the trunks rather than the burning of the whole field after felling and drying out.

Another method, which has been used in Nigeria, employs lighter crawler tractors and avoids gross disturbance of the soil, but it entails the continued use of hand labour. The sequence of operations[31] is as follows:

1. Underbrushing at ground level so that no spiky stumps remain.
2. Lining.
3. Holing with 30 cm cube holes (Acid Sands soil).
4. Felling all possible trees at ground level. Hardwood trees can be sawn, but softwoods, comprising the majority, are felled with axes, the stumps being levelled off after felling. Very large buttressed trees are

felled as usual above the buttresses and the stumps remain (two or three per hectare).

5. Burning when dried out.
6. Cutting logs into lengths of 6 m or less by mechanical saw.
7. Marking each planting hole with a large stick and moving, with tractor and angledozer blade, all logs on to the planting line avoiding the marked planting hole. The tractor moves up the avenue or 'interline' pushing to the left and returns also pushing to the left, thus clearing the whole avenue.
8. Field watchmen continually put fire into the stacked timber, thus much reducing the volume.
9. Timber finally shunted by tractor into place in the lines so that no timber is nearer than 1.2 m to a hole.
10. Repegging holes.

This method has the advantage that all avenues between the palms are cleared and ready for mechanical maintenance while the remaining timber is reduced to a minimum by burning. Moreover, the amount of labour used in planting can be greatly reduced since the plants can be brought by wheeled tractor and trailer into the field and off-loaded at the planting sites. Similarly, tractors can enter the field with sprayers either for applying herbicide to the undergrowth (particularly around the stacked logs) or fungicides to the plants.

One of the difficulties of tractor and bulldozer clearing in a seasonal climate is the need to have a sufficiency of tractors and to ensure that these are maintained in working order. If a breakdown occurs, the forest may not have been felled and dried out in time to obtain a satisfactory burn. The need to carry through all the work during the relatively short period of the dry season means that more equipment per unit area is required. Where the planting season is less restricted, the work can proceed at a slower and more even pace and operations can be under way in one area while planting is proceeding in another. Nevertheless in some areas the use of power saws for felling is now preferred to the use of bulldozers and in Colombia much detailed attention has been given to the methods of felling trees of different sizes.[32] In a forest containing 70 per cent of hardwood species it was found that clearing by mechanical saw could be done at the rate of 5 man-days per hectare where there were 230 trees of diameter less than 50 cm, 24 of 50–100 cm, 14 of 1–1½ m and 10 over 1½ m. This represents a great advance on hand methods.

In the planning of large oil palm development schemes the question of whether to employ fully mechanical methods or to use largely hand methods aided by mechanical saws has been controversial. It has been argued, for example, that the use of the methods described on p. 376 and Fig. 8.2 enables larger areas to be planted in a shorter time thus ensuring the highest return on investment.[30] Experience in Malaysia has shown, however, that equally large areas can be opened per annum through the employment of contractors using mechanical saws and hand labour; and in Africa plantation companies continue to prefer hand methods. Cost comparisons on a per-hectare basis will be affected by prevailing labour rates, and in West Africa labour has been cheap enough to render fully mechanical methods much more expensive than hand clearing assisted by mechanical saws. One argument used against hand methods has been the need to recruit and house large labour forces at an early stage of development; however, in most countries there is a labour surplus eager for employment and a very high proportion of the labour employed for field preparation can later be employed in maintenance and harvesting. At the present time (1975) fully mechanical methods can only be justified in areas where sufficient manpower is not obtainable for clearing.

In spite of its high cost mechanical felling and clearing has been used on a large scale in the Ivory Coast and some other countries. Experience has shown that excellent plantations can be obtained where the terrain is suitable and the work is up to time; there have been cases however where operations have become seriously behindhand and, with the onset of wet weather, soil puddling has followed and the early growth of the palms has been adversely affected.

In some situations a combination of hand (including chainsaw) and mechanical methods has been found most economical. In Malaysia, if felling gets seriously behindhand or if the burn has been less than 30 per cent, it has been found best to stack with bulldozers before reburning; but full mechanization is still considerably more expensive than either hand clearing or partial mechanization.[44] In Ecuador also, hand and chainsaw felling followed by mechanical windrowing before burning has been found very satisfactory for large areas.

Preparation in non-forest areas

There are many areas of the wet tropics where the forest has been felled and the land used for food-cropping and then abandoned. When the intensity of cultivation has not been high and the district is one of low population these food-cropped areas usually revert to forest in stages, shrubs taking hold through the initial ground cover of grass and creepers and the shrubs being succeeded by secondary and later primary forest trees. In these cases preparation would of course be carried out in the same manner as with primary forest or jungle.

Where population density is high or the areas extensive, many factors may prevent a reversion to forest. In Malaysia for instance, areas abandoned to lalang grass (*Imperata cylindrica*) are often regularly, though unintentionally, fired. This tends to kill out the young shrubs and perpetuate the lalang. In West Africa, areas of 'derived savannah', consisting of grasses and fire-resistant trees, are extensive. These are usually to be found in districts with a prolonged dry season where food-cropping has been intensive and been followed by deliberate annual firing of the bush to assist in the capture of game. In other districts (e.g. the 'bad lands' of eastern Nigeria) a poor grassland has even encroached on the areas of palm grove which have become ever-decreasing 'islands' in a sea of grass. Here the palms do not replace themselves and so the grassland obtains entry wherever a patch of palms dies out. The whole succession is often assisted by the roaming of uncontrolled herds of cattle. In the Ivory Coast, areas of this type are, as in the Far East, dominated by *Imperata*, while in both the Ivory Coast and Nigeria strong growths of Siam weed (*Eupatorium ordoratum*) have lately become a menace both in cleared areas and even in recently planted areas. This plant takes fire easily in dry weather.

It was in Malaysia that the recovery of large areas under *Imperata* was first investigated.[33, 34] During the Japanese occupation of 1942–5 large tracts of forest had been felled for tapioca and other food crops while on rubber estates many new plantings were soon swamped by lalang. Hand eradication was not easily carried out owing to high cost and lack of labour. Early experiments showed that disc ploughing followed by several disc-harrowings was the most satisfactory method and that the complete-ness of eradication was much improved by a final cultivation with a straight-tine cultivator. The establishment of covers was an essential part of the sequence and very small quantities of lalang had to be dug out in subsequent patrolling. In coconut and oil palm plantations, where light growths of lalang had become established, rotary cultivation behind wheeled tractors was found to be satisfactory.

Costs of mechanical eradication of lalang over large areas were found to be less than mechanical spraying of sodium arsenite, but two sprayings at weekly intervals *before* rotary cultivation effectively aided eradication.

More extensive work in the Ivory Coast[18] confirmed the Malaysian methods; there, ploughing is preceded by the use of the heavy 'cutaway' disc harrow known as the 'Rome-plow' which breaks up the vegetation. Ploughing with a six-furrow disc plough is then carried out to turn over the soil; this is followed by one or two cross harrowings with the Rome-plow and four to six disc harrowings with a wide and light set of disc harrows at weekly or fortnightly intervals. Lining and the planting of covers follow, and any final eradication needed is done by hand.

On extensive areas the work is rapid as will be seen from the following schedule.

'Rome-plow'	2½ h per hectare	= 1 h per acre
Ploughing (6-furrow)	3 h per hectare	= 1¼ per acre
'Rome-plow'	2½ h per hectare	= 1 h per acre
Disc harrowing (× 6)	25 min (× 6) per hectare	= 10 min (× 6) per acre

The whole sequence therefore takes about 10½ hours per hectare or 13 hours if a second cross-harrowing with the Rome-plow is needed.

In derived savannah areas free from trees and with no dominating weed, cultivation may not be required. The planting sites only need to be cleared and covers planted among the sparse vegetation. Such areas are likely to be very poor and their satisfactory usage for the oil palm has not yet been established. There is no reason, however, why intrinsically suitable areas which have become lalang-dominated through mismanagement should not be reclaimed by the planting and proper management of oil palms and, indeed, several such areas have been successfully brought into bearing in the Far East and Africa; the low capital cost of planting has been an encouragement. Fertilizer requirements will need added attention, however. The procedure in Malaysia has been similar to that in Africa, the disc plough being the main implement, preceded or followed by (or sometimes replaced by) the heavy cutaway Rome-plow. The latter will cut down to 20–23 cm (8–9 in) but disc-ploughing to 30 cm is sometimes required.

No weed killer has yet approached sodium arsenite in effectiveness or cheapness in the destruction of lalang, but as it is highly poisonous its use has been carefully controlled and, in some countries, avoided altogether. Sodium arsenite is a contact weed killer and successive sprayings are required to destroy the leaves and weaken the plant so that it can no longer produce shoots from the runners. The quantity and concentration of sodium arsenite to be sprayed depends on the strength of lalang growth, the weather at the time of spraying, and whether a wetting agent is used. On average a 1½ per cent solution obtained by dissolving 10 kg powdered sodium arsenite (80 per cent As_2O_3) with 250 g of a wetting agent in 700 l of water will be enough for an initial spraying of 1 hectare. Spraying is continued every 8 to 10 days for 7 to 10 rounds, the quantity of solution required to wet the leaves decreasing with each round. After the spraying has been completed, sporadic lalang may be eradicated by digging or by 'wiping' with oil. Several proprietary oils such as Shell Lalang Oil W or A-601, Soracide PY or PYD are on the market. An absorbent cloth is simply dipped into the oil, squeezed of excess oil and then wiped along the blades from the base upwards. Methods of application of both sodium arsenite and oil, and the control and organization needed, are dealt with very fully in Edgar's *Manual of Rubber Planting*.[20]

As alternatives to sodium arsenite, dalapon (sodium 2,2-dichloro-propionate) and Roundup (isopropylamine salt of glyphosate) have been successfully used. Dalapon is sprayed as a 1.5 per cent solution with three treatments at three-weekly intervals which results in the use of about 7–8 kg of the commercial product per hectare per treatment.[46] Roundup has been recommended at 4 kg a.e. (acid equivalent) per hectare, with a 'touch-up' spraying 60 to 90 days later.[47]

In recent years Siam weed, *Eupatorium odoratum*, has become a serious weed in Africa, and it is also to be found in the Far East where, however, it does not appear to grow quite so high or luxuriantly. Its control in established plantations is described on p. 451. If planting is to be carried out in an area where patches of *Eupatorium* exist, these patches should first be eradicated. If eradication can be undertaken and completed before planting, then a mixture of 2,4-D and 2,4,5-T at a rate of 5.5 kg a.i. per hectare may be used. If, however, eradication must continue after planting, it is too dangerous to use these weed-killers since they affect the young palms. The latter are tolerant of substituted triazine and substituted urea herbicides, and of these atrazine at 4.5 kg a.i. per hectare and diuron (DCMU) at 5.5 kg per hectare have been found effective against Siam weed if applied in dry weather.[35]

Inter-row planting of covers or crops

Many areas of oil palms have been established without any planting of inter-row cover plants. The natural vegetation was allowed to come up and was kept under control. In many parts of West Africa *Pueraria phaseoloides* becomes dominant between the rows without deliberate seeding, while occasionally patches of *Centrosema pubescens* arise in the same manner. With the advent of burning the early establishment of a cover became more urgent, and it is now usually regarded as an essential part of any planting programme. As, however, the establishment and maintenance of covers continues into the post-planting period, cover planting will be dealt with in Chapter 10. The sowing of a cover is normally done at the end of the land-preparation operations and before planting although there may be exceptional circumstances where covers are sown after planting. In West Africa, for instance, planting can start in March, well before the rains have fully set in, but covers establish better if planted with the first heavy rain in April or May. In Malaysia covers may be sown from 3 to 6 months before planting.

The inter-planting of food crops is rarely an attractive proposition for large estates since they are not organized for such work and the profit to be obtained is very doubtful. To the smallholder, however, the obtaining of quick-growing cash or food crops while awaiting a return from the palms may be an attractive proposition and sound husbandry. Such cropping should not be confused with a deliberate attempt to combine the cultivation of other crops with the planting of oil palms; such combinations present quite different problems which will be discussed in Chapter 12. *Establishment inter-cropping* is temporary cultivation only and it has no effect on the preparation of the plantation or the spacing of the palms, except of course that a more thorough clearing and burning will be needed.

Several trials of establishment inter-cropping have been carried out in Africa.[36] In Zaire[37] the planting (alone or in succession or combination) of cassava, maize, hill rice and bananas for 1, 2 or 3 years had, in general, a beneficial effect on palm yields and, after 16 years, no deleterious effects had appeared. The area used was in forest, the catch-crops were planted immediately after the palms and the control plots were planted with *Pueraria phaseoloides*. A similar trial planted on forest land near Benin in Nigeria in 1940 included both cropping for 2 years and cropping for as long as crops could be obtained. In practice the latter treatment entailed cropping with maize, yams and cassava until shade made the growing of cocoyams the only possible culture, and this was continued until the twelfth year. In the 2-year cropping plots and the early years of the cropping-to-exhaustion plots, good crops were obtained; these were similar to those obtained in Zaire and showed that food-cropping for 2 or 3 years may, in these circumstances, be an attractive proposition to a smallholder. Some of the crops obtained are shown below:

Crop	Zaire (kg/ha)	Nigeria (kg/ha)
Yams, 1st year	—	12,849
Yams, 2nd year	—	6,148
Cassava tubers (1st year Zaire, 3rd year Nigeria)	17,790	7,385
Maize, dry grains (1st year Zaire, 3rd year Nigeria)	905	1,677
Rice (as dry padi)	1,509	—

In both trials there was a tendency for early bunch yields to be improved by inter-cropping; in the Nigerian trial both leaf production and early yields were significantly increased. In neither trial, after 16 or more years of harvesting, was there any significant fall in yields following the early inter-cropping treatments, though in Nigeria a tendency for yields to fall in later years has been noticed. The cumulative yields over periods of 16 and 19 years' harvesting and the early and late yields are given below as percentages of the control (*Pueraria* cover):

	Zaire			Nigeria		
	3rd year	16th year	16 years cumulative	1st–3rd year cumulative	14th–19th year cumulative	19 years cumulative
Control (*Pueraria*)	100	100	100	100	100	100
Natural cover	—	—	—	103	94	96
Two years' food cropping	125	105	117	108	88	95
Cropping 'to exhaustion'	—	—	—	120	89	98
Three years bananas	124	100	113	—	—	—

The early beneficial effect of planting food crops with the palms as a last 'preparatory' treatment in place of a leguminous cover or a natural cover is perhaps unexpected, but soil studies in the Nigerian experiment have helped to provide an understanding of it. Soil analyses were carried out in 1941, 1945, 1951, 1956 and 1961 and the inter-cropping treatments showed the soil effects which one would expect to follow tillage, namely a general reduction of fertility, as judged by total exchangeable cations, below that of the plots having a natural cover. In this respect the *Pueraria* 'control' showed similar, though not such accentuated, soil effects as the cropped plots, thus indicating that the cultivation needed to establish the cover may be of significance.

The early beneficial effects of establishment inter-cropping have therefore been obtained in spite of a small reduction in topsoil fertility. It is believed[13] that this is because competition is reduced considerably by cropping (and this may apply, in the early stages, to cover establishment), and secondly because cultivation makes some soil nutrients more readily available. Differences of soil composition due to initial cultivation tended to be large after 5 years, but after the palms had been in the field for 16 or 20 years the differences in potassium were insignificant, those of magnesium small, while those of calcium became relatively large. Small differences in exchange capacity and carbon and nitrogen percentages persisted.[38]

The results of the trials so far mentioned were obtained in areas of Zaire and Nigeria opened from heavy forest where, it may be assumed, deficiencies would not immediately arise. In areas already degraded by foodcrop farming or by annual burning, fertility needs to be rebuilt. In experiments on such land in Nigeria[36] even moderate yields could not be obtained without heavy fertilizer dressings, the beneficial effect of establishment inter-cropping was short-lived and regular tillage soon became harmful (pp. 436−7).

Some interest has been taken in Asia in establishment intercropping and in some smallholders' schemes intercropping has been introduced. An experiment in Malaysia on marine alluvial clay soil showed that much depends on the crop used and the amount of tillage entailed. The field was a replanting, and cultivation was carried out to 1.8 m from the young palms. With soya beans planted for 2 years there was no effect, in comparison with a legume cover, on growth or early yield, though there was a suggestion that if deep tillage is undertaken for two annual crops then palm growth is slightly impaired through root damage. Two tapioca crops grown in successive years severely retarded growth and reduced early yield by 14 per cent through competition for both light and nutrients, particularly N and P. However, all intercropping treatments increased in yield relatively to the legume cover treatment with palm age, and by the fourth bearing year differences were very small.[43]

It may be concluded that intercropping can be practised on fertile, water-retentive soils for 1 or 2 years without substantial loss of crop if the crops are not allowed to compete severely for light, nutrients and water, and provided tillage is not too severe. On light soils in Africa and elsewhere, intercropping should only be practised on forest land or where an adequate programme of manuring both intercrops and palms is possible; and the practice should be avoided on very light soils already degraded by bush-rotation farming or periodic burning.

Soil drainage

The oil palm will not grow and fruit in standing water whether the water level be at the soil surface or below it. The majority of plantations are situated either on comparatively light soils of great depth or on undulating land with soils of varying physical composition where the permeability of the soil and the natural drainage courses are adequate. On such plantations drainage problems tend to be localized, i.e. they are confined to small pockets of valley land where the central stream needs to be kept clear and where flood-waters from adjoining rivers may temporarily back up over some planted areas. In these cases the clearing of main drains (canalized streams) and the digging of subsidiary drains from a few low-lying areas is often all that is needed. The bunding of small areas from occasional river floods is expensive and seldom justified. Assistance can sometimes be given to these areas by the digging of 'foothill drains' which lie at the foot of the slopes which rise from a flat valley bottom; these drains are dug to follow the contour and carry seepage water from the slopes direct to the main valley drain at its lower end. They must be deep enough to ensure that the seepage water enters them.

Although the majority of plantings may be on areas which need comparatively little attention to drainage, flat coastal or estuarine alluvial clay soils account for a considerable proportion — and the highest yielding section — of the plantations of Malaysia. These plantations are flat, are sometimes below high tide level, and the clay soil may be only slowly permeable. Initially these clays may be overlain by a layer of muck or peat soil the drainage of which presents special problems.

A drainage system for these soils should be 'built' back from the chosen outlet point or points; A natural outlet should be utilized if possible and the main drain extended from it, but no natural outlet may exist and one has then to be constructed across a main river bank or into tidal waters. The main drain is cut as straight as possible through the area to be drained and inter-field drains or collection drains are then cut at right-angles to meet it. Finally inter-row or field drains are cut parallel to the main drain to meet the inter-field drains.

On areas of heavy coastal clay, whether overlain by peat or not, field drains are commonly dug every fourth row and sometimes even every other row. On the better structured clays, drains may be less frequent, every sixth row or about 47 m, while on lighter soils intervals of 100 m are possible. Inter-field or collection drains are dug at intervals of about 200–400 m. With every fourth row drainage and the palms spaced at 9 m triangular the length of field drains to be dug will be about 320 m per hectare.

In heavy clay soils the drain sides may be sloped at an angle of 70° from the horizontal but angles of 50°–60° are more usual. The main drain should be 3.5–5.0 m wide at ground level and about 2.5 m deep. Collection drains may be 2.0–2.5 m at the top, 1.2–1.8 m deep and 0.6–1.0 m

Pl. 35 Constructing an inter-row drain with an excavator on coastal alluvium in Malaysia.

at the bottom. The most usual size is 2.5 x 1.8 x 0.6 m. Inter-row drains are 1.2 x 1.0 x 0.5 m or 1.2 x 1.2 x 0.3 m. All drains are nowadays dug with mechanical equipment, drag line excavators being used for main and inter-field drains and trencher equipment for inter-row drains (Plate 35).

The drainage system should be designed to interlock with the transport system so that the minimum of bridging is required. Neither the railway nor the road system may pass over the inter-row drains, but excavators and tractors can be driven into each field from one side of the collection road without passing over a bridge. Access to the rows of palms for inspection can be obtained at both sides of a field either directly or over a footbridge. This is illustrated in Fig. 8.3.[39]

When peat overlies clay, care needs to be taken to see that drainage from the peat is not too rapid. If peat is dried out quickly it is found that reabsorption of water is difficult and slow ('irreversible drying') and the palms suffer accordingly. In any part of the plantation which is covered with peat, a system of small water gates must be installed in the collection drains so that in dry weather water may be drained very slowly from the area.

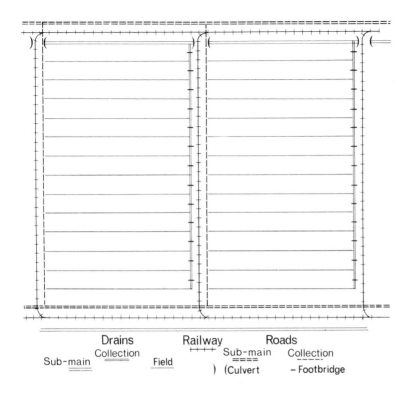

Fig. 8.3 Diagrammatic representation of drainage, rail and road systems for flat low-lying oil palm plantations.

Many flat coastal or estuarine areas must be protected from tidal waters by bunding. This is an expensive undertaking since substantial water gates need to be constructed in the main outlets through the bunds. Hinged automatic gates can be installed which will close under the pressure of the tidal water and open when the tide recedes and water flows from the main drains. Soil should, if possible, be taken from the seaward or river side; bunds should be at least 35 cm above high tide level and be grassed over. The base of the bund should be double the mean height of the tide, while the top should be about 1 m wide. A 3 m tide thus demands a bund 6 m wide at the base, 3.5 m high and 1 m wide at the top. This gives a slope of about 55° to the horizontal. In the Far East crabs bore holes through the bunds, but they can be kept down by applying to each hole one-tenth litre of diesoline in which 70 g of Gamexane dispersible powder has been mixed. Where crabs are likely to be troublesome bunds are now constructed by excavator from borrow pits 14 m from the bund on the landward side.[20]

Irrigation

In Chapter 3 a description was given (p. 103) of the relationships found by French workers between bunch production and water deficit, the latter being calculated for periods of 10 days on the assumption of a soil water reserve of 200 mm and a potential evapotranspiration of 150 mm for dry months and 120 mm for months with more than 10 wet days.[40] At Grand Drewin, where the mean annual water deficit for the 10-year period 1957—66 was 590 mm, an irrigation experiment was conducted on palms planted in 1960 in which each irrigated palm was surrounded by four nozzles each watering a circle of 3 m diameter. The treatments were: A. Sufficient water applied to assure 50 mm rain + irrigation water per 10-day period. B. Water requirements judged by midday stomatal aperture (see p. 153). T. Control. The mean bunch yields for 1966—7 and 1967—8 in tons/hectare/annum were: A. 23.6, B. 22.2, T. 10.6.[41]

Other irrigation experiments have been conducted at La Mé, also in the Ivory Coast, and in Dahomey. At La Mé, where the annual deficit averages 254 mm, water was supplied through a system of spray lines laid along the ground.[42] The cumulative yield over the first 3 years of bearing in the irrigated plots was 32.7 tons per hectare of bunches against 18.5 tons in the non-irrigated plots; the increased yield was mainly due to a higher sex ratio and number of bunches produced per palm. At Pobé in Dahomey where the mean annual deficit is 520 mm, a drip irrigation system, designed to compensate fully for the dry-season evapotranspiration, has given similar results even though the volume of wetted soil represented only 35 to 40 per cent of the total rooting volume. This system is now (1976) being tried on a plantation scale.[48]

These remarkable results indicate that in a climate having several dry months irrigation can raise yields to levels near to those obtained in the Far East. Although the irrigation methods described above may not be practicable or economic on a large scale, in several regions, e.g. south-western Mexico, northern Colombia and south-western Ecuador, there is now an interest in raising yields by irrigation or in planting in areas where the rainfall is clearly insufficient. These areas are largely flat and low-lying, and the provision of water from canals or drains is being attempted. The essential conditions for successful irrigation are (*a*) deficits must be made good by frequent and calculated applications; occasional applications will be largely valueless, (*b*) the method used must supply water to the surface roots, (*c*) areas must not remain waterlogged for long periods. These are guide lines only, from a knowledge of the palm's requirements; much remains to be learnt about practical methods of oil palm irrigation or their costs.

Costs

The costs of preparing the land and planting are given at the end of Chapter 9.

References

1. **Gawthorn, D. J.** (1967) The planting of Elaeis on terraces. *Planter, Kuala Lumpur*, 43, 502.
2. **Ahrens, L.** and **Vanderput, R.** (1951) Contribution à l'étude des travaux d'ouverture d'une plantation en région forestière. *Bull. agric. Congo belge*, 42, 617.
3. **Lebancq, L.** (1957) Mécanisation de l'abattage et du tronconnage en exploitation forestière. *Bull. agric. Congo belge*, 48, 371.
4. **Sheffield, A. F. W.** and **Toovey, F. W.** (1941) Oil Palm Research Station, Nigeria. Second Annual Report, Part A, paras 23–27. Mimeograph.
5. **Toovey, F. W.** (1947) Oil Palm Research Station, Nigeria, Seventh Annual Report, p. 30.
6. **Beirnaert, A.** (1942) Nienwere in de cultuurtechniek van warme landen. *Bull. agric. Congo belge*, Leopoldville, 33, 55.
7. **Vanderweyen, R.** (1943) Quelques directives pour l'établissement d'une palmeraie. *Bull. agric. Congo belge*, Leopoldville, 34, 80.
8. **Rubber Research Institute of Malaya** (1939) Planting without burning. *Plts.' Bull. Rubb. Res. Inst. Malaya*, No. 6, 1.
9. *Anon.* (1957) Jungle clearing. *Plts.' Bull. Rubb. Res. Inst. Malaya*, No. 33, 105.
10. **Sly, J. M. A.** and **Tinker, P. B. H.** (1962) An assessment of burning in the establishment of oil palm plantations in Southern Nigeria. *Trop. Agric. Trin.*, 39, 271.
11. **Bocquet, M.** and **Michaux, P.** (1961) Choix d'une méthode d'ouverture en forêt pour une plantation de palmiers à huile. *Oléagineux*, 16, 149.
12. **Tinker, P. B. H.** and **Ziboh, C. O.** (1959) A study of some typical soils supporting oil palms in Southern Nigeria. *J. W. Afr. Inst. Oil Palm Res.*, 3, 16.
13. **Kowal, J. M. L.** and **Tinker, P. B. H.** (1959) Soil changes under a plantation established from high secondary forest. *J. W. Afr. Inst. Oil Palm Res.*, 2, 376.
14. **Coulter, J. K.** (1950) Organic matter in Malayan soils. *Malay. Forester*, 13, 189.
15. **Gunn, J. S.** (1960) W.A.I.F.O.R. Eighth Annual Report, 1959–60, p. 52.
16. **Barrett, R. G.** (1965) *A method of calculating the optimum 'maximum carry' on oil palm estates.* Sabah Planters Association. Oil Palm Seminar. Mimeograph.
17. **Sankar, N. S.** (1965) *Fruit collection and evacuation by road.* Sabah Planters Association, Oil Palm Seminar. Mimeograph.
18. **Surre, Chr., Fraisse, A.** and **Boyé, P.** (1961) Plantation de palmier à huile sur le sol de forêt et sur savane de *Imperata. Oléagineux*, 16, 91.
19. **Vanderweyen, R.** (1952) *Notion de cultures d'Elaeis au Congo Belge.* Brussels.
20. **Edgar, A. T.** (1958) *Manual of Rubber Planting (Malaya).* Incorp. Soc. of Planters, Kuala Lumpur, Malaysia.
21. **Bevan, J. W. L.** and **Gray, B. S.** (1969) *The organisation and control of field practice for large-scale oil palm plantings in Malaysia.* Incorp. Soc. of Planters, Kuala Lumpur, 166 pp.
22. **Turner, P. D.** and **Gillbanks, R. A.** (1974) *Oil palm cultivation and management.* Incorp. Soc. of Planters, Kuala Lumpur, 672 pp.

23. **Sparnaaij, L. D. and Gunn, J. S.** (1959) The development of transplanting techniques for the oil palm in West Africa. *J. W. Afr. Inst. Oil Palm Res.*, **2**, 281.
24. **Bevan, J. W. L. and Gray, B. S.** (1966) Field planting techniques for the oil palm in Malaysia. *Planter, Kuala Lumpur*, **42**, 196.
25. **Tailliez, B.** (1975) L'aménagement des terrains vallonnés et accidentés pour la plantation de palmiers à huiles. *Oléagineux*, **30**, 299.
26. **Leamey, M. L. and Panton, W. P.** (1966) *Soil survey manual for Malayan conditions.* Min. of Agric. & Cooperatives, Kuala Lumpur.
27. **Dufrane, M.** (1954) Contribution à l'étude de la mécanisation de l'aménagement et de l'entretiens des plantations de palmiers. *Comptes Rendus des Journée d'Etudes sur la Mécanisation de l'Agriculture au Congo Belge*, Oct. 1954, p. 119.
28. **Huguenot, R.** (1963) Coup mécanique des souches et andainage sur sol de forêt. *Oléagineux*, **18**, 623.
29. **Fraisse, A.** (1964) Essais de mécanisation pour l'abattage de la forêt à La Mé. *Oléagineux*, **19**, 73.
30. **Martin, G.** (1970) Le défrichement mécanique pour la création de palmeraies industrielles. *Oléagineux*, **25**, 575.
31. **Sly, J. M. A.** (1963) W.A.I.F.O.R. Eleventh Annual Report, 1962–3, p. 47.
32. **Huguenot, R.** (1965) Utilisation des scies mécaniques pour l'établissement des plantations sur forêt. *Oléagineux*, **20**, 303.
33. **Hartley, C. W. S.** (1949) An experiment on mechanical methods of lalang eradication. *Malay. agric. J.*, **32**, 236.
34. **Keeping, G. S. and Matheson, H. D.** (1949) Mechanical spraying for the eradication of lalang. *Malay. agric. J.*, **32**, 253.
35. **Sheldrick, R. D.** (1968) The control of Siam Weed (*Eupatorium odoratum* Linn.) in Nigeria. *J. W. Afr. Inst. Oil Palm Res.*, **5**, 7.
36. **Sparnaaij, L. D.** (1957) Mixed cropping in oil palm cultivation. *J. W. Afr. Inst. Oil Palm Res.*, **2**, 244.
37. **I.N.E.A.C.** (1955) Division du Palmier à huile. Palmier à huile et plantes vivrières. *Bull. Inf. I.N.E.A.C.*, **4**, 319.
38. **Tinker, P. B. H.** (1963) Changes occurring in the sedimentary soils of Southern Nigeria after oil palm plantation establishment. *J. W. Afr. Inst. Oil Palm Res.*, **4**, 66.
39. **Gray, B. S.** (1965) Private communication.
40. **Anon.** (1969) Recherches sur l'économie de l'eau à l'IRHO. L'eau et la production du palmier à huile. *Oléagineux*, **24**, 389.
41. **Desmarest, J.** (1967) Essai d'irrigation sur jeune palmeraie industrielle. *Oléagineux*, **22**, 441.
42. **I.R.H.O.** (1969 and 1973) *Rapport Annuel 1968*, p. 57, and *Rapport d'Activites 1972–1973*, pp. 67–8.
43. **Chew Poh Soon and Khoo Kay Thye** (1976) Growth and yield of intercropped oil palms on a coastal clay soil in Malaysia. Int. Agric. Oil Palm Conference, Kuala Lumpur, 1976.
44. **Balakrishnan, V. and Lim Cheong Leong** (1976) Jungle clearing by FELDA for planting oil palm. Int. Agric. Oil Palm Conference, Kuala Lumpur, 1976.
45. **Martin, G.** (1976) Méthode d'estimation des temps de travaux pour le défrichement et l'andainage mécanique d'une palmerie industrielle. *Oléagineux*, **31**, 59.
46. **Coomans, P.** (1976) Contrôle chimique de l'*Imperata*. *Oléagineux*, **31**, 109.
47. **Wong Phui Weng** (1976) Use of Roundup (glyphosate) for lalang control prior to planting oil palm and rubber. Int. Agric. Oil Palm Conference, Kuala Lumpur, 1976.
48. **de Taffin, G. and Daniel, C.** (1976) Premiers résultats d'un essai d'irrigation lente sur palmiers à huile. *Oléagineux*, **31**, 413.

Chapter 9

The establishment of oil palms in the field

In planting oil palms in the field, the first object is to bring them into bearing as early as possible and so to reduce the period in which no return on capital outlay is being obtained. Growth to the bearing stage can be influenced at planting time by:

1. The stage of development and the general health of the nursery seedling.
2. The method of transplanting.
3. The time of transplanting.

The second object is so to space the plants in the field that the optimum economic yield will be obtained from the whole period of production.

Planting in the field

Stage of seedling development

In regions with no distinct dry season it is preferable to plant well-developed seedlings which have been growing in the nursery for 10 to 16 months. Experiments in Malaysia have shown that 12- to 18-month-old seedlings (from the two-leaf stage) will come into bearing earlier and give appreciably higher early yields than palms only 6 months old.[1] The growth of older palms is often checked after planting, the first leaves to be opened being sometimes shorter than the last leaves opened in the nursery, but this tendency can be reduced by good transplanting techniques and is reversed soon enough for the older seedlings to maintain a lead over the younger ones. Of more importance, however, is the earlier bearing and the consequently higher bunch yields produced by older seedlings during the first 2 or 3 years of harvesting. The superiority of 12-month over 6-month seedlings may be of the order of 20 to 30 per cent in the latter respect. Seedlings older than 18 months may sometimes, under favourable circumstances, give even higher early yields, but they may suffer severely in transplanting and are of course too large for easy handling.

These findings apply equally to field nurseries and polythene bag nurseries. In two experiments on coastal clay in Malaysia polybag seedlings transplanted at 13 months from the germinated seed stage or older gave significantly higher yields in the first 3 bearing years than seedlings transplanted at younger ages.[2] Other experiments have shown that older seedlings maintain a higher leaf production, bear earlier, and have heavier bunches, a higher fruit-to-bunch ratio and, perhaps most importantly, a higher oil to mesocarp in the first year of harvesting.[31] Estate experience has been that when very young polybag seedlings are transplanted the stand becomes uneven, and planting at not less than 12 months from the germinated seed stage is now generally recommended.

The right age to transplant is really determined by balancing a number of factors. Small seedlings cost less to transplant and do not show so much check in growth; but they will be more uneven owing to their greater susceptibility to pests and diseases and because they have been subjected to less culling, and they will take longer to come into bearing. Older seedlings on the other hand are more costly to transplant and are subject to a greater initial check; but nevertheless they come into bearing earlier. When these factors are all taken into account it will be found inadvisable to transplant seedlings younger than about 10 months or older than about 20 months from the germinated seed stage (about 6 months and 16 months in the nursery where prenurseries are used).

The age at which seedlings are transplanted may, however, depend on special circumstances. The object may sometimes be to bring an area into bearing as soon as possible, when nurseries have not been prepared nor seed purchased. Such circumstances may result from a sudden acquisition of land or a sudden decision, on economic grounds, to change from another crop to the oil palm. The question then is how can the seeds or prenursery seedlings purchased be brought into bearing in the field in the shortest possible time? The answer will be to raise them in polythene bags and to transfer them to the field as soon as possible as young transplants. Where, however, a lengthy planting programme is being undertaken over a number of years or a previous crop is being removed at a steady rate, the question becomes what kind of seedling should be raised to reduce to a minimum the time from transplanting to bearing? The answer in this case will be to prepare nurseries sufficiently far ahead of planting to give large robust seedlings which will come to the bearing stage in the shortest possible time *from transplanting.*

In markedly seasonal climates such as obtain in West Africa, seedlings for planting into the field at the beginning of the rains will have normally been 11 to 13 months in the nursery. This will be clear from what has been said in Chapter 7. If dry-season nurseries (see p. 348) have also been established, seedlings for transplanting from these nurseries will be 7 months old. Experiments have shown that while these seedlings will not come into bearing so soon as the larger seedlings, if seed supplies and the land to be planted are both available, it is better to plant at 7 months from a dry-season nursery than either to keep the seed for the next season or to hold the seedlings for another year in the nursery.[3] Seedlings which pass through a full wet season in West Africa after being forced forward in a dry-season nursery will attain an enormous girth and height by the time they are 18 months old; but apart from the difficulties of transplanting such large seedlings, their contemporaries already planted into the field in the previous year rapidly overtake them in growth and flowering. This can be seen from the following data from a trial in Nigeria:

Nursery	Year of planting in the field	Age at planting (months in the nursery)	Number of leaves produced from May 1960 to Nov. 1962	Number of female inflorescences per palm Nov. 1962
1958 Main nursery (April)	1959	12	60.8	6.00
1958 Dry-season nursery (October)	1959	7	62.5	4.54
1958 Dry-season nursery (October)	1960	18	50.6	1.52

The planting of both germinated seed and of prenursery seedlings direct into the field has been tried. These practices lead to very uneven stands, much supplying and slow progress towards fruiting and are not to be recommended.

Seedling selection and roguing in the nursery

The question of which seedlings should be discarded, or 'rogued' from a nursery is one that has been given much attention. The obtaining of an even stand in the field is of great importance and this has led to the advocacy of a heavy elimination of 'poor doers'. A very careful definition of a 'poor doer' is needed however. Obviously it would be dangerous to transfer to the field any seedlings which are deformed or have had their growth seriously retarded by a non-lethal Blast attack or by the much faster growth of their neighbours.

Selected progenies show very different conformations and rates of elongation in the nursery. This is very obvious to anyone who has examined progeny rows. In a mixed lot of crosses, therefore, many plants will be relatively short because this is a characteristic of their parentage.

Differences in height in the nursery of palms *of the same age* tend to even up when the plants are transferred to the field. In one nursery in Malaysia groups of seedlings averaging from 95 to 150 cm in height (range 55 cm) at planting time showed a group average height range of only 21 cm after 14 months in the field. A group of palms averaging only 67 cm in height in the nursery was still markedly under-developed, but in the following 12 months this group made rapid growth reaching a height with-in 5 cm of the mean height of the other groups. In the first bearing year, significant yield differences between groups were not clearly associated with height in the range 103—150 cm.[1] On an estate in Nigeria, seedling height groups of 1.1—1.5 m and 1.5—1.8 m (length of youngest, fully-opened leaf) gave similar early yields; larger palms, 2.0—2.4 m high, came into bearing earlier but gave slightly reduced early yields, while palms only 0.6—1.1 m in height gave considerably lower yields.[4]

The claim has been made by Devuyst[5] that with groups of palms of the same parentage and age, those groups producing the greatest quantity of leaf matter (measured by number of leaves x average leaf area) will give the highest bunch yields in later years. On the basis of a very low correlation ($r = 0.26$) between leaf number and seedling height, he suggested that height of seedling may be used as a criterion of selection. The system was originally proposed for the purpose of selecting the higher bunch yielders 'belonging to the same variety', but the theory takes little account of environmental factors such as time of germination, relative damage on transplanting from the prenursery, and soil and moisture conditions at the nursery planting site, all of which factors undoubtedly influence the height of individual plants in the nursery even when the highest standards are maintained. Sparnaaij,[6] who discussed all the implications of this theory, showed that, in Nigeria as in Malaysia, differences before transplanting are largely reduced or even reversed after transplanting. Devuyst expected that, under Zaire conditions, eliminating the 50 per cent smaller but *normal* plants would increase bunch production by about 40 per cent. An experiment to test his theory in Nigeria allowed for comparisons of the 50 per cent 'best' seedlings, as judged by height, with both the 50 per cent 'worst' and with normal seedlings selected at random. Mean yields over the first 4 years of bearing were, in kilogram bunches per hectare per annum:

50% Best seedlings	8,654**	50% Best seedlings	8,313
50% Worst seedlings	7,455	Random seedlings	7,987

While the planting of the 50 per cent worst seedlings gave a yield significantly 13.9 per cent lower than that of the 50 per cent best seedlings, the random sample gave a yield which was not significantly lower than that of the best seedlings. Clearly Devuyst's expectations have not been approached and the case for a heavy selection on the basis of height of seedling has not been made out.

Roguing therefore should be of plants whose growth has been obviously retarded or which show abnormalities of habit and of leaflet form.[7] Palms showing habit abnormalities may be (*a*) unusually upright but narrow (dressé), (*b*) flat-topped with successively shorter leaves giving a bunched appearance (ramassé), (*c*) unusually spread out with flaccid, curved leaves also giving rise to a flat-topped appearance (étalé), (*d*) maintaining a juvenile type of growth, although large, so that the leaves do not become fully pinnate (folioles soudées). Leaflet abnormalities may be classified as

(i) those being inserted at an acute angle to the rachis, i.e. 45° or less instead of 60°–90°, (ii) unusually narrow leaflets, rolled longitudinally to give a narrow appearance (étroite), (iii) unusually short but broad leaflets which come to a point abruptly (courte), (iv) leaflets tending to be crowded together and often short and crimped (rapprochée), (v) leaflets much wider apart on the rachis than usual (espacée).

In all the main oil palm growing regions of the world it should now be possible to plant out in the field at least 80 per cent of the nursery seedlings. Of the remaining 20 per cent, 10 per cent may be casualties or deformed seedlings, while the other 10 per cent may be small seedlings which are less than half the average height of the nursery. With a loss of 5 per cent at the prenursery stage (if used) it should therefore be possible to plant in the field at least 75 per cent, and more usually 80 per cent, of the germinated seed provided. This standard is now being achieved on plantations in Malaysia and elsewhere. Where Blast makes an annual appearance, however, the supply of transplantable seedlings is less predictable.

Methods of transplanting

Nursery seedlings cannot be transplanted to the field with bare roots unless they have been given special treatment, and even when such treatment is given, results are variable and establishment is comparatively slow. The failure of many early plantings in West Africa was due to the fact that bare-root seedlings had to be carried over long distances.[8, 9] Transplanting is now most commonly done from polythene bag nurseries; the work of transplanting from a field nursery is more exacting and will first be described.

Transplanting from a field nursery

A nursery seedling needs to be transported to the field and planted there with a substantial ball of earth and for this reason a field nursery needs to be sited as near to the field planting as other considerations will allow.

Methods of digging and transporting the seedling with its ball of earth have been many and various, differences of detail being largely due to differences of soil, situation and personal preference. Any practices which aid the carriage of an intact ball of sufficient size are to be commended.

The effect on the root system of digging a seedling from the nursery and transplanting it has been studied.[10] When planted with a ball of earth young primary and secondary roots continue to grow and after about 4 weeks new secondaries grow out from the old cut primaries. The latter also produce, near their cut ends, one or more thick secondaries which grow in the same direction as the primaries from which they have sprouted. These thick secondaries fulfil the same function as primaries and have been named pseudo-primaries or *racines de substitution*.[11] They are of great importance since, together with the young primaries, they soon begin to produce secondaries (see Plate 36).

Pl. 36   The effect on the seedling root system of root pruning in a field nursery a month before transplanting. Root-pruned plant, bottom plate.

When seedlings are planted with naked roots, many of the longer primaries decay, and establishment depends on the growth of the shorter primary roots and the development of secondaries. Pseudo-primaries are not produced until 2 months after transplanting. If naked-root plants are severely leaf pruned, the decay of the root system is not so extensive and the development of both secondaries and pseudo-primaries is quickened.

The growth of a palm planted with a ball of earth can be improved by causing the root system to develop in the manner described above *before* transplanting takes place. This is done by *root-cutting*, or root-pruning as it is more usually termed. The work is done with a sharp spade in a circle round the seedling at a distance of 15 cm from the base. Such root-cutting checks the growth of the seedling in the nursery and some paling of the leaves may take place. To reduce this effect the work should be done in two stages as shown in Fig. 9.1. Half the circumference of the circle is used for root-cutting 6 weeks before transplanting, while the other half is cut 4 weeks before transplanting.

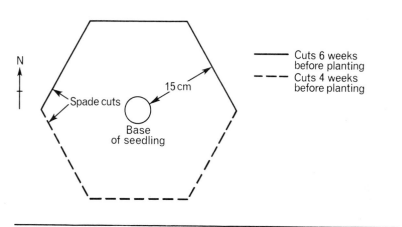

Fig. 9.1   Cutting the roots of a field nursery seedling prior to transplanting.

The blade of the spade must be driven vertically into the ground to its full depth. After the cutting of the roots a light pressing of the soil helps to re-establish contact between the roots and the surrounding soil. One man can treat 300 to 400 seedlings in a day at each stage and the cost of the operation is more than balanced by a reduction in the cost of digging at the time of planting. When digging up the plants care must be taken not to damage the new roots just starting to grow from the old ones; digging is therefore done at a distance rather more than 15 cm from the base of the seedling.

This method was developed in Nigeria and its value demonstrated in experiments. Work in the Far East has confirmed the benefit of root cutting though pruning 8 and 6 weeks before planting may be more effective on Malaysian soils.[12] A further advantage of the method is that the proliferation of the root system in the ball (see Plate 36) helps to hold the soil together so that good balls of earth may be dug in soils from which they are otherwise hard to obtain. Root-cutting has been shown to be advantageous in the transplanting of seedlings of all sizes.[3]

With very large 12- to 18-month-old seedlings some leaf-pruning may be needed to give access for root-cutting and to render the seedlings easier to lift and to handle at transplanting time. Experiments in Nigeria have shown that there is no advantage, as far as early growth and yield is concerned, in pruning the leaves, but that the cutting back of very large seedlings over 1.5 m (5 ft) high to a height of 0.9–1.2 m will not harm them.[3] In general, therefore, field nursery seedlings over 1.7 m in height may be cut back to about 1.2 m to facilitate handling and this may be carried out before root-cutting is started. At the same time any withered or diseased small leaves should be removed. In Malaysia the pruning of large seedlings to a diamond shape is favoured; the older leaves are close-pruned while the upper leaves are pruned more lightly.[2] Seedlings below 1.5 m in height should need very little pruning, only the withered or diseased older leaves being removed.

When lifting seedlings from a field nursery the soil must be damp so that the balls of soil hold together well. In most soils it is possible, with care, to lift a root-pruned seedling without difficulty, but in some very sandy soils and in some very fine-grained inland soils in Malaysia it is difficult to transport intact balls, and special implements must be used. These implements should not be used except in these circumstances since it is difficult to combine their use with the practice of root-cutting and, moreover, the amount of soil transferred by them is usually insufficient.

The most effective implement for lifting a seedling with its ball of earth is a large sharp spade. After digging round a circle slightly larger than that made at the time of root cutting, two men stationed at opposite sides of the palm lever it up with its ball of earth which will be about 40 cm in diameter. The seedling may then be transferred to a pan or wrapped in some material for transportation (Plate 37).

In Africa the commonly used head-pan is useful. Although the ball of earth is of smaller diameter and therefore does not fit tightly into the head-pan, the ease with which the latter can be carried by its two handles makes it a safe form of transport and the balls of earth, even when composed of very sandy soil, rarely arrive at the planting hole in a shattered condition. In the Far East square, sloping-sided pans of the same approximate size as the ball of earth have been used. These have the disadvantage that balls of earth do not fit exactly into them and some scraping off of the outer part of the ball takes place when the seedlings are being lifted into or out of them.

Wrapping the balls of earth in sacking has been widely employed and, when there is any chance of the earth shattering during transport, this is probably the soundest method. The dug seedling is lifted on to a strip of sacking and this is then bound round it and tied up with string. The sacking has to be removed at the planting hole. An ingenious modification of

Pl. 37 Planting a 1-year-old seedling in the field with a substantial ball of earth.

this method is the use of newspaper in place of sacking. The paper does not need to be removed, but is planted with the seedling. A further alternative is to use large and strong leaves from forest plants.

In situations where the soil will not hold together at all on digging, some implement must perforce be used which will contain the soil during lifting and transportation. The most widely employed implement is the Plantoir Socfin or 'frame' invented in Malaysia in 1948 and since used in a number of situations both in Africa and the Far East.[13] The frame measures (externally) 41 cm at the top and 35 cm at the base, being 20 cm high; it is placed over the palm on the ground and four heavy plates are driven into the ground between the inner and outer bars of each of the four sides. These plates cut out and slightly compress a block of soil 28 cm square at the top, and 18–20 cm square at the bottom and some 20 cm deep. The seedling, complete with this block of soil, is then withdrawn from the ground by poles or bars attached to the frame, and the apparatus with the seedling can be transported by two men and placed over a pan of the same dimensions. The seedling is allowed to drop into the pan by the withdrawal, one by one, of the plates from the frame. In the field, the planting hole is made by a Plantoir Socfin so that it has the same dimensions as the block of soil to be introduced, with the seedling, into it.

Other apparatus for lifting field nursery seedlings, e.g. the Plantoir Java and the Watcott cylinder,[14] have been used but with the advent of the polybag nursery the need for any such equipment is disappearing. Moreover experiments have shown that more satisfactory results follow the practice of root pruning and the subsequent lifting of seedlings in good balls of earth. These pieces of apparatus tend to be cumbersome and the extent of the root system transferred is restricted. It is therefore preferable to use polythene bag nurseries or if field nurseries are unavoidable to seek areas where large balls of earth may be dug following root-cutting.

All plants dug from a field nursery should be planted the same day.

The preparation of the planting site has been described in the previous chapter. At the time of planting, a hole is dug of the same shape but slightly wider than the ball of earth in which the transplanted seedling is rooted. The small gap between the ball and the surrounding ground is then filled with topsoil which is firmly pressed in with the feet or with some suitable implement. This is a very important operation as the young primaries and pseudo-primaries must be able to enter the surrounding soil without delay or hindrance. It is important also that the seedling should be planted so that its base is just above the level of the ground. Deep planting has been shown to retard subsequent growth.[15]

Transplanting from polythene bags

The day before transplanting, the bags should be well watered and, if necessary, fungicide or insecticide treatment should be given.

When transplanting seedlings from polythene bags, the base of the bag is cut away and the whole plant in the bag is placed in the hole. The side of the bag may then be slit and the bag is pulled out leaving the soil intact in the ground. If the soil in the bag is very firm, the polythene may be cut off altogether before planting. Filling at the sides is then carried out in the same manner as with a field nursery plant. To make certain that the plant is firm in the ground a soil rammer 1 m long and 7–8 cm in diameter may be used to ram in the soil in successive layers round the ball of earth. Unless seedlings are planted very firmly they are in danger of being blown over by high winds before their root systems are fully established in the surrounding soil.

The time of transplanting

In regions where there is no very distinct dry season it is possible to plant all the year round, though it is more usual to choose those months in which the average rainfall is high. In the south of Malaysia, for instance, some estates start to plant during April (a month with a comparatively high average rainfall) and continue right through the drier months of June and July into the second 'wetter' season of September to December. Experience has shown that, provided planting is done with a large, intact, moist ball of earth, or with a polythene bag plant, the weather at planting time is not of the first importance. If rain falls within 10 days of planting, good establishment should follow; and even if rainfall is still further delayed, seedlings correctly planted will survive and rapidly establish themselves with the onset of a rainy period. A great deal of the 'waiting for rain' which is a common feature of the planting sequence is unnecessary, and it is better for the plant to be in the ground when rain comes than for it to be planted during or just after a period of torrential rain.

Nevertheless in most regions near the equator, e.g. Zaire and Malaysia, where there are two 'wetter' seasons, the one in April and May and the other from August to October or November, is best to choose the *beginning* of one or other or both of these periods for planting out. If a big programme is envisaged the longer season should be chosen with, perhaps, a further small planting at the beginning of the other season. Even in a typical equatorial climate young seedlings may be severely retarded by periods of drought — especially in very permeable soils or in those clay soils which tend to bake hard — and it is therefore best to allow for a period of establishment in wet weather before a dry period is expected. The most severely retarded seedlings will be those planted in wet weather just before the drought period sets in.

In the seasonal climate typical of West Africa the time of planting becomes of much greater importance, the main consideration being to enable the plant to establish itself and grow to the maximum extent possible before the dry season sets in.

The effect of weather on seedlings planted with a good ball of earth and with naked roots was closely examined in Nigeria by planting every day throughout the year except during the months of November, December and January.[10] With satisfactory balls of earth no correlation was found between rainfall during planting and survival or subsequent growth of main-nursery seedlings. With naked-root seedlings however, survival in the early months (before the wet season had fully set in) was largely dependent on the weather at planting time. Such seedlings planted in February or early March only survived if planted within 3 days of rain. Palms planted with balls of earth or with bare roots were of heaviest weight 2 years later when planted in the early months of intermittent rainfall and high number of hours of sunshine. These months were April and May, but ball-of-earth seedlings planted in dry weather in February and March also grew very well.

Palms left over in the nursery through another dry season and planted as 2-year-old nursery seedlings in February to May were heavier after 1 to 1½ years in the field than seedlings planted *late* in the previous year and which had been 2 to 2½ years in the field. This shows that under West African conditions it is better to leave main-nursery seedlings for 2 years in the nursery rather than plant them in the middle of or *late* in, the wet season. It has already been mentioned however (p. 393) that it is better to plant *dry-season nursery* seedlings *early* in the year rather than to hold them over for a further whole year in the nursery and to transplant them as 18-month-old seedlings.

In the Ivory Coast, where the main dry season is much less severe than in Nigeria but the 'short-dry' season of August is more pronounced, planting in October has been practised. In a seed block at La Mé, out of thirteen groups of progenies (planted in 1962, 1963, 1964, 1965, 1966) from each of which some had been planted in May and some in October, the May-planted sub-group was superior in yield in twelve out of the thirteen groups in the first bearing year, in nine out of the thirteen in the second year, in eleven out of the thirteen in the third year and ten out of the thirteen in the fourth bearing year; only by the fifth bearing year had the yields of the two sub-groups become similar.[16]

The work in Nigeria provided some useful data on progeny effects at transplanting. Thirteen progenies were used and the weights of the palms 2 years after transplanting were more closely related to progeny than to weather at transplanting. Five of the thirteen progenies used produced palms which were considerably more vigorous than the average, irrespective of the weather at transplanting. This shows the importance of obtaining planting material which is adapted to local conditions and has been shown to establish itself and grow vigorously in the early months in the field.

As a general axiom it may be said that it is best to transplant seedlings to the field with an intact ball of earth just *before* a period of intermittent rain and sunshine is normally experienced. While it is generally considered that, with polybag seedlings, weather following transplanting is a less critical factor (p. 339) nevertheless the same general principles regarding the timing of transplanting apply.

Cultural practices at transplanting

1. Cultivation and mulching

The site of the planting hole should be levelled and cleared of all vegetation to a radius of 1 m. In some regions mulching of this circle has been recommended and practised. In Zaire mulching with cut vegetation was shown to improve the early growth of very small seedlings (about 60 cm high at transplanting). In the Ivory Coast, where rather severe leaf pruning is practised, mulching with a circle of cut vegetation or with a square piece of black polythene sheeting is recommended.[13] The latter is about 1.3 m across, has a central hole 20 cm in diameter for the seedling, and is perforated with small holes of 1 cm in diameter and 10 cm apart. In

Nigeria two experiments have shown that mulching with cut vegetation and with black polythene improves early growth. The latter appears to have a slightly greater effect in the first year, but if mulching is continued the mulch of cut vegetation seems to gain the advantage.[17]

Elsewhere, the mulching of normal, large transplanted seedlings has not been shown to be advantageous, though if very dry weather were to follow planting the practice might be expected to be of value. Apart from cut vegetation and black polythene, bunch refuse has been used as a mulch.

Hoeing around the plants at the end of the rains showed a small effect on growth in Nigeria but this effect was not as great as that of mulching.[17]

In areas where the soil is very impermeable and young palms suffer severely during periods of heavy rain, improvement through the construction of raised platforms or terraces before planting has been claimed.[18, 19] It should be realized that these measures cannot replace a proper drainage system, though they may be justified in special circumstances. In Colombia, platforms about 1 m in diameter and of varying height have been employed, the small drain around the platform being connected with the inter-line drains, On the heavy 'Massape' soil of Brazil, mounded terraces 2.5 m wide and 0.7 m high have been raised by disc-harrows but this practice has not proved very successful.

2. Manuring at planting

Manuring the oil palm at planting time is fraught with danger since the main need is for nitrogen but the normally employed nitrogen fertilizers have a damaging effect on seedlings wherever part of the root system is exposed. In very carefully managed plantings little or no damage may be experienced when nitrogen fertilizers are placed in the planting holes before or at the time of planting, but in practice there are nearly always some plants with imperfect balls of earth and damage to these plants, particularly where the rainfall during the week after planting is slight, may be serious.

It is a common practice in Malaysia, taken over from the rubber industry, to apply 200 g of Christmas Island rock phosphate in the planting hole; there is as yet no evidence that this practice is of value, however. Phosphorus may or may not be required before the palms come into bearing, and this depends on the soil type; where it is required it is probably best applied after the seedlings have established themselves and in the area of the feeding roots.

Both potassium and magnesium may be required in young plantings. Where required, applications of K have been shown to bring palms earlier into bearing; magnesium is required to counteract a deficiency which is found on certain soils and which is accentuated by N and K dressings.

Compound fertilizers containing unnecessary elements should be avoided. The most usual need is for sulphate of ammonia and sulphate or muriate of potash, and these may be applied in a ring around the seedling at the rate of 200 g each per palm 4 to 6 weeks after planting. Where magnesium is likely to be deficient, it may be added at the rate of 100 g anhydrous magnesium sulphate or 200 g Epsom salts or Kieserite. In areas requiring higher N applications the sulphate of ammonia dose may be repeated after 5 or 6 months.

3. Protection from rodents

In Africa, rodent damage to young transplanted seedlings may be very serious owing to the presence of the large 'cutting grass', *Thryonomys swinderianus*. This animal is usually present in primary or secondary forest adjacent to new plantings and will attack seedlings up to 3 years from planting; it eats through the leaf base to get at the heart of the plant. The single apical bud is therefore almost always destroyed and the plant dies. A new planting may be almost totally destroyed by these rodents in one night. The only really effective measure against this destruction is to surround the young plant *immediately after planting* with a wire collar. Although the collar is open at the top the animal will not normally attack the plant from inside the collar; but if the latter is not properly pegged down it may be lifted off and the plant will then be open to attack. Moreover if the collars are too wide, smaller rodents may enter and cause damage. Experiments have shown that surrounding the plants with wire collars or a fence of sticks packed closely together are the only effective measure; painting the base of the seedlings with tar is not effective.

Owing to the rapidity with which young plants grow it may be necessary to provide wire collars of two different sizes, viz:

1. At planting time: height 45 cm, circumference, 75 cm, diameter 23 cm.
2. After 12 to 18 months: height 60 cm, circumference 120 cm, diameter 38 cm.

With the use of large plants at transplanting time, collars 45—60 cm in height and 120 cm in circumference may be found satisfactory for the whole period. They may be removed between 2 and 3 years after planting depending on the speed of growth of the seedlings.

The transplanting of bare-root nursery seedlings

In areas of peasant production large numbers of small plantings may be undertaken in remote areas and the landowners may be unwilling or unable to plant nurseries themselves. Where possible, small nurseries should be established from central prenurseries by the transportation of seedlings by one of the methods already described. In certain parts of West Africa this proved impossible and fully grown nursery seedlings had to be distributed to these remote holdings from central nurseries. The carrying of these seedlings with a ball of earth is often impracticable and these circumstances gave rise to the development of the root-cutting technique. As already seen, this technique improves the standard of ball of earth plantings; but it often makes the difference between failure and success as far as the planting of bare-root seedlings is concerned.

If field nursery seedlings must be transported without their balls of earth, the seedlings should be root-pruned in two stages as already described. In this way they develop a dense root system which, if treated as described below, will enable the plants to establish themselves in new soil. Immediately after the seedling is uprooted the root system should be dipped in a clay slurry if a suitable clay is available. This slurry will help to prevent drying out. The seedlings should then be placed in tall baskets lined with wet grass and the roots should be kept covered.

If bare-root seedlings are to be sent a long distance by lorry, they can be 'planted' in a layer of soil in the lorry after dipping in the clay slurry. Seedlings can be despatched in this way in a 5-ton lorry, and successful plantings have been carried out as far as 300 miles from the nursery. In such a case all operations need to be carefully timed so that there is no drying out of the root system at either end of the journey. Such a method of transport, carefully supervised, may be superior to the carrying of ball-of-earth seedlings which shatter on the journey while in their pans or sacks and arrive in a dry condition. The method may therefore have application where nursery seedlings are surplus to requirements at site but are urgently needed on a far distant estate.

A special method of planting bare-root field nursery seedlings has been devised.[10] The planting hole is filled and is then opened out again in the form of a cross as shown in Fig. 9.2. The cross has a slight dome in the centre and the base of the seedling is placed on it. The roots are then divided between the four arms of the cross and laid in them and covered in with topsoil which is firmly pressed down. In this way the seedling is held

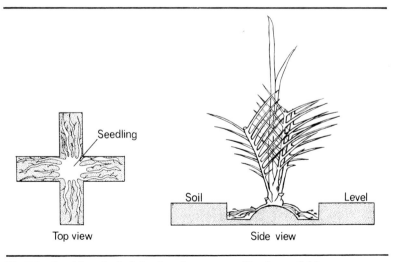

Fig. 9.2 Special method of planting a bare-root seedling from a field nursery.

firmly in the ground and the root system will be less likely to be disturbed by any movement of the plant in the wind.

With the introduction of polybag nurseries the transporting of seedlings in earth has become easier; however, it has been found in Nigeria that bare-root seedlings which are raised in polybags and dipped in clay slurry after freeing of earth can be transported more cheaply and be successfully planted. It cannot be too strongly emphasized however that such measures are palliatives and should only be adopted in the most exceptional circumstances.

Although the growth of the naked-root root-pruned plants will not be as fast as that of ball-of-earth plants without root-cutting, experiments have shown that establishment can be 100 per cent, and after 18 months in the field the height of the plants may be within 90 per cent of the height of ball-of-earth seedlings. Establishment is assisted by cutting back all leaves except the spear to half their length. Such cutting back is *not* advised for normal ball-of-earth transplants, but with naked-root seedlings planted before a dry spell of weather it reduces transpiration and prevents extensive withering. The leaf-cutting is done at the time of lifting.

From what has already been said about fertilizer applications at planting time it will be evident that nitrogen fertilizers are likely to be much more damaging to naked root seedlings if applied before, at or soon after transplanting. This damaging effect is accentuated if the seedlings are planted, as they should be, early in the planting season. In one trial, the production of green leaves in the first 6 months was halved by the prior application of fertilizers in an April planting. With root-pruned bare-root seedlings, therefore, application of fertilizer should be delayed until at least 6 weeks after transplanting.

Supplying

Transplanting techniques have now reached such a high standard that transplanted seedlings, whether from polybags or from field nurseries, should seldom require replacement. Moreover, growth in the field is so rapid that a supply planted 2 years after its neighbours will be able to make little headway, and supplying even after only 1 year from planting is of doubtful utility. In large well-regulated plantings the amount of supplying required after 1 year should be only of the order of 0.5 per cent, though in areas with a very severe dry season rather higher supplying, up to 3 per cent, may be needed.

For these reasons it is best to pay attention to supplying shortly after planting rather than to let a year go by before taking any action. In seasonal climates, if really early planting has been undertaken, it is possible to have all stands inspected about a month after planting; any plants which have clearly not established themselves or have been severely attacked by pests can then be replaced before the season is too advanced for planting. In non-seasonal climates, too, it is best to carry out an inspection 1 or 2 months after planting with a view to supplying any dead or doubtful plants before their neighbours have developed so far that supplies will later become permanently retarded. When there are two 'wetter' seasons it may be best to leave this first supplying until just before the next period of heavy rainfall is expected.[20]

In spite of this early supplying, some further supplying may be required 1 year after planting, particularly if a group of palms has suffered insect or other damage. In this case the plants used should be from a new nursery having seedlings of suitable size for transplanting, supplying should be carried out as early in the season as possible, and the plants should be given special care and attention so that they have the best opportunity of catching up with the other palms.

Spacing the plants in the field

A consideration of the optimum economic spacing of the oil palm in the field raises many difficult questions. In discussing the flowering and fruiting of the palm it was seen that shade had an effect on apparent sex ratio and it would therefore be expected that yield per palm would diminish rapidly with increasing density per hectare not only because of increased competition for nutrients but also because of the mutual shading effect of the palm leaves. This has been shown to be the case. Very closely spaced adult palms always produce a preponderance of male influences and Sparnaaij has shown that the pruning of the leaves of surrounding palms in a close-spaced planting has a marked effect on the apparent sex ratio and hence on the yield of unpruned palms.[21] Apart from the decrease in sex ratio, the weight per bunch decreases with closer spacing, and the usual vegetative effects are an increase in height, a smaller trunk girth, a reduced leaf production and an increased leaf length; the overall leaf area per leaf is, however, little affected (see p. 163).

The optimum yield of a hectare of palms over a given period can only be determined by a knowledge of the potential cumulative yield of palms at different spacings. At a certain density the number of palms per hectare multiplied by the cumulative yield per palm becomes optimal. When plantations were first established, spacing was determined by judging the probable spread of the palms' leaves under plantation conditions. Spacings adopted varied from 7.5−10 m (25−33 ft) and a triangular arrangement was usually employed. Such spacings gave between 115 and 205 palms per hectare. Because the leaves are produced spirally and the palm cannot 'fill in' any gaps in the canopy by production of branches, a very regular arrangement with the largest possible number of similarly spaced surrounding palms is desirable; a triangular spacing in which each palm is surrounded by six of its fellows fulfils this requirement best.

Young palms are little affected by spacing since they are not yet in competition for nutrients or light. Where growth is not rapid, the yield of palms may for several years be unaffected by their spacing. In the Ivory Coast, palms planted in 1926 at 8 and 10 m triangular, gave similar yields per palm until 8 years after planting,[22] while in Nigeria palms planted in 1945 at 6.4 m triangular gave higher yields per palm than palms planted at 12.8 m triangular until 9 years after planting.[23] Close-planted palms may, through their higher proportion of weeding circles and paths, and the higher overall shading of ground vegetation, be better cared for than wider-spaced palms, and so, through reduced competition with surrounding vegetation, be able to give higher yields in the period before mutual shading begins to have the opposing effect. Where growth in the early years is very rapid, however, close spacing effects may be noticeable almost as soon as the palms come into bearing. Palms at double density in a productive field in Malaysia showed a reduced yield per palm from the time of bearing, and at 3½ years old had developed trunks which were appreciably higher than those of normally spaced palms. Nevertheless, the double density palms gave a considerably higher yield per hectare in the first 2 years.

Old palms may be markedly affected by a reduction of density (increase of spacing). Thinning out of a plantation from 124 to 99 palms per hectare at 15 years has been done on fertile land without more than a temporary drop in yield per hectare. Similarly, where stands have been reduced by 10 to 20 per cent by Vascular Wilt disease or *Ganoderma*, yields per hectare have been largely maintained. Heavier bunches per palm are initially obtained and larger numbers of bunches per palm follow. At La Mé in the Ivory Coast there was little difference, from 17 years of age onwards, between the yields of plots with a 'normal' density of 123 palms per hectare and of low density plots of 100 palms per hectare. In two triangular spacings where the density in one case was over 50 per cent higher than the other, the yield in the less dense plots from 17 years of age onwards was more than 20 per cent *higher* than in the denser plots.[22]

In short, high densities can be expected to give high early yields, while lower densities than normal may not reduce yields per hectare in old fields. What, then, is the optimum density for the whole plantation cycle, and what will be the effect of early high densities on later yields in unthinned and thinned plantations?

The early spacing trials which shed light on these problems were carried out in Africa. Prevot and Duchesne showed, on the basis of an experiment with two triangular and three square spacings, that it is possible to calculate the density giving optimum production for any given situation and planting material.[22] This is made possible by the fact that, within the limits usually obtaining, there is a linear relationship between the cumulative bunch yield per palm to any age and the density per hectare. This finding has been confirmed with data from Zaire[24] and Nigeria[23] and is illustrated in Fig. 9.3.

The slope of the straight line relationship between cumulative yield per palm and density gives a competition factor which is calculated as follows:

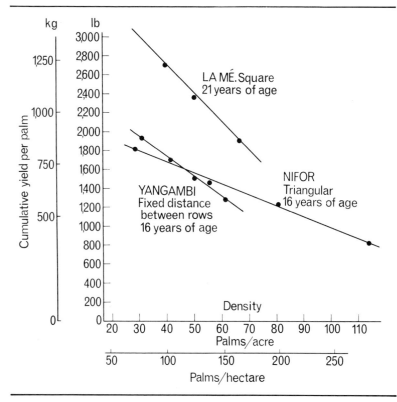

Fig. 9.3 Relationship between cumulative yield per palm and density in the Ivory Coast, Zaire and Nigeria. In each case the relationship is linear.

$$C = \frac{\text{Cumulative yield (wide spacing)} - \text{Cumulative yield (close spacing)}}{\text{Density (close)} - \text{Density (wide)}}$$

This factor, which is an estimate of the amount by which the cumulative yield per palm falls with an increase of one palm in the density, is obtained from the two extreme points of the straight line and is a rough method of calculating the regression coefficient b which is normally calculated from all points available. It should be noted that any particular competition factor applies only to the cumulative yield to the particular number of years for which data has been obtained, and it can be used for finding the optimum density for yield for that period of years. It has already been seen, however, that with older palms yield per hectare does not vary appreciably with density. Provided the data are available for around 12 years, therefore, the optimum density determined is likely to be applicable to a bearing period of 20 to 25 years or the whole life of the plantation. In the La Mé data presented by Prevot and Duchesne the optimum density became constant when the palms were 18 years old and 13 years' data had been obtained.

Production at any density is calculated from the competition factor as follows:

$$P_x = (p_1 \pm (D_x - D_1)C) \times D_x$$

where

P_x is the production per hectare at the density, D_x, concerned, and p_1 is the production per palm at density D_1.

For instance, if the competition factor is 2.14 kg and the cumulative production per palm at the age of 16 years and a density of 70 palms per hectare is 821 kg, then production per hectare at 138 palms per hectare (9.14 m $\triangle$), P_x, is expected to be $(821 - (138 - 70)2.14) \times 138 = 93,216$ kg. The actual accumulated yield in this case was 91,626 kg, the divergence from the expected figure being 1.7 per cent.

Calculation of the density, D_0, which would give the optimum yield per hectare, P_0, over a given number of years is obtained by the formula

$$P_0 = (p_1 - (D_0 - D_1)C) \times D_0 \text{ (as above)}$$

to give

$$P_0 = p_1 D_0 - C D_0^2 + C D_1 D_0 = (p_1 + C D_1)D_0 - C D_0^2$$

hence

$$p_1 + C D_1 - 2 \times C D_0 = 0$$

$$D_0 = \frac{p_1 + C D_1}{2C} \text{ for maximum yield per hectare, } P_0.$$

Taking the previous example, the density for maximum yield up to 16 years of age will then be:

$$\frac{821 + 2.14 \times 70}{4.28} = 226.8 \text{ palms per hectare.}$$

(Calculated from the regression coefficient b, the number of palms per hectare for optimum yield was found to be 228.)

Using these formulae, Prevot and Duchesne showed that a higher cumulative yield per hectare can be obtained from triangular than from square spacing, but that the planting density for optimum yield is lower in the triangular spacings than in the square spacings.[22] This is illustrated by the curves in Fig. 9.4. It was also shown that the decrease in yield per palm with increased density is due both to a reduction in the number of bunches and the weight per bunch, but that the effect on the number of bunches was much the more pronounced.

Linear relationships have also been shown with both square spacing (Ivory Coast) and triangular spacing (Nigeria) between height increment

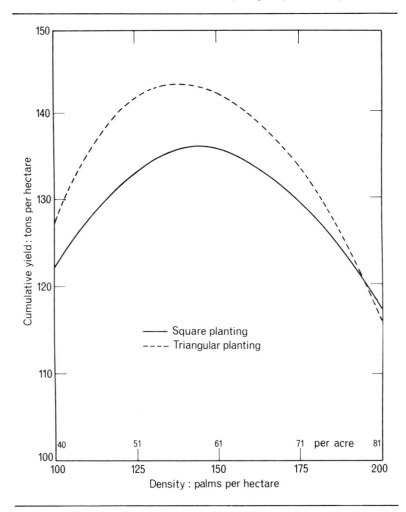

Fig. 9.4 The relationship between cumulative yield per hectare and density in square and triangular plantings (Prevot and Duchesne).

and density, high densities inducing a rapid increase in height. For ease of harvesting therefore, low densities are advantageous.

The most complete experiment including triangular spacings was the one conducted at La Mé in the Ivory Coast and it showed that where the same density is to be maintained throughout the planting cycle, a density of 139 palms per hectare with a triangular spacing of 9.1 m should be used. The Zaire experiment, with 16-year-old palms and using different rectangular spacings but with rows always 8 m apart gave a calculated optimum density of 203 palms/hectare. With rectangular spacing, however, the optimum is always higher than with triangular spacing, and, moreover, it was clear that a constant optimum density had not yet been reached. It is probable therefore that, for triangular spacing and a sufficiency of years, the Zaire optimum would not be far removed from that of the Ivory Coast. In Nigeria a much higher optimum was found (92.3 per acre, 228 per hectare) largely owing to the slowness with which the plants came into bearing. When the calculations were made the palms were the same age as those in the Zaire experiment though 12 years' data were available. But the palms were not competing with each other, i.e. there was no positive competition factor, until the palms were 10 years old. Moreover there were progeny differences which were unfortunately confounded with blocks (see p. 416). Thus several circumstances made it difficult to determine a valid optimum density, though much interesting subsidiary information emerged. From all the information available however it seems that a density of about 143 palms per hectare (9 m triangular spacing) is suitable for Africa if a single density is to be maintained throughout the palms' life.

A number of density experiments have been undertaken in Malaysia (Plate 38) and their early results have become available. In an experiment

Pl. 38   An aerial view of a large spacing experiment in Malaysia.

on inland soils Deli *dura* progenies were confounded with blocks and annual yields were not recorded until 7 years after planting. The results were therefore influenced by lack of early yields and may have been affected by differential progeny responses. The optimum density for 12 years' cumulative yield was calculated to be about 130 palms per hectare, but there were no significant differences between the cumulative yields of densities 96, 114, 138 and 158 palms per hectare. However, as the yield of the four early years not included would have favoured the closer spacings, optimum density for the material employed is likely to have been greater than 130 palms per hectare. This experiment confirmed the linear relationship between density and the cumulative bunch yield per palm, and the competition factor was seen to increase by 1 kg for every year's increase in the age of the palms.[25]

Using the data of four density experiments in Malaysia, Corley *et al.*[26] found that there was little effect of planting density on mean area per leaf (see p. 163), but that the latter factor was directly related to optimum density for current yield and that this relationship appeared to hold good for any level of production. In Fig. 9.5 mean leaf area is plotted against

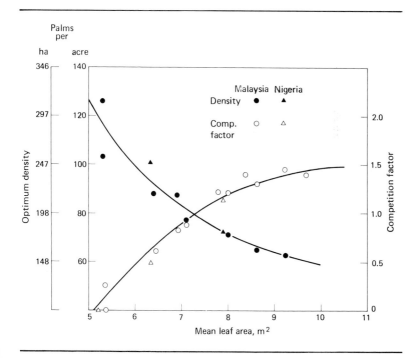

Fig. 9.5   Relationship between mean area per leaf and

(1) optimal density for current yield and

(2) the competition factor for current yield (Corley *et al.*).

optimum densities for current yield and against competition factors calculated for current yield using data from experiments on two soils in
Malaysia and from the Nigerian experiment[23] already referred to. It
should be noted that the competition factor is zero below a certain leaf
area or palm density and tends to become constant when the leaf area per
leaf has reached about 10 m². The current-yield competition factor can be
used for calculating the current yield per palm (Y_D) at any density by the
formula

$$Y_D = Y_x - \frac{C(D-X)Y_x}{100}$$

Where Y_x is the known or expected yield at a standard density, X, and C is
the competition factor as a percentage of the expected yield at the
standard density.

Using competition factors for current yield derived from leaf data and
the yield trends found in experiments conducted under different conditions, Corley *et al.* calculated cumulative yields for 25 and 30 years for
densities varying from about 143 to 180 palms per hectare (58–72 per
acre). They found that optimum densities varied from 150 to 170 palms
per hectare (60–70 per acre) and that the longer the period and the better
the growing conditions the lower the optimum within these limits. However the curves of cumulative yield plotted against density were so flat-
topped that a ten palm per hectare deviation on either side of the
optimum would not reduce cumulative yield by more than 1 per cent.
They concluded that about 158 palms per hectare or 64 per acre was a
good general compromise, but suggested 150 palms per hectare for good
coastal soils, 158 per hectare for good inland soils and 165 per hectare
for poorer soils, using current planting material and with adequate
manuring.

Spacings other than triangular

While the superiority of the triangular over the rectangular arrangement is
established beyond doubt, there are some circumstances, such as mixed
planting or semi-permanent inter-cropping with food crops, in which
avenues of palms, corresponding to rectangular planting, may be
employed. It is therefore useful that a study in Nigeria[23] has shown the
effect on yield of different spacings in the row. Two comparisons were
made:

1. Rectangular spacing with constant interline and varying spacings in the
   row.
2. Various spacings with constant distance within the row.

   The results are given in Table 9.1.
   In the wide avenues of 19.8 m (65 ft), close spacing in the rows failed

Table 9.1 *Effect of spacing on yield (Nigeria)* (cumulative and mean annual yields)

A. *Rectangular spacings with constant interline width*

Cumulative yields	19.8 x 3.7 m 138 palms/ha*		19.8 x 6.4 m 79 palms/ha	19.8 x 9.1 m 55 palms/ha
No. of bunches per palm	53.7	(83.0)	82.7	92.7
Wt. of bunches per palm (kg)	358	(665)	660	810
Mean wt. per bunch (kg)	10.2	(12.0)	12.2	13.8
Mean annual yields (12 years)				
No. of bunches per hectare	618	(954)	544	425
Wt. of bunches per hectare (kg)	4,115	(7,631)	4,331	3,717

B. *Various spacings with constant distance in the row*

Cumulative yields	5.5 x 6.4 m 282 palms/ha	12.2 x 6.4 m 128 palms/ha	19.8 x 6.4 m 79 palms/ha
No. of bunches per palm	60.6	83.1	82.7
Wt. of bunches per palm (kg)	369	646	660
Mean wt. per bunch (kg)	8.8	11.8	12.2
Mean annual yields (12 years)			
No. of bunches per hectare	1,421	887	544
Wt. of bunches per hectare (kg)	8,642	6,893	4,331

* Figures in parentheses relate to same density planting spaced 9.14 m triangular.

to compensate for the heavier and more numerous bunches in the wider spacings. It is of interest to note that with triangular spacing at the same density per hectare as the 19.8 x 3.7 m (65 x 12 ft) planting, the yield was nearly double. The figures in parentheses in the first column of Table 9.1 relate to this triangular spacing. Many of the palms in the 19.8 x 3.7 m planting were spindly and overshaded by their fellows, and there were more dead palms.

The number of bunches and the weight per bunch normally decrease with rising density, but with constant distances in the row, the relationship between cumulative yield and planting density is *not* linear. This is because there is a limit to the advantage to the palms of increasing the distance between the rows. This can be seen in the number and weight of bunches in the 12.2 x 6.4 m and the 19.8 x 6.4 m spacings where reducing the density by two-fifths reduces the yield by about the same amount, whereas in reducing the density by over half, from a spacing of 5.5 x 6.4 m to 12.2 x 6.4 m, the yield was reduced by less than a quarter.

Two conclusions may therefore be drawn:

1. The oil palm cannot be hedge-planted; if for special reasons it is to be grown in avenue formation then the spacing in the rows must be near to the normal spacing for ordinary field planting.
2. Yield per palm cannot be expected to increase further if the palms are

moved more than about 12 m (40 ft) away from each other. Beyond this distance yield per palm or per row will remain constant and the actual spacing of the rows will be determined by other considerations such as the value of inter-crops. These conclusions are supported by the results of an experiment in Malaysia.[25]

Progeny effects

The general conclusions drawn so far may be said to apply to mixed high-yielding material such as is generally available to the grower. The heterogeneity of planting material is such, however, that it would be unlikely that all progenies would require the same spacing for optimum yield. Marked progeny differences have indeed been shown, and when planting material ceases to be a 'collection' and becomes more homogeneous, the optimum spacing for particular material will have to be taken into consideration. As already described in this book (see pp. 163 and 262), the exploitation of variation in growth factors may provide progenies which maintain a high yield per palm at high density.

Table 9.2   *Effect of spacing on selfed* dura *progenies (Nigeria)* (mean height per palm and weight of fruit bunches per hectare in twelfth bearing year)

Spacing	6.4 m △		7.6 m △		9.1 m △		12.8 m △		Diff. 6.4–12.8 m	
Parent	Height (m)	Yield (kg)	Height (m)	Yield (kg)	Height (m)	Yield (kg)	Height (m)	Yield (kg)	Height (m)	Yield (kg)
103.149	4.97	7.679	4.42	12,266	4.30	13,329	4.45	10,696	0.52	−3,017
103.90	5.79	7,255	4.33	10,212	3.99	12,119	3.99	9,676	1.80	−2,421
103.426	4.24	11,297	4.05	12,156	3.38	8,019	3.32	6,120	0.92	+5,852
103.114	6.95	6,448	5.49	8,811	5.27	11,055	4.48	10,367	2.47	−3,919
551.256	6.58	6,080	5.82	9,947	6.40	13,299	4.94	10,019	1.64	−3,939
103.151	4.21	8,345	4.33	9,741	3.50	9,331	4.02	5,327	0.19	+3,018

In the Nigerian experiment already referred to each block was planted with a different progeny. Although this precluded statistical analysis of progeny differences, the latter were so striking that there was little doubt that they were real. Table 9.2 shows the yield in the twelfth bearing year and the height of six selfed *dura* progenies at four different triangular spacings. It will be noted that in the majority of progenies the yield had by this age fallen off seriously in the close spacing of 6.4 m triangular. However, in two cases (103.426 and 103.151), the yield at the latter spacing was still higher than at the widest spacing of 12.8 m and in one case (103.426) was higher than at the 'normal' spacing of 9.1 m triangular. These two progenies, it will be noted, were the shortest and, with the exception of progeny 103.149, were least 'drawn up' by high density. This suggests that progenies showing a short annual height increment may be adapted to high yield at close spacing.

Corley *et al.*[26] have suggested that the progeny differences of optimum density found in this trial might be due to differences in mean leaf area, but that if palms giving a higher yield per unit leaf area can be selected then these would probably not follow exactly the competition factor/ mean area per leaf relationship shown in Fig. 9.5.

The effects of deaths and thinning

It has already been mentioned that thinning, or small reductions of stand following disease outbreaks, may result in only temporary and very small effects on yield per hectare. Bachy[27] has presented some data from 2,211 square-spaced and 1,467 triangularly-spaced stands in fields in the Ivory Coast and, from the yields of palms with different numbers of adjacent palms missing, has calculated coefficients for the 'correction' of their yields. Such coefficients are useful in selection work for reducing to normal the yields of palms favourably affected by the absence of one or more of their six neighbours.

In the case of Deli palms at the usual spacing of 9 x 9 m triangular, Bachy found a linear regression of both the weight of bunches and the number of bunches per palm on the number of adjacent palms missing. Calculated coefficients of correction for weight of bunches per palm were as follows:

Number of missing adjacent palms	0	1	2	3	4	5
Coefficient of correction for wt. of bunches/palm	1.00	0.87	0.77	0.70	0.63	0.58

Though it would naturally be desirable to calculate corrections for each area and spacing, these estimates probably have very general application and are near to those adopted by I.N.E.A.C.

In the case of end palms, Bachy suggests logically that half of these must be treated as having three neighbours missing, while the other half, i.e. those at the end of rows which terminate within the field, have only one neighbour missing. It must be remembered, however, that at the side of a field a whole straight row of palms will be concerned and that each of these palms has two neighbours 'missing'. Moreover, it was noticed in Nigeria that the yield of edge rows depended on how they were oriented. Rows unguarded on one side have their yields increased more when they run N–S than when they run E–W and this may be due to the greater reduction of shading from the sun in the case of the former. Coefficients of correction may therefore need to be adjusted to the orientation of the rows in the case of end palms.

Bachy has also drawn attention to the relevance of these findings to commercial production. If, for instance, a palm is sterile, its removal will enhance the yield of the field, e.g. if the average yield per palm is k, then the removal of a sterile palm will increase the yield of the six surrounding palms from $6k$ to $\dfrac{6k}{0.87} = 6.90k$, an increase of 15 per cent. Thus in fields of very variable material a planter might increase yields considerably by a wise elimination of non-bearing palms.

The optimum economic density

The possibility that the optimum spacing for bunch yield may not be equivalent to the optimum economic density has been considered.[26, 28] Both in the Ivory Coast and, as already mentioned, in Malaysia, changes of density of 10 to 15 per cent have been shown to lead to only very small changes in cumulative yield over periods of 16 to 17 years of harvesting. It has been claimed that small reductions of yield resulting from a sub-optimal density are amply compensated by much larger reductions in cost of establishment, maintenance and harvesting. On the assumption that one density is to be maintained through the planting cycle, it is clear that *establishment* costs will be lessened by reduced charges for planting material, nurseries, lining, holing, planting and disease and pest control. It has been estimated that a density reduction of 10 per cent will reduce the capital cost of all planting operations by 4 per cent. *Maintenance* costs may or may not be reduced; while there will be fewer rings to weed and fewer paths, the wider spacing will, in most fertile areas, induce a heavier growth of interline vegetation which will cost more to control. This may be of particular moment in the early years. If mechanical maintenance is envisaged, however, it is likely that maintenance costs will be reduced because tractors and implements will have easier access to the plantation although cutting a larger area of vegetation outside the rings. Pruning costs will be reduced owing to fewer palms. The relative *harvesting* costs will depend on the method of payment employed. If payment to harvesters is made by number of bunches (or by an allotted number of palms) costs will be reduced as the bunches will be fewer but larger; if however, as is more usual, payment is being made by weight, costs will be unaltered.

In short, a small reduction (up to 10 per cent) from the optimum density will result in a very small reduction in cumulative yield, will reduce the capital required for establishment and may reduce certain of the costs when the plantation is in production. On these grounds it has therefore been claimed that the optimum economic density is always below the optimum density for cumulative yield per hectare. If, however, the yield trend over time is taken into consideration a different conclusion may emerge. Higher density results in higher early yields and under the discounted cash flow method of appraisal these early yields will be worth more than those obtained later on. Although higher densities will produce these higher early yields, they will tend to have somewhat lower yields than those of lower densities in later years. Corley *et al.* found the optima for discounted cash flow differed very little from the optima for cumulative yield.[26]

Variable densities

It has been seen that if a constant density is to be maintained throughout a planting cycle (i.e. from planting a field to replanting) high densities will tend to give high yields per hectare in the early years and low yields per hectare in the later years, while small reductions in densities in mature plantations will have only small effects on yield per hectare. It is logical, therefore, to search for a method of maintaining a high density in the early years and reducing this to a density which will be around the optimum for the middle and later years of the planting cycle. Many fields have in the past been planted at double density, i.e. a planting distance of, say, 9 m triangular having been decided upon, plants are first set at 4.5 m apart in the rows with the intention of taking out every other plant at a later date. The evils of hedge planting have already been noted, however, and the tendency has been, more often than not, to leave the field at double density until the production per palm — and, it should be noted, this includes the palms to be left in — has been seriously reduced and the palms have grown abnormally fast in height. In one such area in Nigeria yield per palm was still well below that in an adjoining normal density area 5 years after thinning although weight per bunch had become the same in the two areas.

Thus if high densities are to be employed with a view to thinning, the latter operation must be carried out well before a serious relative reduction of yield per palm has started.

The lower number of bunches per palm produced at higher densities arises from a higher abortion rate and perhaps also a lower sex ratio. Thinning to a 'normal' density might therefore be expected to restore the production of a palm after about 1 year, with any sex ratio effects taking a further 1½ years for adjustment. Malaysian experiments showed that palms thinned after 2 years of production adjusted their yields after little more than a year, thus indicating that a higher abortion rate was the predominant cause of the lower bunch production per palm at high density. There was a suggestion that if thinning is further delayed recovery will take longer than a year (Fig. 9.6). These experiments also showed that the loss, after thinning, in the recovery year was about 30 per cent and that the overall gain in yield from high density planting and thinning was small. When the higher planting costs and the cost of thinning were taken into account the practice was considered unlikely to be profitable under Malaysian conditions.[26] There is a greater likelihood of finding this practice profitable where growing conditions are slower and double density provides a higher yield for a longer period.

In any policy of thinning, the eventual stand must also be considered.

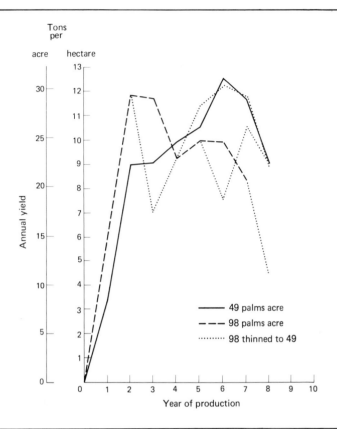

Fig. 9.6   Effects on yields of thinning from high to low density (Corley *et al*).

In an area where high densities can be supported for several years, a relatively low eventual stand, perhaps reached in two stages, might be suitable. However, in areas where high densities lead to a rapid reduction in yield per palm, the final density may be imposed at an early stage and should not therefore be too low.

The simplest form of double density stage is obtained, as already described, by doubling the number of palms in the rows to form long rectangles. This arrangement has the advantage of providing a triangular spacing after thinning, but the early stand takes very poor advantage of the available ground before thinning. A better arrangement in this respect is an hexagonal one[23] which, in effect, is the same as double density in the rows except that every other palm, instead of lying *in* the row, is set in the interline so that it is equidistant from the two adjoining palms in the row *and* the palm opposite in the adjacent row. Every palm is thus surrounded by three palms equidistant from it, and the removal of every other palm in

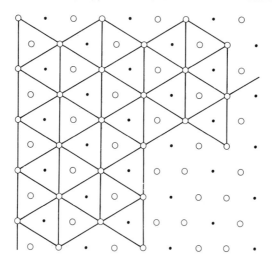

Fig. 9.7 Triangular—hexagonal—triangular planting: thinning from triangular to hexagonal and from hexagonal to triangular spacings. Palms marked · are in the original small triangles only. ○ = palms in small triangles and hexagons; those left in the large triangles are shown in the diagram joined by lines which indicate the eventual spacing.

the zig-zag lines leaves a final triangular formation. This arrangement has the further advantage that, if triple density is possible, this can be planted at triangular spacing by placing one palm in the centre of each hexagon. The sequence:

Triple density: triangular
Double density: hexagonal
Final density: triangular

can be seen in Fig. 9.7.

The relative yields of fields with the same density but with hexagonal and triangular spacings are not known, but even coverage of the ground being in the order triangular, square, hexagonal, and higher yields being obtained from triangular than from square planting of the same density, it is logical to expect that hexagonal planting will give yields inferior to those of triangular plantings of the same density. It would not therefore be advantageous to have planting in hexagonal formation for a lengthy period; this formation could perhaps be best regarded as a short stage either between two triangular spacings or between the onset of bearing and the final triangular spacing.

Thinning schedules should be determined before planting on the basis of a knowledge of the growth rate of palms in the area and any yield figures which may be available per palm for high density plantings and normal density plantings in the area. When reductions of yield per palm take place from the start of bearing, 2 to 3 years at the high density is as much as can be expected. Where two stages are possible, the total period from planting to the final thinning may extend to 5 or 6 years.

Table 9.3   *Table of planting distances and densities*

Triangular planting distance (ft)	Distance between rows (ft)	Density per acre	per hectare	Metre equivalent Planting distance (m)	Between rows (m)
18	15.6	155.1	383.2	5.49	4.75
19	16.5	138.9	343.2	5.79	5.03
20	17.3	125.8	310.9	6.10	5.27
21	18.2	114.1	282.0	6.40	5.55
22	19.1	103.9	256.7	6.71	5.82
23	19.9	95.1	235.0	7.01	6.07
24	20.8	87.3	215.7	7.32	6.34
25	21.7	80.5	198.9	7.62	6.61
26	22.5	74.5	184.1	7.92	6.86
27	23.4	68.9	170.3	8.23	7.13
28	24.2	64.3	158.9	8.53	7.38
29	25.1	59.8	147.8	8.84	7.65
30	26.0	55.9	138.1	9.14	7.92
31	26.8	52.4	129.5	9.45	8.17
32	27.7	49.1	121.3	9.75	8.44
33	28.6	46.2	114.2	10.06	8.72
34	29.4	43.5	107.5	10.36	8.96
35	30.3	41.1	101.6	10.67	9.24
(m)	(m)			Foot equivalent (ft)	(ft)
7.0	6.06	95.4	235.7	23.0	19.9
7.5	6.50	83.0	205.1	24.6	21.3
8.0	6.93	73.0	180.4	26.2	22.7
8.5	7.36	64.7	159.8	27.9	24.1
9.0	7.79	57.7	142.6	29.5	25.5
9.5	8.23	51.8	127.9	31.2	27.0
10.0	8.67	46.7	115.5	32.8	28.4

Notes: To obtain the hexagonal density, with the same spacing, multiply the triangular density by two-thirds.

To convert density per acre to density per hectare multiply by 2.4711.

To convert density per hectare to density per acre multiply by 0.4047.

1 kg per hectare = 0.8922 lb/acre. 1 lb/acre = 1.1209 kg/hectare. 1 ton/acre = 2.51 tonnes/hectare.

Distance between the rows in triangular planting is 0.866 x distance between palms.

To obtain density per acre or per hectare of triangular planting, multiply the square of the planting distance in feet or metres by 0.866 and divide this product into 43,560 or 10,000.

To summarize, the advantages of high density initial planting are:

1. Fuller use is made of the ground from the start of bearing.
2. An earlier return, approaching that obtained from mature areas, is made on capital.
3. Earlier and fuller use is made of capital invested in the mill.
4. The period of no and low yield is reduced when replanting, and thus more frequent replanting with higher yielding material is made possible.

The disadvantages are:

1. Higher initial cost of planting or replanting.
2. Additional field costs, viz. felling half or two-thirds of the stand.
3. Possible reduction in mature yields due to maintaining too high a stand for too long.

Table 9.3 gives densities per acre and hectare at different spacings with some conversion factors. Where the eventual planting distance has been decided, the hexagonal distance, i.e. the distance between palms in the hexagon, is two-thirds of the distance between the rows at the eventual planting distance. For example, with an eventual 9 m triangular planting, distance between rows is 9 x 0.866, or $\sqrt{(9^2 - 4.5^2)}$ = 7.79 m. The hexagonal distance is therefore 2/3 x 7.79 m = 5.20 m. The distance between palms in small triangles (triple density) formed from the hexagons will also be 5.20 m. Densities per hectare would be:

5.20 m triangular	428
5.20 m hexagonal	285
9 m triangular	143

It is unlikely in this case, of course, that the small triangular spacing would be used.

If an initial small triangle or hexagon spacing is chosen, the eventual spacing may be calculated as twice the distance between the rows in the small triangle. For example, with an initial triangular or hexagonal planting distance of 6 m the distance between the rows is 6 x 0.866 = 5.20 m. The eventual triangular planting distance is therefore 10.40 m. Densities per hectare would be: triangular 6 m, 321; hexagonal, 214; triangular 10.40 m, 107.

Costs

The costs of some of the methods of land preparation and planting described in Chapters 8 and 9 are given in Table 9.4 in terms of man-days per hectare. The cost of roads is not included, but is the subject of a special note. The remarks made on p. 357 regarding the variations in man-day costs and the effects of new practices should be taken into account.

Table 9.4   *Man-day costs per hectare*

Africa

From high forest	Man-days per hectare				
	Nigeria	Ivory Coast[14,29]	Zaire (1)[30]	Zaire (2)	Zaire (3)
Underbrushing	19	10	10 ⎫	58‡	44
Felling#	50	50	40 ⎭		55
Beating down	7	30† ⎫	90 ⎫		90 ⎫
Burning	1	30¶ ⎭	⎬	55 ⎬	⎬
Opening rides	26 ⎫	20	30 ⎬		
Lining and realigning	10* ⎭		4 ⎭		19
Planting§	21	12	22	14	25
Cover sowing		5	2	5	5
Fixing wire collars	3				
	137	157	198	132	238

* Lining after underbrushing.
† Includes cutting up.
‡ Light to medium forest.
§ Including lifting from nursery.
¶ Includes heaping, etc.
# Axing. With chainsaws and axes the man-days can be reduced to 5 to 10 per hectare.

Malaysia

From high forest*	Man-days per hectare	
	Manual with chainsaws	Partial mechanization
Underbrushing	6.0	6.0
Felling†	5.0	5.0
Burning	0.25	0.25
Cutting up†, stacking and reburning	12.5	1.35‡
Mechanical stacking	—	0.5
Lining and holing	6.4	6.4
Planting (from polybags)	16.5	16.5
Cover sowing	7.2	7.2
	53.9	43.2

* Data from underbrushing to reburning is from FELDA experience[32]; lining, holing, planting and cover sowing is from the mean of figures of nine estates.
† Units of one chainsaw man and two axemen.
‡ Assuming 5 per cent of the area cannot be done mechanically. If manual assistance is also required with the mechanically stacked areas, then up to a further 5 man-days may be needed.

Notes: Africa.   It will be seen that there is not much variation in felling or in planting costs and that the differences lie largely in the amount of effort expended in the various beating down, cutting, clearing and heaping operations which are undertaken between felling and planting. It should be noted that the felling costs are for felling by axe. With chainsaws the man-days should be reduced to 10 to 20 per cent of the axing figures, thus reducing overall man-day requirements for clearing and planting to around 100 to 150 man-days per hectare. Mechanical felling and windrowing (see p. 376) reduces labour needs by as much as a further 60 man-days per hectare.[29]

Malaysia. It will be noted that clearing costs in terms of man-days are considerably less in Malaysia than in Africa even when the use of chainsaws is taken into account. However, more man-days appear to be expended on lining, holing, planting and cover sowing.

Plantings in Malaysia are commonly on old rubber and coconut land. The work can be done manually or mechanically and costs in the case of old rubber depend largely on the size and number of trees per hectare in the old stand. Man-days per hectare have varied from 40 to 100 for felling, cutting up, clearing and burning. With the need for stumping and burning in old coconut areas (against *Ganoderma* infection) costs may be higher than with rubber.

Planting costs from polybag nurseries have varied widely; the figures given are from a mean of nine estates.

Roads.   Road and initial marking out costs have not been included in the above comparisons as they vary widely with the frequency of roads, the terrain and other factors. Mechanical clearing and levelling of roads is now common practice. However, with a 160 m carry, about 55 m per hectare of road are required. In Africa this entails the employment of 45 to 60 man-days per hectare for road construction.

References

1. **Gray, B. S.** (1963) Annual Report, Oil Palm Research Station, Banting, 1962; **Gray, B. S.** and **Hew Choy Kean** (1964 and 1966) Annual Reports, Oil Palm Research Station, Banting, 1963, 1964 and 1965. Harrisons and Crosfield (Malaysia) Ltd. Mimeographs.
2. **Hew Choy Kean** and **Tam Tai Kin** (1971) Annual Report, Agronomy. Research and Advisory Scheme. 1970. Harrisons and Crosfield (Malaysia) Sdn.Berhad. Mimeograph.
3. **Gunn, J. S.** and **Sheldrick, R. D.** (1963) The influence of the age and size of oil palm seedlings at the time of transplanting to the field on their subsequent growth. *J. W. Afr. Inst. Oil Palm, Res.,* 4, 191.
4. **Unilever Plantations** (1961) Annual Review of Research, p. 36. Mimeograph.
5. **Devuyst, A.** (1954) Selection of the oil palm (*Elaeis guineensis*) in Africa. *Trop. Agric., Trin.,* 31, 133.
6. **Sparnaaij, L. D.** (1955) Seedling selection in oil palm nurseries. *J. W. Afr. Inst. Oil Palm Res.,* 1, (3), 51.
7. **Bénard, G.** (1964) Palmier à huile. Sélection en pépinière. *Oléagineux,* 19, 243.

8. **Toovey, F. W.** (1948) Oil Palm Research Station, Nigeria, Seventh Annual Report, 1946–7, pp. 19–20.
9. **Toovey, F. W.** (1953) Field planting of oil palms. *J. W. Afr. Inst. Oil Palm Res.*, 1, (1), 68.
10. **Sparnaaij, L. D. and Gunn, J. S.** (1959) The development of transplanting techniques for the oil palm in West Africa. *J. W. Afr. Inst. Oil Palm Res.*, 2, 281.
11. **Moreau, C. and Moreau, M.** (1958) Lignification et réactions aux traumatismes de la racine du palmier à huile en pépinières. *Oléagineux*, 13, 735.
12. **Bull, R. A.** (1965) *The production of oil palm seedlings for field planting.* Sabah Planters' Association, Oil Palm Seminar.
13. **Dekester, A.** (1962) La mise en place des jeunes palmiers à huile. *Oléagineux*, 17, 475.
14. **Sly, J. M. A.** *et al.* (1963) W.A.I.F.O.R. Eleventh Annual Report 1962–3, p. 25.
15. **W.A.I.F.O.R.** (1955) Third Annual Report 1954–5, p. 65.
16. **I.R.H.O.** (1974) Rapport d'Activites 1972–1973. Doc. No. 1138.
17. **Sly, J. M. A. and Sheldrick, R. A.** (1965) N.I.F.O.R. First Annual Report 1964–5, p. 27.
18. **Poncelet, M.** (1965) Plantations de palmiers à huile en terrasses sur terrains humides. *Oléagineux*, 20, 431.
19. **Ollagnier, M. and Delvaux, R.** (1966) La plantation du Val d'Iguape. *Oléagineux*, 21, 663.
20. **Lafaille, J.-P. and Daniel, M.** (1965) Soins aux jeunes palmiers et remplacements. *Oléagineux*, 20, 667.
21. **Sparnaaij, L. D.** (1960) The analysis of bunch production in the oil palm. *J. W. Afr. Inst. Oil Palm Res.*, 3, 109.
22. **Prevot, P. and Duchesne, J.** (1955) Densités de plantation pour le palmier à huile. *Oléagineux*, 10, 117.
23. **Sly, J. M. A. and Chapas, L. C.** (1963) The effect of various spacings on the first sixteen years of growth and production of the Nigerian oil palm under plantation conditions. *J. W. Afr. Inst. Oil Palm Res.*, 4, 31.
24. **Beirnaert, A. and Vanderweyen, R.** (1940) Note préliminaire concernent l'influence du dispositif de plantation sur les rendements. *Publs. I.N.E.A.C.* Série Tech., No. 27 bis.
25. **Ramachandran, P., Narayanan, R. and Knecht, J. C. K.** (1972) A planting distance experiment on *dura* palms. In *Advances in oil palm cultivation*, p. 72. Incorp. Soc. of Planters, Kuala Lumpur.
26. **Corley, R. H. V., Hew, C. K., Tam, T. K. and Lo, K. K.** (1972) Optimal spacing for oil palms. In *Advances in oil palm cultivation*, p. 52. Incorp. Soc. of Planters, Kuala Lumpur.
27. **Bachy, A.** (1965) Influence de l'éclaircie naturelle sur la production du palmier à huile. *Oléagineux*, 20, 575.
28. **Surre, C.** (1955) Densité economique de plantation pour le palmier à huile. *Oléagineux*, 16, 411.
29. **Surre, Chr.** *et al.* (1961) Plantation de palmier à huile sur sol de forêt et sur savane à *Imperata. Oléagineux*, 16, 91.
30. **Vanderweyen, R.** (1952) *Notions de culture d'Elaeis au Congo Belge.* Brussels.
31. **Khoo Kay Thye and Chew Poh Soon** (1976) Effect of age of oil palm seedlings at planting out on growth and yield. Int. Agric. Oil Palm Conference, Kuala Lumpur, 1976.
32. **Balakrishnan, V. and Lim Cheong Leong** (1976) Jungle clearing by FELDA for planting oil palm. Int. Agric. Oil Palm Conference, Kuala Lumpur, 1976.

Chapter 10

The care and maintenance of a plantation

Once the plants have been established in the field they need care and protection so that they may come into bearing early and give a high bunch production throughout their bearing life. The palms need protection against competition, soil depletion, diseases and pests, and care must be taken to see that in their pruning and harvesting no damage is done which may affect their future performance. Eventually they must be replaced by another, and generally superior, set of palms by the process of replanting. Some of these factors are interdependent; competition from inter-row plants and soil depletion cannot, for instance, be considered apart from one another nor from nutrition and manuring, while the latter is often intimately connected with disease incidence. For convenience, however, nutrition and diseases and pests will be dealt with in separate chapters.

Cultivation and soil covers

In Chapter 8 it was shown that in opening new areas from forest by burning the last operation was the planting of covers, except where establishment inter-cropping was to be undertaken. Examples were given of the effects of early cultivation on soil nutrient levels and on subsequent yields. The removal of competition was seen to have early advantageous effects on yields but, on already impoverished soils, to reduce soil fertility to a point where later yields were affected. Two of the West African experiments there mentioned also contained treatments in which, although ring-weeding and paths were maintained, the inter-row covers were neglected to the extent that only one round of slashing per year was undertaken as against normal maintenance in which the undergrowth was kept down to about 1 foot by regular slashing rounds. The effects on yields were as follows:

Yields of annually slashed plots as a percentage of the normal maintenance yield

	Forest area	Previously cultivated area
First 8 years of bearing		
Normal maintenance	100.0	100.0
Annual slashing only	79.4	69.1
Later bearing years	(11 years)	(2 years)
Normal maintenance	100.0	100.0
Annual slashing only	102.0	85.4

It will be seen that competition from undergrowth had a severe effect on early yields, this being severest on already impoverished land. In the forest area, however, once the palms had fully established themselves, their shade effected an ascendency over the cover which resulted in yields no longer being reduced by lax maintenance. This effect was less marked in the other area. It is thus clear that in considering the ground cover policy for optimum yield, the effects of competition and cultivation need as much consideration as the type of cover to be maintained.

The type of cover to be maintained

There has been much discussion, but little experimentation, on the most desirable type of ground cover to grow in a plantation. Many plants considered manageable as covers are hard and expensive to establish and the cultivation effect of establishing them may readily be confused with an effect of the cover itself. On the other hand an easily established plant may appear deleterious because it enters quickly into competition with the palms through not being properly maintained. It is not unnatural, therefore, that cover policy has for decades been a matter on which opposing opinions have been firmly held on very little scientific evidence and that fashions for the use or encouragement of certain plants have come and gone.

Historically, the attitudes to ground cover have been different in Africa and in the Far East. This may be ascribed to a number of factors, environmental and accidental. In Africa, particularly in the wet but seasonal parts of West Africa, forest regrowth and the species which follow the felling of forest grow with tremendous rapidity particularly at the beginning and, to a lesser extent, at the end of the rains (April/June and September/October). At the same time the leguminous cover *Pueraria phaseoloides* establishes itself naturally and vigorously and increases the volume of vegetation. The maintenance of the interline covers has therefore consisted of cutting back this adventitious growth to a height of about 1 foot (30 cm). The frequency of cutting depends on the locality, but in areas opened from high forest and subject to a sharp dry season, as many as six 'cutlassing' rounds have been required in the second and third years while ring-cutlassing is needed as or more frequently to prevent the rings around the palms being overrun and the palms themselves covered with fast-growing creepers. As the palms shade the ground, so the frequency of cutlassing and ring-cutlassing decreases. Cutlassing is reduced to two rounds (or even one) per year and ring-cutlassing to two to four rounds.

In the less seasonal climates of Zaire and the Ivory Coast, initial regrowth of natural covers is not so vigorous, but *Pueraria* still establishes itself well; the need for controlling the interline growth still exists as does the protection of the plants by constant ring-weeding. In Zaire, when the forest was not burnt, the frequency of cutlassing rounds was reduced by sowing *Pueraria* in the rows and allowing it to grow over and restrain the forest regrowth in the interline. When the palms shaded the ground the *Pueraria* died out, leaving the regrowth which then needed only very infrequent cutlassing and was, on theoretical grounds, considered a more favourable cover than the *Pueraria*.[1,2] In the Ivory Coast,[3] on the other hand, *Pueraria* has come to be encouraged by low cutting of natural regrowths, and cutlassing is only required in young areas twice a year. In adult areas cutlassing is reduced to one round per year but ring-cutlassing continues four times a year.

By contrast, maintenance in the Far East, through all its changes of emphasis, has been thought of in terms of weeding rather than cutting. The plants held to be desirable, or at least not obnoxious, do not tend to grow rapidly in height, e.g. the leguminous covers and the fern *Nephrolepis biserrata*. A weeding round has therefore largely consisted of chopping out those species not wanted and clean weeding the rings around the palms. One clearly dangerous weed, lalang (*Imperata cylindrica*), must be kept at bay, and this more than anything else has been the cause of maintenance being thought of in terms of keeping out undesirable plants. Lalang is particularly undesirable as it invades the rings and competes directly with the oil palm. Therefore in all countries where lalang is ubiquitous, regular weeding, or oil wiping, patrols are necessary. In Africa, lalang is usually confined to certain locations and seldom presents a problem.

Cover policy has been much influenced in the Far East by current practice in the rubber industry. This has undergone a considerable number of changes in the last four or five decades and these changes have been reflected in similar changes in practice on oil palm estates, especially those where rubber is also grown. Fortunately, the oil palm was established as a plantation crop late enough to escape the full effect of the clean weeding era which afflicted the rubber industry. Nevertheless, the first manual in Malaysia on oil palm growing[4] commended the practice of clean weeding on flat land although conceding that it was costly. Cover plants were, however, recommended for sloping land and on many estates the establishment and maintenance by weeding of cover mixtures became standard practice up to the Second World War. A mixture of two or more of the legumes *Calopogonium mucunoides*, *Centrosema pubescens*, *Pueraria phaseoloides* and *Dolichos hosei*, with the non-leguminous creeper *Mikania cordata* was at that time recommended. In Sumatra the Carilla gourd, *Momordica charantia*, was favoured as was the shrub, *Leucaena glauca*, which was sometimes used as a green manure.

Immediately after the war cover planting fell out of favour and it was reported by research workers of plantation companies[1] that 'there was a great wave of feeling against the use of cover crops in Malaya'. The idea of 'forestry' covers took the place of planted covers and seemed to suit the non-burning procedure then adopted. In some areas young palms were planted in the shade of forest trees which were frill-girdled after 6 months and poisoned with sodium arsenite. The natural cover then consisted of shrubs from stumps, and seedlings which grew from the seed of the forest flora. Plants of secondary forest growth such as *Macaranga* species were common. Slashing rounds were undertaken at 4- to 6-month intervals.

Where areas had been burnt, the 'natural covers' were of different species. Soft weeds were followed by grass species, typically those of *Paspalum* and *Axonopus*, with small shrubs such as *Trema* and *Lantana* and the creepers *Mikania cordata* and *Passiflora* species. Later, the fern *Nephrolepis biserrata* would become dominant, though mixed with grasses and *Mikania*. In Nigeria, unburnt areas not planted with covers would contain many woody species, but *Pueraria* invasion was the rule. Later, *Pueraria* would die down under shade, leaving a shrubby undergrowth.

In the new oil palm plantings of tropical America, grass species are the dominant feature of natural covers and present a serious problem. The presence of ranches, big or small, in almost all parts of the region is responsible for this. Even where plantations are being opened from forest, seeds from grasses in adjoining pastures soon gain access. Most of these grasses are upright and grow vigorously throughout the wet season and many, e.g. *Hyparrhenia rufa* (Uribe or Yaragua), produce tall flowering shoots which seed freely and can easily 'submerge' young plants. *Panicum maximum* and *Pennisetum purpureum* are common. A serious weed of the Pacific coast, particularly in Ecuador, is the Camacho (*Xanthosoma* sp.) the leaves of which grow rapidly to enormous size; the plant requires special control measures.

Where areas in America are opened from forest *Pueraria* establishes well, but can quickly become invaded by grass species. When planting in pasture land the grasses are already there and eradication and their replacement by leguminous covers may be prohibitively expensive. The use of cattle to graze out the grass has been suggested and successfully used in some areas, but heavy soils and high water tables are characteristic of many of the oil palm plantations and the land may become seriously puddled and drainage impeded. Cases of retarded growth have been encountered following puddling of the soil during the rains. If cattle are used, therefore, their grazing has to be most carefully managed.

Cultivation and the sowing of covers is sometimes successful in America; but the original grasses often regain their foothold. Where grasses are dominant therefore it is necessary to arrange for especially frequent and thorough ring-weeding and to control the interline grasses by spraying with herbicides. If ring-weeding is neglected, the grasses soon invade the circles from outside, grow up through the leaves of young palms and enter into competition with them for nutrients and even for light. Once an invasion has taken place, ring-weeding becomes very difficult as the root systems of the palm and the grasses are intertwined. In Honduras a fern cover not dissimilar to the *Nephrolepis* cover of the Far East has been found to establish itself and needs little attention.

With the almost universal practice of burning, and with the replanting of oil palm, coconut and rubber areas with oil palms, the planting of covers has become the general rule in all oil palm growing regions. Recent reviews have suggested however that the controversy of planted legumes *versus* natural covers is not yet dead. Broughton[74] has reviewed Malaysian experiments and showed that legume plots invariably outyielded plots with natural covers or any other managed covers. Moreover palms in the legume plots showed higher leaf levels of N and P and gave higher responses to fertilizers. He found that, in the early years when the interrow cover is not yet competing with the root system of the young palm, the legumes grew faster than other covers, and that later, when the palm canopy closed in, the covers began to die and returned large quantities of their nutrients to the soil. For Nigeria on the other hand Aya[75] considered the initial poor establishment of legumes a hindrance to the control of erosion, and he viewed the subsequent vigorous growth as providing strong competition for water and nutrients. He found no increased yields from legume planting in experiments except in some later years in a long-term experiment at Benin; and, although he agreed with Broughton that this later benefit was likely to be due to residues of the legumes when they died out, he did not consider this an overriding advantage. He drew attention firstly to the strong competitive effects of legumes, particularly in the drier regions of West Africa, and secondly to the work entailed in controlling the cover and preventing the invasion of the weeded circles by runners, and he suggested that the encouraging of a natural cover is preferable. While it is undoubtedly true that in seasonal climates the prevention of undue competition between covers and palms in the dry season is of special importance, nevertheless in most parts of West Africa it has not been found at all difficult either to establish a legume cover quickly or to control it by normal maintenance methods. Moreover in countries such as Ghana or the Ivory Coast *Pueraria* has been shown to be easily managed, to persist for many years and thus to return large quantities of nutrients to the soil. It seems unlikely therefore that the planting of legumes will be discontinued in any appreciable areas of oil palm cultivation.

The important considerations in determining a cover maintenance policy following the burning of forest or replanting are:

1. The need to prevent, by adequate ring-weeding and cover control, competition for both water and nutrients between the palms and other plants.
2. The need for early establishment of suitable plant species which will cover the intervening ground and, while requiring the minimum of maintenance in the early years, will prevent erosion.
3. The need to maintain in adult areas a cover requiring minimal care and attention and not containing species actually harmful to the palms.
4. The need to maintain soil fertility at all stages.

Ring-weeding and path maintenance

Young palms have a small, developing root system and they will suffer greatly if they have to compete in their development with other plants. It is therefore of great importance that the soil in which the young roots are developing should be kept free of weeds. This is commonly done with a cutlass in Africa and America and with a hoe in Asia. With the latter instrument there is a tendency with successive hoeings for the soil to be drawn away from the palm and for a saucer-like depression, in which rainwater may stand, to be formed. An implement for pulling back a creeping cover to the correct circumference is useful. Ring-weeding should be to 1 m radius for the first year and thereafter to 1.5−2 m radius. In the early years the purpose is predominantly the prevention of competition, but later on the ring is needed for bunch and fruit collection and so that the ripe fruit dropping from a bunch may be seen. In many areas carpet grass, *Axonopus compressus*, establishes itself on rings and paths which are frequently cut. This is favoured by some on the grounds that it holds the soil together, preventing saucering, the hard-baking or puddling of the surface and the creation of run-off channels; it is also held that *Axonopus* does not compete with the main root system of the adult palm. However, with the advent of herbicides *Axonopus* is less often seen on the circles.

With hand weeding, maintenance in the early years must be frequent though the actual frequency will depend on the rainfall and soil fertility. In most areas at least six rounds per year will be required for the first few years and this will then be progressively reduced to two to four rounds a year depending on the locality and the degree of shading of the ground by the palms. In non-seasonal climates monthly ring-weeding rounds are normally necessary in the early years and this is gradually reduced to 6-weekly, then 2-monthly rounds and finally to four rounds per year.

With increasing labour costs the maintenance of the rings or circles by herbicide application is becoming more common. There has been some reluctance to use herbicides in the early years for fear of damaging the young palm (Plate 39) or its root system, but while care must be exercised in both the method of application and the choice of herbicides there is no evidence of deleterious effects where such care has been taken. Indeed it has been argued that, with no disturbance of the soil, erosion from the circle will be less and the root system will be undamaged; consequently growth should be improved. Experiments in Nigeria tend to support this conclusion.[5]

With regard to choice of herbicides, it is best to apply contact rather than translocated herbicides and particularly those which are inactivated by soil contact. However, no hard and fast rules can be laid down; paraquat, which has been widely used, is translocated to some extent but is inactivated on contact with the soil. A very full and helpful account of the commoner herbicides and their use in oil palm fields has been provided by Turner and Gillbanks.[6] The herbicides which should not be used near young palms are the phenoxyacetic acids, 2,4-D and 2,4,5-T, or the halogenated aliphatic acids, dalapon and TCA. 2,4-D produces long-necked and bending symptoms in oil palm seedlings with subsequent death, while dalapon and TCA produce extreme fasciation of the leaves and leaflets and

Pl. 39 Damage to young palm by herbicides applied to the circle with insufficient care.

later a flattened rosette appearance.[7] Lesser damage is caused by amino-triazole (amitrole or ATA), individual leaves showing severe chlorosis, but recovery is rapid. The substituted ureas (often used as pre-emergence herbicides), monuron and diuron, and the triazines, atrazine and simazine, although systemic, are also less dangerous, but they can cause temporary damage and the exact effects on the root system are unknown.

Of the contact herbicides sodium arsenite, which has been much employed in the rubber and oil palm industries, is tending to fall out of use because of its human toxicity, but sodium chlorate is still quite frequently used (in spite of the fire hazard in dry weather). However, more widely used now is the sodium salt of methane arsonic acid, MSMA, which is effective against certain grasses. The most complete screening tests for the effect of herbicides on the young oil palm are those of Sheldrick, whose results are summarized in Table 10.1.[7] The effects of MSMA have not been reported.

Herbicide mixtures used for ring-weeding will in practice depend on the above considerations and the current cost of the ingredients. For young palms in Nigeria applications of 2 kg a.i. of paraquat with 3–4 kg atrazine, monuron or diuron per hectare of sprayed ground applied twice a year has been found to give satisfactory weed control.[5] In the Ivory Coast a range of formulations for palms older than 5 years has been evolved.[8] Up to 7 years MSMA at 1.8 kg a.i. per hectare and paraquat at 0.7 l/hectare have been found effective. The latter is applied at the end of the dry season, the MSMA at the end of the main wet season (August) while

Table 10.1   *Symptoms in the oil palm associated with particular herbicides* (Sheldrick)

Herbicide	Type of response				
	Immediate leaf scorch	Subsequent blocking and leaf necrosis	Leaflet chlorosis	Epinastic response	Leaf fasciation
Diquat	***				
Paraquat	***				
MCPA	**	***		***	
2,4-D	**	**		**	
Mecoprop	**	***		**	
Dichlorprop	**	***		**	
2,4-DB	**	***		**	
MCPB	*	**		**	
2,3,6-TBA	***	*		*	
TCA	*	**			**
Dalapon	***	***			***
Monuron	*	*	**		
Diuron	*	*	**		
Mono-linuron			*		
Linuron			*		
Simazine					
Atrazine			**		
Atraton			*		
Prometon			*		
Prometryne			*		
Aminotriazole (Amitrole)	*	**	***		

*** Severe.
 ** Moderate.
  * Slight effects.

MSMA with amitrole at 1 kg a.i./hectare or 2,4-D at 0.7 kg a.i./hectare is applied at the end of the short wet season (December). For older palms various mixtures of MSMA, 2,4-D, TCA, sodium chlorate and/or amitrole are recommended. These mixtures contain herbicides dangerous to the palm but with older palms the danger is much reduced.

In Malaysia circle spraying is usually done three or four times a year. In mature areas a mixture of 6–9 l sodium arsenite liquid + 2–3 kg sodium chlorate per sprayed hectare has been used, but for young palms formulations of paraquat, MSMA and diuron are more common. With young palms it was shown that spraying with mixtures of paraquat, MSMA and diuron can start 3 to 9 months after planting without adversely affecting growth. A satisfactory mixture in terms of a.i. was paraquat 0.28 kg/hectare, MSMA 1.68 kg/hectare and diuron 0.28 kg/hectare.[9]

In West Cameroon MSMA replaced paraquat altogether on grounds of cost, but needed to be sprayed more frequently.[10] Ametryne (a triazine herbicide) was also found effective. With MSMA alone spraying was undertaken twice in the year of planting, monthly in the second year, six times in the third year and thereafter four times a year.

With this variety of experience and practice it is clearly necessary for each plantation to adapt its circle spraying formulations and routines in accordance with prices, climate, weed flora and age of palms.

Circle spraying is most frequently done with knapsack sprayers, a man covering up to 500 palms per day.[10] It should be observed that with 143 palms per hectare the proportion of total area to be sprayed is as follows: 1 m circles, one-twenty-second; 1.5 m circles, one-tenth; 2 m circles, one-sixth.

The maintaining of paths through the palm fields is desirable in the early years for inspection and in the later years for use by harvesters. They can be maintained by any of the methods described above for ringweeding. Harvesters' paths, which in adult areas fall naturally along the rows or on hilly land along the contours, tend to become denuded of vegetation or covered with a 'lawn' of carpet grass (*Axonopus compressus*). The position of inspection paths for the early years has been largely a matter of custom, and maintenance is of course more onerous than in later years. In Zaire[2] paths were maintained between each pair of rows and, where opening by alternate row heaping has been practised, this is convenient. In Nigeria, it was found convenient for inspection to run paths along the palm rows for the first 2 years, but later on, when the palms were still 'on the ground' but the leaves far-spreading, to cut a path between each pair of rows. Later, when the palm crowns were well off the ground, the harvesters' paths again ran along the rows. It is now generally recommended that inspection paths should be maintained from the start between each pair of rows and these later become harvesters' paths. If these are 1.5 m wide they will occupy one-tenth of the area; circles plus paths will then come to occupy about one-quarter of the area.

Cultivation and the early establishment of covers

It has been seen in Chapter 8 that cultivation of the soil in the planting of food crops has a deleterious effect on nutrient status and, if practised on infertile areas, lowers bunch yield. A certain amount of soil cultivation is always required in the establishment of cover plants, the amount usually depending on the ease with which these plants can be established in the locality concerned. Such cultivation consists of the opening of drills or pockets for the planting of seed and the hoeing required for keeping the area free from volunteer competing plants. In some localities the latter cultivation can be very considerable and become very expensive.

It is important, therefore, to know the effect of tillage, for although the amount of cultivation in the establishment of covers may be small in comparison with a full turning or disturbance of the soil, any effects of the latter are likely to be felt in some measure in areas where cover plants are introduced.

Few cultivation experiments have been carried out. In an early trial in Malaysia[11] a triennial slashing of a light weed cover was compared with the amount of cultivation required monthly to keep the ground free of weeds. This resulted in the ground becoming hard and impervious. The palms were already 14 years old and the area had a history of clean-weeding. Cultivation resulted in increased yields for 2 years followed by a heavy fall in yield. In spite of the inadequacy of this experiment its results are in line with those of a trial in eastern Nigeria where regular tillage was carried out with a hoe, monthly for the first few years and later annually. Yields were enhanced by cultivation in the first 3 bearing years in comparison with those of plots maintained normally under a natural cover, but subsequently the yields of the cultivated plots declined as shown below:

Yield of cultivated plots as a percentage of the control plots' yield, Abak (Nigeria)

	First 3 years	4th—7th years	8th—11th years
Burning, normal maintenance	100	100	100
Burning, regular tillage	277	92	82

Similar results were obtained in Malaysia: tillage as if for two annual crops of soya beans but allowing natural weeds other than those known to be harmful to come up between cultivations reduced early growth and yield. With soil maintained bare by herbicide spraying, growth and early yield were initially as good as in plots with leguminous covers, but by the third bearing year leaf length and area and bunch yield were significantly lower.[73]

Soil data from regularly tilled areas are not available except where tillage was combined with the continuous or temporary growing of food crops between the young palms. These results from an afforested area in Nigeria have already been referred to (p. 383), but it should be repeated here that the soil effects in the inter-cropped plots were those which would be expected from tillage, i.e. a general reduction of fertility as measured by exchangeable cations K, Mg and Ca. It was remarked that plots prepared for and planted with a leguminous cover — *Pueraria* — showed soil changes intermediate between those of plots left to natural covers and those tilled for food crops. Plots under natural covers showed the minimum nutrient losses. In the early experiment in Malaysia referred to above, plots in which a selected 'forestry' cover was eventually established showed an initial drop in yield, but after 2 years a large increase in yield (corresponding in time to the drop of yield in the clean-weeded plots) was recorded. It is interesting to note that Wycherley[12] attributes the superiority of rubber growth in natural as opposed to leguminous cover plots in an early trial on unburnt forest land to 'minimal soil disturbance in the coversion of forest to plantation'; but at the same time he realizes that this 'cannot be realized in the majority of replantings' or where heavy burning of the cleared forest has been practised.

On the scanty evidence available, therefore, it is reasonable to suppose firstly that, to avoid soil depletion, covers should be planted as early as possible and with the minimum of cultivation, and secondly that there is some merit in a natural cover which includes deep-rooted plants. Kowal and Tinker[13] believe that the comparatively high nutrient status of plots maintained under such a cover for as much as 10 years after burning is due to the return to the surface of nutrients washed down the profile during the first rains after the burn by the deep-rooted cover which establishes itself.

In regions with a marked dry season, cultivation has been advocated at the beginning of the dry season to reduce competition between palms and cover for the limited moisture which will be available in the coming months. Such treatment in a trial in Nigeria opened from forest had a marked beneficial effect on yield in the early years.[14] The effect was similar to, but greater than, that of establishment inter-cropping and was still substantial after the plants had been in the ground for 7 years. Data from this trial are presented in Table 10.2 and show that effects last for 3 years of bearing, but then wear off. This may well be due to the fact that when the palms are well up they shade the ground and are able to compete adequately for moisture without the aid of cultivation. Under the extremely dry conditions of Dahomey (1,200 mm rainfall) the maintenance of bare soil under young palms to the age of 6 years has been shown to improve vegetative development including the root system and to increase early yields by nearly 100 per cent.[15]

The purpose of sowing an 'artificial' cover or cover mixture after burning is to provide a cover for the soil at the earliest moment and thus prevent erosion and soil nutrient depletion. A secondary purpose is to check the invasion of undesired plants. Because of lack of valid comparisons, the choice of cover plants is usually dictated by experience of their

Table 10.2 *Effect of establishment inter-cropping and dry season soil cultivation* (yield per palm (planted 1957))

Years	Control (kg)	Inter-cropping for 2 years (kg)	Cultivation in the dry season (kg)
1961 + 62	80	90	103
1963	79	84	89
1964	85	89	90
1961−4	244	263	283

Differences between treatments and control are significant in all years except 1964.

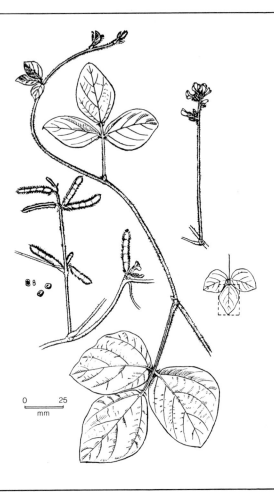

0   25
mm

Fig. 10.1   *Calopogonium mucunoides* (R.R.I. of Malaysia).

establishment in the locality. Leguminous plants are almost always chosen because they form nodules which contain colonies of nitrogen-fixing *Rhizobium* bacteria which live symbiotically on the host plant. The leguminous creeping covers which have, in mixture, stood the test of time and are still widely planted in all parts of the tropics are *Calopogonium mucunoides, Pueraria phaseoloides* and *Centrosema pubescens*[16] (Figs 10.1, 10.2 and 10.3). Much better establishment of these covers than previously is now obtained by (i) seed treatment with concentrated sulphuric acid or scarification or by soaking for 12–24 hours with subsequent heat-treatment for 1 hour at 39°–40°C, (ii) provision of the correct *Rhizobium* species for nodulation, (iii) planting with rock phosphate, (iv)

0   25
mm

Fig. 10.2 *Pueraria phaseoloides* (R.R.I. of Malaysia).

dispersing concentrations of wood ash from the planting lines or pockets, (v) (in West Africa) planting when the rains have set in.[17, 18]

Calopogonium and *Centrosema* seed should be soaked for 15—20 minutes in concentrated sulphuric acid for satisfactory germination. *Pueraria* requires at least 30 minutes and, in Nigeria, 90 minutes has been found to be the optimum time. The acid must be rapidly washed off by repeated changes of hot water after treatment. Mechanical scarification is used in Malaysia as an alternative. The soaking and heat-treatment method mentioned above has been used for *Pueraria* in Nigeria.[18]

Rhizobium inoculum is not always required, and indiscriminate purchase should be avoided. It is only worth obtaining *Rhizobium* culture

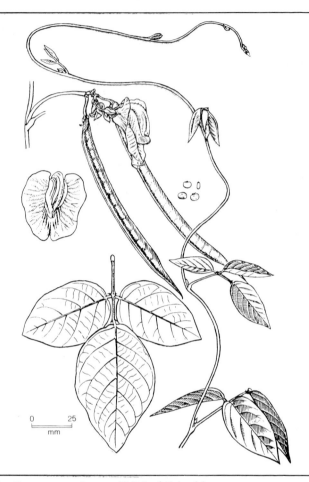

0 — 25
mm

Fig. 10.3   *Centrosema pubescens* (R.R.I. of Malaysia).

if there is an absolute assurance that the correct organism is included and that the culture will be effective. The seed rate of the mixture is 5−6 kg per hectare. The recommended proportions in West Africa are *Calopogonium*: *Pueraria*: *Centrosema* − 2 : 2 : 1. In Malaysia a mixture in the proportions 2 : 2 : 3 is often employed, but a mixture of *Pueraria* and *Centrosema* only in the proportions 2 : 3 has also been recommended for oil palm estates.[19]

The mixture should be planted in three to five drills per interline. No drill should be nearer than 1.5 m to the line of palms. With 9 m triangular planting three rows would be about 2.2 m apart, five rows about 1.1 m apart. Cross drills may also be sown between the palms at the same spacing.

Calopogonium germinates first and grows vigorously but soon dies out to be replaced firstly by *Pueraria*. In West Africa the latter is particularly vigorous. *Centrosema* is not much in evidence for the first few years, but when the palms start to shade the ground it may become dominant.

Failure of establishment of this cover mixture in Africa is rare. Such failures as have occurred have been due to planting before the rains, in too high a concentration of ash or to faulty pretreatment of seed. In the Far East however, establishment has been often more difficult.[16] Creeping covers may, for no obvious reason, become chlorotic and die off and in some cases this is due to lack of appropriate fertilizers. On most Malaysian soils additional applications of rock phosphate at the rate of 60–120 kg per hectare are recommended at 2, 5, 9 and 12 months after planting. Potash, in burnt areas, is usually in sufficient supply but elsewhere a small initial seed dressing of 1 kg muriate of potash to 1 kg seed is recommended with the phosphate seed dressing. When magnesium deficiency is expected or in evidence 120 kg per hectare of magnesium limestone is applied. Legume failure in Malaysia has also been ascribed to the fungus *Rhizoctonia solani* (as well as to *Fomes lignosus* and *Ganoderma pseudoferreum*) and to an aphid, *Aphis laburni*, the ladybird beetle, *Epilachna indica*, which may be particularly destructive of *Centrosema*, the beetle *Pagria signata* (on *Pueraria* and *Calopogonium*), the bug *Chauliops bisontula*, the caterpillar *Lamprosema diemenalis* and a number of other caterpillars and grasshoppers, slugs and snails. Many cover failures in Malaysia are also due to the nematode, *Meloidogyne javanica*, against which there is as yet no economic control.

It is not surprising, therefore, that in Malaysia there should have been widespread trial of and interest in covers other than 'the standard three'. This interest has included bush covers which, besides growing vigorously, have been thought to have the virtues of penetrating and perhaps improving the structure of heavy impervious soils and bringing to the surface nutrients which have been washed down after the forest burning. Though there is little or no direct evidence for the correctness of these theories, there is perhaps support in the work of Kowal and Tinker (see p. 437) for the use of plants which in some measure simulate the growth of the deep-rooting woody species of a natural cover. Among the leguminous plants which have been tried may be mentioned the following (Plates 40–3):

1. *Creepers.* *Dolichos hosei* is by no means a 'new' creeper, having been mixed with the 'standard three' on many occasions as it has the reputation for persisting under shade. It is planted from cuttings or seed. *Indigofera endecaphylla* has also been in occasional use for a very long time as has *Desmodium ovalifolium*. The latter is shade-tolerant. Neither is very successful as a cover though useful for protecting drain banks; planted from cuttings. *Stylosanthes gracilis* is a strong grower in pure culture and usually takes full possession of the land. It has a reputation for being

Stylosanthes sundaica

Fig. 10.4   *Stylosanthes sundaica* (R.R.I. of Malaysia).

'aggressive', i.e. entering into competition for water and nutrients with the oil palm, but this really means that it grows well and therefore needs extra attention so that it does not invade the weeding circles. It has the virtue of forming a pure stand and keeping other plants out, so in other respects maintenance costs are reduced. *Stylosanthes sundaica* is similar but less persistent (Fig. 10.4 and Plate 40). *Phaseolus calcaratus* (Fig. 10.5) is probably the most promising and widely grown of the more recently employed creepers. It somewhat resembles *Pueraria*, but in the Far East it gives a healthier appearance though it is reported to die back early. There are two varieties, *gracilis*, a slender short-lived climber with very hairy pods, and *glabra*, a more vigorous and persistent creeper with thick hairless

Pl. 40 *Stylosanthes gracilis* in pure stand as a cover between young palms in Malaysia.

leaves. A number of shade-tolerant legumes have recently been tried. Of these, *Calopogonium caeruleum* has tended to dominate on Malaysian coastal clays; *Psophocarpus palustris* has also grown well and has been popular in Sumatra for a long period. Seed rates are usually high: 8 kg per hectare.

2. Erect covers. Several species of *Tephrosia*, *Crotalaria* and *Leucaena* have for generations been available in most parts of the tropics as bush covers which can be cut back from time to time to provide material for mulching. Several of these (*T. candida, vogelii, C. anagyroidus, usaramoensis, striata, L. glauca*) have been brought back into use on trial together with the more recently employed *T. noctiflora* and *Flemingia congesta*, the latter being a wild Malaysian plant. The useful species and varieties of *Flemingia* have recently been described.[20] The more common variety of *F. congesta* is *semialata* which is tall and has shiny black seeds. It requires more cutting back than the variety *latifolia*, which is distinguished by its dense covering of brown silky hairs on young stems and petioles. *Flemingia congesta* has trifoliate leaves (Fig. 10.6 and Plate 41). The other species, *F. strobilifera*, has simple leaves and is only likely to be of value in adverse conditions since it grows on sandy soils especially near the sea.

It has been shown that many of these erect covers have tap roots which penetrate to great depths (2–4 m) into soils of good structure and that the amounts of loppings obtained from them may reach 25–40 tons per hectare per annum. The question which needs answering is whether these factors are of advantage in oil palm cultivation or whether these bush

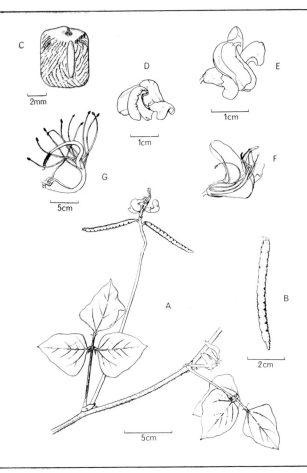

Fig. 10.5   *Phaseolus calcaratus* (R.R.I. of Malaysia).

covers may not enter too far into competition with the young palm for either nutrients or light. A further consideration is how the presence of these covers, perhaps — as is sometimes advocated — with creepers growing over them, will affect the development of the natural cover which eventually establishes itself in adult fields. These questions can only be answered by prolonged study of the effects of different covers and cover mixtures grown under varying management systems over a long period of time. It has been suggested[19] that *Flemingia* may be useful in areas subject to *Oryctes* infestation since it may serve as a barrier to flight.

A few experiments have been laid down in Malaysia to test the effects on newly planted or replanted palms of a few cover species and natural covers. *Stylosanthes gracilis* has been shown to depress early yields, but

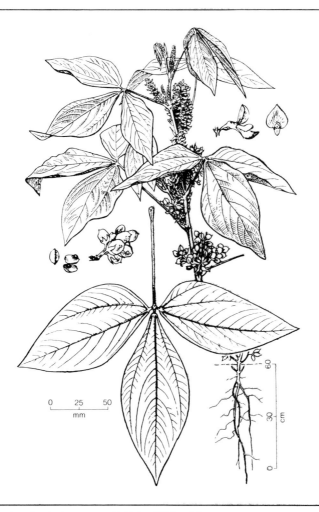

Fig. 10.6   *Flemingia congesta* (R.R.I. of Malaysia).

what has already been said about the control of this cover should be borne in mind. The non-leguminous creeper *Mikania cordata* which is so common a feature of the Malaysian scene has been shown, in pure culture, to depress growth and early yields on both coastal and inland soils.[21] The effect of this plant on rubber is well known and this result was not therefore wholly unexpected. In adult fields, *Mikania* is commonly found in the natural cover mixture. In the Malaysian experiments, the adverse effect of *Mikania* in young palms disappeared after the palms had reached 6 to 7 years of age. *Flemingia congesta* either alone or mixed with *Pueraria* has

Pl. 41   The bush cover, *Flemingia congesta* growing between rows of young palms in Malaysia.

had no deleterious effect when compared with the standard covers, but it must be borne in mind that the maintenance of a pure *Flemingia* stand may entail appreciable cultivation through frequent weeding. Any advantages this cover may have are likely to be seen at a much later date. The effect of grasses and ferns (the other principal constituents, with *Mikania*, of natural covers in adult Malaysian oil palm fields) on young palms is uncertain and maintaining such covers in pure culture presents difficulties.

The establishment of leguminous covers is sometimes assisted by the use of pre-emergence or other herbicides before or around the time of sowing or by spraying the interlines after the cover is visible. As a general rule legumes are sown on ground which has been cleared of other vegetation either by ploughing, hand cultivation or spraying. Successful use of pre-emergence herbicides to prolong the weed-free period and so give the legumes a better start has been reported from both Malaysia and Nigeria.[22, 17] In the former case alachlor, a pre-emergence grass killer, was used, while in Nigeria band-spraying with atraton and simazine at 2 kg a.e. per hectare was undertaken. Damage to covers by low or medium rates of alachlor were negligible, but in Nigeria there was some damage where simazine-laden soil was washed on to the legume rows. Monuron and diuron have also been successfully used at 2 kg a.i. per hectare but these herbicides are not generally used because they may be washed into the seed drills. Recent work has favoured the use of the pre-emergence diphenyl ether herbicide RH 2915 or combinations of the pre-emergence herbicide alachlor with post-emergence application of a paraquat/diuron mixture between the drills.[76, 77]

The maintenance of a suitable cover throughout the palm's bearing life

Attempts to maintain planted covers for a prolonged period raises many problems, both agronomic and economic. In Africa, *Pueraria* will soon become dominant and may be followed by a dominance of *Centrosema* when the palms begin to shade the interlines. Even in Africa, however, grasses, particularly species of *Paspalum*, may invade the legumes at quite an early date. In the Far East, difficulties of establishment may necessitate frequent and prolonged weeding (fortnightly weeding is often recommended during establishment) and, later on, grasses, ferns and *Mikania*, along with many woody plants which will be mentioned later, will crowd out the legumes. There have been advocates of maintaining the leguminous mixture, by constant weeding rounds, for as long as possible; this is expensive work however, and other oil palm growers have favoured the controlled natural replacement of legumes by natural covers. This control consists of keeping out, by regular weeding, plant species which on various grounds are not desired, but allowing other plants to replace the legumes naturally and without hindrance.

The natural covers of rubber plantations have been intensively studied in the Far East and these studies have influenced practice in oil palm plantations. The shade under oil palms is much less than under mature rubber and for this reason a much more permanent natural cover becomes established; with rubber the tendency is for the natural covers, which may be well developed before the canopy closes in, to disappear altogether or be confined to a few very shade tolerant ferns. The study of natural covers was given impetus by the publication by Dr W. B. Haines in 1934, with a completely revised edition in 1940,[23] of a treatise on *The uses and control of natural undergrowth on rubber estates*. An attempt was made to classify plants into Class A: those to be encouraged, Class B: those probably useful but needing to be controlled, and Class C: those undesirable and needing to be suppressed. It was pointed out that the classification was not hard and fast and that where Class A plants were absent, more Class B plants and even some Class C plants must be tolerated. While certain of the Class C plants (e.g. lalang) were clearly undesirable on the grounds that they quickly entered into close competition with the crop, the whole scheme was tentative and was open to criticism on the grounds that the cover plants present in any situation were a reflection of the soil and other conditions obtaining, and that the performance of the rubber was determined by those conditions rather than by the presence of allegedly 'undesirable' plants. This was of course realized by the author whose main object was to point the way to improved cover control under what unfortunately came to be known as the forestry system.

In considering the natural flora of oil palm plantations, it is important to be aware of this background to the subject. Few plants have been shown, experimentally, to be deleterious to the oil palm though many may be considered undesirable owing to their mode of growth and the high expenditure required for their control or elimination. A plant may be considered desirable if:

1. It is not shown to enter into competition with the palm to the extent that yields are reduced.
2. It does not grow so rapidly that it shades young palms and needs to be constantly slashed.
3. It provides a substantial cover of the soil, but needs little attention.
4. It provides a quantity of herbage which rots rapidly to give a mass of decaying vegetation suitable for the growth of oil palm roots. (It has been noted in Chapter 2 that a large part of the root system of the oil palm will grow upwards into the topsoil or any rotting material lying on the soil.)
5. It has a substantial penetrating root system, but is not so woody that it provides little cover and its vegetative parts are resistant to decay.

On almost all plantations in the Far East, the cover which eventually establishes itself is a mixture of the fern, *Nephrolepis biserrata* (Fig. 10.7), with varying quantities of the grasses *Paspalum conjugatum* and *Axonopus compressus* and the non-leguminous creeper *Mikania cordata*. As the latter has been shown to be harmful to young palms, if it has been kept out in the early years it is unlikely to form a substantial part of the undergrowth later on. Grasses are also thought to be more harmful in young areas than in older ones, since they are certainly more aggressive where shade is light. *Nephrolepis* has virtues 3 and 4 above, and rarely grows to such a height that it needs slashing; in situations where it does become bulky it can be given a light slashing and the cut fronds then form a thick decaying mat on the soil surface into which the palm roots grow. Care needs to be taken, however, to see that the slashing is light enough to allow regeneration and does not lead to dying out of the fern. This natural mixture may therefore be considered satisfactory, particularly where the *Nephrolepis* dominates; it is possible that it may be improved by the gradual reduction of *Mikania* through light spraying with herbicides.

Some account may now be given of the plants which may invade the natural association. An excellent series of drawings of plants on rubber estates has been provided by the R.R.I. of Malaysia in its *Planters' Bulletin*; many of the plants illustrated and discussed appear also on oil palm estates.

1. *Obviously undesirable plants.* These are lalang, *Imperata cylindrica*, other strong-growing grasses, *Mikania cordata* and Siam weed, *Eupatorium odoratum*. Control of these plants is best undertaken by regular patrolling, lalang being dug out or wiped with oil and *Eupatorium* being pulled out. Eradication of Siam weed will be discussed later.

2. *Other ferns.* The tropical bracken, *Gleichenia linearis* and maiden-hair creeper, *Lygodium* sp., are common invaders of the poorer and drier soils

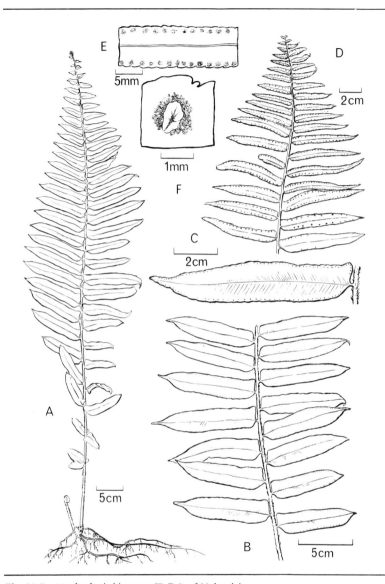

Fig. 10.7   *Nephrolepis biserrata* (R.R.I. of Malaysia).

which will not be improved by their presence. If there is a reasonable chance of *Nephrolepis* taking their place they may be sprayed out with sodium arsenite or other herbicides. Stagmoss, *Lycopodium cernuum*, may obtain a hold in acid, wet soils. It, too, may be sprayed out, though a need for improvement of soil conditions is indicated by its presence. *Blechnum*

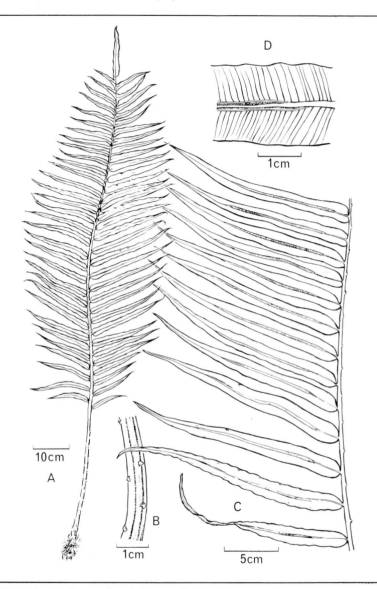

Fig. 10.8   *Blechnum orientale* (R.R.I. of Malaysia).

orientale (Fig. 10.8) somewhat resembles *Nephrolepis* and tends to grow when palm shade is less dense.

3. *Shrubs and young tree seedlings.*   A number of very common shrubs may become widespread in an oil palm plantation. Certain of them are

rated as 'Class C' plants though, in oil palm plantations, there is no evidence that they are harmful. However, they may grow tall and so need constant slashing and many are indicators of poor soils. Among the latter may be mentioned the Straits rhododendron, *Melastoma polyanthum*, *Lantana aculeata* and *Clidemia hirta*. These are usually eradicated in the weeding round by hand pulling or chopping below ground level.

Wycherley[12] has recently drawn attention to one matter which is fundamental to the consideration of cover policy and is of particular importance in the Far East where various schools of thought have arisen. Referring to cover policy in rubber plantations he writes:

> It is tempting to fall back on general experience, but this can be very misleading because cover plant effects are easily confounded with two other factors. Firstly, many plants are indicators of conditions rather than their cause . . . so the mere removal of these plants will effect no improvement. . . . Secondly, the presence or absence of certain plants may testify only to the skill or otherwise of those required to conform with a cover policy based on preconceived verdicts on the nature of the plants concerned. For example, if all grasses are considered harmful, their successful exclusion may owe much to the availability of trained labour under efficient management, whereas the expression of the latter in other aspects of husbandry such as regular manuring may be responsible for any apparent beneficial results of eliminating grass.

These warnings apply with equal force to a consideration of cover policy with the oil palm.

The natural covers in Africa are not so well documented or so stable as in the Far East. Moreover, it is only recently that any serious weed eradication problem has arisen. As already mentioned, it is only in certain areas that lalang is of any importance, while invasion of young areas by Elephant or Napier grass (*Pennisetum purpureum*) is very local. The latter plant must be dug out although, where cattle are available, it may with advantage be grazed.

The arrival of Siam weed, *Eupatorium odoratum*, in West Africa presented a serious problem. If left unchecked it will rapidly become dominant and grow to a height of 3—6 m thus competing with young palms not only for water and nutrients, but also for light. If detected in a plantation early enough it can be kept at bay by regular patrols which pull up the young plants by hand well before they flower and which occasionally spot spray any older ones that may have survived. This was accomplished for more than 15 years at N.I.F.O.R., Nigeria, in an area containing both old and young fields. The expenditure was equivalent to about 2 man-days per hectare per year and control was effective in spite of the whole plantation being surrounded by *Eupatorium* infested areas. Once the weed gets a hold however, it can only be tackled by weed killers or mechanical treatment. Patches can be killed by 2.5 kg a.e. of 2,4-D per hectare, but if spray drift is feared then 4 kg a.i. of atrazine or 5 kg a.i. of diuron per hectare will be effective for about 10 months if spraying is done just before the dry season.[24] Further spraying is then required until the area is sufficiently under control to be taken over by the regular patrol. Mechanical control in Nigeria and in Malaysia (where, however, the weed does not grow so vigorously) has been carried out with the Holt weed Breaker (Mark VIIIB) with fair success.[14, 25] In young areas it is advisable to follow spraying or mechanical treatment by the establishment of a *Pueraria* cover.

Apart from its severe competition with the palm for water, nutrients and light, *Eupatorium* can be dangerous because it easily takes fire in dry weather and young palms may then be destroyed. In this respect it is a particularly serious pest in smallholdings, and in the Ivory Coast the weed presents a peculiarly difficult problem since it has infested village plantings, where there is little control, and from these it invades the large plantations.

The succession from leguminous covers to natural covers in Africa is not the same as in the Far East. In Africa *Pueraria* and *Centrosema* tend to persist and only gradually to be replaced by small and retarded specimens of forest species which are shade tolerant (Plates 42 and 43). Cases are known where *Pueraria* has been maintained in pure stand and without special measures for over 10 years. However, if not dominated by *Pueraria*, the land may become covered with many typical invaders among which

Pl. 42   A *Pueraria phaseoloides* cover still in almost pure stand between 7-year-old palms at the Oil Palm Research Centre, Kusi, Ghana, with good ring-weeding.

Pl. 43 A good mixed *Pueraria* and natural cover in a 'closed in' field in Nigeria.

may be mentioned *Canna bidentata*, which is often found in a thick stand, *Hibiscus seratensis*, *Solanum* sp., *Ipomoea involucrata*, *Sida acuta*, *Ageratum conyzoides*, *Passiflora foetida*, *Momordica* species and other cucurbits, *Aspilia latifolia* and the common grass *Paspalum conjugatum*. Also present in some localities are *Pityrogramma calomelanus* and *Sarcophrynium macrophyllum*.

The succession in the Ivory Coast seems to be rather different to that of Nigeria.[8] After 5 years grasses tend to predominate, chiefly *Axonopus compressus* and *Paspalum conjugatum* with some dicotyledons, viz. *Commelina nudiflora*, *Borreria latifolia*, *Desmodium adscendens* and *Talimum* sp. After 12 years the proportions of grasses and dicotyledons vary but of the latter the following are common: *Commelina* sp., *Ageratum conyzoides*, *Sida carpinifolia*, *Dissotis rotundifolia*, *B. latifolia*, *Urena repens*, *Phyllanthus amaras*, *Oldenlandia affinis*, *Asystasia gangetica*, *Cyathula prostata* and *Mariscus umbellatus*.

Control of these covers is by regular slashing (cutlassing) which, in Nigeria, is required four times a year in areas from 5 to 8 years old, twice a year thereafter, but perhaps only once a year in areas over 15 years old where the palm foliage is dense. In the Ivory Coast rather less frequent slashing is required, one or two rounds being sufficient once the palms are above ground. Similarly, in Zaire, slashing is much less frequent when the methods already described (p. 428) are adopted.

While slashing is usually resorted to in West Africa and there have been no reports of large scale use of herbicides in the avenues (as opposed to the circles) except for control of *Eupatorium* or other special invaders, slashing is not an approved practice in Asia.[6] Previously hand-digging of the undesirable species was the rule, but spot-spraying with herbicides is now more usual practice. The herbicides used must be suited to the vegetation it is desired to eradicate. *Imperata* and *Eupatorium* control has already been discussed. Spot-spraying rounds will usually be monthly in the first year, then 2-monthly, and 3- or 4-monthly by the fourth year after planting. Paraquat is often used for *Mikania* control, but here again the actual herbicide chosen will depend largely on local prices and availability, and new formulations are frequently coming on the market.

One other method of interline cover control, which has been adopted on some estates in Malaysia and is bound up with harvesting methods and mechanization, should be mentioned here since it is entirely different from the standard slashing or weeding methods practised elsewhere. In certain flat coastal estates a strip is cut down the centre of each, or every other, interline by a rotary cutter such as the Servis. This cuts all vegetation near ground level and a lawn of *Axonopus compressus* is formed. Down these lawned paths harvesting carts drawn by bullocks or tractors pass to collect the harvested bunches. The system is clearly suited to mature areas where the crown is well above ground; no trials of its effect on yield have been carried out.

Experiments on close cutting of the interline vegetation have clearly shown that this practice is deleterious to both growth and yield.[26] In a young area the practice depressed both leaf nitrogen and leaf phosphorus. Even in a mature area there was a small yield decrease, leaf size was reduced and leaf-N and leaf-P depressed. It must be assumed that the close cutting of wide paths between rows will have the same deleterious effect at least to the adjoining rows though the effect may not be as severe as with complete close cutting. The practice of sending vehicles into fields with the harvesters has also been employed in parts of Latin America and not only has the cover suffered but severe puddling of the surface soil has been common. There is little doubt that these practices cause considerable yield reduction. The natural covers of Latin America were mentioned on p. 430.

Mechanization of cover control

The mechanization of interline cover control presupposes that mechanical methods have been employed to clear the forest. The advantages of mechanization must be seen to be economic. Trials of mechanical implements must be carried out over a long period to determine the full effects on palm yields, and such trials have not yet provided much information on which definite advice can be based. It is necessary to determine the effect of each implement on the cover and then on the crop. Implements may be divided into the following categories:

1. Slashers — these cut the cover but do not disturb the soil.
2. Rollers — these roll and may also cut the cover with a little soil disturbance.
3. Cultivators — these cultivate the soil.

The slashers may have rotary cutter arms or rotary chains. Rollers may have horizontal blades fitted into them. The cultivators used are disc harrows.

Experience in the Ivory Coast[27] has been that any implement which cuts or cultivates tends to destroy a leguminous cover, and only heavy logs of wood rolled behind a tractor or disc harrows set with the discs absolutely straight maintain the cover satisfactorily. In Malaysia a round hardwood log measuring 183 x 30 cm diameter with four narrow horizontal blades (6.4 cm) attached to it was used in mixed Straits rhododendron (*Melastoma*) and legumes. With this light roll, leading to the minimum of cutting, the legume cover re-established itself satisfactorily. Later work with the Holt weed breaker,[25] which has a much wider blade, showed that legume covers suffered while grasses tended to take their place. In an experiment in Nigeria involving all three types of implement,[14] all treatments discouraged the spread of legume covers. Frequent mechanical slashing or cutter-rolling encouraged grass, the cutter-roller also encouraging annual weeds. Disc harrows used frequently left the soil bare, while any mechanical treatment carried out infrequently encouraged bush covers interspersed with grasses, soft weeds or legumes according to the implement used. Another experiment in Nigeria in which different combinations of implements were used in the early, mid and late rains, showed that a Cambridge ring roller assisted the development and maintenance of a strong *Pueraria* cover. The use, additionally, of disc harrows just before the dry season increased early yields, presumably through its effect in reducing competition for soil water.[28]

Where leguminous covers have already disappeared, small or large rollers with cutting bars (*Charrues landaises* and *rouleaux débroussailleurs*) have been successfully employed in the Ivory Coast, but it is recognized that the continuous use of one implement tends to select one species of plant, and alternation of implements is recommended.[27] An ingenious set of log rollers has been constructed to cover one interline right up to the palms.

In the strong-growing grass areas of America some substantial implement is required owing to the toughness of the grass stems. Rotary cutters and heavy 'Marden' cutter-rollers have been used. The former implements are comparatively slow; the latter, though fast, sometimes fail to chop the tougher stems and leave a flattened sward which soon grows up again. Some estates have even harvested the grasses with cutter and blower and fed to penned cattle; a considerable quantity of nutrients are thus removed from the fields and this method, entailing as it does the gaining of subsidiary produce from the land, can hardly be successful unless the ring-weeding is good and compensatory fertilizer dressings are provided.

To summarize: the close mechanical mowing or slashing of oil palm areas has been shown to reduce yields. Other implements which destroy a leguminous cover or which cause excessive disturbance or puddling of the soil will also be deleterious. Mechanical maintenance can indeed only be considered where light rollers are used and the soil structure is unlikely to be disturbed, though very careful use of disc harrows before a dry season may increase yields in seasonal climates. In very wet climates mechanical maintenance is certainly harmful and spraying herbicides with knapsack sprayers is sounder practice. Spraying with tractor-drawn booms has been practised in Latin America, but it is doubtful if the small reduction in labour usage balances the full tractor costs, quite apart from the damage done to the soil and cover and hence the loss of crop.

Pruning

The practice of pruning leaves from the palm is motivated by the desire of the grower to view the production and ripening of bunches in the crown and by the wish of the harvester to have the burden of his work lightened. The latter wish is particularly dangerous and may have been responsible for the practice of 'pruning up to a bunch', i.e. the cutting of all leaves 'below the next bunch' presumably on the assumption that such leaves are functionless, and that therefore it is advantageous to leave the next bunch to ripen where the harvesters can plainly see it and easily cut it down. Early experiments in Sumatra showed that pruning in excess of this practice was deleterious, but *less* severe pruning was not tried.[29] This system thus remained unquestioned for many years.

The ripe bunch is found to be subtended by leaf 30 to 32. This however is not the 'leaf below the bunch' since the latter leans out from its subtending leaf and lies on one in another spiral on a lower whorl, i.e. on some leaf around number 35 to 37. If pruning is done to leave one whorl of leaves below the bunches then about 35 leaves or just over 4 per spiral will remain; if 2 whorls are left then just over 5 leaves per spiral will remain. But Corley found that on an inland plantation in Malaysia 'minimal pruning' (i.e. pruning only those leaves which harvesters find they need to remove to harvest the bunches) left 35 to 40 leaves with occasional cases of only 27 leaves.[31] It is clearly difficult therefore to leave more than 40 leaves on the palm unless a deliberate policy of insisting on all green leaves being retained is imposed.

All experiments indicate that maximum yields are obtained by retaining as many green leaves as possible, but the increment obtained by leaving more than 40 leaves is usually small. Table 10.3 gives the results of three experiments, one in Nigeria and two in Malaysia.[29, 30, 31] It will be seen from the Malaysian data that a drop of up to 2 tons bunches per hectare can be expected from reducing the canopy from 40 to 32 leaves, and this is through reductions of both number of bunches and weight per bunch.

Table 10.3 *The effect of pruning on yield*

Treatment	No. of bunches per hectare per annum	Mean bunch weight (kg)	Bunch yield per annum (tons/ha)	No. of male inflorescences per hectare per annum
Nigeria*				
1. No pruning	803	10.0	8.01	
2. Dead leaves only, annually	813	10.4	8.51	
3. Pruning to one leaf below the bunch, 6 monthly	726	9.9	7.16	
4. Pruning to flowering inflorescence, annually	699	9.7	6.76	
Malaysia				
Coastal†				
1. Leaving 8 leaves per spiral	2,268	11.2	25.33	1,203
2. Leaving 7 leaves per spiral	2,268	10.9	24.76	1,285
3. Leaving 6 leaves per spiral	2,197	10.1	24.24	1,268
4. Leaving 5 leaves per spiral	2,224	10.8	23.90	1,335
5. Leaving 4 leaves per spiral	2,222	10.0	22.32	1,345
6. Leaving 3 leaves per spiral	2,118	9.0	19.12	1,378
Inland‡				*per palm*
1. Leaving 40 leaves (5/spiral)	1,415	16.5	25.24	9.31
2. Leaving 32 leaves (4/spiral)	1,344	14.5	24.48	10.15
3. Leaving 24 leaves (3/spiral)	1,287	13.9	20.50	9.52
4. Leaving 16 leaves (2/spiral)	1,032	11.2	13.29	10.18
5. Leaving 8 leaves (1/spiral)	212	6.9	4.35	6.00

* Adjusted means for 6 years (pretreatment yields available for 4 years).
† Means for 3 years.
‡ Bunch mean yield for 2 years, other data for the second year.

Sources: Sly[29]; O.P.R.S., Bantang[30] and O.P.G.L., Layang Layang,[31] Malaysia.

Severe pruning has a rapid effect on production. In the inland experiment of Table 10.3 mean bunch weight was reduced within 4 months and bunch number fell suddenly after 8 months. The latter result indicates an immediate increase in the inflorescence abortion rate. Preferential abortion of female inflorescences is shown by the fact that male inflorescence production is not adversely affected until the most severe pruning treatment is reached.[31] It has also been shown that severe pruning reduces the size of developing leaves but does not alter the rate of leaf production.[78]

It will be noted in Table 10.3 that in the Nigerian experiment the removal of dead leaves gave a small increase in yield over the non-pruning treatment. This is difficult to explain though it is possible that no removal of dead leaves or epiphytes reduced pollination efficiency.[29]

If no pruning is done it may become difficult for harvesters to see into the crown and to gain access to the bunches; much loose fruit may also be lost. There is no evidence that the presence of epiphytes reduces bunch yields, but it is of great importance that, during harvesting, no ripe bunches are missed owing to the palms being surrounded by withered leaves and covered with epiphytes. Tidiness in this case has a definite purpose, but the danger of over-zealous tidiness is apparent from Table 10.3.

Time of pruning may be important in seasonal climates. Pruning at the beginning of the wet season should be avoided; in an experiment in Nigeria in which intensive pruning methods were used bunch yield was about 10 per cent higher in the first 4 years when pruning was done at the end of the rains than when carried out at the beginning of the rains.[32] In non-seasonal climates pruning is usually done at times of low crop when labour is available.

On some estates in the Far East it is the practice to carry out a special 'pruning up' when bringing a field into harvest. This is not usual in other parts of the world where the ripe young bunches are simply cut out with a chisel and withered leaves are removed later either during pruning rounds or during later harvests. With palms growing very vigorously, however, the improved access for harvesters is considered to make this practice of early pruning up to a ripe bunch worth while. An experiment in Malaysia showed no effect of this practice on yield when compared with no pruning up.[31]

In Malaysia it is now generally recommended that some pruning of withered leaves be done before assisted pollination begins so as to improve access for this operation, but two whorls of leaves are left below the ripe bunch. Thereafter pruning is at first confined to senescent leaves lying close to the ground. When the palms are well up and the canopy closes in, however, harvesters are allowed to cut leaves during harvesting as may be necessary, but two whorls of fronds are again maintained.[6] This rule is sometimes relaxed for palms over 10 years old, but the criterion of two whorls or a minimum of forty leaves (five per spiral) should be maintained as long as possible. An annual special pruning round to remove excess withered leaves is then all that is additionally necessary.

Pruning may be carried out with cutlasses, axes or chisels. In Africa the cutlass is the usual implement but in Asia the chisel is supplanting the axe for young palms. In both continents the pole and sickle is now being widely used for tall palms. Varying advice is given about the length of the leaf base which should be left on the palm. 'As short as possible' is often advised on the grounds that large leaf bases catch loose fruit, encourage epiphytes and possibly assist in the spread of disease, though for this there is no evidence. There is evidence, however, both in Africa and America that cutting the bases very short may be followed by a serious increase in attack by *Rhynchophorus* species leading either directly to the death of the palm or to Red Ring disease. It is noticeable that attempting to cut the bases very short leads to wounding of underlying leaf bases and this gives a suitable place for the weevil to lay its eggs. The petiole must therefore be cut sufficiently long to give a clean cut and no damage to adjoining parts of the palm.

In some parts of Africa, e.g. the Onitsha area of East-Central State, Nigeria, annual pruning is done to provide leaves for thatching, wall construction and other purposes. At the same time palms producing only male inflorescences are tapped for palm wine. Experiments have shown[29, 33] that these practices, as would be expected, reduce bunch yields, but the reduction of oil and kernels has to be weighed against the other palm products in the village economy of the people.

Bunch production and harvesting

Bunch production may be initially influenced by two practices which in recent times have been commonly though not universally employed. These are ablation (or castration) and assisted pollination. Harvesting methods have been evolved firstly to obtain bunches which will give the maximum quantity of palm oil and kernels which will be of an acceptable quality, and secondly to reduce costs to a minimum while assuring that palms are maintained in high production by retaining as many green leaves as possible.

Harvesting implements

In the early days of plantations in Africa, harvesting was carried out almost entirely with the cutlass and tall trees were climbed either by ladder or by means of ropes. The cutlass is not a handy implement for harvesting, however, and climbing with ladders is slow and with ropes very slow. These methods have largely given place to methods developed in the Far East.

The ideal instrument for harvesting is a chisel on the end of a 0.9 to 1.5 m wooden handle. Suitable chisels for harvesting and pruning have blades 7–12 cm wide and about 25 cm long with overall length including the shaft hole of about 45 cm. Special handled chisels have been designed in Honduras (Plate 44). These instruments can be used for cutting the bunch peduncle without cutting the green leaf which subtends the bunch. Chisels are often used for the first 3 years of harvesting only, but they should be employed for as long as possible as other methods lead to the cutting of too many green leaves. A common procedure in the Far East was for short axes to succeed the chisel when the palms are 'off the ground', i.e. have formed a trunk, but the continued use of the chisel is now preferred. As soon as the bunches can no longer be easily reached from the ground harvesting is carried out with a long curved knife lashed to the end of a bamboo pole (Plate 45). These poles are gradually increased in length and palms of over 9 m in height can be harvested with them. The curved knife or sickle is about 60 cm long (Plate 46).

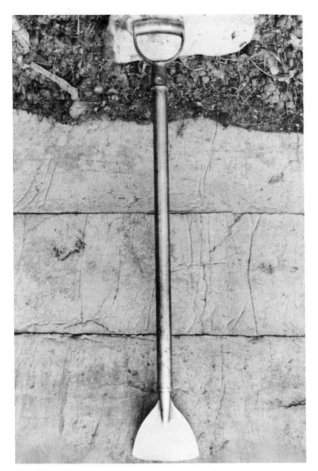

Pl. 44   Harvesting tools: a metal chisel used in Honduras.

When harvesting with either an axe or the knife and pole, the leaf or leaves below the bunches are inevitably cut. There is a tendency, indeed, for more leaves to be cut by the pole-and-knife than by the old method where the harvester climbed with a rope or ladder and used a cutlass. This is because, with a cutlass, the bunch can often be cut out without cutting either the leaf below it or many of the leaves to the side. The pole is, however, much less expensive to use. Moreover a skilled operator can manipulate it so that the minimum of leaves are cut. The blade is first hooked round the peduncle and given a strong downward pull; if the bunch does not fall the blade is applied to the apex of the bunch to ease it out from behind uncut or partially cut leaves.

Recently, many estates have realized the importance of preventing the cutting of green leaves and they have therefore tried to retain the use of

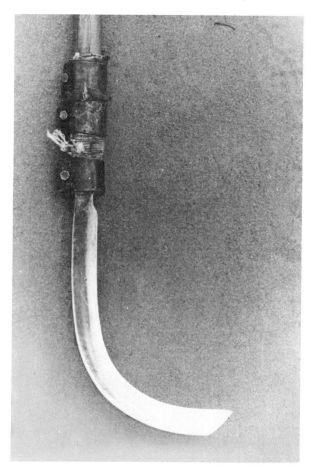

Pl. 45 Harvesting tools: a hooked knife attached to the end of a bamboo pole.

the chisel for as long as possible and to eliminate the axe stage altogether. Harvesting is done with the chisel until the palm is 12 years old and measures perhaps 3.5 m to the crown. The harvesters climb the palm tying a wire rope to a stout leaf base to support them while using the chisel. There is evidence that even on the best soils chisel harvesting with the retention of as many leaves as possible will give a larger number of bunches than axe or knife harvesting (Plate 47).

In short, chisel harvesting should be continued as long as is practically possible and, thereafter, pole-and-knife harvesters should be trained to cut bunches with the minimum removal of green leaves. The curved knives on the ends of the long poles may be a danger to other workers. When being carried along roads from field to field, therefore, they should be guarded by the attachment of a piece of wood.

Pl. 46   Harvesting bunches: the hook and pole method for tall palms.

When to start harvesting

The question of when to start harvesting has always been debated. The first bunches are small and of poor quality both as regards percentage fruit to bunch and oil to mesocarp. On the other hand many young and even fields will nowadays produce a very large quantity of small bunches (5 tons/hectare in Africa and double that amount in the Far East) in their third or fourth year in the field, and this should not be lost. In recent

Pl. 47   Harvesting bunches: a wooden-handled chisel in use on a young palm.

years attempts have been made to eliminate the few bunches of the earliest producing period and thereby to increase the yield of the first harvesting years by the removal, by hand, of the first female inflorescences appearing on the young palms. This has, curiously, been named 'castration' though ablation is a better term. Obviously, if an area can produce more highly in its first harvesting year, field costs are reduced and more economical use is made of the oil mill. This can be attained by high density planting as described in Chapter 9. On the other hand, it must also be the planters' object to reduce to a minimum the period of no crop. All these economic considerations, therefore, must be carefully weighed before a decision is taken on when to start harvesting.

The initial time of harvesting has no effect on later yields, assuming of course that all bunches are allowed to set. In an experiment in West Africa,[34] starting to harvest 2½ years after planting gave a no higher cumulative yield than starting to harvest 3½ years after planting (the normal time). Even when harvest was delayed until 4 years after planting the significant reduction in the number of bunches was partially balanced by the higher average bunch weight and, under the conditions obtaining, it was doubtful whether harvesting earlier than 4½ years after planting would have been worth while. In cases of faster development, and with precocious progenies, earlier harvesting may be expected to be economic, however.

Ablation

There has been a great deal of experimentation on the effect of postponing harvesting by ablation for periods of 6 to 33 months. The general effect is as follows. Palms whose female inflorescences have been removed show increased vegetative growth. The number of leaves produced, their length and weight are all increased as is the palm's girth. There is also evidence of an improved root system. When ablation ceases a high initial yield is obtained; the early bunches are of increased size and the oil to mesocarp is higher than in the earlier fruit bunches of untreated areas, though there does not appear to be any difference in oil content of contemporaneously harvested bunches. The superior vegetative growth does not persist and in the early bearing period the rate of vegetative development may become inferior to that of untreated palms. After initial high production treated palms suffer a drop in production and there is therefore a tendency for high yield fluctuations in the early years. The drop in yield appears to be due to inflorescence abortion. Male inflorescence production is considerably reduced and this may persist for some time; in high-production areas the need for assisted pollination is therefore accentuated, and in low-production areas the need may be created.[35-38]

In view of the fact that no greater yield is obtained by ablation, it must be shown to have other economic advantages if it is to be employed. The advantages which have been claimed are: (*a*) that the same yield is concentrated into a shorter period thus reducing harvesting costs; (*b*) that overall in the early years there will be a higher out-turn of oil to bunch; (*c*) that in practice many small bunches are often imperfectly harvested, are subject to rat damage and, in Asia, may become infected with *Marasmius*; (*d*) that mill investment can be delayed and that the equipment is put to better initial use by the larger early crops which will be obtained. Against these factors must be set the cost of the ablation itself. There is evidence in Malaysia that additional fertilizer applications may in some circumstances be needed to prevent excessive abortion.[39]

It is clear from the above that the decision to carry out ablation must depend on local circumstances. Plantation managers tend to favour the practice partly because of the improved cleanliness and general appearance of the palms. In areas of very high initial production or with very precocious material, however, palms may come into early heavy bearing without ablation and the possible period of ablation is then so short that it is hardly worth while. In Africa if the practice makes assisted pollination necessary then the cost of this must be taken into account. Owners of small or medium-sized plantations are unlikely to favour the practice since they are usually eager to obtain revenue as soon after planting as possible.

In practice ablation usually starts as soon as about 50 per cent of the palms are bearing female inflorescences. This will be at about 14 to 18 months in Asia and up to 2 years in Africa. Ablation will then continue for 6 to 18 months with palms being brought into bearing at 3 to 4 years after planting. The shorter period and the earlier harvesting time is preferable, particularly in areas of rapid growth. Longer ablation would only be undertaken in special circumstances, e.g. where the operation of the only available mill was unexpectedly delayed. It has been suggested that in seasonal climates ablation should be continued until a palm girth of 150 cm has been reached and that the time of discontinuation should be the beginning of the dry season (e.g. November in West Africa). This leads to a delay of harvesting by 1 year if 150 cm girth has not been reached at 2½ years from field planting.[79]

Young inflorescences can be pulled out by hand using a suitable glove. If ablation has to be prolonged beyond 6 months it will probably be necessary to use a narrow-bladed chisel. There is however a great danger of wounding the palm and thus allowing entry of *Rhynchophorus* sp. This has to be particularly guarded against in America where there is danger of Red Ring disease, and it is doubtful if ablation should be undertaken at all in that continent if the use of chisels is necessary. Ablation rounds are usually monthly, but 5 to 6 weekly is permissible in periods of low crop in Africa.

Table 10.4 summarizes the results of three ablation experiments. It will be noted that in the Nigerian experiment harvesting was started 44 months after planting in each treatment whereas in the control in the Ivory Coast experiment harvesting was started after 38 months. From the yields of the former experiment it is fair to assume that a production of some 20 kg per palm could have been harvested in 1964 from the control plots in the Nigerian experiment, in which case the yield to the end of 1968 would not have been significantly less than in treatment 'Ablation 2'. The higher bunch weight induced by ablation was demonstrated in all experiments as was the tendency to greater yield fluctuations. In the Malaysian experiment the earlier and higher yields obtained in Asia can be seen. In the treatment 'Ablation 2' the yield in three successive 6-month periods of harvest was high, low, high.

It has been suggested that the practice of ablation is of more practical utility in regions of low than of high production and particularly so in extremely dry conditions such as obtain in Dahomey. Experiments in the latter country have shown that early maintenance with clean weeding improves vegetative development, including the root system, and that this is further improved by ablation for 20 months. Though the effect of ablation on overall yield to 6 years of age was small, there was evidence of improved drought resistance and the harvesting costs per ton of bunches was much reduced.[15] Special timetables for ablation have been devised.[80]

Corley has suggested that partial ablation could be used to maintain

Table 10.4   *The results of some ablation experiments (Tailliez and Olivin; Obasola; Chew and Khoo)*

Treatment	Planting date	Ablation period (months)	Harvesting dates	Parameter	Production per palm kg and number				
					1966–7	1967–8	1968–9	1969–70	Total
Ivory Coast†									
Control	May 1963	Nil	July 1966	Bunch yield	57	59	81	97	294
				No. of bunches	13.3	9.9	10.0	10.1	43.2
				Wt. per bunch	4.4	6.5	8.4	9.8	
Ablation 1	May 1963	14	Nov. 1966	Bunch yield	70*	65	70*	97	302
				No. of bunches	13.6	10.7	8.2	9.0	41.5
				Wt. per bunch	5.2*	6.4	8.7	10.7	
Ablation 2	May 1963	19	Mar. 1967	Bunch yield	18*	117*	70*	93	298
				No. of bunches	3.1*	18.0*	8.9	9.1	39.1
				Wt. per bunch	6.0*	6.8	8.2	10.4	
					1965	1966	1967	1968	Total 48 mths
Nigeria‡									
Control	May 1961	Nil	Jan. 1965	Bunch yield	48.4	71.1	45.1	61.6	226.2
				No. of bunches	12.2	11.3	7.6	7.8	38.9
				Wt. per bunch	4.0	6.7	6.4	7.6	
Ablation 1	May 1961	10	Jan. 1965	Bunch yield	50.0	63.2	46.0	62.7	221.9
				No. of bunches	12.8	9.0*	7.9	8.1	37.8
				Wt. per bunch	3.8	7.7*	6.1	7.7	
Ablation 2	May 1961	19	Jan. 1965	Bunch yield	61.3*	76.4*	42.5	70.4*	250.1*
				No. of bunches	11.6	10.0	7.0*	8.5*	37.1
				Wt. per bunch	5.4*	7.4*	6.4	8.3*	
					1969–70 (6 mths)	1970 (last 9 mths)	1971 (first 9 mths)	1971–2 (9 mths)	Total
Malaysia §									
Control	Oct. 1967	Nil	Oct. 1969	Bunch yield	9.9	47.3	74.7	113.2	245.1
				No. of bunches	4.4	15.8	16.3	14.3	550.8
				Wt. per bunch	2.3	3.0	4.6	7.9	
Ablation 1	Oct. 1967	12	April 1970	Bunch yield	–	58.7	78.3	111.3	248.3
				No. of bunches		16.0	16.6	14.6	47.2
				Wt. per bunch		3.7	4.7	7.6	
Ablation 2	Oct. 1967	24	Jan. 1971	Bunch yield	–	–	101.2	109.6	210.8
				No. of bunches			14.6	13.1	27.7
				Wt. per bunch			6.9	8.4	

† This experiment covered two types of plantation material. An asterisk (*) indicates that at least the P = 0.05 level of significance between an ablation treatment value and that of control was reached in both halves of the experiment.

‡ In this experiment it should be noted that harvesting started in the same month for each treatment. An asterisk indicates a significant difference at at least the P = 0.05 level, between the ablation treatment and control.

§ Assisted pollination was carried out at ten rounds per month from the cessation of ablation.

monthly bunch numbers at or below an estimated maximum thus flattening yield peaks and perhaps filling in yield troughs. He showed that partial ablation besides increasing the weight of the remaining bunches also increases carbohydrate storage in the trunk, though the overall carbohydrate demand, and consequently the rate of supply, decreases.[81]

Assisted pollination

Pollination and the physiological effects of lack of pollen or of applying additional pollen have been discussed in Chapter 4 (pp. 181–5). It remains here only to describe the practical aspects of assisted pollination. It should be reiterated however that whether to employ the practice or not remains very much a matter of individual judgement, and much work needs to be done on pollination and embryo development before there is a sound quantitative basis for its use and timing. For this reason assisted pollination remains a controversial subject. Even within one country, e.g. Malaysia, some planters will claim that no crop will be set if pollination is not done, while others will assert that it is not needed. There have, however, been well-authenticated cases of large areas showing almost total bunch failure with a low level of male inflorescence production, and under these circumstances it is difficult to believe that inadequate pollination is not the cause; but assisted pollination is a labour-intensive practice and an uncritical attitude towards its prolonged use may prove expensive. Attempts have been made to assess the economics of the practice,[40] but these are based on postulates and are nullified by the plain fact that the actual increase (or decrease) in yield is never known; moreover it is difficult to estimate the inverse effect of increased fruit to bunch on bunch number.

Pollination failure is the probable cause when young inflorescences develop unevenly, i.e. a few rather large fruit are found scattered over the bunch or in one part of it and the remaining fruit are parthenocarpic, small and elongated. Such poorly set bunches can be detected at an early stage and parts of them may rot or they may be subject to complete bunch failure (p. 180). In Asia, however, where early bunch failure is common, it is important to make certain that this is not because of attack by the caterpillar of *Tirathaba mundella* or by *Marasmius*, though in the latter case it is thought that the attack is most likely to be a consequence of poor pollination followed by full or partial bunch failure.

In any area where high yields are expected, and particularly where ablation is practised, the number of male inflorescences will be initially small and assisted pollination may then be carried out from the onset of natural or permitted flowering. It is tempting to rely on male inflorescence counts as a guide to pollen needs, but it has been pointed out[6] that other factors such as rainfall, wind strengths and cover policy are also playing their part. Very roughly, a monthly production of less than twenty-five male inflorescences per hectare may be considered dangerously low while three times that number should be adequate. Gray[41] has shown how male inflorescence production is influenced by age, planting material, density, ablation, pruning and by assisted pollination itself. In Asia there is a tendency for production to be fairly constant after 7 years of age though there are still monthly fluctuations. *Tenera* material tends initially to provide fewer male inflorescences than does Deli *dura*. Pruning increases male flowering while increased density has the opposite effect. In *D x T* plantings in Malaysia Gray found that male inflorescence production fell below fifty per hectare in 66 to 70 per cent of the months during the age period 3 to 7 years. Months of low production were frequent.

Pollination rounds should cover the whole area to be pollinated every 3 days to ensure that each female inflorescence is pollinated when receptive. With very young palms the spathe should be pulled away; after 2 years this is no longer necessary. To 6 years of age a hand puffer is used; later, if pollination is continued, palms up to 4 m to bunch level can be pollinated by a lance pollinator with an aluminium tube about 2.5 m long (Plate 48). Rotary hand dusters and power dusters have also been used. Pollinators using puffers or lances can cover up to 15 hectares per day.

Pollen is taken from inflorescences cut from adult areas. It is shaken out, sun-dried and stored in a desiccator for immediate use. If stocks of pollen are to be held, as is usually the case, the pollen can still be stored in a desiccator for about 3 months, but if it is likely to be stored longer it is better to dry it overnight, after sieving, at not more than 40°C and store it in a deep freezer at about −15° to −20°C. When needed for use the pollen is diluted with four parts of fine talc to one part of pollen. Greater dilution, up to 20:1, can be used if pollen is scarce.

Determination of the time to cease assisted pollination can be assisted by male inflorescence counts or by untreated control plots in the centre of fields; 2 hectares per 40 hectare field has been suggested. These plots may be started after, say, 2 years of pollination and as soon as fruit set is seen to be satisfactory in them pollination can cease in the whole field. It has also been recommended that until pollination finally ceases a new control should be used every 4 months so as to avoid prolonged poor set in any one area. The practical organization of assisted pollination has been discussed very fully by Veldhius[42] and Turner and Gillbanks[6] for Malaysia and by Tailliez and Valverde[43] for Africa.

Time, frequency and criteria of harvesting

Harvesting is carried out with the object of obtaining the maximum quantity of palm oil of a quality, as judged by the free fatty acid content, acceptable to the purchaser. Under-ripe fruit contains less oil than ripe fruit; over-ripe fruit provides oil of a higher f.f.a. content. Ripening takes place unevenly over the bunch, and although the period from anthesis to harvest is shortest in a dry season,[44] the final ripening of fully-developed bunches is hastened by wet weather.

Fruit colour has been discussed in Chapter 2, but it may be mentioned here that *nigrescens* fruit is colourless at the base before ripening, and that it ripens from the top downwards to a reddish-orange colour, the exterior fruits retaining their brown or black caps; *virescens* (green) fruit ripens in the same manner to a light reddish-orange, but it does not retain the cap, only a small ring of green remaining around the persistent stigma at the apex of the fruit.

Bunches usually ripen about 5½ to 6 months after pollination. In Zaire the average maturation period lies, according to age, between 170 and 195 days, though palms are known on which bunches ripen in as little as 4½ or

A

B

Pl. 48   Assisted pollination in Malaysia: **A.** Pollinating with long aluminium tube, pressure bulb and flask. **B.** The end of the tube and the pollen-containing flask.

as much as 7 months.[2] Differences due to season may amount to as much as 12 days.

The ripening of the fruit, which becomes detached or easily detachable when ripe, usually takes place from the tip of the bunch downwards and over a varying period of time. Small bunches may ripen their fruit fully in 11 days whereas large bunches from the same aged palms may take up to 20 days to ripen. Bunches of the same weight take longer to ripen on older palms than on young palms. Before ripening, the mesocarp contains a high percentage of water and carbohydrates and has a low oil content. One week before ripening the oil content may have risen to within 80 per cent of its final amount.[4] The final oil production takes place rapidly with the change of colour of the fruit. When oil content is at its maximum the f.f.a. of the oil is about 0.5 per cent. When the fruit has become detached or the bunch harvested, oil production ceases. Moisture then tends to be lost, and if the fruit or bunches are stored a false impression will be obtained of the oil extraction rate. While the oil to fruit, calculated on a moisture-free basis, remains constant, the mesocarp to fruit may, through loss of moisture, fall by 3 per cent in 4 days and the oil to fruit will rise accordingly. During this post-ripening period, the f.f.a. content of the oil will gradually increase.

Harvesting criteria

Though these facts have been known in broad outline for many years, the criterion for harvesting a bunch, i.e. a stated number of loose fruit per bunch, has remained arbitrary and has varied widely from plantation to plantation. Ideally, fruit should be collected as it drops, fully ripe, from the bunch, or bunches should be examined daily and brought from the palm when the majority of their fruit are fully ripe, but none over-ripe or damaged. Clearly these ideals are in practice impossible, and it has to be recognized that any harvesting arrangement is a compromise: the harvesting round must not be so infrequent that bunches become over-ripe and most of their fruit drop out and are damaged; nor must it be so frequent that the process is expensive and there is a temptation to harvest under-ripe.

Dufrane and Berger,[45] in Zaire were the first (1957) to provide a logical scheme based on the facts of ripening and free fatty acid formation. They analysed 400 bunches averaging about 5 kg in weight and demonstrated a linear relationship between the number of loose fruit and percentage oil to mesocarp. An increase in the number of loose fruit from 5 to 74 (2 to 46 per cent) resulted in an increase of oil to mesocarp of just over 5 per cent. At the same time the free fatty acid content of the fruit rose from 0.5 to 2.9 per cent. The f.f.a. content of the loose fruit was approximately constant at 5 per cent, whereas the f.f.a. of oil from unloosened fruit never rose above 1.2 per cent even for bunches with 46 per cent loose fruits.

These authors obtained, with their 5 kg bunches, an equation

$$Y = 45.2 + 0.083\,x$$

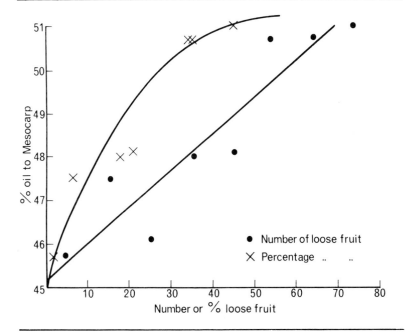

Fig. 10.9   Relationships between percentage oil to mesocarp and number of percentage of loose fruit in bunches averaging 4.8 kg (Dufrane and Berger).

where Y is the percentage oil to mesocarp, x the number of loose fruit (below 50 per cent) and 45.2 the percentage oil to mesocarp calculated to exist in bunches which have not yet loosened any fruit though on the point of doing so. Although linear to 50 per cent loose fruit, this relationship does not appear linear when the number of loose fruit is expressed as a percentage and suggests that a maximum oil percentage of just over 51 per cent was reached at 50 per cent loose fruit (Fig. 10.9). It was suggested that the relationship could be extended to bunches of different average weight by modifying the equation to

$$Y = 45.2 + 0.415 \frac{x}{p}$$

where p = the average weight of bunches in kg. Thus for a given standard of ripeness, as judged by the percentage oil to mesocarp, the number of loose fruit would be greater the larger the bunch size. A similar relationship has been obtained in Malaysia with D x D and D x P material where the maximum oil to mesocarp in 5-, 8- and 11-year-old plantings was reached at 30 per cent detached fruit or less (Fig. 10.10). Free fatty acid at 30 per cent detached fruit lay between 2 and 3 per cent.[46] Rather different results were obtained in the Ivory Coast where it was estimated that

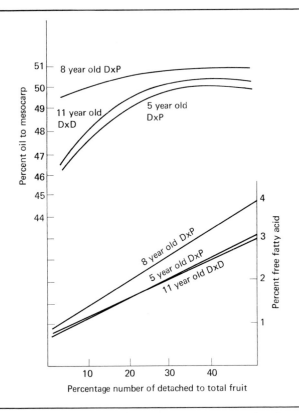

Fig. 10.10   Relationship between percentage oil to mesocarp, free fatty acid and percentage number of detached fruit to total fruit in Malaysia (Ng and Southworth).

there was no further increase in oil to mesocarp beyond 6 per cent detached fruit, after cutting and carrying.[47]

Theoretically, if a criterion of x detached fruit per bunch is used and bunches are harvested daily, all bunches harvested might be represented graphically by a vertical straight line at the x point as shown in Fig. 10.11A, i.e. whatever number of bunches are harvested none have more loose fruit than the criterion. Similarly if the interval is extended to a lengthy period, say 20 days, bunches with all numbers of loose fruit up to the total number of fruit per bunch might be represented by the horizontal straight line shown in Fig. 10.11B. In practice the frequency curves for loose fruit per bunch for 1 day and for 20-day harvesting periods would be as shown in the lower graphs C and D of Fig. 10.11, the asymmetry of the 20-day graph being due to the fact that the curve of bunch maturation is logarithmic.[45]

A curve of maturation showing the mean number of fruit becoming detached over a period of days can be constructed for a given area. From

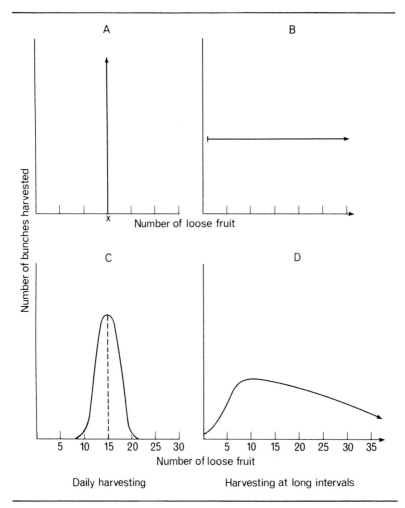

Fig. 10.11 Theoretical and practical representation of the effect of harvesting daily (**A** and **C**) and at long intervals (**B** and **D**) − after Dufrane and Berger.

this curve can be seen, for any given average number of loose fruit taken as a criterion, the upper and lower limits of loose fruit production for any number of days' harvesting interval. For instance, if, using Fig. 10.12, the average number of loose fruit taken as the criterion is 40, then a 4-day harvesting interval will entail the harvesting of bunches with between 22 and 64 loose fruit, i.e. the 'practical criterion' will be 'at least 22 loose fruit'. An 8-day interval would reduce the practical criterion to 8 loose fruit, but bunches with about 90 loose fruit would also be being harvested. If the average criterion is set higher, at 90 loose fruit, then 4- and 10-day

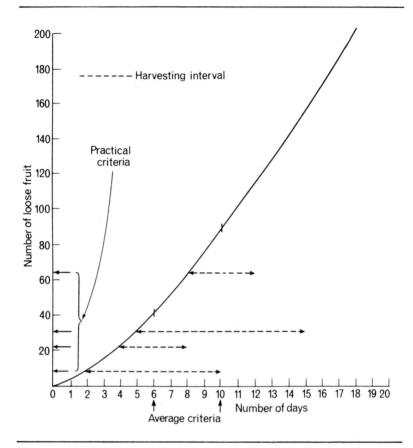

Fig. 10.12 The relation between the harvesting interval, the average number of loose fruit taken as a criterion and the practical criterion − after Dufrane and Berger.

intervals will give ranges of 62 to 116 and 30 to 158 loose fruit, the first number being the practical criterion in each case. From graphs of this kind the effect of the setting of any practical criterion can be seen for different harvesting periods, and the carrying out of this exercise can prevent arbitrary judgements of harvesting standards being made in the field. The increased oil obtained by increasing the practical criterion, in terms of loose fruit per bunch, needs to be balanced against any reduction of price per ton due to increased f.f.a.

When maintaining a given *average* number of loose fruit, reducing the frequency of harvesting may reduce costs but may lead to loss of oil through failure to pick up all loose fruit. Moreover, where harvesting is paid for on the basis of *number* of bunches harvested, real costs may actually be increased.

The translating of these considerations into practical field methods designed to give maximum oil at an acceptable level of f.f.a. is not easy and it is therefore not surprising that different systems have been canvassed. A standard that has been widely used is that there should be not less than two loose fruit on the ground per kilogramme of bunch, i.e. if a 10 kg bunch has any number of loose fruit on the ground less than twenty, then it will be left until the next harvesting round.[48] It is claimed that this system works well with experienced harvesters if used flexibly, e.g. if the round is widened to 10 days or if it is very wet, fifteen fruit on the ground would be acceptable for harvesting a 10 kg bunch.

This system has been criticized on the grounds that it cannot be accurately checked after the bunch has been cut and that the fruit will perhaps be too ripe. Ng and Southworth have shown that, in Malaysia with D x P material, the proportion of detached fruit on the ground to total detached fruit is about 50 per cent when the total of detached fruit is 15 per cent and somewhat more at higher percentages of detachment. A minimum or practical criterion of twenty loose fruit on the ground will give a mean of about sixty with a weekly round (Fig. 10.12), and sixty loose fruit on the ground is equivalent to a total of 120 detached in a 10 kg bunch having 6 kg of fruit. Assuming a fruit weight of 9–12 g, the detached fruit will constitute between 18 and 24 per cent of the total fruit. This will give maximum oil to bunch according to Ivory Coast estimates[47] and a little below maximum oil with Malaysian estimates.[46] It seems therefore that the system is soundly based and it is known to have given good results on estates where good quality oil is produced.[48]

Southworth,[82] who regards 20 per cent total detached fruit as the aim, recommends two detached fruit per pound of bunch weight (4.4 per kg). This is very close to the standard of two detached fruit *on the ground* per kilogramme of bunch discussed above; he points out, however, that although there is the general relationship already mentioned, there is a wide variation from bunch to bunch in the proportions of fruit on the ground to total detached fruit, so that if harvesters can estimate the total detached fruit this is likely to be the most satisfactory measure. In many parts of the world however it may be difficult for harvesters to do this with any accuracy and in this case the criterion of loose fruit on the ground will be the more practical one.

An alternative recommended method is estimating the number of loose fruit by counting the number of loose fruit sockets on a group of bunches at the loading point.[6] It is claimed that by this method it can definitely be seen whether a particular bunch has been correctly harvested or not and that therefore good harvesting discipline is maintained. Under this system five loose fruit per kilogramme of bunches can be used as the criterion.

With young palms it is possible to specify the actual number of loose fruit or fruit easily detachable with the finger and in the Ivory Coast two per kilogramme of bunch has been recommended.[49] This is low by other standards but it has to be remembered that very small bunches ripen their fruit more rapidly than larger bunches and that if a small bunch with loose fruit is left to the next round if may become over-ripe.

In any large organization which commands laboratory facilities it should be possible to make an accurate assessment of harvesting criteria and frequency by following and adapting the methods of Dufrane and Berger and of Ng and Southworth. There is little doubt that the careful computation of the most profitable harvesting methods could be undertaken on a wider, though local, scale than has yet been attempted. Without local information on maturation, oil content, f.f.a. build up and harvesting costs, it is only possible to discuss harvesting procedure in the broadest terms, as follows:

1. *Yield, payment and harvesting organization.*   As yields vary so widely from region to region it is not possible to state a general quantity of bunches which a harvester can bring in. In Malaysia a harvester cuts 100 to 200 bunches per day in mature but not tall areas, this being equivalent to between 1½ and 2½ tons of bunches. Payment is per bunch on a variable scale or per unit of weight at a fixed scale.[50] If the harvester also carries, then the amount harvested will be reduced by about one-third; it is common practice, however, for a cutter-carrier pair, often husband and wife, to work as a team and 1¼−2 tons per day will be a common achievement per team.[51] It is not possible here to enlarge on harvesting organization; a number of useful papers from Malaysia are, however, available on the subject.[48, 51, 52] About 1,100 kg is considered an average task for cutter-carrier pairs in Zaire where a good deal of attention was given to the rate of work of harvesters, i.e. the time taken to cut, walk between palms, carry, collect loose fruit, etc.[45] Similar studies have now been undertaken in the Ivory Coast[53] and Malaysia.[51, 54] While this work will be of interest to those establishing plantations in areas where the oil palm is a new crop, it must be borne in mind that all these components of harvesting time will vary with the age and yield of the field concerned, with the productivity of the region, and with the ability of workers of different traditions.

2. *Frequency of harvesting.*   Harvesting is usually more frequent in Africa than in the Far East. It is not known if this is justified, though there is a suggestion that Deli bunches in Malaysia ripen more slowly than bunches in Africa. In Zaire the calculations of Dufrane and Berger suggest a 6- to 7-day cycle with an increase of frequency to 4 to 5 days at times of peak crops. In the Ivory Coast 4 to 5 days is suggested as normal for peak periods, though this may be reduced to 8 to 10 days during periods of low productivity. In Nigeria weekly rounds have been adopted as standard on some plantations though more frequent harvesting is sometimes practised in the peak period. In Malaysia once-a-month harvesting has been practised for the first 6 months, with twice-a-month for a further year. Thereafter 7 to 10 day rounds are the rule.

3. *Layout of road or rail system; acreage covered per day.* Something has already been said (p. 366) of the importance of plantation layout in assuring the satisfactory harvest of the crop. Roads or rails 320–400 m apart keep the carry of bunches to a maximum of 160–200 m; this has been found satisfactory in Malaysia and Zaire. It has been found that with layouts of this type 4 hectares of mature plantation per day can, under Zaire conditions, be covered by a pair of workers in an area yielding about 12 ton bunches per hectare per annum. In high-yielding areas in Malaysia the weight of bunches harvested is much higher but the area covered nevertheless ranges from 3 to 5 hectares per team per day.[52]

Lastly, mention should be made of the possibility of inducing the bunch, with the aid of growth regulators, to retain its fruit after full ripening. Oil synthesis could then continue in the later-ripening fruit after it would ordinarily have become necessary to harvest the bunch. A number of growth regulators have been tried and of these NAA (d-naphthalene-acetic acid) appears the most promising.[55] Abscission of the fruit is significantly supressed, but the synthesis of oil seems to proceed at the normal rate so that for any given percentage abscission the oil content per bunch will be higher in treated bunches. Application can be done to the ripening bunches 140 days after anthesis and this avoids the inducement of parthenocarpy. Though this method of reducing the problems of harvesting is obviously attractive, the economics of it and the techniques to be adopted require further study and elaboration.

Transport of bunches

Bunches are normally carried from the field to collecting points on the road or railside and then manually loaded into lorries, tractor-drawn trailers or rail cages. Road transport is now the rule in all new plantations and special efforts have in recent years been made to reduce labour costs both in carrying within the field and in loading at the roadside. In these developments the need to maintain f.f.a. at a low level through careful handling of the bunches has been borne in mind.

In South America mules have been used for carrying out the bunches to the roadside (Plate 49), in West Malaysia bullock carts have been employed (Plate 50), while in Sabah the human carriers have on some large estates been replaced by buffaloes drawing sledges.[56] The first of these sytems can be much abused if the mules are allowed to wander uncontrolled through the fields and much puddling of the soil has resulted on many plantations in high rainfall areas. The system is only permissible under strict discipline where the mule is confined to a clear harvesters' path within alternate avenues. Buffaloes and sledges might also be criticized on the grounds of their soil compacting and puddling effects, but here again the effect can be minimized by moving down defined paths in every other avenue. Buffaloes can pull up to 1 ton of bunches.

Pl. 49   Bunches carried out of the fields on mules in Ecuador.

Where animals are employed it is very necessary to remember that they must be kept out of fields where poisonous herbicides have been used, and that the persistence of each such herbicide must be ascertained.

The passage of tractors and trailers through the fields or the use of special small vehicles designed for the job have been tried in many places. The main problem, particularly with tractors, has been the wetness and unevenness of the ground, which leads to much puddling, ponding and sometimes the bogging down of the tractors themselves. In Malaysia a container system of carrying bunches to the factory was combined with the use in the field of a small three-wheeled 3½ hp machine with hydraulic jack (Jackpak) which could lift the containers and carry them to collecting points.[57] In general these methods have not taken on, and the tendency has been to bring roads nearer together rather than enter the fields and damage the soil and cover with vehicles. Where tractors have been continuously used in the fields and much puddling and ponding has resulted it is most probable that yields have suffered. Apart from this, the full costs of running, maintaining and depreciating the tractors must be set against any saving in labour; no study of this has been published.

In Honduras and Costa Rica harvester-carrying booms mounted on track or high flotation wheeled tractors have been used.[58] These bring the harvester to the bunch on palms up to 12 m high and the fallen bunches are loaded into a trailer drawn behind the tractor. Here again the full cost of this operation in comparison with manual work and an efficient system of roadside-to-mill carriage is not known (Plate 51).

Pl. 50   Harvest collection from field to railside by bullock cart on a coastal estate in Malaysia.

There have been several recent developments in transport methods from roadside to mill. Tipping lorries or tipping trailers are now the rule so that off-loading at the mill, either on to storage ramps or into sterilizer cages is rapid. Where a rail system is used with the cages no further storage is required and the cages go direct to the sterilizer. These cages can also be used with low trailer frames drawn by tractors. The use of ramps has been

Pl. 51   Mechanical harvesting with tractor, trailer and hydraulic boom in Honduras.

given as one cause of increased f.f.a., because of the relatively severe impacts on loading and unloading; however, without a rail system and/or direct loading at the fieldside into sterilizer cages, it is difficult to avoid millside ramps in a large plantation where bunches must at times be accumulating at a faster rate than they can be milled. Central storage ramps in the plantation, sometimes used where one mill serves several plantations, should be avoided if possible.

Tractors sometimes draw up to six trailers from the field to the millside. In this case the tractors must be unhitched to tip each tractor in turn or the trailers must be side opening and run on to the tipping platform.

Bunch collection and transport in nets lifted by cranes has been steadily gaining ground in all regions. The 'Kulim System' can be used with lorries or tractors and trailers.[59] The former are more likely to be used where the journey to the mill covers an appreciable distance on a public road. With tractors and trailers it is unnecessary to fit all tractors with cranes since some tractors will be used simply for towing full trailers to the mill. Nets, which are designed to carry bunches and loose fruit, are set out on roadsides at the bunch collection points, and the harvested bunches are carried to them. The lorry or tractor-trailer with attached crane lifts the net and the bunches are discharged into the vehicle. The empty nets are carried to the mill or other collecting points and are redistributed to the fields in a sidecar or light vehicle. There is considerable saving in both time, handling and loading labour and oil quality should thus be improved (Plate 52).

Pl. 52   Lifting bunches from the roadside in nets by tractor-mounted crane in the so-called 'Kulim system' of bunch collection and transport.

With the container system already mentioned (p. 478) the plywood containers on steel frames with a capacity of 0.5 ton bunches are placed along the roadsides and filled by the harvesters. They are loaded by tractor cranes on to trailers and the bunches are eventually off-loaded at the mill into sterilizer cages or into lorries for further transport; this is done by a chain hoist with bridle gear on a gantry.[57]

Replanting

A field of oil palms should be replanted when (i) a large number of palms have become so tall that they cannot be conveniently or economically harvested, or (ii) the yield is so low that replacement with high-yielding young palms would give a much higher return. Tallness has usually been considered the more important factor,[60] but with the change over from *dura* to high yielding *tenera* material, and with the high incidence of disease in some old fields, the second factor may now be considered of equal or perhaps greater importance. Pole harvesting can now be continued until palms reach a height of over 10 m, and replanting may well be justified before the palms have passed that height. Clearly every case must be considered on its own merits, account being taken of the present yield, the cost of harvesting, the bunch yield potential of the area and the planting material available.

Growth in height depends both on environment and on genetic constitution. Though 'dumpy' progenies exist, short-stemmed palms of *E. guineensis* with desirable other characters are not yet available on a large scale. In general it may be expected that palms should be replaced when 25 years old (21 years of harvesting) though many fields have been left for 30 years or more before being replanted. On the grounds that much higher yielding progenies will rapidly become available and that periods of no crop and low crop can be substantially reduced by better planting techniques, selection for precocity and high-density planting, much shorter crop cycles − 20 or even 15 years − are now being suggested.

Historically, it is interesting to note, the first replanting trial was concerned with the possible value of fallows between crop cycles.[60] Later workers were concerned with the problem of underplanting and the maintaining of some crop from the old stand, while recent workers, as mentioned above, have given prime attention to the new crop.

Early experiments on underplanting have been reviewed in a monograph by Ferwerda.[60] An unreplicated trial in Nigeria[61] in which all old palms were retained for a year but were then subjected to a variety of treatments, suggested that if alternate E−W rows were retained for 3 years, the yield from the old palms would more than compensate for the reduced early yield of the young palms. The retention of N−S rows appeared to suppress the early bearing of young palms presumably because of a greater shading effect.

The most complete replanting trial was carried out at Yaligimba in Zaire.[62, 63] An area planted at 9.5 m triangular (128 palms per hectare) was replanted in a replicated trial at the same density and at a density of 160 palms per hectare. Four methods were used:

Proportion of palms felled in each year

Method of felling	Year of replanting			
	1953	1954	1955	1956
3/3	All palms			
2/3	2/3	1/3		
1/3	1/3	1/3	1/3	
0/3	—	1/3	1/3	1/3

Where three fellings were undertaken, the least productive palms were chosen in all the rows for the first felling, then all the palms in alternate rows were removed at the second felling, and the remainder were removed at the third felling.

The results at the end of 11 years are shown in Table 10.5. The bunch yields are as recorded, while the oil yields, though estimated, raise an important consideration. Replanting is often carried out in *dura* and Deli *dura* fields which are being replanted with *tenera* material. Total bunch weights do not therefore give the full picture of yields obtained. While oil-to-bunch ratio will remain steady at a comparatively low figure in the old palms, the oil-to-bunch ratio for the young palms will rise year by year from a low figure, say 10 per cent, to a much higher figure. In this case the oil-to-bunch ratio of the old palms was known to be 17 per cent while in the young palms other determinations suggested a progressive increase from 10 per cent in the first year to 22 per cent in the seventh.

The increases in total yield of the 1/3 plots over the clear-felled 3/3 plots were: bunches 5.35 per cent, oil 5.38 per cent, indicating that, over the limited period, the increased oil percentage of the *tenera* material had not yet affected the comparison.

Table 10.5 *Replanting experiment, Yaligimba, Zaire* (yield of bunches and estimated yield of oil per hectare)

Treatment	Bunches/hectare			Oil/hectare		
	Young palms 1956—63	*Old palms 1953 to time of felling*	*Total*	*Young palms 1956—63*	*Old palms 1953 to time of felling*	*Total*
	(tons)	(tons)	(tons)	(tons)	(tons)	(tons)
3/3	70.9	—	70.9	13.0	—	13.0
2/3	63.5	2.8	66.3	11.7	0.5	12.3
1/3	64.9	9.8	74.7	12.1	1.6	13.7
0/3	57.0	17.2	74.2	10.9	2.8	13.7

Table 10.6   *Replanting experiment, Yaligimba, Zaire* (bunch yield of young palms per hectare)

Treatment	1959−60 (tons)	1960−1 (tons)	1961−2 (tons)	1962−3 (tons)
3/3	11.9	8.9	12.5	13.1
2/3	10.8	9.3	11.4	12.5
1/3	11.1	10.1	12.2	13.2
0/3	9.6	9.6	12.1	13.0

The young palms clearly suffered from the competition of the old palms since their early yields (and their growth measurements) were lower. This effect was not permanent however, as the data in Table 10.6 shows.

The results of this experiment suggested that the method 1/3, in which one-third of the old palms were removed in successive years, starting in the replanting year, would provide the highest overall yield and the shortest period of no crop (some small yield was obtained from the young palms during the last year of felling) and that the young palms would suffer no permanent ill effect. The method, which has come to be known as 'the Congo system', has been used in estate practice and has been incorporated in trials carried out both in Malaysia and in West Africa.

In Nigeria the 'Congo system' was tried against (*a*) cutting out either N−S or E−W rows at replanting and the remainder 2 years later; (*b*) pruning old palms to half canopy at replanting, at 3 months and (to the spear) at 1 year, and felling after 2 years; and (*c*) felling all palms at re-planting.[64] Of the underplanting treatments the pruning treatment was the most effective, giving a significantly higher total yield to the fourth bear-ing year of the young palms. The young palms in the full felling treatment gave significantly higher yields in the first 3 bearing years than in the underplanting treatments, but thereafter there were no differences; so this treatment did not overtake any of the underplanting treatments when the old-palm production was included. To year 8 the mean overall bunch yield of the underplanting treatments was 28 per cent above that of the com-plete felling treatment, while the pruning treatment bunch yield was 18 per cent higher than that of the other underplanting treatments.

In Asia most underplanting tried appears to have been under a full stand[6] so that an exact measure of produce obtainable under various systems has not been obtained even in areas less prone to *Ganoderma* (see below).

In spite of the encouraging results of experiments, doubt remains as to whether the leaving in of any part of the old stand is worth while. Firstly growth of unshaded young oil palms, particularly in the Far East, is now so rapid that they may be brought into bearing at least a year earlier than was previously the case. Experience has shown that underplanting makes it impossible to take advantage of this precocity. Secondly, the incidence of *Ganoderma* Trunk Rot makes it quite impossible to consider such a system in Asia. There are suggestions that species of this fungus are now attacking earlier in Africa and for this reason underplanting is of doubtful wisdom there too. In Malaysia and Sumatra replanting tends to be dominated by the need to combat *Ganoderma* and to reduce *Oryctes* attack (see pp. 635 and 677) and the methods adopted are costly.

To minimize *Oryctes* attack a mechanized system of replanting has been recommended for the Ivory Coast.[65] The main objective is to obtain rapid growth of a *Pueraria* and *Centrosema* cover at the beginning of the wet season, all cultivation and felling having been done early in or during the dry season. It has been shown that if the trunks are covered by the leguminous cover within 4 to 6 months of dry-season felling very little *Oryctes* breeding will occur. The system as recommended allows for planting the palms only after a further year. This would hardly be acceptable with a commercial plantation, and there seems no reason why this general system of mechanical cultivation in the dry season should not be combined with planting in the same year.

Very little is known about cultivation in relation to replanting. By the time the first stand of oil palms is felled, the cover is likely to be largely a natural one, kept free of certain plants considered noxious. The cover may be sparse owing to the shading effect of the palm canopy. Removal of this canopy will lead to a rapid growth of vegetation, and many species not suited to shade conditions will gain access. Many of these will be grasses. The re-establishment of a leguminous cover, as described earlier in this chapter, may be preferable to allowing naturally colonizing weeds to gain access, and the planting of covers may be undertaken in the same manner as in a newly planted area.

Disposal of oil palms

The old palms may be disposed of by 1. felling with axes; 2. felling mechanically; 3. poisoning with chemicals.

1. *Axing of palms* is heavy and tedious work. The number of mature palms which can be felled per man-day is between five and twelve. Directional felling is necessary, particularly if the area has been underplanted. Disposal of the trunks may present a problem in areas where the rhinoceros beetle is common.

2. *Mechanical felling* requires the use of heavy tractors and tackle and will only be possible where large areas are to be cleared. Two methods were tried in Malaysia,[66] (*a*) pulling the palms over with a 1 inch cable drawn by two heavy tractors, one on each side of a row or pair of rows; (*b*) bulldozing the palms with an 85 hp tractor and bulldozer blade. Though the former method may be employed on two rows at once, the cable tended to become fouled; the bulldozer method thus proved quicker and less expensive. Under moist conditions the whole bole of a palm will emerge from the ground on felling, but after a period of dry weather the palms will often snap off at ground level.

Two matters need special consideration in axing or mechanical felling. Firstly, if replanting is carried out before all the palms are felled the felling must be done in a manner which avoids damage to the young palms. Secondly, if it is intended to carry out inter-row cultivation of any kind, the old palms must be felled in such a way, and the new palms be planted in such a position, that tractors can pass between the new rows without being obstructed by palms felled from the old rows. Though this may seem a difficult proposition, felling by bulldozer can be carried out without damage to young palms set as near as 1.2 m (4 ft) to the rows. Felled palms rot rather slowly, but cutting off the leaves and cutting up the trunks into, say, three sections will hasten decay.

3. *Poisoning* the old stand is likely to prove the most satisfactory method of disposal. A preliminary trial in Malaysia before 1940 showed that sodium arsenite was an effective and rapid killer. Later trials in Malaysia showed that doses as low as 28 g per palm were as effective as larger doses. All palms show signs of dying within a few days and 70 to 80 per cent of the fronds wither within 10 days. After 1 month the crowns are completely dead and the stem begins to collapse and disintegrate. Collapse of the trunk takes place at any point between the crown and the base.

The method is as follows: a hole is bored into the trunk by a 1-inch auger to as far as half the distance between the circumference and the centre of the palm. The hole may be 30–90 cm above the ground and has a gentle downward slope. Twenty-eight grammes of powered sodium arsenite is placed in the hole, water is added and the hole is tamped with clay or stopped up with a wooden plug. Crowbars or steel pipes hammered into the trunks have also been used. More than forty trees per day can be bored by one man. The usual precautions in using sodium arsenite are, of course, necessary.

Early trials in Nigeria showed that West African palms could not be poisoned with sodium arsenite quite so readily.[67] Up to 170 g per palm were required and in some palms the rotting portion of the trunk around the injection hole became sealed off by a hard layer of tissue and the remainder of the trunk remained normal. Some exudation of the poison took place and larger holes more firmly plugged were probably needed. In Zaire[68] the introduction of 60–120 g in concentrated suspensions was effective in killing over ninety-five per cent in 70 days but, as in Nigeria, some resistant palms remained apparently healthy. Elsewhere in Zaire[60] it was noted that survivors of a first application were not affected by subsequent efforts to poison them. These palms developed cavities and necrotic areas in the vicinity of the auger hole. In general, sodium arsenite has been shown to be the cheapest and most effective substance for poisoning palms and poisoning is the easiest and least damaging method of disposal. While 28 g per palm is effective in Malaysia, rather larger doses seem to be required elsewhere.

Attempts have more recently been made to find poisons as or more effective than sodium arsenite and with a lower toxicity. In Nigeria the majority of modern herbicides have been screened[69] but only two, the bipyridyl herbicides diquat and paraquat, have shown toxic qualities similar to sodium arsenite. The effect of diquat is as follows:[70] within 2 weeks most of the leaves are desiccated; after 6 weeks they have usually collapsed basally and remain cloaking the stem. In a few palms the leaves remain in their normal position but there is some fracturing along their length. The spear remains upright though dried out. This is in contrast to sodium arsenite poisoning where the spear falls out and the trunk collapses. The trunks of diquat-treated palms collapse much later. The internal progression of change has been described in detail by Gunn and Tatham.[70] Although there is extensive drying out of the leaves, portions of the spear remain green for some time. The stem becomes dry and has a pink discoloration and a wet rot eventually develops in the leaf bases and the growing point. Later a rot develops near the point of injection. Further collapse of the palm normally occurs in high winds or heavy rain.

Diquat at the rate of 20 g a.i. per palm is more effective than a similar quantity of paraquat; 35 g a.i. of the latter is, however, rather more rapid in its action than 20 g diquat. Diquat may be introduced into a single sloping hole in the trunk 20 cm deep and punched 60–90 cm from the ground with a crowbar. The hole is then filled up with water. It has been suggested that *Oryctes* species multiply less rapidly in the stumps remaining after diquat and paraquat poisoning than in stumps remaining after sodium arsenite poisoning. Strong evidence for this is lacking, however; in West Africa, investigations of diquat-poisoned stumps have suggested that termites, ants and mice (*Rattus alleni*) together with other insect predators and parasites severely restrict the breeding of *Oryctes* sp.

In Malaysia it is now common practice to combine poisoning and felling.[6, 71] Felling is done by a bulldozer extension known as a 'de-stumper' a fortnight or more after application of sodium arsenite by which time the leaves have withered.

The importance of replanting methods in the control of *Ganoderma* Trunk Rot is discussed on p. 635.

Costs

Field maintenance

Table 10.7A–B shows representative costs of field maintenance work in terms of man-days per hectare per annum. It must be remembered that these are costs of manual work only and (see Chapter 7, p. 357) are averaged figures from various plantation sources. Quite different costs will be incurred if, for instance, maintenance is with herbicides or by mechanical means, and for these the portions of this book dealing with these subjects and the references given therein should be consulted. The upkeep of roads entails 2 to 3 man-days per planted hectare.

Table 10.7   *Field maintenance work — man-days per hectare per annum*

(a) Africa

Young palms	Nigeria	Zaire (1)[2]		Zaire (2)		Zaire (3)
Years of age:	0–5	0–2	3–4	0–3	3–7	0–4
Ring-weeding	12	17	17 ⎫	20	20	22*
Paths or 'rides'	12	6	6 ⎬			
Slashing	20	12	6 ⎭	20	15	12
	44	35	29	40	35	34

Middle years						
Years of age:	6–11	6–7				4–11
Ring-weeding	7	12 ⎫				10
Paths or 'rides'	10	6 ⎬				
Slashing	15	6				17
	32	24				27

Adult palms						
Years of age:	11+	9–20		8+		12–25
Ring-weeding	5	10 ⎫		12		6
Paths or 'rides'	10	6 ⎬				
Slashing	7	4		6		7
	22	20		18		13

(b) Malaysia

Year in the field	1st	2nd	3rd	9th (adult field)
Ring-weeding and weeding, including spraying if done	100	72	47	20
Pests and diseases	12	17	12	12
Paths, drains, roads, bridges, etc.	15	12	12	15
Manuring	2	5	2	2
	129	106	73	49

* Including mulching.

Maintenance of adult plantations by hand in the Ivory Coast is stated to provide 20 to 30 days work per hectare per annum. Where tractor work is possible, the work of slashing can be done at the rate of one tractor hour per hectare per annum.[27]

The very much higher expenditure on weeding in the Far East compared with that on ring-weeding and slashing in Africa is due to the difficulty of, and heavy expenditure on, the establishment and maintenance of covers in the first few years. Fortnightly weeding at the outset is sometimes undertaken and there may be as many as eighteen weeding rounds in the first year. In addition more attention has to be given to pests and, on coastal clay estates, upkeep of drains and bridges entails much labour though the former is now often mechanized.

In West Africa, maintenance costs in the first 4 to 5 years may be increased by about 40 man-days per hectare per annum by the need to spray every 3 weeks against Freckle (*Cercospora elaeidis*). With the use of motor-operated sprayers the work is reduced by about one-half. If the area has been prepared so that tractors can be driven between the rows then the man-day requirements can be reduced by four-fifths, i.e. to about 8 man-days per hectare, spray-lines being used in conjunction with a tractor-mounted sprayer.[72]

Pruning

	Nigeria		Zaire			Zaire[2]	Ivory Coast[6]	Malaysia
Age (years)	6−10	11+	4−8	9−12	Old	5−20	adult	adult
Frequency	yearly	yearly	8 mths	8 mths	9 mths	yearly	9−12 mths	9 mths
Man-days/ ha./an.	2−5	5−12	3.3	3.9	6	4−10	8.0−10.7	4.4

Ablation

Malaysia
8−12 rounds in 1 year   10 man-days per hectare

Harvesting

Harvesting tasks have been discussed briefly in this chapter. Owing to varying yields, labour expenditure per ton of bunches and per unit area varies widely from region to region and month to month. The figures in Table 10.8 of cutting plus carrying from Zaire and Malaysia are means of several

Table 10.8   *Harvesting: cutting and carrying*

Age	Man-days per ton of bunches	Man-days per hectare/an.	
Zaire			
Year 5	3.7	14	
Year 14	1.5	19	
Old fields	3.2	23	low yields
Malaysia			
Young fields (chisel harvesting)	4.5	33	
Medium fields (axe harvesting)	1.5	19	medium yield
	1.5	30	high yield
Adult fields (pole harvesting)	1.1	22	medium yield
	1.1	33	high yield

estate figures and calculations are based on yields per hectare per annum
on the following scales:

Year of harvest	1	2	3	4	5	6+	Old fields
	Tons bunches/hectare/annum						
Zaire	2.5	3.7	6.2	10.0	10.0	12.5	7.5
Malaysia (medium)	5.0	7.5	12.5	17.5	20.0	20.0	
(high)	7.5	10.0	15.0	20.0	25.0	30.0	

Harvesting costs per ton are found to vary greatly even within regions
and the figures given in Table 10.8 must therefore be taken as very rough
averages.

Plantation labour requirements

From the information given above and in Tables 10.7 and 10.8 it can be
seen that with manual methods the labour requirement on an adult
plantation will vary as follows:

	Man-days per hectare	
	Africa	*Malaysia*
Field maintenance	14–22	49*
Pruning	5–12	4.4
Harvesting	19–23	22–33
Roads	2–3	
Total	40–60	81
	Africa	*Malaysia*
Hectares per worker working 300 days per year	7.5–5.0	3.6

* Including roads.

Figures of 3 to 5 hectares per worker are usually accepted in Malaysia for
the total labour force required, including transport, mill and other labour;
this suggests that the above man-day figures for Malaysia are rather higher
than average, and that, in particular, labour on maintenance can be kept
lower by mechanization of drainage and other work. It should be observed
that these figures are based on maintenance of adult fields and that a high
proportion of young areas on a plantation raises the labour requirement,
though the difference between the upkeep costs of young and adult areas
appears to be much greater in the Far East than in Africa.

The conversion of common weights and measures

1 inch = 25.4 millimetres (mm)   1 centimetre = 0.394 inch
1 foot = 30.48 centimetres (cm)   1 metre = 39.37 inches (in.)
1 yard = 0.9144 metre (m)   1 metre = 3 feet (ft) 3.4 in.
1 chain = 22 yards (yd) = 20.12 m
1 mile = 1.609 kilometres (km)   1 km = 0.6214 miles = 1,094 yd

1 ounce (oz) = 28.35 grammes (g)   1 g = 0.0353 oz
1 pound (lb) = 453.6 g
1 lb = 0.454 kg
1 hundredweight (cwt) = 112 lb = 50.8 kg   1 kilogram (kg) = 2.205 lb
1 long ton = 2,240 lb
           = 1.016 metric tons   1 metric ton = 0.984 long ton
           = 1,016 kg   = 2,204 lb

1 pint* = 34.66 cu. in.   1 litre = 1.76 pints
1 pint = 20 fl. oz*   = 35.2 fluid oz (fl. oz)
         = 0.568 litres (ℓ)   = 1,000 cu. cm
1 cu. in. = 16.39 cu. cm   1 cu. cm = 0.061 cu. in.
1 gallon (gal) = 277.3 cu. in   = 0.0352 fl. oz
               = 4,546 cu. cm
               = 4.546 ℓ   1 ℓ = 0.22 gal
1 cu. ft = 6.23 gal   1 cu. m = 35.31 cu. ft
         = 0.0283 cu. m   = 1.308 cu. yd
1 cu. yd = 0.7646 cu. m   1 cu. m = 1,000 ℓ
1 fl. oz = 28.41 cu. cm   = 220 gal
         = 0.0284 ℓ   1 ℓ = 33.81 US fl. oz
1 gal (imperial) = 1.201 US gal*   = 0.2642 US gal
1 fl. oz (Imp) = 0.9608 US fl. oz*   = 2.113 US pints

1 sq. in. = 6.452 sq. cm   1 sq. cm = 0.155 sq. in.
1 sq. ft = 929 sq. cm   1 sq. m = 1.196 sq. yd
1 sq. yd = 0.836 sq. m   1 hectare = 10,000 sq. m
1 acre = 0.4047 hectare (ha)   1 ha = 2.471 acres
        = 4,840 sq. yd
        = 4,047 sq. m
1 sq. mile = 640 acres   1 sq. km = 100 ha
            = 259 ha   = 247.1 acres
            = 2.59 sq. km   = 0.386 sq. mile

1 lb per acre = 1.121 kg per ha   1 kg per ha = 0.892 lb per acre
1 (long) ton per acre = 2.511 metric   1 metric ton per ha = 0.3982 (long) ton
                          tons per ha   per acre
1 lb per gal = 0.0998 kg per ℓ   1 kg per ℓ = 10.02 lb per gal
1 oz per gal = 6.236 g per ℓ   1 g per ℓ = 0.1604 oz per gal
1 lb per gal = 99.78 g per ℓ   1 g per ℓ = 0.010 lb per gal
1 gal (160 fl. oz) per acre = 11.23 ℓ per ha   1 ℓ per ha = 0.089 gal per acre
                                                            = 14.24 fl. oz per acre
1 fl. oz per acre = 70.2 cu. cm per ha   1 cu. cm per ha = 0.01425 fl. oz per acre
1 lb per sq. in. (psi) = 0.0703 kg per sq. cm   1 kg per sq. cm = 14.22 lb per sq. in.
1 part per million (ppm) = 1 milligram per ℓ

---

* Capacities are in Imperial (British) measures except where otherwise stated. A US fl. oz is one-sixteenth of a US pint. To convert pounds or oz per gal (Imp) into per US gal multiply by 0.8327. One fl. oz per gal (Imp) = 0.8001 US fl. oz per gal = 6.25 cu. cm per ℓ.

References

1. Proceedings of the Conference on Oil Palm Research held at the Oil Palm Research Station, near Benin, Nigeria, Dec. 1949. Mimeograph.
2. **Vanderweyen, R.** (1952) *Notions de culture d'Elaeis au Congo Belge*. Brussels.
3. **Surre, Ch. and Ziller, R.** (1963) *La Palmier à huile*. G.-P. Maisonneuve and Larose, Paris.
4. **Bunting, B., Georgi, C. D. V. and Milsum, J. N.** (1934) *The oil palm in Malaya*, Kuala Lumpur.
5. **Sheldrick, R. D.** (1968) The control of ground cover in oil palm plantations with herbicides. 3. Development of ring weeding techniques. *J. W. Afr. Inst. Oil Palm Res.*, **5**, (17), 57.
6. **Turner, P. D. and Gillbanks, R. A.** (1974) *Oil palm cultivation and management*. Incorp. Soc. of Planters, Kuala Lumpur.
7. **Sheldrick, R. D.** (1962 and 1968) The control of ground cover in oil palm plantations with herbicides. 1. An Introduction and some early investigations. *J. W. Afr. Inst. Oil Palm Res.*, **3**, 344. 2. Screening trials for herbicides suitable for ring weeding. *J. Nigerian Inst. Oil Palm Res.*, **4**, 417.
8. **Coomans, P.** (1971) Entretion chimique des ronds dans les palmeraies adultes de Côte d'Ivoire. *Oléagineux*, **26**, 595.
9. **Seth, A. K., Abu Bakar and Sivarajah, S.** (1970) New recommendations for weed control under young palms. Crop Protection Conference, 1970, p. 85. Incorp. Soc. of Planters, Kuala Lumpur.
10. **Smith, R.** (1973) Notes on chemical weed control in oil palms in Cameroon. *Oil Palm News*, No. 16, 12.
11. **Lucy, A. B.** (1941) A comparison between natural covers and clean weeding on yields of oil palms. *Malay. agric. J.*, **29**, 190.
12. **Wycherley, P. R.** (1963) The range of cover plants. *Plts'. Bull. Rubb. Res. Inst. Malaya*, No. 68, 117.
13. **Kowal, J. M. L. and Tinker, P. B. H.** (1959) Soil changes under a plantation established from high secondary forest. *J. W. Afr. Inst. Oil Palm Res.*, **2**, 376.
14. **Sly, J. M. A. and Sheldrick, R. D.** (1963) W.A.I.F.O.R. Eleventh Annual Report, pp. 28 and 46.
15. **Daniel, C. and de Taffin, G.** (1974) Conduite des jeunes plantations de palmiers à huile en zones seches au Dahomey. *Oléagineux*, **29**, 227.
16. **Edgar, A. T.** (1958) *Manual of rubber planting (Malaya)*. Incorp. Soc. of Planters, Kuala Lumpur.
17. **Sheldrick, R. D.** (1968) Weed control with herbicides during legume cover establishment. *J. Nigerian Inst. Oil Palm Res.*, **5**, 67.
18. **Aya, F. O.** (1973) Germination inhibitors in the seeds of *Pueraria phaseoloides* (Ruxb.) Benth. *J. Nig. Inst. Oil Palm Res.*, **5**, (18), 7.
19. **Bevan, J. W. L. and Gray, B. S.** (1966) Field planting techniques for the oil palm in Malaysia. *Planter, Kuala Lumpur*, **42**, 196.
20. **Anon.** (1962) Species and varieties of Flemingia in Malaya. *Plts'. Bull. Rubb. Res. Inst. Malaya*, No. 61, 78.
21. **Gray, B. S. and Hew Choy Kean** (1968) Cover crop experiments in oil palms on the West Coast of Malaya. In *Oil palm developments in Malaysia*, p. 56. Incorp. Soc. of Planters, Kuala Lumpur.
22. **Wong Phui Weng** (1971) A selective pre-emergence weedicide for grass control in the establishment of legume cover crops. *Planter, Kuala Lumpur*, **47**, 459.
23. **Haines, W. B.** (1940) *The uses and control of natural undergrowth on rubber estates*. Rubb. Res. Inst. Malaya, Planting Manual No. 6, Kuala Lumpur.
24. **Sheldrick, R. D.** (1968) The control of Siam weed (*Eupatorium odoratum* Linn). *J. Nigerian Inst. Oil Palm Res.*, **5**, 7.

25. **Anon.** (1957) Trial of the Holt Weed Breaker. *Plts'. Bull. Rubb. Res. Inst. Malaya*, No. 31, 74.
26. **Hew Choy Kean and Tan Tai Kin** (1970) The effects of maintenance techniques on oil palm yield in coastal clay areas of West Malaysia. In *Crop protection in Malaysia*, p. 62. Incorp. Soc. of Planters, Kuala Lumpur.
27. **Anon.** (1962) L'entretien des palmeraies adultes. *Oléagineux*, 17, 777.
28. **Sheldrick, R. D.** (1968) Mechanical maintenance in oil palm plantations. *Nigerian agric. J.*, 5, 7.
29. **Sly, J. M. A.** (1968) The results of pruning experiments on adult palms in Nigeria. *J. Nigerian Inst. Oil Palm Res.*, 5, 89; quoting Rutgers, A. A. L. *Investigation on oil palms*. A.V.R.O.S., Medan. 1922.
30. **Oil Palm Research Station**, Banting, Malaysia. Agronomy Annual Reports, 1969, 70, 71.
31. **Oil Palm Genetics Laboratory**, Layang Layang, Malaysia. Progress Report, 1970 and 1971.
32. **Gunn, J. S.** *et al.* (1962) W.A.I.F.O.R. Tenth Annual Report, p. 52.
33. **W.A.I.F.O.R.** First Annual Report (1953), pp. 80–1.
34. **Anon.** (1960) When to start harvesting. *J. W. Afr. Inst. Oil Palm Res.*, 3, 187.
35. **Tailliez, B. and Olivin, J.** (1971) Nouveaux resultats expérimentaux sur l'ablation des jeunes inflorescences du palmier à huile en Côte d'Ivoire. *Oléagineux*, 26, 141.
36. **Obasola, C. O.** (1970) N.I.F.O.R. Fifth Annual Report, Nigeria, p. 41.
37. **Hew, C. K. and Tam, T. K.** (1972) The effect of removal of the initial inflorescences on the establishment, growth and subsequent yields of the oil palm on the coastal clays in West Malaysia. *Malay. Agriculturalist*, 11, 13.
38. **Chew, P. S. and Khoo, K. T.** (1973) Early results from disbudding trials on oil palms. In *Advances in oil palm cultivation*, p. 133. Incorp. Soc. of Planters, Kuala Lumpur.
39. **Chan, K. W. and Mok, C. K.** (1973) Castration and manuring in immature oil palms on inland latosols in Malaysia. In *Advances in oil palm cultivation*, p. 147. Incorp. Soc. of Planters, Kuala Lumpur.
40. **Hardon, J. J.** (1973) Assisted pollination in the oil palm: a review. In *Advances in oil palm cultivation*, p. 184. Incorp. Soc. of Planters, Kuala Lumpur.
41. **Gray, B. S.** (1969) The requirement for assisted pollination in oil palms in Malaysia. In *Progress in oil palms*, p. 49. Incorp. Soc. of Planters, Kuala Lumpur.
42. **Veldhuis, J.** (1967) Methods of assisted pollination for oil palms. In *Oil palm developments in Malaysia*, p. 72. Incorp. Soc. of Planters, Kuala Lumpur.
43. **Tailliez, B. and Valverde, G.** (1971) La pollinisation assistée dans les plantations de palmier à huile. *Oléagineux*, 26, 683 and 763.
44. **Broekmans, A. F. M.** (1957) Growth, flowering and yield of the oil palm in Nigeria. *J. W. Afr. Inst. Oil Palm Res.*, 2, 187.
45. **Dufrane, M. and Berger, J. L.** (1957) Étude sur la récolte dans les palmeraies. *Bull. agric. Congo belge*, 48, 581.
46. **Ng, K. T. and Southworth, A.** (1973) Optimum time of harvesting oil palm fruit. In *Advances in oil palm cultivation*, p. 439. Incorp. Soc. of Planters, Kuala Lumpur.
47. **Wuidart, W.** (1973) Evolution de la lipogenèse du régime de palmier à huile en fonction du pourcentage de fruits détachés. *Oléagineux*, 28, 551.
48. **Speldewinde, H. V.** (1968) Harvesting and harvesting methods. In *Oil palm developments in Malaysia*. Incorp. Soc. of Planters, Kuala Lumpur.
49. **Gerard, P., Renault, P. and Chaillard, H.** (1968) Critère et normes de maturité pour la récolte des régimes de palmiers à huile. *Oléagineux*, 23, 299.
50. **Pratt, N. S.** (1963) *Notes on the cultivation of oil palms*. Incorp. Soc. of Planters, Kuala Lumpur.
51. **Sankar, N. S.** (1965) *Fruit collection and evacuation by road*. Sabah Planters' Association. Oil Palm Seminar. Mimeograph.

52. Bevan, J. W. L. and Gray, B. S. (1969) The organisation and control of field practice for large-scale oil palm plantings in Malaysia. Incorp. Soc. of Planters, Kuala Lumpur.

53. Boyé, P. and Martin, G. (1969) Organisation général de la récolte en palmeraie industrielle. *Oléagineux*, 24, 451.

54. Gillbanks, R. A. (1967) Harvesting and fruit transport. A discussion on current practice. *Planter, Kuala Lumpur*, 43, 322.

55. Chan, K. W., Corley, R. H. V. and Seth, A. K. (1972) Effects of growth regulators on fruit abscission in oil palm, *Elaeis guineensis. Ann. appl. Biol.*, 71, 243.

56. Wan, D. (1973) The use of buffaloes in oil palm fruit collection. In *Advances in oil palm cultivation*, p. 432, Incorp. Soc. of Planters, Kuala Lumpur.

57. Cunningham, W. M. (1969) A container system for the transport of oil palm fruit. In *Progress in oil palms*, p. 287. Incorp. Soc. of Planters, Kuala Lumpur.

58. Washburn, R. A. (1973) The African oil palm in Costa Rica. *Oil Palm News*, 16, 1.

59. Price, J. G. M. and Kidd, D. D. (1973) Mechanised loading and transport of oil palm fruit. In *Advances in oil palm cultivation*, p. 415. Incorp. Soc. of Planters, Kuala Lumpur.

60. Ferwerda, J. D. (1955) *Questions relevant to replanting in oil palm cultivation.* Thesis, Wageningen.

61. Gunn, J. S. *et al.* (1961) W.A.I.F.O.R. Ninth Annual Report, p. 58.

62. Unilever Plantations (1961) Annual Review of Research, p. 32. Mimeograph.

63. Green, A. H. (1964) Private communication.

64. Sheldrick, R. D. (1968) N.I.F.O.R. Third Annual Report, 1966–7, p. 48.

65. Boyé, P. and Aubrey, M. (1973) Replantation des palmeraies industrielles. Méthode de préparation de terrain et de protection contre l'*Oryctes* en Afrique de l'Ouest. *Oléagineux*, 28, 175.

66. Hartley, C. W. S. (1949) The felling and disposal of old oil palms prior to replanting. *Malay. agric. J.*, 32, 223.

67. W.A.I.F.O.R. First Annual Report (1953) 1952–3, p. 84.

68. Marynen, T. and Gillot, J. (1957) L'élimination des vieux palmiers par empoisonnement. *Bull. Inf. I.N.E.A.C.*, 6, 167.

69. Sheldrick, R. D. (1963) A note on recent investigations into palm poisoning. *J. W. Afr. Inst. Palm Res.*, 4, 101.

70. Gunn, J. S. and Tatham, P. B. (1961) Diquat as an arboride. *Nature, Lond.*, 189, 808.

71. Stimpson, K. M. S. and Rasmussen, A. N. (1973) Clearing the old stand and some preparation for replanting coastal oil palms. In *Advances in oil palm cultivation*, p. 116, Incorp. Soc. of Planters, Kuala Lumpur.

72. Sheldrick, R. D. (1965) The practical aspects of mechanised spraying in the field. *J. Nigerian Inst. Oil Palm Res.*, 4, 325.

73. Chew, P. S. and Khoo, K. T. (1976) Growth and yield of intercropped oil palms on a coastal clay soil in Malaysia. Int. Agric. Oil Palm Conference, Kuala Lumpur, 1976.

74. Broughton, W. J. (1976) Effect of various covers on the performance of *Elaeis guineensis* Jacq. on different soils. Int. Agric. Oil Palm Conference, Kuala Lumpur, 1976.

75. Aya, F. O. (1976) A critical assessment of the cover policy in oil palm plantations in Nigeria. Int. Agric. Oil Palm Conference, Kuala Lumpur, 1976.

76. Teoh Cheng Hai and Chong Choon Fong (1976) Use of pre-emergence herbicides during establishment of leguminous cover crops. Int. Agric. Oil Palm Conference, Kuala Lumpur, 1976.

77. Tan, H. T., Pillai, K. R. and Fua, J. M. (1976) Establishment of legume covers using pre- and post-emergence herbicides. Int. Agric. Oil Palm Conference, Kuala Lumpur, 1976.

78. **Calvez, C.** (1976) Influence de l'élagage a différents niveaux sur la production du palmier à huile. *Oléagineux,* 31, 53.
79. **Anon.** (1976) Ablation des inflorescences des jeunes palmiers à huile. *Oléagineux,* 31, 9.
80. **Daniel, C. and de Taffin, G.** (1976) L'ablation des inflorescences de jeunes palmiers. Cas particulier des zones sèches. *Oléagineux,* 31, 211.
81. **Corley, R. H. V.** (1976) Oil palm yield components and yield cycles. Int. Agric. Oil Palm Conference, Kuala Lumpur, 1976.
82. **Southworth, A.** (1976) Harvesting − a practical approach to the optimization of oil quantity and quality. Int. Agric. Oil Palm Conference, Kuala Lumpur, 1976.

Chapter 11

The nutrition of the oil palm

Manurial trials with the oil palm were begun in Sumatra almost from the birth of the plantation industry, but serious attention was not given to the use of fertilizers elsewhere until around the time of the Second World War. Nearly all the early plantations were opened from virgin or old secondary forest, and many of the experiments started in such areas, especially in Zaire but also in some parts of West Africa and the Far East, showed little or no response in the early years.

It was not long, however, before it was realized that the oil palm makes heavy demands on the nutrient supplies of the soil, that responses to fertilizers are readily obtained, and that when soils are deficient nutrient need is soon visibly expressed by foliar symptoms.

Nutrient demand and employment

Nutrient uptake and usage is an important physiological process but it was not dealt with in Chapter 4 as it can more usefully be discussed in direct connection with manuring. Attention has already been given to the cumulative production of dry matter in the palm (p. 157) and to the distribution of dry matter production in the various organs. The immobilization of nutrients taken up in the course of this dry matter production will now be considered.

That nutrient requirements can be indicated by the chemical composition of the plant has not been generally accepted, but it has been claimed that analysis can have some value when the crop is bulky, as in the oil palm, and the long-term nutrient supplying power of the soil is poor, as in so many tropical soils.[1] It is an error to think, however, that whatever is removed by the crop must be replaced *pro rata* since all oil palm soils will be continuing to release nutrients at a regular pace from their mineral reserves. Nevertheless, results of analyses in the case of the oil palm are of particular interest in a consideration of replanting since by the time one cropping cycle of 25 to 30 years is completed considerable quantities of nutrients have been removed in the bunches, stored in the trunk, leaves and roots, or returned to the soil in dead leaves and roots. A good idea can therefore be obtained at that time of the proportions in which the various nutrients have been lost in the produce or have been or are to be returned to the soil.

A number of attempts have been made to carry out a complete estimation on these lines. The earliest calculations usually assumed that nutrients in leaves and male inflorescences are permanently lost, and it is probable that exaggerated requirements were thus suggested.[2, 3] Ferwerda combined data from African sources to construct a table of plant nutrients immobilized per hectare by a 20-year-old plantation of *tenera* palms in Zaire. Accounted as immobilized were nutrients in (i) the trunk, (ii) the crown with the leaves and the roots existing on the palms at the time of estimation and (iii) the bunches removed from the palms since they started to bear.[4] The total of these sources constitute the cumulative net nutrient uptake to any given age.

In a study in Nigeria the palms analysed were aged 7, 10, 14, 17, 20 and 22 years and were taken from a forest-felled area where no clear responses to N, P, Ca or Mg in adjoining fields had been obtained and where K had only become deficient in the later years. Bunch yield throughout the palms' life was known and nutrient losses were thence estimated through analysis of bunch stalk and calculations of bunch composition. Analysis was made of the following above-ground parts of the palm separately: leaflets, rachis, apical tissues and trunk. Roots of two 17-year-old palms were excavated to 90 cm and analysed, and on the basis of Ferwerda's work quantities were assumed to be similar for all age-groups except the 7-year-old palms. The 22-year-old palms, though not showing symptoms, were in an area becoming potassium-deficient, and the study also included a group of three 22-year-old palms showing obvious potassium deficiency and two 21-year-old palms in an area of magnesium deficiency. Table 11.1 shows the analyses of these palms. Interest in this data lies largely in the changes that occur with age and with the onset of deficiency symptoms and in the estimations that can be made of the total quantity of nutrients removed by the palms in the course of their productive life.

Examination of the data in Table 11.1 shows that there is very little variation with age in percentage nitrogen or phosphorus content, but that potassium percentage decreases with age of palm while there is a corresponding increase in magnesium and calcium percentages. Potassium deficiency is reflected by lower quantities in all parts of the K-deficient 22-year-old palms and it may be assumed from the very low trunk K content of the 20-year-old palms that they must have been on the verge of potassium deficiency. Perhaps most remarkable of all is the extremely small quantities of magnesium left in any tissues of the Mg-deficient palms.

The cumulative net nutrient uptake, which differs from the total uptake by the exclusion of nutrients recycled to the soil in old leaves, male inflorescences and dead roots, was calculated from the analytical data, estimates of live roots and the actual bunch production from the start of bearing. The means of the 20- and the healthy 22-year-old palms were

Table 11.1   *Dry matter and nutrient content of palms of different ages in Nigeria*

Age	Part	N	P	K	Mg	Ca	Dry matter
A. *Kilogrammes per palm*							
7	Crown	0.61	0.07	0.63	0.15	0.26	68.5
	Trunk	0.33	0.04	0.30	0.14	0.14	64.4
	Total	0.94	0.11	0.93	0.29	0.40	132.9
10	Crown	0.82	0.07	0.69	0.17	0.31	92.2
	Trunk	0.85	0.08	0.65	0.25	0.27	167.5
	Total	1.67	0.15	1.34	0.42	0.58	259.7
14	Crown	0.80	0.08	0.48	0.21	0.36	85.4
	Trunk	1.31	0.17	0.87	0.42	0.36	238.8
	Total	2.11	0.25	1.35	0.63	0.72	324.2
17	Crown	0.89	0.09	0.76	0.19	0.32	96.9
	Trunk	1.53	0.17	1.51	0.66	0.63	280.0
	Total	2.42	0.26	2.27	0.85	0.95	376.9
20	Crown	1.31	0.17	0.68	0.48	0.87	151.5
	Trunk	1.93	0.30	0.83	1.49	1.12	439.1
	Total	3.24	0.47	1.51	1.97	1.99	590.6
22	Crown	1.06	0.11	0.67	0.41	0.64	117.5
	Trunk	1.17	0.20	1.15	0.90	0.50	343.8
	Total	2.23	0.31	1.82	1.31	1.14	461.3
22 K-defic't	Crown	0.86	0.11	0.31	0.44	0.73	99.1
	Trunk	1.26	0.24	0.50	0.93	0.60	311.6
	Total	2.12	0.35	0.81	1.37	1.33	410.7
21 Mg-defic't	Crown	0.44	0.05	0.37	0.02	0.08	47.0
	Trunk	0.56	0.07	0.45	0.03	0.15	88.0
	Total	1.00	0.12	0.82	0.05	0.23	135.0
17	Roots	0.46	0.03	0.60	0.19	0.09	128.0
B. *Percentage of dry matter*							
7	Leaflets	1.64	0.12	0.95	0.27	0.55	
	Rachis	0.41	0.07	0.84	0.16	0.25	
	Apical tissue	2.20	0.38	3.30	0.80	0.82	
	Trunk	0.52	0.07	0.47	0.21	0.22	
14	Leaflets	1.90	0.13	0.70	0.36	0.68	
	Rachis	0.37	0.06	0.43	0.14	0.28	
	Apical tissue	2.18	0.36	2.55	0.91	0.73	
	Trunk	0.55	0.07	0.36	0.17	0.15	
20	Leaflets	1.94	0.14	0.77	0.38	0.78	
	Rachis	0.33	0.08	0.26	0.28	0.47	
	Apical tissue	2.00	0.40	1.75	1.00	1.00	
	Trunk	0.44	0.07	0.19	0.34	0.26	
22 K-defic't	Leaflets	1.82	0.15	0.48	0.46	0.92	
	Rachis	0.38	0.08	0.20	0.43	0.64	
	Apical tissue	2.00	0.48	1.61	0.99	0.87	
	Trunk	0.40	0.08	0.16	0.30	0.19	
21 Mg-defic't	Leaflets	1.63	0.12	0.99	0.07	0.20	
	Rachis	0.45	0.10	0.60	0.05	0.12	
	Apical tissue	3.00	0.47	3.33	0.53	0.53	
	Trunk	0.64	0.08	0.50	0.03	0.17	

Table 11.2   *Nutrients removed from or immobilized in 20- and 22-year-old palms in Nigeria and Zaire* (kg per hectare)

Nigeria	N	P	K	Mg	Ca
Parts — Above ground	390	55	250	230	220
Roots*	70	5	90	30	14
Bunches*	430	90	500	65	76
Total	890	150	840	325	310
Zaire estimate					
Parts — Above ground	713	125	305	174	297
Roots	84	9	86	4	1
Bunches*	564	97	585	82	88
Total	1,361	231	976	260	386

* Based on yields: 1,060 kg/palm, Nigeria, and 1,371 kg/palm, Zaire.

used. These data are shown in Table 11.2 where they are compared with Ferwerda's estimate for 20-year-old Zaire palms.

Considering all the approximations inevitable in both sets of data and the composition from various sources of the Zaire estimate, remarkably similar figures emerge. The high phosphorus figures for the Zaire trunk and crown are no doubt due to the use of a single trunk analysis by Zeller.

The cumulative net uptake of phosphorus, potassium and magnesium are illustrated from the Nigerian data in Fig. 11.1. It will be seen that though the P uptake is much smaller than the K uptake, the proportion of the phosphorus diverted to the bunches is large. Phosphorus also accumulates in the trunk at a steady rate. Nitrogen uptake follows a similar course to that of phosphorus, but a lower proportion is removed in the bunches. Large quantities of potassium are diverted to the bunches and it is apparent that in these palms the K-accumulation in the palm itself is very slow; there is here a suggestion that in the later years the reserves in the trunk were being drawn upon. In contrast, magnesium (where not deficient) accumulates in the trunk most rapidly in the later years and a low proportion is removed in the bunches. This pattern is also followed by calcium.

In drawing any general conclusions from these data notice must be taken of the depletion of potassium which was clearly taking place in this soil under palms, the known antagonisms which exist between the nutrient elements, and the low level of bunch yield — though typical of the region concerned — compared with other producing regions. It is reasonable to suppose that if K dressings had been applied or the soil had had a higher K-content, accumulation in the trunk would have been greater and that the K graph in Fig. 11.1 might have approximated more closely to that of P; bunch yield would have been higher and the proportion of K diverted to bunches might have remained the same.

Phosphorus immobilization is on a much smaller scale, but owing to the

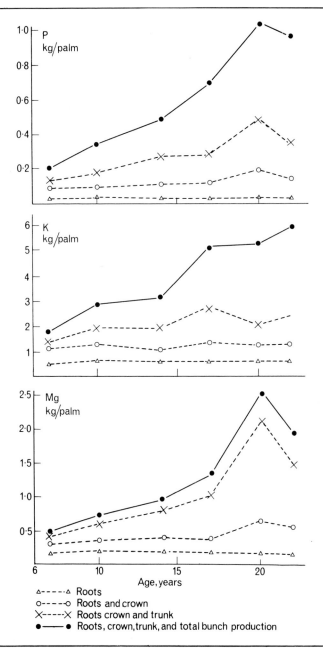

Fig. 11.1   Cumulative net uptake of phosphorus, potassium and magnesium in a field in Nigeria (Tinker and Smilde, 1963).

high proportionate diversion to bunches it is hardly surprising that a phosphorus need often arises when bunch production has been raised by the supply of primarily deficient nutrients.

Owing to the potassium—magnesium antagonism, Mg contents, both absolute and as a percentage of dry matter, tend to be high in the prevailing incipient K-deficiency of the Nigerian palms. As Table 11.1 shows, however, Mg contents can fall to very low levels indeed. In spite of the low demand for magnesium in the bunches, magnesium can become a factor limiting yield in these circumstances.

An extensive study of nutrient uptake has now been undertaken in Malaysia by Ng Siew Kee *et al.*[5] It is not possible to compare the results directly with those from Nigeria since gross annual uptakes were computed in place of net nutrient uptake over the life of a palm. The gross annual nutrient uptake of a group of adult palms growing on marine clay and yielding 25 tons of bunches per hectare is given below together with the gross uptake less nutrients in leaves and male inflorescences recycled to the soil, and the uptake in the bunches.

	Kilogrammes per hectare per annum				
	N	P	K	Mg	Ca
1. Gross nutrient uptake (GNU)	193	26	251	61	89
2. GNU, less leaves and male inflor.	144	15	149	32	23
3. Bunches	73	12	93	21	20

From these figures and from Table 11.2 it may be deduced that cumulative net nutrient uptake is much higher under Malaysian conditions than under conditions obtaining in Nigeria since, for example, the potassium taken up by the palm less K returned to the soil in leaves and male inflorescences in 6 adult years is more than that removed from and immobilized in the Nigerian palms in their whole 20 to 22 years of life. The Malaysian analyses showed figures comparable to those of Nigeria for nutrient concentration in the various organs except in the case of potassium, the percentage of which in the trunk was much higher in Malaysia. In short, the conditions of growth on the Malaysian coastal soils were, in comparison with conditions in Nigeria, conducive to much higher uptake of nutrients for bunch production and storage in the trunk, a greater development of the trunk both in height increment and width, a higher bunch production, but only a small difference in leaf production.

Ng Siew Kee *et al.*[5] have also shown that the proportion of each nutrient in the bunch does not depend on bunch weight since there are linear relationships between the quantities of nutrients and the weight of the bunch. Differences of nutrient content were, however, found between soils, though these differences were not great. From the results obtained the mean quantities of nutrients contained in 10 tons of bunches was estimated as follows:

Per 10 tons of bunches	N	P	K	Mg	Ca	Mn	Fe	B	Cu	Zn	Mo
Kilogrammes	29.4	4.4	37.1	7.7	8.1						
Grammes						15.1	24.7	21.5	47.6	49.3	0.084

Methods of detecting nutrient need

Field experiments with fertilizers are the primary means of detecting and determining nutrient need, but these may be greatly assisted by (a) methods using plants and (b) soil analysis methods. Many of these methods have been tried and found useful with the oil palm, but before giving an account of them it should be stated that no single method gives all the answers to the questions which the practical grower may be asking, though some field experiments may, in a limited sphere, come nearest to answering queries on the nutrients and their rates and frequencies required.

Wallace[6] has stated that the main methods of estimating nutrient needs are qualitative rather than quantitative, are empirical in character and some are obviously suited only for preliminary diagnosis. This is especially the case in the methods depending on the behaviour of the palm itself and these will be considered first.

Deficiency symptoms

The oil palm shows many leaf chloroses and other foliar symptoms attributable to nutrient deficiencies. Symptoms were found most extensively on soils of low nutrient status in Nigeria and Zaire, and the work on deficiency symptoms dates from a classic experiment at Nkwelle in Nigeria[7] in 1940 and the discovery by Hale[8] of the connection between leaf potassium content and 'bronzing' or orange spotting of the leaves. In this experiment at Nkwelle an application of 17 tons of wood ash per hectare applied over a 3-year period restored affected palms to health and increased their yield fivefold. Later, dressings of sulphate of potash in the same area showed that the responses to ash were attributable to its potash content.

These startling results stimulated an interest in deficiency symptoms and in the following 15 years a considerable amount of knowledge was gained both in Africa and Asia. This knowledge was obtained by three methods: (i) leaf injection and spraying of chlorotic leaves, (ii) the correlation of symptoms with leaf nutrient contents and with yield and (iii) the inducement of symptoms in sand culture. In the following descriptions the symptoms induced on seedlings are those described by Broeshart[9] and Bull,[10, 11] while the adult palm symptoms are attributable to general observations in several parts of the world and to accounts by several authors.

Nitrogen deficiency

Symptoms are similar on seedlings of all ages and on young palms in the field. Although adult palm responses to N are not uncommon, N-deficiency symptoms in large bearing palms have not been described; it has been suggested, however, that the characteristic yellowing symptom in the 'plant failure' condition may be attributed to N deficiency.[12]

In the very young seedling the first symptom is the development of a uniform pale green colour over the leaves. Paling is followed by yellowing, green gradually disappears and the production of a deeper yellow colour precedes necrosis. In larger seedlings and in young palms in the field paling and yellowing follows a similar course and, in addition to the necrosis, the leaflet midribs and the rachis take on a bright yellow colour and the laminae tend to be narrow and to roll inwards. Thus N-deficient plants are quite distinctive and when the deficiency is corrected by applications of sulphate of ammonia the plants very rapidly turn a dark green colour (Plates V and VI, between pp. 78 and 79).

Nitrogen deficiency symptoms are often associated with waterlogged conditions and they are soon counteracted by drainage. The leaves of severely waterlogged seedlings, however, often have a dull, olive green and flaccid appearance. Severe grass competition also induces nitrogen deficiency symptoms.

Phosphorus deficiency

Symptoms of phosphorus deficiency can be obtained in pot culture under rigorous systems of sand and water purification.[10] The purification methods used by Broeshart *et al.*[13] were insufficient to reduce phosphorus and calcium contents to levels where symptoms could appear.

In small seedlings the oldest leaves become dull and assume a pale olive green colour. The chlorotic condition increases in severity but the seedlings do not become fully yellow before necrosis of the tips sets in. The necrotic areas are usually of a dark brown colour and in transmitted light the tissues near to these areas are seen to be pale and water-soaked, suggesting rapid cell collapse. The leaves are much reduced in size.

Though responses to phosphorus fertilizers are often obtained, clear symptoms of phosphorus deficiency have not appeared in the field. In an experiment in southern Zaire, however, it was noted that palms not receiving phosphorus showed a high incidence of premature desiccation of the older leaves, a symptom usually found in areas of potassium deficiency.

Potassium deficiency

In very young seedlings in sand culture, leaves begin to show a pale green to white interveinal mottling with minute white or yellowish rectangular spots. Later, the leaves become pale olive or ochre in colour and the tips and margins become necrosed, the necrosis being typically pale grey or silvery.

In larger seedlings the chlorosis is similar, being interveinal with the veins and adjacent tissues remaining a normal green colour. Necrosis of the leaflet tips and margins also proceeds in the same manner as in smaller seedlings, the transition zone between chlorosed and necrotic tissue being very thin and pale brown. Minute clear spots also appear scattered over the laminae. These symptoms are accompanied or followed by marked shortening of the rachis and of the leaflets and this gives the leaf a 'bunchy' appearance, but these symptoms do not necessarily persist in later-formed leaves.

As early as 1941 Thompson,[14] working in Malaysia, obtained similar seedling symptoms in water culture.

A variety of symptoms have been found to be associated with potassium deficiency in the mature oil palm. This should not cause surprise since symptom differences in potassium-deficient plants have not been uncommon and several authors have shown that these differences are due to contemporaneous variations in the concentration of other ions which may be caused by environmental or genetic factors.[12, 14] This suggests that a debate, based on geographically isolated observations, as to what is *the* potassium deficiency symptom is unrewarding; and the removal of any one symptom from the list simply because it has appeared in certain individuals or progenies shown not to be K-deficient is unsound. It is therefore proposed to describe those symptoms which have unquestionably been associated with potassium deficiency as shown by fertilizer responses or leaf analysis and to comment on their extent and variation[15] (Plate VIII, between pp. 78 and 79).

Confluent orange spotting. This name, first used by Waterston, was adopted by Bull to describe a condition in which chlorotic spots, changing from pale green through yellow to orange, develop and enlarge both between and across the leaflet veins and fuse to form compound lesions of a bright orange colour.[16] Under a lens a pale yellow halo can be seen around the spot, but to the naked eye the boundary between spot and green leaf appears sharp. Necrosis within spots is common but irregular, and may be accompanied by fungus invasion which is secondary. Orange spotting described in Malaysia and Sumatra is virtually identical except that in Malaysia the orange spots tend to be more elongated.[12, 17] This tendency is, however, by no means unknown in Africa.

There has been some confusion as to the identity of this symptom owing to previous nomenclatures both in Nigeria and Malaysia and this confusion needs to be dispersed. Chloroses in the oil palm, when not meticulously examined, were often simply termed 'yellowing'. One of the earliest cases of Confluent Orange Spotting was termed 'Nkwelle Yellows' from the place in Nigeria where it was widespread, but the spotted nature of the chlorosis was early recognized in such Nigerian plantation terms as 'speckled yellows', 'speckled bronzing', etc. Significantly, the term 'bronzing' came to be widely used both in Nigerian and in Malaysian areas of Confluent Orange Spotting because the leaves with the most dense production of coalescing spots take on a bronzed appearance when viewed from a distance. This is particularly pronounced in certain coastal areas in Malaysia. Later, Hale, though he obtained rather unsatisfactory descriptions, referred to 'Speckled Bronzing' in his account of some Nigerian material and the majority of the Nkwelle palms which he analysed were described as being yellow-spotted; more recently Coulter and Rosenquist[18] showed that bronzing, as the term was used in Malaysia, included two distinct conditions, firstly the orange spotting similar to that of Nigeria, and secondly a yellowing which was more pronounced on the upper ranks of leaflets. Chapman and Gray[19] distinguished, but did not describe, bronzing and yellowing associated with low leaf K and Mg respectively. The suggestion that the material analysed by Hale[8] was not in fact exhibiting Confluent Orange Spotting is untenable firstly because the descriptions in his Appendix referring to Nkwelle suggest otherwise, and secondly because the Nkwelle condition is well known to so many workers acquainted with that station. It is necessary that these doubts concerning nomenclature should be cleared up since it is on the work of Hale, of Coulter and Rosenquist and on the results of fertilizer trials in Nigeria that the evidence for Confluent Orange Spotting as a symptom of K-deficiency relies.

Hale[8] analysed upper, middle and lower leaves from healthy, mildly and severely affected palms growing at Nkwelle and on two estates. The Nkwelle material is of most interest as all palms were shown to be K-deficient though some did not show the spotting symptom; those with the symptom were, however, more deficient than those without. Middle leaves gave the following K and Mg percentages of dry matter and indicate the extent to which K-deficiency had proceeded.

| | Per cent of dry matter | |
	K	Mg
Healthy	0.37	0.49
Mildly affected	0.17	0.69
Severely affected	0.16	0.63

The outstandingly large responses to K manures at Nkwelle have already been referred to. Equally distinct effects of K applications on Confluent Orange Spotting symptoms were later seen at Umudike and Mbawsi[20] in eastern Nigeria. At the latter place palms exhibiting the symptoms were shown to contain considerably less potassium in leaves 9, 17, 25 and 33 than a palm growing in the same field, but showing no symptoms (Table 11.4). At the same time Tinker found a relationship between Confluent Orange Spotting and the mole fraction of the exchangeable potassium in the topsoil.[20]

The fields of palms in Africa in which these Orange Spot palms appeared were of mixed genetical origin and the symptom, though widespread, was not found on every individual. In recent years it has been shown that individual palms and progenies exhibiting a high degree of Confluent Orange Spotting may have a high leaf potassium content. Six Orange-spotted palms found in a field of generally healthy palms in Nigeria gave a leaf-K percentage of 0.94 against 0.87 per cent for adjacent healthy palms; one selfed progeny showing a high incidence of the condition had a K percentage of 1.05 while adjacent healthy progenies, whether selfed or crossed, gave similar K contents. The palms were not in the areas of severe K deficiency but progenies showing symptoms gave a significantly lower yield than healthy progenies.[21, 22] Forde and Leyritz[22] made a detailed study of these palms and divided their symptoms into three types. They showed that these symptoms could not be attributed to K-deficiency and appeared to be genetic in origin. On the basis of this study it was affirmed that Confluent Orange Spotting is not a deficiency symptom of the major elements; however, in view of all the evidence from elsewhere, it is not possible to accept such an unconditional statement.

In Dahomey, also, individual palms with Confluent Orange Spotting were found which could not be associated with a low leaf-K content.[23]

In Malaysia both leaf-K differences and progeny differences have been studied in a field of palms showing Confluent Orange Spotting.[18] In this study the youngest fully-opened leaf, which was counted as leaf 3 not leaf 1, and leaf 17 on this scale were taken. A very significant correlation was found between the orange-spotting scores and the potassium in the ash and in dry matter of leaf 3, but no correlation was found in the case of leaf 17. A high calcium content was associated with Orange Spotting, and this condition was, as might be expected, associated with a decrease in yield. Marked differences in Orange Spotting incidence were found between progenies. In leaf 3, the less affected progenies had a significantly higher potassium content in the ash than the most affected progenies, but this difference did not appear in leaf 17. Work on nutrient 'gradients', through the analysis of ash from seven leaves of different progenies, was based on single palms and was inconclusive; but there were indications that some susceptible progenies might have a higher percentage of K in the ash of affected than of unaffected palms.

Although common in the highly K-deficient areas of Nigeria, Confluent Orange Spotting is not much in evidence in Zaire and the isolated palms exhibiting the symptom may well be genetically prone to it.

Various attempts have been made to find an association between Confluent Orange Spotting and an excess or deficiency of other elements or to explain its connection with K-deficiency under special conditions of nutrient balance. No definite and consistent trends have come from this work.

It may be affirmed that where Confluent Orange Spotting is found as a widespread condition in a population of mixed genetic origin it is symptomatic of acute potassium deficiency. However, certain individual palms or progenies may be prone to this condition when potassium need is slight, or even absent, and correlations between the symptom severity and potassium leaf-content will not then be obtained. It seems probable that this symptom was the first one to be recognized because it happened to appear in areas where K deficiency was severe, and it has been suggested that it may be regarded as a symptom of premature senescence which in Nigeria and some other countries will often be caused by inadequate potassium supplies.

Mid-Crown Yellowing.   This symptom was first described by Chapas and Bull from highly K-deficient areas in eastern Nigeria,[15] but has since been recognized in old palms in less deficient fields. Leaves around the tenth position on the phyllotaxis became pale in colour and terminal and marginal necrosis follows. A band along the midrib usually remains green. There is a tendency for later-formed leaves to be shorter and the palm has an unthrifty appearance with much premature withering, sometimes referred to as 'grey withering', of older leaves. Leaf analysis of seventy-one palms in a field near Benin where this symptom was prevalent lends support to the belief that this is a primary potassium deficiency symptom, but there is a suggestion that nitrogen deficiency may be a contributing factor (Table 11.3).[24] The different effects of reduced leaf K on the Ca and Mg contents is of particular interest and will be referred to later.

Table 11.3 *Leaf analysis of the seventeenth leaf of palms in a field showing Mid-Crown Yellowing: Benin, Nigeria (Leyritz)*

Mid-crown yellowing:	Percentage of dry matter				
	N (%)	P (%)	K (%)	Ca (%)	Mg (%)
Absent	2.55	0.17	0.63	0.85	0.40
Slight to medium severity	2.36	0.17	0.39	0.87	0.54
Medium to severe	2.17	0.17	0.30	0.90	0.64
	Percentage of total cations				
Absent			33	45	22
Slight to medium			22	48	30
Medium to severe			16	49	35

No very exact descriptions of Zaire symptoms of potassium deficiency in adult palms have been published, but the reports of a general pale chlorosis in older seedlings and marginal chlorosis and desiccation in 7-year-old K-deficient palms are suggestive of a similar condition to Mid-Crown Yellowing.[13] Symptoms described from Dahomey as *décoloration diffuse* also agree with those of Mid-Crown Yellowing and it has been shown that this symptom does not usually appear until the K content of leaf 17 has fallen to about 0.3 per cent of dry matter.[23] The 'shading effect' characteristic of magnesium deficiency is also found in this symptom, i.e. where one leaflet is covered by another the covered portion does not show chlorosis.

Mbawsi Symptom. This distinctive symptom takes its name from a highly K-deficient plot in eastern Nigeria. Large yellow or orange patches appear on affected leaflets. The midrib and a narrow strip on either side of the midrib remain green. The orange patches usually contain a mass of minute orange spots showing little or no necrosis.

Table 11.4 *Leaf analysis of single palms showing Orange Spotting and Mbawsi symptoms: Mbawsi, Nigeria (Bull)* (percentage of dry weight)

Nutrient	N			P			K			Mg			Ca		
Condition of palm	H	COS	MS	H	COS	MS	H	COS	MS	H	COS	MS	H	COS	MS
Leaf No.	(%)	(%)	(%)	(%)	(%)	(%)	(%)	(%)	(%)	(%)	(%)	(%)	(%)	(%)	(%)
1	2.3	2.5	2.5	0.19	0.20	0.18	0.9	0.9	0.7	0.2	0.4	0.2	0.5	0.3	0.4
9	2.2	2.4	1.9	0.15	0.18	0.14	0.5	0.3	0.2	0.6	0.6	0.3	0.7	0.8	0.6
17	2.2	2.1	1.8	0.14	0.16	0.13	0.7	0.4	0.4	0.2	0.9	0.4	0.7	1.0	0.6
25	2.1	2.1	2.0	0.14	0.17	0.14	0.6	0.4	0.3	0.2	0.7	0.6	0.9	1.0	0.9
33	1.9	2.0	1.7	0.13	0.16	0.12	0.5	0.2	0.2	0.1	0.4	0.2	0.9	0.9	0.9

H = Healthy in appearance. COS = Confluent Orange Spotting. MS = Mbawsi symptom with some Confluent Orange Spotting.

Table 11.4 shows the analyses of leaves of a palm without symptoms in a K-deficient area at Mbawsi together with analyses of palms with Confluent Orange Spotting only and the Mbawsi Symptom with some Orange Spotting.[25] The latter two palms also suffered some tip and edge necrosis from the seventeenth leaf.

It will be noticed that the only striking differences are the markedly lower K content of both the 'COS' and 'MS' palms in comparison with the symptomless palm which itself has a rather low K percentage. Magnesium tends to be higher in the affected palms, particularly in the 'COS' palms, and, though the figures have been reduced to one place of decimals for simplicity, it may be said that the healthy palms themselves are bordering on Mg-deficiency and K-deficiency.

The manganese contents of the affected palms were considerably higher than those of the healthy palms. This phenomenon has been noted in another analysis but has not been consistently found with these K-deficiency symptoms.

Magnesium deficiency

In small seedlings in sand culture the older leaves first develop an ochre colour changing later to pale to bright yellow, the deep orange tints of magnesium deficiency in adult palms being absent. Older seedlings develop a similar chlorosis on the distal leaflets of the older leaves and the appearance of magnesium deficiency in a nursery, often induced by unbalanced manuring, is quite characteristic. The symptoms are always most strong on the older leaves which, in a nursery, are still entire or bifurcate. Here the chlorosis is most pronounced in the central part of the leaflet, the paling not having proceeded right to the edge. However, with increasing severity, the chlorosis tends to cover more of the leaf and to be found on the younger leaves also. Eventually, necrosis sets in at the tip of the older leaves and is reddish or chocolate brown in colour. On close inspection it can be seen that the apparently clear chlorotic parts of the leaflets contain a mass of small orange spots (Plate VIIA).

In adult palms and on large seedlings in the field severe magnesium deficiency symptoms are most striking and have been named Orange Frond. (Plate VIIB and C, between pp. 78 and 79). While the lower leaves will be dead, those above them show a graduation of colouring from bright orange on the lower leaves to a faint yellow on leaves of a young or intermediate age. The youngest leaves show no discoloration. In young palms in the field there is often a much smaller proportion of bright orange leaves. The chlorosis proceeds as follows: discoloration begins about 4 to 5 inches from the leaflet tip with an ochre-coloured strip which extends first between and then across the veins until the whole leaflet, except small areas at the tip and the base, becomes first yellow then deep orange. Typically the chlorosis spreads right across the midrib and there is no green band. Later the leaflets are infected by fungi and umber and purplish-coloured areas appear at the tip and extend down the edges. This necrosis is quite distinct from the brown-grey withering of orange-spotted leaves. Most typical of magnesium deficiency symptoms is the strong shading effect of one leaflet lying over another; the shaded portion of the lower leaflet, particularly when in proximity to the upper one, will be found to be dark green.

Magnesium deficiency symptoms as described above have been seen in all three continents where the oil palm is grown. In Nigeria it has been shown, by isolation and pathogenicity tests, that the fungus *Pestalotiopsis gracilis* occurs regularly in the purplish-brown lesions just described and that the organism can be actively parasitic on moribund Orange Frond leaflets.[26] The ultimate stage of this deficiency disease is thus hastened.

That Orange Frond in adult palms was caused by magnesium deficiency was first conclusively shown in 1952 by Bull who obtained rapid responses to magnesium sulphate and magnesium chloride by leaf-tip injection and by spraying. This classic work has been very fully reported[26] and included tissue extraction with Morgan's reagent which showed that the leaf magnesium extracted from Orange Frond palms was around half that of healthy palms. At an earlier date, Thompson[14] had shown in Malaysia that the seedling symptoms of Orange Frond, not yet recognized as such in the field, could be induced in water culture, and Chapman and Gray[19] had reported yellowing of older leaves in palms in a fertilizer experiment where the leaf levels of magnesium were low and potassium had been applied. Subsequent leaf analysis work in Africa and in Malaysia has confirmed beyond doubt that this striking condition is due to magnesium deficiency alone. In the latter country the second type of 'bronzing' described by Coulter and Rosenquist[18] appeared similar to Orange Frond; the full orange colour of heavily affected palms was not seen, but the 'yellowing' score was negatively correlated with the Mg contents of both leaf ash and dry matter and the single yellowed palm taken from all progenies used had a lower Mg content than healthy palms. Further, it was suggested that a species of *Pestalotiopsis* was responsible for the tip and edge necrosis following the yellowing symptom. It is of interest to note that in the Malaysian work leaf magnesium content was correlated with Orange Frond symptoms in leaf 17 but not in the youngest fully opened leaf (counted as leaf 3), whereas potassium content was correlated with Orange Spotting in the youngest leaf but not in leaf 17.

Calcium deficiency

This has only been seen in small seedlings raised in sand culture where rigorous conditions of sand and water purification have been adopted.[10] It has been evident in all pot culture work that oil palms can develop normally even when only minute quantities of calcium are available. The symptoms are therefore of somewhat academic interest: abnormally short and narrow leaves with prominent veins are first produced. Older bifurcate leaves show poor development of one-half, with apical splitting and necrosis. Later leaves are progressively smaller, terminal necrosis increases, and eventually after a progression of malformed blades the youngest leaves merely comprise the basal part of the petiole. There is no chlorosis.

Sulphur deficiency

There is only one record of apparent sulphur deficiency symptoms appearing in a field trial; this was in a missing-element trial in Zaire and, as it has been stated that similar symptoms were induced by excessive applications of borax[13] and the latter was included in all except −B treatments, it is possible that the symptoms were not entirely due to sulphur.

In pot culture with small seedlings the symptoms in the early stages are not unlike those of nitrogen deficiency. Leaves are small and pale green or almost white in colour. Some interveinal streaking occurs. Later, brown necrotic spots appear on the older leaves followed by terminal necrosis. In older seedlings the interveinal chlorosis which follows the general paling is mottled or spotted. These chlorotic areas of spots become orange and brown and then coalesce and become necrotic.

Chlorine deficiency

Although chlorine has been considered to be an essential and important element in oil palm nutrition, no deficiency symptoms have yet been reported.

Minor element deficiencies

Boron

Boron deficiency is perhaps the most elusive of the deficiencies of the oil palm. Seedling deficiency symptoms have been determined, field symptoms have been described and experiments in areas of supposed deficiency have been conducted, but much of the latter work has lacked the final conclusiveness which has been so widely sought.

Ferwerda[27] was the first to report the effect of *absence* of boron in a fertilizer mixture on a leaf and bud condition. In a missing-element trial in the Kasai region of southern Zaire he found that in the minus-boron treatment thirty-four cases of 'Little Leaf' Disease, as described in a previous account by Kovachich,[28] developed, while a maximum of four cases were recorded in the other twelve missing-element treatments and the control. The absence of the Little Leaf condition in the control was ascribed to an overriding deficiency of elements other than boron which limited growth to the extent that boron deficiency did not occur. At that time the several conditions confused in the term Little Leaf disease were not understood and the significance of this trial cannot therefore be determined; Duff,[29] referring to this and subsequent experiments[13] mentioned by these authors, commented that

> it is unfortunate that there is no data on the incidence of Bud Rot from these experiments because the count was masked in the term 'Little Leaf' which in their papers obviously covered not only Bud Rot/Little Leaf but also other foliar symptoms.

Bull and Robertson[30] postulated a Spear (or Bud) Rot-Little Leaf of unknown cause and a Hook Leaf-Little Leaf possibly due to boron deficiency and, somewhat prophetically, added:

> If the above theories are correct it may prove possible to distinguish between boron deficiency and Spear Rot-Little Leaf by dissection of the bud tissues. Where Little Leaf is caused by boron deficiency, leaves will be abnormal prior to emergence and there will be no necrosis of bud tissue in the earliest stages of the disorder. In palms affected by Spear Rot, unemerged leaves will be normal prior to Spear Rot, and the little-leaf condition will be the result of effects developing in unemerged leaves subsequent to the rotting of the spear.

Robertson[31] subsequently proved the latter point by simple surgery and Duff showed that Bud Rot-Little Leaf was caused by a pathogen. Broeshart *et al.*,[13] while claiming that experience showed that a cure of the Bud Rot-Little Leaf condition had been effected by boron, presented evidence from a further missing-element trial which only suggested a cure of the Hook Leaf-Little Leaf symptom.

Symptoms produced in Nigeria with small seedlings in pot culture under rigorous conditions of sand and water purification were as follows:[11] newly emerging bifurcated leaves were much smaller than normal, having a shortened petiole and a truncated leaf blade. Affected leaves were less than one-tenth the size of leaves of the same age in control pots and the continued production of these small leaves produced congestion in the centre of the seedling. Hooking of the apex or corregations of the blade were not seen. However, in similar trials in Malaysia Rajaratnam reported some cases of hooks and flaps on the leaf margins and puckering of the laminae.[32] Some chlorosis of the sixth (bifurcate) leaf was also reported with characteristic wide angles between the two lamina sections. Later leaves, besides being very small, were erect, compact and had necrotic tips.

With 8-month-old healthy seedlings transplanted into sand culture minus boron chlorotic streaking was seen after the sixth leaf to be produced; this was followed by the production of small leaves with no reduction in the number of leaflets but a marked reduction in their size.[32] After fifteen of these leaves both 'incipient little leaves' and, subsequently, 'fish-bone' leaves were found. In the former, the leaflets were much bunched on the shortened rachis and the number of leaflets reduced. The 'fish-bone' leaves were abnormally stiff with leaflets reduced to projections. This symptom was in turn followed by the production of shorter stump leaves and eventual rotting. It is clear therefore that extreme boron deficiency can be followed by a spear or bud rot, but this must not be confused with the spear rot which occurs *before* little leaf production in the palms investigated by Robertson (see p. 643) or described by Duff.

Rajaratnam has also reported other symptoms in the recovery of boron-deficient seedlings after application of the nutrient. These have been described as 'fish-tail' leaf in which the terminal leaflets are fused and basal leaflets absent, and 'hook-leaf', a symptom often reported on oil palms of all ages in which the end of the leaflet is bent over in the form of a hook[33] (Plate 53).

In adult palms the symptoms of boron deficiency have been difficult to substantiate. Ferwerda[27] found 93 per cent of the palms in plots receiving all elements except boron to be showing symptoms of Hook Leaf and a narrowing of the distal leaflets with some leaflets corrugated, reduced in size or even absent. The stunted rachis of an abnormal leaf was often calloused but rotting was limited to some leaflets of the younger spear leaf. There was no chlorosis.

The Fish-bone Symptom has been frequently seen in Malaysia in areas, particularly young plantations, with low boron leaf content. Following the work of Rajaratnam it is reasonable to conclude that the main symptoms of boron deficiency in the field are a reduction of leaf area in certain leaves producing either incipient 'Little Leaf' (defined in this case as leaves smaller than normal with leaflets only half normal width and three-quarters normal length, without reduction in number of leaflets), advanced Little Leaf, with extreme reduction of leaf area and bunching and reduction in number of leaflets, and Fish-bone Leaf (Plate 54). It is important however not to confuse in the field the little leaves produced in this syndrome with those produced following a spear rot (p. 643).

These general conclusions have received some support from an experiment in Colombia[34] where symptoms including reduction in leaf size,

Pl. 53   The 'Hook-Leaf' condition, which is of **widespread** occurrence, seen in Ecuador.

malformations of the leaflets and the temporary halting of leaf production were arrested by boron applications. The leaf malformations included Hook Leaf and a more extreme form of Hook Leaf ('en baionette'); but it was pointed out that these symptoms can have a number of other causes, e.g. faulty application of fungicides or insecticides, or may be connected with seasonal or genetic factors. Hook Leaf is so commonly though sporadically seen that it is undoubtedly an error to attribute it to boron deficiency without other supporting evidence. Chlorosis, if it is a symptom at all, appears to be present only at certain stages of symptom development[32] and there is no evidence that White Stripe (p. 621) is a primary symptom of boron deficiency as has been claimed.[35]

Iron

Cases of iron deficiency in the field have not been detected. In pot sand culture[11] lack of iron is shown by a very marked chlorosis from pale green to pale yellow, spreading uniformly over the whole surface of the seedling's leaves. Necrosis then sets in at the tips, the leaves become yellowish-brown, growth is arrested and the plants die.

Manganese

Seedlings grown in sand culture without manganese assume a flattened appearance and the leaves become dull and pale yellowish-green in colour.[11] Later, a paler longitudinal chlorotic striping appears between the

Pl. 54 Palm showing shortened leaves devoid of leaflets, probably through deficiency of boron (Malaysia).

veins. These stripes necrose and become greyish-red. Apical splitting occurs and necrotic areas expand.

Copper, zinc and molybdenum

Symptoms of molybdenum, copper and zinc deficiencies have not been obtained with seedlings in pot culture in spite of several attempts, and special techniques have not yet proved successful. The determination of symptoms of these deficiencies would be of interest since responses to copper and zinc in fertilizer experiments have been claimed.

The condition known as Peat Yellows in Malaysia has been associated with low leaf levels of both potassium and copper[36] (Plate XVII). In a study in Malaysia this condition was divided into Mid-Crown chlorosis, Peat Yellows and a combination of the two.[37] In this case the mid-crown chlorosis was the symptom associated with very low levels of leaf copper. This symptom (not to be confused with mid-crown yellowing of K deficiency) is characterized by a chlorosis on the tips of the leaflets of the younger leaves and is followed by necrosis or desiccation from the tip downwards. The chlorosis starts with the appearance of small yellowish green specks along the leaflets and these gradually coalesce. Very young palms can be affected.

Multiple deficiencies

It was at one time thought that deficiency of two or more elements simultaneously would lead to distinctive symptoms. Certain symptoms found in the field where both K and Mg were in short supply were thought to be a manifestation of this kind. It has been shown in pot culture, however, that where two elements are lacking the deficiency symptoms of one of them is usually dominant.[38] Thus where nitrogen is deficient in combination with a deficiency of any other major element, the foliar symptoms are identical with those of N deficiency. Phosphorus deficiency symptoms are dominant when the deficiency is in combination with that of K, Ca or S. Magnesium deficiency symptoms dominate when in combination with lack of P and K, but very severe necrosis occurs in the latter case. Exceptionally, calcium produces its deficiency symptoms − short, narrow leaves with prominent veins − in combination with those of K and Mg and, to a lesser degree, of P.

This work indicates that the appearance of a known distinctive deficiency symptom, e.g. that of magnesium or nitrogen, does not necessarily indicate a lack of that element alone, and this fact sets a first limit to the value of visual diagnosis.

Toxicity

Under plantation conditions oil palms have not shown symptoms of any general toxicity due to one element. Fertilizers can be applied in toxic quantities if care is not taken. Sulphate of ammonia applied within 6 weeks of transplanting may damage seedlings seriously. The effect is usually seen on the youngest fully-opened leaf which blackens off, but other leaves become dried up and the palm may die. It is clear that the initial effect in these cases has been on the root system, since similar symptoms follow the transplanting in dry weather of large seedlings with damaged root systems.

Cases of toxic effects following applications of boron to the soil have been known. Growth was severely checked and flowering inhibited. Toxicity symptoms have been induced in 18-month-old seedlings growing in polythene bags;[39] they appeared when the boron level exceeded 250 ppm of dry weight in leaf 3. These symptoms were a terminal chlorosis progressing along the leaflet tips and margins, necrotic spotting on the chlorotic tissue and a drying out and shattering of the necrosed tissue.

The oil palm leaf is extremely sensitive to copper and the common copper fungicides cannot be used. Copper toxicity through the soil has not been encountered, however.

Leaf analysis

'Over the past century the pendulum of analytical diagnosis has swung from soil analysis to plant analysis', writes Smith,[40] though he adds that plant analysis has not entirely replaced soil analysis and that neither procedure, nor the two in combination, can be 'the complete answer' as is sometimes implied.

Leaf analysis for diagnostic purposes has been particularly useful for perennial crops which are relatively slow growing, can provide easily-defined and standard leaf material for analysis, and whose responses to fertilizers will in any case take a longer time than will those of annual crops. One of the dangers of relying exclusively on leaf analysis is that it may fail to show in time the soil depletion that is taking place through poor cultural practices or other causes, and the detection of soil changes, with a crop such as the oil palm, may be an important factor in assuring satisfactory growth and yield in future crop cycles.

Leaf analysis has come to the fore in the last 30 years largely through the work of Lundegardh, and the oil palm has been among the handful of perennial crops on which most of the important work has been undertaken. Following the work of Hale,[8] already mentioned, and of Chapman and Gray[19] in Malaysia, great progress was made through the investigations of Prevot and his co-workers using material of the Ivory Coast, Dahomey and elsewhere.[41-43]

It is important that the present usefulness and limitations of leaf analysis in the oil palm should be thoroughly understood. In much of the early work, attention was necessarily given to the leaf contents of the separate nutrients. It has been found with other crops that the levels of leaf nutrient content can be zoned into a succession of ranges according to their effects. This can be seen in Fig. 11.2 where a growth or yield curve is constructed to show the changes of concentration of a mineral element through a gradual increase in the supply. These zones are as follows:

Zone A.   Decrease in concentration with increased growth through initial increase in supply (the Steenbjerg effect).

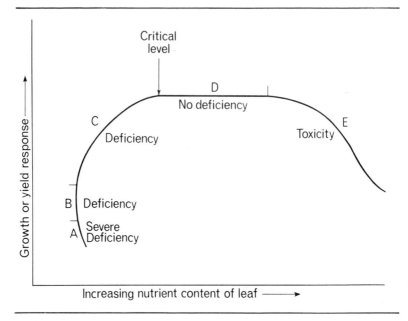

Fig. 11.2   Diagram of deficiency and toxicity in relation to leaf nutrient content and growth and yield (adapted from Drosdoff, Prevot and Ollagnier, and Smith). According to Foster and Chang the luxury uptake represented by D does not usually occur with the oil palm.

Zone B.   Growth response with *no change* in mineral concentration.

Zone C.   Normal growth response with *increasing* mineral concentration reaching the 'critical level' or zone.

Zone D.   No growth response, but continued *increasing* mineral concentration. The work of Foster and Chang[137] suggests however that this further increase does not usually occur with the oil palm; they found on inland soils in Malaysia that maximum yield and leaf nutrient levels were simultaneously reached.

Zone E.   *Increase* of concentration with toxic effect.

   Chapman and Gray,[19] who were the first to show correlations between both leaf P and leaf K and yield in oil palm fertilizer experiments, did not concern themselves with critical levels, but considered that the K_2O/P_2O_5 ratio in the leaf ash was of primary importance. They showed that where this ratio was low potassium increased yields, but where it was high, phosphorus increased yields. An important further observation was that where the ratio was very low and potassium therefore extremely deficient, additions of phosphorus, in disturbing the balance still further, decreased yields. Moreover potassium applications were found to assist the uptake of nitrogen.

Prevot and Ollagnier, in one of the earliest of a large number of papers,[41] defined the critical level as 'the concentration of the element in the leaf (dry matter basis) above which a yield response from the element in the fertilizer is unlikely to occur' and proposed the following critical levels for N, P and K in leaf 17: N − 2.70 (later revised to 2.50); P − 0.15; K − 1.00 per cent of dry weight. Later, Ca and Mg levels were added, viz. Ca − 0.60, Mg − 0.24 per cent, the basis for Ca levels being only that a total of about 2 per cent for the sum of K + Ca + Mg has been observed. Ratios between elements were held to be especially useful in regions where no data from fertilizer experiments were available; correlations, both positive and negative, between the leaf contents of the elements were shown to exist, and the relative constancy of the total leaf K + Ca + Mg content was noted. Prevot and Ollagnier also converted some of the Malaysian data into percentages of nutrient elements in dry matter and showed that the considerable responses to P and K were associated with leaf nutrient increases in leaf 17 in the following manner:[41]

	Phosphorus		*Potassium*	
	P_0	P_1	K_0	K_1
Yield responses (%)	100	193	100	168
Leaf nutrient content, per cent of dry matter	P 0.128	0.153	K 0.56	1.09

Broeshart,[9] working in Zaire and presenting his results in terms of the elements as a percentage of ash, found a reasonable constancy in the sum of K + Ca + Mg, but his introduction of the concept of an 'optimal' leaf composition for nutrients in oil palm leaves requires comment. The idea of critical levels was already generally accepted and if the curve in Fig. 11.2 as presented by many authors has any validity it must follow that, assuming other nutrients are present in sufficient quantity, the 'optimal' composition of a leaf for any given nutrient is the same as the critical level. The former term was said to be 'used to indicate the chemical composition of leaves of palms having regional maximum growth and/or production'. But, by definition, if each of the nutrients is at its final critical level then a yield response to any of the nutrients will not occur, and the maximum growth and/or production will have been reached.

It is necessary to use the phrase 'final critical level' since, to quote Smith,[40] 'the fact that the critical level of an element may fluctuate as it is modified by some other factor does not seem to be generally understood'. The early definition of Prevot is not indeed satisfactory except in a closely defined set of circumstances, and in their later papers the French workers were at pains to point out that the critical level was useful as a first approximation and should not be used in a mechanical manner.[42] The critical level is that at which the supply of the element is barely, but just, above the point of limiting growth or yield; when the critical level for element A has been reached, some other factor may then become limiting for growth or yield. This factor may be element B or it may be any environmental factor such as climatic conditions, or a cultural factor, e.g. pruning. Prevot and Ollagnier[43] have described this factorial method of foliar diagnosis as being based on a modification of Liebig's Law of the Minimum, and they provided data which showed, for instance, that only when the leaf level of N had reached 2.70 per cent was a positive correlation to be found between K levels and yield, i.e.

N < 2.70%.   K: Yield correlation r = +0.239 Insig.
N > 2.70%.   K: Yield correlation r = +0.643***
conversely, K < 1.1%.   N: Yield correlation r = −0.042 Insig.
K > 1.1%.   N: Yield correlation r = +0.655***

This, then, explains why, after many years of leaf analysis, adjustments to the critical levels are still made and why responses to fertilizers applied to palms with a nutrient content below the supposed critical level are sometimes not found. In particular, little is known of the critical levels of minor elements, and where one of them is deficient, then normal responses to major elements may not occur. What is required is, in each oil palm region and in each soil type within each region, for levels to be related to responses in comprehensive fertilizer experiments.

Factors likely to effect critical levels even within geographical and soil areas are the progenies used and the previous cropping. Significant differences have been found between progenies in West Cameroon and between fruit forms (*dura* and *tenera*) in Malaysia.[44] In the latter case however it was pointed out that the differences might be the result of productivity differences varying the demand on nutrient reserves. The number of bunches produced by the *tenera* palms was 20 per cent higher than by the *dura* palms. It was concluded that unless foliar analysis is made more specific in its application it is bound to be less precise than it has often been considered to be.

In areas where the environmental circumstances encourage high production, critical levels may easily be set too low. Initial experimentation may show a response to one element alone following an obviously low leaf content of that element, and the levels of the other elements may be thought to be at or above the critical level. When the deficiency is satisfied however, either the level of another element may fall below the believed critical level or it may remain stationary or rise, but nevertheless be deficient for the new production level. In the latter case the supposed critical level was clearly set too low for the circumstances.

For these reasons it is important that the normal effect of soil applications of nutrients on the leaf composition should be known. If element A is supplied, then elements B, C, etc., can either increase, decrease or maintain their percentages in the dry matter in the leaves. For some crops the effect of the supply of all major and most minor elements

Table 11.5   *Usual effects of applied elements on the mineral composition of the oil palm leaf*

+ = Increase.   − = Decrease.   0 = No marked effect.

Element added	Element measured in the leaves				
	N	P	K	Ca	Mg
N	+	+	−	0	−
P	0†	+	−	+	0
K	0*	0*	+	−	−
Ca	0	0†	−	+	0
Mg	0	−	−	−	+

* Both increases and decreases have been reported.
† Synergism has been reported.

is known, but for the oil palm the effect of major elements only has so far been extensively studied and the general results are shown in Table 11.5.

The effects are described as antagonistic when the leaf content of an element is reduced by the application of another element and synergistic when application increases the leaf content of the element concerned. The latter phenomenon is not common in the oil palm though Chapman and Gray reported an increase of leaf N associated with K applications, and increases in P following N applications have also been reported. Antagonisms are common, the most important being between K and Mg. Ng has pointed out that effects are not often easy to interpret because fertilizers often contain another anion or cation which may also have an effect.[45] Thus the presence of calcium in phosphorus fertilizers suppresses potassium, while a rather longer-term effect would be the increased leaching of potassium or magnesium, and hence their reduced uptake, following application of ammonium sulphate or fertilizers in the form of nitrate or chloride salts.

In recent years the relationships between the elements in the nutrient composition of leaves has been examined statistically by computing partial correlation coefficients (calculated for each pair of nutrients, all others being held constant) and by principal component analysis. Using the latter method Holland[46] showed, from the data of a factorial fertilizer experiment, that some 80 per cent of the total variation could be attributed to antagonism between K on the one hand and Ca and Mg on the other, and synergism between N and P. In a foliar uniformity trial in Malaysia Poon *et al.*[47] showed the same relationships through the computation of partial correlation coefficients from the data of a uniformity trial, although additionally there was a significant synergism between P and K. It is thus possible that the latter is masked in fertilizer experiments by the opposing effect of calcium in the phosphorus fertilizers. In general, when significant, partial correlation coefficients of the same sign were found for leaves 3, 9 and 17, but the correlations in the 9th leaf were closer to those of the 17th than to those of the 3rd, though the coefficients of the latter leaf tended to be the greatest.

Macronutrients

Under each set of environmental circumstances the critical levels for leaf nutrients must be related to the results of fertilizer experiments. Deductions made from leaf analysis data not so based may be misleading. French workers estimated their critical levels from analyses of palms in fertilizer experiments and in areas of obvious deficiency or health, by indirect methods such as the comparison of effects, under different conditions, of the supply of one element on the level of another, and by studies of antagonisms and synergisms.[48] Special attention was given to potassium deficiency and the relation of K levels to the supply of phosphorus and nitrogen.

It was shown that where the leaf-P content in leaf 17 was less than 0.15 per cent of dry matter there was no correlation between K-content and yield, but that when phosphorus deficiency was rectified and the leaf-P content brought above 0.15 per cent, a strong correlation existed and the highest yields were obtained in the leaf-K range 0.9 to 1.1 per cent (Fig. 11.3). Similarly, where K was below 1.0 per cent there was no leaf-P/yield

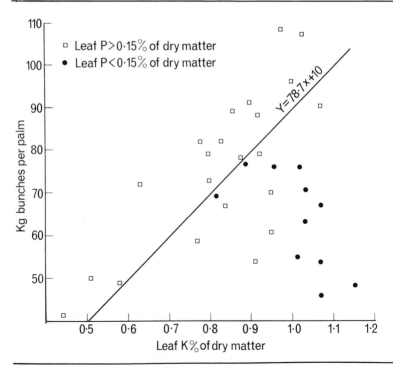

Fig. 11.3   The relation between yield and leaf-K levels with leaf-P content above and below 0.15 per cent of dry matter. Correlations; yield/leaf K with P < 0.15 per cent, insignificant; yield/leaf K with P > 0.15 per cent, $r = 0.739^{***}$ (Ochs, 1965).

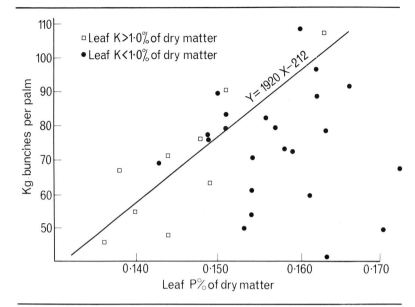

Fig. 11.4 The relation between yield and leaf-P levels with leaf-K content above and below 1.0 per cent of dry matter. Correlations: yield/leaf P with K < 1.0 per cent, insignificant; yield/leaf P with K > 1.0 per cent, $r = 0.832***$ (Ochs, 1965).

correlation but where leaf-K was above 1.0 per cent there was a strong correlation between leaf P and yield[23] (Fig. 11.4). The N–K relationship has already been mentioned on p. 518.

In early work it was estimated that the leaf-N critical level would be about 2.7 per cent of dry matter (p. 517). This was based on the observation that leaf-N fell from 3.0 to 2.8 per cent with potassium fertilization in the Ivory Coast while it rose from 2.6 to 2.7 per cent in Chapman and Gray's Malaysian experiments. Subsequent experience in Africa suggested that a level of 2.5 per cent might be sufficiently high.

The magnesium level was estimated from deficiency areas and experimental results, while a figure of 0.6 per cent was set for leaf calcium. K levels tend to become inflated under conditions of Mg deficiency. In an experiment in Nigeria where K dressings had previously been applied, leaf-17 K was found to rise as high as 1.6 per cent of dry matter with leaf Mg falling as low as 0.07 per cent in control plots.[49] In some experiments in Nigeria where K deficiency had been corrected, yield responses to magnesium were being obtained with leaf-Mg levels as high as 0.28 per cent, suggesting that under certain circumstances critical levels of Mg may move up to 0.35 per cent.

Some responses to nitrogen have been obtained in Nigeria with leaf-N levels of around 2.4 per cent of dry matter and some phosphorus responses with leaf-P around 0.147 per cent. Fertilized plots normally show N percentages of over 2.5 and P percentages of over 0.15. Potassium responses are common, but in old K-deficient palms it is rarely possible to bring the leaf-K content up to 1.0 per cent of dry matter even when very substantial doses of K are applied and large increases in yields obtained;[50] when younger palms respond to K dressings, however, leaf-K often rises to 1.20 per cent. It is therefore likely that the critical level is lower in old bearing palms than in young bearing palms.

In Malaysia, Coulter[51] attempted to set critical levels for coastal clay and 'upland' soils by analysing leaves 1 and 17 in areas of satisfactory production and, in some cases, where no responses had been obtained to fertilizers. The results he obtained did not diverge greatly from the standards of the French workers already mentioned. However, other work in Malaysia made leaf analysis data difficult to interpret and accentuated the need for fertilizer experiments and for the realization of the fact that there may be critical zones rather than exact levels. On inland soils, increases of yield due to N dressings were accompanied by increases in leaf N to 2.7 per cent of dry matter and beyond, while increases of yield due to K dressings followed leaf-K increases up to around 1.25 per cent of dry matter.[52] P levels have been in the range of 0.16 to 0.17 per cent and have tended to be slightly increased by nitrogen applications. The effect of applications of a deficient element on the need for another element was seen in the case of nitrogen and magnesium antagonism. High N dressings leading to leaf-N contents of over 2.8 per cent depressed leaf-Mg to below 0.24 per cent and increased Orange Frond symptoms. On coastal clay soils high-yielding experimental areas, showing little or no response to fertilizers, had high leaf-N contents up to 2.8 per cent, very high P contents up to 0.19 per cent, comparatively low K contents of 0.9 to 1.1 per cent and high Mg contents of 0.3 to 0.4 per cent.[53]

A comprehensive study of leaf levels in relation to yield has been undertaken by Foster and Chang[137] working on the data of ten coastal and ten inland-soil fertilizer experiments in Malaysia. They found that the leaf levels for optimum yield of any specific nutrient differed with the presence or absence of fertilizers providing other elements. For instance, a bunch yield of 24 tons per hectare was obtained in one experiment in the following circumstances:

At zero levels of P and Mg: Leaf N 2.82–2.92% Leaf K 1.07–1.36%

At optimum levels of P and Mg: Leaf N 2.72–2.89% Leaf K 0.83–1.07%

Thus yields can only be related to the leaf level of a specified nutrient if other nutrients are kept constant, and any critical levels proposed must at least assume that the optimum requirement for other elements has been met. Using response functions fitted to bunch yield and leaf nutrient plot data, Foster and Chang found that predicted leaf nutrient levels at optimum fertilizer rates for yield were not the same for inland and for coastal soils. Table 11.6 shows the mean of these predicted levels and the

Table 11.6 *Means of predicted nutrient levels (per cent of dry matter) at optimum rates of all fertilizers, and the range of critical levels for two groups of soils in Malaysia (from Foster and Chang) Leaf 17*

Leaf nutrient	N	P	K	Mg	Ca
Coastal soil trials (10)	2.59	0.169	0.94	0.29	0.45
Inland soil trials (10)	2.87	0.177	1.11	0.26	0.74
Range of critical levels:					
Coastal soils	2.55−2.65	0.165−0.175	0.85−1.05	−	−
Inland soils	2.85−2.95	0.175−0.185	1.10−1.15	−	−

critical ranges they deduced from this work. The question of why there should be such markedly differing levels for the two groups of soils has not been fully resolved. Ng[90] suggested that this was because of differences of moisture status since coastal soils normally have a more regular water supply and this might result in more efficient utilization of nutrients which would require lower levels. Foster and Chang, however, have pointed to the marked difference of calcium status and suggest that as Ca is lower in coastal soils, requirements for K and Mg will also be less, since a balance between the nutrients is likely to be more important than the actual levels. However, the differences of leaf N critical level remains unexplained.

The claim that chlorine is an essential element in the nutrition of the oil palm has so far been based largely on the interpretation of leaf analysis data.[54, 55] In Colombia applications of KCl, while giving small increases in bunch yield, actually decreased leaf-K and increased leaf-Cl from 0.18 to 0.54 per cent of dry matter. Following a world survey of leaf chlorine levels it has been suggested that 0.1 to 0.3 per cent must be considered the acute deficiency range, with 0.5 to 0.6 per cent being regarded as satisfactory.[54] These findings need substantiation through direct comparisons of the effect of KCl and K_2SO_4 on yield. Leaf N−Cl antagonism has also been suggested, with positive N effects being attributed to suppression of leaf-Cl (see p. 579).

Claims for sulphur needs have also been largely based on leaf analyses. In an experiment in the Ivory Coast with young palms the length of leaf 4 was found to be correlated with the S level in the leaves though not with the Mg level in spite of the fact that both the S levels and Mg levels were raised by kieserite application.[55, 56] On the basis of this finding a critical range of 0.20 to 0.23 per cent of dry matter in leaf 17 has been proposed, but, as in the case of chlorine, the effect of sulphur levels on yield needs substantiation through direct comparisons of magnesium application in the sulphate and non-sulphate forms. Rather similar results have been obtained in comparisons of application of ammonium sulphate and urea. Both fertilizers increased the leaf-N percentage but the increase in leaf sulphur levels with the ammonium sulphate application was accompanied by an improved growth of the young palms and reduced *Cercospora* damage.[145]

Micronutrients

In the Ivory Coast no deficiency sumptoms were detected with leaf boron levels of 6 ppm of dry matter or less; in Colombia however, with more rapid growth, Little Leaf symptoms were common in young palms at this level and were reduced by soil application in rings of 30 g borax per palm. Application in the axils of three leaves above the zone of developing bunches was shown to have larger immediate effects on leaf-B, though this effect was short lived.[34] In Sumatra severe symptoms have been reported with leaf-B levels as low as 4—5 ppm where it was pointed out however that such levels could also be encountered on the palms not showing foliar symptoms.[57] In most areas leaf-B levels of 10—20 ppm are encountered.

In a study of the mobility of boron within the palm Rajaratnam[58] has shown that it is relatively immobile and does not move from leaves to the shoot. Boron tends to move to the extremity of the leaflet and is not re-translocated. Some boron is lost by guttation. This work indicates that soil application is likely to be the best method since axillary application will not allow direct translocation to the shoot or younger leaves.

Manganese leaf-content levels vary greatly and experiments have shown both positive and negative responses to this element. Concentrations in leaf dry matter vary from below 100 to around 1,000 ppm.

Molybdenum levels are low, around 1 ppm or less, but can be raised to over 5 ppm by soil applications. When molybdenum levels are very low it is possible that there may be an Mo requirement for the effective use of nitrogen fertilizers.

There have been occasional indications of copper and zinc requirements, but leaf-contents are low and do not vary very greatly. Leaf-copper is usually about 6—15 ppm of dry matter, leaf zinc about 9—39 ppm.[59] Leaf contents of iron are 60—110 ppm of dry matter.[59]

Some indications of the leaf levels of copper required for satisfactory growth and for the elimination of deficiency symptoms (mid-crown chlorosis) were obtained from experiments on peat in Malaysia.[60] With mature palms soil applications of copper sulphate raised leaf 17 levels from 1.4—2.1 ppm to 2—5 ppm and increased leaf length; with young palms foliar sprays of copper sulphate in 200 ppm Cu solution improved the colour of the leaves, increased their length in comparison with untreated palms and gave leaf-3 copper levels of 3.7 ppm and 2.3 ppm at 3 and 6 months after treatment against levels of 2.8 ppm and 1.8 ppm for un-treated palms. It is not possible to gauge a critical level for copper from these results because there is a gap between the normal levels of 6—15 ppm prevailing on non-peat soils and the highest level to which Cu content has been raised by Cu application on peat soils. Ng[60] recommended the use of leaf 3 for the analysis of copper as well as for boron, manganese and iron analysis and suggested that a fall to 2 ppm in this leaf will markedly reduce growth.

Leaf-nutrient content variation

The above account has been based largely on leaf 17 of adult palms and something should now be said about the reason for this and about variations of leaf-nutrient content with leaf number, age, season and other factors.

Early workers made a deliberate choice of either the first fully-opened leaf (usually designated leaf 1), leaf 9, leaf 17 or a leaf making an angle of $45°$ with the horizontal. The choice of leaf 17 was based on convenience and reasoning but not on any exhaustive sampling trials. It was argued that this leaf lay conveniently on the easily-recognized spiral in the succession 1, 9, 17, 25, of the phyllotaxis and was in the middle of the foliage, the lower part of the rachis making an angle of about $45°$. This leaf was fully developed, not yet senescent and carried an inflorescence in its axil. In some, usually older, palms however, leaf 17 is hard to identify or is senescent, and in such a case a leaf round about the leaf 17 position, with its lower rachis making an angle of $45°$, was often chosen; or recourse was had to the easily identified leaf 9.

Chapman and Gray[19] held that young leaves were not in general convenient for routine sampling because the composition might be expected to change rapidly with development. In examining the differences of composition of leaves of different ages they used the central 4 inches of the central two pairs of leaflets. With this material, taken from a fertilizer experiment, they found that there was a downward trend in the percentage P_2O_5 and K_2O of leaf ash with age of leaf, but that with N there was a rise to leaf 5 followed by a decrease with age. From the control plot and the plot receiving the highest PK dressings it appeared that malnutrition showed its greatest effect on the ash composition of leaves of medium age; moreover, although there were no increases in yield due to fertilizers in the experiment except from PK in combination, correlations were found between the nutrient contents of the leaf ash and palm yields. These correlations were high and significant for all four leaves chosen (leaves 1, 9, 17 and 25) for N per cent, but for P_2O_5 and K_2O the correlations were high and significant for leaves 17 and 25 but not significant in the younger leaves. It was largely on this work that the use of leaf 17 has been based. However, calculation of leaf nutrient content on a dry matter rather than an ash basis is now accepted practice and, as described below, for certain elements leaf 1 appears to be less variable than leaf 17, while for some micronutrients leaf 3 is preferred (p. 524).

Some early data were presented on leaf composition variations though these were not based on any established estimate of sampling error. No more than brief mention can be made here of the more important and consistent of these findings. Prevot found decreasing concentrations of N, P and K and increasing concentrations of Ca and Mg with increased age of leaf, though N increased slightly in the younger leaves.[61, 62] The variations in Mg were very small. Small significant differences were found between morning and evening sampling. Many workers followed Chapman and Gray in selecting two pairs of leaflets in the centre of the leaf and taking the centre 4 inches of these leaflets for analysis. Gradients of composition within the leaf have been variable however, and no consistent trend has emerged. Hale took leaflets on alternate sides of the rachis at intervals of ten leaflets and used the whole lamina.

The first comprehensive investigations of the errors of leaf sampling were those of Smilde and Chapas[63] and Smilde and Leyritz.[64] Using leaves 1, 17 and 25 of thirty-two 15-year-old palms in Nigeria they showed that there were significant differences between palms, age of leaf and month of sampling in leaf N, P, K, Ca and Mg. In some cases there were significant interactions between palms and months and months and age of leaf, though not between age of leaf and palms. The results showed that different weight must be attached to the results of analyses of different elements when the same number of palms are used for the analysis of each element. Table 11.7 gives estimates from 2 years' data of the minimum number of palms required in a uniform area of palms such as that investigated in Nigeria to make it possible to detect differences exceeding 5, 10 and 20 per cent of the mean nutrient level at the 5 per cent level of significance; to obtain a proper comparison, the formula $n = 2s^2t^2/D$, where s is the coefficient of variation, t is the reading in the t table for $P = 0.05$, and D is the difference (per cent) of the mean nutrient level to be detected, has been applied to the data of both years.

Table 11.7 *Estimation from two sets of data from thirty-two palms of the extent of leaf sampling necessary to detect significant differences exceeding 5, 10 and 20 per cent of the mean nutrient level* (P = 0.05) *(Nigeria)* (number of palms to sample)

Nutrient measured	N		P		K		Mg		Ca	
Difference of the mean nutrient level to be detected:	1960–1	1962–3	1960–1	1962–3	1960–1	1962–3	1960–1	1962–3	1960–1	1962–3
17th leaf sample										
20%	–	1	–	1	–	14	–	9	–	8
10%	4	3	4	3	56	54	40	33	16	30
5%	12	10	18	11	224	212	156	131	60	117
1st leaf sample										
10%	10	–	8	–	18	–	28	–	28	–
5%	40	–	32	–	68	–	108	–	110	–

The following conclusions may be drawn from this work. Firstly, the necessary degree of sampling may be expected to be about the same from one year to another. Secondly, while the seventeenth leaf is the most satisfactory for N and P determinations and has no disadvantage for Ca, the first leaf gives less variation for Mg and, more especially, for K. Most workers will feel that sampling must be standardized on one leaf and in this case it could be argued that the first leaf should be used since a lesser number of leaves overall would be required. On the other hand the seventeenth leaf has such considerable precision for N and P that its use is justified except where specially precise investigations on K, and to a lesser degree Mg, are being carried out. Thirdly, the first leaf has not been used so extensively in yield/leaf nutrient comparisons in fertilizer experiments and it is not known if it reflects the nutrient status of the palm so accurately as the seventeenth leaf.

One of the main conclusions of this work was that in fertilizer trials all palms in the plots must be sampled; moreover, as plots rarely contain more than sixteen palms an accuracy of 10 per cent of mean values will not be achieved for K, Mg and Ca and differences of less than about 20 per cent of the mean values for these elements cannot be regarded as significant, even when all palms in each plot are sampled. For advisory or confirmatory purposes Smilde and Leyritz recommended the taking of composite leaf-17 samples of 10 per cent (30) of the palms in 5-acre homogeneous blocks to give results accurate to within 20 per cent of mean values for K, 10 to 20 per cent for Mg and Ca, and about 5 per cent for N and P.

Another study of leaf nutrient variation was made on an estate in Nigeria where the first fully-opened leaves of 512 palms in an area of about sixteen times that number of palms were sampled and analysed for N, P, K, Ca, Mg, Mn and B.[65] Thus a much larger number of palms and a different leaf (leaf 1) was used than in the sampling of Smilde and Leyritz. The investigation did not cover leaf-age or seasonal factors. The results are compared in Table 11.8.[66] Variation of leaf N and P was low in both studies, while leaf-K variation was lower in the study using leaf 1 and leaf Ca variation much higher. In spite of the fact that the overall results (except for Ca) of the two studies were not markedly different (Table 11.8) the conclusions drawn were not the same, Ward considering that a 1 per cent intensity of random sampling would give sufficiently accurate results for all major nutrients except calcium.

Table 11.8 *Comparison of deductions from the results of three leaf samplings*
Minimum number of palms to sample in order to detect differences of 5 and 10 per cent of the mean nutrient level with a 5 per cent level of significance (confidence interval). (Smilde and Leyritz, Ward and Poon *et al.*)

Locality	Nigeria*		Malaysia†				Nigeria‡	
Leaf	17		17		3		1	
Differences of mean	10%	5%	10%	5%	10%	5%	10%	5%
N	3	10	4	14	5	17	6	24
P	3	11	4	13	5	17	9	35
K	54	212	22	85	10	39	8	31
Mg	33	131	26	104	22	88	24	94
Ca	30	117	34	137	41	161	615	2,462

* From Smilde and Leyritz.
† From Poon.
‡ From Ward.

Some attention has now been given to this subject in Malaysia, and results from a uniformity trial[47] are incorporated in Table 11.8. It is interesting to compare the data for leaf 17 from Nigeria and Malaysia. The minimum number of palms required is similar for all major nutrients except potassium, the variation of which appears to be considerably less in Malaysia. Ward's leaf 1 corresponds to the Malaysian leaf 3 and again the results are not dissimilar except for the anomalous calcium.

Ng and Walters[67] have compared coefficients of variation from three sites on coastal clays in Malaysia and found these to be as follows:

	Coefficients of variation (per cent)				
	N	P	K	Mg	Ca
Site J*	12.9	14.3	14.1	27.1	22.3
Site B	6.6	7.5	16.1	19.5	21.4
Site SD	4.9	7.0	13.5	20.1	24.1

* Estimated from composite fifteen palm samples.

Using these coefficients they calculated the sample sizes needed for the different nutrients to detect differences of 5 and 10 per cent using three different sampling methods: (1) a compact block within the area, (2) random sampling, (3) random groups of nine palms. The lowest number of palms is required under the completely random system (2), but system (3) only increased the number of palms required by some 50 per cent and the number of points (number of palms divided by 9) was very much reduced. This system was considered to combine practical convenience with satisfactory precision.

The optimal size of the area to be sampled clearly depends on the terrain and soil variations but will lie between 4 and 20 hectares. Poon *et al.*[47] recommended the use of 4-hectare sampling units for a start, with the combining of these areas into larger units if the analysis results indicate that this is feasible. Assuming a sampling area of 10 hectares and using Ng and Walter's system (3), the number of palms required for leaf-17 sampling would vary according to locality from 2 to 5 per cent of all palms to achieve for potassium a precision within 10 per cent of the mean. Five random points of nine palms, as recommended by Ng,[45] is 3.1 per cent of 10 hectares. It is important to realize that with sampling of this order differences of less than 10 per cent are unlikely to be significant with K and Mg though they may be significant with N and P. The viewing of analytical results without these statistical considerations in mind may well lead to unjustified conclusions. Poon *et al.*[47] have attempted to refine the interpretations of results by considering what test is being applied (i.e. whether two means are equal or whether the mean is equal to or not less than a given value) and estimating the types of error involved; however, this greater precision is unlikely in practice to alter to any great extent the sampling intensity used.

A number of sampling methods are currently in use on plantations. The system of taking a block within an area is obviously not to be recommended. The sampling of the tenth palm in every tenth row, giving a 1 per cent sample, is quite common. For this system the unit area must be at least 18 hectares, since not less than twenty-five palms per unit should be sampled. For most areas this low percentage sampling is likely to be insufficient however, and the system of Ng[45] described above is more suitable and will be suitable for areas of 10 hectares which Ng regards as generally most satisfactory when practical considerations and those of palm variation are both taken into account.

Season and time of day.   The detailed study in Nigeria indicated, as might be expected, that the beginning of the wet season is a bad time for sampling particularly on account of the nitrogen flush at that time. Leaving this period out of account there is no time of lowest palm-to-palm variability for nitrogen.[63] Differences in the variability of other elements during the year were marginal, though there was some tendency for leaf K to be less variable in leaf 1 in April to September than in other months and less variable in leaf 17 in October to January. Magnesium levels may vary rather widely between months, but no consistent pattern has emerged from year to year. It was concluded on the basis of 2 years' results that sampling for foliar analysis can be carried out in Nigeria at any time of year except during the April to June period of nitrogen flush. It is probable that this would be a satisfactory rule for any seasonal climate.

For the very dry conditions of Dahomey the end of the dry season is considered the best time for sampling; at this time the P, K and Mg levels were found to be at their highest and differences between ages of palms greatest.[61]

Although differences between months can be appreciable in Malaysia[45] no full study of seasonal variations has been made. It is generally recommended that the periods of drought or very heavy rain should be avoided and that 6 months, or an absolute minimum of 3 months, should elapse between fertilizer application and leaf sampling. For each area it is also best to keep to the same sampling period each year if annual analysis is to be undertaken.

Foster and Chang[137] found low coefficients of variation for seasonal fluctuations varying from 3 to 4 per cent for N and P and 6 to 11 per cent for K, Mg and Ca. They also detected an influence of soil moisture in the month before sampling: in coastal areas P and K leaf levels were as much as 5 and 9 per cent higher after a very wet month than after a dry month. This relationship was not however maintained for K on inland soils.

It is sometimes recommended that sampling should only be done between 6.30 a.m. and noon. Smilde and Chapas[63] considered this to have little justification since the small differences found, besides being insignificant, are unlikely to be of a greater order than day-to-day differences, and the sampling programme always covers many days.

Leaf number (age of leaf).   The leaf concentration of N, P and K decreases with the age of the leaf, with leaf-Mg there is no clearcut trend, but leaf-Ca increases with age. Thus the K + Mg + Ca sum is likely to be constant at about 2 per cent unless there are any marked deficiencies. Critical levels under various circumstances have not been worked out for leaves 1, 9 and 25, but the following suggestions have been made for leaves 1 and 9.

	N (%)	P (%)	K (%)	Ca (%)	Mg (%)	
Leaf 1	2.8	0.22	1.70	0.5	0.25	Nigeria
Leaf 9	2.7	0.19	1.25	0.6	0.24	Malaysia, Inland Soils
Leaf 9 young palms	2.7	0.16	1.25	0.5	0.23	W. Africa (Bachy)

Leaves younger than No. 17 are now rarely used since a leaf-17 sample can usually be obtained in areas of satisfactory growth by the age of 3 years.

Age of palm.   Leaf analysis is likely to be most useful with bearing palms at the commencement or in the middle of their bearing life. Analyses are however sometimes carried out on very young seedling palms in the field. Broeshart[9] compared such palms with older palms growing in the same field and which had reached the bearing stage; he found little difference with leaves 1 and 9 except in the case of magnesium of which the young palm had a lower leaf-content. Using the critical levels for leaf 17 of adult palms and equations relating the levels of nutrients in leaves 9 and 17, Bachy calculated the theoretical critical levels for leaf 9 in young palms of less than 3 years.[68] He found these to accord well with levels in healthy young palms in several countries (see figures above).

Nevertheless it has been general experience that leaf levels of certain nutrients tend to decline with age even where liberal quantities of fertilizers have been supplied. Such a decline is commonly seen with K, usually though not invariably seen with N and Mg, and not commonly with P. Declines of N and Mg can occur at quite an early stage; K declines have been more pronounced on some soils than others. Some illustrations of soil nutrient movements with age in Malaysian soils are given in Table 11.9. In each case a high level of application of the nutrient concerned had been given annually.[69, 70] The periods covered are only the first 3 to 5 bearing years but declines of N and Mg are already apparent. With K a decline is only seen on the marine clay (Selangor series) soil. P movements are irregular. In the twenty trials examined by Foster and Chang[137] declines with age were only found with N, P, Mg and Ca on coastal soils. On inland soils there was some tendency for P to increase; and there was no decline in K levels. These results suggest that the decline of K with age which is often seen may be due to insufficient application rather than age itself. It is also clear that in Asia changes of nutrient leaf levels with age depend to a great extent on soil type.

Table 11.9 *Movement of leaf nutrient levels with age (from Tan*[69] *and Hew et al.*[70]*)* (leaf 17, per cent of dry matter) *Malaysia*

Soil*	Treat-ment†	Nutrient	Years from planting				
			3	4	5	6	7
1. Inland silty clay	N_3	N	2.93	2.97	2.79		
2. Inland sandy colluvial	N_3	N	2.85	2.73	2.68		
3. Inland silty clay loam	N_2	N		2.96	2.77	2.75	3.06
4. Marine clay	N_2	N		2.85	3.00	2.90	2.65
5. Marine/river clay	N_2	N		2.90	2.87	2.74	2.56
1. Inland silty clay	P_3	P	0.160	0.151	0.180		
2. Inland sandy colluvial	P_3	P	0.161	0.177	0.144		
3. Inland silty clay loam	P_2	P		0.168	0.177	0.170	0.156
4. Marine clay	P_2	P		0.169	0.169	0.179	0.176
5. Marine/river clay	P_2	P		0.174	0.183	0.184	0.168
1. Inland silty clay	K_3	K	1.23	1.14	1.22		
2. Inland sandy colluvial	K_3	K	1.69	1.36	1.31		
3. Inland silty clay loam	K_2	K		1.08	1.11	1.20	1.32
4. Marine clay	K_2	K		1.02	0.88	0.98	0.83
5. Marine/river clay	K_2	K		0.90	0.90	1.02	1.02
1. Inland silty clay	Mg_3	Mg	0.403	0.357	0.288		
2. Inland sandy colluvium	Mg_3	Mg	0.353	0.358	0.366		
3. Inland silty clay loam	Mg_2	Mg		0.40	0.38	0.37	0.29
4. Marine clay	Mg_2	Mg		0.31	0.28	0.29	0.26
5. Marine/river clay	Mg_2	Mg		0.36	0.41	0.37	0.35

* Soil series: 1 − Batu Anam/Durian; 3 − Munchong; 4 − Selangor; 5 − Briah.
† In experiments 3, 4, 5 the treatments cited received Nitro 26: 6.4, 3.6, 1.8; Rock phosphate: 3.6, 3.6, 2.8; Muriate of potash: 5.0, 5.4, 2.8; Kieserite: 3.2, 3.6, 1.8 kg/palm/annum respectively. Experiments 1 and 2 received varying but increasing rates from planting.

In West Africa Bachy found declines in K contents with age in all three localities investigated; N and Mg declines were found in one area and an increase of P also in one area only.[68]

The leaf analysis of very old palms may show some peculiar results. For instance, in Nigeria, 38-year-old palms giving a 264 per cent increase in yield as a result of K dressings applied 12 years previously showed leaf analyses as follows (per cent of dry matter):

Treatment	Yield 1960−1 (%)	K (%)	Mg (%)
K_0	100	0.37	0.35
K_1	218	0.62	0.26
K_2	264	0.77	0.25

Sulphate of ammonia depressed yields, though N contents were as follows: N_0 − 2.08 per cent, N_2 − 2.03 per cent.[49]

Environment in relation to yield and critical levels

Such results as those quoted above and anomalous results from other parts of the world have long suggested that critical levels may be influenced by environmental factors and not simply by palm and leaf age, the supply of other nutrients, etc. Ruer[72] has used the Effective Sunshine Factor of Sparnaaij *et al.*[73] (see p. 187) to demonstrate that when climatic conditions are unfavourable the correlation between the leaf-K content and yield becomes insignificant. It was shown, for instance, at La Mé in the Ivory Coast that when effective sunshine over a period of 12 months from 1 March to 28 February was low (below 1,620 h.) there were no significant correlations between the yield in the 12-month period 28 months later and the K or N leaf contents determined 2, 1 or 0 years previously; whereas with effective sunshine above 1,620 h. significant correlations were usually found, particularly between yield and the K or N leaf contents 2 years previously. Through the fitting of yield/leaf-K curves to the data obtained for years of high and average effective sunshine, it was also estimated that much larger yield responses to high leaf-K contents would be obtained in yield years which corresponded to the years of high effective sunshine.

From these results Ruer argued that in climates of even rainfall with regularly high effective sunshine, the critical levels will be higher than where these conditions do not obtain, and he suggested that the following levels for leaf-K might be appropriate:

	Malaysia	*Ivory Coast*	*Dahomey*
Per cent K	1.2–1.3	1.0	0.7–0.8
Effective sunshine hours	1,800	1,600	Dry climate

While this work is a valuable contribution to the understanding of leaf nutrient levels in relation to yield, it is clear from what has been said in the previous section that a percentage of 1.2 to 1.3 for K cannot be used in all circumstances in Malaysia since soil and season must also be taken into account.

Table 11.10 gives examples of leaf analysis data from a number of regions and illustrates the leaf-content percentages typical of certain deficiencies. Some of these are illustrated diagrammatically by pentagons in Figs 11.5 and 11.6. This type of diagram, used by the French workers, is well-adapted to the illustration of the nutrient status of any group of palms and the antagonisms found between nutrient elements, though it presupposes a knowledge of critical levels.

Relation of yield to leaf nutrient levels

A greater study has been made of leaf-potassium levels than of those of any other nutrient and some attempts have been made to judge, in the case of seriously deficient palms, the yield increments that may be expected to

Table 11.10   *Some leaf analysis data of the oil palm* (percentage of dry matter in leaf 17)

Country	Locality or soil	Leaf nutrient content					Notes
		N (%)	P (%)	K (%)	Ca (%)	Mg (%)	
1. Malaysia	Coastal alluvium	2.79	0.185	1.13	0.54	0.33	No apparent deficiency
	Muck soil	2.47	0.113	1.87	0.31	0.12	P and Mg deficiencies. K high through antagonism. N marginal
	Rengam series	2.72	0.172	1.20	0.88	0.21	K fertilizer applied
		2.66	0.172	1.08	0.84	0.27	K Mg fertilizer applied
		2.72	0.177	0.90	0.97	0.25	No fertilizer
2. Dahomey	Pobé	1.93	0.149	0.43	0.81	0.68	Severe K and N deficiency
3. Nigeria	Benin. Acid Sands	2.82	0.20	1.24	0.72	0.44	Leaf 1 ⎰ K
		2.55	0.15	0.69	0.92	0.40	Leaf 17 ⎱ deficient
	Calabar	–	–	1.62	0.50	0.07	Severe Mg deficiency
	Benin	2.29	0.13	0.81	1.20	0.31	P_0 plots ⎰ P effect
		2.28	0.14	0.68	1.33	0.31	P_2 plots ⎱ on K
4. Colombia	Alluvium	1.68	0.13	0.67	0.69	0.31	Extreme NPK deficiency: grass competition

result from the raising of the leaf-K level by a given amount. In the study of nine sets of experimental data Ochs found that yield was raised by various proportions from 4 to 12 per cent for every gain of 0.1 per cent in the leaf-K level; he therefore concluded that other uncontrolled factors made it impossible to relate yield with increases in leaf-K with any precision, but he put forward an approximate average increment of 10 per cent in the yield for a rise of 0.1 per cent in leaf-K.[23] This corresponds with a correlation found in a field in Nigeria between individual palm yields and their leaf-K content.[50] In this case the relationship appeared to hold good with palms showing leaf-K contents up to 1.10 per cent. Correlations between yield and leaf-N and leaf-P have already been referred to (pp. 518 and 521); however, no attempt was made to attribute any generally applicable percentage rise in yield with any given rises in the leaf levels of these nutrients. Prevot and Ollagnier[43] found overall significant positive correlations between N, P and K and yield, but showed that these needed special interpretation owing to the existence of interactions, i.e. in all cases where elements x or y or both x and y are below certain levels, no correlation is found between the level of z and yield. This is illustrated from Ochs's work[23] in Figs 11.3 and 11.4.

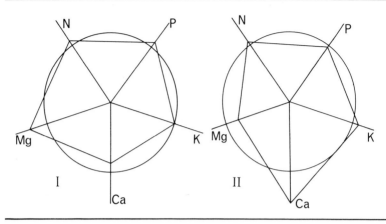

Fig. 11.5   Pentagons of nutrient content of leaves of palms in Malaysia. **I.** Coastal clay soil: essential nutrients in good supply. **II.** Rengam series (inland soil): magnesium deficiency induced by potassium supply.

From their examination of twenty fertilizer trials in Malaysia Foster and Chang[137] have concluded that an increase in bunch yield of 1 ton per hectare is accompanied by increases in leaf nutrient levels of 0.03 to 0.05 per cent of dry matter for N, 0.003 to 0.005 per cent for P and 0.05 to 0.08 per cent for K. In Asia 1 ton may represent a 4 to 5 per cent yield increase, so that in the case of potassium a yield increase of 8 to 10 per cent might be accompanied by an increase of 0.1 to 0.16 per cent in the leaf-K level; this agrees closely with the suggestion of Ochs referred to above. Foster has concluded that, provided factors affecting critical levels are taken into account, leaf analysis is a useful guide to the nutrient status of the palm, and that research directed towards improving the knowledge of critical-level variations will be most productive.

The precise value of leaf analysis with the oil palm has been much debated in recent years. Its ability to show gross deficiencies is not disputed, and as a tool for diagnosing these deficiencies or obvious in-balances it is of undoubted utility. Many plantations, applying substantial annual fertilizer dressings, use annual leaf analysis for monitoring the effect of these dressings on leaf nutrient changes. While this may lead to the detection of some gross errors of fertilizer practice leading to serious leaf nutrient imbalance and consequent lowering of yield, there is a danger-ous tendency to deduce more from small changes than can possibly be justified when regard is had for the variability of the material and the statistical considerations which should be taken into account (p. 528). Moreover many growers expect that annual leaf-nutrient reviews can, with expert examination, provide the correct types and quantities of fertilizers to be applied in that year for maximum production. Green has rightly tried to dispel the idea that these experts exist and concludes that 'the

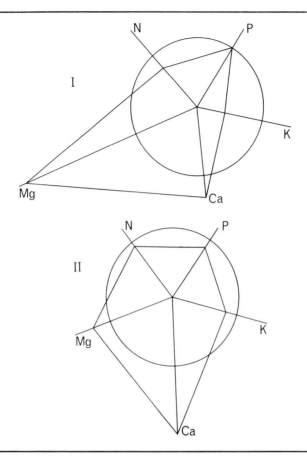

Fig. 11.6 Pentagons of nutrient content of leaves of palms in West Africa. **I.** Highly N and K deficient palms in Dahomey. **II.** N, P, K deficient palms in Nigeria.

unpalatable but inescapable truth is that in the present state of our knowledge once nutrient levels are somewhere around the believed optima, leaf analysis of itself offers no guide to the quantities of fertilizer that ought to be applied' and 'can only tell us, and then retrospectively, when we have gone wrong again and a faulty policy has re-created imbalances or induced new deficiencies'.[74] It is important to note the words 'of itself' in the above sentence since it is in conjunction with the results of fertilizer experiments that leaf analysis data can be most intelligently viewed; and Green's dictum that 'interpretation by experienced and competent advisers is almost always based on their wide familiarity with the subject than on any quantitative interpretation of the data' may be taken to embrace familiarity with relevant fertilizer experiment results.[74]

Sampling practice and analysis

Two methods are commonly used for sampling leaf 17. In one, after the leaf has been cut down, it is cut into three equal parts and six leaflets are taken from each side of the middle section; of these twelve leaflets six will be upper-rank and six lower-rank leaflets.[75] Upper-rank leaflets nearly always contain more K and less Mg than lower-rank leaflets.[76] The leaflets are bulked from each sampling unit and placed in a polythene bag. Another method is to take leaflets from the entire length of the leaf on alternate sides of the rachis at intervals of about ten leaflets.[63]

Leaflet samples should be prepared within 24 hours. The leaflets are first thoroughly cleaned with damp cottonwool. The midribs are then removed. The taking of the middle 20–30 cm has been advocated but there is slightly less between-palm variation when the whole lamina is used. The cutting off of the marginal 2 mm of the laminae is also recommended. The laminae are cut into small pieces and dried for about 5 hours at 65°– 70°C in an air-draught oven. This does not remove all moisture but dries the samples sufficiently for storage or despatch but not so far that alternative methods of analysis cannot still be employed. The samples are placed in polythene bags and hand-crushed.

There are now many analytical laboratories dealing with oil palm leaf material in all three continents and most of the analyses are fully automated. Results have not always been consistent and it is now usual for laboratories to exchange samples regularly as a check on their methods. The distribution of samples of the same material to numerous laboratories has also been organized centrally in Europe. An outline of the modern methods most commonly used for oil palm leaves has been provided by Poon.[75]

Soil analyses

There is a persistent demand by plantation companies, development organizations and others for laboratories to decide from the results of the analyses of soil samples whether an area is suitable for oil palms or not, and what fertilizers should be applied. Soil classification in feasibility studies has been discussed in Chapter 3 (pp. 113 and 132) where it will be seen that it is primarily the physical properties of the soil which determine suitability. Within a particular climatic environment it will be these properties also which will to a great extent determine the general level of yield. Here, however, we are concerned with the determination of the nutrient-supplying power of the soil and its fertilizer needs; and it must be admitted that soil analysis has rarely been of primary value in this respect. There is no set of quantitative standards of soil nutrients comparable to the admittedly imperfect and fluctuating critical levels in the leaf. Some typical profile analyses were given in Chapter 3 and it is difficult to discern from them even the most general relationship between soil nutrient levels and fertility. Soil chemists themselves are not slow to point out their inability to relate fertility to analysis. In the case of the Sabah soils now coming under oil palm cultivation, Paton[77] has written:

> If the analytical figures alone were available the conclusion would be reached that in the soils of Group A there are a number of families of considerable agricultural potential while the soils of the B Group would be considered of very little value. The situation is profoundly altered by the fact that the Table family of soils, in Group B, is actually extremely productive and when last seen in 1959 was growing very good crops of Manila hemp, cocoa and oil palm. On Table Estate itself Manila hemp has been grown more or less continuously without fertilisers for over thirty years, without apparent decline in yield.

He goes on to cite other examples, and the lack of responses to fertilizers in experiments on soils which might be expected to have nutrient requirements.

The great difficulty of correlating soil analysis with 'fertility' is due to the fact that (i) roots extract nutrients from three sources,[78] i.e. the soil solution, the exchangeable ions and the readily decomposable minerals of the soil, which are interrelated in a complex manner, (ii) crops differ considerably in their needs and in their power to extract nutrients from these three sources, (iii) a measure of any of these sources may not indicate the real availability of nutrients for any particular crop, since soils differ in the rate at which they release non-exchangeable ions, (iv) the rooting volume available to the plant varies considerably and (v) the physical condition of the soil effects the power of the root system to exploit it. It should not be surprising therefore that in areas where there are abrupt changes of soil type with marked changes of underlying parent rock, limited use can be made of analytical data, but that in regions where large areas are covered with soil of more or less the same origin some correlations can be found. Although measures of exchangeable cations have been widely used to indicate nutrient status, it has been found in Malaysia, for instance, that differences between soil types in this respect are not very great though nutrient needs are widely different, but on the other hand differences of total cations are sometimes considerable. Rosenquist[79] has shown, for instance, that soils in Malaysia of the Rengam series developed over acid igneous rock have a primary need for potassium while certain soils derived from sedimentary rocks show primary deficiencies of phosphorus and nitrogen. This is not apparent from examination of the levels of the soil nutrients.

Under such circumstances, and in the present stage of knowledge, the soil scientist provides the best service by recognizing, through soil surveys, distinct soil series and, through experience, soil analyses and fertilizer experiments, providing information as to their suitability and requirements. In Malaysia considerable progress has been made on these lines and many plantations have undertaken detailed soil surveys and the mapping of soil series; soil and leaf analysis are carried out using these maps as a basis.[80]

Soil analysis for the detection of fertilizer needs has made most headway in West Africa where Tinker[81] established relationships both on the wide areas of Acid Sands soils and the basement complex soils of Nigeria between certain functions and responses to potassium and magnesium fertilizers. In these areas potassium is usually the dominating need, but no direct relationship was found between exchangeable or total K and yield response. Through the main areas of oil palms in Nigeria a relationship was found, however, between the mole fraction K — Exchangeable K/Cation Exchange Capacity — and yield. There were some exceptions, these being either in areas with unusually high clay content in the sub-soil or on soils of very different origin to that of the Acid Sands. However, in later work on equilibrium ionic activity ratios, Tinker showed a better relationship of yield with a function AR_u,* the unified activity ratio, in which account was taken of the activities of the cations K, Mg and Ca and of aluminium.[82] From this work it was then found that potassium response was closely related to a function of the exchangeable cations, including aluminium, as follows:

$$ER = \frac{K_e}{Ca_e + Mg_e + 2.5\ Al_e} \qquad \text{where } _e = \text{exchangeable.}$$

Since $Ca_e + Mg_e + 2.5\ Al_e$ is correlated with the exchange capacity, the previously found relationship with the mole fraction K (above) becomes clear.[83] This relationship was found to cover palms on Acid Sands, basement complex and a soil probably of volcanic origin, being particularly suited to soils of high acidity with low or varying amounts of calcium.

Correlations have also been found within a field on Acid Sands soil between various functions indicative of the potassium status of the soil and both leaf-K content and individual palm yields.[50, 84] In this case correlations between yield and both exchangeable K and mole fraction K were found in addition to a relationship with the ionic activity ratio without inclusion of aluminium, as the soil was only weakly acid. Such within-field correlations are of interest since they show that oil palm yields vary with relatively small variations in the nutrient status of the soil. In these Acid Sands soils, where exchangeable K falls below 0.10 m eq per 100 g of soil, response to potassium is certain provided no other deficiency is inhibiting a response. Responses in soils with exchangeable K in the range 0.10—0.20 m eq per 100 g are probable. There are strong positive correlations between leaf-K and soil-K. Leaf symptoms of K-deficiency may be expected with leaf-K in the range 0.3 to 0.5 per cent of dry matter which will correspond with soil exchangeable K of below 0.10 m eq per 100 g.[84]

---

* $AR_u = \dfrac{K^+}{\sqrt{[(Ca^{++}) + (Mg^{++})]} + P^3\sqrt{(Al^{+++})}}$ \qquad P is a constant.

A similar relationship between leaf-K and exchangeable K in the top 20 cm of soil has been found in a grove in a grove in Dahomey.[23]

While this work has hardly led to the determination of critical levels for soil potassium parallel with those for leaf-K, there are indications that, on the Acid Sands soils of Nigeria, soils with a mole fraction of below 0.015 are likely to give responses, while values of 0.015 to 0.020 may be marginal. On the West African soils studied by Tinker an AR_u value in (moles/litre)$^{\frac{1}{2}}$ of less than 0.006 is also likely to indicate a potassium response (see Fig. 11.7).

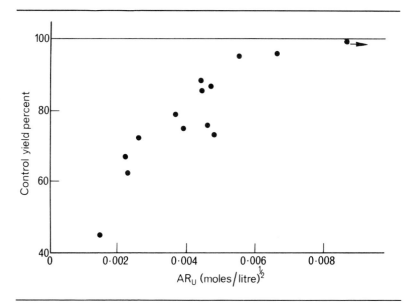

Fig. 11.7 Relationship of the function AR_u in the topsoil of fertilizer experiments with the yield responses obtained to K fertilizer applications (Tinker, 1964).

Ataga[85] has studied the availability and fixation of potassium in the Acid Sands and Basement Complex soils of Nigeria using the b-index (number of successive extractions with 50 ml of 0.005 N $CaCl_2$) and other closely correlated extraction methods. Available reserves were found to be high in Basement Complex soils with a high proportion in exchangeable form. Reserves in depleted topsoils on the Acid Sands were low, with considerable quantities from non-exchangeable sources; but it was shown that with soils newly opened from virgin forest the available reserves could be moderately high for a few years. It was also shown that when K is added to the soils at rates equivalent to about 1 kg per palm, 30 to 65 per cent is fixed, though later released gradually to the palm. This is of course advantageous as it retards leaching. These findings are relevant to fertilizer experiment results which are discussed later (pp. 548 and 550). Another interesting finding was that the clay fraction contributed the bulk of the non-exchangeable K released by these soils, thus providing a further reason for preferring those Acid Sands which contain a higher proportion of clay.

Examination of the potassium reserves and fixation in Malaysian soils[91] gave results rather similar to those in Nigeria.[85] Greater fixation was found in montmorillonitic and micaceous clays than in kaolin clay factions, but the eventual supplying power of the former soils is greater. Some relationship was found between soil extractable K levels and response to K fertilizers.[137] On coastal soils K levels of unfertilized plots varied between trials but results suggested 0.20 m eq per 100 g as a tentative critical level for all unfertilized soils. Difficulties arise, however, where there is a history of K application; on inland soils for instance profitable yield increases were found up to soil K levels of 0.22—0.27 m eq per 100 g where fertilizer had been broadcast, but up to 0.57—0.79 m eq per 100 g where fertilizer was placed in the circles (from where samples were collected).

A recent attempt to utilize the Stanford—De Mont test (which employs barley plants deficient in K) to establish the K status of various soils merely served to demonstrate the usual poverty of oil palm soils in all parts of the tropics and failed to differentiate between them.[86]

Soil deficiencies of magnesium are widespread in Africa but owing to the strong potassium/magnesium antagonism magnesium need tends to be exhibited only where potassium is in good supply, or where the levels of both elements are exceptionally low. In fertilizer experiments, symptom development (Orange Frond) has been common and yield response to magnesium applications occasional. Prevot and Ziller[87] showed a relationship in Congo (Brazzaville) between water-soluble magnesium in the topsoil and both deficiency symptoms and leaf-Mg. Tinker, working with twenty-six Nigerian soils and ten samples from Etoumbi in Congo (Brazzaville) found that symptoms were usually, but not always, related to the level of exchangeable magnesium in the soil.[81] In the worst areas the level of exchangeable Mg was below 0.1 m eq per 100 g but in the Etoumbi soil and some soils in Nigeria symptoms were found where the Mg levels were well above 0.1 m eq per 100 g and in these cases exchangeable potassium would be about 0.2 m eq per 100 g. It was found that a good relationship existed between the topsoil exchangeable Mg : K ratio and symptoms, but the steepness of the curve indicated a very critical ratio level of about 2 below which symptoms appeared.

Later experiments both in the field and with soil in pots showed how easily magnesium deficiency could be induced by potassium applications to the soil, particularly in areas where potassium had, in the course of previous cultivation, been less depleted than magnesium.[88] It was shown that young palms were particularly prone to magnesium deficiency symptoms when K + N fertilizers were applied. This condition appeared to be due not only to the K/Mg antagonism but also to the leaching effect of sulphate of ammonia on Ca and Mg in the soil. At the same time it became clear that, with young palms, symptoms of Mg deficiency could be

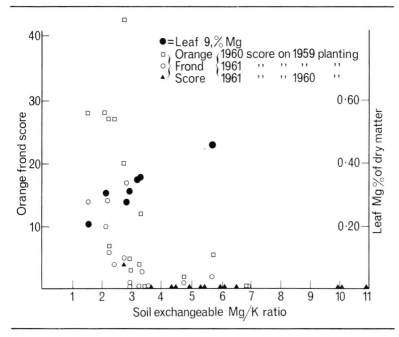

Fig. 11.8 Relation between Orange Frond incidence, soil exchangeable Mg/K ratio and leaf magnesium content (Tinker and Smilde, 1963).

induced where the topsoil exchangeable Mg/K ratio sank below 4; and the critical level in this case seemed to be about 3.5. These results, obtained near Benin in Nigeria, are illustrated in Fig. 11.8; the clear positive correlation between the Mg level of leaf 9 and the topsoil exchangeable Mg/K ratio can also be seen.

The difficulties of assessing phosphorus need from soil total phosphorus or various measures of 'available' phosphorus are notorious. In Malaysia no clear differences can be seen in these measures between soils on which the oil palm responds to phosphorus and on soils on which the palm does not respond. There is perhaps more hope of finding a relationship where the soils have similar genetic properties, and in West Africa some progress has been made with such work. Primary responses to phosphorus have been few, but it is sometimes found that a response will be obtained after the need for K or N has been satisfied.

A number of reference soils from typical areas growing oil palms in West Africa were used in an experiment in which oil palm seedlings were planted in pots treated with complete nutrients with or without phosphorus.[89] With the majority of soils there was a response to added phosphorus in seedling dry matter production and, with some notable exceptions, those soils showing greatest response had given responses in growth or yield in the field. Additionally, certain methods of estimating 'available' phosphorus showed a significant negative correlation between the log percentage response in dry matter of the pot seedlings and the soil values of available phosphorus, the Olsen and Saunder methods giving the highest significant correlations.

Work on phosphorus release from these soils suggested that they could be grouped into three categories: those giving no response with the oil palm to P fertilizers owing to initial high levels of P, those having a low level of P and responding to fertilizer, and those having a low level of P but showing no response owing to high fixation of phosphorus. While this work indicates that some progress may be expected, the problem of relating responses in the field, particularly of adult palms, to measures of soil-P availability remains incompletely solved.

Ng has presented data for P, K and Mg in certain Malaysian soils to indicate their differences and probable needs.[90] In general, soils derived from basic rocks are poor in K though high in P. Sandy soils of various origins are poor in both K and Mg though shale-derived soils can be high in non-exchangeable K; P is generally low. Marine clays are much richer in all three nutrients and this is reflected in their lower initial responses to fertilizers.

No studies of American soils employed for oil palms in relation to their nutrient-supplying power have been reported and there is a primary need in that continent for further soil analyses and fertilizer experiments.

Other analytical methods

The difficulties in interpretation of both soil and leaf analysis have stimulated interest in possible other methods. Growth analysis has been discussed in Chapter 4 and its value in relation to nutrition is considered later in conjunction with fertilizer experiments (p. 558).

Interesting relationships were recently found between leaf contents of K and P and the contents of these elements in mature primary roots of the same palms.[92] With potassium it was shown that there was a good correlation between leaf and root levels particularly where K was deficient. With phosphorus there was a large difference in root content between palms which had received and had not received a dressing of dicalcium phosphate 3 years previously, whereas the leaf levels were only marginally, though significantly, altered as shown below:

Per cent of dry matter	Without P	With P	Increase (%)	Without K	With K	Increase (%)
Leaf content	0.170	0.177	4	0.578	0.907	57
Root content	0.051	0.114	124	0.346	0.739	114

These results indicate that whereas the root does not act as a storage organ for K it appears to be capable of storing P. The large percentage differences in the roots also suggest that this analysis method merits much further investigation. Root analysis does not however appear likely to be so promising for N or Mg. With the former, the values are low compared with leaf values, while with Mg there is little variation even when leaf-Mg varies considerably.[93] The range of root nutrient values (per cent of dry matter) is approximately as follows: N, 0.30—0.45; P, 0.03—0.16; K, 0.20—0.95; Ca, 0.05—0.15; Mg, 0.10—0.15.

Soil changes during the life of a plantation

Most plantations of oil palms are planted from forest, pass through 25 to 35 years of life under one stand of palms and are then replanted. This cycle of growth and replanting is then likely to be repeated. It would be natural to suppose therefore that studies of soil changes throughout these processes would provide information relevant both to nutrition and plantation practice, and would assist in assuring that desirable soil characters were maintained. It is probable that soil analysis is of greater value than leaf analysis in indicating long-term trends and in signalling advance warnings.

Long-term changes have only been reported from Nigeria, fortunately from an oil palm field containing a cultural experiment.[94, 95] The field was planted in 1940 on Acid Sands soil, Benin fasc (see p. 125) and sampled in 1941, 1945, 1951, 1956 and 1961. Shorter studies were made of an experiment planted in 1951 on the more leached Calabar fasc of the Acid Sands, and of another planting on the Benin fasc.

The results of these studies showed that where there was no burning there was a rapid release of K from the vegetation followed by substantial increases in Mg and Ca in the topsoil. In burnt areas there was a larger early build-up of K, but by the fifth year these differences had largely disappeared and there was then, to a soil depth of 46 cm, a gradual reduction of all three cations over the next 15 years. Organic matter is surprisingly little affected unless there has been tillage or inter-cropping, and the changes in carbon and nitrogen are small. The most serious steady losses during the life of the palms were found to be in potassium, but magnesium and calcium losses were also substantial as can be seen from the following data which apply to areas not cultivated or inter-cropped, but carrying a natural or leguminous cover throughout the period:

Depth (cm)	K		Mg		Ca	
	0—38	*38—117*	*0—38*	*38—117*	*0—38*	*38—117*
	m eq/100 g					
Highest level, 1945 or 1951	0.153	0.091	0.85	0.58	4.36	2.29
Lowest level, 1961	0.045	0.025	0.58	0.30	2.95	1.67
Loss	0.109	0.066	0.27	0.28	1.41	0.62
Loss in kg per hectare	99	115	73	150	645	549

Estimating that the loss of exchangeable K from the top 3 m of soil in 15 years would be around 335 kg per hectare, Tinker pointed out that, to fulfil a requirement of 670 kg uptake, the other 335 kg would be provided by only a very small proportionate release of K from the non-exchangeable sources, and a manuring recommendation of 335 kg KCl or K_2SO_4 per hectare every 3 years thus seemed an adequate supply.[95]

Long-term studies of this nature in other oil palm regions would be of great interest and value. One of the practical difficulties, however, is the finding of areas which have not been subjected to fluctuating fertilizer policies.

Fertilizer experiments and application

Some account must now be given of the knowledge that has been gained of nutrient requirements through fertilizer experiments. It will already have been noted how important such experiments are in relation to the correct interpretation of both leaf and soil analysis, but they have, of course, a particular value in themselves since not only do they show which nutrients are needed but, if properly designed, they can assist in determining the most economic rate and frequency of application. For this reason a quantitative discussion of fertilizer application has been left to this section.

One attempt has been made to relate fertilizer experiments to a theoretical concept. Homès,[96-98] following other workers, attempted to demonstrate with oil palm seedlings — and other plants — the importance of the anion—cation balance, and the independence of the separate cationic and anionic balances. By measuring height and other growth factors of seedlings grown in sand culture and their degree of health, he deduced that the six major elements should be supplied to oil palm seedlings in certain proportions. Later he suggested an adaptation of the method applicable to field trials, having regard for the fact that each soil would provide in different proportions the six major elements with which he was dealing and that therefore the proportions of added nutrients needed in each case would differ from his 'optimum balance' worked out in sand culture. The method has been applied to other crops, but although the first work was done with the oil palm and was later repeated and reported in much detail[99] it was not pursued; indeed, the facts of nutrient balance seem to show themselves in a more practical manner in factorial fertilizer experiments combined with leaf analysis or soils data.

Early experimentation

Although some fertilizer experiments were conducted in the between-wars period, extensive trials were not begun in most producing areas until after the war. Reviews of experiments up to 1955 have been provided by

May[100, 101] and the early knowledge of manurial needs is summarized below.

In most areas opened from undisturbed or very old secondary forest little or no responses were initially obtained to fertilizers. In Zaire many experiments were conducted without positive results and in Malaysia the same lack of response was experienced even on the poorer soils. In Nigeria, areas opened from old secondary forest showed no responses until the palms were well into bearing, though when improved planting techniques and selection began to induce more rapid growth in the early years responses in growth and early yield to nitrogen fertilizers appeared.

Thus in typical forested country the nutrients accumulated in the top-soil together with those supplied following burning or during the decay of the forest debris seemed sufficient or nearly sufficient to provide a healthy stand of young palms; this satisfactory state of affairs tended to be short-lived, however, and on all but exceptionally suitable soils some nutrients began to be required soon after bearing started.

Where an area had a history of any kind of cropping, fertilizers were soon shown to be required from the start. The early experiments on bearing palms showed responses typically to potassium in Africa and on some soils in the Far East; to nitrogen in the early years, but less commonly with bearing palms; to phosphorus on certain soils in the Far East; to magnesium on very poor soils in West Africa, Congo (Brazzaville), Zaire and Malaysia; to organic manures and wood and incinerator ash in West Africa and Zaire; and, in addition, there were some hints of trace element responses.

Undoubtedly the most startling responses were obtained to potassium, particularly in the case of old plantations which had never received fertilizers and where yields were declining and leaf chloroses appearing. Lack of K responses in early experiments in Malaysia was due to the fact that experiments were largely confined to soils derived from sedimentary rocks. Soil applications of magnesium fertilizers in Nigeria were spectacular in their elimination of the Orange Frond chlorosis.

Rates of application were largely guesswork, being based on some experience with other permanent crops, and applications were either annual or once-and-for-all. Thus dressings of potassium sulphate, for instance, varied from 1–9 kg per palm for bearing palms, this being the equivalent of about 150–1,250 kg per hectare. Young seedling palms in the field might receive anything between 0.25 and 1 kg per palm. Some of the negative results may have been due to the very small dressings, although in other experiments results were obtained from applications which were two to four times as large as were required. Comparatively few of the early experiments allowed for varying rates or different methods of application. In one set of experiments in Nigeria arrangements were made for increasing the dosage by 0.45 kg of fertilizer per palm for each year of age until a ceiling dosage was seen to have been reached. As early responses were not obtained in forested areas, however, there was a tendency to allow these applications to reach very large and unprofitable amounts before they were discontinued.

However, this period saw the carrying out of three important sets of experiments, the Unilever 'Crowther Experiments' and the early W.A.I.F.O.R. and I.R.H.O. experiments which threw some light on fertilizer rates and frequencies applicable to African soils.

The Crowther experiments

These experiments, designed by Dr E. M. Crowther, were 3^n NPK, NPKMg and NPKCaMg factorial experiments started in 1940 on young 1932 to 1940 plantings on four soil types in Nigeria and West Cameroon and one soil type in Zaire; they have been described in detail by Haines and Benzian.[102] Early fertilizer applications were at low rates and irregular intervals, but from 1947 annual dressings were as follows:

		Usual application per palm per annum			Estates	Soil
		Rate	(kg)	(lb)		
					Cowan, Nigeria	Acid Sands, Calabar fasc.
N	Sulphate of ammonia	1	1.5	3.3		
		2	3.0	6.6	Cowan, Nigeria	Acid Sands, poor Benin fasc.
P	Ground rock phosphate	1	1.5	3.3	Ndian, Cameroon	Gravel over basalt
		2	3.0	6.6	Ndian, Cameroon	Sand over gneiss basement complex
K	Sulphate or muriate of potash	1	1.0	2.2		
		2	2.0	4.4	Kangala, Zaire	Sandy
Mg	Magnesium sulphate	1	0.25	0.55	On Cowan Experimental blocks 1, 2 and 3 only, from 1951	
		2	1.0	2.2		
Ca	Limestone	1	10	22	On Kangala experiment only	
Mg	Limestone	1	10	22		

The results of the Cowan experiments where treatments were continued in the 1950s are given in Table 11.11. They show a predominating response to K, with a suggestion that, with adequate K dressings, phosphorus and perhaps nitrogen might have some effect on higher yielding areas. At the end of a 10-year yielding period it was clear that an *annual* dressing of 1 kg was as much as was required on the Calabar fasc, but the higher yields of the K_2 treatment in the earlier years on the Calabar fasc and throughout the period on the poorer Benin block suggested that higher dressings at the start might have given a better return.

At Kangala and Ndian the last applications were in 1949 and 1951 respectively, and results have been presented up to 1954 and 1951.[102] At Kangala there was a response to potassium and it appeared that the higher

Table 11.11　*Responses to annual potassium dressings on Acid Sands Soil (Crowther Expts. Cowan Estate)* (kg per palm per annum and percentages)

	K_0 (kg)	K_1 (kg)	K_2 (kg)		K_0 (kg)	K_1 (kg)	K_2 (kg)
Modified Calabar fasc soil, 1948−59							
P_0	49.7	64.1	66.0	N_0	46.6	64.3	66.9
P_1	47.1	65.7	71.8	N_1	46.7	66.9	70.1
P_2	48.4	68.6	71.5	N_2	51.9	67.3	72.2
	48.4	66.2	69.8		48.4	66.2	69.8
	(%)	(%)	(%)				
1948−59	100	137	144				
1961	100	146	146				

	K_0 (kg)	K_1 (kg)	K_2 (kg)
Poor Benin fasc soil			
1948−59	38.6	46.8	49.2
	(%)	(%)	(%)
1948−53	100	114	116
1954−9	100	127	141
1948−59	100	121	128

level was required, but dressings were only applied for three consecutive years. The results from Ndian, with its two distinct soil types, were of greater interest. Here the brown lateritic clay soil derived from basalt gave strong responses to potassium though no additional yield was obtained from the higher level. By contrast, the sands derived from basement complex material showed sharp responses to phosphorus, also at the lower level. There was also a small response to potassium, particularly on plots receiving phosphorus. In short, these Crowther experiments, while showing that areas in Africa existed where phosphorus was the primary need, gave evidence for a K or P requirement where a sufficiency of the other element had already been provided; at the same time there was evidence that the use of higher rates of P and K fertilizers than 1.5 and 1.0 kg/palm per annum might be beneficial in the early years.

The early W.A.I.F.O.R. and I.R.H.O. experiments

The early experiments of the West African Institute for Oil Palm Research (now N.I.F.O.R.) which provided useful preliminary information on the Benin, Calabar or intergrade soils of the Acid Sands are referred to as the Main Station, Akwete and Umudike Experiments respectively. In addition to these, the Nkwelle Potash Experiment already mentioned (p. 501) included two rates, 7½ and 20 lb (3.4 and 9.1 kg) per palm of potassium sulphate applied to four-palm plots in 1944−5. The results in this area of

Confluent Orange Spotting were startling, the two rates leading to increased bunch yields of 94 and 184 per cent respectively in the period 1945 to 1948, but thereafter poaching of fertilizer caused a rapid rise in the yield of the control plots and further comparison was not possible. This experiment indicated, however, that in extreme cases of soil K depletion doses of over a ton per hectare may be required to give maximum response.

In the Umudike Experiment (604—4), on intergrade Acid Sands soil, a 3^4 layout allowed for application in November 1948 of 628 and 1,255 kg per hectare (5 and 10 cwt per acre) respectively of sulphate of ammonia, superphosphate and potassium chloride, and 1.25 and 2.50 tons per hectare (10 and 20 cwt) of lime on 23-year-old palms.[15] The higher dose of the latter caused an early yield decrease of 20 per cent which soon disappeared; and a positive effect of nitrogen was equally short-lived. The superphosphate dressings had no effect. The effect of potassium was small in 1949 but thereafter substantial, as can be seen below:

Potassium levels	Kilogrammes bunches per hectare per annum		
	1949—52	*1953—6*	*1957—60*
K_0	3,682	3,283	2,993
K_1 628 kg/hectare	5,909	7,122	6,739
K_2 1,255 kg/hectare	6,001	7,488	7,524

Percentage increases over control, reaching 169 per cent in 1960 for the K_2 level, were somewhat misleading since continuing soil-K depletion led to declining control yields as can be seen in Fig. 11.9. It was evident that, on these old palms, the lower level of application was sufficient to sustain yields for some 8 years and that thereafter there was no falling off in the yields of the K_2 plots. Soil analysis 10 years after application showed that differences in exchangeable potassium in the topsoil between treatments had completely disappeared, but that the fertilized plots had maintained a higher *total* potassium at all depths down to 1.2 m, thus indicating that part of the original applications had been fixed in a form which was not immediately available but may be assumed to be released slowly over a long period.[81, 85] It is also probable that the original application so improved the general health and hence the root system of the palms that the dry season effect on growth and yield was not so severe.

It was shown at the I.R.H.O. Stations at Pobé in Dahomey[103] and at Dabou in the Ivory Coast[23] that even small potassium dressings could have a marked residual effect. The lower dressing in the Umudike Experiment is equivalent to nearly 4.5 kg KCl per palm at normal spacing. In the Dabou Experiment (DA—CP2) plots receiving 1.5 kg KCl per palm still showed an increased yield of over 100 per cent of control 5 years after application, though the leaf-K content of the treated palms had by that time reverted to the same level as that of the control palms. These experiments did not,

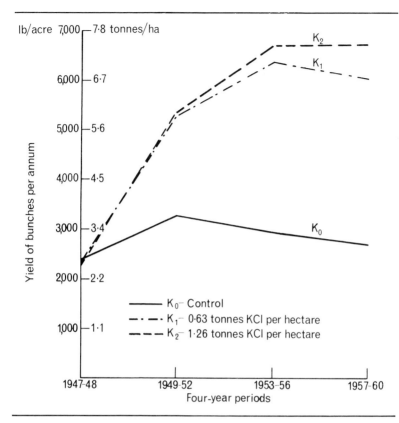

Fig. 11.9 The effect of single applications of potassium chloride on the yield of 23-year-old palms at Umudike, Nigeria.

however, allow for different levels of application, and few responses other than to K were obtained.

The Akwete Experiment (651—1) lay on an area of poorly developed palms on highly K- and Mg-deficient Calabar fasc Acid Sands soil in Nigeria. Clay contents were very low. The presence of Mg deficiency was not at first recognized and an experiment was designed to determine the rate of K dressing required, whether the fertilizer should be applied annually, biennially or triennially, and whether it was best to apply the dressing at the beginning of the wet season or to distribute it in three or six doses through the wet season. The design of the experiment was not wholly satisfactory and it was later seen that magnesium deficiency was limiting the response to K, but it was nevertheless clearly shown that there was no advantage in distributing the potassium sulphate in small doses through the season, and there was no evidence that annual applications were more productive than applying every 3 years.

Table 11.12   *Responses to NPK fertilizers and farmyard manure on Acid Sands soil, Benin fasc., Nigeria (Expt 34—1) (yield of fruit bunches per hectare per annum)*

Fertilizer and manure	Rate of application Per palm per year of age (kg)	1952—5 (kg)	1956—9 (kg)	1960—3 (kg)
N_0	—	2,446	7,024	—
N_1	0.30	2,718	7,759	—
N_2	0.61	2,798	7,553	—
P_0	—	2,694	7,008	6,587
P_1	0.30	2,618	7,642	7,530
P_2	0.61	2,651	7,687	7,599
K_0	—	2,734	7,109	6,787
K_1	0.30	2,565	7,428	7,297
K_2	0.61	2,663	7,800	7,634
FYM_0	—	2,283	6,600	6,478
FYM_1	19	2,731	7,871	7,656
FYM_2	38	2,949	7,807	7,581
L.S.D., P = 0.05		394	654	621

N = Sulphate of ammonia. P = Superphosphate. K = Sulphate of potash.

The 'Main Station' experiments of importance in this context were a 3^4 NPK and farmyard manure experiment in which placement application methods were adopted, and a time of application experiment including both fertilizers and bunch refuse.

The 3^4 NPK, FYM Experiment (34—1) was planted in 1948 in conjunction with a 'presence and absence' experiment on a poor area of secondary forest. The latter experiment showed significant responses to K throughout 8 bearing years and occasional responses to P either broadcast or placed. Experiment 34—1 had the treatments shown in Table 11.12, and the following points are of particular interest: (i) early small responses to N were maintained and 0.3 kg per palm per year of age appeared more than sufficient, (ii) higher doses of K were required in the second 4 years, but sufficient K seemed to have accumulated from the annually increasing doses to satisfy the need later on and (iii) P requirements only became apparent after 4 years of bearing, the lower dosage being sufficient. Farmyard manure, seldom obtainable and costly to transport, could supply the same needs as the fertilizers.

The fertilizer and farmyard manure in Experiment 34—1 was applied by a placement method in sectors of circular trenches and applications ceased in 1958 by which time over 16 kg of each fertilizer had been applied at the lower rate, i.e. about 1 kg per palm per year over their whole life. The maintenance of responses to both K and P since application ceased is of interest and suggested that more effective and economic responses might be obtained by larger doses applied less frequently. As the 'presence and absence' experiment suggested that broadcasting potassium fertilizers was a superior method to placement, it is probable that larger responses would have been obtained to K if broadcasting had been used in Experiment 34—1.

A number of small 'Main Station' experiments investigated the effect of applying fertilizer at the beginning and end of the wet season and the general conclusion was that applications early in the season are the more effective. Experiment 3—3 included both applications of bunch refuse and a complete NPK mixture. Applications were made from the third year at 25.4 kg bunch refuse and 1.4 kg fertilizer mixture per palm per annum and increased to 203 kg and 5.4 kg respectively after 6 years of application. The combined applications gave the following yields:

Bunch refuse and fertilizer mixture	Yield of bunches per hectare per annum (kg)	
	1951—6	1957—62
April applications	6,566*	10,156**
October applications	3,893	7,619

Production and application of bunch ash.   It is interesting to note in connection with bunch refuse applications that pre-war experiments in Malaysia demonstrated that similar yield increases are obtained from bunch refuse applied in equivalent amounts as raw refuse, composted refuse or bunch ash and that, for reasons of costs, application in the form of ash is to be recommended.[104] The ash obtained constitutes 2.4 per cent of the refuse or 0.6 per cent of the original bunches and provides 30 to 35 per cent K_2O which is about half to two-thirds the K_2O equivalent of muriate of potash.[105] Ash applications should therefore be about double the quantities that would be applied as KCl. The ash has an MgO content of 3 to 5 per cent.

If an area is yielding 20 ton bunches per hectare about 120 kg ash per hectare or 0.84 kg per palm will be produced in a year, equivalent to 0.42 kg KCl per palm. If dressings of, say, 3 kg KCl per palm were required, ash would provide one-seventh of requirements.

The information which emerged from this early work in Africa may be summarized as follows:

1. The need for potassium is widespread; where old plantings are very deficient, dressings up to 4 kg per palm may be needed for full responses, but the latter will be maintained for many years. In areas opened from good secondary forest there may not be a potassium requirement for several years and annual dressings of the order of 1 kg per palm will be sufficient. Broadcasting appears more satisfactory than placement. K applications induce magnesium deficiency on poor Acid Sands soils.
2. Nitrogen dressings are of importance in the early years and may be of

the order of 225 g per palm in the planting year and thereafter 450 g per palm per year of age for 3 or 4 years.
3. Phosphorus is required on basement complex soils and on Acid Sands soils, though in the latter case not until after several years of bearing.
4. Applications of fertilizer early in the rains are most advantageous and there is no advantage in dividing the dressing into several small doses.
5. Yield responses can be obtained to organic manures in the form of farmyard manure or bunch refuse, but refuse incinerated to form ash is just as effective, less expensive to apply and can supply a small but significant portion of potassium needs.

Recent fertilizer experiments and their interpretation

The foregoing may be said to represent the published knowledge of the manuring of oil palms up to about 1955, by which date very few results had been published in the Far East though many experiments were under way. At this time, however, the common need for magnesium, the dependence of responses of one nutrient on the adequacy of supply of another, and the possibilities of minor element requirements were beginning to be seen. From about 1955 a very large number of factorial experiments covering different rates of application were laid down both in the Far East and in Africa by governmental research organizations and commercial research units. It is clearly impossible here to deal fully with the extensive data which have now accumulated. In the following pages the advances made in the interpretation of experiment results will first be discussed and this will be followed by some selected examples of results from various parts of the tropics.

Factorial experiments

Factorial designs have been employed for the majority of fertilizer experiments in all parts of the tropics. Green, in a stimulating review of a number of such experiments, has drawn attention to two important aspects of their interpretation.[106] Firstly, these experiments usually rely on the pooling of second and third order interactions as error, and there is a great need to distinguish between those interactions which are genuinely small or non-existent, and so can be legitimately pooled, and those which should be taken into account.

Secondly, each of the main effects of N, P, K, etc., only show the response to these elements in the presence of the mean level of all the other elements (or other treatments) in the experiment; the effect of each nutrient *alone* or in *specific combinations* is not shown in the summarized data or two-way tables; but it is exactly this information which is required for an economic appraisal of the results, since fertilizer recommendations must depend on the gains in bunch weight, oil and kernels believed to be obtainable from specific formulations applied alone and not along with varying quantities of additional nutrients. Green has now provided a method of estimating the expected yield value of all eighty-one treatment combinations in 3^4 factorial experiments and has applied this to his experiments. Using these values, he calculated the profit or loss expected to result from the application of each single nutrient or specific combination.

From three experiments in West Cameroon and one each in Nigeria, Zaire and Malaysia it was shown that (*a*) maximum yield was usually obtained from a balanced fertilizer mixture of three or four nutrients; (*b*) applications of the 'wrong' fertilizers or fertilizers in inappropriate proportions could lead to substantial reductions in yield and considerable financial loss; (*c*) the increased yield was usually accountable predominantly to one or two nutrients and the highest profit was thus obtained from applying one or two nutrients only, the addition of others resulting in financial loss; (*d*) only in one instance, in Malaysia, did application of all four major nutrients give a maximum profit, and in this case the profit was not significantly different from that obtained from two or three nutrients only; (*e*) foliar analysis data from unfertilized plots indicated that this data could not have shown the requirements for maximizing yield let alone the mixtures for highest profit. Gross deficiency was only apparent in two cases. In spite of substantial yield increases due to P and K, leaf-P and leaf-K were only increased significantly in one case each by the optimum treatment. Leaf-Mg followed the manurial applications more closely.

In general 3^4 factorial experiments using plots of twelve to sixteen palms have given very useful results in many countries, but more recently experiments with four or more levels of certain nutrients have been used. It has been claimed that for obtaining a sound production function for the economic assessment of results four to five levels of nutrients are needed and, to reduce the number of plots, incomplete factorial designs have been advocated.[107]

The economics of fertilizer application

When manuring results in increased bunch yield and hence in increased oil and kernel outturn, there are no increases in general overheads and calculations of profitability are relatively simple.[108] If the cost of harvesting, transporting and milling a ton of bunches and the extraction rate of oil and kernels are known, then the net value of any additional ton of bunches produced will be

$$V_{nt} = a + b - c$$

where a and b = value of oil and kernels extracted respectively from a ton of bunches, and c = the cost of harvesting, transporting and milling a ton of bunches. In this latter figure should be included, of course, charges for maintenance, etc., of vehicles and milling overheads, but it can be noted that where a mill is running below capacity the milling of a larger crop may increase costs very little and so reduce the overall milling cost per ton.

The cost of fertilizer and its application in the field now has to be set against the expected bunch weight increase at this net value per ton, V_{nt}.

This is best done by first calculating the weight of bunches of equivalent net value to the cost of fertilizer and its application:

$$x V_{nt} = C + A,$$

where C = cost of fertilizer/hectare and A = cost of application/hectare. Thus if C and A are respectively 8 and 0.5 monetary units and V_{nt} is 8 units, then $x = \dfrac{8.5}{8} = 1.0625$ tons. If the actual response is expected to be greater than x, or in this case greater than 1.0625 tons per hectare, then the fertilizer application will be profitable.

While these calculations are very simple for adult palms responding directly in bunch yield to applied fertilizers, it is more difficult to assess the profitability of fertilizing young palms. Any assessment must be somewhat arbitrary since the object of manuring at this early stage is to establish strong healthy plants which will come into bearing early and, if thereafter supplied with nutrients sufficient for optimum yield, have a productive bearing life. It is probable that palms insufficiently fertilized when young will never be able to produce at the high rate of others properly manured. To a certain degree, therefore, early manuring is an establishment cost. Moreover the quantities required during the early years are so small on a per hectare basis that yield increases of 1 to 5 per cent over the first 3 years of bearing would amply cover the costs of prebearing applications on areas of fair production. It has, however, been suggested that preharvesting applications may be set against the yield of the first 6 years of harvest.[108]

Another important point to note is that the quantity, x, given above for the weight of additional bunches required to balance the cost of fertilizers and their application, is an absolute amount; therefore, if the *basic yield*, i.e. the yield expected without the use of fertilizers, is low, then the *percentage increase* in yield must be high to be profitable. On the other hand if high yields can be obtained *without fertilizers* then very small percentage responses are required for their profitable use. The proposition, therefore, that a high yielding area requires no fertilizer needs careful examination; increases as low as 5 per cent can often be profitable, though such increases can only rarely be shown to be statistically significant in a fertilizer experiment. As Gunn has pointed out,[108] the corollary of this *for high-yielding areas*, is that any response shown in a fertilizer experiment to be significant is almost bound to be profitable.

Studies of the profitability of fertilizer applications have recently been carried out in Malaysia. Hew *et al.*[70] presented the discounted accumulated gross profit to be obtained from N, P, K and Mg applications with yields similar to those obtained in five experiments on different soils. The calculations allowed for the time lag of response by a discounted cash flow procedure; certain constants were used for costs with three levels of oil and kernel prices. Three- or four-year yields from the start of bearing were employed. As might be expected, results differed widely between soils. The only two consistent results were that nitrogen was uneconomical to apply and that the higher rates, e.g. over 3 kg per palm, of muriate of potash or rock phosphate were not justified. The lower rate of muriate of potash was the most profitable treatment on Munchong, Serdang, Rengam and Briah series (see pp. 114–20), but was unprofitable on the Selangor series marine clay; the lower rate of rock phosphate was the most profitable treatment on Munchong soil only. Magnesium, as Kieserite, was uneconomical to apply on all the three soils where it was used. The main effects in this study appear to have been taken straight from the experiments and not adjusted to 'expected values' as advocated by Green;[106] nevertheless the results do indicate that fertilizers applied before bearing or in the early bearing years will only give economic returns if used with discrimination through knowledge obtained from fertilizer experiments.

Lo and Goh[107] applied a response surface analysis to an experiment on Rengam series (p. 117) with adult palms using varying fertilizer and oil price levels. They concluded that optimum fertilizer rates depend on yield response, price of produce and price of fertilizer in that order of importance, but that there are difficulties of adjustment to prices in view of the time lag between application and response. Rather than making continuous changes to fertilizer rates with prices, managements are recommended to estimate average prices of both produce and fertilizers for several years and to adhere to these estimates. Another factor is the variation in the estimates of yield response. If the profit margin is in any case small and the 95 per cent confidence limits of the experimental results are wide (i.e. the error is large) management may be inclined to reduce costs to conserve the cash flow. All these considerations point to the conclusion that the economics of manuring the oil palm is not a simple or straightforward matter, that decisions can only be taken after a painstaking appraisal of all the factors involved and that the oil palm grower should be on his guard against short cuts. A computer system has recently been proposed for making fertilizer recommendations for individual fields through computer 'reports' based on information on age, ground cover and leaf analysis fed into the computer. The objectives are stated to be the duplication of what an agronomist would recommend and the education of managers and others in the importance of plant nutrient status.[138] It must be at once apparent from the problems discussed in the preceding pages that such a scheme, unless it includes a proper regard for soil series differences, fertilizer experiment results, and appraisals of the actual returns of oil and kernels known to be obtainable from the computer's recommendations, could be both wasteful and misleading, and far from educating managers might lead them to believe that manuring the oil palm is a much simpler matter than in fact it is.

Leaf analysis in fertilizer experiments

Much has already been said about the uses, and danger of mis-interpretation,

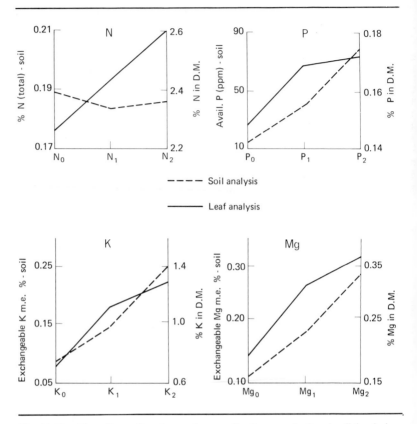

Fig. 11.10 The effect of main nutrient applications on leaf and soil levels in a fertilizer experiment on Kulai series soil in Malaysia (Marriar and Piggott).

of leaf analysis and of the need to adjust critical levels following appraisals of the results of experiments. Numerous examples could be given of leaf analysis data from experiments, but two examples will suffice to show that where deficiencies are pronounced good correlations between leaf data and yield responses can be found.

With adult Deli *dura* palms in Malaysia on Kulai series soil derived from volcanic tuff a 3^4 NPKMg experiment showed, over a 7-year period, yield responses to N, P and K.[109] Figure 11.10 shows the effects of these three elements and Mg on both leaf and soil analyses, and Table 11.13 gives the correlation coefficients between these functions and yield. It will be seen that where the nutrients had not been applied the leaf contents reached very low levels in the cases of N, K and Mg and a rather low level with P. The very considerable yield increases obtained (Table 11.13) are not therefore unexpected. Palms without fertilizers were yielding less than 7 tons

Table 11.13 *Total and partial correlation coefficients between analysis data and bunch yield†* (Warriar and Piggott)

	Total correlations			Partial correlations‡			Bunch yield, 1969–70, high levels as per cent of nil levels §
	Soil and leaf	Leaf and yield	Soil and yield	Soil and leaf	Leaf and yield	Soil and yield	
N	−0.119	0.640***	−0.240*	0.046	0.634***	−0.215	155
P	0.561***	0.492***	0.279*	0.507***	0.421***	0.004	123
K	0.648***	0.449***	0.056	0.698***	0.543***	−0.345***	140
Mg	0.531***	−0.395***	−0.153	0.518***	−0.375***	0.073	101

† Yield of year 1970; analyses carried out in Feb. 1970.
‡ The partial correlations are calculated by keeping the third variable fixed.
§ Kilogrammes of fertilizer applied over 7 years: N, 5.54; P, 4.81; K, 11.53; Mg, 2.45.

bunches per hectare in 1969, whereas those with complete dressings at the higher rates were producing over 25 tons. In some plots leaf levels as low as N − 2.00 per cent, P − 0.129 per cent, K − 0.319 per cent and Mg − 0.094 per cent could be found and K and Mg deficiency symptoms were visible. A curious feature of the results was that, although soil and leaf analysis were well correlated, yield and soil analysis showed little relationship, and under these circumstances of extreme deficiency leaf analysis therefore seemed to be a better guide. The negative correlation between Mg leaf levels and yield is of interest here and simply indicates the strong K deficiency.

Following the results of this experiment it was found possible to rehabilitate the deficient plots within 2 years by giving differential dressings of N, P, K and Mg to the plots according to the previous treatment and leaf analysis.[109]

A very different situation existed in an experiment at La Mé in the Ivory Coast where potassium was the principal deficient element.[110] Here 2 kg per palm per year each of sulphate of ammonia, superphosphate, muriate of potash and magnesium sulphate were applied to clayey-sand soil derived from tertiary sandstone in a 2^5 experiment. On an area planted in 1946 applications were made from 1954 with the regular dressings of 2 kg from 1956 (N, K and Mg) and 1959 (P). The effect of K became significant in 1957–8 and over 11 years the plots without K gave 62 kg bunches per palm per annum and the K plots 98 kg. This 58 per cent increase was believed to be exaggerated by some 20 per cent owing to unusually low yields from some of the P without K plots which were highly deficient at the beginning of the experiment. Leaf analysis data computed as the mean of the 1961, 1963 and 1969 determinations are shown in Table 11.14.

It will be noted that the effect of K on both leaf content and yield were very great and that K altered the leaf content of the other nutrients, though not sufficiently to bring Mg into the range of deficiency. N and P had only minor effects on leaf contents and no effect on yield, and the P

Table 11.14   *Leaf analyses from an* NPKMg *experiment at La Mé, Ivory Coast (Bachy)* (per cent of dry matter – mean of 1961, 1963, 1967 data)

Plots	With N	Without N	With P	Without P	With K	Without K
N	2.64[*1]	2.60	2.61	2.64	2.66[*3]	2.58
P	0.173[*1]	0.170	0.175[*3]	0.168	0.170[*2]	0.173
K	0.644[*1]	0.679	0.634[*1]	0.689	0.849[*3]	0.474
Ca	0.868	0.876	0.919[*3]	0.824	0.791[*3]	0.953
Mg	0.402	0.396	0.401	0.397	0.338[*3]	0.461

Significant differences in 1 year only ([*1]), 2 years ([*2]), or in all 3 years ([*3]), at P = 0.01.

effects on K and Ca were no doubt due to the Ca content of the fertilizer. In the Ivory Coast the yield limiting factor is water and it is not improbable that if this impediment were removed nutrients other than K might have an effect in spite of their present leaf levels.

While these experiments show useful correlations between leaf nutrient contents and yield, and others will be quoted later in this chapter, such correlations may, as already mentioned, be entirely absent. Among Green's experiments[106] (see p. 553) for instance the leaf analysis data for the experiment in Nigeria were rather similar to those of La Mé; maximum yield was obtained by K applications with the addition of N and P but significant differences were not seen in the N and P leaf analyses between fertilized and unfertilized plots. In the Cameroon and Zaire experiments N and P leaf differences were also absent although these nutrients featured in the optimum treatment. K leaf values were anomalous, showing reductions (two significant) with K applications, while Mg, although featuring in all optimum dressings, showed a significant leaf increase in only one case. In the Malaysian experiment only a very small leaf-P increase was shown although the NPKMg mixture increased yields by 32 per cent.

Growth and density in fertilizer experiments

The anomalous and variable results of leaf analysis in fertilizer experiments stimulated interest in growth factors in their relationship to yield. Corley and Mok[111] examined the net assimilation rate (E), crop growth rate (CGR), leaf area index (L), vegetative dry matter production (VDM) and bunch index (BI) in relation to yield in the plots of two 3^4 NPKMg experiments on Rengam series (p. 117) in Malaysia. On this soil the main response is to K, but responses can additionally be obtained to N, P and Mg.[79] VDM, L and CGR were increased by N, P and K with the increases in yield. E was increased only by N, which also increased the number of leaves per palm. In one of the experiments the responses to N were dependent on the application of K and vice versa.

The main point of interest in these findings was that bunch index (BI), which is the ratio of bunch yield to bunch yield plus VDM, was unaltered by manurial treatment, suggesting firstly that this is an inherited character (see p. 262), and secondly that VDM might be used as a tool for diagnosing nutrient need. The highest values of *VDM* found in the two experiments (139 and 135 kg/palm/annum) were similar to those found on the highly productive marine clays in Malaysia. Corley and Mok suggested that only where annual values of *VDM* fall below 120 kg per palm might leaf analysis be undertaken. Lo *et al.*[112] have compared the correlation coefficients between *VDM* and bunch yield in the same two experiments with those of N, P, K and Mg and yield and found that the former were of a greater magnitude and less fluctuating.

There are a number of factors which complicate this approach. Firstly, it appears likely that environmental factors are of the greatest importance in the determination of *BI* and that available assimilates are first used for vegetative growth,[113] and that therefore variation in availability will have a greater effect on yield than on *VDM* (see p. 161). Secondly, density has an effect on the growth parameters and complicates the interpretation of data. For instance, the two experiments cited above were planted at 114 and 138 palms per hectare respectively and Corley and Mok believed that this accounted for the lower net assimilation rate, higher leaf area index and the lower bunch index in the denser planting. Moreover Beure showed that bunch yield responses to fertilizers may be markedly affected by density. He obtained a 40 per cent yield increase at a density of 110 palms per hectare compared with only 6 per cent at normal density for the same medium rates of application and no response at all at higher density.[143] Obviously the differential effect of fertilizers with density needs much more study and economic appraisal. Thirdly, the N effect on leaf number also complicates interpretation. In practice leaf number in adult plantations is affected by the need to prune leaves during harvesting. A higher leaf number might reflect increased longevity or the ripening of the bunches at a later leaf number through a higher rate of leaf production.[111] Lastly, the relationship between *VDM* and yield at the higher levels of yield may be as indefinite as those of the leaf nutrients.

Effect of fertilizers on bunch composition

Fertilizer effects are usually measured only in terms of bunch yield and very little work has been done to examine the possibility that the percentage oil to mesocarp, or even mesocarp to bunch, might be altered by the fertilizer being applied. Corley[113] has drawn attention to three experiments carried out in south Malaysia in which the oil-to-bunch percentage was found to decrease with increasing doses of potassium fertilizers, as follows:

Percentage oil to bunch

K level		0	1	2
Expt. PF 73	*Dura*	20.0	18.6	18.2
Expt. PF 73	*Tenera*	23.5	21.8	20.8
Expt. PF 75	*Dura*	19.0	17.4	17.0
Expt. PF 78A	*Tenera*	23.5	22.2	22.3

It was pointed out that this decrease in oil percentage was more than compensated for by increased bunch yield, but nevertheless it is important to know more exactly the effects on bunch composition of the main nutrients supplied. In the above cases it is not known whether the fall in oil content was due to decreases in the mesocarp oil or to lower fruit-to-bunch or mesocarp-to-fruit ratios. Furthermore the kernel must also be taken into account, and it is thus clear that much more work on this subject needs to be done.

Rates, frequencies and methods of application

Rates of application have featured in the majority of experiments though rarely with more than two actual levels of the fertilizers concerned. 3^n experiments have been most common and the quantities used for adult palms in a representative sample of experiments are shown in Table 11.15. Where only one rate has been used it has been placed in the two-level column. Where three rates have been used the highest rate is shown in parentheses. In general rates tried in Malaysia have been higher than in Africa, but the lower rates, usually about 1.5—2.5 kg per palm, have often been sufficient for maximum yield. Rates of magnesium fertilizers tried tend to be lower than with the other nutrients. In plantation practice also, the Malaysian rates tend to be higher than those in Africa,[74] and this reflects, though it is not necessarily justified by, the much higher general level of yields resulting from climatic factors.

Very few trials have been carried out on *frequency* of application on adult palms. The long-lasting effect of the potassium application in the Umudike experiment in Nigeria and in Dabou experiment DA—CP$_2$ (pp. 548—9) suggested that applications on a triennial basis might be as satisfactory and cheaper than annual applications and several fertilizer experiments have been laid down on this assumption in Nigeria; it has been contended however that large triennial dressings may induce magnesium deficiency when the smaller annual ones would not,[84] and, in general, annual applications are more convenient for management and costing.

In Malaysia the tendency has been to apply nutrients more frequently rather than less frequently than once a year. The effect of nitrogen on growth and colour in the first few years of the palm's life has often been so rapid but short-lived that frequent N applications have been strongly indicated. Application of 200 g sulphate of ammonia 4 to 6 weeks after planting, with a similar amount of muriate of potash, has been successfully employed in Nigeria, with one further similar dressing in the same year towards the end of the rains. In Malaysia dressings of N at 3, 6 and 12 months have been advocated with P and K or PKMg at the two latter times.[71] Magnesium is often required in young plantings and can be applied as 200 g Kieserite at the same time as the N or NK.

In Malaysia, application twice a year has also been advocated until bearing or even later,[71] and where there are two wetter seasons this might be justified. However, no evidence has been provided to support this practice;

Table 11.15 *Rates of fertilizer application used in some experiments* (kg per palm per annum)

Country	Place	Soil	N_1	N_2	P_1	P_2	K_1	K_2	Mg_1	Mg_2
1. Nigeria	Benin	Acid Sands	0.68	1.36 (2.04)†	0.68*	1.36* (2.04)	0.38*	0.76* (1.14)	0.38*	0.76* (1.14)
2. Cameroon	La Dibamba	Clayey sand	—	—	—	—	1.00	2.00	0.50	1.00
3. Ivory Coast	La Mé	Clayey sand	—	2.00	—	2.00	—	2.00	—	2.00
4. Ivory Coast	Dabou	Clayey sand	—	—	—	3.00	0.8	1.2 (1.6)	—	—
5. Malaysia		Batu Anam/Durian	1.36	2.72 (5.44)	1.13	2.27 (4.54)	1.36	2.72 (5.44)	0.68	1.36 (2.72)
6. Malaysia		Kulim	2.27	4.54	2.27	4.54	2.27	4.54	0.91	1.82
7. Malaysia		Rengam	3.63	7.26	7.26	14.53	3.63	7.26	2.72	5.45
8. Malaysia		Selangor	1.80	3.60	1.80	3.60	2.70	5.40	1.80	3.60

* Application made every 3 years, but quantities shown on an annual basis.
† Figures in parentheses represent a third rate of application in the experiment.

Fertilizers: N: Sulphate of ammonia −1, 3, 5, 6, 7. Nitre 26−8.
P: Superphosphate −1 to 4 and 6; rock phosphate −5, 7, 8.
K: Potassium chloride or sulphate. Mg − Kieserite or other forms of magnesium sulphate.

much will depend on the intrinsic nutrient-supplying power of the soil, and on good soils the ideas that 'maintenance dressings' or the quantitative replacement of nutrients removed by cropping are required at frequent intervals cannot be accepted without experimental evidence. The interpretation of leaf analysis in the early years is particularly difficult and this accentuates the importance of having fertilizer experiments laid down in the first planting in any region where oil palms are being planted for the first time.

As already seen (p. 551), in seasonal climates fertilizer applications should be made shortly before the onset of the wet season and not before or during the dry season or in the middle of the rains. Where rainfall is better distributed the drier periods and periods of very heavy rainfall should also be avoided and, if this is possible, those months should be selected in which a fair amount of light or medium rainfall occurs, preferably preceeding a rainy season rather than a dry season.

The *methods of application* of fertilizers has been a subject of some controversy[71, 74] but few experiments. With adult palms in Nigeria the concentrated placement of fertilizers in sectors of trenches showed less yield response than broadcasting treatments.[84] Experiments in the Ivory Coast and Dahomey have failed to show differences between broadcasting and application in circles round the palms.[114] Ng quotes one case in Malaysia where there were no differences in uptake of N and K, as judged by leaf analysis, between broadcasting and application in the weeded circle and another case in which there was a slightly greater uptake of N from broadcasting.[142] The development of the root system was fully described on pp. 55−9. The relationship of the mode of development to nutrient uptake has been discussed by Ollagnier et al.[114] Since absorption is through the quaternaries and the tips only of the larger roots it is reasonable to supply nutrients to all areas where quaternaries are found in quantity. Although the density of primary and secondary roots tends to decrease with distance, the total quantity of absorbing roots increases in successive surrounding rings at least to 3.5−4.5 m.[115] Furthermore, roots show a positive tropism to areas where water and nutrient supply is good. The argument that fertilizer should be applied only within the weeded circle because there is a concentration of roots of all kinds near the base of the palm is unsound since it ignores the facts that there are larger quantities of absorbing roots at a distance from the palm, that the roots are deeper and more lignified in the circle, and that better conditions are provided in the interline for the encouragement of positive tropism.

Ruer[115] has recommended increasing the circle of application from 1.5 m at 1 year to as far as 4 m at 5 years of age provided the doses being applied are not too small (i.e. giving less than about 40 g per square metre); and broadcasting on strips in the interline or in wide circles to 3.9 m (halfway across the interline), for adult palms. However, the fertilizer should not be thrown on the bare ground near the base of the palm, and application in a broad band to a little beyond the extremity of the leaves corresponds approximately with Ruer's recommendations for young palms. Once the palms have closed in the fertilizers should be broadcast in the interline or in a very broad band around the palms.

It has also been shown that development of the root system is much greater under a good cover of *Pueraria* than under grass[116] or under the paths along the rows. Therefore when broadcasting between or around adult palms small isolated areas of grass should be avoided if possible and the fertilizer should not be scattered on the trampled ground of paths.

Individual nutrient effects

Before citing some representative fertilizer experiments something will be said about the common effect and need for the major nutrients and the fertilizers which supply them.

Nitrogen. Nitrogen is commonly required for the rapid growth of young palms in the field. Sulphate of ammonia should be used unless, on the basis of nitrogen content, urea is found to be appreciably cheaper. This is because up to 25 per cent of the nitrogen of urea can be lost through volatilization as ammonia. The need for nitrogen in the production years is not very common[114] but it varies according to the situation and needs to be ascertained by experiment. For instance, at La Mé in the Ivory Coast the effects of annual dressings of sulphate of ammonia were positive and significant with young palms, but there were no effects with adult palms in the presence of adequate K supplies. Where K was not supplied, N could actually reduce yields (see also p. 579):

	Young palms[114]		Adult palms[92]	
	Indices of: Colour (at 7 mths, scale 0–4)	Vigour (dm³)	15 years: bunch yield (kg/palm/an.)	
			With K	Without K
Without N	1.14	2,120	96	72
With N	2.41	2,370	99	63

By contrast, the Malaysian experiment quoted on p. 556 showed a 55 per cent yield response to N.

Nitrogenous fertilizers containing calcium must be avoided because of the antagonistic effect of calcium on potassium uptake. Ammonium phosphate can be used where there is a need for both N and P and this fertilizer has been found particularly valuable in prenurseries and nurseries.

Phosphorus. Rock phosphate (from Christmas Island, 36 per cent P_2O_5) has been the traditional source of P in the Far East while the super-phosphates, single, double or triple (18, 38 and 48 per cent P_2O_5), have been the most available forms of P fertilizer in Africa and tropical

America. On a very P-deficient soil in northern Brazil a large initial application of rock phosphate spread over the whole ground had a superior effect to annual doses of triplesuperphosphate;[92] rock phosphate may therefore be a more suitable source for oil palms than the superphosphates.

Phosphorus is required on specific soils in all three continents and in these cases often produces very large yield increases. On other soils it may provide a further yield increment when K requirements are satisfied, but on its own it may actually reduce yields, possibly through the antagonistic effect of the calcium content of the phosphatic fertilizers. The indiscriminant application of phosphorus 'maintenance dressings' is therefore inadvisable. Ollagnier *et al.*[114] believe that there is a direct synergistic relationship between leaf-P and leaf-N levels and that this relationship largely determines responses to P and N.

Potassium.   Potassium is the most commonly required element for adult palms and with nursery or young palms it may reduce the incidence of leaf diseases such as *Cercospora elaeidis.* It is usually applied as the chloride (containing the equivalent to 60 to 62 per cent K_2O) or sulphate (48 to 52 per cent K_2O) according to availability and price; but it has been suggested (see p. 523) that the effects of some KCl applications can be attributed to Cl rather than K.[54] Potassium is sometimes not required where there is a strong P or Mg deficiency and in the latter case its application can produce or increase intense Orange Frond. Potassium has also been applied as Patent Kali, sulphate of potash magnesia, containing 26 to 30 per cent K_2O and 9 to 12 per cent MgO, and as wood ash or bunch ash (see p. 551). The latter is particularly useful for very acid soils such as the Acid Sulphate soils of Malaysia.

Magnesium.   The need for Mg is commonly exhibited through its deficiency symptoms which are often induced by potassium manuring. Mg is required on the poorer soils of West Africa and on many of the American soils derived from recent volcanic material whether sedentary or alluvial. However, its application rarely results in very large increases in yield. Magnesium is usually applied as hydrated magnesium sulphate (Epsom Salts, 46 to 48 per cent $MgSO_4$ or 16 per cent MgO), crude magnesium sulphate (kieserite, 26 per cent MgO) or anhydrous magnesium sulphate (96 to 98 per cent $MgSO_4$, 30 to 32 per cent MgO). Magnesium limestone is not usually suitable owing to the K/Ca and Ca/Mg antagonisms, but may be useful where magnesium is required in very acid conditions as on the Malaysian Acid Sulphate soils. It is not uncommon for Mg deficiency to be exhibited by young palms in areas where it is not seen on adult palms and where dressings of Mg fertilizers are not needed to maximize the crop. In such cases a small dressing will correct the symptoms but they will also disappear on their own without fertilizers as the palm develops. Mg deficiency is much commoner in replantings than new plantings.[141]

The use of compound fertilizers of fixed formulae are not advisable with the oil palm except perhaps where very small quantities are needed in prenurseries or nurseries. Most compound fertilizers contain, for the oil palm, a low amount of Mg which, when Mg is not needed, will be detrimental or simply wasted, or, when required, will be insufficient. Oil palm requirements with young or adult palms are usually so specific that it would be very rare for a compound to meet these requirements, and some ingredients are almost always wasted or actually harmful. Costs per unit of the nutrients supplied are also higher with compounds.

Asia

Indonesia

Some information on past work and fertilizer practice has been provided by Werkhoven.[117] Rock phosphate became widely used in Sumatra owing particularly to its effect when applied to 'liparitic' latosols and to some soils derived from sedimentary rocks, but Dell and Arens showed that the indiscriminate use of P might be dangerous. In areas shown by leaf analysis to be deficient in K, rock phosphate applications either depressed yields or had no effect, while on the liparitic soils where leaf analysis showed K to be in good supply a yield response to P of 24 per cent over 4 years was obtained.[118] The yield depression due to P was most marked on marine soils where already low leaf-K levels were further depressed.

On the liparitic soils (referred to as Yellow Podzolic in Indonesia) experiments demonstrated a positive interaction of N and P; 4 kg ammonium sulphate per palm with 3 kg triplesuperphosphate nearly doubled yields while application of one of these fertilizers alone only gave a 40 per cent increase (Table 11.16). A distinctive feature of these soils are the very low Mg levels. In one experiment a correlation was found between leaf-Mg and yield responses to kieserite or magnesium chloride and substantial yield increases were had only where leaf-Mg fell below 0.10 per cent of dry matter. Responses to K were not found on these soils in spite of leaf levels falling to between 0.8 and 0.9 per cent.[140]

On latosols derived from Tertiary sandstones experiments have shown strong P and K requirements with the need for corrective dressings of Mg.

Table 11.16 *Effect of sulphate of ammonia and triple superphosphate on adult palms on liparitic latosols (yellow podzolic) in Sumatra*[140] (kg per palm per annum — mean of 3 years)

N levels		N_0	N_1	N_2
P levels	Fertilizer rates (per palm per an.)	0	2 kg	4 kg
P_0	0	86	98	127
P_1	1.5 kg	97	134	142
P_2	3.0 kg	121	124	167

Malaysia

Early experiments in Malaysia were largely on soils derived from sedimentary rock and the few positive responses were to P. One case of a response to Patent Kali was reported on soil derived from quartzite parent material,[119] suggesting a need for K or Mg or both on these soils once the overriding P requirement has been satisfied.

A series of experiments on soils of the Rengam series (derived from granite) and the Pamol series (derived from shale) gave results which may be taken as a first guide to the fertilizer requirements of inland soils derived from acid igneous and sedimentary rocks.[79] The experiments allowed for three levels of each nutrient, but unfortunately at each site one replicate of a 3^3 or $3^3 \times 2$ experiment had the fertilizers applied only once, in 1949, while the other received the fertilizers annually from 1949 to 1953 on the Pamol soil and from 1949 to 1956 on the Rengam soil, after which, in the latter case, annual dressings were given to both replicates. The results were presented for the 1956 to 1959 period for the Rengam soil and the 1950 to 1953 period for the Pamol soil. The experiments are therefore not fully annual-rate experiments, though nearly so. Rates of application were as follows, in kg per palm:

Sulphate of ammonia		Rock phosphate		Muriate of potash		Kieserite	
N_0	0	P_0	0	K_0	0	Mg_0	0
N_1	0.91	P_1	1.36	K_1	0.91	Mg_1	0.91
N_2	1.82	P_2	2.72	K_2	1.82		

The results on the sedimentary derived soil, where Mg was not included, were straightforward: K gave no significant response even at the high rate of P; the higher rate of N was required for a response to that element and this was only obtained in the presence of added P; and there was a response to P but no significant advantage in the higher rate. These results are illustrated in Fig. 11.11.

On the Rengam soils the primary deficiency is potassium, and the experiments showed that other elements are of little value, or may be actually harmful, if a primary deficiency remains undiscovered (Tables 11.17, 11.18). In this case there were significant negative responses to P in the absence of K and to N where P was high and K low.

The experiments covered two sites of rather dissimilar Rengam series soil, site A being on a typical sandy clay while site B had a coarser soil of lower fertility as judged by yield. The negative response to P in the absence of K was found on the typical soil of higher general fertility (Table 11.17). Positive responses to P in the presence of K were more marked on the coarser soil. Responses to Mg were only obtained in the presence of K and the positive response was similar at both levels of K.

Responses to N were also dependent on an adequate supply of K but the presence or absence of P fertilizer also had an effect. When K was not

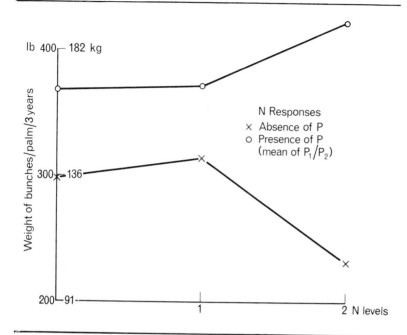

Fig. 11.11 Responses to nitrogen in the presence and absence of P fertilizer on soil derived from sedimentary rock, Malaysia (Data, Rosenquist, 1962).

supplied or, as in one replicate, was inadequate owing to discontinuity of applications, high P applications caused the N effect to be negative, although in the absence of P a positive N effect was obtained. Where K was sufficient, however, N responses were positive at all levels of P and the highest yields were obtained from N_2P_1 and N_1P_2 with K_1, and from N_2P_2 with K_2 (Table 11.18). In experiments with this type of response pattern a simple response to K may show a high degree of curvature, indicating that a low level of application is all that is required; however, when the right balance of other elements is supplied the response to K may be linear, and it is then difficult to say whether the higher rate is sufficient or not (Fig. 11.12). Experiments of the design 4^n with really high top rates of application may therefore be needed.

These experiments indicated the importance firstly of determining the primarily deficient element, secondly of determining the other elements needed for maximizing responses to the primary element, and thirdly of providing rates of application sufficient to discover the optimum level of application for any combination used. On both soils it was seen that the application of certain elements without satisfying the primary requirement could actually reduce yields below a 'control' level which was itself already low. In certain circumstances N applications may be dangerous in the

Table 11.17   *Responses to N, P, K and Mg on Rengam soil series (granite derived), sites A and B, Malaysia (Rosenquist) (weight of bunches per palm per annum, 1956—9 (kg))*

Major response	K_0	K_1	K_2
Both sites	66	89	94

Secondary responses — both sites

	Nitrogen		Phosphorus			Magnesium	
	K_0	K_1/K_2		K_0	K_1/K_2		K_0	K_1/K_2
N_0	62	88	P_0	69	86	Mg_0	69	89
N_1	72	89	P_1	66	91	Mg_0	64	94
N_2	65	96	P_2	63	96			
Linear effects	Insig.	Sig.*		Insig.	Sig.**		Insig.	Sig.*

Site A				Site B		
	K_0	K_1/K_2			K_0	K_1/K_2
P_0	77	94		P_0	61	79
P_1	69	97		P_1	63	86
P_2	62	101		P_2	64	90
Linear effects	Sig.*	Insig.			Insig.	Sig.*

For rates of application see text.

absence of K or P, P applications may be dangerous in the absence of K, while K applications may be dangerous in the absence of Mg or vice versa.

More recent experiments on soils derived from sedimentary rock have confirmed the need for N and P but have suggested that in some circumstances K is additionally required for maximum yields and Mg may be needed for early growth.[69, 70] The mean effects over 3 years in an experiment on Munchong series soil were as follows:[70]

	N_0	N_1	N_2	P_0	P_1	P_2	K_0	K_1	K_2
Tons bunches/ ha./an.	24.8	27.7	28.5	23.1	28.6	29.3	25.3	27.6	28.0
Leaf 17 N, P or K per cent of d.m.	2.73	2.94	3.06	0.137	0.152	0.156	1.17	1.26	1.32

It will be seen that P gives the highest responses and shows the most obvious deficiency. N and K leaf levels are well above the supposed critical levels but these elements gave substantial yield increases. An NK experiment on Serdang series, where all the plots had received applications of P and Mg, also showed substantial responses to K.[70] It is difficult however to interpret the results of trials receiving regular and perhaps excessive overall

Table 11.18   *Interactions between* N *and* P *at different rates of* K *on Rengam series soil derived from granite, site B (annual dressings), Malaysia (Rosenquist)* (weight of bunches per palm per annum, 1956−9 (kg))

Potassium	Nitrogen	Phosphorus			
		P_0	P_1	P_2	Mean
K_0	N_0	50	75	73	66
	N_1	71	73	65	70
	N_2	66	67	58	64
	Mean	63	72	65	67
K_1	N_0	85	99	92	92
(Muriate of potash	N_1	73	97	107	92
0.9 kg/palm/an.)	N_2	96	104	99	101
	Mean	84	101	99	95
K_2	N_0	86	77	96	86
(Muriate of potash	N_1	90	99	103	98
1.8 kg/palm/an.)	N_2	99	101	113	104
	Mean	92	93	104	96

L.S.D. between means, P = 0.05 − 14

Annual rates of application N_1, N_2 − 0.9, 1.8 kg S/ammonia; P_1, P_2 − 1.4, 2.7 kg rock phosphate.

applications of one or two nutrients particularly when these may have antagonistic and synergistic effects.

Several further experiments have been carried out on the common Rengam series soil derived from granite. The mean effects in two of these experiments were:[112]

	N_0	N_1	N_2	P_0	P_1	P_2	K_0	K_1	K_2
Tons bunches/ha./ an., 5 years Leaf 17, N, P or K per cent of	19.04	22.9	22.9	21.1	22.2	21.5	18.6	22.7	23.5
d.m.	2.60	2.79	2.86	0.175	0.179	0.184	0.766	1.090	1.169
Tons bunches/ha./ an., 4 years Leaf 17, N, P or K, per cent of	18.9	23.4	25.1	22.0	22.0	23.3	19.7	23.6	24.2
d.m.	2.53	2.83	2.92	0.168	0.176	0.178	0.629	1.017	1.096

Lack of P responses may have been due to the lasting effect of considerable quantities of rock phosphate applied in pretreatment dressings. There were no responses to Mg. The lower levels of N and K appeared to be sufficient. These levels were 3.6 kg per palm per annum of ammonium sulphate

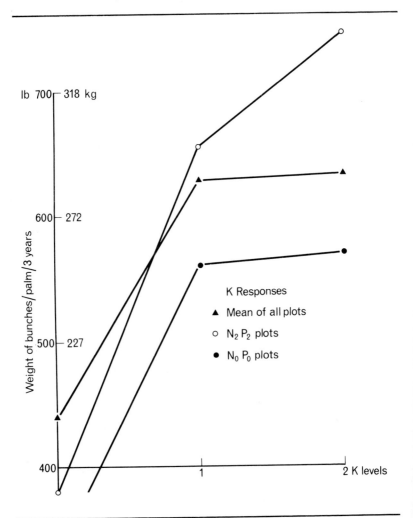

Fig. 11.12 Potassium responses on Rengam series soil, Malaysia (Data, 1962; Rosenquist, 1962).

and potassium chloride in the first experiment and 4.5 kg in the second. On another area of Rengam series soil where all plots had received P and Mg, N responses did not appear until the third year of bearing, while higher levels of K than 2.7 kg KCl per palm per annum gave no further return.

On the fertile marine clays (Selangor series) in Malaysia many fertilizer experiments have now been undertaken. Early trials gave responses only to K and even this was not achieved until after several years of bearing and did not amount to more than about 5 per cent or 1—1.5 tons of bunches. Dressings of under 3 kg per palm per annum of KCl are usually sufficient. More acid phases of the marine clays such as Sedu series or the Acid Sulphate soils respond more sharply to potassium particularly when applied as bunch ash; in an experiment on Sedu series 2.7 kg KCl per palm per annum gave nearly a ton of bunches less than an equivalent rate (K content) of bunch ash.[120] Magnesium also appears to be deficient on Acid Sulphate soils and leaf levels are raised by application but yield responses have not always been obtained.

The Briah series is derived from mixed marine and riverine alluvium and oil palms normally respond to both K and N,[120, 70] dressings of about 2.5 kg each of sulphate of ammonia and KCl being sufficient. As with Selangor series, responses are not large and the N effect particularly may take some years to establish itself, e.g.:

	Tons bunches per hectare		
	N_0	N_1	N_2
4th year from planting	13.9	13.9	13.0
5th year from planting	21.3	22.0	21.7
6th year from planting	23.8	24.1	24.8
7th year from planting	22.7	23.8	24.3

N_1, N_2: 1.8 and 3.6 kg Nitro 26 per palm per annum.

With mixed alluvial and colluvial soils on the west coast of Malaysia P is needed and an experiment exhibited the positive interaction so often seen between P and K; the highest rate of P application only gave a significant increase in the presence of K (mean quantities, rock phosphate 3.8 kg and muriate of potash 1.62 kg per palm per annum).[121]

The peat soils of Malaysia present special problems, and the Peat Yellows symptoms have already been described (p. 514). Very low leaf levels of copper are usual (1.4—2.5 ppm) with low leaf levels of N and K. Ng *et al.*[122] have obtained considerable changes in the growth of palms suffering from Peat Yellows or mid-crown chlorosis. Foliar sprays with 200 ppm Cu solution of copper sulphate resulted in large increases in leaf and leaflet lengths and raised leaf-Cu levels. Soil applications of 2.5 kg per palm copper sulphate increased leaf length in adult palms after a 3-month delay. Spear drenching with copper sulphate had a quicker effect. Evidence of yield responses to soil applications have also been presented.[144]

Of greatest importance on peat is potassium. A 2^5 NPKCaMg experiment showed marked yield increases with K, small increases with Ca and Mg and yield decreases with N and P.[123] These effects became more pronounced with time and in the ninth year after planting bunch production in kg per palm and leaf nutrient levels (leaf 17) were as follows:

N_0	N_1	P_0	P_1	K_0	K_1	Ca_0	Ca_1	Mg_0	Mg_1
184	137**	175	148**	140	181**	153	168*	156	165
2.542	2.64**	0.154	0.208**	0.594	1.022**	0.537	0.589*	0.454	0.490*

** Sig. at $P = 0.01$. * Sig. at $P = 0.05$.

Nutrient applications were 2.7 kg per palm per annum of sulphate of ammonia, rock phosphate, muriate of potash and limestone dust and 3 kg of kieserite. The 10 per cent increase in yield due to Ca is the only known positive response to this element with the oil palm.

Foster and Goh[139] fitted response functions to yield data of twenty fertilizer experiments, ten on coastal and ten on inland soils. They found that the yield without fertilizers was appreciably higher on the coastal than on inland soils and that responses to individual fertilizers, in the absence of other limiting factors, declined with increasing yield level. Thus responses to P and K were less on coastal soils and this was considered to be due not only to higher soil supplies but also to their higher buffer capacities. Nevertheless P application was found to be profitable on all soils except river alluvium (Briah series) and peat; and, provided P requirements had been satisfied, K and N fertilizers gave profitable responses on all inland soils and on Briah series. Profitable responses to Mg were rare, thus confirming the study of Chan and Rajaratnam[141] who found Mg responses on inland forest soils only after 23 years growth and calculated that applications of kieserite of more than about $1-1.5$ kg per palm per annum had no further effect on yield.

Foster and Goh also found, as so frequently indicated earlier in this chapter, that responses to individual fertilizers were dependent on the adequacy of other fertilizers. There were also indications that responses could be inhibited on coastal soils by limited rainfall and on inland soils, in the case of N, by leaching. These authors charted responses on isoquant, or yield contour, maps showing both the range of dressings of two fertilizers giving positive responses and the most profitable combination at different yield levels: for this interesting method their paper should be consulted.[139] Their work also demonstrated the importance of having nil levels in all fertilizer experiments.

Africa

Nigeria

The most recent phase of fertilizer investigations in Africa, particularly West Africa, started around 1959 with the realization that much more information was required on rates of application, on the effect of the supply of one nutrient on the need for others, and on frequency of application. Experiments in Nigeria and Sierra Leone have, in general, been of a 4^n design though four levels have not always been retained for each

Table 11.19 *A $4^3 \times 2^3$ fertilizer experiment undertaken in Nigeria* (kg per palm)

Expt. 508-1 Fertilizer	Level	Year 0	1	2	3	4	5	6	7	8	Total (kg)
Urea	N_0										0
	N_1	0.11	0.11	0.23	0.34	0.34	0.34	0.34	0.34	0.34	2.50
	N_2	0.23	0.23	0.45	0.68	0.68	0.68	0.68	0.68	0.68	5.00
	N_3	0.34	0.34	0.68	1.02	1.02	1.02	1.02	1.02	1.02	7.49
Super-phosphate	P_0										0
	P_1	0.23	0.45	0.91	3.41	—	—	4.09	—	—	9.08
Sulphate of potash	K_0										0
	K_1	0.11	0.23	0.57	0.91	—	—	1.14	—	—	2.95
	K_2	0.23	0.45	1.14	1.82	—	—	2.27	—	—	5.90
	K_3	0.34	0.68	1.70	2.72	—	—	3.41	—	—	8.85
Magnesium sulphate (anhydrous)	Mg_0										0
	Mg_1	0.06	0.18	0.34	1.02	—	—	1.14	—	—	2.72
	Mg_2	0.11	0.34	0.68	2.04	—	—	2.27	—	—	5.45
	Mg_3	0.18	0.51	1.02	3.06	—	—	3.41	—	—	8.17
X Trace element	X_0 X_1										
Y Trace element	Y_0 Y_1										

Note: Year 0 is the year of planting.

nutrient. The different levels have allowed for rising total quantities of fertilizer applied, over the 8 or 9 years, annually in the case of nitrogen, but triennially with other elements.[124] The nutrients and the number of levels to be employed were chosen after a consideration of both the soil type and the earlier, less exact, information on nutrient needs. For instance, on an area of K-deficient Acid Sands soil where N responses had sometimes been obtained, Mg responses were probable and P responses in the presence of K were not unlikely, the $4^3 \times 2^3$ design shown in Table 11.19 was arranged.

On the Acid Sands soils of Nigeria interest has centred on K and Mg needs with N and P as subsidiary requirements. On young palms N has usually been needed. Applications of NK can sometimes induce magnesium deficiency on these soils and dressings of anhydrous magnesium sulphate at about half the rate of the minimum potash dressings are then advisable.

On the *Benin fasc* of the Acid Sands an experiment of the same general design as that shown in Table 11.19, but with only two rates (0 and 1) of N and Mg, has shown that responses to K are obtained up to the K_2 rate, that P also increases yield and that there is a possible boron requirement, though the latter is only shown in the presence of Mg. N and Mg responses are absent, at least in the early years. On the *Calabar fasc*, with treatments as in Table 11.19, the first 4 years' yields showed that responses to K were significant up to the K_3 level and Mg responses to the Mg_1 level (Table 11.20). Here also there was a suggestion of a boron requirement but only in the presence of P.[125] In the early years there were severe Mg deficiency symptoms in the Mg_0 plots and K applications increased these symptoms though improving growth:[126]

	K_0	K_1	K_2	K_3
Mg-deficiency symptom score:	78	93	106	136
Mean height per palm (cm):	161	165	166	171

The presence of P also increased Mg-deficiency symptoms though slightly improving initial growth. N dressings enhanced early growth on both Benin and Calabar fasc soils, but this effect was not reflected in enhanced early yields; in fact on the Benin fasc there was small initial negative response. It is a curious feature of some experiments that in spite of the better growth that is obtained from early applications of N fertilizers, this effect does not always persist, either in later development or in bunch yields. Nevertheless on older palms N responses have been obtained on Acid Sands soils.

In an impoverished area of old palms on Benin fasc at Mbawsi, mean annual yields in K_0 plots were around 5.6 tons bunches per hectare. Single dressings of P, K, Mg and Ca had been applied in a 2^4 layout in 1956. The K response averaged about 50 per cent. In 1961 second, lower applications of P, K and Mg were made and annual dressings of urea in one replication and sulphate of ammonia in the other were introduced. A response to urea was immediate and a response to sulphate of ammonia followed. In addition, by 1963, a year of generally high yield, K responses reached nearly 100 per cent as shown below:

Years	K application (per palm)	N application (per palm)		Tons bunches per hectare per annum			
	Sulph. Pot. (kg)	Urea (kg)	S/Am. (kg)	K_0	K_1	N_0	N_1
1956	4.5	—	—				
1957–8	—	—	—	5.17	7.54	—	
1961	2.7	0.9	1.8				
1961–2	—	1.8	3.6	5.54	9.17	7.03	8.03 (Urea only)
1963	—	0.9	1.8	6.16	11.93	8.46	9.63

Small responses to P in the presence of K were also found in this experiment; the 1956 application of superphosphate was 9.1 kg per palm, followed by 5.4 kg in 1961.

Experiments on widely separated fields of older palms on the Calabar

Table 11.20 *Responses to potassium, magnesium, phosphorus and boron on Acid Sands Soils, Nigeria (N.I.F.O.R.) (kg per palm per annum in first 4 or 5 years of bearing)*

Expt.	K_0	K_1	K_2	K_3	P_0	P_1	B_0	B_1	L.S.D. $P = 0.05$
Benin fasc.* (2-15)	44.2	50.1	52.4	52.5	48.2	51.4	44.3	46.2	
					Mg_0	Mg_1	Mg_2	Mg_3	
Calabar fasc.† (508-1)	29.4	39.2	42.1	46.1	36.6	40.4	39.3	40.5	3.1

* First 5 years for K and P, 4 years for B.
† First 4 years of bearing.

fasc showed yield responses to K and Mg. Many of the experiments were on old fields of poorly grown palms and, although Mg deficiency symptoms were largely abated, yield responses to Mg were small and irregular. At Itu, where duplicate experiments were sited on adjoining areas showing, and not showing, Orange Frond symptoms, even combined KMg dressings did not raise the yield of old palms to the comparatively high figures obtained at the non-Mg-deficient Mbawsi area. It was evident that the Mg requirement was as high where there were no symptoms as where symptoms were shown. Applications of Mg to old palms on very poor sandy soil at Akwete gave similar positive responses, but yields remained generally low. The fact is that many soils which are extremely Mg-deficient and are bearing old palms in south-eastern Nigeria are intrinsically very poor sands with a history of long usage for annual crops. Besides the exceptionally low exchangeable Mg level they have a much lower level of exchangeable K than, for instance, the Mg-deficient area at Etoumbi in Congo (Brazzaville) and their clay plus silt content is exceptionally low (only 12 per cent at Akwete) even at 60 cm depth.

Information on the fertilizer requirements of palms growing on *basement complex soils* remains meagre. In the Kwa Falls area, where soils derived from mica and quartzose schists predominate, responses to P have not been as great as expected. Highest yields were obtained in one experiment from ammonium phosphate with K applications, each fertilizer being applied at 2.27 kg per palm every 3 years; in another experiment an economic 'response' was obtained to both broadcast and placed single applications of 4.54 kg per palm of superphosphate but the 11 per cent increase in yield over 6 years was not significant.

In the Calaro area young palms on soil derived from granite gneiss have shown a 24 per cent bunch yield increase to half the P_1 levels of phosphorus shown in Table 11.19 and a 7.5 to 13.5 per cent increase to the higher rates of magnesium in the first 4 years of bearing. In the same area, on soils derived from mica schist, early yields suggested P and N requirements.[127]

Ghana

Fertilizer experiments laid down by N.I.F.O.R. in *Ghana* have been extensively reviewed by Van der Vossen.[128] These experiments, of 2^4 or 2^5 design, were on soils derived from Tertiary sands contiguous with those of the Ivory Coast and on soils derived from granites and phyllites (Lower Birrimian formation). All fertilizers were applied at the rate of 0.23 kg per palm in the planting year and then, for 6 years, 0.45 kg per palm per year of age, i.e reaching 2.72 kg per palm. Thereafter, P, K, Mg (and Ca if included) were applied triennally at 4.54, 2.27 and 2.72 kg of single super-phosphate, muriate of potash and magnesium sulphate per palm respectively (and 3.18 kg lime if included), with sulphate of ammonia applied annually at 1.82 kg per palm.

On the silty clay soil of the Lower Birrimian phyllites, in Ghana's main cocoa area, very good yields were obtained through a 12-year period even on unfertilized plots. Apart from an early indication that the highest yields might be obtained from KMg applications, leaf analysis indicated that, later on, N, P and K may all be required. On the sandy and silty clays with some concretions overlying granites there was a P and K requirement with a positive PK interaction at Pretsea (Table 11.21 (2)) and a negative KMg interaction at Assin-Foso (Table 11.21 (1)). N responses were also recorded at the latter place, but Mg was clearly not required on these soils.

The results of the experiment on Tertiary Sands soil are of particular interest since they are in marked contrast to those obtained in the Ivory Coast on soils of the same derivation. At Aiyinasi high early yields were obtained from the superphosphate dressings with a later positive PK inter-action giving up to 16 ton bunches per hectare in certain years. These yields fell away rapidly from the seventh year of bearing (Table 11.22) some 4 years after the triennial dressings started, and this decline continued until yields were only one-third of what they had been in the palms' early adult years. Before this time foliar analysis had indicated low P even in the P_1 plots and low N. Van der Vossen[128] has considered the possible causes of this unusually severe decline and concludes that, although in part it could be accounted for by a nearly continuous fall from 1962 to 1968 in effective sunshine (see p. 187), the main and later factor was a depletion of organic matter and available nutrients, particularly P, in the soil, through root uptake, leaching and P-fixation. He suggests a reversion to annual dressings, with at least $2-2.5$ kg superphosphate per palm with addition of $1-1.5$ kg of muriate of potash, and applications in August instead of May. From the leaf analysis it was seen that K content was severely reduced by P applications and the K applications were insufficient to bring the leaf level in the P_1 plots above 0.9 per cent of dry matter.

As will be seen below, soils developed over Tertiary Sands in the adjoining Ivory Coast are predominantly K deficient with little or no demand for P. Van der Vossen[128] suggests that the Aiyinasi soils are not typical of the

Table 11.21 *Responses to fertilizers on soils derived from Pre-Cambrian rocks in Ghana (Van der Vossen)* (tons bunches per hectare per annum)

	N_0	N_1	P_0	P_1	K_0	K_1	Mg_0	Mg_1
Birrimian phyllites								
First 3 years	9.44	9.80	9.46	9.78	9.27	9.71	9.51	9.73
Next 9 years	12.29	12.85	12.59	12.54	12.30	12.84	12.41	12.73
Leaf 17 per cent of dry matter	N 2.41	2.37	P 0.142	0.145	K 0.83	0.93	Mg 0.39	0.41
Granite								
(1) First 3 years	2.83	3.48	2.71	3.60*	2.81	3.51*	3.17	3.14
Next 6 years	8.61	9.96*	8.06	10.51*	8.80	9.77	9.30	9.26
Leaf 17 per cent of dry matter	N 2.68	2.71	P 0.134	0.155*	K 1.25	1.34	Mg 0.30	0.32
(2) First 3 years	8.76	8.53	8.21	9.09**	8.56	8.74	8.80	8.49
Next 6 years	9.02	9.26	8.74	9.54	8.57	9.70*	9.40	8.87
Leaf 17 per cent of dry matter	N 2.35	2.42	P 0.151	0.159	K 0.61	0.74	Mg 0.48	0.51

Interaction

P_0K_0	P_0K_1	P_1K_0	P_1K_1
Last 6 years			
8.83	8.66	8.32	10.76*

For fertilizer dressings see text. * Significance at least at P = 0.05. ** Significance at P = 0.1.

Table 11.22 *Yield declines and responses to fertilizers on sedentary oxisols over Tertiary Sands in Ghana (Van der Vossen) (tons bunches per hectare per annum)*

	N_0	N_1	P_0	P_1	K_0	K_1	Mg_0	Mg_1	Ca_0	Ca_1
First 3 years	8.19	8.04	6.82	9.40**	8.24	7.98	8.13	8.10	7.99	8.23
4th–6th year	12.04	12.23	10.56	13.70**	11.84	12.42	12.30	11.96	12.11	12.16
7th–9th year	7.76	8.21	6.95	9.02**	7.57	8.40	8.01	7.96	8.02	7.95
10th–12th year	3.82	4.38*	3.56	4.64**	3.97	4.24	4.23	3.97	3.92	4.28

Leaf 17 per cent of dry matter

	N		P		K		Mg		Ca	
5th bearing year	2.79	2.91	0.134	0.150*	0.82	1.07*	0.36	0.36	0.93	0.95
8th bearing year	2.25	2.27	0.133	0.139*	0.87	1.09**	0.31	0.34	0.95	0.97
12th bearing year	2.42	2.55**	0.138	0.147*	0.75	1.01**	0.41	0.42	0.67	0.68

P_0K_0 0.81 P_1K_0 0.67 P_0K_1 1.11 P_1K_1 0.90

Leaf K

8th bearing year	P_0 1.08	P_1 0.88**	
12th bearing year	P_0 0.96	P_1 0.79**	

* Significance at $P = 0.05$.
** Significance at $P = 0.01$.

Tertiary Sands soils as a whole, being comparatively thin and underlain by granite-derived material.

Ivory Coast

In the *Ivory Coast* experiments up to 1968 reviewed by Bachy[129] showed K to be the predominant need on the Tertiary Sands soils. In areas of derived savannah plots yielding 17 kg per palm were raised to a mean production of 82 kg per palm (11.7 tons/hectare) by the application of 1 kg KCl per palm per annum, it taking some 4 years to establish the new level of yield. Also in the savannah region a small increase in yield following magnesium sulphate applications was attributed to sulphur on the grounds that there was a small positive correlation between yield and leaf S.[130] On an experiment at La Mé (LM—CP7) in the forest zone 2 kg of KCl per palm per annum raised the mean yield over 8 years by 59 per cent to 13.9 tons per hectare compared with a mean of 8.8 tons for plots not receiving K. Leaf-K was 0.90 per cent of dry matter in the plots receiving K and only 0.48 per cent in the K_0 plots. P has been largely ineffective on these soils whether the area has been opened from forest or grass. However at Grand-Drewin in the Sassandra region, where soil phosphorus is reported as unusually low, responses to 2 kg per palm per annum of dicalcium phosphate were obtained with high rates of K (1.6 kg per palm) as follows:[129] (kg bunches per palm per annum)

	1957—60		1961—5	
	K_0	K_3	K_0	K_3
P_0	65	75	51	59
P_1	73	91	49	69

Nitrogen has been found to depress yields on the Tertiary Sands through, it is believed, its depressing effect on leaf-K; in the presence of K there is a slight increase in yield due to N.[92] Ollagnier and Ochs[131] believe this to be the most common NK effect, but that negative interactions can also occur through the intervention of a third factor such as Mg.

Recently there has been some anxiety that the effect of repeated KCl applications will be deleterious through the raising of leaf-Cl levels above the believed optima (0.5 to 0.6 per cent). These levels are attained in the Ivory Coast without K application. Some depressive effects of KCl have been recorded, and a change is being made to potassium sulphate in some experiments. However, it is also believed that positive N effects, where K is sufficient or excessive, may be due to an N—Cl antagonism and consequent reduction of the leaf-Cl content.[92] In one experiment (LM—CP14), in which there was a reduced yield with higher K dressings in the absence of N, application of sulphate of ammonia restored the yield of the K_2 plots to the control plot level.

Yield data have not been presented for the soils derived from granite-gneiss in the Ivory Coast but leaf analysis suggests that K is in sufficient supply but that responses to P may be expected.[130]

Cameroon

In *East Cameroon*, oil palms respond to both K and Mg on Tertiary Sands soil, experiments indicating an application of 1 kg KCl and 1 kg kieserite per palm per annum as probably optimal.[130, 92] In *West Cameroon* on poor sandy soil a strong K requirement has also been demonstrated. In an experiment on poor lateritic soil derived from basement complex there were P and K responses with a positive PK interaction and a further yield increment provided by N and Mg. Rates giving optimum yield appeared to be 3 kg each of sulphate of ammonia and rock phosphate and 1 kg each of KCl and kieserite.[106] In another experiment, on soils developed over basaltic flows, an early P response was later supplemented by responses to K and Mg.[106]

Dahomey

In *Dahomey*, on the 'terre de barre' soils (p. 127), K deficiency is common, but responses are limited by the water deficits experienced in this dry part of the West African coast. In one experiment significant responses were only obtained when the water deficit was below 400 mm.[92] Leaf-K levels for optimum yield appear to lie between 0.75 and 0.85 per cent according to the current water deficit. An average of 0.75 kg KCl per palm per annum has been recommended.[132] N is not recommended for young palms beyond their second year. Low production in some areas supplied with KCl suggested a possible excess of chlorine and in certain experiments leaf-Cl levels have reached more than 0.9 per cent of dry matter. In a comparison between fertilizing with KCl and K_2SO_4 the following leaf levels were attained (mean of 1972 and 1973 determinations):[92]

	K	Cl		K	Cl
KCl	0.755	0.715	K_2SO_4	0.837	0.589

Sierra Leone

In *Sierra Leone* there have been N.I.F.O.R. experiments on river alluvium and on lateritic soils, but the sedimentary soils have not yet been covered. The latter are similar to the Acid Sands of Nigeria.

Symptoms of magnesium deficiency are widespread in Sierra Leone and it was expected that this deficiency would be found on both lateritic and inland alluvial soils. In the experiment on the former soil, with treatments as in Table 11.19 but including four levels of P and two of K instead of the reverse, there were early Mg deficiency symptoms on K and NK plots, but the main requirements for yield have been shown to be P and N; yield levels are poor on this soil in which both rooting volume and water-holding capacity are low: [133]

Tons per hectare, first 4 years			
N_1	N_2	P_0	P_2
12.73	15.51	11.55	14.48

On river alluvium yields are considerably higher and the experiment allowed for six rates of magnesium sulphate and two rates of sulphate of potash (0, 1.36, 2.72, 4.09, 5.45 and 6.81 kg per palm of Mg sulphate and 0 and 4.09 kg of sulphate of potash over the first 6 years). Mg increased yields up to the Mg_2 level while there were also responses to K and a positive KMg interaction. The differences were small though significant, viz: [133]

Tons per hectare, first 4 years			
Mg_0	Mg_2	K_0	K_1
32.88	35.58	32.78	33.88

Zaire

In *Zaire* there was a history of lack of response to fertilizers. In the Crowther Experiment at Kangala (p. 546) a response to potassium was obtained. Descriptions are available, however, of numbers of prewar and postwar experiments at the Yamgambi station of I.N.E.A.C. and elsewhere which showed no responses whatever. Examination of the treatments of these experiments [134] suggests a reason. There appears to have been a general expectation, perhaps as the results of early Sumatran experience, that phosphorus would be the primary requirement on Zaire latosols. Most experimental treatments allowed for fairly liberal doses of phosphorus fertilizers with none or only small applications of potassium; even when included, dressings of the latter rarely rose above 1 kg of chloride or sulphate of potassium. In none of the earlier experiments reported were substantial potassium applications made either alone or with small supplements of other nutrients or as part of a factorial experiment. Indeed, in one experiment, dressings of 1.8 and 3.2 kg per palm were considered to be 'shock' applications. Even the conclusions drawn from these negative results seem to have been dominated by a preoccupation with phosphorus, for it was thought that fixation, primarily of the latter nutrient, was the first cause of failure of response. However, instances have already been given of negative responses to P in the absence of K fertilizer. The one element which gave any yield response in these experiments was magnesium when added to an NPK dressing.

A strong indication that palms in northern Zaire would in fact readily respond to fertilizers was later given by an experiment at Binga where application of 12 kg per palm of an NPKMg plus Zu, Cu, B, Mn fertilizer mixture distributed over 4 years gave rise to a 34 per cent yield increase over a 5-year period.[135] Later, small responses to N were reported from northern Zaire, to P from both north and south, and to K in northern Zaire.[106] A recent experiment at Yaligimba on the northern latosols (p. 122) has shown responses primarily to P and K but with additional yield increments from N and Mg so that relatively high rates of all four elements (2 kg per palm per annum of Calnitro, Fertiphos and KCl and 1 kg of kieserite) maximized yields. However, owing to the high cost of fertilizers to inland Zaire plantations only 1 kg per palm of Fertiphos appeared to be economic, this treatment by itself giving a calculated yield increase of 12 per cent.[106]

America

Very few fertilizer experiments have been carried out in America and the areas of planting are so widely separated that no general conclusions can be drawn from the few results available. The need for P in the area around Belem in Brazil has already been referred to (p. 564).[92] The apparent need of N and Mg in the Adean region has been mentioned in Chapter 3 and there have been responses to N in Central America. In Peru an experiment on young palms with magnesium sulphate and magnesium chloride suggested that yield responses may be attributable to the chlorine supplied. MgCl applications increased bunch yields significantly and also increased Cl leaf levels while failing to increase Mg leaf levels. Chlorine leaf levels are exceptionally low in this area, lying between 0.03 and 0.06 per cent of dry matter.[136] The effect of applications of KCl in Colombia were described on p. 523.

With the large new areas being planted there is at the present time (1976) an urgent need for the laying down of fertilizer experiments by research institutes and plantation companies.

This chapter should not close without a warning that many of the responses from fertilizers which have been quoted from factorial experiments are not necessarily economic, and they must also be considered with the reservations alluded to in the discussion of factorial experiments on p. 552.

References

1. Tinker, P. B. H. and Smilde, K. W. (1963) Dry-matter production and nutrient content of plantation oil palms. II. Nutrient content. *Pl. Soil*, **19**, 19.
2. Anon. (1957) *The oil palm, its culture, manuring and utilisation*. International Potash Institute. Berne.

3. Jacob, A. and Uexküll, H. von (1958) *Fertiliser Use. Nutrition and manuring of tropical crops.* Verlagsgesellschaft für Ackerbau GmbH, Hanover.

4. Ferwerda, J. D. (1955) *Questions relevant to replanting in oil palm cultivation.* Thesis presented to the Agric. University of Wageningen, Holland.

5. Ng Siew Kee (1967 and 1968) Nutrient contents of oil palms in Malaya. I. Nutrients required for reproduction: fruit bunches and male inflorescences. II. Nutrients in vegetative tissues. *Malay. agric. J.,* 46, 3 and 332.

6. Wallace, T. (1957) Methods of diagnosing the mineral status of plants. In *Plant analysis and fertiliser problems.* p. 13, I.R.H.O. Paris.

7. Toovey, F. W. (1948) Seventh Annual Report, Oil Palm Research Station, Nigeria, 1946–7.

8. Hale, J. B. (1947) Mineral composition of leaflets in relation to the chlorosis and bronzing of oil palm in West Africa. *J. agric. Sci., Camb.,* 37, 236.

9. Broeshart, H. (1955) *The application of foliar analysis in oil palm cultivation.* Thesis presented to the Agricultural University of Wageningen, Holland.

10. Bull, R. A. (1961) Studies on the deficiency diseases of the oil palm. 2. Macronutrient deficiency symptoms in oil palm seedlings grown in sand culture. *J. W. Afr. Inst. Oil Palm Res.,* 3, 254.

11. Bull, R. A. (1961) 3. Micronutrient deficiency symptoms in oil palm seedlings grown in sand culture. *J. W. Afr. Inst. Oil Palm Res.,* 3, 265.

12. Bull, R. A. (1957) Techniques for visual diagnosis of mineral disorders and their application to the oil palm. Proceedings of the Anglo-French conference on the Oil Palm, Jan. 1956, *Bull. agron. Minist. Fr. outre mer,* No. 14, p. 137.

13. Broeshart, H., Ferwerda, J. D. and Kovachich, W. G. (1957) Mineral deficiency symptoms of the oil palm. *Pl. Soil,* 8, 289.

14. Thompson, A. (1941) Notes on plant diseases. *Malay. agric. J.,* 29, 241.

15. Chapas, L. C. and Bull, R. A. (1956) Effects of soil application of nitrogen, phosphorus, potassium and calcium on yields and deficiency symptoms in mature oil palms at Umudike. *J. W. Afr. Inst. Oil Palm Res.,* 2, 74.

16. Bull, R. A. (1954) A preliminary list of the oil palm diseases encountered in Nigeria. *J. W. Afr. Oil Palm Res.,* 1, (2), 53.

17. Anon. (1952) Notes on current investigations, Oct. to Dec. 1951: Oil palm. *Malay. agric. J.,* 35, 41.

18. Coulter, J. K. and Rosenquist, E. A. (1955) Mineral nutrition of the oil palm. *Malay. agric. J.,* 38, 214.

19. Chapman, G. W. and Gray, H. M. (1949) Leaf analysis and the nutrition of the oil palm. *Ann. Bot.,* N.S., 13, 415.

20. Bull, R. A. (1957) W.A.I.F.O.R. Fifth Annual Report 1956–7, p. 98 and Tinker, P. B. H. ibid., p. 79.

21. Smilde, K. W. (1962 and 1963) W.A.I.F.O.R. Tenth and Eleventh Annual Reports 1961–2, 1962–3, pp. 74 and 75 respectively.

22. Forde, St C. M. and Leyritz, M. J-P. (1968) A study of Confluent Orange Spotting of the oil palm in Nigeria. *J. Nigerian Inst. Oil Palm Res.,* 4, 372.

23. Ochs, R. (1965) La fumure potassique du palmier à huile. Oil Palm Conference, London, 1965, and *Oléagineux,* 20, 365, 433 and 497.

24. Leyritz, M. J-P. (1964) W.A.I.F.O.R. Twelfth Annual Report 1963–4, p. 79.

25. (1956) W.A.I.F.O.R. Fourth Annual Report, 1955–6, p. 82.

26. Bull, R. A. (1954) Studies on the deficiency diseases of the oil palm. 1. Orange Frond disease caused by magnesium deficiency. *J. W. Afr. Inst. Oil Palm Res.,* 1, (2), 94.

27. Ferwerda, J. D. (1954) Boron deficiency on oil palms in the Kasai region of the Belgian Congo. *Nature, Lond.,* 173, 1097.

28. Kovachich, W. G. (1952 and 1953) Little Leaf disease of the oil palm (*Elaeis guineensis*) in the Belgian Congo. Parts 1 and 2. *Trop. Agric., Trin.,* 29, 107 and 30, 61 respectively.

29. **Duff, A. D. S.** (1963) The Bud Rot Little Leaf disease of the oil palm. *J. W. Afr. Inst. Oil Palm Res.*, 4, 176.
30. **Bull, R. A. and Robertson, J. S.** (1959) The problems of 'Little Leaf' of oil palms – a review. *J. W. Afr. Inst. Oil Palm Res.*, 2, 355.
31. **Robertson, J. S.** (1960) W.A.I.F.O.R. Eighth Annual Report 1959–60, p. 112.
32. **Rajaratnam, J. A.** (1972) Observations on boron-deficient oil palms (*Elaeis guineensis*). *Expl. Agric.*, 8, 339.
33. **Rajaratnam, J. A.** (1972) 'Hook Leaf' and 'Fish-tail Leaf': boron deficiency symptoms of the oil palm. *Planter, Kuala Lumpur*, 48, 120.
34. **Ollagnier, M. and Valverde, G.** (1968) Contribution à l'étude de la carence en bore du palmier à huile. *Oléagineux*, 23, 359.
35. **Tollenaar, D.** (1969) Boron deficiency in sugar cane, oil palm and other monocotyledons on volcanic soil in Ecuador. *Neth. J. Agric. Sci.*, 17, 81.
36. **Turner, P. D. and Bull, R. A.** (1967) *Diseases and disorders of the oil palm in Malaysia*, p. 167. Incorp. Soc. of Planters, Kuala Lumpur.
37. **Ng, S. K. and Tan, Y. P.** (1974) Nutritional complexes of oil palms planted on peat in Malaysia. I. Foliar symptoms, nutrient compositions and yield. *Oléagineux*, 29, 1.
38. **Smilde, K. W.** (1961) W.A.I.F.O.R. Ninth Annual Report 1960–1, p. 82.
39. **Rajaratnam, J. A.** (1973) Boron toxicity in oil palms (*Elaeis guineensis*). *Malaysian Agr. Res.*, 2, 95.
40. **Smith, P. F.** (1962) Mineral analysis of plant tissues. *A. Rev. Pl. Physiol.*, 13, 81.
41. **Prevot, P. and Ollagnier, M.** (1954) Peanut and oil palm foliar diagnosis interrelations of N, P, K, Ca and Mg. *Pl. Physiol.*, 29, 26.
42. **Prevot, P. and Ollagnier, M.** (1957) Méthode d'utilisation du diagnostic foliaire. In *Plant analysis and fertiliser problems*. I.R.H.O. Paris.
43. **Prevot, P. and Ollagnier, M.** (1961) Law of the Minimum and balanced mineral nutrition. In *Plant analysis and fertiliser problems, Am. Inst. Biol. Sci.*, p. 257.
44. **Poon Yew Chin, Varley, J. A. and Ward, J. B.** (1970) The foliar composition of oil palms in West Malaysia. III. Differences in leaf composition between fruit types. *Expl. Agric.*, 6, 335.
45. **Ng Siew Kee** (1972) *The oil palm, its culture, manuring and utilization.* International Potash Institute, Berne, Switzerland.
46. **Holland, D. A.** (1967) The interpretation of the chemical composition of some tropical crops by the method of component analysis. *Oléagineux*, 22, 307.
47. **Poon Yew Chin, Varley, J. A. and Ward, J. B.** (1970) Foliar composition of the oil palm in West Malaysia. I. Variation in leaf nutrient levels in relation to sampling intensity. II. The relationships between nutrient contents. *Expl. Agric.*, 6, 113–21, 191–6.
48. **I.R.H.O.** (1962) Vingt ans d'activité. *Oléagineux*, 17, 249.
49. **Smilde, K. W.** (1963) W.A.I.F.O.R. Eleventh Annual Report 1962–3, pp. 70–2.
50. **Forde, St C. M., Leyritz, M. J-P. and Sly, J. M. A.** (1965) The importance of potassium in the nutrition of the oil palm in Nigeria. *Potash Review*, Subject 27, 46th suite, June 1966.
51. **Coulter, J. K.** (1958) Mineral nutrition of the oil palm in Malaya. *Malay. agric. J.*, 41, 131.
52. **Bull, R. A.** (1964) Chemara Research Station, Oil Palm Division, Annual Report for 1962. Mimeograph.
53. **Gray, B. S. and Hew Choy Kean** (1964) Annual Report, Oil Palm Research Station, Banting, 1963, Harrisons and Crosfield (Malaysia) Ltd. Mimeograph.
54. **Ollagnier, M. and Ochs, R.** (1971) Le chlore, nouvel élément essential dans la nutrition du palmier à huile; *and* La nutrition en chlore du palmier à huile et du cocotier. *Oléagineux*, 26, 1–15 and 367–72.
55. **Ollagnier, M.** (1973) La nutrition anionique du palmier à huile. Application à la

détermination d'une politique de fumure minérale à Sumatra. *Oléagineux,* **28**, 1.

56. **Ollagnier, M. and Ochs, R.** (1972) Les déficiences en soufre du palmier à huile et du cocotier. *Oléagineux,* **27**, 193.

57. **Purba, A. Y. L. and Turner, P. D.** (1973) Severe boron deficiency in young oil palms in Sumatra. *Planter, Kuala Lumpur,* **49**, 10.

58. **Rajaratnam, J. A.** (1972) The distribution and mobility of boron within the oil palm, *Elaeis guineensis.* I. Natural distribution. II. The fate of applied boron. *Ann. Bot.,* **36**, 289−97, 299−305.

59. **Ng Siew Kee** *et al.* (1969) Nutrient contents of oil palms in Malaya. IV. Micronutrients in leaflets. *Malay. agric. J.,* **46**, 1.

60. **Ng Siew Kee, Tan Yap Pau and Cheong Siew Park** (1974) Nutritional complexes of oil palms planted on peat soil in Malaysia. I. Foliar symptoms, nutrient composition and yield. II. Preliminary results of copper sulphate treatment. *Oléagineux,* **29**, 1−14, 445−56.

61. **Scheidecker, D. and Prevot, P.** (1954) Nutrition minérale du palmier à huile à Pobé (Dahomey). *Oléagineux,* **9**, 13.

62. **Prevot, P. and Montbreton, C. Peyre de** (1958) Étude des gradients en divers éléments minéraux selon le rang de la feuille chez le palmier à huile. *Oléagineux,* **13**, 317.

63. **Smilde, K. W. and Chapas, L. C.** (1963) The determination of nutrient status by leaf sampling of oil palms. *J. W. Afr. Inst. Oil Palm Res.,* **4**, 8.

64. **Smilde, K. W. and Leyritz, M. J-P.** (1965) A further investigation on the errors involved in leaf sampling of oil palms. *J. Nigerian Inst. Oil Palm Res.,* **4**, 251.

65. **Ward, J. B.** (1966) Sampling oil palms for foliar diagnosis. *Oléagineux,* **21**, 277.

66. **Ward, J. B.** (1966) Private communication.

67. **Ng Siew Kee and Walters, E.** (1969) Field sampling studies for foliar analysis in oil palms. In *Progress in oil palm,* p. 67. Incorp. Soc. of Planters, Kuala Lumpur.

68. **Bachy, A.** (1964) Diagnostic foliaire de palmier à huile. Niveaux critiques chez les arbres jeunes. *Oléagineux,* **19**, 253.

69. **Tan, K. S.** (1973) Fertilizer trials on oil palms on inland soils on Dunlop estates. In *Advances in oil palm cultivation,* p. 248. Incorp. Soc. of Planters, Kuala Lumpur.

70. **Hew, C. K., Ng, S. K. and Lim, K. P.** (1973) The rationalisation of manuring oil palms and its economics in Malaysia. In *Advances in oil palm cultivation,* p. 306. Incorp. Soc. of Planters, Kuala Lumpur.

71. **Turner, P. D. and Gillbanks, R. A.** (1974) *Oil palm cultivation and management.* Incorp. Soc. of Planters, Kuala Lumpur.

72. **Ruer, P.** (1966) Relations entre facteurs climatiques et nutrition minérale chez le palmier à huile. *Oléagineux,* **21**, 143.

73. **Sparnaaij, L. D., Rees, A. R. and Chapas, L. C.** (1963) Annual yield variation in the oil palm. *J. W. Afr. Inst. Oil Palm Res.,* **4**, 111.

74. **Green, A. H.** (1974) Book Review. *Planter, Kuala Lumpur,* **50**, 242.

75. **Poon Yew Chin** (1969) An outline of the technique of oil palm foliar analysis. *Planter, Kuala Lumpur,* **45**, 452.

76. **Bull, R. A.** (1960) W.A.I.F.O.R. Eighth Annual Report, 1959−60, p. 104.

77. **Paton, T. R.** (1963) A reconnaissance soil survey of soils of the Semporna Peninsular, North Borneo, Dept. Tech. Cooperation. Colonial Research Studies, No. 36. H.M.S.O.

78. **Russell, E. W.** (1961) *Soil conditions and plant growth.* Ninth Edn. Longmans, London.

79. **Rosenquist, A. E.** (1962) Fertiliser experiments on oil palms in Malaya. Part 1. Yield data. *J. W. Afr. Inst. Oil Palm Res.,* **3**, 291.

80. **Ng, S. K. and Selvadurai, K.** (1967) Scope for using detailed soil maps in the planting industry in Malaysia. *Malaysian Agric. J.,* **46**, 158.

81. **Tinker, P. B. H. and Ziboh, C. O.** (1959) Soil analysis and fertiliser response. *J. W. Afr. Inst. Oil Palm Res.,* 3, 52.
82. **Tinker, P. B. H.** (1964) Studies on soil potassium. IV. Equilibrium cation activity ratios and responses to potassium fertiliser of Nigerian oil palms. *J. Soil Sci.,* 15, 35.
83. **Tinker, P. B. H.** (1961) W.A.I.F.O.R. Ninth Annual Report 1960–1, p. 105.
84. **Forde, St C. M., Leyritz, M. J-P. and Sly, J. M. A.** (1968) The role of potassium in the nutrition of the oil palm in Nigeria. *J. Nig. Inst. Oil Palm Res.,* 4, 331.
85. **Ataga, D. O.** (1974) Release and fixation of potassium in some soils supporting the oil palm (*Elaeis guineensis* Jacq.) in Nigeria. In *Potassium in tropical crops and soils.* Proc. 10th Colloquium Int. Potash Inst., p. 131.
86. **Ochs, J. and Quemener, J.** (1972) Application aux sols de palmiers à huile de la technique de Stanford et De Ment pour l'extraction de potassium. *Oléagineux,* 27, 127.
87. **Prevot, P. and Ziller, R.** (1958) Relation entre le magnésium du sol et de la feuille de palmier. *Oléagineux,* 13, 667.
88. **Tinker, P. B. H. and Smilde, K. W.** (1963) Cation relationships and magnesium deficiency in the oil palm. *J. W. Afr. Inst. Oil Palm Res.,* 4, 82.
89. **Forde, St C. M.** (1965) *The phosphorus status of some soils of West Africa.* Nigeria Institute for Oil Palm Research, Conference Paper.
90. **Ng Siew Kee** (1970) Greater productivity of the oil palm (*Elaeis guineensis* Jacq.) with efficient fertilizer practices. In *Role of fertilization in the intensification of agricultural production.* Proc. 9th Congress Int. Potash Inst., p. 357.
91. **Ng Siew Kee** (1965) The potassium status of some Malaysian soils. *Malaysian Agric. J.,* 45, 143.
92. **I.R.H.O.** (1974) *Rapport d'activites, 1972–73.* Document 1138, Paris.
93. **I.R.H.O.** Private communication.
94. **Kowal, J. M. L. and Tinker, P. B. H.** (1959) Soil changes under a plantation established from high secondary forest. *J. W. Afr. Inst. Oil Palm Res.,* 2, 376.
95. **Tinker, P. B. H.** (1963) Changes occurring in the sedimentary soils of Southern Nigeria after oil palm plantation establishment. *J. W. Afr. Inst. Oil Res.,* 4, 66.
96. **Homès, M. V.** (1949) L'alimentation mineral de palmier à huile. *Publs. I.N.E.A.C.* Série Sci., 39.
97. **Homès, M. V.** (1955) A new approach to the problem of plant nutrition and fertiliser requirement, Part I. *Soils Fertil.,* 18, 1.
98. **Homès, M. V.** (1955) Part II. Field experiments. *Soils Fertil.,* 18, 101.
99. **Homès, M. V.** (1959) Études complémentaires sur l'alimentation minérale et le fumure du palmier à huile. *Publs. I.N.E.A.C.* Série Sc., 79.
100. **May, E. B.** (1956) The manuring of oil palms – A review. *J. W. Afr. Inst. Oil Palm Res.,* 2, 6.
101. **May, E. B.** (1956) Early manuring experiments on oil palms in Nigeria. *J. W. Afr. Inst. Oil Palm Res.,* 2, 47.
102. **Haines, W. B. and Benzian, B.** (1956) Some manuring experiments on oil palm in Africa. *Emp. J. exp. Agric.,* 24, 137.
103. **Prevot, P. and Ziller, R.** (1957) Étude d'une carence en potasse et en azote sur palmier à huile au Dahomey. *Oléagineux,* 12, 369.
104. **Arokiasamy, M.** (1969) Investigation on the best method of using the oil palm bunch waste as a fertilizer. *Commun. (Agron) Chemara Res.,* 7, 1–7.
105. **Uribe, A. and Bernal, G.** (1973) Incinerateur de rafles des régimes de palmier à huile. Utilisation des cendres. *Oléagineux,* 28, 147.
106. **Green, A. H.** (1972) *Annual Review of Research, 1970.* Unilever Plantation group, Unilever, London. 140 pp. Mimeograph.
107. **Lo, K. K. and Goh, K. H.** (1973) The analysis of experiments on the economics of fertilizer application on oil palms. In *Advances in oil palm cultivation,* p. 338. Incorp. Soc. of Planters, Kuala Lumpur.

108. **Gunn, J. S.** (1962) The economics of manuring the oil palm. *J. W. Afr. Inst. Oil Palm Res.,* **3**, 302.
109. **Warriar, S. M. and Piggott, C. J.** (1973) Rehabilitation of oil palms by corrective manuring based on leaf analysis. In *Advances in oil palm cultivation*, p. 289. Incorp. Soc. of Planters, Kuala Lumpur.
110. **Bachy, A.** (1969) A propos d'un cas typique de carence potassique du palmier à huile en Côte d'Ivoire. *Oléagineux,* **24**, 533.
111. **Corley, R. H. V. and Mok, C. K.** (1972) The effects of nitrogen, phosphorus, potassium and magnesium on growth of the oil palm. *Expl. Agric.,* **8**, 347.
112. **Lo, K. K., Chan, K. W., Goh, K. H. and Hardon, J. J.** (1973) Effect of manuring on yield, vegetative growth and leaf nutrient level of the oil palm. In *Advances in oil palm cultivation*, p. 324. Incorp. Soc. of Planters, Kuala Lumpur.
113. **Corley, R. H. V.** (1973) Oil palm physiology: a review. In *Advances in oil palm cultivation*, p. 324. Incorp. Soc. of Planters, Kuala Lumpur.
114. **Ollagnier, M., Ochs., R. and Martin, G.** (1970) The manuring of the oil palm in the world. *Fertilité,* **36**, March–April 1970, 63 pp.
115. **Ruer, P.** (1967) Répartition en surface du système radiculaire du palmier à huile. *Oléagineux,* **22**, 535.
116. **Taillez, B.** (1971) Le systéme racinaire du palmier à huile sur la plantation de San Alberto (Colombie). *Oléagineux,* **26**, 435.
117. **Werkhoven, J.** (1965) *The manuring of the oil palm.* 51 pp. Verlagsgesellschaft für Ackerbau GmbH. Hanover.
118. **Dell, W. and Arens, P. L.** (1957) Inefficacité du phosphate naturel pour le palmier à huile sur certains sols de Sumatra. *Oléagineux,* **12**, 675.
119. **Hartley, C. W. S.** (1950) The effect of a potassium–magnesium fertiliser on the yield of oil palms on hill quartzite soil. *Malay. agric. J.,* **33**, 38.
120. **Hew, C. K. and Poon, Y. C.** (1973) The effects of muriate of potash and bunch ash on uptake of potassium and chlorine in oil palms on coastal soils. In *Advances in oil palm cultivation*, p. 239. Incorp. Soc. of Planters, Kuala Lumpur.
121. **Mollegaard, H.** (1971) Results of a fertilizer trial on a mixed colluvial alluvial soil at Ulu Bernam in West Malaysia. *Oléagineux,* **26**, 449.
122. **Ng Siew Kee, Tan Yap Pau, Chan, E. and Cheong Siew Park** (1974) Nutritional complexes of oil palms planted on peat soil in Malaysia. II. Preliminary results of copper sulphate treatment. *Oléagineux,* **29**, 445.
123. **United Plantations,** Seventh Annual Report, Research Department, Teluk Anson, Malaysia.
124. **Gunn, J. S.** (1960) W.A.I.F.O.R. Eighth Annual Report 1959–60, p. 57.
125. **Aya, F. O.** (1972) N.I.F.O.R. Fifth Annual Report (1968–69), pp. 28–33.
126. **Sly, J. M. A.** *et al.* (1963) W.A.I.F.O.R. Eleventh Annual Report, 1962–3, p. 35.
127. **Sheldrick, R. D.** N.I.F.O.R. Third Annual Report (1966–67), pp. 38–9.
128. **Van der Vossen, H. A. M.** (1970) Nutrient status and fertilizer responses of oil palms on different soils in the forest zone of Ghana. *Ghana J. Agric. Sci.,* **3**, 109.
129. **Bachy, A.** (1968) Principaux résultats acquis par l'I.R.H.O. sur la fertilisation du palmier à huile. *Oléagineux,* **23**, 9.
130. **I.R.H.O.** (1972) *Rapport d'Activites 1971*, pp. 56–7.
131. **Ollagnier, M. and Ochs, R.** (1974) Interaction entre l'azote et le potassium dans la nutrition des oléagineux tropicaux. Proceedings of 10th colloquium of the Int. Potash Inst. Berne. pp. 215.
132. **De Taffin, G. and Ochs, R.** (1973) La fumure potassique du palmier à huile au Dahomey. *Oléagineux,* **28**, 269.
133. **Aya, F. O.** (1970 and 1972) N.I.F.O.R. Fourth and Fifth Annual Reports (1967–8 and 1968–9), pp. 32 and 31 respectively.
134. **Vanderweyen, R.** (1952) *Notions de culture d'Elaeis au Congo Belge.* Brussels.

135. **Anon.** (1962) Modalités de replantation du palmier à huile avec ou sans apport d'engrais. *Bull. Inf. I.N.E.A.C.,* **11,** 113.
136. **Daniels, G. and Ochs, R.** (1975) Amélioration de la production des jeunes palmiers à huile du Pérou par l'emploi d'engrais Chlore. *Oléagineux,* **30,** 295.
137. **Foster, H. L. and Chang, K. C.** (1976) The diagnosis of the nutrient status of oil palms in West Malaysia. Int. Agric. Oil Palm Conference, Kuala Lumpur, 1976.
138. **Tan Kok Yeang and Chaplin, M. H.** (1976) The use of a computer in fertilizer extension and management for oil palm. Int. Agric. Oil Palm Conference, Kuala Lumpur, 1976.
139. **Foster, H. L. and Goh, H. S.** (1976) Fertilizer requirements of oil palm in West Malaysia. Int. Agric. Oil Palm Conference, Kuala Lumpur, 1976.
140. **Umar Akbar,** *et al.* (1976) Fertilizer experimentation on oil palm in North Sumatra. Int. Agric. Oil Palm Conference, Kuala Lumpur, 1976.
141. **Chan Kook Weng and Rajaratnam, J. A.** (1976) Magnesium requirement of oil palms in Malaysia: 45 years of experimental results. Int. Agric. Oil Palm Conference, Kuala Lumpur, 1976.
142. **Ng Siew Kee** (1976) Review of oil palm nutrition and manuring — scope for greater economy in fertilizer use. Int. Agric. Oil Palm Conference, Kuala Lumpur, 1976.
143. **Beure, G. J.** (1976) Preliminary results from an oil palm density fertilizer experiment on young volcanic soils in West New Britain. Int. Agric. Oil Palm Conference, Kuala Lumpur, 1976.
144. **Cheong Siew Park and Ng Siew Kee** (1976) Copper deficiency of oil palms on peat. Int. Agric. Oil Palm Conference, Kuala Lumpur, 1976.
145. **Calvez, J., Olivin, J. and Renard, J. L.** (1976) Study of a Sulphur deficiency on young oil palms in the Ivory Coast. *Oléagineux,* **31,** 251.

Chapter 12

Mixed cropping, rearing livestock among oil palms and tapping for wine

The desire to combine either the cultivation of a subsidiary crop or the raising of livestock with the growing of a plantation crop does not usually arise where the latter is being cultivated on a large scale. Those in control normally wish to concentrate all their efforts and resources upon the opening of new land and the successive establishment of large fields, well-planted and designed solely for the crop concerned. It is to the small farmer that a subsidiary means of livelihood is more usually attractive since in the first place he often needs an income for the few years before the main crop comes into bearing, and secondly his acreage may be so small that he is forced to consider a more intensive use of the land throughout the life of the crop.

With the oil palm these latter considerations are further influenced by the small stand per hectare, though this is counteracted in some measure by the rapidity with which the palm comes into bearing. A farmer depending largely on his own resources is likely to wish not only to engage in establishment inter-cropping, which was discussed in Chapter 8, but to use his land to provide himself each year with food crops and perhaps livestock. Furthermore, he may wish to use some of his palms for the production of palm wine. In Africa, particularly where population pressure is high, the growing of food crops among oil palms is already standard practice, and any small farmer who planted an area with palms would almost certainly wish to cultivate the interlines with such crops in rotation with a natural cover.[1]

The low stand per hectare has also had some influence on the large grower, and the inter-planting of such crops as cocoa or coffee has been tried or advocated on many occasions and has sometimes met with success.

There is very little tradition for cattle raising in the wet tropical lowlands, except in Latin America, but in recent years improvements in the control of cattle diseases and an increasing demand for animal products has encouraged an interest in cattle where this only existed previously on a very small and primitive scale. For these reasons the grazing of cattle or sheep among the palms has been given serious consideration in certain regions, while on some of the new American plantations it is almost taken for granted.

Inter-cultivation of food crops: planting palms and farming

In those parts of Africa where the oil palm is the cash crop of the peasant population, the farmers are also engaged in subsistence farming. The relationship of the palm groves to farming has already been discussed in Chapter 3. In these regions palms and food crops have to share the same areas permanently and the question arises as to how they should be arranged. The majority of food crops, e.g. maize, cassava, yams (*Dioscorea* sp.), are not suited to growing under shade, though cocoyams (*Colocasia* and *Xanthosoma* sp.) are adapted to shade conditions. Where food crops are grown haphazardly under palms, yields suffer through competition both for nutrients and light, and if both cultures are to be organized on the same holding they must either have separate areas allotted to them or the spacing of the oil palms must be such that competition between the crops is reduced to a minimum. As the growing of food crops demands a resting period of several years, the farmer will normally wish to cultivate them over as wide an area as possible, and the allotment of separate areas for food crops and palms may not be to his taste.

These considerations have given rise to several experiments in Nigeria where this problem is most acute. In an early spacing experiment, already described in Chapter 9, rows were spaced 65 feet apart to allow sufficient room for food-crop cultivation or for grazing. It was shown that the palms could not be crowded in the rows to compensate for the wide spacing. Palms spaced at 19.8 x 3.7 m (65 x 12 ft) gave a lower yield than palms spaced at 19.8 x 6.4 m (65 x 21 ft). Provided both the palms and the food crops received satisfactory manurial treatments, individual palms spaced normally *within* the widespaced rows yielded slightly more than palms at normal spacing both within and between the rows. Food crops kept 3.7 m away from the palm rows had no effect on the palms' yield. These results suggested that the combination of planting palms with farming should be studied in more detail and the effects of fertilizers, establishment intercropping and cultivation should be taken into account.[2, 3] Experiments laid down on forest and old grove areas in Nigeria and on previously cultivated land in Sierra Leone had the following treatments:

Factor	Level 0	Level 1	Level 2
A. Fertilizer (to palms)	None	In April (beginning of rains)	In October (end of rains)
B. Establishment intercropping or cultivation	None	Establishment intercropping for 2 years	Dry season cultivation for 2 years*
C. System of planting and farming	Normal density: No farming	Half density; rotational farming in wide interlines	Two-thirds density; rotational farming in wide interlines†

* Omitted in the old grove area in Nigeria, and in Sierra Leone.
† In Sierra Leone this was replaced by half density with *grazing* in the wide interlines.

Pl. 55   Establishment inter-cropping: young palms growing between lines of yams (*Dioscorea* sp.) in Nigeria.

In view of the poor growth of palms crowded in the rows, all palms were planted at the basic spacing of 8.83 m (29 ft) triangular; the half density plots represented areas in which every other pair of rows had been omitted, and the two-thirds density plots represented areas with one row omitted in every three, these two arrangements giving a wide interline for farming of 15.24 m (50 ft) in the first case and 7.62 m (25 ft) in the second, no farming being permitted within 3.8 m (12½ ft) of the rows. Establishment inter-cropping (Plate 55) covered the whole area except the weeded circles of the palms and produced the expected results: increased early yields of the palms. Dry season cultivation also increased yield per palm in the first 4 years.[4] Fertilizers were immediately shown to be necessary in the areas not opened from forest.

In these experiments the palm rows were set N–S since it was expected that the high incidence of sunlight thus obtained might increase the sex ratio and thus induce a higher yield per palm; a reduction of stand by one-third or one-half to enable food-cropping to take place would not then result in reductions of crop of the same magnitude. In the event it was seen that the leaving out of one row in three did not provide enough space for satisfactory food-crop cultivation and the best means of combining the two cultures at present seems to be the alternating of normally spaced pairs of rows with food-crop cultivation in the lanes which would have been occupied by other pairs of rows, the foodcrop cultivation being kept about 4 metres away from the palm rows. Even with these wide lines there is still, of course, some shading in the morning and evening of the sides of the food-crop strips. A considerable period must elapse before the long-term returns from each treatment combination can be assessed.

Experience of replanting palm groves has shown that there is often the greatest difficulty in persuading farmers not to plant food crops right up to the palms and to damage the palms in so doing. Moreover cultivation tends to continue in the vicinity of the palms for very many years after the latter are planted. As a result, the palms come late into bearing and their subsequent yield is reduced. It is therefore best to confine even establishment inter-cropping to strips of 4–5 m width down the avenues between the rows and not to allow the whole area to be covered by annual crops. There is no doubt whatever that if mixed-cropping of palms and food crops is to be successful, the work must be carried out under disciplined control; if this is not practicable, then palm areas and food-crop areas are better separated entirely one from the other.

In Malaysia it appears that the effect on the palms is likely to be a lesser problem in inter-cropping; there the main effort has been directed towards improving the standard of cultivation of the food crops themselves.[5]

Mixed cropping with other perennial crops

Wherever a main crop is a large tree leaving plenty of space between individuals, there has been a desire to make fuller use of the land by planting some smaller economic crop in the intervals. In considering this question from the viewpoint of the rubber industry, Allen[6] stated that the second or subsidiary crop:

1. should not grow as tall as the main crop and its root system should exploit different soil horizons;
2. should be tolerant of partial shade;
3. should not be more susceptible than the main crop to diseases which they may have in common;
4. should not demand harvesting or other operations which would damage the main crop or induce soil erosion or damage soil structure;
5. should not have an economic life longer than that of the main crop.

To these characteristics may be added the more positive ones that the soil shall be suitable for both crops, that the combined yield of the two crops shall be greater in monetary terms than that of the main crop when grown without the subsidiary crop, and that, if and when the subsidiary crop comes to the end of its bearing life, the yield of the main crop shall continue at an economic level unaffected by the previous presence of the subsidiary crop.

It is not easy for a subsidiary crop to fulfil all these exacting requirements and this is why, by and large, mixed cropping has seldom been successfully practised on a large scale. Some mention may be made here, however, of those crops which have been suggested or tried.

Rubber and oil palms were planted together in the early years of oil palm cultivation in Indonesia. These crops are obviously incompatible, however, since rubber will overtop the oil palm and its canopy will spread out and shade the palms. Competition for nutrients will be fierce, and while the rubber must be widely spaced and so yield poorly per hectare, oil palm fruiting and yields will be seriously suppressed. Systems of inter-planting rubber and oil palms have been suggested, but they have nothing to commend them; moreover, on plantations, harvesting and transporting costs would be much increased.

Coffee and cocoa are small trees which can be planted among oil palms. Both these crops, but particularly cocoa, are more exacting in their soil requirements than is the oil palm. In West Africa satisfactory yields of these crops cannot be obtained on the great areas of Acid Sands soils where oil palms are found in greatest number. On the cocoa soils of Ghana and western Nigeria, cocoa is likely to be more profitable as a sole crop than when mixed with palms, even though the latter will yield well on these soils. *Robusta* coffee will produce quite well on the soils of the Congo river basin, on the Tertiary sands and other soils of the wet coastal belt of south-western Ghana, the Ivory Coast and Liberia, and in some parts of Sierra Leone. In the Far East cocoa has been a rather uncertain crop but is proving suitable on volcanic soils in East Malaysia and on coastal clays in West Malaysia. *Robusta* coffee is also a fair yielder in many parts of the Far East, but the return to be obtained does not make it such an attractive proposition as cocoa.

Work in Zaire[7] confirmed that considerations of soil type were the most important for inter-planting with cocoa and that, though cocoa is shade tolerant, shading with the oil palm presented certain difficulties; while cocoa benefits from greater shade when young, the palm provides increasing shade as the plantation matures. Reducing shade by leaf pruning is impracticable because of the mode of growth of the oil palm and the adverse effect of pruning on yield, while planting at abnormally wide spacing might reduce oil palm yields to a point where the cocoa yields were insufficient compensation. Vanderweyen therefore recommended the planting of cocoa when the palm trunks are about 6 feet in height (7 to 8 years old) either in two rows between palms normally spaced at 9 m triangular, or in three rows between pairs of palm rows approximately 10.5 m apart, the palms being spaced 7.5 m apart in triangular formation in the pairs of rows.

In Malaysia productive stands of cocoa have been established under mature palms on coastal alluvium soils and some promising yields have been obtained. Cocoa production is supplementary to the main enterprise and, as such, appears to be worth while. In Sumatra,[8] cocoa has also been successfully established under mature palms, but the fields concerned supported old stands which were thinned out by 40 per cent before inter-planting with cocoa and shading with *Gliricidia* was undertaken. The economics of this procedure has not been reported.

In an experiment in Zaire on the inter-planting of *Robusta* coffee among oil palms, the cultivation of coffee had no overall effect on oil palm yields, but in one treatment it appeared that the cultivation applied to the coffee may have had an effect similar to that of cultivation for establish-ment inter-cropping with food crops, since the bunch yield was slightly enhanced. Coffee yields soon declined, however, and it was concluded that it was better to plant the two crops separately. In this experiment all treat-ments allowed for an almost full stand of both palms and coffee bushes per hectare, so the result was not unexpected. Coffee can, of course, be planted at the same time as the palms; there is no need, as has appeared to be the case with cocoa, to await the palm's shade. More recently it has been suggested that coffee can be grown with oil palms in Zaire for 6 or 7 years after which it must be cut out.[9]

A more informative experiment has given interesting results on an area of river alluvium in Sierra Leone. In this experiment large plots of 0.34 hectare have compared the two crops planted separately (30 ft (9.1 m) triangular for palms and 10 ft (3 m) square for coffee) with coffee inter-mingled in different ways among incomplete stands of palms. The treat-ments are given below and the results for the first 4 years are shown in Table 12.1.

Treatments

(A) Control. Pure stands of coffee and oil palms, the plots being divided between the two crops. Oil palms spaced 30 feet (9.1 m) triangular, coffee 10 feet (3 m) square.

(B) Oil palms at 30 feet (9.1 m) triangular but with one row in three omitted. In the wide interline thus formed, coffee planted at 10 feet (3 m) square.

(C) Oil palms 30 x 40 feet (9.1 x 12.2 m) rectangular; coffee at 10 feet (3 m) square.

(D) Oil palms 30 feet (9.1 m) square, but alternate plants in alternate rows omitted thus leaving squares into which coffee at 10 feet (3 m) square was planted.

It should be noted that in treatments (C) and (D) in which the crops are intermingled there is a higher number of palms plus bushes per unit area than there is in the treatment (A) where half the area is planted with palms and half with coffee. It would appear from the early results that a higher total of produce may be obtained in the first years by arrangements such as (C) and (D) than by having separate areas of palms and coffee, and this order of total production was maintained to the sixth year. The arrange-ment in treatment (B) has too low a stand of coffee to compete with treat-ment (C). It cannot, however, be assumed that a higher economic return will be obtained from mixed cropping, since this will depend on the rela-

Table 12.1 Inter-planting oil palms and coffee: Sierra Leone

Treatment	Stand			Yield of coffee, 1961–4 per annum (kg)		Yield of oil palm bunches, 1961–4 per annum (kg)		Total production per hectare (kg)
		per acre	per hectare	per bush	per hectare†	per palm	per hectare†	
A. Separate stands within the plots	O.P.	55.9 (27.9)*	138 (69)	1.85	995	33.0	2,265	3,260
	C.	435.6 (217.8)*	1,076 (538)					
B. Coffee in wide interlines	O.P.	37.3	92	1.64	528	32.2	2,950	3,478
	C.	130.1	321					
C. Wide oil palm spacing, coffee interplanted	O.P.	36.3	90	1.39	983	36.3	3,249	4,232
	C.	286.7	708					
D. Coffee in hollow squares	O.P.	36.3	90	1.66	1,186	31.0	2,779	3,965
	C.	289.2	714					

* Stands per acre in treatment A were 55.9 and 435.6 respectively but, for comparative purposes, the 'stands per acre of plot' are taken as half these figures, i.e. 27.9 and 217.8 (69 and 538 per hectare).

† The figures for treatment A are 'per acre or hectare of plot' since it is this that must be compared with the other per acre or hectare figures.

tive value of the two crops and on the ability of the coffee to continue to produce under the stand of palms. There was evidence from the experiment that coffee yields were in fact declining; and with low prices of coffee it would clearly be better for the *whole* area to be under palms.

Experiments involving the underplanting of oil palms with cocoa are also under way in western Nigeria but there have been difficulties in establishing the two crops together on an experimental scale in this area.[10]

Both cocoa and *Robusta* coffee have been grown commercially with oil palms on estates in the Congo basin. Coffee is grown as a catch crop and planted at the same time as the palms, and three crops are taken. A total yield of about 3 tons per hectare is expected and this provides revenue during the mainly unproductive years of the palms' life. Cocoa, on the other hand, is planted under the shade of mature oil palms. With two rows of cocoa trees between the rows of palms a yield of around 135 kg of dry cocoa is obtained and, provided there is no reduction in the palms' yield, this forms a useful cash increment when prices are satisfactory.

Recently Patchouli (*Pogostemon cablin*) has been suggested as a catch crop in Malaysia since it will tolerate some shade and the essential oil extracted from the shoots has long been marketable in the Far East.[11]

The grazing of livestock under oil palms

Oil palm fields provide shade which is a good deal lighter than that of other plantation crops, e.g. rubber and cocoa, and even in fully mature fields there remains a considerable quantity of undergrowth which has to be kept under control. This undergrowth is, of course, much increased by the use of slightly wider spacings. It is not unnatural therefore that attention has been given to the raising and rearing of livestock on oil palm plantations as a subsidiary source of income.

In Africa, oil palm plantations lie in areas where livestock are subject to trypanosomiasis, but immune breeds of cattle exist and sheep and goats are also to be found. In the Far East it is not uncommon for herds of cattle (and sometimes buffaloes) to roam through plantations, but these are rarely owned by the estate and are usually regarded by managers as a nuisance.

The keeping of cattle and sheep in an oil palm plantation may entail: (i) provision of protection for the young palms, (ii) a change in the cover policy so as to provide suitable herbage for the stock and (iii) a change in the spacing of the palms.

Livestock will damage young palms in the field so either all young fields must be excluded from the areas to be grazed or individual palms must be protected. The latter is only possible on an experimental scale so that livestock rearing can only take place, in practice, in areas where the palms are 'off the ground'. The stock can then most easily be controlled by electric fences (Plate 56).

Pl. 56 Ndama cattle grazing Elephant (Napier) grass in an experiment in Nigeria; leguminous cover plot in the foreground.

Neither sown leguminous covers nor natural covers are suitable for intensive grazing; the former tend to be grazed out and the latter to provide an insufficiency of material and the flora changes in composition. Some grass or grasses suitable for grazing are required. There is no information on trials among oil palms in the Far East, but in those areas *Axonopus compressus* (carpet grass) and some *Paspalum* sp. graze well, and *A. compressus* grows well under shade. Napier (Elephant) grass, Guinea grass and Guatemala grass (*Pennisetum purpureum, Panicum maximum* and *Tripsacum laxum*) are often regarded as fodder grasses and are considered not to graze well. In West Africa, however, *Pennisetum purpureum* grazes well if properly controlled and in America *Panicum maximum* is also grazed.

In the early years of a plantation at normal spacing (about 9 m triangular) grasses will grow well, and in West Africa a strong stand of Elephant grass can be maintained. If this grass is allowed to become too luxuriant and encroaches on the circles, however, reduced early yields may be expected. Later, the palms tend to shade out the grass in their immediate vicinity and the effect on yield disappears. This is shown by the following figures of bunch yield from a spacing–inter-cropping–grazing experiment in Nigeria:

9.14 m (30 x 30 ft) spacing:	Pueraria plots		Elephant grass plots	
Bunch yield:	tons/ba./an.	per cent	tons/ba./an.	per cent
1st—3rd year of bearing	3.18	100	1.78	56
4th—6th year of bearing	7.85	100	8.89	113
7th—9th year of bearing	10.00	100	10.89	109

If Elephant grass is to be planted at the same time as the palms, therefore, special measures need to be taken to see that it does not encroach on the circles in the early years. Up till about the eighth year it will grow fairly satisfactorily among normally spaced palms in West Africa, but experiments have shown that after 10 or 12 years it is completely shaded out.[12]

In the spacing—inter-cropping—grazing experiment in Nigeria, plots were planted with Elephant grass or *Pueraria* or allowed to grow natural covers. Grazing was started 8 years after planting (the palms had established slowly) in plots of the following spacings:

	Triangular			Rectangular		
Metres:	9.14	12.8	9.14 x 19.8	6.4 x 19.8	3.66 x 19.8	6.4 x 12.8
Feet:	30	42	30 x 65	21 x 65	12 x 65	21 x 40
Density						
(per acre)	55.9	28.5	22.3	31.9	55.8	51.9
(per hectare)	138	70	55	79	138	128

The rows were arranged E—W so that the interline obtained a minimum of shading and this assisted the growth of Elephant grass. As already mentioned, the Elephant grass was soon completely shaded out in the 30 feet triangular plots but in other spacings growth was satisfactory. From the point of view of palm yields, however, the wide 65 feet lines and the very low density plots were quite unsatisfactory, the overall yield for both Elephant grass and other cover plots being as follows:

	Fruit bunches per hectare per annum (12 years)					
Spacing:	Triangular			Rectangular		
Metres:	9.14	12.8	9.14 x 19.8	6.4 x 19.8	3.66 x 19.8	6.4 x 12.8
Feet:	30	42	30 x 65	21 x 65	12 x 65	21 x 40
Density						
(per hectare)	138	70	55	79	138	128
Yield (tons)	7.63	4.80	3.72	4.33	4.12	6.89

It will be noted that the adoption of the rectangular spacing 21 x 40 feet (6.4 x 12.8 m), where the Elephant grass grew quite satisfactorily in the wide interlines, resulted in a yield only slightly lower than that of normal triangular spacing. Half of each Elephant grass plot was used for grazing while the remaining half was normally maintained. The yield of bunches was unaffected by the grazing and the carrying capacity of the plots growing Elephant grass was found to be as high as one beast per 0.8 hectares (2 acres) in a herd of small upgraded Ndama cattle. Strict control of the grazing was required to maintain a steady growth of the grass. The general conclusion was that with a very small reduction in yield owing to wider spacing, a considerable contribution might be made to meat supplies in palm-growing areas where the diet of the people is deficient in protein, and that at the same time the cost of maintaining the palm fields would be reduced.[13] Strict management would be needed and trials on a larger scale are required, but there is no denying that under conditions obtaining in Nigeria the combining of oil palm planting and cattle rearing is an attractive prospect.

A study has been made in the Ivory Coast of running cattle in large plantations.[14] The method found most satisfactory was to leave the cattle on the plantations throughout the day and night. During the day the animals follow a strict rotation under the control of herdsmen while at night they are confined by electric fencing into an enclosure giving 0.25 hectare per animal. It was found possible to employ a stocking rate of 125 kg cattle per hectare, equivalent to half a head per hectare of the Baoulés breed, and to obtain a live weight gain of about 500 g per day. Under this relatively light stocking rate it is claimed that the vegetation is not in general degraded nor is the soil poached and that the cost of herd management is met by the saving in field maintainance expenditure which the system makes possible.

In Latin America, many of the new oil palm plantings are in areas where ranching is already an established occupation. These plantings may be laid down in old pasture land or may be adjacent to pastures so that strong growing grasses soon gain access.[15] There has been some grazing in young plantations, but the palms in these cases usually suffered damage. On heavy soils in high rainfall areas grazing caused puddling of the soil and adversely affected the root system of the palms which demand a good surface structure. In many plantations in Colombia and Ecuador *Panicum maximum, Pennisetum purpureum* or *Hyparrhenia rufa*, often mixed with other grasses, can grow very strongly. Manual control is costly and control by herbicides, though increasingly employed, is becoming more expensive. Controlled grazing among young palms was successful in Colombia; this success was however only achieved by restricting the herd size to one animal per 2 hectares and carefully rotating the grazing so that regeneration of good covers, including *Pueraria*, was encouraged.[16] If overgrazing was allowed to take place the cattle soon began to eat the leaves of the young palms and to damage the drains. Most plantation owners in Latin America do not yet have staff available to exercise proper control of such grazing and, with heavy soils and high rainfall, the use of cattle in this manner cannot yet be generally recommended.

The production of palm wine

Mention was made in Chapter 1 of the tapping of oil palms in the vicinity of African towns and villages for the production of palm wine. The industry is of considerable economic and nutritional importance in Nigeria and other parts of West Africa. Tuley studied production from palms in eastern Nigeria and described tapping methods;[17, 18] Bassir investigated the composition and fermentation of palm sap from the male inflorescence stalk.[19] From samples taken from one palm he found the chemical constituents of fresh palm sap to be as follows:

	g/100 ml palm sap
Sucrose	4.29 ± 1.4
Glucose	3.31 ± 0.95
Ammonia (NH_3)	0.038 ± 0.015
Lactic acid	Present
Amino acids	Present

Many yeast cells and two main types of bacteria were found to be present after 8 hours fermentation in the open air at about 25°C. Experiments suggested that fermentation proceeded in two stages: in the first, bacterial action was responsible for the production of organic acids; in the second, sucrose was inverted by yeast and ethyl alcohol and more organic acids were produced.

The production of the so-called 'down-wine' from the terminal 'cabbage' of a felled palm is entirely destructive of the crop. It is said to be 'preferred' in Ghana,[20] but this is of course simply because the rural economy of that country has not forced the people to preserve their palms as has been the case in eastern Nigeria. The flow of sap is induced from the stem apex by an incision in the region of the unopened leaf bases and it continues for about a month. The fresh wine appears to be of different composition from the inflorescence wine as it is reported to contain glucose, sucrose, fructose, maltose and raffinose. However, the presence of the latter three sugars might be due to post-felling changes since even with inflorescence wine they are found after 24 hours' fermentation.[21] Another difference, however, is in the alcohol content. Fermented inflorescence palm wine is reported to contain only ethanol whereas 'down-wine' contains some methanol and propanol and this is another undesirable feature of 'down-wine' production.

Stem tapping[18] of standing palms is damaging to the soft tissues around the growing point and may kill the palm or provide entry for injurious insects, bacteria or other organisms. The only acceptable and manageable form of tapping is through excising the male inflorescence and drawing off the sugary solution which will exude from the cut surface.

The leaf subtending an immature male inflorescence is removed to

Pl. 57   Palm wine flowing from the tapped immature male inflorescence.

obtain access to the inflorescence enclosed in its spathes. An incision is
made near the apex of the inflorescence and the top of the tissue inside
the spathes is removed. A piece of the front spathe is removed and the
main stem of the spadix is cut horizontally to form a 'tapping panel'. The
cut is covered with a piece of 'felt' composed of the fibrous leaf sheath
fabric, and a new slice is taken daily until the wine begins to flow. A
funnel of bamboo is inserted in the felt cover which is then set in position
and the wine allowed to flow into calabashes or bottles. It is collected
morning and evening and a new slice taken from the tapping panel at each
collection. The spathes have been described as forming a cylindrical casing
for the collection of the wine[18] (Plate 57).

Wine production depends on the number of male inflorescences available and on the quantity produced per inflorescence. In the tapping of
sixty palms over a 4-year period in eastern Nigeria[17] it was found that
yields were highest at the beginning and end of the rains in the periods
March to April and October to November. It appeared that high production in the latter period was due to the large number of male inflorescences available for tapping, whereas the high yield in March to April was
due to high production per inflorescence, this presumably being associated
with the increased rate of development of the organs of the palm at the
beginning of the rains. The mean monthly production and the mean
number of palms in tapping are shown in Fig. 12.1. The mean annual yield
per palm was 26.3 l of wine, which, at Nigerian internal prices, was estimated to have a value more than double that of the oil and kernels from
similar palms and some 60 per cent higher than oil and kernels from high
yielding palms in eastern Nigeria.

Oil palm wine has a milky appearance due to high concentrations of

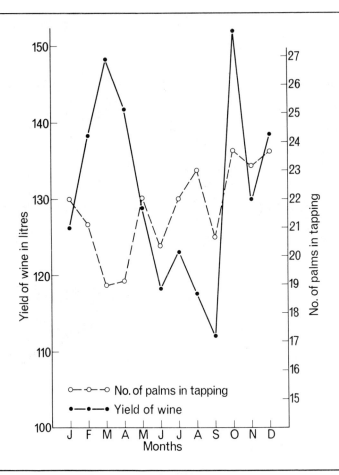

Fig. 12.1   Mean monthly production of palm wine and number of palms in tapping, 1959−62, at Umudike, Nigeria (Tuley, 1965).

yeast. It has a slightly sulphurous smell because of its sulphur-protein content. As it is not collected under sterile conditions fermentation is quite rapid, but if this has not proceeded too far palm wine forms a nutritious drink which provides an important source of the vitamin B complex.

The pH of fresh wine is 7.4; it falls to 6.8 in the first stage of fermentation and then to 4.0 in the second stage when fermentation virtually comes to an end.[21] The wine is usually drunk at pH 5.5−6.5 after about 12 hours' fermentation. In Nigeria the organic acid content includes acetic, lactic and tartaric acids and thirteen amino-acids have been identified. Vitamins B_1, B_2, B_6 and vitamin C are present. In Zaire inflorescence wine was found to contain seven organic acids, twenty-five amino-acids and vitamin B_{12}.[22] The microbial flora causing fermentation appeared to differ from that found in Nigeria.

The sealed bottling of palm wine is now being undertaken in West Africa.[21] In Cameroon a process of double pasteurization and controlled fermentation has been used with *Raphia* palm wine.[23] The distilling of a spirit or 'gin' from palm wine has long been practised in crude distilleries. The raw spirit was later used in Ghana for the industrial production of branded spirits.[20]

References

1. Sparnaaij, L. D. (1957) Mixed cropping in oil palm cultivation. *J. W. Afr. Inst. Oil Palm Res.*, **2**, 244.
2. Gunn, J. S. and Sparnaaij, L. D. (1957) W.A.I.F.O.R. Fifth Annual Report 1956–7, p. 55.
3. Sly, J. M. A. *et al.* (1963) W.A.I.F.O.R. Eleventh Annual Report 1962–3, p. 28.
4. Sly, J. M. A. and Sheldrick, R. A. (1965) N.I.F.O.R. First Annual Report 1964–5, p. 30.
5. Cheng Yu Wei (1970) Improving the performance of catch crops in Malaysia. In *Crops diversification in Malaysia*, p. 66, Incorp. Soc. of Planters, Kuala Lumpur.
6. Allen, E. F. (1955) Cultivating other crops with rubber. *Plts'. Bull. Rubb. Res. Inst. Malaya*, 16.
7. Vanderweyen, R. (1952) *Notions de culture d'Elaeis au Congo Belge.* Brussels.
8. Garot, A. and Subadi. (1960) Tanaman tjoklat dibawah kelapa sawit dikebun Marihat dan Pebatu di Sumatera-Utara. *Menara Perkebunan*, **29**, 7, 123.
9. Marynen, T. (1960) Précis de phytotechnie des principales cultures industrielles. *Publs. I.N.E.A.C.* Hors Série.
10. Sly, J. M. A. and Sheldrick, R. D. (1966 and 1968) N.I.F.O.R. First and Third Annual Reports, 1964–5 and 1966–7, pp. 31–2 and 33–4 respectively.
11. Chin Hoong Fong and Rashid bin Ahmad, M. (1970) Preliminary observations on a potential catch crop – Patchouli. In *Crop diversification in Malaysia*, p. 107, Incorp. Soc. of Planters, Kuala Lumpur.
12. Gunn, J. S. *et al.* (1962) W.A.I.F.O.R. Tenth Annual Report 1961–2, p. 39.
13. Gunn, J. S. *et al.* (1961) W.A.I.F.O.R. Ninth Annual Report 1960–1, p. 44.
14. Rombaut, D. (1974) Étude sur l'élevage bovin dans les palmeraies de Côte d'Ivoire. *Oléagineux*, **29**, 121.
15. Hartley, C. W. S. (1965) Some notes on the oil palm in Latin America. *Oléagineux*, **20**, 359.
16. Van den Hove (1966) Utilisation du bétail pour la lutte contre les graminées dans les plantations de palmiers à huile en Colombie. *Oléagineux*, **21**, 207.
17. Tuley, P. (1965) Studies on the production of wine from the oil palm. *J. Nigerian Inst. Oil Palm Res.*, **4**, 284.
18. Tuley, P. (1965) How to tap an oil palm. *The Niger. Fld.*, **30**, 28.
19. Bassir, Olumbe (1962) Observations on the fermentation of palm wine. *W. Afr. J. biol. Chem.*, **6**, 2, 20.
20. Sodah Ayernor, G. K. and Mathews, J. S. (1971) The sap of the palm *Elaeis guineensis* Jacq. as raw material for alcoholic fermentation in Ghana. *Trop. Sci.*, **13**, 71.
21. Bassir, Olumbe. (1968) Some Nigerian wines. *W. A. J. biol. Chem.*, **10**, 2, 42.

22. **Van Pee, W. and Swings, J. G.** (1971) Chemical and Micro-biological studies on Congolese palm wines. (*Elaeis guineensis*). *E. Afr. agric. J.*, **36**, 311.
23. **Fyot, R.** (1973) Industrial processing of a traditional beverage: pasteurized palm wine. *Techniques and Développement*, No. 8, 10.

Chapter 13

Diseases and pests of the oil palm

Until the time of the Second World War it was true to say that the oil palm was largely free from serious diseases and pests although the temporary condition Crown disease was known and a number of bud and stem rots had been reported. Since that time however there have been serious, and at times devastating, outbreaks of disease in several parts of the world. Of greatest importance have been the devastation caused by *Fusarium* Wilt and a bacterial Bud Rot in southern Zaire, the considerable losses sustained through Dry Basal Rot (*Ceratocystis*) in Nigeria and through *Ganoderma* Trunk Rot in old and replanted areas in Asia, and the sudden and devastating attacks of bud rots and Sudden Wither on new plantations in Colombia, Peru and Central America.

Diagnosis and cure of the more recent of these diseases has proved extremely difficult, firstly because the size and manner of growth of the palm makes investigation difficult and time-consuming, and secondly because there has been a lack of plant pathologists to apply themselves single-mindedly to the diseases in question. Partly for these reasons there has been a tendency to avoid the usual plant pathological approach and to search for resistance to or tolerance of the diseases both within *E. guineensis* material and in inter-specific hybrids with *E. oleifera*, and some progress has been made with this work. Arnaud and Rabechault[1] have compared the roots of the two species and postulate that resistance to pathogens gaining access through the roots will be greater in *E. oleifera* and the inter-specific hybrid because of the greater lignification of the hypodermis and the external cortical parenchyma and the presence of tannins in the endoderm and phloem. The exact relevance of these differences to specific diseases has yet to be worked out however; the most outstanding resistance yet shown by *E. oleifera* and its hybrids has been to undiagnosed bud rots in Colombia and it is not thought that the pathogen in this case enters through the roots.

Attacks by one pest, the Hispid *Coelaenomenodera elaeidis*, have been growing more serious in West Africa while sporadic defoliations have also been caused by caterpillars and bagworms of various species in Malaysia and South America. With the greatly increased areas under oil palms there has been a general increase, particularly in Asia and America, in the incidence of pests of several natural orders, but particularly of the Lepidoptera and Coleoptera. The prediction that owing to the vast assembly of palm species in that continent oil palms in America would soon be troubled by many species has been amply fulfilled.

Although no comprehensive works on diseases or pests have yet been written, useful and practical handbooks have now been compiled on diseases and disorders by Robertson *et al.*[2] for West Africa and by Turner and Bull[3] for Malaysia, and on pests in Malaysia by Wood.[4] Descriptions in this chapter must necessarily be condensed, but reference is made to original papers where greater detail can be found. Nutritional disorders have been described in Chapter 11 and this chapter will therefore deal with conditions caused by pathogenic organisms, with important diseases of unknown cause, and with insect and other animal pests causing more or less serious damage to the palm.

Diseases

It will be most convenient to deal with diseases according to the stage of growth at which the palm is attacked and the organs affected.

1. Germinating seed

Brown Germ. Cause: *Aspergillus* sp. (probable)

Distribution.   This disease is common in Africa but is known in Malaysia and is probably universal.

Symptoms.   Brown spots appear on the emerging 'button'. These spread and coalesce as the embryo develops, and the tissues become slimy and rotten.

Control.   The fungus grows best under moist conditions at a temperature of 38°–40°C; use of the wet heat treatment for germination therefore encourages its development and spread. Though sanitary measures in the germinator may reduce incidence, the best method of control is to adopt the dry heat treatment method of germination (p. 323), since the seeds are dry when being heated at 39.5°C, and when germinating they are at around 27°C, a temperature which does not encourage the growth of the organism. In Malaysia it has been the practice to immerse seeds, after depulping, in a fungicidal and bactericidal solution of T.M.T.D. and streptomycin,[5] but there is no evidence that this is a necessary procedure and seed treatment is normally only needed to fulfil phytosanitary requirements for despatch to other countries (see p. 326).

Other seed infections

Where removal of the mesocarp in cleaning the seed has been incomplete a species of *Schizophyllum* may invade the shell. Infection can spread to the inner shell surface and kernel. However, Turner and Bull state that where kernel rotting occurs its association with the fungus is obscure. The condition is not of great importance since it is unlikely to occur where removal of the mesocarp is complete and the seeds have been properly dried.[3]

2. Seedling diseases

There are certain leaf diseases which may attack seedlings in the prenursery, nursery or when they are young palms in the field. They will be dealt with together here even though some of them are characteristic of very small and some of larger seedlings.

(1) Leaf diseases

Anthracnose. Cause: Species of *Botryodiplodia, Melanconium, Glomerella*

Distribution. Anthracnose is a disease characterized by limited lesions, necrosis and arrested development typically caused by one of the Melanconiales, e.g. *Melanconium, Pestalotiopsis,* but also by other fungi. It appears as dark necrotic lesions on the leaves of seedlings, usually at the prenursery or young nursery stage. The disease has a wide distribution and several causal organisms; the appearance of the necrotic area depends on the organism concerned (Plates IX and X, between pp. 78 and 79).

Symptoms. The pathogenicity of three Anthracnose organisms has been confirmed by tests in Nigeria.[6] These are:

1. *Botryodiplodia palmarum.* Small translucent spots, typically near the top or edge of the leaf or where the leaf is damaged, change to dark brown and are surrounded by a yellow halo or transition zone. The lesions enlarge and their centres turn grey while drying out; pycnidia develop here and liberated spores give rise to further lesions.
2. *Melanconium* sp. (probably *M. elaeidis,* also found in Malaysia). Development rather similar to that of *B. palmarum,* but the lesion is lighter brown with a pale yellow halo; dries out more rapidly and therefore has a proportionally larger grey and dry area. Acervuli develop with spherical spores which cause further infection.
3. *Glomerella cingulata* produces long lesions between, but not crossing, the veins. The necrosed tissue is brown or black and at first has a water-soaked appearance. The transition zone is yellow. The fungus produces acervuli with conidia and, on the older dried-out lesions, flask-shaped perithecia containing asci each with eight ascospores.

Anthracnose in Africa is typically a disease which follows transplanting from the prenursery to the nursery though it originates in the former. In Malaysia it is more a disease of the prenursery.

Control. The fungi concerned appear to be weakly parasitic since the disease rarely appears if agronomic practices are sound. The latter are for this purpose considered to include sufficient space — at least 3 inches — between plants in the prenursery, regular watering to prevent wilting and fertilizing to induce steady growth, transferring plants to the nursery or bag with a block of soil to prevent leaf desiccation, and careful planting. As the disease may be carried from the prenursery to the nursery an added control measure is to spray the seedlings with a fungicide weekly in the prenursery from the two-leaf stage and for 6 weeks after transplanting to the nursery. Captan (Orthocide M50) and ziram at 1 kg and 0.5 kg respectively per 500 l water have given good protection in West Africa but thiram and Dithane M45 are likely to be equally effective. In Malaysia it is considered that nursery standards of cultivation are high enough to enable prophylactic spraying to be dispensed with.[7]

Corticium solani has been recorded as a cause of Anthracnose in Zaire; this pathogen is mentioned on p. 612.

Freckle, or Cercospora Leaf Spot (Cercosporiose). Cause: *Cercospora elaeidis*

Distribution. This *Cercospora* Leaf Spot, first described by Steyaert in Zaire, is widespread throughout Africa but has not been reported in Asia or America (Plate XI, between pp. 78 and 79).

Symptoms. Freckle is a disease of nursery seedlings which sometimes starts in the prenursery and is frequently carried to field plantings where it survives for many years. It has been noted during the last decade that there has been a tendency for Freckle to survive in the field as a troublesome condition for an increasingly lengthy period.

The youngest leaves of nursery seedlings become infected and minute translucent spots surrounded by yellowish-green haloes enlarge and become dark brown. Conidiophores emerging through the stomata in the centre of the spots, mainly on the undersurface of the leaf, produce conidia which give rise to further, surrounding spots. This results in a freckly appearance, but later the lesions coalesce and the tissue dries out to become greyish-brown and brittle. The disease tends to become aggressive as the leaves age, and the process described above may proceed very rapidly at certain periods of the year. In West Africa this is usually the middle or end of the wet season, and in the following dry season the drying out of the older leaves is much hastened by *Cercospora* incidence. Proof of pathogenicity was obtained by Kovachich[8] in Zaire and Robertson in Nigeria.[6] For details of growth and reproduction of *Cercospora* in the host the papers of these authors and of Weir[9] should be consulted.

Control. The gradual increase in the severity of *Cercospora* invasions, particularly in the field, has caused anxiety. Even moderate attacks materially reduce the green leaf area and are therefore likely to affect early bunch production. The obvious course is to try to eradicate the disease in the nursery and to prevent reinfection of the young seedlings in the field. Copper-based fungicides will control the infection, but unfortunately they have a toxic affect on palm leaves and so cannot be used. Organic fungicides have been extensively tested[10, 11] and captan and ziram, at the same strengths as when used for Anthracnose, have given a 60 per cent control of the disease. These fungicides have now however given place to Dithane M45 and benomyl (Benlate). The former has been shown to be considerably more effective than captan even when applied at only 0.5 kg per 500 l of water.[12] Benomyl is a systemic fungicide and is taken up by oil palm seedlings both through soil application and by spraying on the leaves.[13] Penetration is higher when application is to the undersurface of the leaf. Spore germination of *Cercospora* is inhibited by concentrations of 0.2 ppm of benomyl and growth is arrested at 0.1 ppm. Concentrations of this order are rapidly attained with the leaves of treated plants. Although no data have been published of comparative trials on *Cercospora* control with this fungicide, it is used in the Ivory Coast in polybag nurseries.

That spraying has not yet been fully effective has been shown by the attention given to pruning affected leaves. In the nursery, pruning is relatively simple and regular rounds are organized for the purpose of pruning off all old dry leaves and any others that may be badly infected. In the field, however, pruning becomes an anxiety since on the one hand removal of green leaf will reduce growth, delay flowering and induce cycles of male inflorescences; on the other hand failure to prune and destroy prunings may increase the severity and prolong the incidence of the disease. The practical planter has to find some middle way which does not affect growth, even though this may not substantially reduce the incidence of *Cercospora*. A pruning standard suggested is: any leaf which shows dead or badly necrosed areas over more than one-third of its total surface should be cut off, removed and burned.

It has been shown that nitrogen manuring may cause a small increase in the incidence of Freckle in the nursery, but that potassium substantially reduces it. Small favourable effects of phosphorus have also been noted.[14] When *Cercospora* is prevalent potassium manuring is justified for its *Cercospora*-reducing effect alone.

There are significant differences between *E. guineensis* progenies in *Cercospora* susceptibility[15] and, since serious loss of crop through *Cercospora* attack in field plantings has been demonstrated, Duff has suggested that breeding for tolerance would be worth while.[16] With better control being obtained from the newer fungicides however, it is doubtful if such work will be included in breeding programmes. Some *E. oleifera* progenies are very susceptible.

Seedling Blight or Curvularia Leaf Spot. Cause: *Curvularia eragrostidis*

Distribution.   This disease has been recognized in Malaysia since 1952 and in 1959 was reported to have spread to seedlings in several states in that country.[17] It is not present in Africa.

Symptoms.   Like other Leaf Spot diseases, this disease originates in a small, translucent yellow spot, but its subsequent development is distinct. The spots tend to become irregularly elongated along or between the veins. They consist of a well-defined yellow halo with a narrow raised greenish-brown rim within it; inside is a reddish-brown region with concentric ridges, while the centre of the spot is thin and light brown in colour. Spots may reach 7—8 mm in length.

An attack may be light, with just a few spots, or heavy; in this case the leaves dry out, curl downwards and may disintegrate. The disease progresses inwards and in severe cases all the leaves are affected and the plant dies. Nursery plants which recover show retarded growth and are unlikely to be transplantable. Attacks at the prenursery stage are not usually serious; the most serious attacks occur when the plants are a few months old in the nursery. Serious attacks do not usually occur in the field, though they have been reported.[7]

Pathogenicity tests confirmed that a *Curvularia*, resembling *C. eragrostidis*, was causing the disease. A *Gleosporium* was often isolated in association with the disease but tests proved that this organism, probably saprophytic, could not cause the condition. The *Curvularia* produces tri-septate spores in dense whorls on long erect conidiophores.

Control.   Early experiments showed the value of organic fungicides both on the incidence of the disease and on seedling growth, which was considerably improved. Treatment similar to that for *Cercospora* may therefore be applied. Thiram is favoured in Malaysia.

Helminthosporium Leaf Spots

Various leaf spottings in Zaire, Malaysia and America have been associated with *Helminthosporium* species. In one case, named Eye Spot, the fungus, *H. carbonum*, spread from the plant *Sarcophrynium arnoldianum*. A Helminthosporium leaf spot due to *H. halodes* has been recorded both in Zaire and in Malaysia where it is associated with drought.[3]

Other leaf conditions of seedlings

In the prenursery malformed seedlings are not uncommon and are variously attributed to the after-effects of Brown Germ or incorrect orientation of the germinated seed at planting. Some nursery diseases without known cause have been constant enough in their symptoms to acquire distinctive names.

Infectious chlorosis. There is no evidence that this condition is infectious, and its name is therefore misleading. The tips of the apical leaflets become chlorotic interveinally and later turn dull or bronze. The condition was first recognized in Zaire, but is also found in Malaysia.[3] In Africa, leaf chlorosis may be associated with malformations.

Bronze Streak. This is a very curious condition which appears, until recently, to have been confined to West Africa. Typically, the condition is found on nursery seedlings shortly before transplanting time when they are seen to have a diffused orange or bronze streaking along the leaflets. This originates in a white chlorotic streaking or spotting at the tips of newly opening spear leaves. The bronze streaks on the upper surface of the leaves sometimes appear water-soaked and they may cover as much as 80 per cent of the leaf surface. The condition may appear alarming when seen to affect a large group of seedlings but, curiously, the affected seedlings are often large ones eminently suited, in other respects, for transplanting, and almost as soon as they are transplanted to the field the condition disappears. There is no evidence that spraying or pruning is advantageous (Plate XII, between pp. 78 and 79).

Ringspot. A condition in which yellow or white rings with dark green centres are found on the leaflets has been a feature of nurseries in Nigeria. The rings may coalesce in severe cases and cause some stunting of the seedling but in general the disease has not been serious. Similar symptoms have been described in Malaysia.

Leptosphaeria Leaf Spot. *Leptosphaeria elaeidis*, of which the imperfect stage is *Pestalotiopsis*, was isolated from anthracnosed tissue by Booth and Robertson in Nigeria.[18] In West Africa it is considered to be saprohytic or only a weak parasite; in Sabah the fungus has been found in several nurseries associated with circular, dull-orange leaf spots surrounded by a yellow halo. The spots enlarge and elongate and necrosis of the tissue follows.[19]

Freak conditions. These occur in young nursery seedlings at the bifurcate leaf stage and have been described as (i) *Leaf Crinkle*, in which the lamina between the veins is folded in lines across the leaf, (ii) *Leaf Roll*, in which the lamina is rolled under the leaf giving it a spiky appearance, (iii) *Collante*, in which the lamina between the veins becomes laterally compressed at a band about halfway along the leaf so as to form a constriction there.[20] Suggestions that these conditions are due to deficiencies have not been substantiated. The fact that they rarely occur in shaded nurseries has suggested that factors connected with water supply to the plant may be concerned (Plate 30, p. 345).

(2) Spear and bud rots

Cases of spear and bud rots occasionally occur in nurseries; two distinct diseases have been described in Zaire and will be briefly mentioned here.

Phytophthora Spear Rot. Cause: *Phytophthora* sp.

Distribution.   This disease has only been reported by Kovachich from Zaire.[22]

Symptoms.   The median leaflets of the spear leaf are affected by a rotting which varies greatly in intensity. When the leaf opens the rotted portions become desiccated though bounded by a well-defined orange-brown transition zone. Adjacent tissue is chlorotic. Sporangiophores of the *Phytophthora* species are produced on the diseased leaflets under moist conditions. Pathogenicity has been established. Attention has been drawn to the similarity between this disease and Crown disease of older palms.

Control.   The disease has not yet been sufficiently virulent to require special control measures.

Corticium Leaf Rot. Cause: *Corticium solani*

It is not clear whether this disease should be classified as a leaf disease or a spear rot since Kovachich,[22] while mentioning that in nursery palms the leaflets rot in the spear leaf, also states that its attack on prenursery seedlings is similar to Anthracnose as described by Bull.[21] Pathogenicity was established, but the disease, though affecting many thousands of seedlings in southern Zaire, was not thought to be of great economic importance. The fungus has also been reported in Malaysia[3] and has many other hosts.

More recently the disease has been troublesome in high rainfall areas of Sabah and New Britain. Here seedlings are attacked at a young age. The rot appears at the base of unexpanded leaves which, when opened, show transverse rows of lesions. These are dark brown, become grey and brittle and may fragment leaving holes in the leaves. Captan and ziram are reported to be ineffective and Thibenzole (80 per cent a.i. used at 0.1 per cent) has been recommended where the disease is severe.[7]

A *Nursery Bud Rot* has been reported from Zaire as causing about 6 per cent deaths in some nurseries through a rot at the base of the spear leaf proceeding into the bud.[22] The roots are healthy in the early stages of the disease. No pathogen has been discovered.

(3) Root diseases

Blast Disease. Cause: *Rhizoctonia lamellifera* and *Phythium* sp.

Distribution.   The term Blast, coined by Trueblood in 1944, only gradually received acceptance for a nursery root disease in Africa characterized by sudden withering and death of the plant with extensive rotting of the root cortex. Attention was drawn to the disease on the Unilever plantations in the early 1940s and it was soon discovered that it was widespread wherever the oil palm was being planted in Africa.

Sudden deaths of seedlings have in recent years been found to occur in nurseries in Malaysia accompanied by cortical rotting,[23] but pathogenicity of the associated fungi has yet to be established. Sclerotia of *Rhizoctonia lamellifera* have been found on the stele of affected plants, but no *Pythium* species has been isolated. Williams[19] has therefore preferred to distinguish the Malaysian condition as Root Rot. Attacks have so far been more serious in polybag than in field nurseries. A report of a rather similar condition has also come from Brazil.[24]

Symptoms and causes.   The symptoms of the disease have been described in great detail by Bull and Robertson.[21, 25] Affected seedlings lose their normal gloss and become dull and flaccid, the leaf colour changing successively to olive green, dull yellow, purple or umber (at the tips) and finally, with full necrosis and drying out, to a brittle dark brown and grey. Necrosis of the central spear is usual and death occurs in a few days. Incipient cases of Blast can be picked out in a nursery by a practised eye and, after a few days, these plants will be seen to be dead (Plate XIII, between pp. 78 and 79).

Necrosis of the spear leaf may proceed from the tip or the base and, in the latter case, though the exposed part of the spear may remain green, the basal rot can destroy the growing point and affect the whole of the spear. In about 5 per cent of all cases the rot may not reach the growing point nor may the leaves become completely withered; the seedling then survives to make a partial recovery as a weakly and unacceptable plant.

The roots of diseased plants will be found to include large numbers whose parenchymatous tissue within the hypodermis has been rapidly destroyed from the tip towards the 'bulb', the stele remaining loose within the hollow cylinder. At the point of advance of the rot the transition zone between rotten and healthy cortical tissue is not well defined. When the rate of cortical rotting becomes greater than the rate of production of new absorbing roots desiccation and death follows very rapidly. There is also, however, evidence of a toxin being formed in the roots by the fungi.

Before 1954 many fungi had been isolated from diseased roots in West Africa but they were all free-living soil saprophytes. In Zaire *Thielaviopsis basicola* was isolated together with a *Fusarium* species, but pathogenicity was not established. In 1954—5 *Rhizoctonia lamellifera*, and a *Phythium*, probably *P. splendens*, were isolated.[26] The former was present in decaying cortical tissue behind the transition zone and its small black sclerotia were to be found on the naked stele. The *Pythium* species was only isolated from primary infections of the root tips where there had been no secondary infection by saprophytes or *Rhizoctonia*; under these circumstances it was shown to penetrate the cells and cause their collapse. Sporangia were found within infected cells just behind the transition zone.

In laboratory experiments Robertson showed that the *Phythium* may be parasitized by *R. lamellifera.*[26] In inoculation experiments, mixed inoculum of *Pythium* and *Rhizoctonia* produced more extensive root rotting and the leaf symptoms were more pronounced than with individual inoculations. Inoculations with *Rhizoctonia* alone were successful only when the roots had been artificially damaged. In *Pythium* inoculations, damage was confined in the root tips. In all these cases pathogenicity was established by reisolation of the organisms. Typical leaf symptoms rarely appeared in any of the experiments which were carried out on seedlings which were younger than the normally recognized susceptible stage. One set of experiments was carried out with seedlings at this stage and in this case typical symptoms appeared. It was concluded that *R. lamellifera* plays a very important part in Blast disease in the destruction of cortical tissues and that it gains access either through a prior invasion of *Pythium* which it parasitizes or through root damage from some other cause. *Pythium* is thought to be of importance through its role as a primary invader and its ability to penetrate the parenchyma cells and develop within them. For a further discussion of these relationships Robertson's paper[25] should be consulted. Moreau and Moreau have studied the anatomy of roots suffering from Blast.[27]

Entirely new light has now been shed on the etiology of the disease in the Ivory Coast where the Blast problem has always been peculiarly severe. It was noted in 1973 that plants grown in metal cages covered with mosquito netting showed only 0.75 per cent Blast in comparison with 15.5 per cent outside in unshaded areas. In 1974 a polythene bag nursery trial compared a completely closed cage with very fine netting (to give the minimum shading effect), an open-top cage, plots treated twice weekly with parathion (40g a.i./hl), and unshaded control plots which had natural grass between the bags.[159] The results were as follows:

Treatment	1. Completely caged	2. Open-top cage	3. Parathion	4. Control
Per cent Blast by end of Dec.	0	6	27	46
end of Jan.	2	9	35	63

These results gave rise to the hypothesis that an insect vector is concerned in promoting the disease. At the same time treatments 3 and 4 above illustrate the peculiar severity of the disease in the Ivory Coast in comparison with other parts of West Africa and account for the strong advocacy of shade as a control measure in that country.

Further trials showed that plants covered during the whole night had a low Blast incidence (6 per cent) as against 48 per cent for uncovered plants, and that plants covered between 6 a.m. and 10 a.m. suffered an intermediate attack. Covering from 4 p.m. to 6 p.m. was less effective.[159] It was noted that night covering prevents dew deposition and it was postulated firstly that this could account for the effect of normal shading practice where dew, which may well attract insects, does not form, and secondly that the lesser effect of covering during a morning period would be due to the fact that this would be the daylight period when dew would normally be on the plants.

It will be of great interest to see firstly whether an insect vector can be identified and secondly whether the results can be repeated in other parts of West Africa, e.g. Nigeria. It is noteworthy that in the Ivory Coast it has not been possible, as in Nigeria, to reproduce the disease with *Phythium* and *Rhizoctonia* inoculations[159] and that while nurseries without shade but with sufficient irrigation are successful in Nigeria, shade has always been considered an obligatory control measure in the Ivory Coast. It will be noted that the 6 per cent Blast in the night covering treatment described above is similar to the percentages obtained by the cultural measures used in Nigeria and described in the following section.

Control. One of the most interesting features of Blast disease is the importance of time of attack. It has been shown both in the Ivory Coast[28] and Nigeria that a relationship exists between Blast incidence and the age of the seedlings at the time of attack. If the seedlings are either very young (1 to 3 months) or old (10 to 12 months) at the beginning of the Blast season, the casualties are very small. Results from the Ivory Coast were as follows:

Number of months in the nursery to 1 November	1	2	3	4	5	6	7	8	9	10	11	12
Blast (%)	0	1	3	5	18	22	23	20	21	21	12	5

And in Nigeria:[25]

Date of planting out in nursery	Early Nov.	Late Oct.	Early Oct.	Mid. Sept.	Late Aug.
Blast (%)	0	4.7	11.2	18.5	19.5

Plants left in a nursery beyond 12 months or young plants in the field are only very rarely attacked by Blast.

The precautions which are taken in nurseries against high Blast incidence were briefly discussed in Chapter 7 (p. 356) as they form an integral part of the agronomy of the crop at this stage. The effect of shade in reducing Blast incidence has been established, but the provision of shade for large plants nearing the end of their nursery life has disadvantages, and in Nigeria has generally proved unnecessary. A significant negative correlation was found between Blast incidence and rainfall for August and September. This period covers the 'short-dry' season, which is very variable in severity and which precedes the onset of the Blast season which normally extends from October to January. Experiments confirmed that the provision of irrigation during the short-dry season in August, and the extension of this as needed into September, substantially and significantly reduced Blast incidence. The following results were obtained:

		1	2	3	4
Treatments during:	August and September	Overhead irrigation. Up to 100 cm per week rain and irrigation	Hand-watering. Up to 100 cm per week rain and irrigation	Nil	Nil
	Dry-season, Oct.–Jan.	Normal irrigation from Oct. 1	Normal irrigation from Oct. 1	Normal hand watering from Oct. 1	Nil
Blast incidence (%)	1959–60	4.9	8.9	12.3	13.8
	1960–1	7.3	11.0	14.0	17.5

Later experiments confirmed the effect of short-dry season irrigation except in years where rainfall is sufficiently high during this period;[29] they also showed that for shade to have its maximum effect it must be maintained from planting time. This of course leads to considerable etiolation and cannot be generally recommended.

It must be emphasized that for satisfactory control of Blast a high standard of nursery cultural practice and the observance of correct planting dates are essential. Experience has shown that insufficiently developed seedlings, late-planted, cannot be protected from Blast simply by short-dry-season irrigation. Also, some progenies are highly susceptible and may need special shading. The control measures for Blast in West Africa may therefore be summarized as follows:

1. Plant well-developed prenursery seedlings early in the rainy season and ensure their rapid growth.
2. Irrigate during the short-dry season making sure that there is no severe drying out of the nursery before irrigation begins. Polythene bag nurseries must be assured of a sufficient though not excessive water supply throughout the nursery period.
3. Provide bent-over shade for dry-season nurseries and shade from planting until January for any known susceptible material or in areas of high Blast incidence.
4. Breed for resistance (see Chapter 5, p. 304).

3. Diseases of the adult palm

In this section will be included all those diseases known to attack the palm after it has been established in the field. Some of the conditions mentioned are characteristic of the palm's early life, e.g. Freckle or *Cercospora* Leaf Spot, which is primarily a nursery disease, some are characteristic of early bearing life, e.g. Dry Basal Rot, and some are characteristic of senescence, e.g. *Ganoderma* Trunk Rot; but there is much overlapping and it is not therefore practicable to allot the diseases to particular stages of growth. Moreover some diseases, e.g. *Ganoderma*, may be characteristic of one age in some regions and of a different age in other regions, while other diseases may have different manifestations at different ages.

(1) Leaf diseases

Patch Yellows (Déchiqueture). Cause: *Fusarium oxysporum*

Distribution.   This disease appears to be confined to Africa where it is widely distributed, though sporadic (Plate XV, between pp. 78 and 79).

Symptoms.   Wardlaw[30] reported that following the discovery of *Fusarium oxysporum* associated with Vascular Wilt disease, a second strain of *F. oxysporum* which closely resembled the first had been shown to be associated with a leaf disease in Zaire known as Patch Yellows. Kovachich[31] later proved the pathogenicity of the organism and described the symptoms and extent of the disease.

Infection takes place in the unopened spear leaf and for this reason the lesions at the sites of infection appear opposite each other on the leaflets when the leaf opens. When this stage is reached the lesions are more or less circular or oval with rings of pale yellow surrounding straw-coloured or brown centres where conidiophores can be found; some of the lesions, however, are simply chlorotic with no brown centres. The patches may appear all along the lamina. Later the centres of the patches dry out and drop away giving rise to the typical 'shot-hole' appearance, or, if the patches are towards the edge of the leaflets, to a raggedly indented appearance. The purely yellow patches persist and darken and can be seen to have small orange spots within them.[21] Wither Tip disease of Malaysia, probably a variant of Patch Yellows, is also caused by a *Fusarium* species.[3]

Susceptibility and control.   The disease affects between 0.2 and 1.8 per cent of the palms in areas where it is found, and evidence for genetic susceptibility was provided in Kovachich's pathogenicity tests. In these only three out of forty and later nine out of forty inoculated 6-year-old palms showed any symptoms. Inoculation of nursery seedlings and field palms from 18 months to 6 years of age suggested that nursery palms were immune and field-palm susceptibility decreased from a peak of 50 per cent at 18 months to 7.5 per cent at 6 years old. Kovachich suggested that seed palms might be sprayed with inoculum to eliminate very susceptible lines which might otherwise be multiplied, but the disease does not seem to have reached sufficient proportions for this precaution to be taken, and increased incidence has not been reported.

A condition in Malaysia known as Wither Tip has been described by Turner and Bull[3] who suggest that it is allied to Patch Yellows both in symptoms and cause. However, in Wither Tip, which is not common, rotting of the unopened leaflets seems to be confined to the tip of the leaf which, when it opens, presents a ragged appearance at the distal end with stumps of affected leaflets and adhering wisps of dead laminae. It is claimed that a *Fusarium* is the causal organism, but the pathological work on which this is based has not been published.

Crown Disease (Arcure défoliée). Cause: Unknown

Distribution.   The disease is found in all oil palm areas but is largely confined to palms of Deli origin. The significance of this distribution is discussed below.

Symptoms.   A young 2- to 4-year-old palm suffering severely from Crown disease has many of its leaves bent downwards in the middle of the rachis; at this point the leaflets are absent or small and ragged. These symptoms originate in the spear leaf where the folded leaflets begin to show a rot of their edges or centre.[22] This rot is brown and spreads throughout the central portion of the leaf so that when the leaf unfolds the leaflets of this section are missing. The bend of the leaf at the point where the leaflets are absent is a curious symptom and the real cause of it has not been discovered though Thompson stated emphatically that it had been established that the 'decreased rigidity' was due to insufficient lignification of the parenchymatous tissue.[32] The leaves, however, tend to be quite rigid, though bent. In severe cases all the leaves surrounding the spear may be bent down, and the spear itself may have a rot of its terminal portion which turns brown and hangs down. Under these extreme circumstances Crown disease may have a severe effect on early development and yields. The disease normally affects palms in the second to fourth year in the field, but instances have been reported in the nursery and up to 10 years of age.

Causes and control.   Much attention was given to this disease in the Far East where it was prevalent in the early plantations, but in the absence of a pathogen it was assumed that the disorder was physiological and might be inherited. This latter assumption has been shown to be correct; with regard to the former it has been suggested that palms suffering from the disease have low leaf-Mg contents and there is some supporting evidence for this. It has also been suggested that the incidence of the disease may be affected by Mg and K manuring,[33] but the evidence for this is weak.

The most valuable work on Crown disease is that of de Berchoux and Gascon[34] who showed that pure Deli progenies in the Ivory Coast were highly susceptible, that La Mé material, free of Crown disease, gave crosses with Delis which were also free of Crown disease, but that Zaire material, which showed several cases of the disease, gave Deli x Zaire crosses with a quarter to a half of the palms showing the disease. The authors postulated that susceptibility to Crown disease is due to a monofactorial recessive character. In twelve La Mé x Deli crosses the disease was either absent or amounted to less than 2 per cent. In one Deli x Deli cross the disease was present in 99.5 per cent of the progeny. In Zaire x Deli crosses, or (La Mé x Deli) x Deli crosses, 3 : 1 ratios were expected and often nearly obtained except with the presumed homozygous recessive Deli, where 1 : 1 was expected. Some examples from de Berchoux and Gascon's data are given below. Crown disease susceptibility is assumed to be genetically aa, and from the results obtained by these workers it seems practicable to select palms which will not throw susceptible individuals in their progeny; in particular it would be valuable to have *pisifera* shown to be homozygous for absence of Crown disease (AA), as the Zaire (Sibiti) palm S 127 P appears to be.

Presumed type of cross	Cross	Progenies No. of palms		Expected segregation	
		Without Crown disease	With Crown disease	AA or Aa (%)	aa (%)
aa x aa	Deli D 115 D Selfed	1	205	0	100
AA x aa	L 10 T (La Mé) x D 115 D	130	0	100	0
Aa x aa	L 219 T (La Mé x Deli) x D 115 D	116	87	50	50
Aa x Aa	L 219 T (La Mé x Deli) x D 10 D (Deli)	104	25	75	25
aa x AA	D 115 D x S 127 P (Zaire)	217	0	100	0
Aa x aa	S 7 T (Zaire) x D 115 D	68	62	50	50
Aa x Aa	L 236 T (Zaire) x L 269 D (Deli)	105	25	75	25
Aa x Aa	L 239 T (Zaire) x D 128 D (Deli)	149	61	75	25

There remains one curious feature of the distribution of this disease. Kovachich[22] stated that Crown disease had only been observed in Zaire on palms originating from imported seed of the Deli type; Moreau did not mention the disease in his contribution to Vanderweyen's *Notions de Culture de l'Elaeis au Congo Belge*. It is strange therefore that some of the Zaire material travelling to the Ivory Coast by way of Sibiti in Congo (Brazzaville) should be susceptible. De Berchoux and Gascon consider the fact of susceptibility in some Zaire palms at La Mé as supporting evidence for a Zaire origin of the Deli palm. The immunity of African material in the Far East was early noted by Thompson, but in Nigeria cases have occurred in individual palms not thought to have any Deli parentage. In

Sierra Leone the disease has also been reported on palms not of Deli origin.[21]

These apparent anomalies may perhaps be explained by Blaak's work in West Cameroon.[35] He found that with four palms in crosses and selfs the expected inheritance occurred, assuming that each was Aa. However, the results of crossing and selfing three other palms and crossing them with the first four gave segregations which could only be explained by the presence of an inhibitor gene which, in homozygous condition, prevents the expression of the disease in the aa genotypes. Blaak points out that the presence of an inhibitor gene complicates selection for absence of Crown disease since detection of a palm of aa genotype (susceptibility) is only possible by test crossing with a palm which is known not to have the inhibiting gene, and this will require 3 years.

Leaf Wither. Cause: uncertain, possibly *Pestalotiopsis* sp.

Distribution.   A virulent type of leaf withering has recently been troublesome in parts of Colombia and Honduras and has caused much defoliation. It is also commonly seen in Colombia on *Elaeis oleifera* palms. It is not certain whether this disease should be considered as the same as the disease variously named *Pestalotiopsis* Leaf Spot and Grey Leaf Blight in Malaysia.[3] This latter condition is not considered of economic importance since it is rarely severe.

Symptoms.   The first symptom is the appearance of small brown spots with yellowish halos. These spots soon coalesce into brown necrotic areas which spread over the leaflet tissue and later become grey and brittle. There is a sharp line between the brown and grey areas and in the latter a species of *Pestalotiopsis* is found, black specks indicating the location of spore-bearing acervuli[36] (Plate XVI, between pp. 78 and 79).

The disease has caused considerable defoliation on certain plantations and, as would be expected, this has been followed by serious yield decline; falls in bunch production from 18–20 to 12–15 tons per hectare in adult areas, and from 11 to 7–8 tons per hectare in young plantings have been reported.

Cause, susceptibility and control.   Although *Pestalotiopsis* sp. have been found on moribund tissue in both Colombia and Malaysia it is not certain that this fungus is the primary pathogen. Turner and Bull[3] point out that although several species are associated with leaf spots of mature palms these are normally weak parasites only attacking senescent tissues or those showing nutrient deficiency symptoms particularly of magnesium. In Colombia, leaf analysis does not suggest that magnesium deficiency is an underlying cause, and a number of possibilities have been canvassed. Firstly it is possible that an aggressive strain of *Pestalotiopsis* has developed in the region. Secondly, another and more aggressive parasite may be giving entry to the *Pestalotiopsis*. A further possibility is entry through insect attack. A *Leptopharsa* species, a sucking bug of the Tingidae family, has been noted as producing minute necrotic spots on oil palm leaflets in the area where Leaf Wither is severe.[36] Much pathological work remains to be done before these issues are fully elucidated.

In Malaysia control is recommended simply through the improvement of the growth of the palm by correct manuring.[3] In Colombia, where more immediate measures are needed, spraying with Dithane M45 is being tried on an experimental scale. Aerial spraying in Honduras has not been successful. Control of the Tingid has also been suggested (see p. 657).

White Stripe. Cause: Unknown

Distribution. This condition is sporadic and, in Asia, is said to be more common on alluvial soils, particularly organic clays or mucks; it has however appeared on almost all the main soils supporting oil palms though it is less noticed in Africa than in either Asia or America.

Symptoms. Narrow white stripes are found on each side of the leaflet midrib and extend its whole length. The stripes are at any point from the midrib to the margin and are sharply divided from the adjoining green (often dark green) tissue. The course of the disease has been insufficiently studied. It is certainly believed that many affected palms recover, and Rajaratnam[37] reports that the chlorotic tissue may turn green after about 7 months and that the symptom is more prominent in young leaves than in old. He also showed that chlorosis was due to failure of the palisade mesophyll cells to elongate and that apparent recovery was through an increase in the chlorophyll content of the spongy mesophyll and not through development of the palisade cells. Turner and Bull[3] believed that the condition is associated with an abnormally erect habit and slender leaflets, but White Stripe has often been seen on palms of otherwise healthy appearance.

Causes and control. The cause of the disease has been thought to be nutritional: either boron deficiency or a high leaf N/K ratio. These claims have not been substantiated.[37] It has also been suggested that the disorder is of genetic origin. A certain *tenera* x *dura* cross showed similar percentages of White Stripe when planted in the Ivory Coast and in East Cameroon; also certain Deli selfs in the Ivory Coast showed the symptoms while others did not.[38] In Malaysia, however, associations between incidence and progeny could not be found.[37] The cause of White Stripe therefore remains unknown, though a genetic origin seems not unlikely.

It is difficult to assess the loss of crop resulting from this disease since its incidence varies so greatly and it is often merely sporadic. Information from Malaysia has been contradictory. In one area palms previously suffering from White Stripe actually yielded more than those which had not been affected.[39] However, in an area of moderate and severe incidence yields of affected palms were compared with adjacent healthy palms; mean bunch yields were from 16 to 47 per cent lower in affected palms.[37] No specific control measures can be recommended, but where the disease is prevalent it is clearly wise to ensure optimum growth and development through good cultivation and wise manuring.

Minor leaf diseases

The oil palm leaf is unusually susceptible to patchy discoloration and necrosis from minor pathogens and to surface covering by epiphytic and saprophytic organisms. These often cause the older leaves to appear far from healthy and the area of actively photosynthesising leaf may be seriously reduced. Brief notes will be given on the more important of these conditions. Identifications have proceeded much further in Africa than in the Far East.

Necrotic Spot.   A common disease of the older leaves in Africa, caused by *Cercospora elaeidis*, though of a different strain to that causing Freckle. Brown necrotic spots on the leaves are surrounded by a narrow greenish yellow halo which later expands and becomes bright orange. The disease sometimes becomes aggressive and causes much premature withering, leaflets tending to die off from the tip and the margins. Proof of pathogenicity was obtained in Zaire by Kovachich who considered Necrotic Spot to be the most common foliar disease of adult palms. The fungus *Oplothecium arecae* is found in the old spots in Zaire but there is no evidence that it is a pathogen.[40]

Crusty Spot (Croûtes noires).   This is a disease of senescent leaves in which circular or oval orange spots appear on the leaflets with a blackened crust at the centre of each. Commonly the spots are at the bases of the leaflets. Around the solid crust the tissue becomes grey and is bounded by an orange halo which merges into the green of the leaf. Only rarely does the disease become aggressive, in which case large areas of the leaflets die off.[21] The crusts have been found to contain the stromatic tissues of the Ascomycete *Parodiella circumdata*, and pathogenicity has been established.

Orange Leaf Blotch (*Bigarrure*).   This is a disease of older leaves in which large irregularly-shaped orange patches appear on the lamina; in advanced cases necrosis occurs in the centre of the patches giving rise to dark brown mottling. Descriptions of the disease have tended to be confusing and it could be mistaken for *Cercospora* Leaf Spot. The cause has not been established though *Pestalotiopsis* species[21] and *Helminthosporium* species[41] have been suggested (Plate XIV, between pp. 78 and 79).

Algal Leaf Spot.   Leaf damage caused by an alga, *Cephaleuros virescens*, is now common. Pin-point yellow spots develop on the upper surface of the leaflets and on the rachis. Later, the production of sporangiophores gives the lesions an orange tinge. In Zaire the organism has been regarded as mildly parasitic on senescent leaves, but Weir[42] has shown that even on vigorously growing palms up to 10 years old leaves which are only entering the second half of their life span may be infected and thereafter the infection may increase so that there may be as many as twenty algal spots per centimetre length of the entire leaflet. It has proved difficult to assess the real effect of the alga in hastening senescence, since it is often found in company with such other conditions of the ageing leaf as Crusty Spot and Freckle.

Epiphytic and Saprophytic Moulds and Lichens. Black 'Sooty Mould' is often found to grow on the older leaves of adult palms and occasionally spreads over a large proportion of the leaf surface giving the palms a blackish-grey appearance. No exhaustive study has been made of the organisms concerned or their relative prevalence. In modern works the term 'sooty mould' is used for the complete epiphytic flora which often becomes covered with a sooty layer of spores and mycelium. Although mainly epiphytic, this flora may harm the plant by blocking stomata and screening the leaves from light. That this is particularly undesirable in the oil palm will be clear from what has been said of the flowering of the adult palm, though it is usually the older leaves which are most affected.

Several of the commonest fungi to be found in Africa as constituents of the epiphytic flora appear in Turner's 'Micro-organisms associated with oil palm'.[43] Among these the Ascomycetes *Apiospora* sp., *Meliolinella elaeidis* and *Meliola elaeis* may be mentioned. *Meliolinella elaeis* is recorded as also being found in America (Costa Rica) on *Elaeis oleifera*. Sometimes the epiphytic flora consists largely of algae. A widespread red 'rust' of oil palm leaves found in all continents belongs to the genus *Cephaleuros*. Species of *Trentepohlia* were found in Sabah.

Epiphytic flora may appear on the upper or lower surface of the leaves. In West Africa the black mould usually found on the upper surface consists of discrete circles of about 5 mm diameter; on the lower surface the black mould is in irregular patches of less dense material. In Malaysia, sooty moulds of *Brooksia*, *Ceramothyrium*, and *Chaetothyrium* sp. develop on insect secretions on the leaves.[3, 19] *Brooksia tropicalis* is common in Africa.

Lichens are often found among the epiphytic flora on oil palm leaves.

(2) Root and stem diseases

The serious diseases of the adult palm are either root and stem diseases or bud rots. The investigations of many of these diseases have proved lengthy and arduous tasks and many lethal disease conditions of unknown cause are still being found. The external symptoms of all these diseases tend to be variable and Wardlaw has told how the investigation of oil palm diseases in particular had shown him how little confidence could be placed in these symptoms alone.

In general, however, root and stem diseases are characterized by fracture and drying out of fully-developed leaves, leaving the spear leaf and some surrounding leaves standing erect until the disease has proceeded much further. These early symptoms may be accompanied by a change of colour, drying out or wilting of one of the more erect younger leaves. Bud and spear rots on the other hand tend to be characterized by symptoms in the centre of the crown. The spear leaf may be directly affected or the surrounding leaves show a sudden chlorosis. Successive spear leaves may be shortened, have peculiar 'little leaf' formations, or cease to develop, leaving a palm with an empty centre.

These general symptom differences between the two groups of diseases give a first rough guide when deaths occur or alarming disease symptoms appear; but dissection of the palm must follow to see exactly where the site of *destruction* is. While the root and stem diseases kill the palm by a process of denial of water and nutrients to the crown, the bud rots eventually kill the palm by growing towards and reaching the single growing point. The site of decay in the first case may be expected therefore to be in the bole, trunk or roots, but with bud and spear rots the changes will be found in the 'funnel' or 'cabbage' of the crown.

Root and stem diseases have been divided into those which depend for their development on large infection foci such as are provided by old stumps and trunks, and those which are caused by common soil saprophytes and where infection seems to depend on predisposing factors.[44] This division is of great significance in evolving methods of combating the diseases.

Dry Basal Rot. Cause: *Ceratocystis paradoxa* (imperfect stage = *Thielaviopsis paradoxa*)

Distribution.   The distribution of this disease has proved unusual. It appears to be confined to West Africa and, though the pathogen is a common soil inhabitant found in all parts of the tropics and was recorded in Ghana in 1924, the disease was not discovered in epidemic form until 1960. One estate in Nigeria was devastated and thereafter minor outbreaks occurred in several parts of Nigeria, West Cameroon and Ghana. In the first epidemic deaths were common, but more recently recovery has been the rule and no more serious outbreaks have occurred. A mention has been made of the disease in Sabah but evidence of pathogenicity from the Far East has not been published.[19]

Symptoms.   The foliar symptoms are preceded by extensive bunch and inflorescence rot. The rachis of certain leaves then becomes fractured submedianly, though the leaflets remain green for a considerable period before they eventually die. Occasionally a young leaf high up in the centre of the crown becomes necrotic and dries out, and this precedes the necrosis of the older leaves. It is quite common for a complete ring of leaves to exhibit the submedian fracture while the upper leaves are still erect, and this gives the newly-affected palm its characteristic appearance. Later, the upper leaves and the spear will be similarly affected and the palm dies, or it may make a recovery at any stage. A recovered palm will take several years to come back into bearing (Plate 58).

One interesting feature of the epidemic at Akwukwu in mid-western Nigeria was the dying of the *Pueraria* cover. Inoculation experiments showed that *Pueraria* seedlings could be infected and killed by the fungus *Ceratocystis paradoxa* following root-dipping in an inoculum.[45]

The characteristic internal symptom of the disease is a dry rot at the base of the trunk. This rot is well established by the time the primary leaf symptoms are apparent. In the transition zone between rotted and healthy material many vascular bundles are necrotic, and it is possible to trace infection from an infected root or leaf base into the base of the trunk.

The majority of palms attacked have been palms which have recently come into bearing, but 10-year-old palms have also been affected. Symptoms usually appear at the end of the dry season.

The cause and spread of the disease. The cause of Dry Basal Rot was shown in pathogenicity tests conducted by Robertson to be due to the Ascomycete *Ceratocystis paradoxa* of which the imperfect stage is known as *Thielaviopsis paradoxa*.[45, 46] *Ceratocystis paradoxa* is a soil inhabitant widely distributed throughout the tropics of Africa and Asia and causes diseases of several other crops. Its sudden appearance in West Africa as the cause of a serious condition was unexpected and has given rise to investigations on conditions conducive to its spread. The epidemic at Akwukwu occurred on Acid Sands soils with an unusually low clay content at depth (15 to 17 per cent at 2 m), and minor outbreaks at the N.I.F.O.R. Main Station also occurred on fields with less clay in the profile. This led to the belief that incidence might be connected with soil—climate relationships. A further outbreak at N.I.F.O.R. in 1967 followed a severe dry season. Incidence varied between fields from 0.1 to 10.0 per cent.

Two features of the spread of the disease are important. In the outbreak at Akwukwu many deaths occurred in the first 2 years, amounting to about 30 per cent in one area. Thereafter very few deaths occurred and there was considerable recovery; though new infections occurred, these did not give rise to many further deaths.[47] The results of surveys in one area of 586 palms planted in 1954 illustrate this.

Date of survey	Healthy palms (%)	Slightly infected (%)	Severely infected (%)	Dead palms (%)
April 1959	32	21	22	25
March 1960	33	28	11	29
February 1961	68	1	—	31
March 1962	56	7	3	34

All the palms which recovered were bearing bunches by 1965.[48]

A   B

Pl. 58   Dry Basal Rot, *Ceratocystis paradoxa*: **A.** A severely infected palm showing sub-median fracture of the lower leaves. **B.** A palm showing external symptoms of the disease dissected to expose the dry rot at the base of the trunk.

The other feature of the spread of this disease has already been referred to in Chapter 5 (p. 304). At the N.I.F.O.R. Main Station incidence in one field of 9-year-old palms was mainly confined to progenies having the same female parent. It was also shown in this field that affected palms had given yields significantly below the average of the field during the 3 years prior to the attack.

Control.   Although in the pathogenicity tests all seedlings were infected by inoculation through a root-dipping technique, the suggestion of progeny tolerance obtained from the N.I.F.O.R. Main Station survey mentioned above appeared to be worth pursuing. Inoculated seedlings of chosen progeny lines planted out in a nursery showed marked differences of disease incidence and there is evidence that tolerance is inherited monofactorially.[49] In regions where the disease is liable to occur, all progenies for distribution should be screened against the disease.

Vascular Wilt Disease (Fusariose or **Tracheomycose).** Cause: *Fusarium oxysporum.* Schl. f. sp. *elaeidis* Toovey

Distribution.   Since its description by Wardlaw in 1946[50] in Zaire this disease has always been considered the most menacing of all oil palm diseases. This view has been somewhat mitigated in course of time by the absence of serious epidemics in West Africa and elsewhere, but its early reputation has died hard and there has been a tendency to assume an outbreak of disease in the adult palm to be Wilt from external symptoms and before the pathogen has even been isolated.

The early history of the disease was briefly stated by Wardlaw as follows:

> During a visit to the Belgian Congo in 1946 I observed a wilt disease of the oil palm (*Elaeis guineensis*), and isolated *Fusarium oxysporum* from the necrosed vascular strands. In 1947, Messrs. S. de Blank and F. Ferguson, in a private report, announced the presence of this disease in Nigeria and submitted cultures to me for identification; and in 1948 I was able to confirm their diagnosis during a visit to the affected plantations. Substantial proof of the pathogenicity of *F. oxysporum* is now being obtained by workers in the Belgian Congo.[51]

Pathogenicity was confirmed by Fraselle in 1948.[52] Thereafter Vascular Wilt was found on several plantations in Nigeria and West Cameroon and in the Ivory Coast and elsewhere in West and West-Central Africa, but the disease has not been detected in Asia or America although a *Fusarium* sp. has been isolated in association with a disease in Brazil. In West Africa it has been almost entirely confined to plantations, the great areas of palm groves being almost free of it. Some cases have occurred among planted palms at the N.I.F.O.R. Substation at Abak in the centre of a district thick with palm groves but only one case of Wilt was found in the groves themselves. The greatest devastation has occurred on replantings on estates in southern Zaire.

Symptoms. The external symptoms of Vascular Wilt are difficult to define and most of them have been observed in other diseases of the adult palm. In the plantation areas of southern Zaire wilt is commonly found in young palms which have recently come into bearing; this is also so in Nigeria in replantings. However, in West Africa the disease has attacked older palms which have been in production for 10 years or more. A full account has been given by Prendergast.[53]

In the most usual, chronic form of the disease in mature palms the older leaves become desiccated and the rachis breaks near the base or at some distance from the base, the ends of the leaves hanging downwards. This feature has been used to distinguish the disease from *Ganoderma* Trunk Rot in which the leaves collapse at the base and closely cloak the stem. The disease usually proceeds gradually along several leaf spirals with younger leaves becoming successively affected. The erect and still green leaves in the crown are now much reduced in size and are often chlorotic and the palm may stay in this state for several years before the crown eventually collapses.

Occasionally a mature palm suffers a rapid death through an acute attack. The leaves dry out and die rapidly while still in an erect position and then snap off several feet from the trunk, usually during strong winds. The remaining leaves die quickly. All stages between the acute and chronic forms are encountered (Plates 59 and 60).

Symptoms of the disease in young palms of up to about 6 years of age in which no trunk has yet been formed are somewhat different. In these palms the 'Lemon Frond' symptom is frequent; a leaf somewhere in the upper middle part of the crown (fourth to fifteenth leaf) develops a bright lemon yellow colour before drying out from the tip to the base. Leaves at about the same level then turn yellow and dry out to be followed by some of the younger leaves which will die while many of the older ones remain green. Newly-developed leaves become successively smaller, and death of the whole palm usually takes less than a year. It should be noted that the striking 'Lemon Frond' symptom is not always to be seen, and in southern Zaire a general yellowing of the leaves before death was more usual, but it is interesting to note that this symptom, later attributed to Vascular Wilt, was the subject of a report made in 1940 in Nigeria some 5 years before the pathogen was isolated.[53]

Nursery plants suffering from Wilt often recover. Shortening of the leaves, causing a bunched appearance, is followed by browning and drying off of the older leaves and the condition progresses inwards.

The pathogen is soil-borne and usually enters the palm through the roots, growing along the stele which becomes blackened. Infection can take place through wounds in the stem base and through uninjured roots.[54]

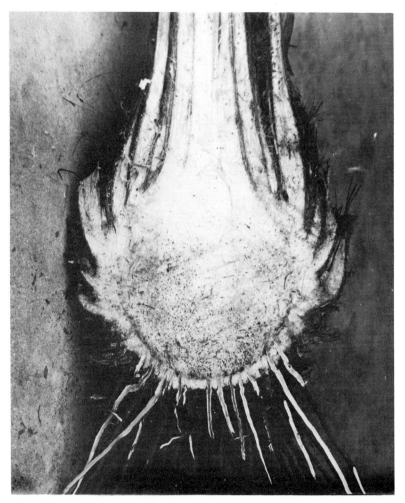

Pl. 59 Vascular Wilt — longitudinal section of a seedling, showing continuation of necrosed vascular strands from roots to stem.

Renard, who conducted many inoculation experiments with seedlings, considered that entry of the mycelium was much impeded by lignification even with wounding and that rapid infection was mainly through the transmission of spores in the vascular system.[55] Locke, also working with seedlings, showed that the pathogen is confined to the conducting elements of the xylem and can reach the stele from the tip of a lateral root or the damaged cortical tissue of a pneumathode.[56] He considered that the plant had little defence against serious infection in spite of resin formation and tyloses.

Pl. 60   Vascular Wilt disease, caused by *Fusarium oxysporum*, of a 16-year-old palm.

From the roots the mycelium penetrates into the wood vessels of the vascular strands where conidia and chlamydospores are also found. The vascular bundles are normally pale yellow or whitish, but when diseased they become brownish-grey or black, and a cross section of the trunk therefore shows a speckled appearance. Discoloration, which is associated with the presence of gum, is confined to the wood vessels; blackened fibre strands do not indicate Vascular Wilt. Such blackening often occurs in older palms and sometimes in other conditions and the inexperienced observer can therefore be misled into a wrong field diagnosis. Moreau points out that though discoloration of the vessels is normal in palms over 20 years old, in these palms the blackening decreases towards the top of the palm instead of becoming accentuated as in the case of Wilt. Drying-up of the leaves and death of the palm is caused partly by the destruction of the roots and partly by the blocking of the wood vessels by gum. Diseased vessels may at first occur in only one section of the stem base and this probably accounts for leaf symptoms being confined at first to certain spirals. Vessels in the centre and at the top of the stem then become diseased and the symptoms spread across the stem so that a large proportion of the vessels at the top are affected. In young palms up to 6 years old diseased vessels are usually widely dispersed throughout the base.

Incidence, susceptibility and spread. Prendergast[53] claimed that healthy vigorous palms in good soil suffer little from the disease and he showed that in areas of K deficiency incidence is substantially reduced by applications of potassium fertilizers. This finding has been substantiated in potassium experiments in areas of Vascular Wilt in both the Ivory Coast and Dahomey.[152] However, the absence of the disease in highly K-deficient palm groves is noteworthy and unexplained.

It has been shown that although, where prevalent, the outbreaks are often sporadic, new cases do tend to develop next to old ones. Perhaps the most difficult feature of the disease is the tendency for many young palms in a replanting to become infected when they are near to the sites of previous cases of Vascular Wilt.

Data on the incidence of the disease in Zaire is lacking, but in southern Nigeria incidence reached 23 per cent in one-half of an estate and 12 per cent in the other half with 17-year-old palms. In fertilizer experiments incidence was over 30 per cent in plots not receiving potassium but usually below 20 per cent when potassium had been applied.

Control. Direct control of the disease is unlikely to be practicable. In Zaire the destruction by fire of all diseased palms and their neighbours was recommended and the replanting of areas where Vascular Wilt had been prevalent was not advised.[41] It was suggested that such areas could be left fallow for many years before being replanted, but as this pathogenic form of *F. oxysporum* is a soil-inhabiting fungus which has been shown to colonize competitively dead organic matter in soils,[57] it is doubtful if even a lengthy period of fallow could prevent recurrence of the disease.

The most promising method of control is by the breeding of resistance lines or the screening of selections for tolerance to the disease. Prendergast[58] was the first to develop a technique for testing seedling resistance at the nursery stage and his methods were adopted with very little modification by Renard *et al.*[59] For detailed descriptions these authors papers should be consulted, but in essence the method consists of applying a mixed inoculum of isolates of the causal organism from different sources to nursery seedlings by pouring 10 ml of the inoculum on to the bulb of the seedlings or the exposed roots around the collar. Later, prenursery seedlings were used. The work showed clearly that no seedlings are resistant to invasion and that differences are really ones of tolerance to the fungus.

Different and empirical methods of division of progenies into 'resistant' or 'susceptible' have been used. Prendergast divided progenies into (1) good, (2) probably resistant, (3) possibly resistant and (4) poor. (1) and (2) had a lower percentage loss in a given test than the mean of that test; (1) had a lower percentage loss than the mean of (1) and (2). Renard *et al.* calculated a Wilt Index which is the percentage of wilt-infected plants in a progeny as a percentage of the wilt percentage of all the plants in a trial. Another Index, Ig, was attributed to a parent by taking the mean of the indices of the progenies in which it figured. Those with indices above 100 were called 'susceptible', those below 100 'resistant'. Special attention was given to *pisifera* parents.

It has been shown[60] that these methods lead to much the same selection of so-called 'resistant' material; but it will readily be appreciated that the degree of tolerance conferred is very difficult to evaluate, that differences between progenies adjacent in order of susceptibility are not significant and that the tests themselves leave unanswered the question of what casualties may be expected in field plantings, On the latter point some evidence has been presented that progenies shown to be more susceptible in seedling tests do give rise to palms which, on average, will have a higher percentage incidence in the field.[59, 60] At the present time (1976) it cannot be said that these methods will do more than effect an indeterminate reduction of Wilt incidence in areas where the disease is prevalent. Nevertheless it is of importance that, where progenies are to be planted in areas where *Fusarium* attack is expected, the moderate protection afforded by such methods should be employed.

Both the inexact nature of these methods and difficulties of reproducibility from test to test have led to work being started on more refined techniques of testing. The methods referred to above do not ensure that the dose of inoculation per seedling is uniform. Locke and Colhoun developed a method (of which full details are given in their papers[56, 61]) of inoculating very young seedlings grown in compost with two known levels of inoculum, and then comparing their growth with that of seedlings grown in uncontaminated compost. Determinations were made of the number of propagules in the soil so that subsequent inoculations could bear a relation to normal soil levels. A technique was devised for providing talc—chlamydospore mixtures for storing and for use in inoculation experiments. The fungus was recovered from progenies showing both large and small reductions either in weight per plant or in 'leaf area product'. No progenies were immune. It was demonstrated that some progenies were tolerant of infection in lightly contaminated compost only, some in both lightly and heavily contaminated compost; others showed high susceptibility at both levels. A high level of repeatability was attained.

The method has not yet been employed on a large scale, but it is expected to be useful for testing large numbers of progenies in a relatively short period of time; forty seedlings per treatment should be planted in uncontaminated compost and in compost contaminated at two levels and grown for 4 months before they are removed for dry weight determinations.

Fusarium oxysporum f. sp. *elaeidis* has been shown to be seed-borne, and there is also evidence that in spite of some routine seed disinfection methods, this and allied species of fungi can easily be transmitted by seed from one country to another.[62]

Ganoderma Trunk Rot (Basal Stem Rot). Cause: *Ganoderma* sp.

Distribution. Species of *Ganoderma* are soil-borne fungi of widespread occurrence, causing disease in a great many economic crops. In Africa the species attacking the oil palm are found throughout the palm groves, killing the old palms and occasionally younger ones which have not yet dropped their leaf bases. In planted areas it is also a disease of senescence, being rarely found except in fields which should already have been replanted. In Asia, though it is also a disease of old palms, it is of much greater economic importance since in certain areas it has attacked palms of 5 years old and upwards, particularly in replantings. Disease caused by *Ganoderma* has not yet been reported from America.

Symptoms. In old palms the onset of the disease is shown by the collapse of one or more of the lower leaves which do not snap, but hang vertically downwards complete with their petioles. This is followed by the drooping of younger leaves which turn a pale olive green or yellowish colour and die back from the tip. The leaflets roll back around the rachis and the top of the stem becomes heavily cloaked by the desiccated leaves (Plate 61).

Later, the base of the stem blackens, gum may be exuded and the well-known fructifications of *Ganoderma* appear. The whole head of the palm may then fall off, or the trunk collapse.

Bull[21] has described and illustrated the internal symptoms of old palms exhibiting *Ganoderma* Trunk Rot. Briefly, it is found that the peripheral tissues are hard and unaffected by the rot, the black fibres in this zone being normal. Within the stem at the base of the palm there is usually a large peg of dark brown infected material; between the peg and the outer part of the stem are narrow dark brown bands. In this zone the majority of the tissue is yellow-coloured and breaks up easily; mycelium can be found extending through the tissue.

Roots are also found to be infected, the cortex being brown and decaying, the stele black. In Malaysia, white mycelium of *Ganoderma* has been found surrounding the cortex which becomes spongy. Large numbers of

Pl. 61 *Ganoderma* Trunk Rot of an old palm on coastal alluvium in Malaysia.

fructifications, the sporophores, may be formed, the early ones being small and rounded, the later ones being typical brackets. The upper surface of the sporophore is reddish-brown and rather uneven, though shining. There is a clear white band at the outer edge; the underside is pale yellow-brown and covered with minute pores.

The symptoms of young palm infection in Asia are variable but a common first symptom is that of drought conditions, i.e. a failure of the young leaves to open so that a number of fully elongated but unopened 'spears' are seen in the centre of the crown.[63] The internal symptoms are similar to those of infected old palms, but the sporophores often appear at an early stage, thus confirming that the failure of leaf opening is a disease-induced symptom.

Cause and control. The taxonomy of the genus *Ganoderma* has been in a confused state. Three species, *G. miniatocinetum, chalceum* and *tornatum*, have recently been identified on oil palms.[64] Other species commonly recorded are *G. applanatum* and *G. pseudoferreum.*[7] Their geographical distribution is not fully known. *Ganoderma lucidum* is a temperate climate species.

Ganoderma is of importance in the African palm groves through its ability temporarily to reduce the stand in an old productive grove and thus to impair the yield of the grove as a whole for the time being. However, as much higher yields could be obtained on the same area from planted palms which do not normally contract the disease until they are overdue for replacement, there is no economic justification for undertaking or considering control measures within the groves other than through the act of clearing and planting.

In planted areas in Asia the position is quite different and recent studies by Navaratnam[65] and Turner[66] are relevant. The former investigated methods of spread and obtained successful inoculations, with mycelium, of both roots and stems of 40-year-old palms. It seems likely that infection under natural conditions is mainly by root contact, but whether spread by spores is significant is not known. The disease is much more prevalent on the coastal clay soils than on inland soils and it is on the former that the serious attacks on young palms have occurred. Turner, examining attacks on young palms, showed that incidence on areas where the preceding crop was coconuts was much higher than where planting followed forest or rubber (Table 13.1). Data from replantings of oil palm fields on the coastal

Table 13.1 *Incidence of* Ganoderma *infection in young oil palm fields following forest, rubber and coconuts (Turner)*

Previously under:	Forest		Rubber		Coconuts	
Age (years)	Acres	Ganoderma (%)	Acres	Ganoderma (%)	Acres	Ganoderma (%)
14	208	0.1	76	0.6	27	42.1
13	997	0.2	1,285	0.5	366	24.4
12	1,154	0.2	962	0.4	120	14.4
11	671	0.2	310	0.5	–	–
10	516	0.2	312	0.3	31	16.7
9	166	–	373	0.2	78	7.7
8	140	0.5	468	0.1	58	3.9
7	199	0.0	915	0.1	184	1.8
6	–	–	1,610	0.1	125	1.3
5	161	0.2	1,725	0.1	–	–
4	29	0.0	1,135	0.0	–	–

soils are few, but surveys in two areas suggested that nearly as high an infection as after coconuts may be expected.[66] Devastating attacks have been experienced in replanted coastal areas of Sumatra where underplanting was practised or the old stand not treated.

The occurrence of the disease on old coconut sites has been as high as 50 per cent after 13 years, incidence rising steadily from the seventh or eighth year. There is circumstantial evidence that coconut and oil palm stumps act as infection foci for young palms. On four sites where contiguous areas of rubber and coconuts had been simultaneously felled and planted with oil palms, *Ganoderma* incidence was high on the coconut areas, reaching 39 per cent by the fourteenth year in one case, and very low on the rubber areas. Two years after coconut felling almost every stump exhibits *Ganoderma* sporophores and production of these continues for several years. Sporophores develop even earlier on felled or poisoned trunks. Turner has pointed out that whether infection is by spores or root contact the main source of infection is the stump. He therefore recommended the removal of old coconut or oil palm stumps and trunks for destruction by burning.[66] Though costly, this at present seems the only practical method of reducing incidence; an advantage of the method is that inspection of stumps for *Oryctes* will no longer be required. Dead coconut stumps and felled trunks become colonized by *Ganoderma* almost to the exclusion of other organisms and sporophores soon begin to appear. Dead oil palms are not colonized so exclusively and they rot faster, but they are still an important source of infection and if poisoned palms are left standing the amount of colonization is increased.[7]

Felling of both old coconuts and oil palms is nowadays done by bulldozer, sometimes with the assistance of a special 'de-stumper' projection. Old oil palms are usually poisoned with sodium arsenite some 2 weeks before felling (see p. 485). Both palms have subsequently to be cut up and burnt. Palm stems are notoriously difficult to burn; but there has been much written on the methods adopted[3, 4, 7] and these will not be described in detail here. Stimpson and Rasmussen[67] have provided a valuable account of practical methods of replanting which entail burning or, if this is not possible, cutting up, splitting the boles and windrowing the old oil palms so that they rot rapidly. The systems adopted include prior poisoning of the palms and subsequent root raking and ploughing to bring up and dispose of pieces of palm base and other material which may form a focus for *Ganoderma*. The operations are costly but are considered essential in coastal areas. In inland areas where the hazard of *Ganoderma* is not so great, poisoning and felling may be followed by any method which encourages the rapid rotting of the old palms.

There is some doubt as to whether, or how far, poisoning with sodium arsenite encourages rotting. It is often stated that rotting is hastened,[4] but in Sumatra it is claimed that applications of urea are more effective, not only accelerating decay but inhibiting the development of *Ganoderma*.[68]

Because the disease develops through infection from massive sources of inoculum it was believed that breeding for resistance was an unpromising line of control. However, it has been shown that marked differences of progeny incidence can exist. In a trial of four progenies in Sumatra one progeny from Zaire showed only a 3.5 per cent incidence after 14 years against 10.5 to 26.2 per cent with the other progenies, and the rate of incidence increase was also less.[69, 153]

Following reports of successful use of systemic fungicides with other perennial crops, including control of *Ganoderma* in coconuts,[154] a start has been made on trials of their use with oil palms in Malaysia.[155] Though promising, these trials are at present (1976) only in a preliminary stage.

Some soil and leaf analysis studies have been made in relation to *Ganoderma* incidence, and these indicate that nitrogen and magnesium may have special roles in combating the disease.[69]

The difference in incidence in fields on coastal soils in Asia and inland soils remains unexplained. One curious feature of the disease is its low incidence on clay areas where there is a covering of peat.

Ganoderma infection is sometimes found to be confined to a portion of the trunk well above ground level. In palms of 10 to 15 years, surgical removal of isolated lesions in the trunk may sometimes be successful.[70]

Armillaria Trunk Rot. Cause: *Armillaria mellea*

Distribution. This disease became conspicuous in the northern oil palm areas of Zaire in the late 1940s. Although *A. mellea* was recorded from the oil palm in Ghana in 1927, the disease as such has not been recorded as present in West Africa nor in any other oil palm region of the world. In Zaire the disease can occur on all soil types. The fungus attacks a number of other crops, including tea and tung, as well as forest trees.

Symptoms. The symptoms have been described by Wardlaw[71] and Moreau.[41] Palms 4 to 12 years of age are usually affected. External symptoms are similar to Vascular Wilt disease but, in addition, the lower leaf bases become rotten and easily fall from the trunk. Palms often break near ground level and fall over.

Internally, the trunk shows regions of mass infection in a soft rot which may progress across the base of the trunk or in an upward direction. The disintegration becomes so complete that the trunk collapses. Infected tissue has, at first, a yellowish-brown colour and the fibre strands, not the wood as in Vascular Wilt, become dark. Occasionally, the palm seals off the rotted portion of the trunk, produces new roots and survives. The fungus produces light brown mushroom-like sporophores near the base of the trunk during the wetter parts of the year.

Invasion takes place through the roots, only a few of which may be attacked; the cortex is destroyed by the matted white mycelium. Rhizo-morphs spreading round the base of the palm cause the loosening of the leaf bases.

Incidence and control. Although the incidence of *Armillaria mellea* has varied considerably it can become locally important. The fungus may sometimes be found in the palm in conjunction with *Fusarium oxysporum* and this makes it difficult to measure its incidence and to determine its pathogenicity.

A number of control measures such as the regular inspection and disposal of infected palms, removal of infected leaf bases and the removal of infected forest stumps have been suggested, but the disease has not proved sufficiently widespread for investigation to proceed very far, and its incidence seems to have decreased considerably during the 1950s.

Sudden Wither (Marchitez Sorpresiva). Cause: Unknown

Distribution. This disease has been serious on one plantation in Colombia and has been reported on other plantations in that country and in Ecuador and Peru. A similar disease has been prevalent in the Taperoa area of Bahia, Brazil.

Symptoms. The disease is characterized by a sudden rotting of all developing bunches, a reddish discoloration of the top of the petioles and a rapid drying out of the leaves from the oldest ones upwards. This drying out is preceded by the appearance of reddish-brown streaks at the ends and centres of the lowest leaflets. The leaf then becomes successively pale green (like nitrogen deficiency), yellow, reddish-brown and ash-grey. The palm dies in 2 to 3 weeks and as soon as the external symptoms appear the root system will be found to have rotted and to a large extent dried out. Similar symptoms though proceeding at a slower rate are sometimes seen, and in Colombia this has been referred to as 'Marchitez progresiva'. Descriptions of the disease in the literature have varied; this seems to be due to the disease having appeared in some areas already being attacked by bud rots.[72] In the typical Marchitez symptoms, however, the spear is initially unaffected.

The root rot is cortical. The cortex liquifies in wet weather but in the dry season tends to become necrosed and to detach itself from the stele. The rot starts to develop from the extremities and moves towards the trunk and towards the lower roots. The trunk itself usually remains healthy but cases are reported where the base is rotted sufficiently to form a cavity.[73-75]

Palms have been attacked by Sudden Wither from the age of 4 years onwards.

Cause and control. Sudden Wither has the appearance of an adult Blast attack and it is clear that, as in Blast, the destruction of the roots quite suddenly exceeds their replacement to such a degree that the palm no longer absorbs sufficient water for survival. The normal balance of death and replacement of roots has been upset, and it is of course possible that this may be due to failure of replacement as much as to death or destruction of the roots.

In some areas the disease has been found mainly near to rivers or on the periphery of plantations near forest. At first it was attributed to an effect of heavy, compacted, poorly-drained soils; on the plantation in Colombia suffering greatest devastation the soils had been compacted by cattle and tractors. However, the appearance of the disease in other situations suggested that soil conditions cannot be a primary cause. The discovery of the widespread occurrence of a root-eating caterpillar, *Sagalassa valida*, altered current views on the disease. Although the biology of this insect has been thoroughly studied,[76] its role in the disease remains uncertain. It is not considered to be a primary cause, but the hypothesis that it is either a vehicle for the real pathogen or that by causing a permanent debility it can provide conditions for the entry of a pathogen has been canvassed.[77] While some zones of high incidence have shown a relatively high *Sagalassa* larval population, other such areas have not shown a similar incidence although higher populations do tend to be found near to the forest or rivers. Furthermore healthy palms or those just beginning to show symptoms often exhibit a root rot near the palm base in the area where larval damage is frequently found.[77]

There is other circumstantial evidence for a role for *Sagalassa* in Sudden Wither. Treatment with endrin at the base of the palm was followed by a progressive reduction in incidence in Peru. However, this may have been fortuitous since there is evidence that the disease can be seasonal and show sudden decreases in incidence. Moreover controlled experiments did not give similar results.[78] It has also been observed that on the most devastated plantation in Colombia it has not been possible to find *Sagalassa* in the quantities expected, and that evidence of *Sagalassa* attack has been seen on oil palms in several parts of Colombia, Ecuador and other countries where Sudden Wither has not appeared.

On the plantation in Colombia most seriously affected by Marchitez an incidence of 53 per cent was reached on one plot within 4 years of planting. The interline cover was predominantly *Panicum maximum*. In adjoining plots (a) weeded with herbicides, (b) with palms and soil treated with malathion insecticide, and (c) with both herbicide and insecticide treatment, the percentage casualties were respectively 12.8, 34.6 and 2.3. The adult of the bug, *Haplaxius pallidus*, was found on the leaves of the palms and the nymphs on the roots of the guinea grass. It was postulated that both killing the grass by herbicides and the application of insecticides, though predominantly the former measure, had been responsible for reducing the *Haplaxius* population and the Marchitez incidence, and that therefore there was a strong probability that the insect was a vector of the disease pathogen.[79]

In view of the uncertainty of diagnosis it is difficult to recommend definite control measures. On the grounds that *Sagalassa valida* is likely to be playing a part in the transmission of the disease, applications of endrin around the base of the palms at rates of 2 l of solutions of between 0.75 and 1.5 per cent endrin (19.5 per cent a.i.) have been used to suppress the insect and have been strongly recommended in Colombia, Ecuador and Peru.[77, 163]

Cases of Sudden Wither in *E. oleifera* or the interspecific hybrid have not been definitely recorded and it is possible that the planting of hybrids may be a method of avoiding the disease. Certain *E. guineensis* palms on a plantation devastated by this disease have remained healthy; it may therefore be possible to select resistant progenies within the species.

Upper Steam Rot. Cause: *Fomes noxius*

Distribution.   Thompson described a lethal trunk rot attributable to *Fomes noxius,* which was serious only on deep peat and inland valley soils.[80] This disease has however appeared on other soils in both Malaysia and Indonesia.

Symptoms.   Fructifications of *Fomes noxius* only appear on palms where the leaf bases are extensively decayed. The brown decay appears to proceed slowly inwards from the leaf bases and in many cases a typical collapse of the stem at one point occurs, this usually following high winds. Pathogenicity was proved by inoculation experiments.

Investigations of this condition have remained largely in abeyance for nearly 30 years, but recently Navaratnam and Chee Kee Leong[81] have returned to a study of *Fomes noxius* in the same area, i.e. Serdang, Malaysia. Examination of affected palms showed that the disease was confined to the stem and did not enter the roots. Typically, the lower leaves first become yellow and this symptom gradually extends to the middle leaves and then to the spear. It was evident that spore infection of leaf bases takes place and that from these the fungus gains entry to the peripheral tissues of the stem. The rot thence spreads upwards and downwards in the stem, eventually killing the palm by invading the crown. Two forms of fruiting bodies (normal and resupinate) appear later; these are small greyish-brown bodies with velvety-brown margins and are inconspicuous among the leaf bases.

Control.   As there is usually much penetration of the stem by the time sporophores appear it is desirable to detect the disease at an earlier stage. This can be done on palms of 10 years or older by a sonic method, i.e. striking the leaf bases with a wooden pole to detect the dull sound of an infected base. Incidence is insufficient to justify surveying palms below 10 years. When the diseased leaf bases are cut away the extent of the infection can be explored. The lesion is exised from the stem with a harvesting chisel and the cut surfaces are treated with a preservative.[82] Coal tar has been reported to give the best overall results. Treated palms give as high a yield as untreated palms, so the measures are considered well worthwhile wherever incidence is likely to be significant. If palms are allowed to collapse with this disease they become a focus for *Ganoderma*.

In a fertilizer experiment containing different progenies, there was evidence firstly that fertilizers containing potassium reduced incidence, and secondly that progeny resistance and susceptibility existed.[81] The area was a poor one and the palms generally retarded; the disease has not yet been of great importance in large commercial plantings, but a survey has shown that incidence in some fields may reach 5 per cent.[82]

Other stem or trunk rots

Basal Decay. Known in Malaysia as Stem Wet Rot, it is found sporadically in Africa. In Nigeria palms 4 to 8 years old are affected and very occasionally older palms. For a stem-rot the disease is unusual in that the centre of the crown is first affected, the spears being shorter than normal so that the palm shows a central depression. Soon, however, the lower leaves are affected and then the younger ones. Some of them are subject to curious twisting and all of them eventually collapse and the palm dies. The deaths tend to be so sporadic and occasional that they are often not noticed until all characteristic symptoms have passed; if noticed early enough, however, the base of the young palm's trunk will usually be found to be destroyed internally by a wet, putrid rot which, on dispersion, leaves a large cavity with a surrounding zone in which yellow-brown fibres are found although the cortex has been destroyed. Incidence is not often more than 2 per cent.

In Malaysia this disease seems to be more sudden in its onset, all unexpanded leaves and a few expanded ones dying suddenly at the same time and complete death of the crown takes 7 to 14 days.[3] The lower leaves do not then appear to become affected before the younger ones. The condition has been wrongly equated with Marchitez (Sudden Wither), a disease which has quite different symptoms.[7]

The cause of Basal Decay is unknown. The symptoms in Africa suggest infection through the roots while in Malaysia a severe bud infection has been suggested. No control measures can be recommended; fortunately incidence is low and is confined to young palms. Waterston[83] isolated a *Ganoderma* species from a case of Basal Decay, but no similar isolation has since been made and no *Ganoderma* fruiting bodies have been found on palms suffering from the disease. Moreover the symptoms do not resemble those of *Ganoderma* attack on young palms in Malaysia.

Charcoal Base Rot. A minor disease of Malaysia attributed to the fungus *Ustulina zonata* or *U. deusta*,[3, 7, 84] but its pathogenicity is uncertain. A black dry rot is formed at the base of palms whose foliage, particularly the older leaves, have become chlorotic. After the rot has advanced across the base the palm may fall over. Incidence is low and the disease so far unimportant.

(3) Diseases of the bud or stem apex

Under this heading must be grouped any disease condition occurring in the emerging spear and younger leaves inside the crown. Such diseases normally move towards the growing point through the enclosed, developing leaves of the 'cabbage' and when they reach it the palm is killed. Bud rots have been of wide occurrence in all three continents and provide, perhaps, the most difficult problems of oil palm pathology.

Investigation is difficult owing to the position of the transition zone, usually in the heart of the palm, the rapid entry of secondary organisms into any rot within the cabbage and the multiplicity of confusing symptoms, some of which may be similar to those of deficiencies or genetic abnormalities.

Slight differences of symptom often cause doubt as to whether there is one or several conditions; but Thompson's early description from Malaysia is wide enough to have general application:

> This disease is found in both young and mature palms. The central spear of young unfolded leaflets collapses, even while still green, and can be pulled out from the crown of the palm. A fresh-smelling bacterial rot develops in the bases of these leaves, the bud tissue being similarly affected. . . . The majority of affected palms recover when the decayed leaves are pulled out from the bud cavity. Usually the new growth is twisted and stunted, but eventually becomes normal.[32]

It is interesting to compare the above with the description given by Duff of the only bud rot disease so far diagnosed.[85]

Bud Rot-Little Leaf Disease. Cause: Bacterium of the genus *Erwinia*

Distribution.   The disease has caused serious losses in the oil palm areas of southern Zaire where deaths exceeding 30 per cent have been common. In northern Zaire it is of occasional occurrence while in West Africa cases rarely exceed a few per cent and are often confined to certain progenies; deaths occur but are rare. Symptoms would suggest that the disease in West Africa is the same as that in Zaire, but this has not been proved. Bud rots in the Far East have been insufficiently described and studied; it is therefore not possible to say if this specific disease exists in that region. Spear and bud rots in America have different and varying symptoms.

Symptoms.   The confusion resulting from the descriptions of 'Little Leaf' encountered under various circumstances has already been discussed in Chapter 11 (p. 510) and will not be dealt with again here. Almost every conceivable cause was assigned at one time or another to this symptom and its place in the symptomatology of Bud Rot and of contrasting, probably nutritional, conditions was not made clear until Bull and Robertson reviewed the position in 1959;[86] for a full understanding of the history of investigations of the little leaf symptom their paper should be consulted. The Spear Rot-Little Leaf of these authors is the disease under consideration here.

The first sign of attack is a wet, brown rot on the lower part of the unopened spear leaf. Robertson, working in Nigeria on palms of a susceptible progeny having regular cycles of infection, showed that the cause of Bud Rot-Little Leaf disease was an active pathogen since the appearance of little leaves and bud rot could, he found, be prevented by cutting off the rotting spear below the rotted portion.[87] Although it is not known how infection of the spear takes place, this spear-rotting is the primary symptom and the disease cannot emanate from any other infection site. Duff has described how in very mild cases only the leaflets may be affected and the leaflet rot is passed from spear to spear until it either develops further or the palm grows out of the attack.[85] Normally, however, the rachis becomes infected and the spear collapses and hangs down; it is not uncommon to find a spear leaf, in which the infected portion has rotted away altogether, lying on the ground where it has fallen.

The spear rot grows downwards and becomes a bud rot but it is only if this reaches the growing point that the palm dies; in other cases a callus layer is formed and the palm produces a varying number of 'little leaves' according to how many unemerged, undeveloped leaves have been partially destroyed by the rot. The first leaves to emerge after the spear rot are stumps consisting of the malformed basal portion of the rachis. Subsequent leaves are very short with a few corrugated shortened leaflets, but each successive leaf will be longer, and the leaflets less abnormal, until fully normal leaves are again produced. Little leaf is therefore a recovery symptom and does not precede rotting (Plate 62).

Cause and susceptibility.   A bacterium of the genus *Erwinia*, similar to *E. lathyri*, was consistently isolated in Zaire from young lesions and from tissue in advance of visible rotting.[85] This bacterium has been found on the surfaces of spears and opened leaves of healthy palms, and inoculation experiments have shown that Bud Rot-Little leaf Symptoms can be induced by it.

Susceptibility seems to be genetic, physiological and seasonal. Bachy[88] found that there were genetical differences of Bud Rot susceptibility in the Ivory Coast though it is not certain that he was dealing with the same disease. In a field in Nigeria the disease was confined to one progeny. Genetic differences were also found in Zaire where there was an association between rate of growth and disease incidence. The former was judged by the rate of elongation of spears and in susceptible palms the growth rate fell below normal levels 2 or 3 weeks before an attack of the disease. It was believed that these circumstances, encountered in 'unhealthy' palms, allowed susceptible tissues to be exposed to infection for longer periods than normal. Palms whose growth rate was artificially reduced by root or leaf cutting showed greater than normal susceptibility.

Pl. 62   **A 'Little Leaf' with pronounced lamina contortion.**

In southern Zaire the disease was more prevalent during the dry season and early rains than at other times. In most cases the onset of the disease occurs towards the end of the dry season when growth rate is at its minimum. In Nigeria, susceptibility was also found to be distinctly seasonal.

Control. Duff has provided growth and health records showing that the more vigorous progenies suffer less from the disease and he infers from this that anything interfering with vigorous growth increases susceptibility. While, therefore, the disease is not likely to be serious enough for control measures to be taken in areas where growth conditions, particularly those of water and nutrient supply, are good, in marginal areas the breeding of particularly vigorous progenies will be necessary.

Lethal Bud Rot. (Pudrición del Cogollo) Cause: Unknown

Distribution. A lethal bud rot with variable symptoms, but not usually including the typical 'little leaf' progression, has caused serious damage on some plantations in Central and South America. Some plantations have been totally devastated while others have suffered serious losses with many palms remaining in a moribund, unproductive condition for long periods.

Symptoms. This description is taken from a plantation in northern Colombia where the disease killed almost all palms in many of the fields.[73, 89, 90] In many cases about four to six young elongated leaves remain unopened and stuck together; this is the 'baton' effect. On separation these spears are found to be sticky. Whether this symptom is seen or not, spear decay starts some distance from its base. The rot, which is reddish-brown at first, spreads downwards and the spear eventually collapses, this process taking from 1 to 9 weeks. The youngest leaves become yellowish, typically with yellow stripes on each side of the leaflet midribs, and this is often the first symptom noticed. Some of these leaves later collapse and hang down in the same manner as the spear. Fruit rot does not appear at this stage. In some cases the rot may affect the surrounding leaves before it affects the spear. The early stages of the disease have also been accompanied by leaf splitting.

The later stages of the disease are more rapid. After spear collapse the bud tissue and developing leaves show various stages of decomposition into a wet, putrid mass. The rot may advance in a broad zone towards the growing point or it may be found to be in a narrow strip between the spear and a younger leaf running down towards the bud. This rot is of a light orange-brown colour and wet. The spear leaflets also rot and become dark grey. When the rot reaches the growing point the palm collapses and dies. Death may be rapid or the palm may remain moribund for some time before death. Spontaneous recovery occasionally occurs and in some cases 'little leaves' may be produced, but the regular succession found in Bud Rot-Little Leaf is not usually encountered.

In Panama the pattern of Lethal Bud Rot in young palms has been a little different; chlorosis is often absent and spear rotting takes place in the 'funnel' well above the growing point. The rot spreads to adjoining rachises and further spears do not emerge though some rotted tips can be seen; this gives the palm a spreading appearance with an empty centre.[91] The existing

Pl. 63   Young palm suffering from spear-rot, with no central leaves, Panama.

leaves remain green for a long time. Internally, the rot can be found in the funnel where it continues to attack any new spears starting to elongate. The palm can remain in this state for perhaps a year without the rot penetrating far towards the growing point. Little leaves have occasionally been seen, but complete recovery without little leaf has been noted in other cases (Plate 63).

In Nicaragua, Bud Rot has occurred as a lethal disease on tall adult palms. The disease starts as a rot of successive spear leaves leading to the production of smaller, usually chlorotic, leaves and finally to the rotting away of all leaves as they emerge; the centre of the crown is thus markedly reduced in size although there are no little leaves. On dissection, the rot is found to be in the funnel, and even when the external symptoms are far advanced rotting has only reached within about a foot of the growing point. Sudden recoveries occur, but the disease tends to be recurrent and eventually a damp rot reaches the growing point and kills the palm.[91]

Cause and incidence.   The cause of this serious disease remains obscure. Insects, fungi and bacteria have been suspected, while unfavourable environmental circumstances such as poor drainage, compacted soils and unbalanced nutrition have been put forward as predisposing factors. No direct evidence of infectious spread was found, and it was thought that the development of large infected areas might be related to local growth factors. In Colombia a relation was found between former land usage and incidence, areas with compacted pasture soils suffering the highest casualties; however, this did not prevent the disease spreading over most of the plantation irrespective of the prior vegetation. A low potassium/magnesium ratio was also suspected, but corrective manuring did not stem the spread of the disease.

Fungi which have been isolated from diseased spears include *Furarium oxysporum*, *F. moniliforme* and *Botryodiplodia* sp. Invasion of bud tissue by bacteria species, including *Erwinia*, appears to follow the spear rot and bacteria are also present in diseased spears. Pathogenicity of these organisms has not yet been established.

Lethal Bud Rot can attack palms of any age, but in Colombia cases were noted at 30 months and incidence was at its highest on palms which had been in bearing for 2 to 5 years. However, on replanting devastated areas the disease appeared on palms as young as 3 to 8 months.[92]

It is sometimes claimed that bud rots are initiated by insect attacks, and in Colombia an unidentified caterpillar was often found in diseased palms. Evidence for insect attack as a primary cause has not been forthcoming however. In Central America *Rhynchophorus palmarum* has been found in the rotting matter of a few infected palms but there was no evidence that this species was a usual companion of the disease.[91]

Control. Without a knowledge of the causal organism it has not proved possible to initiate direct control measures. Speculative prophylatic applications of mixed fungicides and insecticides to the palm's crown have not been successful. In view of the early appearance of the disease in very wet situations where soils are compacted or poorly drained and manuring is insufficient, it is clearly of importance to maintain the highest standard of cultivation and nutrition and to avoid areas of pasture in regions where the disease is prevalent.

By far the most promising measure is the introduction of resistant progenies. It is possible that, as with Bud Rot-Little Leaf disease, there may be resistant lines within *E. guineensis*, but the fortunate discovery on La Arenosa (Coldesa) plantation in Colombia that specimens of *E. oleifera* and the inter-specific hybrid were likely to be immune or resistant to the disease has led to the replanting of devastated areas with the hybrid and the establishment of the first commercial plantation of this cross in the world. On this plantation hybrids raised from *E. oleifera* parents growing in Surinam and crossed with six African *E. guineensis pisifera* had been laid down in 1963 in a number of fields where the *E. guineensis* population was subsequently devastated.[92] From 1968, crosses of Colombian *E. oleifera* with *pisifera* were also planted in small plots. All these hybrids survived while the surrounding *E. guineensis* palms died off in large numbers (Plate 64). In 1970 *E. guineensis* and hybrids were planted in a field in alternate rows; by 1974 several *E. guineensis* were dying while the hybrids were still unaffected. The plantation also had areas where wild *E. oleifera* were growing. Further specimens were inter-planted to give a regularly planted field and their production under conditions of improved

Pl. 64   *Elaeis guineensis* x *E. oleifera* **hybrids growing healthily in a field previously** having a full stand of *E. guineensis* which was devastated by Lethal Bud Rot (Coldesa plantation, Colombia).

drainage and maintenance was recorded. These palms have not yet been subject to the disease.

The breeding of the inter-specific hybrid and its production have been discussed in Chapter 5 (p. 299); there seems little doubt that in areas subject to Lethal Bud Rot the planting of this cross is the best known method of combating the disease.

(4) Diseases of the bunches and fruit

The occasional bunch and fruit rots which are encountered have not been extensively studied. Bunch-End Rot has been associated with the Deli palm, particularly in Malaysia.[32]; it has been postulated that this may be a physiological reaction to 'overbearing',[3] i.e. that the full development of the bunch in cases of heavy female inflorescence production is more than can be sustained by the palm's processes of assimilation. This is however an unsupported hypothesis. In this condition the unaffected part of the bunch matures normally.

A Stalk Rot has been connected with an undiagnosed condition in West Africa known as Leaf Base Wilt.[21] The leaves bend down towards the ground and the stalks of bunches in the leaf axils also bend and may then begin to rot. The disease seems to be of purely mechanical origin and provided the rot is not so extensive that the bunch falls, the majority of fruit will develop. The small splits that appear in the stalk are invaded by a variety of saprophytic bacteria and fungi.

Marasmius fruit rot

Distribution. A number of species of *Marasmius* have been recorded on the oil palm in Africa and Asia[43] but only one, *Marasmius palmivorus*, has been associated with rotting of fruit in the bunch and this condition has been confined to the Far East.

Symptoms. Marasmius palmivorus is commonly found on the cut petioles and on the decaying debris between these and the trunk.[93] The white or pinkish strands or rhizomorphs can easily be seen in association with the fructifications. The latter have white caps, 5—8 cm in diameter, which are upturned when fully developed; on the undersurface are the spore-producing white gills. The fructifications are produced in greatest abundance in wet weather, and in drier weather they tend to be smaller and pinker.[94] Both the rhizomorphs and fructifications extend from the leaf axils to rotting or apparently healthy bunches under conditions which will be discussed below.

Susceptibility and parasitism. Marasmius palmivorus is a saprophyte on the leaf bases and in the leaf axils of the oil palm and its standing as a parasite has been much debated. The fungus has been stated to 'grow up over mature fruit bunches and render them useless for oil production' and to 'invade living tissue' and it has thus been accounted a facultative parasite.[94] It has been described as forming discs of mycelium on the ripe fruits beneath which the mesocarp is usually rotted.[32] There are no published records, however, of pathogenicity tests. It has been a matter of common observation that heavy infection with *Marasmius* has been associated with the presence of large numbers of rotting immature bunches which were presumed to have 'failed' through lack of pollen, and the most acceptable theory has therefore been that maturing or ripe fruit may be penetrated whenever climatic conditions are suitable and the inoculum potential has been increased by saprophytic invasion of large quantities of debris in the leaf axils and of numbers of rotting inflorescences and bunches.[94]

Control. If the above postulate is accepted, then the most obvious means of control is to reduce as far as economically possible, through sanitary measures, the media on which the fungus grows on the palm. Where bunches are rotting and *Marasmius* is seen to be prevalent, infected bunches, dead male flowers and as much of the axil debris as possible should be periodically removed; burying or burning are of doubtful value and fungicide spraying only justified in special cases.[3] The severing of leaves as near to the trunk as possible has been advocated, thus reducing the space where debris may collect; but this practice should certainly not be followed where there is danger of attack by *Rhynchophorus* species. Lastly, fruit set should be assured through assisted pollination wherever an insufficiency of pollen is suspected.

Other abnormal conditions of the palm

The oil palm is subject to many abnormal conditions of growth and development the causes of which are not known. Usually, though not always, these abnormalities are encountered where conditions are in some way adverse: impoverished sandy soils, long dry seasons, excessively wet conditions, intermittent waterlogging, pockets of unusual valley-bottom soils, etc. In the more severe conditions bunch yield is usually negligible.[95] Only a few can be mentioned here.

1. *Plant failure* (*Boyomi*).   This term was used by Wardlaw for palms which ceased to grow. Owing to a marked lowering in the rate of root and spear production the number of green leaves decreases and those that remain are erect and crowded (the 'fasciculate' habit) through lack of heavy leaflets and bunches in their axils. This in turn leads to a tapering of the trunk and progressive deterioration of the leaves which are subject to various kinds of chloroses, dry out prematurely and become brittle. There has been much speculation on the reasons for such palms being found dotted about among normal ones.[3] The condition rarely occurs in Asia. In Africa it is considered either to be of genetic origin or may be associated with severe potassium and magnesium deficiency.[3]

2. *Dwarfed Crown.*   This condition is encountered in fields in America suffering from Red Ring Disease (see p. 652), but does not appear to have the same cause.[96] It has been referred to as 'hoja pequeña' (little leaf) but this is unsuitable as it is important that the term 'little leaf', in oil palm symptomatology, should be reserved for the recovery symptom of Bud Rot-Little Leaf. In this condition all the leaves are smaller than normal, green, erect, bunched together and twisted with varying amounts of atrophy or corrugation of the leaflets. A sudden recovery from the condition is frequent, a tall bunch of normal new leaves being produced in the centre of the deformed ones giving the palm a two-tier appearance. This type of deformity is not unknown elsewhere and the term 'choke' has been used in Malaysia to describe a similar condition (Plate 65).

3. *Genetic chloroses.*   Some forms of chlorosis are genetic. An orange spotting has been confined to certain progenies in Africa and Asia. It can be distinguished from Confluent Orange Spotting (p. 503) firstly by the fact that it is not general in the field but is confined to a progeny and secondly by the absence of marginal necroses on the affected leaflets of older leaves.[3]

Pl. 65 Sudden recovery from an abnormal 'choked' condition on coastal alluvium of low pH in Malaysia.

Another chlorotic condition thought to be genetic consists of leaves which may be wholly or partially pale or bright yellow and which continue to be produced on the same palm.

4. *Fomes lignosus* and *Poria ravenalae.*   Fructifications of these fungi are occasionally seen at the base of young palms in Asia, *Poria* invading the leaf bases.[3] Neither fungus appears to be pathenogenic on the oil palm.

5. *Leaf Base Wilt.*   This condition is said to be widespread in Malaysia,[3] but is less common in Africa. The leaves appear prematurely to loose their normal full attachment to the stem and therefore are set at a more obtuse angle than is normal for their age. With older palms they often bend over

sufficiently to touch the ground or, if the palm is tall, to hang down to cloak the trunk. The effect on the bunch has already been mentioned (p. 648). The cause of this condition is unknown; there is some evidence that heavily bearing palms are most likely to be affected.[3]

6. *Rachis Internal Browning.*   This condition has recently been studied in Malaysia and has been described in detail by Turner and Bull.[3] Damage to the palm as a whole appears to be slight and the cause is unknown.

7. *Lightning.*   As with the coconut palm, the oil palm is occasionally killed by lightning strike. The frequency of such deaths in different parts of the world has not been recorded, and it is possible that they are not so frequent as with the coconut. Young palms collapse rapidly and wither; in older palms the trunk base is often charred. Surrounding palms show scorching on the side facing towards the strike.[3]

Pests

The injurious pests of the oil palm include nematodes, mites, insects, birds and mammals. Most attacks by pests have been localized and often short lived, so they have only been the subject of study, if at all, in certain localities or at certain times when locally alarming outbreaks have occurred. The published work on pests is still rather scanty, but recently a comprehensive work has been provided for Malaysia by Wood, and much of the information he gives on ecology and control will be found applicable in other oil palm countries.[4]

Nematodes

Nematodes have often been thought to be responsible for diseased conditions in the oil palm — at one time it was thought they might be responsible for Blast disease. The nature of the symptoms of nematode attack inclines one to describe them with the diseases, but they may more correctly be considered among the pests.

Red Ring Disease (Anneau rouge, Anillo rojo). Cause: *Rhadinaphelenchus cocophilus*

Distribution.   'Red ring' of oil palms has been found in Venezuela, Surinam, Brazil and Colombia where the similar disease of coconuts is prevalent. It has been studied on an estate in Venezuela where it has done very considerable damage.

Symptoms.   The symptoms of this disease have been described by

Malaguti.[96] The centre of the crown takes on a dwarfed appearance and the newly opened leaves become bundled together into an erect compact mass, the leaflets being corrugated, twisted and sometimes adhering to the rachis. Gum is exuded. Later this crown of leaves turns slowly yellow and dries out, the rachis being a light brown colour with yellow spots. One or two of the intermediate leaves become bronzed and after 2 to 5 months all the leaves gradually become yellow or bronzed, though remaining erect. Developing bunches rot away and inflorescences fail to set fruit.

The most striking interior symptom is the brown (not really red) cylindrical ring found in the trunk, 7–8 cm from the periphery and 1–2 cm broad. This ring is most distinct towards the base of the palm but the infection proceeds upwards into the petioles and rachis of the leaves in the crown in which, on cross-sectioning, necrosed areas or spots can be found (Plate 66). This infection does not, however, invade the tissues of the stem apex or surrounding very young leaves. The effect of the invasion of the stem and rachis therefore appears to be the prevention of normal water and nutrient supply to the foliage.

Incidence and cause.   It has been shown that in unprotected areas incidence can become high with some rapidity; Malaguti cites a group of 100 palms showing only 16 doubtful cases in January which by August had 22 deaths, 9 doubtful or affected cases and only 69 palms remaining healthy. On this Venezuelan estate about one-third of the original stand is believed to have been affected.

The coconut nematode, then named *Aphelenchus cocophilus*, seems first to have been recorded on oil palms by Freeman in Trinidad in 1925.[97] Proof that the nematode was the cause of the condition was obtained by Malaguti, who undertook inoculations with inoculum both from the oil palm and the coconut. The disease appeared in inoculated palms 2 to 10 months after inoculation. The vector of the nematode is usually, but perhaps not always, the weevil *Rhynchophorus palmarum*. Maas, who found six cases of Red Ring in coconuts and four cases in oil palms in Surinam, noted that about 7 per cent of weevils were contaminated with the nematode. The low incidence of Red Ring appeared to be connected with the fact that there were many recently felled trunks of many palms in which both the nematode and its vector could breed in preference to live palms.[98]

Control.   Incidence on the affected estate in Venezuela has been greatly reduced by the taking of regular sanitary measures. Any diseased palm is poisoned with sodium arsenite and, if possible, felled and burned later. The whole estate is covered every 2 months in a search for diseased palms. Most important is the protection of the palm against the type of wounding which will provide sites for *R. palmarum* to lay its eggs. A sudden attack of Red Ring in Brazil was preceded by very close leaf pruning which had resulted in underlying petioles being wounded. Care must be taken when

Pl. 66 Longitudinal (A) and transverse (B) sections of an oil palm suffering from 'Red Ring' in Venezuela.

removing leaves to make a clean cut sufficiently far up the petiole to avoid wounding other parts of the palm. Ring weeding with herbicides instead of with hand tools may also help to prevent wounding. Regular disinfection of tools has been suggested together with treatment of the cut leaf and bunch-stalk surfaces. Providing wounding is avoided, however, it is doubtful if the latter precautions, which are expensive, are necessary.

Arachnids

The arachnida include both spiders and mites, but only mites have been pests of the oil palm.

Germinator Mites

The embryo and part of the kernel may be destroyed by mites in the course of germination particularly if charcoal boxes or the wet heat treatment are used. The seeds will be found to have their germ-pores filled with fibrous material below which the translucent mites are found. They have been described as appearing under a lens to be like elongated drops of water,[99] but there are no records of identification.

Germinators may be fumigated with methyl bromide, but with the dry heat treatment there is little chance of damage from these mites.

Red Spider Mite

The true spider mites belong to the family Tetranychidae and the spider mite prevalent in Malaysia has been identified as a species of *Oligonychus.* These are large enough (0.5 mm) to be seen moving on the leaf which they cause to turn a bronze colour. Webbing and cast skins of eggs and young stages can also be seen as white flecks. Badly infected leaves die prematurely and a widespread attack, which is favoured by dry weather, may have severe effects on a nursery (Plate XVI, between pp. 78 and 79).

Control.   Tetradifon (Tedion) will control the pest without killing the natural enemies, and its use can be combined with rogor (Rogor 40) which has systemic properties. The latter as a 0.1 per cent high volume spray is followed 10 days later by Tedion in either a 2 per cent emulsion or 1.0 per cent wettable powder.[7]

Insects

A large number of insects have been recorded on oil palms, but only a few have become major pests. It is proposed here to pass briefly over those orders of insects from which only minor damage has been reported and to deal more fully with the half-dozen pests and allied species which have at some time and place caused serious harm to the palm.

Many oil palm pests have been reported from the newly developing areas of America. This region is a vast museum of palm species all with their insect fauna. It is only to be expected that many insect species will turn their attention to the oil palm and that some may for a time become troublesome pests.[162]

Orthoptera

Acridoidea (Grasshoppers, locusts and crickets)

Grasshoppers can be a nuisance in prenurseries and nurseries in all oil palm regions and they are normally controlled by hand collection. The stink locust, *Zonoceros variegatus*, has damaged young and adult palms in West Africa following the slashing of overgrown covers. In Malaysia, the grass-hopper *Valanga nigricornis* has been found to eat large pieces of the leaf of young palms. Crickets and mole crickets are also occasionally troublesome in Malaysia.

Isoptera (Termites)

While termites of the genus *Coptotermes* have been reported[4] as damaging the oil palm in Malaysia, they cannot be regarded as potentially serious pests.

Hemiptera

Aphididae (Aphids)

Aphids are occasionally found on nursery seedlings. In Zaire they have been reported as being found near the base of the leaves under earth cover-ings constructed by black ants.[99] Here the dark blue aphids of less than 2 mm width puncture the epidermis and may cause a slowing down of growth rate. Several species are found in Malaysia both on nursery plants and on the leaves of mature plants.[4] The severity of the attack does not usually justify the use of insecticides.

Coccidae (Mealybugs, scale insects)

The species *Planococcus* (*Pseudococcus*) *citri* is reported as damaging nursery seedlings in Zaire by attacking the plant just under the surface of the soil.[99] The damage to the epidermis leads to the formation of a small 'canker'. The insect, with its white mealy covering, measures about 4 mm. Control measures are rarely needed where prenursery or nursery plants are growing vigorously, but if necessary the systemic insecticide Parathion may be added to the irrigation water at the concentration of 0.5 to 10,000.[99]

Smaller Coccids are sometimes found on the leaves and more often on the developing fruit of bearing palms. In Zaire *Aspidiotus destructor* and *A. elaeidis* cover leaves and fruit while *Pinnaspis marchali* is not un-common on bunches in Africa, causing discoloration of fruit and inhibiting ripening. In Malaysia mealybugs are found on leaves, unopened spears and fruit (*Dysmicoccus* sp.). A species of *Geococcus* has been recorded on seedling and young palm roots[4] causing some root distortion and weakening of the palm.

Mealybugs and aphids are naturally controlled by many predators of which the ladybird beetles are the most important. They are often protected by ants. Although control measures are rarely required, control of the tending ants is sometimes helpful. Dieldrin may be used as a spray on nests or as granules on the ground.[4]

Other Hemiptera

The presence of adults of *Haplaxius pallidus* (Cixiidae) in the foliage of oil palms has been mentioned (p. 639) in connection with Sudden Wither (Marchitez sorpresiva). However, no description has been given of the life history of this bug or of its behaviour on the palm.[79] The *Leptopharsa* species (Tingidae) found on plantations also suffering from Leaf Wither (p. 620) is probably the same species as that described as *Gargaphia* by Genty *et al.*[100] The insect spends its whole life on the under surface of the leaf and pierces the tissue on each side of the midrib of the leaflets. The piercing depth is so great that the effect can be seen on the upper surface of the leaflets as bleached necrotic spots. The underside of the leaflet is spotted with excrement. The highest populations are found in the middle leaves (17—25) but there are also considerable numbers in the younger leaves. It has been postulated that the damage done by the insect gives entry to *Pestalotiopsis* sp. and other weak parasites or saprophytes which may become aggressive.[100] Treatment with trichlorfon (Dipterex) at 1.2 kg a.i. per hectare from the air is reported to have given good results.

Lepidoptera (Moths and butterflies)

The caterpillars which do damage to oil palm leaves are mainly those called slug or nettle caterpillars belonging to the Cochlidiidae or Limacodidae family, or bagworms of the Psychidae family. *Pimelephila ghesquierei* belongs to the allied family Pyralidae. Caterpillars found in Latin America belong to several families including the above. A feature of recent serious attacks has been the mixing of many leaf-eating species on the same palms and leaves.

Pyralidae

Pimelephila ghesquierei — *The African Spear Borer*

Distribution. This pest is found damaging leaves of young palms in all African territories. Damage is most common between the second and fifth year in the field, but both nursery seedlings and older palms are sometimes affected. The moth was first described by Tams from specimens obtained in 1929 by Ghesquière in Zaire.[101]

Incidence, life cycle and damage. This moth cannot be described as a common pest, but on occasions the damage done has been severe. It has perhaps been more troublesome in Central than in West Africa. Shade encourages attack in nurseries.[102]

The eggs are laid by the moth at the base of the spear leaves and even one larva hatching can do considerable harm. Usually two or three are found on young palms or up to a dozen on older ones and, typically, they penetrate the rachis and leaflets of the growing, unopened spears, forming galleries through them. The attack may proceed downwards towards the growing point and the rachis may be so damaged that later, in a strong wind, several young leaves may snap near the base. When unbroken spears open, the holes left by the caterpillar are seen to be symmetrically placed on either side of the rachis. The caterpillar does not kill the palm, but may be followed by weevil larvae, e.g. those of *Temnoschoita*, or a bacterial rot which may finally prove lethal.[100]

The caterpillars reach a length of 3—4 cm before pupating in a cocoon of fibrous debris. The colour of the caterpillar changes from dark red to yellowish as it develops. The olive to brown moths are not long in emerging from the pupae and the whole life cycle takes 35 to 45 days; attacks can therefore be made at frequent intervals.

Light attacks can be dealt with by removal of infected leaves and collection of the caterpillars and pupae. In the case of more serious infestation, or in the vicinity of infested areas, sprays of 1 per cent DDT, 0.02 per cent parathion or 0.1 per cent endrin have been found effective in Zaire when directed at the base of the spears and allowed to penetrate. In the Ivory Coast the DDT solution has been sprayed at intervals of 2—3 weeks in nurseries and the first 2 years in the field. Older palms are less vulnerable to attack and are not usually treated.[103]

Tirathaba mundella — *The Oil Palm Bunch Moth*

Distribution.   Occasionally found on other palms, *T. mundella* is widespread in Malaysia and Indonesia and can reach epidemic proportions especially in young areas.

Life cycle and damage.   Eggs are laid in the bunches, especially those overripe or rotten, and in inflorescences or bunches lying on the ground. Caterpillars bore into developing fruit or feed on the surface of ripening fruit. They are sometimes found tunnelling into the base of a spear leaf. They are light to dark brown and grow to 4.0 cm before pupating as dark brown pupae inside the bunch.

Control.   Tirathaba mundella is thought to be heavily parasitized. Dipterex at 0.55 kg in 370 l per hectare gives good control. A wetter added at 0.25 per cent with white oil at 0.1 per cent improves penetration and effectiveness.[4] More recently endosulphan (Thiodan) has been preferred.

Sufetula *species — The Oil Palm Aerial Root Caterpillar*

Distribution.   Species of *Sufetula* feeding on the aerial roots have only recently been investigated.[76] *Sufetula sunidesalis* is found in Asia, *S. nigrescens* in West Africa and *S. diminutalis* in Colombia.[156]

Life cycle and damage.   Eggs are laid on palms having a supply of roots at the base of the trunk; they are inserted among the roots or at their base.

Sufetula diminutalis lays fifty to eighty eggs. The young caterpillars feed on the tips of the growing aerial roots and the older ones can hollow out these roots for a few centimeters. Although attack is normally above ground it can continue when the root has already penetrated a short distance into the soil. The caterpillars grow in five larval stages from 1–1.5 mm to 1.2–2.0 cm. From the fourth larval stage they acquire both dorsal and lateral plates. Pupation takes place in the soil at a depth of a few centimeters and about 50 cm from the trunk, or among the aerial roots. The pupa, which is straw-coloured, is protected by a silk cocoon. The whole life cycle lasts only 28 to 31 days.[156]

Experiments in Colombia suggest that aerial root development is severely curtailed by *S. diminutalis* attack. Palms treated with endrin have shown a tremendous development of these roots, which would otherwise be systematically destroyed.[156] It remains to be seen, however, how far the maintenance of a supply of aerial roots by control of the pests will enhance growth and yield.

Other Pyralids

In Colombia a species of *Tiquadra* has been found invading the spear and rotting bunches. This latter pest appears to be the American counterpart of *Tirathaba*.

Limacodidae

Parasa viridissima *and other species — West African slug caterpillars*

Distribution.   Species of *Parasa* defoliating oil palms have been found in West Africa but have not been serious pests in Zaire, although Lespesme recorded the oil palm as a host plant for *Parasa carnapi* in that country. *Parasa* species have also been found on the coconut in West Africa. Specimens from Cameroon, Nigeria, Liberia and Uganda have been identified as *P. viridissima*.[104] *Parasa pallida* has also been a pest of oil palms in the Ivory Coast.[105]

Incidence, life cycle and damage.   *Parasa* attacks in West Africa have been infrequent and localized on planted areas; serious attacks on grove palms have not been reported. In Nigeria attacks were mostly on adult palms and the larvae spread rapidly from palm to palm; in one field 138 palms became infected in the course of a month by the spread of the caterpillars from a single palm.[106]

In heavily infected palms the caterpillars start feeding from the tip of the leaflets and may consume the whole of the lamina, or they may transfer to another leaf after consuming the terminal half of the laminae only. The spear and youngest leaves are not consumed when in that stage. Feeding continues for about a month. The youngest larvae of *P. viridissima* do not consume the vascular network and upper epidermis of the leaflets.

Eggs are laid in groups on the under surface of the leaflets. The young caterpillars, which are brown, then feed in colonies. After a time the caterpillars disperse and gradually change to a green colour as they grow to a final length of 3—4 cm. There are rows of orange-red bristle tufts along the back. These insects pupate on the under surface of the leaves and on the rachis. *Parasa viridissima* cocoons are spherical. 1.0—1.3 cm in diameter, and are gummed to the leaf and covered with a whitish secretion and upright brown stinging bristles; those of *P. pallida* are oval, 1.5—2.0 cm in length and covered over and held to the leaf by a network of stinging bristles. The adult moth of *P. viridissima* is green and hairy, *P. pallida* is white with black markings except for the posterior wings and the abdomen which are pale yellow.[105]

An unidentified species, probably of *Parasa*, which caused a serious attack on adult palms at Awka, Nigeria, in 1953, had green caterpillars turning yellow and reaching 3.0—3.5 cm. Pupation took place under the soil surface around the palms; the cocoons were dark brown, hard and brittle. Adults have not been bred.[106] (Plate XIX, between pp. 78 and 79.)

It appears that these insects have a fairly short life cycle and serious attacks may succeed each other at intervals of 4 to 9 months. Normally, fungi and natural predators keep the populations in check by attacking the larvae and pupae. For serious attacks in the Ivory Coast Dipterex at 0.7 to 1.0 kg a.i., DDT at 1.0 to 1.5 kg a.i. and carbaryl at 0.8 to 1.2 kg a.i. per hectare have been recommended for application 3 weeks after the appearance of the first caterpillars.[105] As the later larval instars do the greatest damage, treatment should not be delayed.

Setora nitens

Distribution.   This nettle caterpillar is found throughout South-east Asia. It also feeds on the coconut and nipa palms.

Incidence, life cycle and damage.   Severe infestations of *S. nitens* and other nettle caterpillars may occur very rapidly as the life cycle is about 6 weeks and reproduction rate high. The eggs of *Setora* are deposited on the under surface of the leaflets near the tip. The caterpillars usually feed on the under surface and eat away the whole lamina leaving the midrib. They then drop to the ground to pupate in cracks in the soil, the pupal stage lasting about 25 days.[107]

The caterpillars are yellowish-green when young, becoming green. They have a longitudinal purple band and four prominent tufts of spines at the corners of their rectangular bodies (Plate XVIIIA, between pp. 78 and 79). The caterpillar reaches about 3.5 cm in length. The moth is brown with a wing span of 3.5 cm.

Control.   Outbreaks of *Setora* damage have been attributed to the prior use of contact insecticides[108] which kill the parasites. Numerous parasites and predators of *Setora* have been recorded by Wood[4] including five species of wasp, four parasitic flies and a bug. Lead arsenite at 4 kg per hectare has been preferred because of its minimal effect on the natural enemies, but trichlofon (Dipterex) at 2 kg per hectare may also be used. However, results with these insecticides have been variable and Wood *et al.*[161] have tested a range of chemicals and *Bacillus thuringiensis* insecticides. In general organochlorines and organophosphates gave good kills but are not sufficiently selective, though monocrotophos by trunk injection was promising. The carbamate, aminocarb, was effective and specific. (For details, their paper should be consulted.)

Darna *species*

Distribution. *Darna* (formally *Othocraspeda*) *trima* (Plate XVIIIB, between pp. 78 and 79) is a common pest of South-east Asia. *Darna metaleuca* is troublesome on plantations in Colombia.

Incidence, life cycle and damage. *Darna trima* is widespread in Malaysia where it consumes the margins of leaves giving a serrated appearance; but total defoliation also occurs. *Darna metaleuca* is among the more serious of the now numerous South America caterpillar pests. The biology of *D. metaleuca* has been studied by Genty who showed that the life cycle was 44 to 56 days.[160] The life cycle of *D. trima* is 51 to 60 days.[157] Pupation takes place on the under surface of the leaves. The caterpillars of *D. trima* are light brown with orange markings and smaller than *Setora*. The moth is dark brown with a wing span of 1.8 cm.[4]

Control. In East Malaysia (Sarawak) spectacular progress has been made with the control of *D. trima*. A virus inoculum was prepared from diseased late-instar larvae and from healthy larvae confined with them. The larvae were naturally infected by the virus and the inoculum was prepared by simple maceration, straining and dilution to 0.1 per cent with untreated clean soft water. This was sprayed with mist blowers repeatedly until the larval census showed that resurgence was not taking place. There was a high mortality within 8 days in comparison with unsprayed areas, and resurgence such as is common after 5 weeks with chemical insecticide spraying did not take place.[157]

Darna metaleuca has several important parasites, including a wasp, *Casinaria* sp., which is itself parasitized, and it is recommended that the parasite population be carefully examined before control measures are decided upon. However, good results have been obtained with a mixture of Dipterex and carbaryl applied twice at the very young larval stage, and with chlorfen amidine.[160]

Sibine fusca *and other species*

Distribution. *Sibine* species have a prominent place in the destructive caterpillar populations on oil palms in South America. *Sibine fusca* is the most common, but *S. apicalis*, *S. modesta* and other unidentified species have also been recorded.[109]

Life cycle and damage.   Genty has described in detail ten stages of development of the caterpillars of *S. fusca* from 1.6 mm in length when they are pale yellow to 2.5—3.5 cm when they are green. They tend to feed in groups until reaching the last stages. The pupa is oval, 1.2—1.6 cm long, brown and fixed to the base of the rachis. The young caterpillars eat the under surface of the leaflets, but later the whole lamina is consumed. Young and developed caterpillars and cocoons can be found on the palm simultaneously. The moth is brown, the female having a wing span of 4.0—4.5 cm, the male 3.0—3.5 cm.[109]

Control.   *Sibine fusca* is parasitized by several wasps and attacked by several bugs. It is also subject to a virus disease developing in intestinal cells; 20 g of infected larvae ground down and immersed in 220 ml of water and applied at 50 ml per hectare has spread the disease over the whole population within 18 days.[110] If this treatment is not possible then carbaryl at 1.0 to 1.5 kg a.i. per hectare gives satisfactory results. Treatment should be at least 3 weeks before pupation because the three last caterpillar stages are responsible for 95 per cent of the damage; and treatment is considered necessary with more than an average of fifteen to twenty larvae per palm.[110]

Thosea asigna *and other species*

Distribution.   Outbreaks of *Thosea asigna* have been reported in two areas in Malaysia,[4] while *T. bisura* caused heavy damage on one plantation in Johore state.[111] *Thosea vetusta* is occasionally encountered.

Life cycle and damage.   *Thosea asigna* is green with a characteristic longitudinal purple-grey band of uneven width. *Thosea bisura* is also green with a straight narrow bluish band. The life cycle of the former species is said to resemble that of *Setora nitens*. The larval stage of *T. bisura* lasts 28 to 35 days. The caterpillar then moves down the palm and pupates in the soil and plant debris. The cocoon is ovoid, 10—1.2 cm long, dark brown to black. The moth is brown, the female larger than the male.

Control.   The Assassin bug, *Sycanus dichotomus*, is a major predator. Ants cause deaths of the pupal stage. Parasitic wasps and flies have been seen.[111] The serious damage by *T. bisura* came in dry weather and high natural mortality followed the onset of rain. Fair control was obtained with Dipterex at 0.3 kg commercial product per 100 l water and good control with Gusathion at 0.15 kg commercial product per 100 l. Hand collection of cocoons was also effective.[111]

Other Limacodidae

In Malaysia outbreaks of *Ploneta diducta* sometimes occur.[112] Other occasional leaf eaters are *Susica pallida*, *Cania robusta*, *Birthamula chara*, *Cheromettia sumatrensis* and *Trichogyia* sp. All these Limacodidae pests of oil palms in Asia have been well illustrated by Wood.[4] In Colombia the

Limacodidae are further represented by *Natada pucara, Phobetron hipparquia* and other species. These may form part of the general leaf-eating population but are not yet individually serious.

Psychidae

Bagworms: Species of Metisa, Cremastopsyche *and* Mahasena

Distribution.   Bagworms have been pests of the oil palm in Asia since the start of the plantation industry, but the prevalent species have changed. In the between-war period *Mahasena corbitti* was extensively studied, but since the war first *Cremastopsyche pendula* and latterly *Metisa plana* have been the common species in Malaysia. In Indonesia *Mahasena corbetti* is still reported to be the principal species[113] (Plate 67).

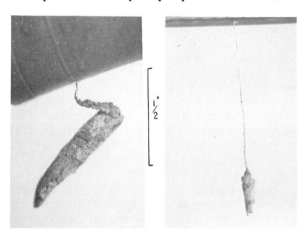

Pl. 67   Bagworms in Malaysia: Left, *Metisa plana*; right, *Crematopsyche pendula*.

Incidence, life cycle and damage.   The larvae, encased in bags constructed of pieces of leaf bound with silk, feed on the upper surface of the leaf, the scraped portion first becoming dried out and then forming a hole. Further damage is done by the removal of pieces of leaf to make the case. Badly damaged leaves soon dry up and this gives the lower and middle part of the crown its characteristic grey and upstretched appearance, the only green leaves being the younger ones. *Mahasena corbetti* feeds on the under surface of the leaf.[4]

Surviving caterpillars migrate to the under side of the leaves where they pupate in their cases. The size and form of the bags and the manner in which they are attached to the leaf help to distinguish the species. *Metisa plana* has a short hooked attachment and the bag is about 13 mm long. The case of *C. pendula* is about 6 mm long, is rather rough and hangs on the end of a long vertical thread. *Mahasena corbetti* is much larger and more ragged. The winged male moths fly from their cases, but the wingless females remain in them and each lay 100 to 300 eggs. On hatching, the caterpillars acquire their own cases and feed in groups. The life cycle takes about 3 months and it has been pointed out that theoretically one female of *Metisa plana* could give rise to 100 million caterpillars in the course of a year. Actually, the caterpillars die in large numbers from parasitic and predatory attacks and other causes, but explosions of population do occur locally from time to time when the natural balance is disturbed for one reason or another. It is believed that the increase in bagworm incidence in Malaysia has largely been built up by the use of contact insecticides sprayed both against other minor pests and against the bagworms themselves[114] thus destroying their natural enemies which are wasps, a tachinid fly, a beetle and an assassin bug. This hypothesis was substantiated in experiments done by Wood, who sprayed a block of about a hectare recurrently with dieldrin and showed that *Metisa plana* gradually increased in and spread from the sprayed block.[115]

Control.   Owing to the danger of contact insecticides, hand-picking is now recommended for very small attacks and stomach poisons such as lead arsenate (4 kg per hectare) or Dipterex (1—1½ kg per hectare) for larger outbreaks. It is interesting to note that lead arsenate was successfully used against *Mahasena corbetti* on coconuts as early as 1927.[116] Aerial spraying over wide acreages has been successful with Dipterex at 1—2 kg in 12—25 l water per hectare.[4]

Other Bagworms

Wood[4] mentions species of *Pteroma, Clania* and *Amatissa* as showing limited increases in Malaysia from time to time. The life cycles are similar to that of *Metisa plana*. In Colombia, *Stenoma cecropia*, which belongs to the Stenomidae, causes minor damage and is remarkable for having a fixed bag and eating by journeys from it while remaining attached by fibres.

Glyphipterigidae

Sagalassa valida. *Oil Palm Root Miner*

Distribution.   The caterpillar of this moth has been found mining in the roots of oil palms in several South American countries including Colombia, Ecuador, Peru and Brazil.

Life cycle and damage.   The female moth has a wing span of 2.1 cm, the male 1.8 cm. They live in the undergrowth and among the cut palm leaves in the interline and their dull colour blends with that of the withered material. The moths are frequently captured on the wing by dragonflies. The position of egg-laying has not been observed, but it is presumed that it is in humid material such as lichens, mosses or humus at the base of the palm. The egg is 0.55 x 0.35 mm and the young larvae are at first 0.8—0.9 mm long growing eventually to 2 cm. The larvae penetrate the primary roots immediately after hatching but can also move through the soil to attack roots some distance from the point of hatching. The young caterpillars at first eat the external part of the root leaving the central cylinder intact; this partial destruction stimulates the production of new roots. Larger larvae cause complete destruction of the tissues of the roots in which they mine.[76]

In areas of attack it has been noted that the number of caterpillars increase with the age of the palm, and it is stated that there have been palms in which 50 to 80 per cent of the root system has been destroyed, including old and recent damage.[76] The method of estimating percentage damage has not however been described. It has also been found that the amount of damage is highly variable but tends to be greater on the edge of a plantation near the forest or near to rivers and streams.

Effect of damage, and control. The effect of this caterpillar damage in relation to Sudden Wither (Marchitez sorpresiva) has already been discussed (p. 639). Genty considers that the extent of the damage done to the roots in itself justifies treatment.[76] The application around the base of the palm of 2 l of solutions of 0.75 to 1.5 per cent endrin (a.i. 19.5 per cent) has been recommended.[77, 163] By analogy with other insect pests of the oil palm it is not impossible that a sudden increase in the incidence of *Sagalassa*, and hence a weakening or even death of the palm, might be due to a temporary reduction in the numbers of a natural parasite or predator. In this case the use of endrin might, in the long term, be disadvantageous. Much still needs to be learnt about the natural enemies of *Sagalassa* and the incidence of, and damage done by, the insect in different situations before general recommendations can be made for its control.

Other caterpillars

Damage by caterpillars of other families is also reported from time to time. In Africa, the Zygaenid caterpillar *Chalconycles catori*, which has a life cycle of about 30 days, has been reported as doing serious local damage in the Ivory Coast.[117] Less common are *Pteroteidon laufella* and *Dasychira* sp. *Pyrrhochaleia iphis* (Hesperiidae) and *Epimorius adustalis* (Psychidae) have been reported from Zaire.[99]

In Malaysia, Wood[4] has recorded a large number of leaf-eating caterpillars, few of which, however, have done any significant damage. These include *Odites* sp. (Xylorictidae); *Elymnias hypermnestra*, and *Amathusia phidippus*, butterfly caterpillars of the Nymphalidae; *Cephrenes chrysozona* and *Erionota thrax* (Hesperiidae); *Turnaca* sp. (Notodontidae); woolly bear caterpillars of *Asota*, *Asura*, *Diacrisia* and *Amsacta* genera (Arctiidae); cutworms (Noctuidae), usually *Agrotis*, sp., which can do damage in prenurseries and *Spodoptera litura* which strips the leaf epidermis in nurseries; and tussock moths (Lymentriidae) *Dasychira mendosa*, *Laelia venosa* and *Orgyia turbata* which occasionally consume significant quantities of leaf in nurseries. There are also small moths of various families which develop in the detritus in and around bunches, but appear to do no damage.

In America, colonies of mixed species of Lepidoptera have characteristically developed on some plantations and the method and timing of control has influenced the balance between species. *Opsiphanes cassina* (Brassolidae) did much damage as a leaf eater and was reported to be encouraged by carbaryl spraying but reduced by using lead arsenate or *Bacillus thuringiensis.*[162] In the mixed colonies there have been species of *Megalopyge* (Megalopygidae), species of Dalceridae and Hesperiidae genera, and *Herminodes insula* (Noctuidae) in the spear leaf, and some Psychidae.

Coleoptera (Beetles)

The beetles which damage the oil palm belong to three distinct groups: (i) the Chrysomeloidea, of which the Hispid leaf miners are of the greatest importance, (ii) the Curculionidae or true weevils and (iii) the Scarabeoidea of which the Dynastid 'Rhinoceros' beetles form the most damaging group.

Chrysomeloidea: Hispidae

Coelaenomenodera elaeidis — *The West African Oil Palm Leaf Miner*

Distribution.   This pest is found on oil palms and, to a much lesser extent, on the coconut and *Borassus* palms throughout West and Central Africa; but the more serious attacks, causing widespread defoliation, have taken place in the drier, more marginal areas of West Africa, e.g. parts of Ghana, Dahomey and the western side of western Nigeria, notably attacks occurring throughout the palm groves of Dahomey in 1950 and 1955. Recently, however, the pest seems to have become of greater importance in more favoured areas and serious attacks have been reported from the Ivory Coast and West Cameroon.

Incidence, life cycle and damage.   For a long period this pest was only reported from Ghana where its method of feeding and life history were recorded by Cotterell.[118] Very detailed studies have now been made by Morin and Mariau[119] of the biology of this insect and of its control. The life history in days is as follows: eggs, 20; larvae, 44; pupae, 12; adult to egg laying, 18; total, 94. The adults continue to live on the under surface of the leaf for 3 to 4 months during and after laying eggs. The length of the life cycle accounts for the pest damage re-appearing in some cases every 3 or 4 months.

The larvae, which grow to about 6.8 mm in length, are brown and their heads are squeezed into the thorax. Their flattened bodies are transversely divided by deep furrows and they have no feet. They mine under the upper epidermis of the leaflets of palms of all ages except, normally, those below 3 years old in the field. The paths of the miners are longitudinal, and in a severe attack the greater part of the leaf tissue will be destroyed.

A single passage or gallery mined by a larva to attain its full development measures about 15 cm in length and is 1 cm broad.[120] Severely attacked palms have a typical appearance; the young leaves are green, being little attacked, while the remainder are grey-brown and withered with desiccated rolled-in leaflets. Later, the withered laminae shatter, leaving the leaflet midribs only.

The pupae are found in the dead tissue of the leaves, and the adults, which are 4–5 mm long, emerge after about 12 days. The pupae are mobile and are found in the centre of the galleries. The adult emerges through the upper epidermis and shows a preference for migrating to the higher leaves. These adults are pale yellow with reddish wing cases; they make grooves about 1 cm long on the leaflets and the female lays her eggs in a small cavity on the under side of the leaf.

It has been observed that natural parasites normally reduce the severity of successive attacks and Cotterell reported Hymenoptera parasites of both the eggs and the larvae as well as fungal parasitism. In the drier parts of the West African palm belt where leaf miner damage has in some years been serious, the attacks seem ordinarily to have been controlled by the natural predators, and resurgence has not occurred again until, for some reason, the parasite population has fallen below normal; in the Dahomey outbreak of 1955–6, for instance, no parasites were found. It is reported that maxima of larvae invasion have been found in March, June, September and December and these attacks have caused successive set-backs to the growth of the palms.

Control. The parasitism of *Coelaenomenodera* has now been studied in detail.[119] The eggs are parasitized by the chalcid fly, *Achrysocharis leptocerus*, and by *Oligosita longiclavata* (Trichogrammatidae). Three larval parasites attack towards the end of development; these are Eulophid flies *Dimmockia aburiana* and *Pediobius setigerus*, and, less frequently, *Cotterellia podagrica*. A rare attacker of young larvae is another Eulophid fly, *Closterocerus africanus*, and another, unidentified, species. Attempts are being made to introduce further parasites from other regions.

In the Ivory Coast a method of assessing the level of attack through an *Intensity Index* has been adopted as follows.[120] Counting of adults and larvae is done on a leaf between the twenty-fifth and thirtieth, small and large larvae, nymphs and adults being recorded separately. As counting is done on one palm per hectare the index on an area basis is obtained by dividing the total insects counted by the number of hectares. The palms selected for counting are changed on each count. Counting is done 3-monthly when the larval index is below 10 and the adult index below 1; monthly when the indices are 10–20 and 1–3; and weekly when more than 20 and 3. When the latter stage is reached treatment is considered necessary; penetrating insecticides then have to be used.[121]

In mature plantations the most effective insecticide against adults was found to be 25 per cent BHC applied as a dust at the rate of 15 kg per hectare and directed to the under surfaces of the leaves. Application at 8-day intervals in dry, still weather is advocated. It was suggested that if this treatment, regarded as preventative, is carried out efficiently, then measures against larvae may be unnecessary.[122]

Control at the larval stage is more expensive and more difficult. The spray used must reach the upper surface of the leaves. The most effective insecticide in trials was Lindane (gamma BHC) in vegetable oil emulsion; this was applied as 3 l of emulsion containing 12 per cent active ingredient in 650 l of water to cover 1 hectare of palms. It was suggested that treatment against larvae could start some 50 days after the first treatment against adults. Two, or at most three, sprayings would be needed.[122] Parathion and malathion have also been advocated.

In a recent attack on plantations in West Cameroon, parasites were found in the early surveys, but specimens of the fly *Cotterellia podagrica* were found after aerial spraying had been carried out.[123] Control of both the larval and adult stages was attempted from the air and the latter was the most successful. Trials confirmed that Lindane was the most effective chemical for adult as well as larval control, a Lindane/oil emulsion at the rate of 11 l per hectare with an active ingredient rate of 0.5 kg per hectare was found to be suitable and a complete kill was obtained; adults covered the ground 24 hours after treatment.

It is clear that in the control of *Coelaenomenodera* care must be taken not to harm the natural parasites, but when a serious attack occurs owing to a shortage of these parasites, chemical control must be used. In any spraying programme areas of low incidence should be omitted. Attempts have been made to correlate large scale swarming of the pest with climatic factors and there is a suggestion that below average temperatures in the dry season are conducive to swarming.

Hispoleptis elaeidis — *The South American Oil Palm Leaf Miner*

Distribution.   This little Hispid has recently been found damaging the leaves of oil palms in Ecuador. Its distribution is at present unknown.

Description.   No work has yet been done on the life history, incidence, etc., of this leaf miner, but the larval damage has much the same appearance as that of *Coelaenomenodera* and it is a potentially serious pest. The adults are slightly larger than those of *Coelaenomenodera*, being 8—10 mm long, pale yellow, with the body and wing cases similarly slightly enlarged at the rear. The ends of the wing cases are, however, dark brown and there is also a dark brown line in the centre towards the thorax.

Alurnus humeralis — *The Alurnus Beetle*

Distribution.   This large Hispid was first reported on the oil palm in Ecuador but is to be found in other parts of tropical America. Allied species are common pests of the extensive palm flora and *A. humeralis* was collected from the coconut palm near Cojimíes on the Pacific coast before it was found to be damaging an oil palm plantation some 50 miles inland.[124]

Description, incidence and damage. This hispid beetle is much larger than the leaf miners already mentioned and the leaf damage is different. The eggs are laid in typical rows of seven to ten joined by mucilage and adhering to the leaflets or rachis. They hatch after 29 to 43 days. The light brown larvae are 7—8 mm long on emergence but reach 40 mm before pupating. In a severe attack the laminae are almost entirely eaten away leaving the rachis and midribs bare, and this gives the palm its characteristic appearance. In the centre of the crown of young palms, larvae in all stages of development can be found between the leaflets of the unopened spears. The larval period lasts 7 to 8 months and there are seven stages. The pupae are to be found on the surface of the petioles of the younger leaves, and the pupal stage lasts 44 days. Although larval damage is more severe, the adults also damage the leaflets by consuming longitudinal strips several centimetres long and about 1 mm wide. The adult has an average life of 113 days.[125]

Incidence can be heavy; in one group of 410 palms 198 were found to be infested, and the larval foliar damage over a whole plantation was estimated to be about 30 per cent.

Control. The natural predators are mites and wasps and it is to be supposed that severe attacks will occur when the populations of these are temporarily suppressed. Several insecticides have been tried against the larvae in the crown of the palm and heptachlor (0.1 per cent), toxaphene (0.5 per cent) and metoxychlor (0.4 per cent) applied at 4-monthly intervals are reported to have given fair control, reducing damage by 70 to 82 per cent. DDT, diazinon and dieldrin were also found to cause a high mortality of the larvae when sprayed at strengths of 0.34 per cent, 0.16 per cent and 0.1 per cent respectively,[124] but the dangers of contact insecticides must not be overlooked. Fortunately the number of serious outbreaks has been limited.

Cassididae

Pseudimatidium neivai *and* P. elaeicola

Distribution. *Pseudimatidium neivai* was discovered in Brazil and named in Bondar's large work on the insects of Bahia[126] as *Himatidium neivai*. Bondar transferred the genus to the Hispidae. The insect was reported as attacking young coconuts when still green, but later it was said to be common on palms native to the area, particularly *Desmoncus* and *Bactris* species and on young fruits of the oil palm, and it was considered to be responsible for the failure of several plantations. Following the extended planting of the oil palm in Colombia, a species of *Pseudimatidium*, possibly *P. neivai*, was reported from the Magdalena valley, and it has become a pest of the oil palm in all parts of the continent. A new species, *Pseudimatidium elaeicola*, has been discovered on the Pacific coastal plain near Calima damaging oil palm fruit, and the genus has now reverted to the Cassididae[127] though some authors still refer to the insect under the name *Himatidium neivai* in the Hispidae.[128, 129]

Description and damage.   The adult of *P. neivai*, which is of flattened shape, measures 5 × 3 mm and is at first white but rapidly becomes a shiny brown with fine longitudinal lines along the wing cases. Single eggs are laid. The larvae are more flattened than the adult and their feet are short and withdrawn; they are at first translucent, later turn dull red and reach 7 mm in length and 4 mm in breadth. The pupae are brown and otherwise resemble the larvae.

The insect is found on the under surface of leaves, but the main point of attack is the fruit. Most of the damage is from the larvae which nibble the exocarp beginning at the apex. A fungus then develops at the point of attack and the exocarp becomes lignified and grey.[129]

Extent of damage, and control.   It has been estimated that a heavy attack leads to a loss of 7 to 9 per cent oil and that losses from the more usual lesser attacks are, in spite of the alarming appearance of the bunches, negligible.[129] Young plantations are more vulnerable. *Pseudimatidium elaeicola* is reported to do similar damage to developing fruit so that the latter becomes dry and hardened.[130]

The pupae are parasitized by two flies, *Terrastichus* sp. and *Psychidosmiera* sp., but the amount of control exercised by them appears small. Ant species are considered to play a more important role in limiting the population.[129] Only if the attack becomes severe (a general attack of more than 70 per cent of the palms, or more than 10 per cent attacked heavily) is insecticidal treatment thought necessary. Endrin in 0.15 per cent solution of active ingredient has been found effective and, once applied, treatment is not necessary for a year.[129] The toxicity of endrin must also be taken into account and time of application should not be near to harvesting rounds.

Other Chrysomeloidea

A little Cassid, *Calyptocephala marginipennis*, has been found doing minor damage to oil palm leaves in Honduras. In Malaysia, the Hispid *Promecotheca cumingi* is occasionally explosive on coconuts and has been noted on oil palms though not in great numbers.[4] It is rather larger than *Coelaenomenodera* to which it is closely related.

Curculionidae (The weevils)

Temnoschoita *species*

Distribution.   These weevils are to be found damaging the oil palm throughout Africa but appear to be more commonly encountered in Zaire than in West Africa. The commonest species is *T. quadripustulata* (*quadri-maculata*); *T. delumbata* is less common.

Life cycle and damage.   The adults of *T. quadripustulata*, which are 8–10 mm long, are dark brown with the thorax spotted with indentations. The light brown wing cases have four reddish blotches and do not fully cover the abdomen. The females lay their eggs on cuts and wounds on the leaf petioles of young palms, both those recently transplanted and palms in early bearing. Nursery plants may also be infested. The young larvae tunnel their way through both dead and living tissue towards the heart of the palm or, in the case of older palms, they move into the inflorescence and cause rotting; they pupate in the tunnels so formed. In bearing palms the adults are attracted to the inflorescences where eggs are also laid. The damage is sometimes severe, and young palms may be killed through penetration of the crown and growing point (Plate 68).

Control.   The following measures have been suggested.[99] In areas where the weevil has been noted, care should be taken to avoid wounding the palms by excessive leaf pruning, particularly just before transplanting. Unfortunately this injunction may conflict with control measures against *Cercospora* Leaf Spot. With bearing palms the collection and destruction of rotted bunches and scattered fruit is also recommended as these may

Pl. 68   *Temnoschoita* damage to leaves in West Cameroon. Note typical 'windows'.

contain eggs, larvae and pupae. When harvesting begins it may be advantageous to undertake a general cleaning of the crown followed by dusting twice with an insecticide such as BHC at 3-weekly intervals, the dust being applied in the crown from the centre to the base, not on the leaves. Traps for the adults have also been constructed from recently cut and split petioles or banana trunks. Banana plants are an attractive host and should not be grown near nurseries or young plantations where infection with *Temnoschoita* is feared. For nursery attacks a 0.3 per cent solution of dieldrin has been recommended.[131]

Rhynchophorus *sp.* — *Palm weevils*

Distribution.   Species of the large *Rhynchophorus* weevil are to be found attacking palms in all parts of the tropics. As pests of the oil palm the distribution of the more important species is as follows:

R. phoenicis	Africa	
R. palmarum	America	The Gru-gru beetle
R. ferrugineus	Asia	The red palm weevil
R. schach		The red-stripe weevil

(*R. papuanus* is found in the Celebes and New Guinea.)

Description.   The larvae attain a length of some 5 cm and are ovoid or rounded, legless, yellowish-white sacks with small brown heads. The last abdominal segment is flattened and has brown edges carrying bristles (Plate 69).

The cocoons of the pupae, constructed of concentrically placed fibres, extend to 8 cm in length and 3.5 cm in breadth.

The adults show distinct specific differences but are usually about 4—5 cm long and 2 cm broad. *Rhynchophorus phoenicis* is black but on the thorax there are two narrow longitudinal dark brown bands. The wing cases have about a dozen longitudinal grooves. The underside of the body is light brown with diffuse black spots.[99] *Rhynchophorus ferrugineus* is the common red palm weevil of the Far East, while *R. schach*, the red-stripe weevil, is the more dangerous for the oil palm, since it is the species most commonly found in Sumatra and Malaysia. The oil palm has, however, proved far less liable to attack than the coconut palm. *Rhynchophorus ferrugineus* is rather variable in length (2—5 cm) and is red-brown with a few irregular black spots on the thorax. *Rhynchophorus schach*, previously considered a variety, is black with a longitudinal red-brown line down the centre of the thorax.[132] The American species, *R. palmarum* is entirely black with velvety thorax slightly prolonged at the base and shiny grooved wing cases.

Life cycle and damage.   *Rhynchophorus* weevils are wound parasites laying their eggs, which are 2—3 mm long, in the cut or damaged surfaces of many palms. In Malaysia the order of preference in the case of *R.*

Pl. 69 *Rhynchophorus* larva found in a 'spear rot' palm in Nicaragua.

schach is the sago palm, the nibong, the coconut, the oil palm, the sugar palm, the areca or betel nut palm.[133]

The adults may survive for 3 months. The eggs hatch in 3 days and the larvae tunnel into the crown and trunk. The tissues around the growing point then begin to decay and the palm may be killed. The external symptoms of attack have been described as similar to those of *Fusarium* Wilt, i.e. the leaves show a gradually increasing chlorosis, and fracture in strong winds.

The larval stage lasts about 2 months and pupation then occupies about 25 days, the larvae moving towards the periphery of the trunk to pupate. The whole life cycle lasts less than 3 months. The weevil more commonly breeds in the stumps of a felled palm field, newly cut stumps being preferred. *Oryctes* (see p. 675) and *Rhynchophorus* are often present in a plantation at the same time, wounds made by *Oryctes* giving a means of *Rhynchophorus* infection, while *Rhynchophorus* damage will provide conditions suitable for *Oryctes*. *Rhynchophorus* is also sometimes present in an oil palm as a secondary organism in cases of bud rot (see p. 647). The connection between *Rhynchophorus* attack and Red Ring disease has already been mentioned (p. 653).

Through its lethal effect, *Rhynchophorus* is a potentially serious pest, but in Asia and Africa it cannot be said that its incidence on the oil palm is very high. Deaths have however been noted in Africa where leaves have been cut abnormally short and wounding of adjacent leaf bases has resulted. In America, incidence may well be higher. Deaths from *Rhynchophorus palmarum* attack have been noted in young plantings within the grove areas in Bahia, Brazil, and the pest is quite frequently encountered on oil palms in other parts of the continent. In Malaysia the seriousness of *R. schach* as an oil palm pest has come to be doubted.

Control.   Effective control of *Rhynchophorus* attack is not easy. In the first place, however, the wounding of the palm must be avoided and the petioles must not be cut close to the trunk. Secondly, all dead or felled palms should be destroyed within the period of the life cycle. In some areas collection of adult beetles has been resorted to. Measures for the control of *Oryctes* and other large beetles will help to reduce the incidence of *Rhynchophorus.*

In areas of infection, cut portions of the petioles should be inspected, and treated if found to be infested. Though the labour involved is considerable, it is possible to hook out the larvae from their tunnels with the aid of a wire. The tunnels can then be treated with disinfectant and stopped up with clay or putty. Control measures have been described in detail by Mariau.[134] Several parasites of *Rhynchophorus* species are recorded in entomological literature.

Other weevils

Several other weevils have been reported as doing damage similar to that of *Rhynchophorus*. *Rhinostomus (Rhina) barbirostris* (the bearded weevil) is found in company with *Rhynchophorus palmarum* in Brazil and elsewhere in America while *Rhinostomus (Rhina) afzelii* has been described in Sierra Leone.

Two species of *Prosoestus, P. sculptilus* and *P. minor*, are found in Zaire and live in the female inflorescences; they are only some 6 mm long.[99] The flower stigmas are punctured and become prematurely brown. The eggs are laid and the young larvae tunnel in the stile working their way towards the ovary which naturally fails to develop. The amount of damage is small, however, the adjoining undamaged fruit tend to grow larger to compensate for the smaller number of fruit developing.

The presence of small weevils in the debris around cut leaf bases and bunch stalks is not uncommon. In Malaysia *Diocalandra frumenti*, which is 8 mm long, is common; the larvae develop in tissue around wounds but do not penetrate deeply.[4] In Brazil and other parts of South America *Metamasius hemipterus* is also not uncommon and develops in rotting leaf bases. *Leurostenus elaeidis* and *Pseudostenotrupis filum* have been reported on oil palms in Zaire, and the former species also in Sierra Leone. There is no record of damage.

Scarabeoidea: Dynastidae

Oil palms in all stages of growth are sometimes attacked by large beetles, though the attacks are rarely as serious as with the coconut palm. The principal culprits are the 'Rhinoceros beetles' which on the continents of Asia and Africa are species of *Oryctes*, but in America belong to the genus *Strategus*.

Oryctes *species – The Rhinoceros beetles*

Distribution.   *Oryctes* are to be found throughout the palm-growing areas of Africa, Asia and the Pacific. Goonewardena states that there are forty-two species of which four are found in South-east Asia.[135] The following species may be mentioned:

O. rhinoceros. The common Rhinoceros beetle of the Far East and which has spread to the Pacific islands (Plate 70).

O. gnu (*trituberculatus*). Asia, less common.

O. boas ⎫
O. monoceros ⎬ Africa. *O. boas* is probably the most common.[99]
O. owariensis ⎭

The Asian species are primarily pests of the coconut palm, but they are to be found attacking many of the other palms of the continent, whether cultivated or wild. The African species attack the coconut and *Borassus* palms but, owing to its ubiquity, the largest population is to be found on the oil palm. Three other species are recorded on oil palms by Jerath.[136]

Description.   The male adult has the characteristic rhinoceros horn; in the female the horn is smaller or, in the African species, is reduced to a triangular protuberance. The beetle is black and measure 4–6 cm long and 2–3 cm broad according to species, *O. trituberculatus* being larger than *O. rhinoceros*, and the African species *O. owariensis* being larger and *O. boas* being smaller than *O. monoceros*. The horn of *O. boas* is particularly long and curved.

The eggs are white, 3–4 mm in diameter and easily observed on the breeding grounds. The young larva is white at first but its head soon becomes brown and its body blue-grey, then yellowish or greenish-white; it reaches a length of 4–10 cm (Plate 70).

Life cycle, incidence and damage.   The eggs are laid on rotting vegetable matter, logs or cow dung and compost. On an oil palm estate decaying palm trunks and bunch refuse are common breeding grounds. About twenty eggs are usually laid but higher numbers have been recorded; they hatch after 11 to 13 days. The length of the larval stage varies considerably, ranging from around 100 to 200 days. Similarly, the insect's life as an adult may last for a few months or extend to over half a year. Before pupation there is a short prepupal stage of a week; the adults emerge after a further 3 weeks.

The oil palm is damaged by the adult beetle which burrows into the cluster of developing spears in the crown and bores its way through the

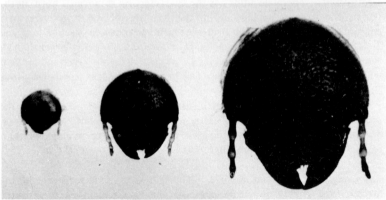

Pl. 70 *Oryctes rhinoceros* in Malaysia: **A.** Adults, l. male; r. female. **B.** Larval instars. 1st, 2nd, 3rd (early, late, prepupal). **C.** Head capsules of larval instars; l. to r. 3rd, 2nd, 1st.

petioles into the softer tissues of the younger unopened leaves. The effect can be seen when these leaves develop and open, but the regularity of wedge-shaped cuts so characteristic with the coconut palm is not always so clearly seen in the oil palm. Where the rachis has been penetrated, leaves may later snap off. Previous attacks may be detected by the presence of holes in the petioles of older leaves.[4] An attack is most dangerous in young palms since the growing point may occasionally be reached or a Bud Rot may develop which will kill the palm. As already mentioned the bore holes also give *Rhynchophorus* weevils access to the palm. In other cases the attack may be so severe that recovery is slow and the new leaves formed are small and contorted. Wood has described and illustrated the kinds of damage that the palm may suffer from the Asian species.[4]

Control. The majority of work on the control of the *Oryctes* beetle has been done in connection with coconut cultivation, but these measures are largely applicable to the oil palm. Overriding all other means is the institution of through-going sanitary measures for the elimination of the larval stage. All rotting vegetable matter, dung heaps and composts should be dispersed and rotting palm trunks and timbers should be disposed of. This is often more easily said than done, but with proper heaping and firing it is possible to burn both old coconut and oil palm trunks and stumps. In certain Asian countries the disposal of breeding grounds for *Oryctes* can be made compulsory by orders issued under the law.

Poisoned palms appear to rot down more rapidly than felled un-poisoned palms, and the period over which the trunks are suitable breeding grounds may thus be reduced.[4] However, this has been disputed and in Indonesia urea has been advocated as a rotting agent.[68] In replanting oil palms in Asia the measures to be taken against *Ganoderma* (see p. 636) are likely also to reduce the incidence of *Oryctes*.

Unless the palm trunks or other timbers such as rubber have been fully disposed of by fire, it will be necessary in a replanting to carry out regular inspection by workers who will break up the rotting material and collect the grubs. Thorough inspection by workers armed with large hoes and axes is advocated in Malaysia.[4]

Measures taken against the adult, whether by trapping or treating the palms with insecticides, have been largely unsuccessful or too costly. In very severe infestations of young palms hand collection may be resorted to, the beetles being extracted from holes by means of a hook. However, larval searches and destruction are considered preferable.[137] It has recently been found that ethyl chrysanthemumate is a strong attractant of *Oryctes* and traps, at the rate of twenty-five per hectare, using 0.2 ml per trap of this substance have been suggested.[138]

It has been noticed in Malaysia that palms along or near roadsides may be heavily attacked while those within the field escape injury, and it is thought that inter-row vegetation may form a barrier[4] and, in young areas, may blur the palm silhouette which is believed to attract the beetle. In areas of young palms (*a*) kept bare, and (*b*) sown with a mixture of erect and creeping covers *Oryctes* damage was considerably higher on the bare areas. It is now considered that rapid covering of timbers with a dense ground cover, usually leguminous, is by far the most effective way of suppressing *Oryctes* attack[138] and the encouragement of an early cover is therefore an important part of planting or replanting schedules.

Many parasites and predators of *Oryctes* have been known for a long time and larvae in particular are kept in check by rodents, birds, lizards, ants and termites. Certain mites feed on the eggs. In recent years much attention has been given, again in connection with coconut cultivation, to the introduction and spread of insect parasites or fungi. Of the former, the Scoliid wasp, *Scolia procer*, parasitizes the larvae. In Malaysia there is a regular, though small, larval mortality from the virus, *Rhabdionvirus oryctes*, and the fungus, *Meterrhizium anisopliae*,[137] and work on the propagation of the virus has been continuing.[139]

In the large new plantings in West Africa the incidence of *Oryctes* has been found to be relatively low, and rapid covering of the felled forest timbers or old palms by leguminous covers is found to be the most effective measure.[140, 158] This can be augmented by the use of ethyl chrysanthemumate traps at the rate of four per hectare.[158]

Strategus aloeus − *The Strategus beetle*

Distribution.   This beetle somewhat resembles *Oryctes* and is distributed throughout tropical America where it has been troublesome in several oil palm plantations.[141]

Incidence and damage.   The adults attack young palms in the field or nursery by digging a hole in the ground near the palm from which they bore their way into the plant just above the roots. Often, in a young palm, the growing point is reached and the plant killed. Eggs may be laid in the palm which is then consumed by the developing larvae. However, eggs may also be laid in rotting stumps, trunks and vegetation, and the measures to be taken against the larval stage are therefore the same as for *Oryctes* species.

Control.   In view of the lethal attack on young palms by the adult, additional control measures are required in areas where the beetle is common. During wet weather fortnightly inspections are recommended. Into all holes 2 l of a 0.2 per cent solution of endrin is poured; this treatment has been found effective in Colombia.[141] In dry weather *Strategus* attack is rare.

Other large Dynastid beetles

Certain other Dynastid beetles are occasionally troublesome in oil palm plantations. In Africa the large horned beetle *Dynastes* (*Augosoma*) *centaurus* can attack and sometimes kill young nursery or field plants. Large numbers may be captured in light traps in the dry season. In Malaysia the rather similar Gideon beetle, *Xylotrupes* (*Dynastes*) *gideon*, is found on coastal estates feeding on the leaf rachis, sometimes causing it to break.[4] The males only of these beetles have large thoracic horns. The breeding habits are similar to *Oryctes* and the grubs can be confused with *Oryctes* larvae, though they are much more hairy. The Atlas beetle, *Chalcosoma atlas*, is also found with *Oryctes*.

Cetoniidae

Platygenia barbata

Distribution.   This African beetle has long been reported as damaging the oil palm in Zaire.[99]

Life cycle, incidence and damage.   The eggs are laid at the base of the petioles and the larvae develop in the debris in the leaf axils. They then bore holes 2—3 cm deep into the petioles which weaken them so that in a wind the leaf will snap low down and hang parallel with the trunk; this gives an attacked palm a characteristic appearance. The larvae also tunnel into the peduncles and bunches thus fail to ripen. Penetration into the terminal bud, with consequent death, is rare, but the holes made allow other insects such as *Rhynchophorus* and *Temnoschoita* to enter, and bud rots may ensue. The adult is a black beetle, about 30 mm long, with rounded wing cases. The larvae reach 50 mm in length, have a brown head and are distinctly hairy. The pupae are enclosed in a cocoon of palm fibres.

Control.   Control measures have been confined to hand collection and the clearing of debris from the leaf axils.

Scarabaeidae (Cockchafers or Night-flying beetles)

Adoretus *and* Apogonia *species* (Plate 71)

Distribution.   Cockchafer beetles of these genera can be a considerable

Pl. 71   Common Malaysian Cockchafers which damage oil palm seedlings: l. to r. *Apogonia expeditiones, Adoretus borneensis, Adoretus compressus.*

nuisance in nurseries in Malaysia by chewing holes in the leaves. Such beetles do not appear to have done significant damage elsewhere. *Adoretus* is in the Rutelinae sub-family and *Apogonia* in the Melolonthionae.

Incidence and damage. The four common species in Malaysia are *Apogonia expeditionis* and *A. cribricollis*, which are black, and the brownish *Adoretus borneensis* and *A. compressus* which are slightly larger and about 12 mm long. They make holes in the leaf at night, *Apogonia* being responsible for the larger incursions.[108] During the day they rest in the soil at the base of the seedlings.

Control. The night-feeding habit makes control by collection difficult. Dipterex sprayed as 0.1 per cent (high volume) has given good control.[4] Dicrotophes (Bidrin) is also effective.

Leucopholis rorida

The larvae of this beetle (Melolonthinae) have fed on the roots of nursery seedlings in Malaysia and more recently cockchafer grub damage has been reported in young field areas.[139] In nurseries control by digging aldrin into the soil is recommended.[4]

Another cockchafer, *Psilopholis vestita*, has been reported as damaging young palms and the inter-row vegetation by root feeding by the larvae. Attacks are nearly always confined to forest boundary areas.[142]

Other Coleoptera

Many other beetles are encountered on oil palms but few do serious damage. In Malaysia many of them are co-inhabitants of *Oryctes* breeding sites and they have been listed and illustrated by Wood.[4] Some of them like *Aegus chelifer* (Lucanidae) are large beetles with prominent mouth parts. Other Lucanids and the Malaysian Cetonids are smaller and the latter are brightly coloured. The other Malaysian species are of several different families.

Hymenoptera

Formicidae (Ants)

Leaf-cutting ants have been reported as doing damage to palms in Zaire;[99] in Costa Rica they have been seen to remove considerable quantities of the leaf laminae to their nests below ground. In Malaysia ants are regarded mainly as a nuisance, particularly the aggressive Kerengga (*Oecophylla smaragdina*). The little ant *Crematogaster dohrni* is also common. These ants attend honeydew-producing scale insects. If the nuisance of these ants becomes excessive then treatment may be given (see p. 656).

Apidae

Bees are reported seriously to have reduced the supplies of pollen on some estates in Malaysia.[143] In one attack the bee identified was the common *Megapis dorsata*. Pollen was removed systematically from one male inflorescence after another. Control has been effected by tracing out the nests and burning them.

The Hymenoptera, together with the Diptera (flies), provide the chief parasites of many oil palm pests. The Hymenopterous parasites are mainly ichneumen, chalcid, scelionid and scoliid wasps.

Insect control on oil palms

It will have been noted from the previous pages that more and more attention is being given to what has been called 'integrated control', that is the timing and use of insecticides in a manner that helps to preserve the natural enemies of a pest. This subject has been dealt with very fully by Wood[4, 115] and cannot be expanded here. It is worth mentioning however that the evil of indiscriminate application of contact insecticides has been clearly seen in Malaysia with such pests as bagworms and nettle caterpillars,[144] and there is considerable scope for biological control through the spreading of natural enemies whether they be other insects, fungi or viruses.

Methods of insecticide application are also described in some detail by Wood[4] and much attention has been given to the use of tractor-drawn powered sprayers and to aerial application over adult plantations. For use in nurseries or on young palms in the field shoulder-mounted knapsack sprayers, manually operated, are normally adequate and may be used with taller palms with an extension lance. Motorized knapsack sprayers may be used in areas where palms are large but tractors cannot pass easily through the fields. In the Ivory Coast very efficient use has been made of tractor-drawn dusters and sprayers for control of the leaf miner, *Coelaenomenodera elaeidis*, on both medium-aged and tall adult palms.[121, 145] Where large areas have to be covered aerial application has been successful. The Piper Pawnee aircraft has been employed both in Asia and Africa for oil palm pest control[4, 145] and 400 hectares per day can be treated. Detailed instructions for the layout, marking and organization of an aerial spray programme have been given by Wood.[4]

Vertebrates (birds and mammals)

Vertebrates, especially birds and rats, have assumed much greater importance in recent decades, particularly in Asia and America. This was only to be expected since oil palm plantations have moved into new areas in these continents and have presented opportunities for certain species to adapt to, and even increase in, a new environment.

Birds

Parrots

The long-tailed parakeet (*Psittacula longicanda*), the blue-rumped parrot (*Psittinus cyanurus*) and the Malay lorikeet (*Loriculus galgulus*) have all been troublesome in Malaysia. Most destructive is the long-tailed parakeet, which feeds in flocks of up to thirty birds and carries away ripe fruit from the bunch and tends to scatter it about half eaten. Such damage can be distinguished from rodent damage by the single beak groove in the fruit. The other species feeds close to the bunch and do not scatter the fruit. Shooting is the only control known and with the long-tailed parrot this does not seem to have been very effective.[4]

Vultures

The American Black Vulture (*Coragyps atratus*) has become a serious pest. These birds are scavengers and eaters of carrion and flock around slaughter houses. In Brazil, Colombia, Honduras and elsewhere in America they have migrated to nearby oil palm plantations and established themselves among the palms, gorging themselves on the fruit as it becomes ripe. Much localized loss is suffered; and, once established, the birds tend to distribute themselves through the plantations and losses then become more general. In most countries the birds are protected by law as useful scavengers, and special permission must be obtained to shoot them. This course has been adopted in Colombia. In Africa, the palm-nut vulture, *Gypohierax angolensis*, does some damage in the groves.

Other birds

In West Africa the weaver bird (*Ploceus cuculatus*) may be locally troublesome through stripping the leaflet laminae from a wide area to make nests in adjoining trees. It is usually necessary to fell the nesting trees to disperse the birds. Birds reported as causing local damage to fruit in Asia include the common mina, *Acridotheres tristis*, the crow, *Corvus macrorhynchus*, and the Philippine glossy starling, *Aplonis panayensis*.[146]

Rodents

The 'cutting grass' — *Thryonomys swinderianus*

This large rodent is common in Africa and young areas planted near to the forest are particularly at risk from its devastating attacks. Protection against this pest is described under transplanting practices on p. 404.

Rats — *Rattus* sp. and other genera

Rats have rather suddenly assumed considerable importance as pests of the expanding plantations of Asia. As a result, extensive studies have been made of their biology and habits and control measures have been devised.[4, 7, 146, 147] Populations of 250 to 500 per hectare are not uncommon and may rise to 1,000. Early infestations by the rice field rat, *R. argenti-venter*, give place to *R. tiomanicus*, the Malayan Wood rat, which is found in virtually all oil palm plantations. Although young palms are sometimes attacked (and these can be protected with wire-netting collars as against the 'cutting grass', though the collar must be turned in at the top[4]) the main damage is to the fruit, both ripe and unripe. Much of this damage may pass unnoticed. Fruit is often carried from the bunch to the interline vegetation where it is not easily noticed, and loose fruit is stolen. Inflorescences may also be damaged or even destroyed. Hunting and trapping are not considered to have much effect on the populations and poison baiting is the main method of systematic control, using anticoagulent chronic poisons. Very detailed instructions for baiting have been published by Turner and Gillbanks[7] and in other papers[4, 148] and cannot be given in full here. Paper-wrapped baits of broken rice, prawn powder and 1.5 per cent a.i. coumachlor (Tomorin or Ratafin) in the ratio 16:2:1 have been widely used. This mixture is used in 15 g lots in greaseproof paper. Wax baits are also commonly employed and consist of maize or rice bran, 7 kg; fish heads 0.7 kg; palm oil 2.5 l; paraffin wax 0.7 kg; coumachlor 0.4 kg. If the anticoagulant used is warfarin (0.5 per cent a.i.) then about double the quantities given for coumachlor are needed. Wax baits require careful preparation.[4, 7] Timing of baiting depends on population and damage assessment.[7, 148]

Other rat species found in Asian oil palm fields are *R. exulans* and the bamboo rat, *Rhyzomis sumatrensis*.[4]

Other rodents

Rat populations and damage in African and American oil palm fields have not been so fully studied as in Asia. In West Africa the populations of *Dasymys incomtus, Lemniscomys striatus, Lophuromys sikapusi* and *Uranomys ruddi* were investigated on Ivory Coast plantations.[149] In fields surrounded by derived savannah the populations tended to increase suddenly and then decrease to a stable population. Details of control measures used in the savannah areas of the Ivory Coast have been given by Bredas.[150]

Squirrels (*Callosciurus* sp.) are occasionally troublesome in Asia, eating the mesocarp and sometimes attacking nursery plants. Porcupines (*Hystrix brachyura*) attack young palms on forest margins, gnawing through to the bud. Wire collars are not effective and zinc phosphide baits with palm oil in cassava roots have been used.[4]

Large mammals

Wild pigs damage or kill young palms. Strong fencing may be necessary for nursery protection, otherwise baiting or shooting are the main control measures.

Elephants have done great damage to young plantings in Sabah, systematically uprooting rows of newly planted seedlings. Barriers formed by ditches, electric fencies, tangle-felled forest trees and scaring devices have been tried, though not very successfully. Automatically firing carbide guns firing through the night at irregular intervals are valuable.[4] Elephants are usually protected so the advice of game departments needs to be obtained.

Lastly, monkeys very occasionally pull up seedlings and may even eat the growing point of older palms. Shooting is the only practicable control measure.

Monitoring disease and pest incidence

With the establishment of large plantations in new areas the need to monitor disease and pest incidence, i.e. to warn oneself of approaching danger before it is too late, has been increasingly apparent. Various census systems have been proposed, usually with particular pests in mind.[4] One of the best methods in use is one devised in Sabah[151] and which can easily be adapted for diseases as well as pests. The system consists of two steps, detection and enumeration. Monthly or bi-monthly *detection* rounds by experienced detecting staff are recommended. The intensity of the detecting work can be varied by taking either every harvesting (or inspection) path or alternate or every third path. The presence of pests or a disease symptom is recorded on special cards, and the work is summarized at the end of each day. Rough levels of infestation (high, medium or low) are recorded. *Enumeration* follows detection and is carried out within a day or two with the object of determining the density of pest attack and so deciding on the control measures to be taken. Diseases detected are of course examined if necessary by plant pathologists and appropriate action taken. In the enumeration of pests about twelve palms are taken per hectare of the affected area and six leaves per palm are used for counting larvae, pupae, etc.

This system has been described as an 'extensive method' in contrast to a 'sampling intensive method' in which one point is chosen per 3 hectares and pest enumeration is done on six leaves, two from each of three adjacent palms.[4] As a general system for disease and pest monitoring in a new area the Sabah system is to be preferred since the whole area is covered. It is relatively inexpensive, and a detection worker can usually cover 10 hectares a day.

References

1. **Arnaud, F. and Rabechault, H.** (1972) Premières observations sur les charactères cytohistochimique de la résistance du palmier à huile au 'dépérissement brutal'. *Oléagineux*, **27**, 525.

2. Robertson, J. S., Prendergast, A. G. and Sly, J. M. A. (1968) Diseases, disorders and deficiency symptoms of the oil palm in West Africa. *J. Nigerian Inst. Oil Palm Res.,* 4, 381.

3. Turner, P. D. and Bull, R. A. (1967) *Diseases and disorders of the oil palm in Malaysia.* Incorp. Soc. of Planters. Kuala Lumpur, 247 pp.

4. Wood, B. J. (1968) *Pests of oil palms in Malaysia and their control.* Incorp. Soc. of Planters, Kuala Lumpur, 204 pp.

5. Bevan, J. W. L. and Gray, B. S. (1966) Germination and nursery techniques for the oil palm in Malaysia. *Planter, Kuala Lumpur,* 42, 165.

6. Robertson, J. S. (1956) Leaf diseases of oil palm seedlings, *J. W. Afr. Inst. Oil Palm Res.,* 1, (4), 110.

7. Turner, P. D. and Gillbanks, R. A. (1974) *Oil palm cultivation and management.* Incorp. Soc. of Planters, Kuala Lumpur, Malaysia.

8. Kovachich, W. G. (1954) *Cercospora elaeidis* leaf spot of the oil palm. *Trans. Br. mycol. Soc.,* 37, 209.

9. Weir, G. M. (1968) Leaf Spot of the oil palm caused by *Cercospora elaeidis* Stey. Aspects of atmospheric humidity and temperature. *J. Nigerian Inst. Oil Palm Res.,* 5, 41.

10. Robertson, J. S. (1957) Spraying trials against Freckle, a leaf disease of oil palm seedlings caused by *Cercospora elaeidis* Stey. *J. W. Afr. Inst. Oil Palm Res.,* 2, 265.

11. Bachy, A. (1956) Essais de divers produits anti-cryptogamiques contre la cercosporiose du palmier à huile. *Oléagineux,* 11, 231.

12. Rajagopalan, K. (1968 and 1970). N.I.F.O.R. Third and Fifth Annual Reports, pp. 102−3 and 77−8 respectively.

13. Renard, J. L. (1973) Transport et distribution du bénomyl dans les palmiers à huile au stade de la pépinière. *Oléagineux,* 28, 557.

14. Robertson, J. S. (1960) W.A.I.F.O.R. Eighth Annual Report 1959−60, p. 107.

15. Robertson, J. S. (1963) W.A.I.F.O.R. Eleventh Annual Report. pp. 77−8.

16 Duff, A. D. S. (1970) *Cercospora elaeidis,* Stey., and the oil palm. *Oléagineux,* 25, 329.

17. Johnston, A. (1959) Oil palm seedling blight. *Malay. agric. J.,* 42, 14.

18. Booth, C. and Robertson, J. S. (1961) *Leptosphaeria elaeidis* sp. nov. isolated from anthracnosed tissue of oil palm seedlings. *Trans. Br. mycol. Soc.,* 44, 24.

19. Williams, T. H. (1965) *Diseases of the oil palm in Sabah.* Sabah Planters' Association. Oil Palm Seminar. Mimeograph.

20. Gunn, J. S. *et al.* (1961) The development of improved nursery practices for the oil palm in West Africa. *J. W. Afr. Inst. Oil Palm Res.,* 3, 198.

21. Bull, R. A. (1954) A preliminary list of oil palm diseases encountered in Nigeria. *J. W. Afr. Inst. Oil Palm Res.,* 1, (2), 53.

22. Kovachich, W. G. (1957) Some diseases of the oil palm in the Belgian Congo. *J. W. Afr. Inst. Oil Palm Res.,* 2, 221.

23. Turner, P. D. (1966) Blast disease in oil palm nurseries. *Planter, Kuala Lumpur,* 42, 103.

24. Cardoso, R. M. E. (1961) Podridao de raizes em dendezeiro. *Biologico,* 27, 246.

25. Robertson, J. S. (1959) Blast disease of the oil palm; its cause, incidence and control in Nigeria. *J. W. Afr. Inst. Oil Palm Res.,* 2, 310.

26. Robertson, J. S. (1959) Coinfection by a species of *Pythium* and *Rhizoctonia lamellifera* Small in Blast disease of oil palm seedlings. *Trans. Br. mycol. Soc.,* 42, 401.

27. Moreau, C. and Moreau, M. (1958) Le 'Blast' de jeunes palmiers à huile. Observation sur le système radiculaire de l'hôte et sur ses parasites. *Revue. Mycol.,* 23, 201.

28. Bachy, A. (1958) Le 'Blast' des pépinières de palmier à huile. *Oléagineux,* 13, 653.

29. **Rajagopalan, K.** (1968–74) N.I.F.O.R. Third, Fourth and Fifth Annual Reports, p. 105, pp. 87–8 and 78–9 respectively; and Influences of irrigation and shading on the occurrence of Blast disease of oil palm seedlings. *J. Nig. Inst. Oil Palm Res.,* 5, (19), 23.

30. **Wardlaw, C. W.** (1946) *Fusarium oxysporum* on the oil palm. *Nature, Lond,* 158, 712.

31. **Kovachich, W. G.** (1956) Patch Yellow disease of the oil palm. *Trans. Br. mycol. Soc.,* 39, 427.

32. **Thompson, A.** (1934) in *The oil palm in Malaya*, by Bunting B., Georgi, C. D. V. and Milsum, J. N. Chapter 7. Kuala Lumpur.

33. **Hasselo, H. N.** (1959) Fertilising of young oil palms in the Cameroons. *Pl. Soil,* 11, 113.

34. **Berchoux, C. de, and Gascon, J. P.** (1963) L'arcure défoliée du palmier à huile. *Oléagineux,* 18, 713.

35. **Blaak, G.** (1970) Epistasis for Crown disease in the oil palm (*Elaeis guineensis* Jacq.). *Euphytica,* 19, 22.

36. **Hartley, C. W. S.** (1974) *Oil palm research and development in Colombia.* Min. Overseas Dev., London, Mimeographed report.

37. **Rajaratnam, J. A.** (1972) White stripe disorder of oil palm (*Elaeis guineensis*) in Malaysia. *Expl. Agric.,* 8, 161.

38. **Ollagnier, M. and Valverde, G.** (1968) Contribution à l'étude de la carence en bore du palmier à huile. *Oléagineux,* 23, 359.

39. **Harrisons and Crosfield** (Malaysia). Oil Palm Research Station. Annual Report for 1974.

40. **Kovachich, W. G.** (1956) Necrotic spotting of the oil palm by *Cercospora elaeidis,* Steyaert. *Trans. Br. mycol. Soc.,* 39, 297.

41. **Moreau, C.** (1952) in *Notions de culture de l'Elaeis au Congo Belge*, by Vanderweyen, R., Brussels.

42. **Weir, G. M.** Algal spot of the oil palm. *Cephaleuros virescens* Kunze, in reference 2.

43. **Turner, P. D.** (1971) Microorganisms associated with oil palm (*Elaeis guineensis,* Jacq.) Commonwealth Mycological Institute, Phytopathological Papers, No. 14, 58 pp.

44. **Turner, P. D.** (1970) Some factors in the control of root diseases of oil palms. In Toussoun, t.a. *et al., Root diseases and soil-borne pathogens.* California University Press.

45. **Robertson, J. S.** (1962) Dry Basal Rot, a new disease of oil palms caused by *Ceratocystis paradoxa* (Dade) Moreau. *Trans. Br. mycol Soc.,* 45, 475.

46. **Robertson, J. S.** (1962) Investigations into an outbreak of Dry Basal Rot at Akwukwu in Western Nigeria. *J. W. Afr. Inst. Oil Palm Res.,* 3, 339.

47. **Robertson, J. S.** (1963) W.A.I.F.O.R. Eleventh Annual Report 1962–3, p. 79.

48. **Rajagopalan, K.** (1965) N.I.F.O.R. First Annual Report 1964–5, pp. 88–9.

49. **Robertson, J. S.** (1962) W.A.I.F.O.R. Tenth Annual Report 1961–2, p. 82.

50. **Wardlaw, C. W.** (1946) A Wilt disease of the oil palm. *Nature, Lond.,* 158, 56.

51. **Wardlaw, C. W.** (1950) Vascular Wilt disease of the oil palm caused by *Fusarium oxysporum* Schl. *Trop. Agric., Trin.,* 27, 42.

52. **Fraselle, J. V.** (1951) Experimental evidence of the pathogenicity of *Fusarium oxysporum* Schl. to the oil palm. *Nature, Lond.* 167, 447.

53. **Prendergast, A. G.** (1957) Observations on the epidemiology of Vascular Wilt disease of the oil palm (*Elaeis guineensis,* Jacq.). *J. W. Afr. Inst. Oil Palm Res.,* 2, 148.

54. **Kovachich, W. G.** (1953) Report quoted in reference 53 above.

55. **Renard, J. L.** (1970) La Fusariose du palmier à huile. Role des blessures des racines dans le processus d'infection. *Oléagineux,* 25, 581.

56. **Locke, T.** (1972) *A study of vascular wilt disease of oil palm seedlings.* Thesis, University of Manchester.

57. **Park, D.** (1958) The saprophytic status of *Fusarium oxysporum,* Schl. causing Vascular Wilt of oil palm. *Ann. Bot.,* N.S., **22,** 19.
58. **Prendergast, A. G.** (1963) A method of testing oil palm progenies at the nursery stage for resistance to Vascular Wilt disease caused by *Fusarium oxysporum,* Schl. *J. W. Afr. Inst. Oil Palm Res.,* 4, 156.
59. **Renard, J. L., Gascon, J. P. and Bachy, A.** (1972) Recherches sur la fusariose du palmier à huile. *Oléagineux,* 27, 581.
60. **Green, A. H. and Ward, J. B.** (1973) *Unilever Plantations Group Annual Review of Research, 1971.* London. Mimeograph.
61. **Locke, T. and Colhoun, J.** (1973) Contributions to a method of testing oil palm seedlings for resistance to *Fusarium oxysporum* Sch. f. sp. *elaeidis* Toovey. *Phytopath Z.,* 79, 77.
62. **Locke, T. and Colhoun, J.** (1973) *Fusarium oxysporum* P. sp. *elaeidis* as a seed-borne pathogen. *Trans. Br. mycol. Soc.,* 60, 3, 594.
63. **Turner, P. D.** (1966) Infection of oil palms by *Ganoderma* in Malaya. *Oléagineux,* 21, 73.
64. **Steyaert, R. L.** (1967) Les *Ganoderma* Palmicoles. *Bull. Jard. bot. nat. Belge.,* 37, 465.
65. **Navaratnam, S. J.** (1961) Successful inoculation of oil palms with a pure culture of *Ganoderma lucidum. Malay. agric. J.,* 43, 233.
66. **Turner, P. D.** (1965) The incidence of *Ganoderma* disease of oil palms in Malaya and its relation to previous crop. *Ann. appl. Biol.,* 55, 417.
67. **Stimpson, K. M. S. and Rasmussen, A. N.** (1973) Clearing the old stand and some preparations for replanting coastal oil palms. In *Advances in oil palm cultivation,* p. 116. Incorp. Soc. of Planters, Kuala Lumpur.
68. **Parnata, J.** (1972) Suatu tjara untok mempertjepat pelapukan dari bekas bowl dan batang sawit. *Bull.* III, 1. Balai Penelitian Perkebunan (R.I.S.P.A.) Medan.
69. **Umar Akbar, Kusnadi, M. and Ollagnier, M.** (1971). Influence de la nature du matérial végétal et de la nutrition minérale sur la pourriture sèche du tronc du palmier à huile due à *Ganoderma. Oléagineaux,* 26, 527.
70. **Turner, P. D.** (1968) The use of surgery as a method of treating Basal Stem Rot in oil palm. *Planter, Kuala Lumpur,* 44, 302.
71. **Wardlaw, C. W.** (1950) *Armillaria* Root and Trunk Rot of oil palms in the Belgian Congo. *Trop. Agric., Trin.,* 27, 95.
72. **Corrado, F.** (1970) La maladie du palmier à huile dans les Llanos de Colombie. *Oléagineux,* 25, 383.
73. **Sanchez Potes, A.** (1972) Dos enfermedades de importancia económica que afectas la palma Africana de aceite en Colombia. Inst. Col. Agropecuaria, Typescript.
74. **Martin, G.** (1970) Le déssechment des feuilles. Maladi du palmier à huile dans la région du Nord-Santander en Colombie. *Oléagineux,* 26, 22.
75. **Van den Hove, J.** (1971) Un type de pourriture des racines du palmier à huile. *Oléagineux,* 26, 153.
76. **Genty, Ph.** (1973) Observations préliminaries du lépidoptère mineur des racines du palmier à huile, *Sagalass valida,* Walker. *Oléagineux,* 28, 59.
77. **Genty, Ph.** (1973) *Informe de misión sobre una enfermedad similar a la 'Marchitez sorpresiva' en varias plantaciones de palma Africana en El Ecuador.* Ass. Nac. de Cultivadores de Palma Africana, Quito, Ecuador, Mimeograph.
78. **Vito Varingaño, C. and Julio Vera, M.** (1973) Avances logrados en el estudio de la Marchitez sorpresiva en palmera (*Elaeis guineensis* Jacq.) Paper presented at VI Congreso de Nematologos de los tropicos Americanos.
79. **Mena Tascon, E., Cardona Mejia, C., Martinez Lopez, G. and Dario Jimenez, O.** (1975) Efecto del uso de insecticidas y control de malezas en la incidencia de la marchitez sorpresiva de la palma africana (*Elaeis guineensis* Jacq.) *Rev. Colombiana Ent.,* 1, 1.

80. **Thompson, A.** (1937) *Observations on Stem Rot of the oil palm.* Dept. of Agriculture, Malaya, Scientific Series No. 21.
81. **Navaratnam, S. J. and Chee Kee Leong.** (1965) Stem Rot of oil palms in the Federal Experiment Station, Serdang. *Malay. agric. J.,* **45**, 175.
82. **Turner, P. D.** (1969) Observations on the incidence, effects and control of Upper Stem Rot in oil palms. In *Progress in oil palm.* Incorp. Soc. of Planters, Kuala Lumpur.
83. **Waterston, J. M.** (1953) Observations on the influence of some ecological factors on the incidence of oil palm diseases in Nigeria. *J. W. Afr. Inst. Oil Palm Res.,* **1**, (1), 24.
84. **Thompson, A.** (1936) *Ustulina zonata* on the oil palm. *Malay. agric. J.,* **34**, 222.
85. **Duff, A. D. S.** (1963) The Bud Rot Little Leaf disease of the oil palm. *J. W. Afr. Inst. Oil Palm Res.,* **4**, 176.
86. **Bull, R. A. and Robertson, J. S.** (1959) The problems of 'Little Leaf' of oil palms — a review. *J. W. Afr. Inst. Oil Palm Res.,* **2**, 355.
87. **Robertson, J. S.** (1960) W.A.I.F.O.R. Eighth Annual Report 1959—60, p. 112.
88. **Bachy, A.** (1954) Contribution de l'étude de pourriture du coeur du palmier à huile. *Oléagineux,* **9**, 619.
89. **Turner, P. D.** (1970) Spear rot disease of Plantació 'La Arenosa'. Mimeographed report. Harrison Fleming Advisory Services.
90. **Turner, P. D.** (1970) Oil palm diseases on Plantación 'La Arenosa' Mimeographed report. Harrison Fleming Advisory Services.
91. **Hartley, C. W. S.** (1965) Some notes on the oil palm in Latin America. *Oléagineux,* **20**, 359.
92. **Anon.** (1974) Replanting oil palm areas with *Elaeis oleifera* x *Elaeis guineensis* hybrids. *Oil Palm News,* No. 18, 1.
93. **Sharples, A.** (1928) Palm diseases in Malaya. *Malay. agric. J.,* **17**, 313.
94. **Turner, P. D.** (1965) *Marasmius* infection of oil palms in Malaya — a review. *Planter, Kuala Lumpur,* **41**, 387.
95. **Courtois, G.** (1968) Arbres anormaux chez *Elaeis guineensis.* Leur production comparée à celle des arbres normaux. *Oléagineux,* **23**, 641.
96. **Malaguti, G.** (1953) 'Pudrición de Cogolla' de la palmera de aciete africana (*Elaeis guineensis,* Jacq.) en Venezuela. *Agronomia Tropical,* **3**, 13.
97. **Freeman, W. G.** (1925) Report of the Department of Agriculture, Trinidad and Tobago, 1925. Abstract in *Rev. appl. Ent.,* **14**, 546.
98. **Maas, P. W. Th.** (1970) Contamination of the palm weevil (*Rhynchophorus palmarum*) with the Red Ring nematode (*Rhadinaphelenchus cocophilus*) in Surinam. *Oléagineux,* **25**, 653.
99. **Fraselle, J. and Buyckx, E. J. E.** (1962) *Maladies et animaux nuisibles du palmier à huile. In Publs. I.N.E.A.C.* Hors Série.
100. **Genty, Ph., Gildardo Lopez, J. and Mariau, D.** (1975) Dégâts de *Pestalotiopsis* induits par des attaques de *Gargaphia* en Colombie. *Oléagineux,* **30**, 199.
101. **Tams, W. H. T.** (1930) A new moth damaging oil-palm in the Belgian Congo. *Bull. ent. Res.,* **21**, 75; (1930) abstract in *Rev. appl. Ent.,* **18**, 426.
102. **Jover, H.** *et al.* (1954) Sur la biologie des chenilles de *Pimelephila ghesquierei* Tams, parasite des palmiers à huile en pépinière. *Revue Path. vég. Ent. agric. Fr.,* **30**, 149. Full abstract in *Rev. appl. Ent.,* **42**, 277.
103. **Mariau, D. and Morin, J. P.** (1971) La Pyrale du palmier à huile. *Oléagineux,* **26**, 379.
104. **Tams, W. H. T.** Private communication.
105. **Mariau, D. and Julia, J. F.** (1973) Les Parasa. *Oléagineux,* **28**, 129.
106. **Allen, J. D. and Bull, R. A.** (1954) Recent severe attacks on oil palms by two caterpillar pests belonging to the Limacodidae. *J. W. Afr. Inst. Oil Palm Res.,* **1**, (2), 130.

107. **Miller, N. C. E.** (1929) Notes on *Setora nitens*, a 'nettle caterpillar', etc., *Malay. agric. J.,* **17**, 315.
108. **Wood, B. J.** *Insect pests of oil palms in Malaya.* Annual Report, Johore Planters' Association, 1965.
109. **Genty, Ph.** (1972) Morphologie et biologie de *Sibine fusca*, Stoll, lépidoptère défoliateur du palmier à huile en Colombie. *Oléagineux,* **27**, 65.
110. **Genty, Ph. and Mariau, D.** (1973 and 1975) Les Limacodidae du genre Sibine. *Oléagineux,* **28**, 225; Utilization d'un germe entomopathogène dans la lutte contre *Sibine fusca* (Limacodidae). *Oléagineux,* **30**, 349.
111. **Hertslet, L. R. and Duckett, J. E.** (1971) *Thosea bisura* — A new pest of oil palms. *Planter, Kuala Lumpur,* **47**, 398.
112. **Leitch, T. T.** (1966) *Ploneta diducta* — A pest of oil palms. *Planter, Kuala Lumpur,* **42**, 433.
113. **Wijbrans, J. R. and de Weille, G. A.** (1958) Rupsenbestrijding in de Oliepalmcultuur. *Bull. Res. Inst. S.P.A.,* No. 15; (1960) abstract in *Rev. appl. Ent.,* **48**, 499.
114. **Conway, G. R. and Wood, B. J.** (1964) Pesticide Chemicals — help or hindrance in Malaysian agriculture? *Malay. Nat. J.,* **18**, 111.
115. **Wood, B. J.** (1972) Integrated control: critical assessment of case histories in developing economies. *Planter, Kuala Lumpur,* **49**, 367.
116. **Corbett, G. H.** (1927) Insect pests of the oil palm. *Malay. agric. J.,* **15**, 338.
117. **Genty, Ph.** (1968) Deux lépidoptère nuisables au palmier à huile. *Oléagineux,* **23**, 645.
118. **Cotterell, G. S.** (1925) The Hispid leaf miner (*Coelaenomenodera elaeidis* Maul) of oil palms (*Elaeis guineensis* Jacq.) on the Gold Coast. *Bull. ent. Res.,* **16**, 77.
119. **Morin, J. P. and Mariau, D.** (1970–74) La biologie de *Coelaenomenodera elaeidis* Mlk. Parts I, II, III, IV and V. *Oléagineux,* **25**, 11; **26**, 83 and 373; **27**, 496; **29**, 233 and 549.
120. **Mariau, D. and Bescombes, J. P.** (1972) Méthode de controle des niveaux de population de *Coelaenomenodera elaeidis. Oléagineux,* **27**, 425.
121. **Bescombes, J. P.** (1968) Essais de traitments des palmeraies adultes avec le B.S.E. Bangui Special. *Oléagineux,* **23**, 715.
122. **Ruer, P.** (1964) Les conditions de lutte contre un prédateur du palmier à huile. *Oléagineux,* **19**, 387.
123. **Shearing, C. H.** (1964) *A serious attack of Hispid leaf miner* (Coelaenomenodera elaeidis Maul.) *at Mpundu Palms Estate.* Cameroons Development Corporation, Ekona Research Unit. Mimeograph.
124. **Merino, M. G. and Vasquez, A. V.** (1963) El 'Gusano cogollero' *Alurnis humeralis* (Rosenberg) como plago de la palma Africana aceite y su combate químico en Ecuador. *Turrialba,* **13**, 6.
125. **Santos, J. V.** (1968) Algunas caracteristicas biologicas y etologicas del *Alurnus humeralis* Rosenburg 'Gusano chato o cogollero' de la palma Africana. *Oléagineux,* **23**, 159.
126. **Bondar, G.** (1940) Notas entomologicas da Bahia. Parts V, VI, VII. *Revta Ent., Rio de J.,* **11**, 199 and 842; (1941) **12**, 268. Also (1941) abstract in *Rev. appl. Ent.,* **29**, 9 and 468; and (1942) **30**, 425.
127. **Aslam, N. A.** (1965) On *Hispoleptis* Baly (Coleoptera, Hispidae) and *Imatidium* F. (Coleoptera, Cassididae). *Ann. Mag. nat. Hist.,* Ser. 13, **8**, 687.
128. **Figueroa, A. and Van den Hove, J.** (1967) Contributión al estudio de la 'escoriación' de los fontos de la palmera de aceite en Colombia, el *Himatidium neivai. Oléagineux,* **22**, 15.
129. **Genty, Ph. and Mariau, D.** (1973) Le genre *Himatidium. Oléagineux,* **28**, 513.
130. **Arens, F. P.** (1965) Private communication.
131. **Dubois, J. and Gerard, Ph.** (1968) Temnoschoites dans les pépinières de palmier à huile. *Oléagineux,* **23**, 571.

132. **Dammerman, K. W.** (1929) *The agricultural zoology of the Malay Archipelago.* Amsterdam.

133. **Corbett, G. H. and Ponniah, D.** (1923) Summary of observations on *Rhynchophorus Schab,* Oliv., the Red Stripe Weevil of coconuts. *Malay. agric. J.,* 9, 4, 79; (1923) abstract in *Rev. appl. Ent.,* 11, 389.

134. **Mariau, D.** (1968) Méthodes de lutte contre le Rhynchophore. *Oléagineux,* 23, 443.

135. **Goonerwardena, H. F.** (1958) The Rhinoceros beetle (*Oryctes rhinoceros* L.) in Ceylon. *Trop. Agric. Mag. Ceylon agric. Soc.,* 114, 39.

136. **Jerath, M. L.** (1968) A list of insects found on palms in Nigeria and their known parasites and predators. *J. Nig. Inst. Oil Palm Res.,* 4, 411.

137. **Barlow, H. S. and Chew Poh Soon** (1971) The Rhinoceros beetle, *Oryctes rhinoceros,* in young oil palms planted after rubber on some estates in West Malaysia. In *Crop protection in Malaysia,* p. 133. Incorp. Soc. of Planters, Kuala Lumpur.

138. **Turner, P. D.** (1973) An effective trap for *Oryctes* beetle in oil palms. *Planter, Kuala Lumpur,* 49, 488.

139. **Wood, B. J.** (1972) Developments in oil palm pest management. *Planter, Kuala Lumpur,* 48, 93.

140. **Boyé, P. and Aubry, M.** (1973) Replantation des palmeraies industrielles. Méthode de préparation de terrain et de protection contre l'*Oryctes* en Afrique de l'Ouest. *Oléagineux,* 28, 175.

141. **Bachy, A.** (1963) Insects et animaux nuisibles au palmier à huile. *Oléagineux,* 18, 15—18 and 173—6.

142. **Wood, B. J. and Ng, K. Y.** (1969) The cockchafer, *Psilopholis vestita,* a new pest of oil palms in West Malaysia. *Planter, Kuala Lumpur,* 45, 577.

143. **Gray, B. S. and Turner, P. D.** (1966) Removal of pollen from oil palms by *Megaspis dorsata. Pl. Prot. Bull. F.A.O.,* 14, 37.

144. **Wood, B. J.** (1971) The importance of ecological studies to pest control in Malaysian plantations. In *Crop Protection in Malaysia.* Incorp. Soc. of Planters, Kuala Lumpur.

145. **Bescombes, J. P.** (1972) Le matériel de traitements insecticides en plantation de palmier à huile. *Oléagineux,* 27, 479.

146. **Wood, B. J.** (1969) The extent of vertebrate attacks on the oil palm in Malaysia. In *Progress in oil palm.* Incorp. Soc. of Planters, Kuala Lumpur.

147. **Wood, B. J.** (1969) Population studies on the Malaysian wood rat (*Rattus tiomanicus*) in oil palms, demonstrating an effective new control method and assessing some old ones. *Planter, Kuala Lumpur,* 45, 510.

148. **Wood, B. J.** (1971) Sources of reinfestation of oil palms by the wood rat (*Rattus tiomanicus*). In *Crop protection in Malaysia.* Incorp. Soc. of Planters, Kuala Lumpur.

149. **Bellier, L.** (1965) Evolution du peuplement des rongeurs dans les plantations industrielles de palmier à huile. *Oléagineux,* 20, 735.

150. **Bredas, J.** *et al.* (1968) La lutte chimique contre les petits rongeurs en jeune palmeraie. *Oléagineux,* 23, 15.

151. **Syed, R. A. and Speldewinde, H. V.** (1974) Pest detection and census on oil palms. *Planter, Kuala Lumpur,* 50, 230.

152. **Ollagnier, M. and Renard, J. L.** (1976) Influence du potassium sur la résistance du palmier à huile à la fusariose. *Oléagineux,* 31, 203.

153. **Akbar, U. and Kusnadi, T. T.** (1976) Relationship between tolerance to *Ganoderma* basal stem rot and the origin of oil palm planting material. Mimeograph, Medan.

154. **Rao, A. P. Subrahmanyam, K. and Pandit, S. V.** (1975) *Ganoderma* Wilt disease of coconut and control. Information pamphlet No. 32, Andrah Pradesh Agric. Univ., India.

155. **Loh, C. F.** (1976) Preliminary evaluation of some systemic fungicides for

Ganoderma control and phytotaxicity to oil palm. Int. Agric. Oil Palm Conference, Kuala Lumpur, 1976.

156. **Genty, Ph., Desmier de Chenon, D. and Mariau, D.** (1976) Infestation of the aerial roots of oil palms by caterpillars of genus *Sufetula* (Walker) (Lepidoptera: Pyralidae). Int. Agric. Oil Palm Conference, Kuala Lumpur. 1976.

157. **Tiong, R. H. C. and Munroe, D. D.** (1976) Microbial control of an outbreak of *Darna trima* (Moore) on oil palm (*Elaeis guineensis* Jacq.) in Sarawak (Malaysian Borneo). Int. Agric. Oil Palm Conference, Kuala Lumpur, 1976.

158. **Julia, J. F. and Mariau, D.** (1976) Recherches sur l'*Oryctes monoceros* Ol. en Côte d'Ivoire. Parts I, II and III, *Oléagineux*, 31, 63, 113 and 263.

159. **Renard, J. L., Mariau, D. and Quencez, P.** (1975) Le Blast du palmier à huile: rôle des insectes dans la maladie. Résultats preliminaires. *Oléagineux*, 30, 497.

160. **Genty, Ph.** (1976) Etude morphologique et biologique d'un lépidoptère défoliateur du palmier à huile en Amérique latine *Darna metaleuca* Walker. *Oléagineux*, 31, 99.

161. **Wood, B. J., Hutauruk, Ch. and Liau, S. S.** (1976) Studies on the chemical and integrated control of nettle caterpillars (Lepidoptera: Limacodidae). Int. Agric. Oil Palm Conference, Kuala Lumpur, 1976.

162. **Rojas-Cruz, L. A.** (1976) Insect pests of oil palms in Colombia. Int. Agric. Oil Palm Conference, Kuala Lumpur, 1976.

163. **López, G., Genty, Ph. and Ollagnier, M.** (1975). Contrôle préventif de la "Marchitez sorpresiva" de l'*Elaeis guineensis* en Amérique latine. *Oléagineux*, 30, 243.

Chapter 14

The products of the oil palm and their extraction

Before the extraction of oil palm products can be considered, attention must be paid to the nature and characteristics of these products and the manner in which they are held in the harvested organ, the bunch.

Harvesting of the bunch has already been described (p. 469). When the bunch arrives at the place of extraction, a portion of the fruit will already be loose; but the remainder will be held on the bunch for some time. Complete loosening of fruit from a bunch may take more than a week, and if natural loosening is awaited the quality of oil, as indicated by the free fatty acid (f.f.a.) content, will deteriorate, since a proportion of both the loose fruit and fruit on the bunch will inevitably be bruised and lipolysis, which is the splitting of fat molecules by hydrolysis into glycerol and fatty acids through enzyme action, will proceed.

It is recognized, therefore, that the rapid extraction of fruit from the bunch* is an important part of the product extraction process and that provision must be made to arrest the deterioration of palm oil quality even before the fruit has been removed from the bunch. This is to some degree recognized in non-mechanical processes through the careful cutting up of the bunches into sections and the hastening of natural loosening by sprinkling heaps of these sections with water and covering them up with leaves. The value of even this crude expedient is shown by the fact that low f.f.a. oil can be produced, after hand picking, by non-mechanical extraction methods, or by mechanical methods not including equipment for dealing with bunches.

Apart from the overall retention of the fruit in the bunches, the palm oil is held in the cells of the fibrous mesocarp of the fruit and the kernel is firmly encased in the hard shell of the nut.

Palm oil is solid at ambient temperatures in a temperate climate. At tropical temperatures it is a fluid with certain fractions held in crystalline

---

* As some mill engineers use the word bunch to indicate that part of the bunch which is left over after stripping, it is necessary to define the terms used in this chapter as follows:

Bunch = the whole ripe bunch with loose fruit as cut from the palm.
Fruit = the palm fruit as detached from the bunch before or during stripping and sometimes containing a proportion of perianth segments ('the calyx leaves').
Bunch refuse = that part of the bunch which is discarded in the stripper.

form. On settling, there is a clear liquid section and a crystalline fluid base. The prerequisites for the release of palm oil from the fruit are a physical breakdown of the mesocarp sufficient to rupture the cells, and a temperature sufficient to aid in this rupturing and fully to homogenize the fat constituents.

For the satisfactory release of the kernels from the fruit the requirements are that the oil-bearing mesocarp shall be removed and the shells cracked without damage to the kernels.

Palm kernel oil is not usually extracted on the plantations, though occasionally mills contain presses designed for this purpose. The conditions for the release of palm kernel oil, which is liquid at tropical day temperatures, are different from those of palm oil, but similar to those of copra and hard oil-bearing seeds. Very small quantities are extracted in producing countries by primitive means, but the great bulk of the palm kernels produced are subjected to industrial processes. As the matrix is solid and hard it must be crushed to a meal before the oil can be extracted under pressure. Though hammer or attrition mills can be used, roller mills are usually employed to grind the material so fine that a high proportion of the oil-containing cells are ruptured. 'Cooking' in stack cookers under steam pressure releases the oil still further and expression of the oil is then undertaken in various types of hydraulic batch presses or in continuous screw-press expellers,[1] or the oil may be extracted by solvents.

Oil palm products, their formation and characteristics

The three commercial products of oil palm fruit are palm oil, palm kernel oil and palm kernel cake. Chemically, the word 'fat' is coming to be used to cover vegetable oils and fats whether they are in the solid or liquid state, though in normal parlance the word oil is applied to a fat when it is in the liquid state.

Fats have been defined as the esters of fatty acids with the trihydric alcohol glycerol, and they must be distinguished from other simple lipids like the waxes which are esters of fatty acids with high molecular weight, straight chain alcohols. The triglyceride fats, which predominate in plant and animal fats, have the following general formula:

$$
\begin{array}{l}
\overset{\displaystyle H}{|} \quad \overset{\displaystyle O}{\|} \\
HC\!-\!O\!-\!C\!-\!R1 \\[2mm]
\quad\quad\ \overset{\displaystyle O}{\|} \\
HC\!-\!O\!-\!C\!-\!R2 \\[2mm]
\quad\quad\ \overset{\displaystyle O}{\|} \\
HC\!-\!O\!-\!C\!-\!R3 \\[2mm]
\overset{\displaystyle |}{H}
\end{array}
$$

where R1, R2 and R3 represent the hydrocarbon chains of fatty acid radicals.

Fatty acids may also combine with glycerol to form mono- or di-glycerides when only one or two of the hydroxyl groups of the glycerol will be in fatty acid combination, but these occur naturally only in fats which have become partly hydrolysed. When more than one fatty acid radical is involved, as in tri- or diglycerides, these may be alike or different. There being many different naturally occurring fatty acids, there will therefore be a multiplicity of fats formed from them.[2, 3, 4]

Naturally occurring vegetable fats are mixtures of fats and their characters are taken largely from the fatty acids which predominate in them and from the arrangement of these fatty acids in the triglycerides. The fatty acids are hydrocarbon chains in which two hydrogen atoms are attached to all or the majority of carbon atoms within the chain. The carbon atom at one end of the chain has three hydrogen atoms attached to it and the one at the other end is attached to a carboxyl group to give the general structure:

$$
\begin{array}{cccccc}
\text{H} & \text{H} & \text{H} & & \text{H} & \text{O} \\
| & | & | & & | & \| \\
\text{HC} & \text{--C} & \text{--C} & \cdots & \text{C} & \text{--C--O--H} \\
| & | & | & & | & \\
\text{H} & \text{H} & \text{H} & & \text{H} &
\end{array}
$$

Fatty acids of this general formula are saturated fatty acids as they have the full number of hydrogen atoms attached to the carbon atoms of the chain. In unsaturated fatty acids there are one, two or three double bonds between carbon atoms which then have only single hydrogen atoms attached to them, in the following manner:

$$
\begin{array}{cccc}
\text{H} & \text{H} & \text{H} & \text{H} \\
| & | & | & | \\
\text{--C} & \text{--C} & \text{=C} & \text{--C--} \\
| & & & | \\
\text{H} & & & \text{H}
\end{array}
$$

Double bonds can occupy different positions in the chain, thus giving rise to different isomers. Furthermore, there are also geometric isomers (the *cis-* or the *trans-* forms) according to whether portions of a molecule joined by a double bond extend in the same or opposite directions.

Chemically, the most important reaction of the fats from the producer's point of view is *hydrolysis*, i.e. the formation of free glycerol and free fatty acid through a splitting of the fat molecule and the addition of the elements of water which may be partly represented by the equation

$$CH_2OOCR + H_2O = CH_2OH + HOOCR$$

where CH_2 represents one-third part of the glycerol radical and R the hydrocarbon chain of the fatty acid radical. Hydrolysis can be either auto-catalytic in the presence of water, be catalysed by metals, or be brought about by the action of the enzyme lipase. The latter is, of course, the fat-splitting enzyme of animal digestion, but it is also found in palm fruit and in fungi and other organisms which gain access to fats. One of the most important tasks of the palm oil producer is to prevent hydrolysis by reducing to a minimum the amount of water and impurities present in the oil and by the desctruction of the enzyme. This is discussed in detail on p. 704. Hydrolysis is alkalis is distinguished as *saponification* and gives rise to soaps and glycerol.

The second important reaction of fats is *oxidation*. Unsaturated fats are commonly oxidized at the double bonds and the oxidation products, the first of which are hydroperoxides, lead to rancidity with the loss of palatibility due to obnoxious flavours and odours, and may effect the bleachability of the oil, a subject which is discussed on p. 708. In oil production the substances most likely to promote oxidation (pro-oxidents) are free atmospheric oxygen and traces of metals; the process is accelerated by light. Oxidation and consequent rancidity does not, however, proceed so fast in vegetable as in animal fats owing to the presence of naturally occurring protective materials or anti-oxidants. Oxidation to hydroperoxides is measured as the *peroxide value* of a fat; this represents the reactive oxygen content, and is estimated through the liberation of iodine from potassium iodide in glacial acetic acid and recorded in terms of milliequivalents of peroxide-oxygen per 100 g fat.

The third important reaction of the fats is *hydrogenation*, which is a process of manufacture and therefore does not directly concern the subjects treated in this book. However, it is generally acknowledged that hydrogenation or 'hardening' of fats has contributed more to the interchangeability of fats and fatty oils than any other process and therefore is a factor in the maintenance of stable economic conditions in the production of all fats. Broadly, hydrogenation processes add hydrogen atoms at the double bonds of unsaturated fats converting these into the higher melting point saturated fats. Thus the oleic ester triolein (the fatty acid of which is a primary constituent of palm oil and is present in palm kernel oil) can be directly hydrogenated to the stearic ester tristearin, the respective melting points being $-32°$ to $5.5°C$ and $54°$ to $73°C$. Fatty substances which are liquid at ordinary temperatures can therefore be made suitable for products which must be sold in the solid state. Apart from this main feature of hydrogenation processes, selective hydrogenation now makes possible the production of products containing almost any desired range of fats. These processes are particularly applied to marine oils and to subsidiary parts of liquid oils and palm oil.[4]

Fourthly, mention may be made of the reaction which is used as a measure of the proportion of unsaturated constituents present in a fat. This is *halogen addition* to the double bonds of the unsaturated fatty acids and the quantity of halogen taken up is expressed in terms of iodine as the *iodine value*, which is the number of grams of iodine absorbed per 100 g fat.

Lastly, with much larger quantities of palm oil coming on the market and fears, so far unfounded, that there might be difficulties in disposing of these quantities, there has been an increasing interest not only in providing a standard product of very high quality but also in the possibilities of improving overall income through *fractionation* of the oil in the countries of production.[5] The liquid fraction would then be marketed as cooking oil for local consumption while the solid fraction would be sold as a component of frying fats, margarine and other products. Various fractionation methods have been compared, and with solvent fractionation a production of 75 to 85 per cent liquid oil and 15 to 25 per cent solid fat is achieved.[6, 7] In a fractionation process by transesterification a liquid fraction can be produced with 59 per cent tri-unsaturated and 34 per cent mono- or di-unsaturated glycerides giving a composition not far removed from those of soya bean or corn oils. This oil has a high stability and can be used for superior quality salad oil. It has an iodine value of 80; the residual solid fraction has an iodine value of 5 and a melting point of 62°−63°C.[7]

Much use has been made of the physical properties of fats, particularly optical properties, in identification of constituents. Solid fats were found to have several melting points. This was shown to be due to polymorphism, i.e. the triglycerides can crystallize to give different structural forms which in turn have different melting points as well as differing in other physical properties. The solid forms all revert to the same liquid when heated to the melting point of the highest melting form. Vegetable fats such as palm oil therefore melt gradually over a range of temperatures not only because they contain a mixture of fats, but also because any solid fat of the same chemical composition may have several crystalline forms or *cis-* and *trans-* isomers.

It is outside the scope of this book to describe or discuss in detail the modern techniques used for the identification and estimation of individual fats and fatty acids, their isomers and their geometric isomers. The methods now in use rely mainly on spectroscopy, thin-layer and gas-liquid chromatography and X-ray techniques.

Fat formation in the ripening fruit

The crude physical changes accompanying ripening have been referred to in Chapters 2 and 10. Development of fats in the kernel precedes that of the mesocarp. At 8 weeks from pollination the content of the seed is liquid. At 10 weeks it becomes semi-gelatinous, and it does not become really hard until the fifteenth week. At 10 weeks from pollination the amount of fat is very small and the composition is different from that of mature kernels. It is believed that at this stage the oil is present only as basal protoplasmic fat. Unsaturated fatty acids preponderate and this is indicated by high iodine values (*c*. 85). As much as 67 per cent oleic and 14 per cent linoleic acids have been found at this stage with comparatively small quantities of saturated acids.

From this stage there is a slow accumulation of fats until about the twelfth to the thirteenth week when fat formation becomes more rapid; the fats laid down are largely saturated, principally lauric (C12) which rises to 46 to 50 per cent and myristic (C14) which reaches 18 to 20 per cent by the twentieth week. The biggest accumulation occurs around the fourteenth to sixteenth week. The fact that considerable quantities of oleic acid are also formed has led to the belief that the laying down of the highly saturated fats is a separate physiological development. There is no evidence of conversions of one fat to another or of free fatty acid accumulation prior to esterification.

Fat formation in the mesocarp takes place very late in fruit development. From the eighth to the sixteenth week after pollination fats constitute less than 2 per cent of the dry weight. There is in fact very little addition of any kind to the dry weight of the mesocarp from the eighth to the nineteenth week when, just prior to ripening, dry weight increases by 300 to 500 per cent and fats rather suddenly come to constitute 70 to 75 per cent of dry matter. During the long period of low oil content palmitic and linoleic acid esters predominate. Oleic esters are present only in very small quantities. During the final week of ripening all the fatty acids in combination increase in quantity and there is no evidence of conversions. Oleic acid, however, increases in greatest proportion and becomes second only to palmitic acid in quantity.

Some data from the work of Crombie are given in Table 14.1.[8,9] The results of this work confirmed and expanded those of Desassis[10] in Dahomey who found in Deli *dura* fruit an oil to wet mesocarp content of 1.4 per cent 40 days before maturity, i.e. 15 to 16 weeks from pollination; oil began to increase rapidly from about 30 days before maturity, replacing water weight for weight so that there was very little increase in

Table 14.1   *Changes in weight and composition of developing fruit of Palm No. 6—173 in Nigeria*

Weeks after pollination	Dry weight	Oil extracted*	Saturated acids						Unsaturated acids		
			C6 + C8	C10	C12	C14	C16	C18	Oleic	Lino-leic	Lino-lenic
	(g/nut)	(g/nut)	(%)	(%)	(%)	(%)	(%)	(%)	(%)	(%)	(%)
A. Kernel											
10	0.09	0.01	1.6		8.5	4.3	28.2	7.8	35.2	14.4	
11	0.12	0.03	6.4		9.2	4.3	20.0	4.6	48.0	7.5	
12	0.30	0.05	2.3	2.0	34.4	13.4	13.7	5.0	28.1	1.1	
14	0.54	0.20	3.3	0.3	48.4	14.2	7.6	2.9	18.7	4.6	
16	0.93	0.44	3.3	3.6	48.3	20.7	9.6	1.5	12.4	1.0	
19	1.11	0.51									
20	1.46	0.57	2.8	2.5	48.3	19.8	6.8	2.4	16.5	0.7	
B. Mesocarp	g/fruit	Per cent of dry weight					+C20				
8	0.33	1.3									
12	0.26	1.4				tr.	57.5	7.0	0.9	29.6	5.2
16	0.33	1.3				2.8	67.0	8.6	0.0	20.3	1.5
19	0.34	22.4				0.2	55.7	6.1	28.2	9.7	0.0
20	1.70	70.5				0.4	45.5	7.8	34.0	11.8	0.0

* A. In the kernel analysis, weights were on a per nut basis. Oil was estimated as light petroleum extract. B. In mesocarp, oil was estimated as liquid extract as per cent of dry weight.

total mesocarp or fruit size after about 100 to 110 days from pollination. Similar studies in Malaysia have confirmed these results with *dura* x *tenera* material, though the acceleration of oil production seemed to take place a little earlier.[11]

The composition of palm oil and palm kernel oil

1. Palm oil

Although palm oil has a high proportion of the saturated palmitic acid (C16) it also contains a high quantity of unsaturated fats, principally those derived from oleic acid. About three-quarters of the glycerides are mixed saturated and unsaturated triglycerides. The oil melts over a range of temperatures from 25°–50°C.

Eckey quoted the fatty acid composition of twenty-one samples of commercial palm oils, seven from plantations and fourteen from countries producing oil mainly from the groves.[3] The ranges and the means of the percentages of the constituent acids are given in Table 14.2 together with analyses from other sources. Loncin[12, 13] found traces of linolenic and lauric acids in his Zaire oils and stated that oil from grove palms had almost the same composition as plantation oil.

Table 14.2   *Fatty acid composition of palm oil* (E. guineensis) (per cent: means and ranges)

	Lauric C12	Myristic C14	Palmitic C16	Stearic C18	Palmitoleic C16:1	Oleic C18:1	Linoleic C18:2	Linolenic C18:3
Eckey[3]								
Range	–	0.6–5.9	32.3–45.1	2.2–6.4	0.8–1.4*	38.6–52.4	5.0–11.3	–
Sierra Leone, Liberia and Ivory Coast samples	–	1.8	34.8	5.4	–	50.4	7.4	–
Far East, Zaire, Nigeria, Cameroon and other samples	–	2.3	41.2	4.3	–	42.5	9.6	–
Loncin and Jacobsberg[13]								
Zaire plantations	0.2	1.2	43.7	6.2	–	36.5	11.8	–
Malaysia[14]								
Plantations†	–	0.6	49.9	2.8	–	40.6	6.2	–
Range	–	0.1–1.0	45.2–58.5	0.6–5.4	–	34.6–44.0	4.3–9.2	–

* In three samples only.
† Oil from four *dura* and four *tenera* bunches.

The generally higher palmitic percentages and lower oleic percentages found in Eckey's plantation samples were attributed by him to 'varietal differences'; this is untenable, however, as the Zaire plantations owed their origin to recent grove descendence while the Far Eastern plantations, with very similar compositions, were Deli plantings. Moreover, Nigerian samples were not significantly different in composition from Zaire or Far Eastern samples except that two Nigerian samples had high myristic contents (4.5 and 5.9 per cent). Sierra Leone, Liberia and the Ivory Coast had unusually high oleic and low palmitic contents, however, but whether this was due to a natural geographic variation or to some alteration in the oil during extraction or subsequent treatment it was not possible to say. Similarly, there is no basis for a belief that 'plantation oils' as such have different proportions of fully saturated triglycerides or unsaturated glycerides than extracted grove oils. Iodine values average 54.9.

However, Bienaymé and Servant[15] examined oil from twenty places, the majority in the African palm belt, but some in Malaysia and Sumatra. They confirmed, through iodine value determinations, that from the Ivory Coast westwards there was a higher proportion of unsaturated acids. At the same time they found that when plantations were planted with Deli material the iodine value was slightly lower than for Dahomey, Nigeria and Zaire.

With the considerable range in fatty acid composition it is obvious that progenies will differ significantly. Some significant differences were found in Malaysia between fruit forms and types, but as the ranges of composition overlapped the differences were probably ones of progeny within the forms and types.[14, 105]

Loncin and Jacobsberg[12] showed that in Zaire the triglycerides were to be found in the following proportions:

Triglycerides S-S-S about 6 per cent
Triglycerides S-S-U about 48 per cent   where S = saturated,
Triglycerides S-U-U about 43 per cent   U = unsaturated fatty acids.
Triglycerides U-U-U about 3 per cent

The low quantity of fully unsaturated triglycerides of high melting point is said to make the oil particularly suitable for margarine manufacture.

Thus palm oil is firstly a fat containing a very high proportion of palmitic acid to which may be attributed its value in soap-making. Secondly, the high quantities of oleic and linoleic acids give the fat a much higher unsaturated acid content than that of coconut or palm kernel oils which are essentially lauric oils giving a hard soap with greater lather.

The most important minor constituents of palm oil are the carotenoids; among these the carotenes are so conspicuous that the fat was termed 'red palm oil' in the Far East. Both the total carotenoid content of the oil and the proportions of the constituents vary. The carotenoids includes α, β, γ and ζ carotene together with lycopene and lutein (xanthophyll). Vitamin A is derived from the carotenes, β-carotene having twice the 'vitamin A activity' of either α-carotene or γ-carotene. The carotenes, which are among the most highly pigmented substances in nature, are hydrocarbons with long chains of conjugated double bonds; they have the formula $C_{40}H_{56}$. When the carotene molecule is split with the addition of the elements of water to both halves, vitamin A is produced. In β-carotene the two halves are identical, but in the other carotenes they are not, only one-half being of the constitution necessary for vitamin A formation.[16]

The carotenoids are easily separated chromatographically as they are absorbed from solution in light petroleum by several powders, e.g. aluminium oxide. In chromatographic columns the different affinities are made use of, lycopene being absorbed preferentially to the carotenes and therefore appearing at the top of the column. β-carotene appears above α-carotene.

Differences in fruit colour have already been noted (p. 65), and apart from the *albescens* type which contains very little carotene, there are big differences in the carotene content of the mesocarp. For instance, two fruit samples which were 'red' and 'orange' when ripe and gave almost identical quantities of oil to dried mesocarp had the following carotene contents:[17]

Fruit	Red	Orange
Carotene in dry mesocarp (mg/100 g):	207	89
Total carotene in oil (ppm):	2,560	1,100

The carotene content of the red sample is exceptionally high; oils taken direct from individual palms rarely contain more than 2,000 ppm. The carotene content of the orange sample is quite usual for palms in West Africa. Samples of Sumatran oil, however, were found to have carotene contents of between 320 and 475 ppm and commercial Malaysian oils are reported as having between 300 and 500 ppm.[18] There is thus evidence that, on average, Deli palms produce a lighter coloured oil than the general run of African palms. In the examination of progenies in Nigeria it has been shown that carotene content may vary widely from bunch to bunch.[19] In one progeny examined, thirty-eight out of fifty-one bunches had carotene contents in the low range 250–500 ppm, and ten were in the range 501–700. In another progeny, in which sixty-two bunches were examined, there was a much greater scatter, viz:

Carotene range (ppm)	250–500	501–600	601–700	701–800	801–900	901–1,000	over 1,000
Number of bunches	12	7	3	9	11	9	11

Bienaymé and Servant[15] have suggested that carotenoids are in greater concentration in oils from Togo, Dahomey and Guinea than elsewhere in Africa and that in the palms of Angola carotene contents are as low as in the Deli palm.

The proportions in which the different carotenoids are found are said to be variable. In the 'red' and 'orange' samples referred to above[17] the proportions were as follows:

	Red (%)	Orange (%)
α-carotene	28	34
β-carotene	54	50
Lycopene	<3	<3
Lutein and ζ carotene	<1	<1

The red fruit thus had a higher proportion of the higher 'vitamin A activity' β-carotene. The above examples are typical of many other analyses. The association of carotene and bleachability is discussed on p. 708

Of the other minor constituents of palm oil, the tocopherols are found in a number of forms. They are anti-oxidants and may be found in quantities as high as 800 ppm in well-prepared plantation oil, though in oil coming from the groves the quantity is usually around 500 ppm. Palm oil contains only very small quantities of phospholipids and sterols.

2. Palm kernel oil

Palm kernel oil resembles coconut oil with which it is readily interchangeable. Both fats have a preponderance of saturated fatty acids, but palm kernel oil has a lower quantity of the low molecular weight acids, caprylic and capric. The usual ranges of fatty acid constituents, per cent, are as follows:

	Saturated						*Unsaturated*	
	Caprylic *C8*	*Capric* *C10*	*Lauric* *C12*	*Myristic* *C14*	*Palmitic* *C16*	*Stearic* *C18*	*Oleic* *C18:1*	*Linoleic* *C18:2*
Range (%)	3–4	3–7	46–52	14–17	6–9	1–2.5	13–19	0.5–2
Typical sample (%)	3.0	6.0	50.0	16.0	6.5	1.0	16.5	1.0

Traces of the saturated caproic and arachidic acids, C6 and C20, and of the unsaturated palmitoleic and linolenic acids, C16:1 and C18:3, are also found.[20]

With such a high proportion of saturated acids it is not surprising that saturated triglycerides constitute over 60 per cent and monooleic disaturated triglycerides more than 25 per cent of the total glycerides. The iodine value is about 17 and the melting point 25°–30°C.

3. Oils of *Elaeis oleifera*

The mesocarp oil of the American oil palm, *Elaeis oleifera*, has an iodine value of 78 to 88, much higher than that of the oil of *E. guineensis*, showing that it contains higher percentages of oleic and linoleic acids and lower percentages of palmitic and other saturated acids.[3] Similarly, the kernel oil of *Elaeis oleifera* has an iodine value of 25 to 32, indicating that saturated acids do not predominate to such an extent as in palm kernel oil. In these respects the American oil palm is similar to quite a number of other American palms which can yield both pulp and kernel oils. For instance, *Jessenia batana*, the Patana or Seje palm of Brazil and Venezuela, has a pulp oil with 84 per cent of unsaturated fatty acids.

Table 14.3 shows analyses of the mesocarp and kernel oils of *E. oleifera* and of its hybrids with *E. guineensis*. The unsaturated fatty acid content of the mesocarp oil of the hybrid is intermediate between those of the

Table 14.3 Fatty acid composition of the oils of Elaeis oleifera and of the hybrid E. guineensis × E. oleifera (per cent)

	Caproic C6	Caprylic C8	Capric C10	Lauric C12	Myristic C14	Palmitic C16	Stearic C18	Palmitoleic C16:1	Oleic C18:1	Linoleic C18:2	Linolenic C18:3	Iodine value
Mesocarp oil												
E. oleifera												
(1)*				0.1	0.3	25.0	1.2	1.4	68.6	2.1	0.9	81.3
(2)				0.5	0.2	22.5	0.6	2.4	55.4	18.0	0.4	81.8
(3)				—	tr	22.9	1.0	1.3	54.8	20.0	—	81.5
(4)				tr	1.2	24.6	0.6	1.0	56.6	15.9	0.1	—
Hybrid, E. oleifera × E. guineensis												
(1H)*					0.5	31.7	4.1	0.1	49.5	13.4	0.5	70.7
(2H)*				0.1	0.9	32.5	3.4	0.2	48.0	13.8	0.4	62.0
(3H)				0.1	0.8	27.3	6.1	0.5	52.5	11.4	1.3	69.8
(4H)				tr	0.6	33.4	3.4	—	51.8	10.9	—	63.0
(5H)				—	0.4	35.8	1.4	—	55.0	7.5	—	63.0
(6H)				—	0.7	30.8	3.6	—	51.8	13.1	—	67.0
(7H)				tr	0.2	31.5	3.6	0.3	55.1	9.2	0.1	—
Kernel oil												
E. oleifera												
(1K)	2.9	0.9	0.8	29.4	23.8	9.9	2.1	—	25.1	5.1	—	30.3
(2K)	0.3	0.4	0.5	25.1	27.8	12.7	2.2	—	24.9	6.1	—	31.9
Hybrid, E. oleifera × E. guineensis												
(1HK)	tr	2.8	3.0	48.9	17.7	8.2	1.7	—	16.0	1.6	—	16.6
(2HK)	tr	1.8	2.1	43.1	21.3	9.4	2.2	—	16.3	3.7	—	20.4

* Traces also of Arachidic acid, C20.

Sources: Colombia — (1), (3H)[21]; (3), (6H), (2K), (2HK)[22]; Nigeria — (2), (4H), (1K), (1HK)[22]; Zaire — (1H)[21]; Malaysia — (2H)[21], (5H)[22]; France (source of material not stated) — (4), (7H)[23]. (4H), (5H), (1HK) and (2HK) are means of determinations of *E. oleifera* × *E. guineensis*, *dura*, *tenera* and *pisifera*.

parent species. It has been postulated that if the hybrid is back-crossed with either of the parents then half the progeny will have hybrid-type fatty acid analysis and half will have the analysis of the parent; support for this was obtained from two back-cross analyses on Malaysian material which showed both a hybrid-type and an *E. guineensis*-type analysis.[22] There is, however, considerable overlapping in analyses and it will be noted that Eckey's data in Table 14.2 for *E. guineensis* oils from the western part of West Africa are not unlike those of some hybrid oils.

The fatty acid content of the hybrid kernel oil approximates more to that of the *E. guineensis* parent.

4. Palm kernel cake

Palm kernel cake contains around 18 to 19 per cent of protein and is thus the lowest of the oil cakes in protein value. The amino acid composition has been given by Babatunde *et al.* and shows high proportions of arginine and glutamic acid.[106] The cake or meal is more suitable for use in the rations of ruminants than of pigs; for the latter its gritty and rather fibrous nature, together with its low level of lysine, makes it less effective than other available protein sources.[106] An average analysis is as follows:

	Carbohydrates	Oil	Proteins	Fibre	Ash	Water
Per cent	48	5	19	13	4	11

The protein content of *E. oleifera* and hybrid meals is lower than that of *E. guineensis* though the amino acid composition is similar.[107]

Producing countries with developing livestock industries are increasing their palm kernel processing plants and much larger quantities of cake and meal are now available in African and Asia.

The quality of palm oil and kernels

The processing of oil palm bunches has as its object not only the extraction of the maximum quantity of the products, but also the production of palm oil and kernels of high quality. Before considering extraction methods, therefore, it is important to know the meaning of high quality and what factors will promote or prevent its attainment. User requirements have been fully discussed by Pritchard.[24]

1. Palm Oil

A poor quality palm oil may be defined as having any or all of the following characteristics:

1. A high free fatty acid (f.f.a.) content.
2. Contamination with water and other impurities.
3. Poor bleachability.

These characteristics are not entirely independent, high water and dirt contents being contributory causes of high f.f.a.; factors causing high f.f.a. may also contribute to poor bleachability. It should also be noted that, for local consumption in Africa, oil of a low f.f.a. content is not always desired. Dishes made with oil of 10 to 20 per cent f.f.a. are often preferred.

The formation of free fatty acid

The fatty acids of palm oil may be formed by autocatalytic action, by the action of the lipolytic enzyme lipase from the palm fruit, or by microbial lipases. In practice the main 'set-up' of f.f.a. is caused by the action of lipase before processing.

Palm fruit contain a very active lipase which can only effect the breakdown of the fats into fatty acids and glycerol when the cellular structure of the fruit has been disturbed. The lipase is active even when the temperature has been reduced below 15°C. It is believed that the fat is protected from the lipase in the fruit by the membranes of the vacuoles, and that these may be ruptured either mechanically or by low temperature.[13] The enzyme is completely inactivated by high temperature.

Oil from fresh ripe fruit contains as little as 0.1 per cent fatty acid (estimated as palmitic acid), but in bruised and crushed fruit the f.f.a. may increase to 50 per cent in a few hours. Fickendey, who was the first to draw attention to enzyme lipolysis, showed that oil from mesocarp peeled from the fruit attained a free fatty acid content of 67 per cent after 24 hours, but that if the fruit was heated to 90°–100°C before depulping, little rise in f.f.a. took place. Later is was shown that a temperature of 55°C was sufficient to prevent enzyme action, that with crushed mesocarp the speed of fatty acid formation was remarkably fast (Fig. 14.1), and that partial bruising of fruit prior to sterilization may be followed by only a small overall rise, to around 5 per cent, in the f.f.a. of the oil of the fruit.[25] In all this work the fat splitting process, as indicated by the f.f.a. content, reached its peak in 30 to 60 minutes. With careful harvesting and carriage, the bruising of fruit need not be very extensive and these results show how it is that on well-regulated estates low f.f.a. oil can often be obtained from bunches kept overnight.

Fruit which has been kept for several days before processing, or which has been allowed to become over-ripe on the palms, may be covered and invaded by a number of moulds. Usually these fungi invade the base of detached fruit or wounds on the fruit surface. Many years ago an *Oospora* was claimed to have been responsible in the Congo for causing lipolysis but more recently lipolytic species of *Rhizopus, Aspergillus, Penicillium, Trichoderma, Circinella, Cunninghamella, Fusarium* and *Phoma* have been identified in Nigeria.[26] Under ordinary plantation practice, however, this cause of f.f.a. rise is likely to be insignificant in comparison with lipase activity originating in the fruit itself.

In Malaysia Turner examined the microbial flora which causes lipolysis prior to processing.[27] He drew attention to the invasion of fruit by

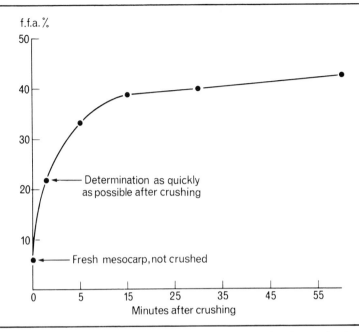

Fig. 14.1   Free fatty acid set up in mesocarp following crushing. (Determinations by Van Heurn quoted by Desassis.)

Marasmius palmivorus and species of *Sclerotium, Diplodia,* and *Glomerella* before the bunch is removed from the palm, and to invasion of loose fruit by seventeen different fungal species, fourteen of which showed lipolytic activity, together with yeasts, bacteria and nematodes. Of the fungi, *Aspergillus* species were particularly abundant. However, oil degradation through these agents was only thought to be very slight on plantations with good harvesting standards.

Turner has also examined the possibilities of microbial degradation of oil during processing and has shown that lipolytic fungi can be found in all parts of a mill both in the atmosphere and on the floors, walls and apparatus.[27] Sterilized fruit if left for more than 24 hours before further processing is invaded by micro-organisms, particularly *Neurospora sitophila* which grows rapidly over both bunches and fruit. However, temperatures are normally so high at all stages of processing that the potentially lipolytic organisms have little chance of significant activity except when a forced close-down of a mill takes place.

From the above it will be clear that for the *production* of low f.f.a. oil the major requirements are: (i) minimal bruising of the fruit during harvesting, carriage and movement at the mill side and (ii) minimal time between harvesting and sterilization. To these may be added (iii) the processing system must be such that the fruit or extracted oil does not cool down and come into contact with apparatus or materials which could cause a recommencement of lipolysis.

While enzyme action is the major cause of rises in f.f.a. between harvesting and processing, the work of Loncin has shown that, after destruction of the enzymes, oil in *storage* can deteriorate through auto-catalytic hydrolysis.[12, 28] The fatty acids already present in small quantity act as catalysts in the reaction between the triglycerids and water according to the formula

$$kt = 2.3 \log \frac{A}{A_0}, \text{ or } \log A = \log \left(A_0 + \frac{kt}{2.3} \right)$$

where k = the velocity coefficient, t = the acidification time, A_0 = the initial f.f.a. content, A = the f.f.a. content after time t. If t is given in 10-day units, $k = 0.12$ at 60°C and is approximately doubled for every 10°C temperature increase.

The reaction velocity is independent of the water concentration so long as enough water is available to saturate the oil. Below the water saturation level water concentration has to be taken into account, but in practice hydrolysis stops almost completely when the moisture content is kept below 0.1 per cent; increases in f.f.a. between purification centres and destination should therefore be negligible. Figure 14.2 is Loncin and Jacobsberg's representation of spontaneous hydrolysis at 70°C, yielding diglycerides and monoglycerides and glycerol.

There appear, however, to be two modifying factors to Loncin's formula. Firstly, some experiments have shown that at high temperatures the water concentration has an effect on hydrolysis, i.e. it increases the rate of f.f.a. formation. Vanneck and Loncin[29] obtained the following figures:

Time and temperature	f.f.a. percentage (as palmitic acid)	
	Oil with 0.25% water (%)	Same oil with addition of 5% water (%)
At beginning of experiment	2.8	2.8
After 62 days at 18°C	3.3	3.3
After 55 days at 55°C	3.5	4.1
After 36 days at 75°C	4.6	9.7

Secondly, there has been increasing evidence of the action of lipolytic micro-organisms in stored oil. Desassis showed that with oil stored in drums in Dahomey the f.f.a. rise was greater than could be expected from autocatalytic hydrolysis.[25] Loncin reported that *Geotrichium candidum*, possibly the same organism reported many years previously as a kind of *Oospora* causing acidification on palm fruit, has been found in many mills in Zaire to be responsible for lipolysis.[13] More recently, Coursey has

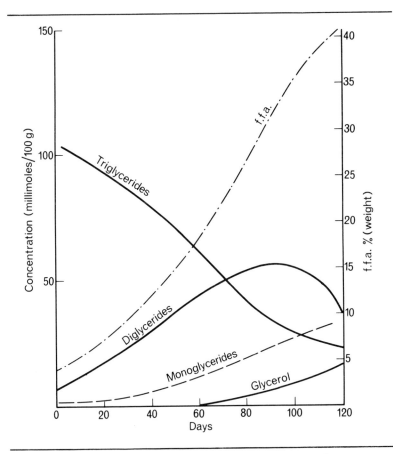

Fig. 14.2 Spontaneous hydrolysis of water-saturated palm oil at 70°C (Loncin and Jacobsberg, 1965).

clearly demonstrated that increased hydrolysis can be brought about by infection with lipolytic micro-organisms.[26] Examination of samples of oil in Nigeria showed that the more important identified lipolytic fungi were species of *Paecilomyces, Aspergillus, Rhizopus* and *Torula*. Inoculation experiments showed that *additional* rises of f.f.a. of the order of 1 to 3 per cent could be induced during 8 weeks' storage. The sources of infection are palm fruit, bunch refuse, oily films on drums and other receptacles. Rises in f.f.a. due to lipolytic fungi are likely to occur wherever oil is produced under generally dirty conditions and where the means of drum sterilization are inadequate. With drum storage, variations from the auto-catalytic reaction formula have been most common at the lower temperatures and it has been suggested that the biochemical factor is unlikely to be important where storage at high temperatures in bulk tanks is the rule.

It has further been suggested that the effect of impurities, including water, on the rates of hydrolysis may also be explained by supposing that they contain, or provide nutrients for, micro-organisms. When sterile conditions do not obtain, therefore, the rate of f.f.a. formation will become dependent on the amount of dirt and water present in the oil. The blending of oils containing lipolytic organisms is likely to increase further the f.f.a. production in storage.

It should now be clear that in order to provide an oil which will maintain a low f.f.a. content, the final milling operations must be directed primarily to attaining a water content of less than 0.1 per cent, since below this figure little autocatalytic hydrolysis is likely to take place. Secondly, the dirt content must be reduced to a minimum and clean and sterile conditions must be maintained in order to avoid the invasion of the oil by lipolytic micro-organisms.

The bleachability of palm oil

Although some of the materials for which palm oil is used, e.g. margarine, are coloured, in manufacturing practice oils must be bleached to definite specifications for the various uses to which they are put. They are blended following the refining process and the quantities of each oil used depend on availability and price. The composition, soft texture and plastic range of palm oil make it satisfactory for blending in fair quantities in margarine and shortening fats and for use in commercial baking and biscuit manufactures. But the oil must first be transformed into a bland, colourless, stable, edible product, and colours in the range 5.0 red to less than 1.0 red with the 5¼ inch Lovibond cell are needed.[30]

Any difficulty in bleaching palm oil and any additional expenditure involved must, therefore, militate against its use in manufacture; the bleachability of the shipped product has thus become an important factor particularly since, in the 1950s, Nigerian oils became in other respects suitable for use in edible fat manufacture.

Manufacturers found that the bleachability of oil received from plantations in the Far East was superior to oil received from the west coast of Africa, but later it became apparent that easily bleached oil could also be obtained from plantations in Zaire and other parts of Africa. It was at first thought that the carotene content of the oil would be the primary factor in bleachability, but later there was evidence that oxidation of the oil and of the carotenoids was of more importance than the absolute carotene content.

Oil from the Deli palm estates of Malaysia and Indonesia is generally found to have a fairly uniform carotene content of around 500 ppm. Wide variations exist in Africa, however, both in carotene contents and bleachability. One of the first surveys carried out of oils from fruit and from small scale and industrial extraction plants gave some remarkably diverse figures.[31] Certain fruits in the extreme west of Africa were found to contain over 3,000 ppm carotene while Deli palms growing in the same region gave, as expected, oil with only 400–600 ppm. Some African fruits, however, also gave low carotene contents and mills provided oils varying in carotene content from 600 to 1,600 ppm.

Of particular interest was the discovery that oil from the residual fibre was much higher in carotene content than oil as normally expressed, suggesting that the oil first squeezed out from digested fruit is unlikely to show the full carotene content of all the available oil. Subsequent work showed that hand-squeezed oil has a lower carotene content than press-extracted oil. It is thought that the exocarp fibres, which lie in that part of the fruit having the highest pigmentation, are less completely separated in the course of pounding or digestion than those of the remainder of the pulp (i.e. the true mesocarp).

In a survey of Nigerian oils expressed in a laboratory hand press from fruit samples in different producing areas, carotene contents from 400 to 1,800 ppm were obtained.[19] All samples with carotene contents below 1,200 ppm gave a low residual colour after bleaching with 5 per cent Fuller's earth at 105°C for 1 hour. Progeny differences in carotene content were also found to exist and it is clear that where less colour and superior bleachability is claimed in Africa for 'plantation' oils as against 'grove' oils extracted by the same processes, such differences are due to the chance selection for plantation use of grove material with a lower than average carotene value.

While, therefore, it is recognized that Deli palm estates produce oils of both low carotene content and good bleachability and certain Zaire estates produce oils with almost similar properties, it is not thought that, in general, carotene content is a major cause of poor bleachability, and investigations have been directed to handling and processing methods.[32]

Studies carried out at the Tropical Products Institute, London, suggested that poor bleachability of West African oil, which was largely produced by smallscale processing methods, might be due to (i) oxidation of lipoxidases in bruised fruit stored for various periods before processing, (ii) oxidation, catalysed by iron, during processing and bulking and (iii) mixing of oils of good and poor bleachability.[18] It was also suggested that high carotene oils are more likely to deteriorate than oils of low carotene content. More recent work has helped to determine the relative importance of these factors.

Fruit storage for a few days is quite common in West Africa but provided care is taken no rise in peroxide values or deterioration of the bleachability of the oil occurs.[19] If fermentation is allowed to take place, as in one of the traditional extraction processes, bleachability will be affected, since oxidation of the unsaturated fatty acids gives rise to compounds responsible for colour fixation in fats. However, these oils, which are also characterized by a high f.f.a. content, do not normally enter into international trade and fruit storage cannot therefore be considered as an important cause of poor bleachability.

No significant deterioration in bleachability occurs during the milling process in a well-regulated mill of standard design; oxidation can be reduced to a minimum by the avoidance of prolonged heating in the presence of air at any stage, by the careful regulation of the sterilization process and by the elimination of copper parts. However, in palm grove areas where crude or improperly regulated extraction methods are used, bleachability may be impaired even though the oil is relatively resistant to oxidation through its low poly-unsaturated fatty acid content and the presence of tocopherols.

The handling of oil in metallic containers at various stages of village processes is likely to be of some importance. Metals and metallic soaps formed from the fatty acids have pro-oxidation effects. Copper is a strong oxidation catalyst and zinc, tin and aluminium form metallic soaps. Vessels of these metals are commonly used.

Of similar or perhaps greater importance is the storage, transport and blending of crudely extracted palm oils in metal drums. Surface contact in this case is increased by rolling the drums during handling. Moreover drums in continual use, even though steamed out at bulk oil establishments, become encrusted with a chocolate-coloured, tarry, rancid oil. Eight samples of this oil gave a mean f.f.a. of 4.4 per cent, a mean peroxide value of 22.3 and a mean carotene content of 232 ppm, indicating considerable carotene breakdown. This drum residue was unbleachable.

If drums are not properly cleaned out and steam-sterilized before filling, considerable deterioration takes place. In one trial in Nigeria with oil drum-stored for 6 weeks, the residual colours after bleaching expressed as 10 (R + Y) Lovibond units, was 256 against 25 units in drums cleaned out, burnt and steamed. It was further shown that purifying the drums with an inert gas, CO_2, reduced the peroxide value and residual colour after bleaching of oil stored for 2 months. As neither steam-sterilization nor an inert gas are readily available in areas of smallscale production, the collection and marketing of an easily bleached oil in several African countries presents many difficulties. Furthermore, no easily-conducted bleachability test, on which price differentials could be based, exists.

The manner in which mill processes are operated has also been shown to effect bleachability.[33] In the sterilization process both failure to eliminate air and the application of too high a pressure (see p. 739) impair bleachability. Digestion with steam injection and clarification with centrifugal separation help to maintain the bleachability of the oil.

Quality standards

Palm oil is traditionally bought on a 5 per cent f.f.a. basis by importing countries, with penalties for exceeding this figure. To keep within these standards producers must achieve an oil of about 3½ per cent f.f.a. and most mills in the Far East and elsewhere have been satisfied with the production of oils with f.f.a. ranging from 2.5 to 4.0 per cent. Although palm oil is still bought on the basis of its f.f.a., water and dirt contents alone (with the exception of SPB oil from Zaire) consumers are becoming much more quality conscious of, and have been demanding, good bleaching qualities.

In Nigeria, from whence the bulk of the exported oil produced by small operators was previously derived, the internal requirement for edible oil purchased by marketing authorities has been 3½ per cent of f.f.a. for many years (SPO grade). Oil with higher f.f.a. contents are bought at substantial discounts and this policy, together with the production of relatively low f.f.a. oil by Pioneer mills, was responsible for the remarkable improvement in the quality of oil exported from Nigeria. Whereas in 1950 only 0.2 per cent of Nigerian exports had an f.f.a. content below 4½ per cent, by 1963 94.7 per cent of the oil had an f.f.a. content of 3½ per cent or less on leaving the producer.

Water and dirt levels have been variable. In Nigeria the small producer has been permitted to sell oil containing as much as 0.4 per cent water and, in the non-sterile conditions obtaining, it is believed that this has been partly responsible for later rises in f.f.a. and perhaps in reduced bleachability. The maximum dirt content has been 0.1 per cent which again compares unfavourably with normal mill performance. Most mills in the Far East, Zaire and elsewhere achieve moisture contents of 0.1 per cent or below, though occasionally samples with as much as 0.3 per cent are to be found. Dirt contents are often as low as 0.005 per cent. Results tend to reflect the efficiency of the mill and manager as much as the marketing requirements. Thus one mill of high efficiency in Nigeria was

	f.f.a. (%)	*Moisture* (%)	*Dirt* (%)
consistently producing oil of these qualities:	2.2	0.4	0.02
against a marketing requirement of:	3.5	0.4	0.1

Under efficient management the mill would inevitably produce oil of around 2 per cent f.f.a. and dirt contents well below requirement, but there was no incentive to reduce the moisture content below the legal limit.

Arnott[34] has given the results of some 1,500 analyses done by the Department of Agriculture in Malaysia between 1949 and 1962. Histograms obtained showed considerable variability in f.f.a. There were two peaks, at 2.6 per cent and 3.6 per cent, while the single peaks for moisture and dirt were 0.15 per cent and 0.01 per cent respectively. Quality was categorized as follows:

Factor	Very low	Low	Medium	High	Very high
f.f.a.	<2.0	2.0 to 2.7	2.8 to 3.7	3.8 to 5.0	>5.0
Moisture	<0.1	0.1 to 0.19	0.2 to 0.39	0.4 to 0.6	>0.6
Dirt	<0.005	0.005 to 0.01	0.011 to 0.025	0.026 to 0.05	>0.05

It was observed that most of the samples in the high and very high categories were obtained in the years of rehabilitation after the war (1949—54) and that thereafter almost all samples were in the medium and low categories.

Certain large producers have for very many years been marketing oil of very low f.f.a. content, sometimes as low as 1.5 per cent and usually around or under 2 per cent. Investigation of the chemistry and biochemistry of palm oil extraction led to the production of a special high quality oil in Zaire. The characteristics of this oil (SPB) in comparison with what may be termed 'ordinary' plantation oil are given below: [13, 33]

	SPB	Ordinary
f.f.a., as palmitic acid (%)	1—2	3—5
Moisture (%)	<0.1	>0.1
Dirt (%)	<0.002	0.01
Iron (ppm)	<10	>10
Copper (ppm)	<0.5	>0.5
Iodine value	53 ± 1.5	45—56
Carotene (ppm)	500	500—700
Tocopherol (ppm)	800	400—600
Bleachability (Bleaching Standard: Lovibond 5¼ in.)	2.0R, 20Y	3.5R, 35Y

While commending the above standards, there is an inherent danger in producing oil of around 1.5 per cent f.f.a. which should not go unnoticed. The handling of fruit before milling, and in particular the loose fruit, plays an important part in the f.f.a. 'set-up' achieved. Attention has been drawn in this chapter to the very rapid increase in oil content of the mesocarp during ripening, and the relation of harvesting criteria to oil yield and f.f.a. set up has been discussed in Chapter 10. If a very low f.f.a. oil is insisted upon, managers and harvesters may be forced into the cutting of underripe bunches instead of concentrating on the careful handling of fully-ripe bunches, and the loss of production may then be far greater than any advantage to be gained by a 1 per cent lower f.f.a. content. This tendency has, indeed, often been noted in the field.

Although users of palm oil have been increasingly demanding oil which is easily bleached, the measurement of bleachability and the devising of standard bleachability tests still presents problems. This is because the damage which palm oil may have suffered can be due to the formation of oxidation products or to a decrease in the natural anti-oxidants or to the presence of oxidation promoters.[35] No single test has been found satisfactory. Until recently the peroxide value was commonly determined, but a 'Totox value' (2 x peroxide value (milli-eq/kg) + 1 x anisidine value (E_{350})) has been used by Swedish importers.[36] A recent study[35] has shown a good correlation between various analytical oxidation figures and bleachability. Peroxide value alone gives insufficient information since for highly oxidized oils this value is not well correlated with the oxidation process.

UV 233 nm and UV 269 nm absorbance have also been tested. Various direct bleaching tests with active earth or a combination of active earth and heat bleaching have been put forward. The International Union for Pure and Applied Chemistry has organized parallel tests in a number of countries and are publishing a choice of methods with notes on their limitations.[37] The standards now (1976) being adopted have been summarized by Cornelius.[99]

2. Palm kernels, palm kernel oil and palm kernel cake

The quality of the eventual products palm kernel oil and cake depend primarily on the quality of the kernels. It is therefore important to know what qualities are required in the products and what kernel characteristics are needed to provide these qualities.

Palm kernel oil is required to be of a low f.f.a. content and a light yellow colour, easily bleached.

Palm kernel cake is required to be relatively light-coloured and its nutritive value, particularly in respect of the constituent amino acids of the proteins, must not be impaired.

It is not surprising, therefore, that palm kernels are themselves judged mainly on the f.f.a. content of the oil they contain, on the amount of external and internal discoloration, and on factors such as moisture content and presence or absence of mould which are likely to effect the eventual f.f.a. content and colour. An average sample of palm kernels in a producing country has a composition as follows:

	Oil	*Moisture*	*Protein*	*Extractable non-nitrogen*	*Cellulose*	*Ash*
Per cent	47–52	6–8	7.5–9.0	23–24	5	2

There appears to be some variation in the composition of the non-oily and non-protein solids. The portion given as 'extractable non-nitrogen' has been found to contain variable quantities of sucrose, reducing sugar and starch, but in some samples no starch has been detected though mannose has been present. As to their moisture content, kernels eventually come to an equilibrium according to the prevailing relative humidity.[38] This being, on average, in the region of at least 80 per cent in producing countries, the equilibrium moisture content is around 7 per cent, though in very wet periods this might rise to 8 or 9 per cent. It is believed that a safe moisture content for seed storage is 14 per cent of the non-oily portion of the seed, and it would therefore appear that the equilibrium moisture content is on the border line of what has been termed 'safe'. Therefore, until the moisture has been reduced to its equilibrium level kernels are liable to reactions or activities for which moisture is needed.

These reactions are similar to those of palm oil, namely autocatalytic hydrolysis and lipolysis by fat-splitting enzymes in the kernels and from lipolytic moulds. At ordinary tropical temperatures the more important of the latter are four species of *Aspergillus* and species of *Paecilomyces, Syncephalastrum* and *Penicillium* identified in Nigeria,[39] and a species of *Rhizopus* and a lypolytic yeast, *Petasospora rhodanensis*, found in Zaire.[40] In stacks of palm kernels which heat up to $50°-60°C$ a different fungal flora, which is thermophilic, appears. The most prominent species are a *Thermomyces, Chaetomium thermophile*, and a *Penicillium*, probably *P. duponti*. All are actively lipolytic and their optimum temperature for growth lies between $42°C$ and $52°C$. Only *C. thermophile* grows below $30°C$.[41] Turner has detected lipolytic micro-organisms at various stages in the milling process but particularly in hydrocyclones and clay baths during shell and kernel separation.[27] Disturbing features of infection with these organisms are their ready carriage by insects and the continued activity of their lipases even when the moisture content of the kernels has been brought well below the equilibrium value after several hours heating at $100°C$. Lypolysis may be reduced by sterilization and a reduction in the number of cracked and broken kernels; but, as will be seen later, overheating produces discoloration of the kernels. It is for these reasons that the highest quality kernels are to be obtained by skilled hand-cracking of nuts which, under village processes, have not been subjected to high temperatures for long periods. West African kernels produced by these means and properly dried and bagged are usually of superior colour to mill-produced kernels. It has been reported that oil from the testa of the kernel has an f.f.a. content three times as high as that of oil from the enclosed endosperm, indicating that access to the oil-bearing tissue is all-important and that the amount of breakage will inevitably effect quality.

For the provision of kernels giving a low f.f.a. oil, therefore, low moisture contents and low breakage are required. This has been well illustrated in Sumatra for whole and broken kernels of average and high moisture content and the results obtained are shown in Fig. 14.3.[42] Even though precautions may be taken to reduce f.f.a. production by proper drying and minimum breakage, there still remains, however, the problem of the heat-resistant enzymes already mentioned. In Zaire[13] kernels are found to have in their oil an f.f.a. content, as lauric acid, of 0.5 per cent after extraction from the nut and this rises to 1.0 to 1.5 per cent after drying. During storage the f.f.a. may rise to any point between 2 and 10 per cent and this has been shown to be due to the variable action of microbial lipases, undestroyed by drying or high temperature. Free fatty acid contents of 3 to 7 per cent or more are common in the oil of kernels arriving in importing countries, and it is not possible to prevent this except by preventing the development of micro-organisms before and during storage. This has been effected by successive injections of steam and cold air, thus sterilizing and drying the surfaces of the kernels before introducing them into the drier. By this method the f.f.a. of the oil in the kernels has been maintained below 1.5 per cent for 6 months of storage.

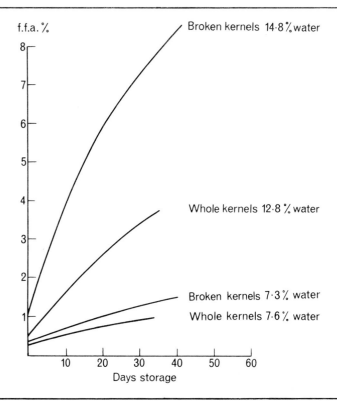

Fig. 14.3 Rises in f.f.a. of oil of old kernels of average and high moisture content stored at about 28°C (*Stork Palm Oil review*).

In Malaysia steam sterilization at temperatures exceeding 90°C for 5 to 6 minutes has been shown to inhibit rises in f.f.a. and to give rise to kernels which are stable during storage.[43, 44]

It has long been realized that the browning of the endosperm and darkening of the testa of palm kernels extracted in mills fitted with steam-pressure sterilizers was to some degree inevitable. Further discoloration in the digester is normally slight and in the nut bin and kernel drier negligible.[45] Kernels which show only moderate browning do not provide a more coloured oil than white kernels, but when severe browning has occurred bleaching is impaired or may even become impossible. Any browning effects the colour of the cake.

A correlation which was found in Malaysia between the percentage of kernels classified as 'off-colour plus mouldy' and the f.f.a. of the kernel oil was no doubt due mainly to the lipolytic activity of the moulds.[34] A somewhat similar relationship shown for kernels of differing colour and mouldiness arriving in the United Kingdom seemed to have a similar cause since the appreciable rises were nearly always associated with the presence of mould.[38] However, small f.f.a. rises are also obtained on browning kernels in the laboratory.

Heating, with consequent discoloration, can also take place in heaps of uncracked nuts[46] and in kernels stored under unsuitable conditions.

The chemical reasons for browning are not fully understood but are likely to be (i) the reaction of free amino groups of proteins with aldoses to give brown polymers and co-polymers, and (ii) the reaction of mannose (among the carbohydrates) with amino groups to form brown compounds.[38]

Some attention has been given by mill manufacturers to the colouring of kernels since this is so intimately bound up with the bunch sterilization process. In Indonesia with sterilization at 130°C for 2 hours, it was observed that by the end of the milling process the kernels were coloured and it was concluded that at 120°C the risk of discoloration was much smaller.[47]

The proportions of white, discoloured and broken kernels, and of shell, in samples of commercial palm kernels received in the United Kingdom, are shown in Table 14.4. These figures give a general idea of the extent of the browning problem.[38] The moisture content of the samples varied very little, being between 4.3 and 5.9 per cent. Oil contents lay between 47 and 54 per cent, leaving broken kernels out of account.

Table 14.4 *Colour of samples of palm kernels received in the United Kingdom* (percentage by weight)

Sample	White	Off-white	Brown	Brown and mouldy	Broken	Shell
Malaysia	62.6	12.7	11.7	4.7	7.3	1.0
Ghana	31.8	24.7	24.0	4.8	12.6	2.1
Sierra Leone	65.1	13.1	9.3	6.2	4.7	1.6
Nigeria (1)	69.4	11.1	4.9	4.9	6.7	3.0
(2)	54.7	22.0	8.3	7.1	5.5	2.4
(3)	62.2	13.0	9.9	5.5	6.8	2.6

(From J. A. Cornelius, Oil Palm Conference, London, 1965.[38])

Non-mechanical traditional methods of extraction of oil and kernels

The satisfactory extraction of palm oil requires specially designed machinery, whether hand or machine operated, and the provision of ancillary equipment of correctly calculated capacity for the prior preparation of the fruit and for the subsequent preparation of the products for sale.

Before the advent of machinery, oil was extracted in Africa by crude means to give a product of generally poor quality; and kernels were extracted from the nuts one by one by hand cracking. These processes still continue to be used in many parts of Africa, owing partly to lack of capital and partly to the unchanging organization of the rural communities. Though they vary considerably in their details, their essentials are those briefly described below.[48-50]

'Soft oil' production

In the great areas of Eastern Nigeria from which the bulk of the early export supplies were brought, the process employed gave rise to the so-called 'soft oil', thus named because the greater part of the oil was liquid at tropical temperatures. The harvested bunches are cut up into sections and kept in heaps for 2 to 4 days. The heaps are sprinkled with water and covered with leaves.

1. *Boiling.* The fruit is picked from the bunch sections and boiled in large pots for about 4 hours; 44-gallon petrol drums are commonly used. The fruit may be left in the vessel for up to 3 days.

2. *Pounding.* The boiled fruit is pounded in a wooden mortar with a wooden pestle until a mixture of nuts and crushed pulp of more or less even consistency is obtained.

3. *Separation.* The oil is separated from this mass of pulp by immersing the latter in water. The initial stage may be carried out either in a special pit with its sides coated with cement or mud or in some large vessel. The whole mass is stirred and first the crude oil which has risen to the surface is skimmed off into another vessel, then the fibre is sifted out of the water and finally the nuts, now largely free of fibre, are picked out and laid out to dry. Later they are hand-cracked. The crude oil thus obtained is boiled in smaller vessels where any fibre it contains sinks to the bottom. The now purer oil is again skimmed off and is then 'fried' in a shallow pot to get rid of the last traces of water. When this condition is reached, drops of water sprinkled on the oil will rapidly evaporate with a crackling noise – hence the use of the term 'frying'.

The amount of oil extracted depends largely on how far heat has been maintained throughout the process and how assiduous the women are in skimming and in teasing out the fibre. Usually the mass of pounded pulp is allowed to cool off and extraction rates are low. Such figures as exist for extraction rates in the soft oil process, i.e. 6 to 10 per cent oil to fruit of low mesocarp content, suggest that a normal efficiency (extracted oil to total oil in the fruit) would be 40 to 45 per cent, occasionally rising to 50 per cent, but sometimes falling to as low as 30 per cent. Average f.f.a. content used to be about 7 to 12 per cent, but lower f.f.a. oils can be produced by this method.

'Hard oil' production

In some parts of Africa the people are not accustomed to pounding, and a less efficient method, in which the fruit is trodden, is the rule. This is the typical method of the people of the Niger Delta. The sequence of events is:

1. *Fermentation.* After hand picking from the stored, chopped up bunches the fruit is placed in a pit or in a long wooden canoe and covered over with leaves. Fermentation caused by microbial and enzyme action takes place with the generation of heat and the fruit thus becomes softened.

2. *Treading and separation.* After some days the fruit is in a condition where it can be vigorously trodden in a canoe. After the first treading the oil is allowed to drain for 3 days from the lower end of the canoe. Water is then added and a second treading is done. When the work is considered to have proceeded far enough and the remaining separated oil has risen to the surface it is skimmed off and boiled in vessels for final preparation as in the soft oil process, though usually with less care. The oil thus prepared solidifies rapidly. Considerable hydrolysis and oxidation takes place during the process and the proportion of low melting point constituents is reduced. As would be expected, the extraction rate is very low, about 4 to 6 per cent oil to low mesocarp *dura* fruit, with a correspondingly low efficiency of 20 to 30 per cent; and the f.f.a. content, owing to the high initial set-up during fermentation, is usually between 30 and 50 per cent. The changes taking place in the fruit and oil during the process of fermentation in the canoe have not been fully studied.

The hard oil process continues in use because it makes low demands on labour and firewood and much carrying and boiling of water is avoided. In some areas it is claimed that there is still a taste for high f.f.a. oil in cooking.

Hand-operated presses

With an increasing demand for palm oil from the west coast of Africa it was clear even by the turn of the century that traditional methods of extraction could not satisfy either demand or quality requirements and that they gave a very meagre return to the producer. Consideration was therefore given to providing oil extraction methods which could be easily adopted by the small or large producer.

Before the First World War two hand-operated machines were designed in which the fruit was placed in cylinders with hot water and submitted to the action of beaters, the oil and water subsequently being run off through a sieve. The 'Gwira' machine was the better known and was given serious trial on the west coast, but never made much headway.[51] After the war attention was first given to improving traditional methods without introducing any machinery but, although some improvement in quality was obtained, extraction efficiency remained low.[52] Attempts were therefore made to introduce pressing into the soft oil process which already provided for fruit boiling and mashing in a mortar. The process was for some time known as the cooker-press system[53] because special steam or water cookers were introduced. These never became popular, and boiling in a 44-gallon drum became the general practice which has continued to this day.

'Depericarping' before pressing was at first thought to be necessary and various types of hand depericarpers were tried in combination with a 'Cully-Ducolson' hand press of the 'bridge' type which consisted of a cylindrical steel press-cage and vertical ram screwed down manually from the bridge. A hand operated centrifuge was also tried,[54] and fair numbers supplied. The system included a steampressure fruit sterilizer, a hand-operated digester and the hand-operated centrifuge itself.

The curb press

The press which was eventually found most satisfactory and captured the market was a curb type of press similar to those used for the extraction of juice from soft pulp fruit in the wine and cider industries. The first machines to be used were modified wine presses of the type developed by André Duchscher, founder of Duchscher & Cie of Luxembourg (later Usine de Wecker), but others were later produced by a number of firms and known by various trade names. Adapted for use as oil palm extractors they gave efficiencies of 55 to 65 per cent with a maximum efficiency of around 70 per cent which was achieved only under very favourable conditions of fruit digestion. These machines were tested both in Nigeria and Malaysia and in the former country they gradually came to be widely used in the eastern part of the country where soft-oil production had been the rule. In places where the people were averse to pounding, the use of the curb press was not so popular. No survey of the extent or performance of curb-press extraction has been carried out, but it is believed that there have been as many as 10,000 operating in eastern Nigeria.

The press consists of a screwed steel shaft, fixed in the centre of a base plate, and a cage, composed of strips of stout wood set vertically about 3 mm apart, and looped externally with two iron bands (Fig. 14.4). The cage, which is in two halves, can be opened and lifted off so that the pressed fibre and nut mixture can be easily removed. Pressure is applied by a ram which is worked downwards on the shaft by a crosshead turned manually by two long iron bars on which two to four men work. The bars, which are detachable, are fitted into the screw crosshead which lies above the ram. The base of the press is wooden and is surrounded by a metal trough fitted with one spout. The oil is squeezed out of the digested material between the wooden strips into the trough and runs through the spout into any collecting vessel. The press is often known as the cage or screw press. The first term is unobjectionable, but the latter may cause confusion with other types of screw press (Plate 72).

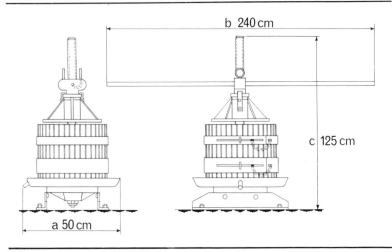

Fig. 14.4   The Duchscher curb press (model NG 1, Usine de Wecker).

In Africa, sterilization and pounding proceeds as described for the soft-oil process and inferior extraction rates are often attributable to the cooling of the pounded mixture before and during transfer to the press, as well as to variations in the length of the sterilization period and the efficiency of pounding. In a trial in Malaysia, Deli fruit was boiled in a 40-gallon 181 l) drum for 5 to 6 hours and immediately pounded with a wooden pestle in smaller drums for 10 to 15 minutes and transferred hot to the press.[55] A recovery of 72 per cent of the oil contained in the fruit was obtained by double pressing which involved removing the whole charge after the first pressing, separating the nuts and repressing the fibre from the first charges; this fibre had been kept warm and partially dried by placing it on iron sheets over a small fire. It was noted that with mixed fruit there was difficulty in obtaining fully mashed material from the pounding drums if small undeveloped fruit were present. It seems probable, therefore, that a better extraction rate can always be expected from good Deli fruit than from the mixed lots of fruit usually obtained from the African groves.

The Malaysian trials suggested that of the 72 per cent oil extracted 5 per cent was attributable to the second pressing, a maximum of 67 per cent efficiency being obtained from one pressing. In Africa double pressing is done by adding fresh material to a charge already pressed.

Presses of various sizes have been used, but an average charge of the Duchscher N.G.I. press is about 70 kg fruit, equivalent to about 115 kg of bunches.

The average rate of production of a curb press as used in Nigeria is not known since the system is usually operated on a cooperative or family basis, sometimes with employed labour, and production depends on availa-

Pl. 72   The Curb Press, extensively used in eastern Nigeria for palm oil extraction in the villages.

bility of fruit and many other factors. Trials in Malaysia suggested that seven charges could be handled in a day of 10 hours, by two men doing all the ancillary work.[55] The quantity of fruit treated would be around 450 kg, equivalent to 770 kg of bunches. It was pointed out that the press was not the limiting factor and that if more men were available at least twice the number of pressings could be undertaken, giving the equivalent of 1½ tons bunches per day.

Hand-pressing was adopted in Malaysia in the very early days of estate cultivation. The fruit was hand stripped from the bunches and sterilized by boiling for half an hour in water. It was then pressed in hand-screw presses consisting of perforated steel cylinders and a screw ram. Pounding of the 14 kg charge was undertaken in iron mortars *after* this pressing and the nuts were separated by hand. The oil-bearing fibre was then heated over boiling water and pressed, wrapped in sacking, in the same screw press, heating and pressing sometimes being repeated. Clarification was done, as in West Africa, by heating and skimming. An extraction efficiency of between 60 and 70 per cent was obtained and the quality was surpisingly good, moisture being 0.3 to 0.4 per cent and f.f.a. 3 to 4 per cent.[56]

Hydraulic hand presses

While the use of the curb press was a substantial advance on traditional methods of extraction and it was soon shown that curb-press operators were able to produce low-f.f.a. oil, the loss of extractable oil was still considerable and it was therefore not surprising that sooner or later a hand-operated press of much greater efficiency would come on the market. In 1959 the Dutch firm of Gebr. Stork began to manufacture, and to test in cooperation with the West African Institute for Oil Palm Research (now N.I.F.O.R.), an hydraulic hand press having a cage capacity of 45–55 kg fruit, equivalent to some 90 kg of bunches, and capable of a higher extraction efficiency than the curb press. The extraction of the crude oil from a single charge was rapid and the problems of introduction of this press, if it were to be used economically, were therefore many. Apart from modification of the prototype press itself, these problems were defined as follows:[57]

1. The determination of the rate of operation and efficiency of the press itself.
2. The design and construction of ancillary equipment for preparing the fruit for pressing and for clarifying the crude oil; the adjustment of this equipment to the rate of work of the press.
3. The organization of workers within the system and modifying as necessary; appraising the economics of the operations.

The hydraulic hand press.   This press is shown in Plate 73; its development has been described by Nwanze.[58] A ram, which is the cylinder of the hydraulic mechanism, moves downward into a perforated press cage when hydraulic fluid pressure is increased by the hand operation of a two-piston pump; both pistons are operated until considerable force is required and thereafter the small piston only is operated until full pressure is reached. On the release of pressure the press ram is withdrawn upwards by springs.[59]

The cages are filled and emptied on a table in front of the press ram and with skilled operation one pressing, including the insertion and withdrawal of charges, takes 6 to 10 minutes. As two cages are provided, and one can be emptied and filled with hot digested material during the pressing of the charge in the other, it is possible to complete six to ten pressings per hour and thus to press 270–450 kg fruit per hour, equivalent to 0.45–0.75 tons bunches per hour.

The cages are surrounded by covers or oil guards while pressing is proceeding. The crude oil, which is heavily laden with sludge, is channelled into a spout and runs off into buckets. As there is seepage of oil on to the press table this should be perforated with weep holes and be fitted underneath with an oil catch.[58] Nwanze showed that the efficiency of the press alone, which must not be confused with full process efficiency, can exceed 95 per cent. The efficiency of this machine led to the construction of hydraulic hand presses by a number of other manufacturers. In some cases pressure was exerted from below, but the mechanism was generally similar. Later, small motors were used by some operators to activate the press, and presses of this type with motor drive can now be obtained.

Pl. 73 The Stork hydraulic hand press.

A hand press of the type described is suitable for use by a small holder or an operator buying in fruit from groves or smallholdings; such an operator will not normally have much capital for expensive ancillary equipment. Nevertheless, it is desirable on economic grounds that the press should be kept in continuous operation at least throughout the day. Most early operators of hydraulic hand presses were surprised at its rate of operation and were unable to supply sufficient equipment and men to deal with the amount of fruit required by the press. This situation, as already mentioned, was not uncommon even with the curb press.

Suitable cheap ancillary equipment which can be made locally has been designed for use with hydraulic handpresses and this, with the quantities needed, is described below.[58]

1. *Sterilization.*   If loose fruit is being dealt with (some operators buy in loose fruit) boiling is best carried out in 44-gallon (200 l) *sterilizing drums* set in a special tipping frame (Fig. 14.5). If sufficient loose fruit is being supplied to keep the press in operation during the daytime then about four drums will be needed. Fruit boiling is carried out for 1 hour after steam

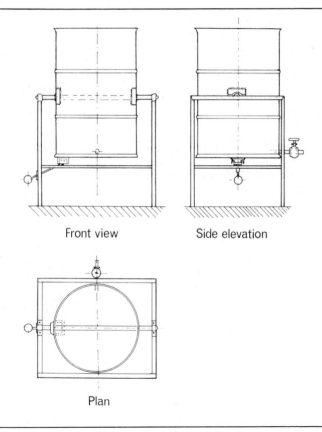

Front view   Side elevation

Plan

Fig. 14.5   Fruit sterilizer for use with the hydraulic hand press.

Pl. 74 Bunch sterilizers for the hand press mill.

appears at the top of the drum. While the drums can also be used for sterilizing cut up bunches, even with as many as seven of these it is difficult to supply sufficient fruit for the press, and fuel consumption (wood, fibre and shell) and labour usage is excessive.

Large *bunch sterilizers* for use with hand presses have been designed to hold about 1 ton of quartered bunches. These are locally constructed from cut-up oil drums welded together to form a cylindrical vessel 2.5 m tall by 1.1 m diameter and fitted below with two boiler compartments surrounding a fire space. Two or three steam pipes rise from the boiler compartments to distribute the heat (Plate 74).

2. *Stripping.* Stripping in a hand-operated mill can be by beating out the fruit from the sterilized bunches by means of wooden batons or pronged forks, or by the use of a hand-turned octagonal slatted drum. The latter has proved very efficient (Plate 75).

Pl. 75   Bunch stripper or thresher for the hand press mill.

3. *Digestion.* Digestion is by pounding stripped fruit but, in the hand press system, the fruit tends to get cold while waiting its turn for pounding. For this reason two *fruit reheating drums* of similar design to the bunch sterilizers are used for storing the stripped fruit until ready for pounding. The pounding is best done in a large concrete mortar of 90 cm diameter and 50 cm deep; six to eight men using wooden pestles pound the fruit until a 'mash' of uniform consistency is achieved (Plate 76).

The pounded material is then placed in special *pounded-fruit reheating drums* which are fed at the top, and from which the material can be continuously removed through a door near the bottom for transfer to the press. This drum has a water compartment with perforations for steam to rise into the drum. At least two preheating drums are needed.

4. *Clarification.* With occasional operation of the press, 44-gallon drums may be used for clarification. For continuous milling, however, two large drums measuring 90 x 100 cm are employed. These are fitted with a 44-gallon drum (200 l) as an inner compartment, and a funnel of at least 25 cm diameter attached to a tube of 5 cm or more which forms a feed pipe and extends downwards near the side of the large drum to within 7.5 cm of the bottom. The inner 44-gallon drum is supported in the centre of the large drum by bars or an attachment to the feed pipe. After the drum has been one-quarter filled with water which is brought to the boil, the oil-and-water mixture obtained from the press is poured through the funnel. The clean oil rises through the water and the dirt tends to fall to the bottom. As more oil or water is added the level rises and eventually the oil overflows into the inner drum and can be drawn off through a pipe

Pl. 76 Pounding sterilized fruit.

leading from the bottom of the inner drum to the exterior. Various arrangements may be made for linking the two drums together so that oil partly purified in the first drum may pass over into the second drum for final separation and clarification (Fig. 14.6 and Plate 77).

Oil exudations at various stages of the process, e.g. reheating, preheating, may be caught and transferred to the clarification drums and, conversely, drainings of heavy unclarified sludge from the first clarification drum may be used for wetting the pounded material in the preheating drum. The final extraction rate may thus be kept at a maximum.

An efficient system of hand milling requires a substantial opensided building for housing the equipment and giving space for the cutting up of the bunches.

The operation of this equipment and the number of vessels required for continuous operation have been extensively studied and described by Nwanze,[58, 60, 61] who showed that with one press, three large sterilizing drums, one stripper, two fruit preheaters, two fruit reheating drums and two large clarifiers, throughput could be raised to nearly 100 tons of bunches per month on double shift working (3.8 tons per working day) and over 130 tons if a three-shift system was introduced. The pressing rate was about 0.5 tons per hour *pressing time*, but when working with two shifts in weekly stretches (i.e. starting operations at the beginning of a

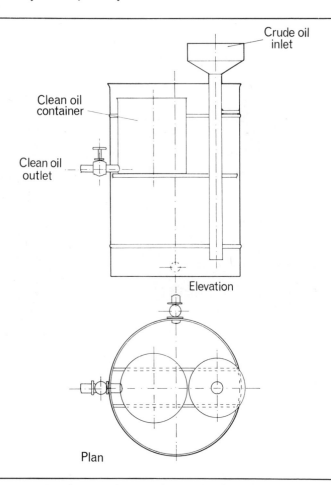

Fig. 14.6   Clarification drum for use with the hydraulic hand press.

week and closing down at the end) overall working time was still two or three times the pressing time though daily throughput could be gradually raised to 4 tons. The extraction efficiency of the whole process varied from 75 to 85 per cent; it is believed that this could be raised to 93 per cent if pounding were as efficient as conventional digestion.[60] Over a period of 7 months the following results were obtained:

No. of working days	Weight of bunches milled (tons)	Weight of bunches per milling day (tons)	Weight of oil produced (tons)	Extraction rate (%)	Extraction efficiency (%)
183	648.4	3.54	88.9	13.7	80.4

Pl. 77 Clarification drums for the hand press mill.

Theoretically the press could deal with 12—16 tons of bunches per 24 hours, but in a hand-press mill the practical operation time of the press itself is unlikely to exceed 8 or 10 hours per day, or a throughput of 4—7 tons of bunches. Only with continuous operation is this likely to be achieved or pressing time to approach overall working time.

Nwanze has demonstrated the considerable gains to be had from a change from the curb press to the hydraulic hand press and has shown how, in countries where the bulk of the rural population are not wage earners, the economics of hand-press operation cannot be satisfactorily assessed by reference to the high statutory or large-enterprise wage rates which may exist alongside the very much lower income level of the farmer.[58]

With one press and the equipment described it is clearly possible to achieve annual throughputs of well over 1,000 and perhaps nearly 2,000 tons of bunches. Thus, depending on locality, areas of 80 to 200 hectares might be served. Areas of this magnitude might be able to command the capital for small mechanized mills, and thus the hydraulic hand press is best suited to the grove areas or to groups of very small holdings agreeing to mill in cooperation.

In spite of this potential and its relatively low costs and high efficiency the hydraulic hand press system has not taken on in the West African grove areas. It has even been reported that the number of the curb-press operators has decreased while employers of primitive soft oil production methods have increased. The reasons why primitive methods have persisted in spite of the availability of relatively inexpensive and efficient apparatus has been the subject of some controversy, it being maintained on the one hand that small operators have only family, unpaid labour and a better choice of fruit,[62] and on the other hand that these factors are spurious or insufficient explanations and that much more social and economic research is required.[63]

Kernel extraction.   In Africa it has been common practice in hand-operated mills to give back the nuts to the women folk for hand-cracking. There are on the market, however, a number of nutcrackers driven by small petrol engines and these can with advantage be introduced into an otherwise hand-operated mill. Motors of around 3 hp are required to turn the cracker at between 2,000 and 2,500 rpm and about 0.75 ton of nuts may be cracked per hour.

Power-operated palm oil mills

The mechanical milling of oil palm fruit started in Africa before the First World War. The first mills were of very imperfect design and construction and were on a small and experimental scale. Progress was somewhat delayed by the belief that successful large scale extraction would depend on the invention of an efficient depulping or 'depericarping' machine for the removal of the mesocarp from the nuts prior to oil extraction. Much attention was therefore given to the design of depulpers. Though small-scale extraction was carried out in Sumatra and, later, in Malaysia as soon as estate planting began, oil was not produced from a large power-operated mill in Sumatra until 1919.*

The first power driven machinery for cracking nuts is said to have been introduced into West Africa in 1877 by Mr A. C. Moore of Liverpool, being devised by Mather and Platt Ltd of Salford.[51]

An incentive to the production of palm oil extracting machinery was given by the offer of a prize in 1901 by the German colonial Wirtschaft-lichen Committee; this was awarded to the firm of Haake of Berlin for a

---

* Palm oil extracting plants were termed mills in Africa from the beginning of their development. In the Far East they were and still are called factories, possibly because of the common use of the word for latex-processing plants on rubber plantations. The earlier and perhaps more accurate word is used in this book.

set of small machines including depericarpers. Progress was slow, but in 1909 Haake erected plant at Mamfe and Victoria in the West Cameroon. A French firm, Fournier, erected a small plant at Cotonou at about the same time and this firm was the first to develop the idea of pressing the fruit whole. Thereafter a number of rather crude bunch strippers was invented and the central problem of how to extract the oil from the pulp began to be tackled in earnest.

Two pressing methods were tried. In the first, the fruit was sterilized, after removal from the bunch, by steam-heating for 15 to 30 minutes and was then transferred to hydraulic presses. This was the principle of the Cotonou mill of French design. Two pressings were carried out, with heating in between, and the nuts were then separated from the fibre in a rotating drum and cracked in a centrifugal cracker. The process sounds thoroughly modern except for the omission of digestion. The design of the presses is attributed to Paulmier, and the nut cracker was by Poisson. The firms of Louis Labarre and Fournier of Marseilles constructed the plant.

Digestion was first undertaken in a stamping mill in a plant of German design in Togoland. The entire spikelets seem to have been cut off the bunches and subjected to digestion, so that fruit-loosening and pulp-crushing were simultaneous. The whole mass was treated in hydraulic presses, the nuts were separated in a rotating drum and cracked in a Haake centrifugal cracker.

Pre-war attempts by Hawkins to extract oil in rotating pans where the fruit was crushed and pressed by rollers were not further pursued.

The earliest machinery of Haake of Berlin followed the second method of press extraction and was designed to separate the mesocarp before pressing, but it is difficult to see why this process was persisted with for so long in view of the relatively early success of hydraulic pressing of sterilized fruit. Numerous firms, German, British and French, designed and tried out depericarping machines. The Haake depericarper consisted of a shaft carrying triangular metal blades rotating inside a cylinder which was itself rotating in the same direction but at a different speed. Other machines were supplied by the French companies already mentioned, by Olier & Co. of Argentuil, and by a number of British firms of which the most noteworthy were Culley Expressors Ltd, a firm much concerned in the early development of small processing, and Manlove, Alliott of Nottingham. No really satisfactory machine for continuous working was invented, however.[64]

During this period experiments were also being carried out by Lever Brothers in their Zaire and other concessions in Africa, but little information became available until after the First World War. Apart from the Lever Brothers mills the only pre-war mill of any size was that built at Maka in the Cameroons for the Syndikat für Oelpalmen Kultur by German firms. The mill consisted of six hydraulic presses for pressing before de-pulping, four depulpers and two presses for pressing after depulping. Stripping fruit from the bunches was done by hand. About 525 tons of grove bunches could be processed in a week with day and night work and the extraction rate was 18.5 per cent oil to fruit and 6.9 per cent kernel to fruit.[65] Krupp presses were used in the mill.

Centrifugal extraction was first used in Zaire in 1916[66] and was the main feature of the Lever Brothers mills. This innovation was taken up by Manlove, Alliott and Co who, through Nigerian Products Ltd, exhibited a complete mill of that type at the British Empire Exhibition at Wembley in 1924. Centrifuges proved very suited to the African grove fruit with its thin layer of mesocarp, but they also became the standard extracting equipment in Malaysia during the first 10 to 15 years of the industry's life in that country. The first Manlove mill was erected at Tennamaram Estate in 1925. In Sumatra the first mills followed the Maka pattern and the hydraulic press, developed by Konrad Loens, Krupp and later by Gebr. Stork of Amsterdam, underwent further improvement for palm oil extraction. The first large mill was erected at Pulu Radja Estate in 1921 and owed its design to Dr Fickendey and its construction and installation to Krupp. By the early 1930s the hydraulic press was beginning to displace the centrifuge in Malaysia. It should be realized that the centrifuge was an entirely new modern method of extracting oil from fruit or seeds and as such it was highly successful. Although the most modern mills now employ hydraulic or screw presses there are centrifuge mills still operating in Asia and America.

Milling began to take its present shape soon after the First World War. Depericarping fell out of favour and, as mentioned above, centrifuges and hydraulic presses became efficient oil extractors in combination with rotary-arm digestors. In Africa much of the early machinery in the inter-war period was installed by Lever Brothers, and their successor, Unilever Ltd, was responsible for the design of the small mechanical Pioneer mills which used centrifugal extractors (Plate 78) and were introduced into the palm grove areas of Nigeria during and after the Second World War. Later, this firm erected several very large mills of its own design both in Africa and in the Far East.

Oil mills are normally powered by steam engines and the boilers use fibre and shell in their furnaces. Transmission in small and medium sized mills was, from early days, by transmission shaft and belting, and this method is now being superseded by the generation of electricity for driving electric motors of suitable sizes.

The choice of a plantation mill

In their essential features oil mills are standardized in that they consist of sections, sometimes grouped as 'stations', for (i) *sterilization* of bunches, (ii) *stripping* of bunches, (iii) *digestion* and mashing of fruit, (iv) *extraction* of mesocarp oil, (v) *clarifying* the oil, (vi) *separation of fibre* from the nuts, (vii) *nut drying*, (viii) *nut grading and cracking*, (ix) *kernel separation* and discarding of the shell and (x) *kernel drying* and bagging.

Pl. 78 The centrifuge of a Pioneer mill.

The central machines of the mill are the extracting machines. It is the rate of operation of these machines, in terms of weight of bunches entering the mill, that determines the capacity of the mill. All other machinery is, or should be, adjusted to the rate of operation of the extracting machines. This to some extent simplifies the procedure of choosing a suitable mill, but there are many other considerations to be taken into account and these will be discussed below.

In the first place it is necessary or desirable to know the area to be planted, the rate of planting per annum and the expected yield at maturity. From this data estimates can be made of the probable crop for each year until the plantation is fully mature. If there are to be possible extensions, then it must be decided whether to allow for these extensions or to adjust the mill to the original area and to erect further mills for any extensions which may eventuate.

Secondly, it is necessary to estimate the probable distribution of the mature crop throughout the year. In countries with an even climate it is usually reckoned that in the peak month about 12.5 per cent of the total annual crop will be harvested. In certain countries, however, as much as 18 to 19 per cent of the crop may be harvested in 1 month. The mill's capacity must allow for this.

Thirdly, it must be known what form of fruit is to be produced. This will effect the extraction and kernel sections and will be discussed under these headings. The fuel supply will also be affected. In addition, it may be that the mill will lie in an area where it is profitable to include palm kernel oil extracting machinery and to sell kernel oil and cake in addition to palm oil.

Table 14.5   *Example of calculation of milling requirements*

Year of planting	1975	1976	1977	1978	1979	1980	Total				
Area planted (hectares)	400	400	800	800	800	800	4,000				
Year of production	1st	2nd	3rd	4th	5th	6th	7th	8th	9th	10th	11th
Estimated tons bunches/hectare	2.5	5	10	14	18	20	20	20	20	20	20
Estimated production (tons)	1978	1979	1980	1981	1982	1983	1984	1985	1986	1987	1988
1st year's planting	1,000	2,000	4,000	5,600	7,200	8,000	8,000	8,000	8,000	8,000	8,000
2nd year's planting		1,000	2,000	4,000	5,600	7,200	8,000	8,000	8,000	8,000	8,000
3rd year's planting			2,000	4,000	8,000	11,200	14,400	16,000	16,000	16,000	16,000
4th year's planting				2,000	4,000	8,000	11,200	14,400	16,000	16,000	16,000
5th year's planting					2,000	4,000	8,000	11,200	14,400	16,000	16,000
6th year's planting						2,000	4,000	8,000	11,200	14,400	16,000
Total production	1,000	3,000	8,000	15,600	26,800	40,400	53,600	65,600	73,600	78,400	80,000
Tonnage in peak month*	125	375	1,000	1,950	3,350	5,050	6,700	8,200	9,200	9,800	10,000
Capacity needed in peak month†	0.25	0.75	2.00	3.90	6.70	10.10	13.40	16.40	18.40	19.60	20.00

* 12½ per cent of annual crop.

† Capacity in tons hourly throughput, assuming 500 hours' operation in the peak month.

An example is given in Table 14.5 of the calculation of the capacities required for an area of 4,000 hectares expected to give an eventual yield of 20 tons of bunches per hectare. The milling capacity is based on the assumption that three shifts will be worked in the peak month with a total operation of 20 hours per day giving 500 hours per 25 working-day month. Obviously if conditions are such that three-shift work cannot be done then the calculations must be based on a lesser number of hours (including overtime) but this will lead to a larger capital outlay.

Usually, after calculations of this kind have been done, decisions have to be taken on (i) whether to have one mill or several smaller mills, (ii) the stages of instalment of the equipment, particularly the extracting equipment, so as gradually to expand the mill or mills to their maximum capacity, and (iii) the method of dealing with the small crop of the first few years.

There are various methods of dealing with the latter problem. In the first place harvest may be delayed and a larger first harvest obtained by the practice of ablation, i.e. the removal of young inflorescences during the first few months of flowering. In this way the need for a small plant may be obviated and early crops may be accommodated in the mill.

Secondly, an hydraulic hand press or presses may be used and these may be mechanized. This must, of course, be combined with the provision of sterilizing and digesting equipment and, if considered worth while, a simple nut-cracker. With a press capacity of over half a ton of bunches per hour it will be seen that four mechanized hand presses could deal with production in the Table 14.5 scheme for 2 to 3 years and that the heavy outlay of capital on the large mill could be delayed. Such a scheme is very suitable in areas which are expected to come slowly into production.

In some cases pilot plants within the mill building have been offered for processing the first few years' crop, it being assumed that the equipment will then be sold off or employed elsewhere. Unless the mill is to be very large and the rate of coming into production slow, this method is not nowadays considered a very attractive proposition. In general, attempts are now made to plant larger acreages per year so that full capacity may be reached as early as possible and the capital be fully employed.

Milling machinery can be supplied and installed in several stages. The stages will depend on the rate of plantation development, on the machinery chosen and on the number of lines of machinery to be installed. In the imaginary scheme in Table 14.5 an eventual peak throughput of around 20 tons per hour must be allowed for. Six or seven hand-operated hydraulic presses of 3 tons, four automatic presses of 4½–6 tons or two screw presses of 10 tons bunches per hour rating could therefore handle this crop. Thus the choice of stages of mill development will be bound up with the type and individual capacity of the units of extracting equipment chosen. When units of 3 tons capacity are being used, three stages might be adopted; sterilizers, strippers, digesters and much other equipment would then be provided in stages to serve the extracting units as they are supplied. It will be seen that the choice of the extracting unit is in more ways than one the most important exercise in the choice of mill.

In the development of very large areas for oil palms, e.g. 8,000 to 12,000 hectares, it is sometimes questionable whether it is not better to have two mills rather than one very large installation. With the single large mill there are, of course, all the usual economies of large scale operation : reduced overheads, lower capital outlay (though this economy varies in magnitude according to the size comparisons being made), lower labour costs, lower fuel requirements, etc., per unit of throughput. Against this must be set certain factors which, owing to the multiplicity of circumstances, cannot be dealt with in any great detail here. In the first place if the area is composed of discrete portions of land at a distance one from another the extra transport costs may outweight the advantages of a single central mill, and in that event mills of appropriate size must be erected in the centre of any large discrete areas. Secondly, the same considerations will apply if the area is very long and thin. Thirdly the water supply at any one point may be insufficient for a very large mill. Fourthly, if many years are to elapse between the first and last plantings and the details of full development are uncertain, then there is the possibility of technological advance in the interim period; advantage cannot be taken of this if one large mill of a certain design has already been constructed or is in course of development.

In Malaysia and other areas of large-scale development the installation

Pl. 79   A large mill serving about 20,000 acres of oil palms in Malaysia.

of single large mills (Plate 79) serving several plantations has been shown to be more economic than two or more smaller mills even where the plantations are as much as 60 km apart.[67] The determining factors in this case were the high standard of the existing road systems, the lower capital outlay per ton of throughput, lower staff and other running costs and the provision of sufficient storage space on a ramp at the mill side to ensure continuous factory operation (with the addition of field storage ramps if necessary). However, in recent years the cost of larger mills involving the installation of all the most modern equipment has shown an unprecedented rise and this has cast doubt on the value of large-scale operation in all circumstances. In South America, where road systems are often inferior and the financing of large mills more difficult, the installation of small, less sophisticated, locally-manufactured mills with only a few imported parts has become popular. These mills have hourly capacities of ¾–6 tons per hour.

The siting of a mill

To reduce transport costs to a minimum a mill should be in the centre of a plantation. Other considerations may prevent this however. These are:

1. The mill must be close to a suitable water supply sufficient for its needs. As a rough guide a mill of about 18–20 tons/hour capacity will require about 40,000 l or 40 m³ per hour. The exact water require-

ments will be specified by the manufacturers. The water, or at least the part required for boilers, must be of specified purity and if necessary a filtering and purifying plant must be installed.

2. The mill must be as near as possible to the point of despatch of the palm oil, particularly if stored in large tanks. The latter are with advantage placed on a slight rise.

3. The site must not be subject to flooding, must have sufficient space for overnight storage of bunches and for the necessary workshops and other subsidiary buildings.

4. It must be possible to discharge sludge into a fast flowing river or stream or to effect disposal in some other manner (see p. 755).

Certain of these factors may prevent the siting of the mill anywhere near the centre of the plantation. Several plantations in Malaysia obtain their water supplies from large rivers and ship their produce directly on to seagoing tankers. Under these circumstances the mill may have to be at one end of a long plantation. Economies of despatch compensate to some extent for higher internal transport costs.

The milling process

It would be beyond the scope of this book to describe the milling process in large factories in their full engineering detail. Although a comprehensive volume on the subject has yet to be written, there are several useful and detailed accounts of much of the equipment used, and these will be referred to where appropriate. It will be the aim here to describe the component processes briefly with a view to drawing attention to their main features and their effects on the products. Where alternative processes exist, their comparative value will be discussed and attention will be drawn to any deficiencies of knowledge in this respect.

Sterilization

Vertical bunch sterilizers. These are only suitable for small mills where the cages on rails for horizontal sterilizers cannot be afforded. They are usually of 2 or 3 tons capacity, but can be constructed to take up to 6 tons. In one type the bunches are forked at ground level into a bucket elevator which drops them into the sterilizer through a circular door at the top. After sterilization at pressure the bunches are discharged through a rectangular hinged door at the foot of the sterilizer. The discharge, though manual, is aided by the presence of a sloping perforated plate at the bottom of the sterilizer. Nevertheless, discharge may take 30 to 40 minutes.

An alternative type of vertical sterilizer containing an immersion basket involves less labour and time in charging and discharging. The bunches are tipped from lorries into the basket which is then lifted by monorail and

Pl. 80   Small cages, rails and weighing scales for short horizontal bunch sterilizers in a small locally-designed mill in Colombia.

lowered into the sterilizer which is closed by a specially locked lid. After sterilization, the basket is hoisted out of the sterilizer and moved to the platform above the stripper; there the load is discharged automatically on lowering.

Horizontal sterilizers.   Horizontal sterilizers were previously employed only in large mills, but in recent years many small mills in South America have been supplied with short horizontal sterilizers and cages on rails (Plate 80).

The immediate advantage of the horizontal sterilizer is that it does not require hand emptying and that the bunches are sterilized in cages in which they have been packed and are therefore not subject to further bruising before sterilization. It is also claimed that, largely owing to the way the bunches pack in a vertical sterilizer, there is less loss of oil in the condensed steam and in the bunch stalks coming from the horizontal type of sterilizer.[68]

Horizontal sterilizers are long cylinders placed at ground level into which cages of standard design can be pushed on rails through a door which then closes the whole of one end of the cylinder. Their capacity depends, therefore, mainly on their length. A large sterilizer would hold about 15 tons of bunches in six cages. Several of these would be required for mills of high capacity. A sterilizer with a 'bayonet-type' door is shown in Plate 81.

Pl. 81 Horizontal bunch sterilizers with bayonet-type doors.

Mention has already been made of the importance of sterilization in reducing the f.f.a. of the oil and the danger, during sterilization, of causing the kernels to become brown. One of the primary purposes of *bunch* sterilization, however, is to loosen the fruit on the bunch so that stripping is not difficult. 'Hard bunches' are often a problem, particularly in Africa, and much attention has been given to methods of injecting and 'blowing off' the steam and raising and lowering the pressure. The physical conceptions behind the trials carried out are beyond the scope of this book, but a few of the important considerations in practical sterilization may be mentioned:

1. The air has to be swept out of the sterilizer to attain the temperature value corresponding to a given pressure, and to avoid oxidation. This takes about 5 minutes.
2. Increasing the steam pressure above 2 kg/cm^2 (30 lb/in^2), which corresponds to a temperature of approximately 130°C, leads to kernel discoloration and may affect the colour of the oil.
3. To avoid zones of very high temperature due to the presence of superheated steam, particularly near the inlets, it is important to work up to full pressure slowly.

With regard to 1, it is common practice to introduce steam slowly from a high point and to allow the air to flow out from the bottom, continuing this procedure for a short time after steam is seen to be coming from the outlet. Venting systems have also been used. A single venting system entails bringing the pressure to 2 kg/cm^2 for 5 minutes and then venting off the steam and recharging to the same pressure. Double venting entails a second 'blow-off' after a further 5—10 minutes.[69] These systems employ a greater amount of steam.

Sterilization of bunches is carried out for 60—75 minutes of which 5 minutes is taken up by replacing the air and 15—20 minutes in working up to full pressure which is maintained for 40—55 minutes. The higher figure is usually employed as reduction of sterilizing time has been shown to increase the number of hard bunches. With horizontal sterilizers 5 minutes is occupied in blowing off and 10—12 minutes in discharging and reloading so that the whole cycle occupies about 96 minutes. If the mill is large enough to have four sterilizers some steam may be saved by a system of blowing over from one sterilizer to another. The system must be organized in such a way that there is no overall delay in the sterilizing process.

The required sterilizing time is affected by the size of bunches and the degree of ripeness. If a harvest consists solely of bunches from 4- or 5-year-old palms, then sterilization for half an hour may be quite sufficient to loosen the fruit. Similarly if the harvest consists of only very ripe bunches with a high proportion of loose fruit sterilizing time can be reduced.

Sterilization usually results in a loss of weight of about 10 per cent of the weight of *fruit*, due to evaporation. This dehydration is important, as, particularly with *tenera* fruit, it reduces the volume of digested fruit in the presses and improves oil extraction.

In small mills, such as the Pioneer, in which fruit, not bunches, are sterilized, the time taken is much shorter; a sterilizing time of 12 minutes has been found sufficient in an autoclave designed to hold 200 kg of fruit.

Stripping of fruit from bunches

There are two kinds of stripper (or thresher), the beater-arm type and the rotary-drum type.[70] The former is smaller and cheaper and is suitable for small mills with a capacity of up to 5 tons bunches per hour. It consists of an inclined cradle of curved bars between which the beater arms attached to the stripper shaft can pass. Bunches falling into the top of the cradle incline are knocked and turned by the tips of the beater arms and pass gradually down the cradle. The empty bunches pass out of the cradle at the end and the fruit and 'calyx leaves' which have been knocked out of the bunches fall through the bars to be picked up by a bucket conveyer and raised to the top of the digester.

The rotary drum stripper has a diameter of about 1.8 m and a length of 2.7 m to 5 m. The longer the drum the more complete the stripping should be. Even feeding is very important. Bunches falling on the platform above the stripper are usually pushed manually into a shute carrying them into the drum, and loose fruit fall through a grill on the platform to be carried straight to the digester without entering the stripper. Automatic feeders can, however, be employed. These consist of large hoppers into which the sterilized bunches are tipped. At the bottom of the hopper is a moving belt with slats which feeds the bunches at a uniform rate into the stripper. The stripper drum is composed of horizontal metal bars with sufficient space between them to allow the stripped fruit to fall through onto a conveyer; this carries the fruit to the bucket conveyer which takes it to the digester.

The centrifugal force exerted on the bunches is sufficient to raise them nearly to the top of the revolving drum whence they fall back onto the bottom and scatter their fruit. The bunches, while repeating this motion many times, gradually work their way along the drum aided by guide strips attached to the inside of the drum, and they fall out at the end and may be carried by a conveyer to a special bunch refuse incinerator outside the mill. It will be obvious that the speed of the drum is important; if too fast the bunches will be simply carried round and round, if too slow they will remain rolling gently at the bottom. Speeds suitable for large heavy bunches will be too high for small light ones. Speeds of between 21 and 23 rpm are usual with drums of 1.8 m diameter. The maximum capacity of a rotary drum stripper is about 20 tons of bunches per hour.

The rotary drum stripper is favoured not only for its higher capacity, but also for its smoother and more efficient running. Bunches occasionally get jammed in beater-arm strippers and this may increase loss of oil in the bunch stalk.

In the use of any stripper a check has to be kept visually on the completeness of fruit removal from the bunches and, in the laboratory, on both the fruit retention and the oil loss in the bunch refuse. Bunches seen to be 'hard', i.e. to have retained their fruit through the stripping process, must be returned for resterilization. Both losses of fruit and losses of oil in bunch stalk are much reduced by making sure that large heaps of sterilized bunches do not accumulate on the platform and that the bunches are fed into the strippers at a regular pace.

Incinerators are designed to operate by autocombustion, after initial firing, with a natural draught. The resulting ash is a useful fertilizer containing 30 to 35 per cent K_2O and 3 to 5 per cent MgO (see p. 551).

Digestion

The treatment of fruit in digesters prior to oil extraction is a very important process and it is partly through the examination of the action of digesters over a relatively long period of time that the proportion of oil extracted from the digested material is now so high. The purpose of digestion, as with pounding, is to break up the pulp physically and, further, to liberate the oil from the cells in which it is contained. The fault of simple pounding, and of some small-mill digesters, is that heat is not also provided, for heat assists in the loosening of the cells from the fibre.

Modern digesters are steam-jacketed cylindrical vessels with a vertical rotating shaft in the centre driven from the top or, less commonly, from the bottom. The fruit is mashed by pairs of stirring arms attached to the shaft. Various devices are used for preventing the mash going round with the arms. Baffles may be set vertically between liner plates on the wall of the digester or bars may be inserted across the vessel. Sufficient steam pressure is maintained in the jacket to ensure that the mash leaves the digester at about 90°C. There are perforations at the bottom to allow seepage of up to 50 per cent of the oil direct to the crude oil tank. For poor *dura* fruit there should be no perforations. Digesters are sometimes provided with a steam-jacketed hopper in which a quantity of fruit is kept hot before passing into the main cylinder where the stirring arms are revolving.

The digester should usually be of sufficient size to provide fruit for an hour's pressing, and the size must therefore be adjusted to the rate of work of the press or centrifuge to be used. Fruit will be fed at the top as charges are removed from the bottom through a special discharging hopper (Plate 82).

The important points to be checked in the digestion process are (i) the temperature must be maintained at around 95°C but the water in the mash must not be allowed to boil, (ii) the mash as produced must be homogeneous without any undigested fruit, (iii) the perforations must be kept cleared so that crude oil may escape at the bottom, (iv) the arms must be replaced when they show wear and (v) the digester must be kept at least three-quarters full at all times. [71, 72, 73]

If digesters are being used with centrifugal extractors they are slightly modified (see p. 744).

Extraction of oil

Although early trials were made with hydraulic presses, the earliest *efficient* machines for extracting oil from the digested fruit were centrifuges. It was mainly in Nigeria and Zaire that pressing difficulties were encountered and these were attributable to the fact that the bulk of the fruit treated had a very thin layer of pulp. For a long time it was considered that presses would inevitably crack the nuts and that therefore a depulper must first be used. Even as late as 1925 the opinion was expressed that the day of the hydraulic press was past. [54] The advent of the centrifuge (which did not need a depulper) in African territories was acclaimed as the answer to all previous difficulties and for a time it held sway as *the* modern method of extraction. [74] However, it was later found that hydraulic presses working at relatively low pressures could be used and they were first adopted for large estates in Sumatra in 1928. The first hydraulic press in Malaysia started operation at Elaeis Estate, Johore, in 1930. Pressing was still in two stages; in the first the digested fruit was pressed and in the second stage, after the nuts had been removed and the fibre dried, the latter was pressed alone. This practice was later abandoned in favour of a single pressing or the 'pre-pressing' of the digested material followed by the addition of further material at the top of the press for

Pl. 82   A set of six digesters with hand-operated Stork hydraulic presses.

final pressing. In postwar years screw presses were developed, the larger models of which have a much higher throughput than any single centrifuge or hydraulic press. The three types of extraction are briefly described below.

The centrifuge.   Centrifuges can be driven from the bottom or top but the latter type has become standard in oil palm factories (Plate 78). A basket is rotated at high speed, between 950 and 1,250 rpm, by a belt-driven shaft. The charge of digested material depends on the size of the basket which is usually between 91 and 125 cm in diameter with capacities, in terms of bunches, of 1—2 tons per hour. Charging, spinning and discharging takes from 20 to 25 minutes and three spinnings should normally be completed in the hour. The oil is expelled through the mass by centrifugal force and passes through the perforations of the basket into the space between the basket and the stationary casing from where it is pumped to the clarification section. A typical schedule for centrifuging in a small centrifuge would be:

Filling   3 minutes
Accelerating   2 minutes
Spinning   6 minutes at full speed
Spinning   2 minutes at full speed with steam injection
Spinning   2 minutes at full speed
Stopping   2 minutes
Discharging   3 minutes

The following characteristics of centrifugal extraction should be noted. The oil obtained is comparatively sludge free, and clarification losses are small. Steam injection is important, but should not be used until towards the end of spinning. At the very end of spinning it is desirable to cease steam injection and then dry out the centrifugal matte to assist subsequent fibre separation. The centrifuge should be covered with a metal cover during spinning. Discharging the matte which has caked around the circumference of the cage is an arduous job which must be done manually through the bottom of the cage after the steel ring around the shaft, which closes the bottom, has been raised. Though covered digesters with steam inlets to the interior were used with centrifugals, open-type digesters, as already described, are now usually employed, but they are fitted with steam injection tubes at the bottom to ensure that the digested material has not dried out too far for satisfactory centrifuging.

The hydraulic press.   The hydraulic hand press has already been described. Standard presses for large mills may be hand-operated or automatic. The former have capacities from 1.5 to 3.0 tons bunches per hour. The material is introduced into the perforated press cage from the digester shute through a hole (of the same diameter as the press) in a table-like sliding door which is operated by a hand-wheel. The door is closed on the interior of the press by sliding it forward so that the hole no longer lies above the press cage. Plate 82 shows the sliding doors in both the charging and pressing positions. This type of press works at a cake pressure of 75 kg/cm^2 and six or seven pressings can be done in an hour. Hydraulic pressure is applied at the base of the press to a ram which moves upwards to press the mash and returns by its own weight. The mash is introduced in portions with up to six circular steel plates inserted between the portions. This assists the operator by enabling him to push off the press cake in manageable slices after pressing. If prepressing or double pressing is adopted a plate is held back to cover the material already in the press before material is added for the second pressing. A typical schedule for pressing at the rate of six pressings per hour would be:

Filling	2 minutes − from digester.
Pressing	6 minutes, including double pressing if employed.
Emptying	2 minutes − pushing press cake into conveyor.

The time can be controlled either from a clock or by a system of lights, and the pressing cycles are recorded from a pressure gauge onto a drum recorder chart.

Hydraulic presses may be of the revolving type comprising two cylinders swivelling on a central column. Discharging and filling is carried out in one press cage while pressing is proceeding in the other. Ejection of the press cake, loading the digested fruit, insertion of the plates and some prepressing is carried out with the aid of an auxiliary piston. These presses may have capacities up to 5 tons of bunches per hour.

An automatic hydraulic press has been evolved in which the digester forms one unit with the press.[75] In this unit the pressure is applied by a ram from above the press cage. When the ram is at its highest position mash passes freely into the cage in which are retained the press cakes of four or five previous pressings. Thus by a succession of pressings the mash is moved slowly downwards through the press-cage cylinder. This is made possible by the presence of a cone at the bottom of the cylinder which closes on the cylinder when the ram moves downwards and presses the mash, but opens after pressure has been applied for a certain time. When the cone moves downwards to open the cylinder the ram continues on its way pushing out a portion of the press cake. The ram then moves swiftly upwards, opening the digester door and closing the cone which in its motion breaks up the press-cake exuding from the cylinder. At the same time the cone pushes back the press cake in the cylinder causing some friction which is considered to assist oil expulsion (Fig. 14.7).

The speed of action of this automatic press can be controlled and its capacity thus increased from the usual 4.5 tons bunches per hour to 6.5 tons. The oil to dry fibre will of course vary with the pressing rate.

It has been shown that the performance of an hydraulic press depends to a great extent on the composition of the press cake. Trials conducted by press manufacturers suggest that there is an optimum proportion of

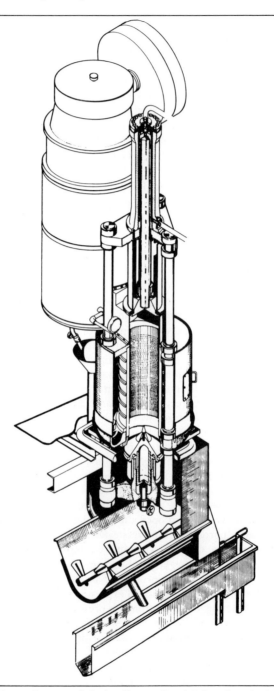

Fig. 14.7   The Stork automatic hydraulic press (*Stork Palm Oil Review*).

nuts to fruit for maximum oil-extraction, i.e. there is an increased loss of oil in the fibre when the proportion of nuts to fruit exceeds or falls short of a certain percentage. Where there is an excess of nuts, they bear on each other and full pressure is not exerted on the oil-bearing fibre.[76] No very clear reason has emerged for the low extraction obtained when there is a shortage of nuts but it is thought that the addition of nuts to *tenera* digested fruit may reduce friction resistance within the press cake and help to ensure a better distribution of pressure through the press cake.

Another measure which improves distribution of pressure is the insertion of the metal plates between sections of the introduced digested material. If this is not done the press cake will tend to be more compact and drier in the middle than at the circumference and more fully extracted at the bottom than at the top. The insertion of plates reduces this tendency by redistributing the forces bearing on the cake.

It has also been shown that the presence in the cake of small pieces of trash such as the 'calyx leaves' (perianth segments) from the bunch, far from hindering pressing, actually improves oil extraction from the cake.

To obtain the most satisfactory expression of oil, therefore, (i) fibre should be added to the sterilized fruit if the fruit has a high nut/mesocarp ratio, e.g. above 50/50, (ii) nuts should be added if the nut/mesocarp ratio is low, e.g. below 25/75 and (iii) no attempts should be made to remove calyx leaves. The actual quantities of nuts or fibre to be added usually need to be found by trial.

Screw presses. Two types of continuous screw press have been used for the expression of oil from digested material and it is claimed that with these presses the digestion period may be shorter. Both types of press allow for the gradual passage of the mashed fruit through a cylinder while under compression by screws against the walls of the cylinder and against cones at the end of the cylinder.

The *single-shaft screw press* has a gear box and a pressing section consisting of two screws − a feed screw and a press screw. The screws are placed end to end on the same shaft but are of opposite thread and rotate in opposite directions. The feed screw in large screw presses of this type rotates at 10 rpm while the opposing press screw rotates at 6 rpm. The digested fruit is fed to the press screw by the action of the feed screw and the tapering design of the former gradually reduces the volume available around the screw, thus increasing the pressure until the matte is expelled around the end cone, which is adjustable. Oil is expelled outwards through the perforated cage around the screws and inwards through a short cage around the shaft at the end of the press screw (Plate 83).

This type of screw press can be obtained in small sizes (about 3 tons bunches per hour), but very large presses capable of dealing with 13 tons of bunches per hour are also in operation.

The *double-shafted screw press* also consists of a gear box and pressing section, but the latter contains twin shafts with screws turning at the same

Pl. 83   The opposing screws and perforated shaft of the Colin 6 B.I. Screw Press.

speed but in opposite directions. The digested fruit enters a hopper below which lie the screws and a strainer through which some of the oil will pass. The material then passes into the perforated press cylinder through which oil is expelled; finally the matte is screwed further along the perforated cylinder where the shaft tubes are also provided with holes through which further expelled oil passes. The outlet at the end of the cylinder is regulated by adjusting the cones by means of an hydraulic pressure control system[73] (Fig. 14.8).

Another make of double-shafted screw press expels the oil only through the press cage and not through the shaft tubes. It is equipped with a feed screw between the digester and press and incorporates a variable speed drive which regulates the throughput and an automatic cone adjustment system which depends on the power absorbed by the main screws.[77]

Comparison of extraction methods It is rarely possible to make satisfactory comparisons of milling processes or individual milling machines. This is because two parallel plants of different construction are hardly ever found on one plantation to enable crops from the same fields, treated in all other respects identically, to be fed to the machines to be compared. This is, however, the only way in which a valid comparison can be made. Articles extolling this or that system and presenting results of processing have been written, but these, as often as not, have ignored the fact that either no comparison is being made or the comparisons, by neglecting the numerous other variables, are invalid.

Until recent years the only published scientific comparison between a standard centrifuge and a standard hydraulic press was that of Georgi, carried out at the Federal Experiment Station at Serdang in Malaysia in 1932–3.[78, 79] The factory was of small size and fruit extracted from the bunches by hand was used. All processes were identical except that digesters suited to the particular extractor were naturally used and, in clarifying the press extracted oil, the crude oil was diluted with water. Oil to dry fibre percentages were by today's standards high. The following results were obtained from four trials of each process:

	Centrifuge	Press
1. Weight of fruit treated (lb)	40,578	31,767
2. Palm oil recovered (lb)	11,097 (27.3%)	8,117 (25.5%)
3. Losses, Total (lb)	1,442 (3.6%)	1,191 (3.8%)
4. Total oil accounted for (lb)	12,539 (30.9%)	9,308 (29.3%)
5. Distribution of losses, Cake	87.4%	78.3%
Nuts	6.1%	6.8%
Clarification	6.5%	14.9%
6. Efficiency (2/4 x 100)	88.5%	87.2%

Figures for oil to dry fibre in the actual tests were not given, but previous trials showed that for the centrifuge the percentage was about 23 to 25 and for the press 21 to 22.

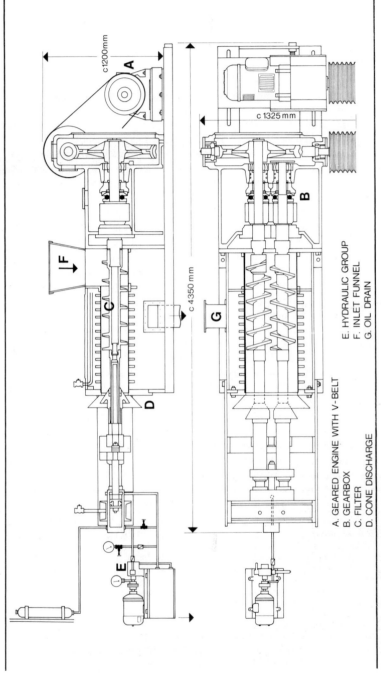

A. GEARED ENGINE WITH V-BELT
B. GEARBOX
C. FILTER
D. CONE DISCHARGE
E. HYDRAULIC GROUP
F. INLET FUNNEL
G. OIL DRAIN

Fig. 14.8   The double-shafted De Wecker screw press.

These results showed the general differences in the performance of centrifuges and presses which may be stated as follows: (1) Centrifuges provide a clearer crude oil, but give a higher oil content in the residual cake. (2) Oil from a centrifuge is more easily clarified and losses in sludge are not so high.

A valuable comparative study of the performance of centrifugals, hydraulic presses and screw presses has been published by Bek-Nielson.[73] He showed that the type of extraction unit chosen will influence not only the oil losses, but also kernel breakage and quality, iron absorption during processing, and wear and tear of the extraction machinery, including the digesters (Table 14.6). His studies confirmed that the loss of oil in the fibre was higher with centrifugals than with presses, though the loss in sludge was lower. It was also shown that losses in the fibre were lower with screw presses than with hydraulic presses, but the losses in sludge were higher. Controversy about the relative merits of the two types of press at one time centred on the non-fatty pressing quotient (NFPQ) which is the ratio of non-oily solids (n.o.s.) in the sludge to the total of n.o.s. in press cake and sludge.[80, 81] This quotient tended to be lower with hydraulic presses. It will be seen from Table 14.6 that the overall extraction efficiencies are only marginally in favour of screw presses; and the view has been expressed[73] that efficiency has now reached a level which, with any kind of pressing, is unlikely to be exceeded.

The advantages and disadvantages of screw presses in comparison with hydraulic presses may be summarized as follows:

Advantages: (i) Higher throughput, (ii) lower capital outlay per ton of throughput, (iii) lower power consumption, (iv) lower labour requirement.

Disadvantages: (i) Higher kernel breakage and loss of kernel in fibre, (ii) more difficult and costly clarification. In the past maintenance costs have been higher with screw presses, but it is now claimed that this is no longer the case.[73, 77]

It should also be mentioned that although screw presses have been successfully employed with mixtures of *dura* and *tenera* material, they are certainly best adapted to the plantations of pure *tenera* material which are now usual, particularly in new areas. Another advantage claimed for screw presses is flexibility of throughput. Presses with a nominal throughput of 10 tons bunches per hour can be run down to 6 tons or up to 14 tons when necessary.

Solvent extraction has often been suggested and tried on a small scale, but the technical and economic difficulties have not been entirely overcome. In the early days of the Sumatran industry solvent extraction of the residual fibre after the first pressing (containing over 20 per cent oil to dry matter) was put into practice on a number of estates.[64] The solvent used was naphtha (boiling point 85°−105°C) which was a product of local refineries. The solvent extraction of residual oils in fibre after normal

Table 14.6   *Loss of oil, analysis of residues, nut breakage and iron contamination when using different extraction equipment (Bek-Nielson[69]) (material: 85 per cent Deli dura, 15 per cent tenera)*

	(1) Centrifuge	(2) Hydraulic press	(3) Automatic hydraulic press	(4) Screw press A	(5) Screw press B
Oil loss, kg per 100 kg bunches, in:					
Sterilizer condensate	0.040	0.040	0.040	0.040	0.040
Empty bunches	0.450	0.450	0.450	0.450	0.450
Press residue	1.400	0.874	0.701	0.606	0.614
Nuts	0.148	0.124	0.150	0.152	0.146
Sludge	0.054	0.107	0.192	0.231	0.271
Total	2.092	1.595	1.533	1.479	1.521
Oil to dry fibre (%)	14.91	9.82	8.32	7.27	7.58
Oil to total dry matter in sludge (%)	3.35	5.09	7.48	8.67	9.32
*n.f.p.q.**	14.6	19.1	23.3	4.2	6.4
Iron loss, kg per ton bunches†		0.034	0.027	0.013	0.012
Nuts in press residue (%)					
Unbroken	99.54	94.48	84.66	76.11	78.01
Smashed	0.28	2.95	5.61	8.33	7.56
Split	0.00	1.62	4.43	5.29	5.28
Free shells	0.12	0.81	4.11	8.14	7.07
Free kernels	0.05	0.14	1.19	2.14	2.08
Extraction efficiency					
Deli *dura* material	89.3	91.8	92.2	92.4	92.2
Deli with 20% *tenera*			93.5	93.8	

* Non-fatty pressing quotient = non-oily solids in sludge/n.o.s. in press cake + n.o.s. in sludge.
† From recorded losses in weight of press plus digester after 1,000 hours' pressing.

Notes: Manufacturers: (1) Manlove Alliott & Co., (2) and (3) Stork Apparatenbouw, (4) Usine de Wecker, (5) Speichim.

pressing or centrifuging has also been suggested and much discussed. Various objections have been responsible for this idea not being pursued. The quantity of solvent required would be large and the loss, under tropical conditions, considerable. Oil extracted from the residual fibre is known to contain a high proportion of carotenoids and is contaminated with an unacceptable quantity of non-oil components such as waxes[100] and this would introduce bleaching problems.

The method of *impulse rendering* (the Chayen process)[82] used for the extraction of oil and proteins from some seeds has also been suggested for the extraction of oil, and small-scale tests have been carried out. The principle of the process is as follows. If any cellular matter is subjected, in a fluid such as water, to impulses or shock waves of sufficient frequency and intensity, the cells are burst and their contents mix freely with the fluid. Oil produced from oil seeds by this method is of good quality and light colour, though complete separation from the protein fraction is not obtained. If the process could be adapted for the extraction of oil from oil palm fruit, the product might be expected to be of both low f.f.a. content and good bleachability.

Clarification

Oil coming from the presses or centrifuges is crude oil, that is to say it is accompanied by water, dirt and cellular matter from the mesocarp. Oil from centrifuges will contain only about 40 to 50 per cent water and comparatively small quantities of dirt and other matter. Oil from hydraulic presses contains about 55 per cent water and from screw presses about 60 per cent with considerably more impurities, particularly cellular matter.

Crude oil is largely an oil–water mixture in various phases of oil dispersion in the water. The majority of the oil settles out on top of the water easily enough but, if care is not taken, emulsions may be formed. Impurities tend to prolong emulsification and temperatures of over 100°C tend to produce emulsions.

Crude oil flows by gravity straight to the crude oil tank where it passes through screens to get rid of the larger impurities. In a small plant the screens may be stationary and set in the crude oil tank, but in larger plants vibrating screens of stretched gauge are used above the crude oil tank and the material which does not pass through is automatically returned to the digester. Sand sinks to the bottom of the tank. Crude oil pumps direct the screened crude oil to the clarification tanks.

In small mills the oil is separated from the water and dirt (the sludge) in simple settling tanks containing water and heated to about 95°C by steam coils. Pure oil is drawn off from the top and sludge passes to another settling tank for further separation after addition of more water. The quantity and kind of clarification equipment depends on the size of the mill and the market requirements.

What are often termed continuous clarifiers are really semi-continuous since although crude oil may be supplied at a regular pace to them and oil and sludgy water will emerge from two separate pipes, the flow, and sometimes the internal piping, need regulating and, moreover, the really heavy particles must be specially drawn off from the bottom from time to time. In these clarifying tanks (see Fig. 14.9) the crude oil is passed via a heat exchanger (A) at 85°–90°C to near the bottom of the tank, and the oil collects and rises in a thick layer until it reaches the top of a pipe (B) through which it will be automatically or intermittently drawn off. The sludge water syphons off at a slightly lower level since the specific gravity of the oil plus sludge is lower than the specific gravity of sludge alone which rises in the side pipe (C). The liquid fractions X + Y in the tank and Z in the pipe are in balance. Temperature is maintained by a steam coil in

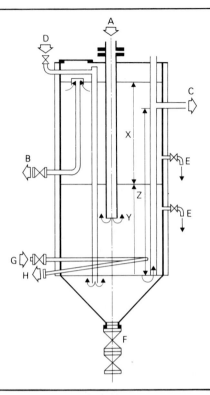

Fig. 14.9   The Stork continuous clarification tank (for explanation see text).

the tank or a steam jacket. The quantity of crude oil being fed and the volume of the oil layer together determine the residence time of the oil in the tank and if this becomes too short for satisfactory clarification then an adjustment must be made to the height of the oil exit pipe. A slow supply leads to too long residence and tends to increase the f.f.a. set up.[83] When supply ceases, the separated oil must be drawn off by allowing entry of water into the sludge layer through pipe D, and the sludge water and the heavy material must be run off by manipulation of the various external taps. It will be seen, therefore, that the control of clarification, though continuous for much of the time, needs careful handling by an experienced operator.

The remaining clarifying equipment depends on the scale of operations and the results required. Further settling tanks may be employed and vacuum driers or centrifugal separators. A number of alternative systems have been advocated. In one system a new type of centrifugal separator, designed to prevent air entry with consequent oxidization, is used, and the oil is then passed to a vacuum drier which reduces moisture to below 0.05 per cent.[84] In another system a nozzle-type centrifugal is used and a method of recycling the water is included. The oil is sent direct, or through a drier, to the storage tanks.[85]

The old and usual method of dealing with the sludge is to pass it through a series of fat pits. Sludge is objectionable, malodorous material and it should be the object of every mill manager to discharge from his care the very minimum of solids and water contaminated with rancid oil. This task will be much assisted if the bunches enter the mill in as clean a condition as possible. The fat pits are in series: the heavier water and dirt go to the bottom of the first and succeeding pits, and pass to the next pit in order under a partition which itself holds back the lighter oil. The partitions can usually be let into the pits in slots and are easily removed. In Africa it is often possible to dispose of the sludge oil for local soap-making rather than risk contaminating the pure oil with it. Otherwise it must be returned to the clarifiers.

An alternative method of treating sludge is to conduct it to a tank for boiling and then to a special sludge centrifuge of which several types are available. In one type a star-shaped rotor has four nipples with 2 mm orifices. Sludge water enters through the hollow halfshaft and is thrown by centrifugal force towards the four nipples at the extremities of the star. The heavier particles and water pass out, but the oil builds up in the centre and is forced back through a small diameter outlet shaft inside the inlet shaft and returns to the continuous clarifier.[71]

Effluent disposal

However effective the system of recapturing oil from the sludge may be, the effluent discharged from an oil mill is still objectionable and will pollute streams, rivers or surrounding land. While mills were comparatively few and mostly on large fast-flowing rivers, the problem was not a serious one, but the situation in many countries is now quite different and much attention has recently been given to the subject of effective disposal. Apart from the sludge water itself, which amounts to 250–325 kg per ton of bunches milled, there are also about 175 kg of sterilizer condensate and 140 kg of effluent from the hydrocyclone or clay bath separators per ton of bunches.[86] In milling 20 tons of bunches per hour, more than 200 tons of effluent may be discharged over 24 hours and this may contain up to a ton of oil and 9 tons of dissolved or suspended solids. This effluent has a biochemical oxygen demand (BOD) of about 20,000 ppm at 20°C for 5 days which is extremely high.[86, 87]

Apart from (*a*) discharge into large rivers and the sea, the disposal methods which have been considered are (*b*) biological treatment and (*c*) dehydration. With regard to discharge into large rivers, their size and flow should be sufficient for a dilution of about 10,000 times the discharge. Olie and Tjeng examined the possibilities of both aerobic and anaerobic biological treatment and concluded that the costs would be too great.[86] Regarding dehydration, they examined soil filtration, filtration under a vacuum or by centrifuging and drying by evaporation and concluded that the latter method is likely to be the most practical. They suggest sprinkling the effluent onto a surface of 1.8 m² per ton of effluent discharge per day by means of pipes and upward sprinklers, with recycling through another pipe and sprinkler system. After concentration from 5 per cent to 20 per cent dry matter content the sludge would be discharged by pumps and dried in a rotary drying kiln, the solid product having possible use as a fertilizer or fuel.

Methods of effluent disposal which have been investigated have been discussed by Stanton and for the biochemical factors involved his paper should be consulted. He concluded that combined aerobic and anaerobic treatment is indicated, with special techniques of design and temperature and pH control to hurry the process.[88] Research has recently been undertaken into a system of absorption of concentrated sludge slurries into tapioca or palm kernel meals or a mixture of these meals. Under the Centrifugal Solids Recovery or Censor system the mixture is then dried using rotary driers and surplus heat from the boilers. The resultant material is expected to have a good value as animal feed. The more the water content of the sludge is reduced the less absorbing material is required so the system includes concentration by centrifugals as used in many mills. This is followed by anaerobic fermentation using specific bacterial populations and with the production of methane gas and concentration of the solids. There is also a final liquid-phase effluent which can be used as a fertilizer.[89, 101] This system is open to criticism on the grounds that sludge water may need to be taken by tanker to central points and that the amount of tapioca required may be unobtainable. Palm kernel cake is insufficient on its own.

A perhaps more promising method under development entails flotation of oil and solids by a flocculent chemical with carefully controlled aeration. The resultant flotate is then skimmed off and is either treated by anaerobic digestion to form an innocuous digested material rendered suitable as a fertilizer following breakdown of the oil and other organic material, or is evaporated and dried to provide an animal food.[102]

Although the above disposal methods are under investigation at the present time (1976) no standard method has so far been adopted. Some mills have been ponding the effluent or spraying it on areas of up to 10 hectares, but these are not satisfactory or permanent solutions. One important factor in considering evaporation methods is that in large modern mills there is now a surplus of fuel to the extent of some 3 tons shell plus fibre per hour in a 40 tons bunches per hour mill. Thus surplus steam is available for aiding evaporation.

Kernel extraction

The matte coming from the presses consists of nuts and moist fibre with some residual oil. To extract the kernels it is necessary to (i) extract the nuts from the fibre, (ii) crack the nuts and (iii) separate the kernels from the cracked shells. In large mills built in the 1950s and later the kernel extraction plant occupied a great deal of space and consisted of a large fibre separator, nut drying silo, nut screens, hydrocyclones or mechanical clay baths, with at least four elevators and numerous screw conveyors for moving nuts, shell or kernels. In comparison with the oil extraction plant the kernel recovery equipment was very expensive, it being estimated that while the investment and operating costs of the oil extraction equipment was only twice as great as that of the kernel equipment, the value produced in the oil extraction section of the mill was nearly nine times as great as the kernel value produced by the kernel extraction plant. This costly and space-occupying equipment has been simplified by using air only to separate the fibre and to blow the nuts first to the silo and later to the hydrocyclones. A pneumatic transport system is also used for conveying kernels to the kernel silo and shell to the boiler platform.[90]

Separation of nuts from fibre

This process has unfortunately been called depericarping, a term already used for removal of the fresh mesocarp prior to oil extraction, and the machines used are often called depericarpers. The term 'fibre separator' has also been used and is more accurate.

Fibre separation may be hydraulic, pneumatic or mechanical. The hydraulic separator performs a washing process in which the fibre floats and the nuts fall to the bottom; this process is not used today.

Pneumatic fibre separators[91] which are partly mechanical have been the rule in nearly all larger modern mills. The press cake has a high temperature as it comes from the press and if it is immediately pushed into a 'breaker conveyer' — i.e. an open-topped, but steam-jacketed conveyer with a shaft carrying blades which cut up the press cake as well as forcing it along the conveyer — it will dry out appreciably under its own heat. This assists the loosening of the fibre from the nuts. The effect may be enhanced by arranging for the air which is to pass through the separator to be heated by a steam-operated air-heater.

In the most commonly used type of fibre separator the fibre-nut mixture passes into a large drum rotating at about 15 rpm. This drum has baffles mounted on the inside which carry the mixture upward, and allow it to drop. The current of air passing through the drum is sufficient to carry the partially dried fibre to the exit tube and blow it to the boiler platform. The nut and fibre mixture is gradually moved by the rotating motion towards the end of the drum and in the course of this motion is further dried by the hot air current. The nuts will fall into a smaller, lower rotating drum where they are polished by friction. Some air passes through this drum also, carrying any light particles upwards to join the main flow. Various releases can be arranged under the second drum to allow the exit of polished nuts and the passing over to the end of the drum of heavy pieces of bunch stalk and other debris.

The currents of air are provided by a fan placed above the separator and the air-fibre mixture is blown to a cyclone which is situated above the boiler house and serves the purpose of separating the fibre from the air, the former falling onto the boiler platform or some convenient place whence it can be easily employed as fuel.

The separator is efficient but, besides being noisy, it makes comparatively heavy calls on the power supply of the mill.

Purely mechanical fibre separators have been used for a long time, particularly in small mills or in mills, such as the Pioneer, where a low capital cost was imperative. One type consists of a screened drum which is rotated and allows the separated fibre to fall through the screen. A second type is a modification of a cotton ginning machine; a revolving shaft is fitted with studs which tease off the fibre from the fibre-nut mixture fed into the machine. A third type has a rotary cage bounded by rollers revolving in opposite directions in pairs which remove the fibre to the outside of the cage but retain the nuts inside.

Rotating drum separators are now being replaced by stationary direct air separation columns in which the velocity of the upward current of unheated air which removes the fibre is adjustable. Separators of this type occupy less space and reduce power consumption.[90, 92] They can be combined with simplified cracking and kernel separating systems in which pneumatic transport replaces chain elevators and screw conveyers.[90]

Nut screening and cracking

The clean nuts may be dried in a nut silo or, if the drying in the breaker-conveyer and in the fibre separator has been sufficient for cracking, they may be conveyed straight to screens for grading according to size before cracking. It is generally considered that in a mill with a large throughput the additional drying in the silo is necessary, and the opinion has been voiced that the moisture content of the kernels must be less than 16 per cent if they are to be sufficiently shrunk away from the shells for easy cracking. Nevertheless, quite satisfactory cracking is often obtained in mills without nut silos, and if heating in the latter is too high kernel quality will suffer. It should be remembered, however, that small nuts are more difficult to crack than large ones and that proper drying is therefore likely to be of greater importance on all *tenera* plantations. Too prolonged drying leads to a high percentage of broken kernels.

For satisfactory nut cracking it has been usual to send manufacturers samples of the nuts produced or to be produced on the plantation. The calculation of the correct, or most nearly satisfactory, screens can then be carried out and a decision taken on how many nut screens, crackers and cracked-mixture screens are required, and of what size and type they should be. The cracking section usually consists of revolving screens to grade the nuts, the crackers themselves, and screens to separate the uncracked nuts from the mixture. In old mills this section is often difficult to improve especially when nuts of widely different sizes are being produced, and if an incorrect installation has been made at the start it may be difficult ever to rectify it completely. More modern systems avoid these difficulties.

The larger the number of nut grades used, the less will be the number of kernels returned to the nut-cracker. This is because the frequency curve or histogram of nut size usually overlaps the frequency curve of kernel size.[93] If no grading is done, therefore, *some* large kernels will pass over the cracked mixture screen and be returned to the cracker where they may be broken and partially lost as dust. If the perforations of the cracked mixture screen are made wider, then the return of kernels to the cracker will be less, but more small nuts will pass through the screen and appear in the cracked mixture for separation into kernels and shell. Cracking in at least two fractions is usual and in three fractions if the size of the plant warrants it. This reduces the overlap of kernel and nut size; with large nuts there will be no overlap, though with small nuts a small overlap remains. In grading into three fractions calculations have to be made as to the proportions of each grade passing through the cracking section of the mill. If, then, five crackers are being used there may need to be one small, one medium and three large graders or any other suitable combination of five screens of three sizes. The object here is to distribute the work as evenly as possible over the five crackers while bearing in mind that, there being the greatest overlap of nut and kernel size in small nuts, more than one small-nut screen will be undesirable. With *tenera* nuts, which are often small, the overlap of nut and kernel size is inevitably larger than with *dura*, particularly Deli, nuts and it is partly for this reason that kernel recovery is less satisfactory with the former fruit.

Nut screening is not as simple as might be supposed. If hole size is correct for the desired screening, many small nuts may still pass over the screen if it is operated too fast, fed too much or is too small. There is thus a relationship between capacity and hole size and if high efficiency is required more screens of lower capacity must be installed. Design of the screen is also important, and *dura* grading is always easier than *tenera* grading; it is therefore not surprising that methods of avoiding screening altogether have now been sought.[90, 92]

Modern nut-crackers are centrifugal, the nuts fed into the inlet falling into and then being thrown out of slots on the face of a shaft rotating at high speed and being hurled against a cracking ring. The vertical crackers (horizontal shaft) previously employed are suitable for large *dura* or Deli nuts, but in predominantly *tenera* plantations cracking is best done with horizontal crackers (vertical shaft) which have a larger diameter. The theory behind the use of the latter crackers is that, firstly, *tenera* nuts, having a distinct rounded head and tapering fibre-covered tail, are enabled to take up a 'head-foremost' position in the extra distance covered and so be cleanly cracked on hitting the plate. If the fibre-covered tail hits the plate the nut may not be cracked. Secondly, in the confined space of a vertical nut cracker the *tenera* nuts may be deflected by rebounding shell fragments. Thirdly, the larger the distance between the cracker rotor and the cracking ring the more obtuse will be the angle at which the nut strikes the ring and there will then be less likelihood of the nut glancing off the ring surface. The speed of nut crackers varies from 800–2,500 rpm according to the diameter of the rotor.

Not all the nuts will be cracked when they are first fed into the cracker. A 90 per cent cracking is usually achieved and this may rise to 95 per cent with efficient crackers. The object of the cracked mixture screens is to allow dust and small shell particles to be first released and then fed by conveyer to the boiler house, to allow the cracked mixture of shell and kernels to pass to the kernel separators and to return the uncracked nuts to the cracker. The cracked mixture screens have in some modern mills been replaced altogether by vertical separating columns (see below) in which dust and small pieces of shell are blown away; uncracked nuts are recovered *after* kernel and shell separation.[73, 92]

Kernel and shell separation

The task of kernel and shell separation can be lightened, particularly with *tenera* nuts or with *tenera* and *dura* mixtures, if a winnowing system is placed between the cracked mixture screens and the kernel separating machine. Fragments of light shell and small nuts with long fibres often have a specific gravity differing little from that of kernels, and if they can be first blown away a more perfect separation will be effected. Normally the winnowing device consists of a rectangular section vertical tube in which the cracked mixture meets an upward current of air. The fragments and dust carried up by the air current are separated in a cyclone.

In the early days of milling the mixed shell and kernels were placed in a salt bath of such a specific gravity that the shells sank and the kernels floated. The kernels were then skimmed off and dried for bagging. Later a mixture of water and clay to give a specific gravity of 1.17 was found more suitable. The shells have a specific gravity of 1.3 to 1.4 and sink to the bottom; the kernels, which have a specific gravity of about 1.07, are skimmed off, washed and dried. However shell of *tenera* material has a lower specific gravity than *dura* shell, and a figure of 1.17 has been quoted.[94] This makes separation based on specific gravity differences more difficult.

A fine clay must be used which does not contain any appreciable quantities of sand or organic matter. It has been estimated that about 2 cubic yards of clay is required per 100 tons of bunches. Simple perforated tanks have been devised which are suspended in clay bath tanks and assist in separation and in clearing away the shell after removal of the kernels. The cracked mixture is placed in the perforated tank and after the kernels have been skimmed off the tank can be raised manually or on hinges and the shell tipped out.

In one type of mechanical clay bath there is a shaft running longitudin-

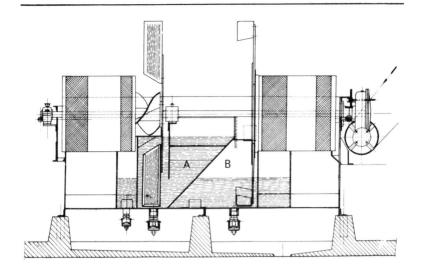

Fig. 14.10 Mechanical clay-bath kernel and shell separator (Gebr. Stork & Co.). (For explanation see text.)

ally above or at the surface of the clay-water mixture and carrying two discs or wheels with lifting buckets and, on either end, washing drums (Fig. 14.10). The bath tank consists of two compartments separated by an inclined baffle. From one compartment (B) clay water is continuously scooped by the buckets of one wheel. This clay water is tipped out of the buckets and runs into the other compartment (A) so that the latter is continually overflowing back into the first compartment (B). The cracked mixture is introduced into the overflowing compartment and the floating kernels pass out with the overflow into one of the washing drums while the clay water is running back into compartment B. At the other end of the bath the shells fall along the inclined baffle to the bottom of the overflowing compartment A and are picked up by the second set of lifting buckets and dropped into the other drum where they are washed and pass out of the machine. The wear and tear on these machines is naturally high and efficiency is limited owing to the tendency for floating kernels to get picked up by the shell buckets.

In another type of mechanical clay bath the clay-water mixture is circulated by a pump from a ground-level tank into a large cone-shaped funnel above it. The mixture is introduced into this funnel. The floating kernels pass from the surface through a pipe on to one side of a vibrating screen while the shells sink to the neck of the funnel and are syphoned over with the clay water on to the other section of the screen. The clay water from both pipes falls through the screen back into the ground-level tank and is recirculated. Both kernels and shell are washed by a sprayer on

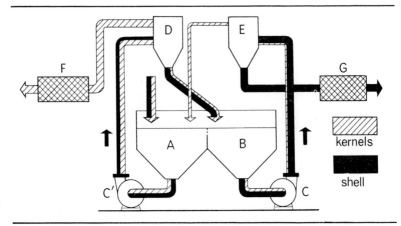

Fig. 14.11   A representation of the operation of the hydrocyclone — see text (*Stork Palmoil-review*).

the screen and are discharged. Kernel separators of this type have capacities of 2—4 tons of nuts per hour.

The advent of the hydrocyclone has reduced the use of clay baths. The hydrocyclone's main advantage is that it is compact and requires only water. Clay and the manual labour that goes with its collection is now eliminated from the mill; however the hydrocyclone has a higher capital cost than the clay bath and its efficiency may not be higher.

A hydrocyclone unit consists of two cyclones, two tanks, two washing drums and two pumps with connecting pipes.[42] The mixture is introduced into one tank (A — see Fig. 14.11) which is filled with water, and is forced with the water by a pump (C^1) to a cyclone (D). Here the water is subjected to two forces; firstly, a centrifugal force, and secondly a non-tangential force directing the water to the vortex in the centre of the cyclone through which the majority of the water flows (Fig. 14.12). A particle lighter than water and/or with high flow resistance tends to leave the cyclone by the vortex with large quantities of water. Heavy particles and/or particles with a low flow resistance follow the centrifugal force and are discharged at the apex at the lower end of the cyclone with the smaller quantity of water. It appears that with *dura* nuts flow resistance plays little part and thus kernels are discharged from the vortex and shell at the apex; but with *tenera* material, separation, as in the clay bath, is less efficient. In practice the greater part of the kernels pass out through the washing drum (F) from which the water is allowed to flow back into the tank (A); the shell, with some kernels, passes into the other water-filled tank (B) and is pumped to another cyclone (E) where separation continues. In this case the kernels going out at the vortex re-enter the first tank (A) while the shell from the apex goes to the other washing drum (G) and so leaves the separator.

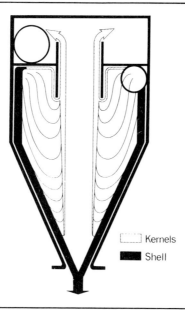

Kernels
Shell

Fig. 14.12   The movement of kernels and shell in the hydrocyclone.

Kernel drying

The washed shell passing out of the clay bath or hydrocyclone is carried by conveyer to the boiler house where it serves as fuel. The kernels have to be dried before they are fit for bagging and if the separation is not sufficiently complete they must be picked over by hand. This is usually done by passing the kernels slowly in front of a picker on a conveyor belt either before or after the kernels have been dried.

Kernels can be dried on trays passing slowly over hot air, but the kernel silo is now the most common method of drying. The wet kernels are admitted at the top and move down the silo in a continuous flow as kernels at the bottom are discharged through a shaking grid and bagged. The silo incorporates a drier which blows heated air through the kernels at different levels and is thermostatically controlled.

Mill layout

In small or medium mills all the equipment is usually under one roof, but in very large mills the boilers, engines and electricity generating equipment may occupy a separate building. A separate power house assists cleanliness and is made possible when all the plant is electrically driven.

When shaft and belt drive is used the boiler room is partitioned off from the rest of the building and the furnaces face outwards; the steam engine is then best positioned in an adjoining room for belt drive on to the middle of the long shaft which drives all the milling equipment in the main part of the building. Next to the engine room will come the clarification plant and next to that a small laboratory and office. Bunches are brought to one end of the building for loading into vertical or horizontal sterilizers while kernels are bagged at the other end. Oil may be drummed at the side entrance to the clarification room or pumped to storage tanks.

Figure 14.13 represents the layout of a mill with horizontal sterilizers and a capacity of 6 tons bunches per hour extensible to 12 tons. The figure gives a good idea of the general scheme of an oil mill and the shape of its components but does not, of course, represent the ideal layout for all sizes or conditions.

In a large installation the reception area must be arranged so that lorries may tip bunches on to a loading ramp for temporary storage. The bunches are then pushed into the cages for sterilization. An electric hoisting crane is required for hoisting the cages of sterilized bunches on to the stripper platform.

The bunch refuse usually leaves the mill by conveyer, but it may go to an incinerator or to a cutter so that it may be used in the boiler, or it may go to the fields. Incineration is now the usual practice.

A layout for a large electrically driven factory of over 20 tons per hour capacity usually has a separate power house with boilers, steam engines and alternators. Clarification is carried out in the main building of the mill though well separated from the presses. The laboratory, offices and workshops are separated from the main building.

Small-scale mechanical milling

For a long time palm oil production was considered either as a very large-scale business or as a small-scale peasant undertaking. For this reason there has been much technical progress in the installation of very large mills and, at the same time, attention has been given to providing for the non-mechanical needs of the small producer. However, in some parts of the world, notably America, the planting of medium-sized holdings of a few hundred hectares has become common and is increasing. Under these circumstances it might be thought that several plantation owners would combine to build a medium-sized mill of high efficiency; such a course is not always possible however, owing to the isolation of small estates and other causes. For this reason there has been an increased interest in small plants of high efficiency. Such plants need to have a capacity of between 1 and 3 tons of bunches per hour and a low installation cost.

These considerations are quite different to those which gave rise to the introduction of the very inexpensive Pioneer mills into Nigeria in the late 1940s and early 1950s, but the history of the Pioneer mill scheme is relevant. At that time it was desired to increase the production and quality of the oil from the palm groves. Capital was limited but labour was

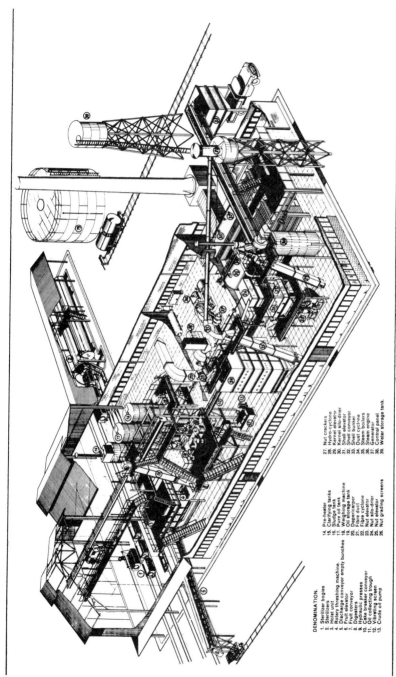

DENOMINATION.

1. Sterilizer bogies
2. Sterilizers
3. Holst unit
4. Rotary threshing machine.
5. Discharge conveyor empty bunches
6. Fruit elevator
7. Fruit conveyor
8. Digesters
9. Hydraulic presses
10. Cake breaker conveyor
11. Oil collecting trough
12. Vibrating screen
13. Crude oil pump

14. Pre-heater
15. Clarifying tanks
16. Sludge tank
17. Pure oil tank
18. Weighing machine
19. Oil storage tank
20. Depericarper
21. Fibre duct
22. Fibre cyclone
23. Nut elevator
24. Nut silo-drier
25. Nut elevator
26. Nut grading screens

27. Nut crackers
28. Hydro-cyclone
29. Kernel elevator
30. Kernel silo-drier
31. Shell elevator
32. Shell conveyor
33. Shell bunker
34. Dust cyclone
35. Steam boilers
36. Steam engine
37. Generator
38. Control panel
39. Water storage tank.

Fig. 14.13   Layout of medium capacity mill.

extremely cheap. Thus it was hoped that these mills would provide a good deal of wage-work but, at the same time, be economic. The usual equipment was provided and consisted of a fruit autoclave, an open unheated digester, a centrifuge, a cotton-ginning type fibre separator, a nut-cracker and screen, an open clay-bath, a crude oil tank and three settling tanks. The power was supplied by an upright boiler and vertical steam engine with shaft and belting, and the capacity is 600 kg fruit (equivalent to 1 ton of bunches) per hour. The Pioneer mill was not supplied with any elevators so gravity was not made use of except for the flow of sterilized mash from the autoclave to the digester. Fruit was lifted into the autoclave and digested fruit lifted into the centrifuge. Thus labour usage was high and, probably owing to the relatively low temperature of digestion and centrifuging, efficiency of oil extraction was only 85 per cent. Wages rose much faster than was expected in Nigeria and this, together with poor siting, competition from hand press operators, etc., rendered the Pioneer mill uneconomic in most grove areas though it could be operated profitably on plantations when worked near to full capacity.

Reorganization of the Pioneer mill machinery and the use of pulleys and gravity feeds much reduced the labour costs of operation, and a number of minor technical improvements such as steamjacketing the digester and covering the centrifuge increased the efficiency.[95] However, in the main, these improvements and the proper training and supervision of mill operators came too late to be put into general practice. Many of the alterations could not be effected without considerable injection of capital when costs were already much inflated.

In most of the older oil palm countries, and in the new ones, wages are already high, and the fate of the Pioneer mill scheme, which in many respects was of great indirect value to Nigeria, is a warning that while low capital cost may be a necessity, labour saving must be a primary consideration in the design of a small mill. Furthermore, small mills will not be operated economically unless there is skilled supervision and spare parts are readily available. A plan of a small mill for ¾ to 1 ton bunches per hour and incorporating the hydraulic hand press with added mechanical drive is shown in Fig. 14.14.

Mills of ¾ to 3 tons capacity using hydraulic presses have been installed in several parts of South America. These mills have been locally designed and incorporate short horizontal bunch sterilizers with a compact rail system which enables the sterilized bunches to be moved easily from the sterilizers to the stripper (Plate 80). In many cases the mills have been constructed on the side of a hillslope so that gravity can be utilized for moving the bunches to the sterilizers and to the strippers and the fruit to the digesters.

The worst feature of small-scale milling in America has been the lack of provision for milling control. Weighbridges or other means of bunch weighing are not provided as standard equipment and the owners do not know either the extraction rates or efficiencies being obtained.

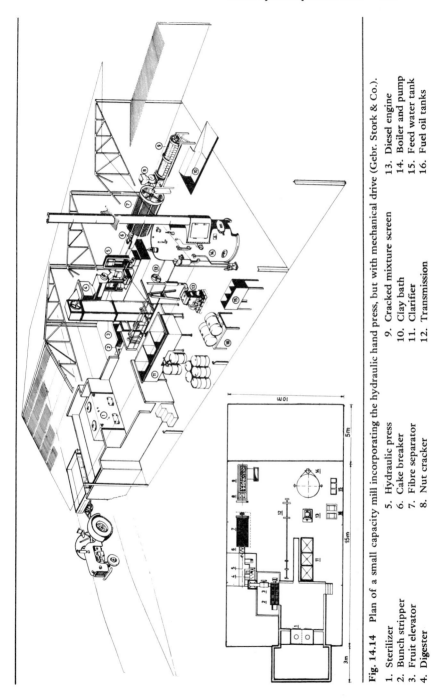

Fig. 14.14 Plan of a small capacity mill incorporating the hydraulic hand press, but with mechanical drive (Gebr. Stork & Co.).

1. Sterilizer
2. Bunch stripper
3. Fruit elevator
4. Digester

5. Hydraulic press
6. Cake breaker
7. Fibre separator
8. Nut cracker

9. Cracked mixture screen
10. Clay bath
11. Clarifier
12. Transmission

13. Diesel engine
14. Boiler and pump
15. Feed water tank
16. Fuel oil tanks

Milling control

The processing of bunches for the extraction of oil and kernels cannot be properly controlled unless what is going out of the mill is known in relation to what is entering it.

Coming into the mill are:
1. Bunches
2. Water
3. Steam
4. Dirt

Going out of the mill are:
1. Bunch refuse
2. Palm oil, containing varying quantities of free fatty acids, water and dirt
3. Palm kernels, containing varying quantities of shell, dirt and moisture
4. Sludge water
5. Sterilizer condensate

In addition press fibre and shell is going to the boilers to be used as fuel.

Control of quantity

The calculation of the efficiency of oil and kernel extraction would be much simplified if the composition of bunches was relatively constant or if sufficient of them could be sampled and analysed to give a representative analysis of the daily or weekly intake. Figures of efficiency for oil and kernel extraction could then be obtained by dividing the quantity going out of the mill by the calculated quantity coming in. Such efficiency calculations can be made in a mill such as the Pioneer, which has an intake of stripped fruit, not bunches, since samples of unsterilized fruit can be readily analysed at intervals and a very fair estimate can be made of the average quantity of oil and kernels contained in the weight of fruit which has been milled.

With bunches, the difficulties of a direct calculation of oil and kernel content are (i) the varying fruit-to-bunch ratio and (ii) the difficulty of obtaining a representative sample of fruit *before* stripping; a sample of fruit can be obtained after stripping, but by this time its moisture content has appreciably decreased and losses have begun to occur.

A direct method of estimating the oil and kernel content of batches of bunches has, however, been proposed by Natta. This entails clearing the stripper before the test begins, determining the total bunch weight of the test bunches and the weight of sterilized fruit obtained from them.[96] The oil and kernel contents are then determined by a system of sampling. Such a method requires the separation of different lots of bunches for sampling and treatment and it will not in itself show the cause of any sudden drop in efficiency.

For these reasons the majority of large mills are 'controlled' by a system of analyses for *losses*. All possible sources of loss are examined and analyses for oil on the one hand and kernel on the other are done. The system is admittedly imperfect and the total of oil produced plus oil losses are rightly recorded in some mill control forms as 'Total Accounted for' but wrongly recorded in others as 'Total Oil present in Bunch'. With oil there is inevitably some loss on machinery and some leakage or spillage which cannot or does not get recorded. Some of the kernel present may disappear as dust. The real total of oil and of kernels coming into the mill is in fact never known and never fully estimated under the loss system, but an approximate figure is known and, in a well-managed mill, the various losses will be fairly constant or improving, and any aberrent figure will soon point to a cause of inefficiency. The impossibility of obtaining a true and complete balance of incomings and outgoings has its accounting disadvantages, however, and may give rise to a certain complacency. The method does not constitute a complete check on events and the need for supervision of all possible points of leakage remains.

In the quantitative control of the mill the following absolute weights are obtained:

1. The weight of bunches entering the mill.
2. The weight of clarified oil produced.
3. The weight of kernels bagged.

From these figures the true *extraction rates* of the mill as a whole can be obtained, e.g.

Month	*January*
Bunches milled	3,120.65 tons
Oil produced	576.03 tons
Kernels bagged	151.76 tons
Extraction rate: oil	18.46 per cent oil to bunch
Extraction rate: kernels	4.86 per cent kernels to bunch

Extraction rates depend on type of fruit, age of palms, season, etc., so although they are exact and very useful for estimation of returns, they do not in themselves give any information on how the mill is operating.

Most of the other data relating to quantity analysis are obtained by sampling; attention must therefore be given to the adequacy of sampling. There is a great need to submit mill sampling methods to statistical examination but until recently there was no published work on this subject at all.[103] Most sampling methods, though sensible and possibly adequate, are empirical, and if a proper statistical examination of sampling could be undertaken it might be possible to reduce some of the work while at the same time judging the limits of each determination. Loss measurements are made as follows:

Oil losses

1. *Sterilizer waste or 'condensate'.* The quantity of sterilizer waste is

measured, and samples taken at regular intervals from the discharge line are analysed for oil and hence the total quantity of oil lost over a period is estimated.

2. *Bunch refuse* (sometimes termed empty bunches, though they are not entirely empty of fruit) must be weighed *in toto* or sampled and analysed for (*a*) *unreleased fruit* and (*b*) *oil absorbed in the stalk and empty spikelets.* The weight of bunch refuse is commonly obtained by weighing daily about 10 per cent of the cages of refuse going to the incinerator or field. Average weight x total number of cages then gives the total weight of bunch refuse. One method of analysis is to check over one random truck per shift and weigh the quantity of unreleased fruit, analysing a sample for oil and kernel content. Fixed percentages may be used for unreleased fruit if the crop is reasonably uniform, but they must be checked from time to time. For oil absorbed in the refuse one stripped bunch can be taken hourly at random from the stripper outlet conveyer and weighed. All the samples from one shift are then reweighed, quartered, one-quarter of each being taken after removal of loose fruit; the bulk is then mixed and quartered down to two samples of 100 g. The total dry weight of the bunch refuse coming from the mill is calculated from the dry weight obtained in the course of these analyses. Percentage oil and kernel losses on dry weight can then be used to calculate total losses. This avoids inaccuracies due to different rates of evaporation.

Chan has studied the variation of oil content of bunch refuse.[104] He found there was more oil in the stalks of overripe than underripe bunches and that more oil was absorbed by the empty spikelets than by the stalk, though there was a positive and significant correlation between the oil contents of stalk and spikelets. As a result of his studies he recommended the sampling of three bunches every half-hour and quartering longitudinally to take one-quarter of each sampled bunch.

Hard bunches are usually returned to the sterilizer and stripper for a second treatment, so bunch refuse weight estimations must be made during the final exodus of stripped bunches from the mill.

3. *Fibre.* The oil loss in fibre leaving the mill must be obtained from estimates of the mesocarp fibre entering the mill in the bunches and the oil to dry fibre leaving the extraction plant. The oil-to-dry fibre estimation is simple and may be done at any stage after the matte has left the centrifuge or press, but in view of the difference in oil content of fibre from different parts of the press cake the sample is perhaps best taken at the fibre cyclone. Then the quantity of nuts and kernels being loose in the fibre may also be analysed and estimated. Samples of about 200 g are taken throughout the day or shift and mixed and quartered down to duplicate 100 g samples.

Two methods of estimating the amount of mesocarp fibre entering the mill have been commonly employed. In one the quantity of fibre is estimated by sampling the sterilized fruit travelling in the fruit elevator. Through a knowledge of the ratio of sterilized fruit (calculated as the difference between the weight of sterilized bunches and bunch refuse) to bunches entering the mill and of the quantity of nuts entering the nut drier an estimation of fibre in the total weight of bunches may be made.

An alternative method, in which the weight of sterilized bunches is not required, is to carry out trial runs during a period of, say, 4 hours and measure the total quantity of all products including the press cake in its component parts, and hence through sample analysis to obtain an average figure of dry fibre to bunches entering the mill.

In sampling for oil to dry fibre, Southworth showed that variations between samples of a bulk sample were small compared with hourly variations, and he found that duplicate tests from a bulk sample made of hourly samples gave a result within 0.6 per cent oil to dry matter of the true figure in 95 out of 100 cases. He recommended bulking hourly samples to give a daily sample and duplicating the analysis, but he admitted that further work on sampling technique is required.[103]

The figure of oil to dry fibre being the largest among the accounted losses and this loss obvious to the eye and touch, a great deal of notice has been taken of it. Moreover gradual improvements in digestion and extraction methods have tended to reduce the average figures over the years from a little under 20 per cent to well under 6 per cent. There has been a tendency even for the oil to dry fibre figure to be taken as a kind of measure of efficiency; and this may lead to a neglect of other and more easily rectified sources of loss. Two points should be borne in mind concerning the oil to dry fibre estimations. In the first place they will depend on the solvent used in the Soxhlet apparatus and will tend to decrease if the fibre is stored for any length of time before analysis. It is necessary, therefore, to adhere to the use of one solvent and to carry out the analysis as soon as possible, Secondly, it should be realised that a difference of 1 per cent in the oil to dry fibre results in a very small difference in estimated milling efficiency. For instance, in a mill where the total accounted losses were estimated to be 1.3 per cent of the weight of bunches entering the mill and the total oil produced by the mill was 17.2 per cent of the weight of bunches, an increase of oil to dry fibre from 9 per cent to 10.5 per cent only led to a reduction of estimated efficiency from 93.0 per cent to 92.4 per cent. In general, with other losses unchanged, a decrease of 1 per cent in oil to dry fibre will only increase efficiency by about 0.4 per cent. Thus, while this loss should clearly not be treated as of no account, it is important that those in charge of mills should not concentrate their attention on oil to dry fibre to the exclusion of other losses which may be occurring, though less obviously, in other parts of the mill.

4. *Sludge.* The number of sludge tanks filled and discharged is counted or some other method is adopted for measuring liquids. Analysis of

samples for oil and for solids not fat are carried out. Regular samples are taken and mixed and the analysis is by day or by shift.

5. *Nuts.*   There are various ways of estimating the small amount of oil lost on nuts. In some mills the nuts are weighed automatically entering the nut drier so that samples are taken at the point and analysed for shell, kernel, moisture and oil on shell and these estimations are related to the input of bunches. If the nuts are not weighed, however, then samples must be taken of nuts coming out of the fibre separator and, after cracking, the oil and moisture content of the shell must be estimated in the laboratory. The calculation of the oil loss will then depend on measurement of the quantity of shells produced and determination of the quantity of shell per unit of bunches entering the mill, and on estimation of the percentage of oil and solids not oil in the shell.

Kernel losses

Losses of kernels are incurred (1) in bunch refuse, (2) among the shells going to the boiler, (3) sticking to shell, (4) in uncracked nuts among the shell and (5) in nuts in fibre.

(1) and (5) have already been mentioned when oil losses in bunches and fibre were discussed; (2), (3) and (4) can be estimated by sampling the shell going to the boiler house and separately they give an indication of the working of the cracking section of the mill.

In relating these losses to the total weight of bunches entering the mill, account is taken of the quantity of nuts, if this is measured, and of the quantity of shell, fibre and bunch refuse.

Efficiencies

The efficiency of *oil extraction* is obtained from the estimations described, as follows:

$$\frac{\text{Oil produced} \times 100}{\text{Oil produced} + \text{oil losses accounted for}} = \text{estimated oil extraction efficiency}$$

The efficiency of *kernel extraction* is similarly obtained:

$$\frac{\text{Kernels produced} \times 100}{\text{Kernels produced} + \text{losses accounted for}} = \begin{array}{l}\text{estimated kernel} \\ \text{extraction efficiency}\end{array}$$

or, if nuts are automatically weighed entering the nut drier and the dry weight of kernels estimated in nut samples, the efficiency of cracking and separating may be estimated as follows:

$$\frac{\text{Dry weight of kernels produced} \times 100}{\text{Dry weight of kernels entering nut drier}}$$

This, however, leaves out of account losses in bunch refuse and in fibre which take place before the nut drier stage.

The control of quality

A great deal has already been said in this chapter about oil and kernel quality and, for completeness, it remains only to record here that the quality analyses required to be done in the mill laboratory are:

(*a*) *Oil.* (1) Moisture content. The sample is weighed to constant weight in a drying oven at 105°C.
(2) Dirt content. This is determined by processes of filtration.
(3) Free fatty acid content. This is normally estimated by titration against caustic soda.

(*b*) *Kernels.* This is entirely a physical process, pieces of shell, dirt, and broken kernels being separated from weighed samples of bagged kernels and weighed separately. Moisture content is determined by grinding and drying to constant weight in an oven.
Determination of the amount of discoloration is a somewhat subjective process. Usually it is considered sufficient to divide the sampled kernels, which have been bisected with a sharp knife, into white, slightly discoloured, badly discoloured or mouldy.

(*c*) *Other determinations.* The majority of other determinations in the mill laboratory are either moisture or oil determinations. The former are carried out by drying in ovens to constant weight while the latter, e.g. of fibre, bunch refuse, is usually done in the Soxhlet apparatus.
Several publications are available on laboratory methods for milling control.[34, 97]

Milling report forms

Milling report forms or control sheets will vary according to the measurements actually made in the mill and the analyses carried out in the mill laboratory. They will be made up from the many books which have to be entered up daily.
The forms should have the sections shown in Table 14.7 where imaginary figures are entered for a mill processing predominantly Deli *dura* bunches. In this scheme there is no repetition of figures and the form can thus be kept in manageable proportions. The tables can be extended to the right to allow for the recording of past or cumulative data for comparison.
Operating data are also recorded for the proper physical and mechanical

Table 14.7   *Milling report forms*

A. Throughput, output, and intermediate weights

Throughput and output	Weight (tons)	Extraction rate (%)
1. Bunches milled	4,462	–
2. Palm oil produced	794	17.8
3. Palm kernels produced	187	4.2

Intermediate + other products		As per cent of bunches milled
4. Sterilized bunches	3,998	89.6
5. Loss of wt. on sterilization	464	10.4
6. Bunch refuse, wet	1,258	28.2
7. Bunch refuse, dry	415	9.3
8. Sterilized fruit (4–6)	2,740	61.4
9. Fruit in bunch refuse	12	0.27
10. Fibre, dry	339	7.6
11. Nuts	897	20.1
12. Kernel in nuts, dry	170	3.8
13. Shells	678	15.2
14. Sludge	830	18.6
15. Sterilizer condensate	647	14.5

control of a mill. The following records (among others) are usually kept:

Sterilizing
Maximum steam pressure
Average sterilizing time
Maximum temperature
Number of sterilizing cycles
Average sterilizer load, tons
Capacity per milling hour, tons
Condensate, in kg per ton of
   bunches

Presses
Number of pressings per pressing hour
Number of presses in use
Quantity pressed per hour, in terms of
   bunches

Kernel cracking
Cracking hours
Quantity of nuts cracked per hour

Clarification
Data on throughput and sludge/oil
   ratio

   In some mills measurements of the degree of ripeness of the bunches are attempted. This is in the nature of an overall check on the harvesting. The figures are open to misinterpretation since the amount of fruit loosening will depend, apart from ripeness, on the number of hours between harvest and the recording of the number of detached fruit, and on whether the bunches are dry or have become wet in the rain. In some mills the bunches are sampled on receipt and are classified according to the number of detached exterior fruit. In other mills the quantity of loose fruit delivered is measured and related to the total bunch plus loose fruit weight. In spite of the difficulties and admitted inaccuracies of these

Table 14.7 — *continued*

B. Oil and kernel quantitative analyses — losses and efficiency

Oil in:	Per cent oil	Loss tons	As per cent of bunches milled	Notes
16. Sterilizer condensate	0.32	1.7	0.04	Calculated from No. 15 above
17. Bunch refuse:				
Fruit	33.3	4.0	0.09	Calculated from No. 9 above
Dry trash	2.10	8.7	0.19	Calculated from Nos 6 and 7 above
18. Dry fibre	9.92	33.6	0.75	Calculated from No. 10 above
19. Nuts	0.33	3.0	0.07	Calculated from No. 11 above
20. Sludge	0.25	2.1	0.05	Calculated from No. 14 above
21. Total accounted losses	—	53.1	1.19	
22. Mill efficiency	93.7%			(No. 2 x 100) ÷ (No. 2 + No. 21)

Kernels in:	Per cent kernels			
23. Fruit in *bunch refuse*	5.00	0.6	0.01	Calculated from No. 9 above
24. Losses among *shell*	1.11	7.5	0.17	Calculated from No. 13 above
25. Sticking to *shell*	0.31	2.1	0.05	Calculated from No. 13 above
26. In uncracked nuts (*to shell*)	0.36	2.4	0.05	Calculated from No. 13 above
27. In fibre, dry	0.02	0.8	0.02	Calculated from No. 10 above
28. Total accounted losses	—	13.4	0.30	
29. Mill efficiency	93.3%			(No. 3 x 100) ÷ (No. 3 + No. 28)

C. Laboratory qualitative analysis

Oil	Per cent	Previous periods
30. F.F.A.	2.1	
31. Moisture	0.072	
32. Dirt	0.006	

Kernels		
33. Moisture	6.4	
34. Shells	0.68	
35. Dirt	0.25	
36. Broken	4.2	
37. White	85.0	
38. Slightly discoloured	12.0	
39. Badly discoloured	3.0	
40. Mouldy		

measurements it is certainly advantageous from the point of view of the mill engineer to obtain some picture of the state of the produce he is receiving from the field so that he may be able to note any variations from his usual supply. In one scheme devised for a group of estates bunches are sampled each day and divided into three ripeness classes: (1) Ripe – more than 1 loose fruit per pound of bunch weight (2 per kg); (2) Over-ripe – less than one-eighth of the fruit retained in the bunch; (3) Under-ripe – bunch with less than 1 loose fruit per pound weight. Bunches are further classified and enumerated if found unsatisfactory in the following respects: (1) empty bunches; (2) poor fruit set; (3) diseased; (4) damaged by pests; (5) dirty; (6) bruised; (7) long stalks.

Milling costs and bunch purchase

Very little has been published on either the capital costs or the running costs of mills of different sizes or designs. Such studies as have been made have usually demonstrated that very large mills have a lower capital cost per ton of throughput and a lower operating cost than smaller mills. Costs of large mills have, however, been rising steeply. In 1969 a 15-ton-per-hour mill was quoted as costing about US $900,000 as a turnkey job;[81] at the present time (1976) a similar mill might cost well over $3 million, and the raising of sufficient capital has become a problem in some countries. This has been an additional reason for the popularity of smaller, less sophisticated mills in some American countries, and such mills, locally constructed, now appear to be little more expensive on a ton of throughput basis than large mills, though operating costs on this basis are still higher.

It has been observed that mill operating costs have often been approximately equal to the revenue from kernels. In a 1970 case study of operating a 40-ton-per-hour mill in Malaysia the operating cost per ton of bunches including all overheads and other charges was estimated to be M$17 against a kernel revenue on the same basis of M$15.75.[98] This mill was expected to purchase bunches from nearby plantations, and graphs were drawn up showing the linear relation between the f.o.b. oil price and the fair purchase rate per ton of bunches with different oil extraction rates and assuming a profit of M$10 per ton for the mill.[98] With the extension in many countries of the practice of selling bunches as a commodity, *tenera* bunches will obtain a current market price of their own which should be directly related to the prevailing oil and kernel prices.

References

1. **Perry, J. H.** (ed.) (1950) *Chemical engineers' handbook.* McGraw-Hill, New York.

2. Bailey, A. E. (1945) *Oil and fat products.* Interscience Publishers, Inc., New York.

3. Eckey, E. W. (1954) *Vegetable fats and oils.* Reinhold Publishing Corp., New York.

4. Devine, J. and Williams, P. N. (eds.) (1961) *The chemistry and technology of edible oils and fats.* Proc. Conf. arranged by Unilever Ltd, Port Sunlight. Pergamon Press.

5. Martinenchi, G. B. (1972) Traitments de l'huile de palme. 1. Séparation en fractions liquide et solide. *Oléagineux,* 27, 267.

6. Bernardini, E and Bernardini, M. (1975) Oil palm fractionation and refining using the C.M.B. process. *Oléagineux,* 30, 121.

7. Koslowsky, L. (1975) Chemical fractionation of palm oil by transesterification. *Oil Palm News,* 19, 14.

8. Crombie, W. M. (1956) Fat metabolism in the West African oil palm (*Elaeis guineensis*) Part 1. Fatty acid formation in the maturing kernel. *J. exp. Bot.,* 7, 181.

9. Crombie, W. M. and Hardman, E. E. (1958) Fat metabolism in the West African oil palm (*Elaeis guineensis*), Part III, Fatty Acid formation in the maturing exocarp. *J. exp. Bot.,* 9, 247.

10. Desassis, A. (1955) Le détermination de la teneur en huile de la pulpe de fruit *d'Elaeis guineensis. Oléagineux,* 10, 739 and 823.

11. Thomas, R. L. *et al.* (1971) Fruit ripening in the oil palm *Elaeis guineensis. Ann. Bot.,* 35, 1219.

12. Loncin, M. and Jacobsberg, B. (1963) Studies in Congo palm oil. *J. Am. Oil Chem. Soc.,* 40, 18.

13. Loncin, M. and Jacobsberg, B. (1965) *Recherches sur l'huile de palme en Belgique et au Congo.* (*Research on palm oil in Belgium and the Congo.*) Paper presented at the Tropical Products Institute Oil Palm Conference, London, 1965, p. 85. Ministry of Overseas Development.

14. Oil palm genetics laboratory, Malaysia. Progress report 1972.

15. Bienaymé, A. and Servant, M. (1958) Variation des characteristiques des huiles de palme et notamment de leur carotênoides. *Qualitas Pl. Mater. veg,* 1, 3/4, 336.

16. Booth, V. H. (1957) *Carotene, its determination in biological material.* W. Heffer, Cambridge.

17. Purvis, C. (1957) The colour of oil palm fruits. *J. W. Afr. Inst. Oil Palm Res.,* 2, 142.

18. Ames, G. R., Raymond, W. D. and Ward, F. B. (1960) The bleachability of Nigerian palm oil. *J. Sci. Fd. Agric.,* 2, 194.

19. Nwanze, S. C. (1961, 1962, 1964) W.A.I.F.O.R. Ninth, Tenth and Twelfth Annual Reports, 1960–1, 1961–2 and 1963–4, pp. 98, 90 and 97 respectively.

20. Bezard, J. A. (1971) The component tryglycerides of palm-kernel oil. *Lipids 6* (9), 630.

21. Hardon, J. J. (1969) Interspecific hybrids in the genus *Elaeis.* II. Vegetative growth and yield of F_1 hybrids *E. guineensis* x *E. oleifera. Euphytica,* 18, 380.

22. Macfarlane, N., Swetman, T, and Cornelius, J. A. (1975) Analysis of mesocarp and kernel oils from the American oil palm and F_1 hybrids with the West African oil palm. *J. Sci. Fd. Agric.,* 26, 1293; and *Oil Palm News,* 19, 12 & 20, 1.

23. Naudet, M. and Faulkner, H. (1975) Compositions et structures glycéridiques comparées des huiles *d'Elaeis guineensis, Elaeis melanococca* et *d'hybride guineensis—melanococca. Oléagineux,* 30, 171.

24. Pritchard, J. L. R. (1969) Quality of oil palm products — User requirements, *Trop. Sci.,* 11, 103.

25. Desassis, A. (1957) L'acidification de l'huile de palme. *Oléagineux,* 12, 525.

26. Coursey, D. G. (1963) The deterioration of palm oil during storage. *J. W. Afr. Sci. Ass.,* 7, 101.

27. **Turner, P. D.** (1969) The importance of lipolytic microorganisms in the degrada-
    tion of oil palm products in Malaysia. In *The quality and marketing of oil palm
    products*, p. 53. Inc. Soc. of Planters, Kuala Lumpur.
28. **Loncin, M.** (1952) L'hydrolyse spontanée autocatalitique des triglycérides.
    Oléagineux, 7, 695.
29. **Vanneck, C. and Loncin, M.** (1951) Considération sur l'alteration de l'huile de
    palme. *Bull. agric. Congo belge.* 42, 57.
30. **Jasperson, H. and Pritchard, J. L. R.** (1965) *Factors influencing the refining and
    bleaching of palm oil.* Paper presented at the Tropical Products Institute Oil Palm
    Conference, London, 1965, p. 96. Min. of Overseas Development.
31. **Bienaymé, A.** (1954) Les huiles de palme et leur richesse en carotène.
    Oléagineux, 9, 603.
32. **Hartley, C. W. S. and Nwanze, S. C.** (1965) *Factors responsible for the produc-
    tion of poor quality oils.* Paper presented at the Tropical Products Institute Oil
    Palm Conference, London, 1965, p. 68. Ministry of Overseas Development.
33. **Jacobsberg, B.** (1971) La Production d'une huile de palme de haute qualité.
    Oléagineux, 26, 781.
34. **Arnott, G. W.** (1963) The Malayan oil palm and the analysis of its products. Min.
    Agriculture and Cooperation, Fed. of Malaya, *Bull.* 113.
35. **Jacobsberg, B. and Jacqmain, D.** (1973) Palm oil quality. Appreciation and
    forecast of stability of crude palm oil during transport and storage. *Oléagineux,
    28*, 25.
36. **Johansson, G. and Persmark, U.** (1971) Evaluation and prediction of oil palm
    quality. *Oil Palm News*, 10 and 11,2.
37. **Cornelius, J. A.** (1973) The assessment of palm oil quality — International
    collaborative work. *Oil Palm News*, 15, 1.
38. **Cornelius, J. A.** (1965) *Some technical aspects influencing the quality of palm
    kernels.* Paper presented at the Tropical Products Institute Oil Palm Conference,
    London, 1965, p. 105, Min. of Overseas Development.
39. **Coursey, D. E., Simmons, E. A. and Sheridan, A.** (1963) Studies on the quality
    of Nigerian palm kernels. *J. W. Afr. Sci. Ass.*, 8, 18.
40. **Loncin, M. and Jacobsberg, B.** (1964) *Study on palm kernel acidification during
    storage.* Int. Soc. for Fat Research Congress, Hamburg.
41. **Eggins, H. O. W. and Coursey, D. G.** (1964) Thermophilic fungi associated with
    Nigerian oil palm produce. *Nature, Lond.*, 203, 1083.
42. *Stork Palmoil-review.* Kernel Recovery, 3, 4—5, (1963).
43. **Bek-Nielson, B.** (1969) Quality aspects of oil palm kernel production. In *The
    Quality and marketing of oil palm products*, p. 161. Inc. Soc. of Planters, Kuala
    Lumpur.
44. **Clegg, A. J. and Teh, Y. C.** (1972) Production de palmistes hydrolytiquement
    stables. *Oléagineux, 27*, 101.
45. **Thieme, W. L. and Olie, J. J.** (1969) Discoloration of oil palm kernels in
    relation to processing temperature and time. In *The quality and marketing of oil
    palm products*, p. 144, Inc. Soc. of Planters, Kuala Lumpur.
46. **Coursey, D. G.** (1961) Quelques observation sur l'altération de la couleur des
    palmistes. *Oléagineux, 16*, 385.
47. *Stork Palmoil-review.* Sterilization, 1, 2 (1960).
48. **Gray, J. E.** (1922) Native methods of preparing palm oil. *First A. Bull. Dep.
    Agric. Nigeria*, p. 28.
49. **Faulkner, O. T. and Lewin, C. J.** (1923) Native methods of preparing palm oil. II.
    Second A. Bull. Dep. Agric. Nigeria, p. 3.
50. **Manlove, D. and Watson, W. A.** (1931) Press extraction of palm oil in Nigeria.
    Tenth A. Bull. Dep. Agric. Nigeria, p. 19.
51. The African Oil Palm Industry. II. Machinery, *Bull. imp. Inst., Lond.*, 15, 57 (1917).
52. **Barnes, A. C.** (1924) A modified process for extraction of palm oil by natives.
    Third A. Bull. Dep. Agric. Nigeria, p. 3.

53. **Barnes, A. C.** (1926) An improved process for the extraction of palm oil by natives. The Cooker-Press process. *Fifth A. Bull. Dep. Agric. Nigeria*, p. 33.
54. **Barnes, A. C.** (1925) Mechanical processes for the extraction of palm oil. *Second Special Bull. Dep. Agric. Nigeria.*
55. **Milsum, J. N. and Georgi, C. D. V.** (1938) Smallscale extraction of palm oil. *Malayan Agr. J.*, 26, 53.
56. **Bunting, B., Georgi, C. D. V. and Milsum, J. N.** (1934) *The oil palm in Malaya.* Dept. Agric., Malayan Planting Manual No. 1, Kuala Lumpur.
57. **Nwanze, S. C.** (1963) W.A.I.F.O.R. Eleventh Annual Report 1962–3, p. 88.
58. **Nwanze, S. C.** (1965) The hydraulic hand press. *J. Nigerian Inst. Oil Palm Res.*, 4, 290.
59. **Gebr. Stork & Co.** *Smallscale palm oil processing with the aid of Stork hydraulic hand presses* – Mimeograph.
60. **Nwanze, S. C.** *Semi-commercial scale palm oil processing.* Paper presented at the Tropical Products Institute Oil Palm Conference, London, 1965, p. 63. Min. of Overseas Development.
61. **Nwanze, S. C.** (1964) W.A.I.F.O.R. Twelfth Annual Report 1963–4, p. 96.
62. **Kilby, P.** (1967 and 1968) The Nigerian Palm oil industry. *Food Res. Inst. Studies*, 7, 2, 177; and 8, 2, 199.
63. **Purvis, M. J.** (1968) The Nigerian palm oil industry: a comment. *Food Res. Inst. Studies*, 8, 2,191.
64. **Blommendaal, H. N.** (1927) *De Fabricage van Palmolie.* A.V.R.O.S. Algemeene Serie No. 33.
65. **Van Heurn, F. C.** (1921) *Considérations sur l'installation de fabriques d'huile de palme.* A.V.R.O.S. Com. Gen. Ser. No. 10.
66. **Dyke, M. F-M.** *in* Leplae, E. (1939) Le palmier à huile en Afrique, son exploitation au Congo-Belge et en Extrème-Orient. *Mém. Inst. r. colon. belge Sect. Sci. nat. méd.*, 7 (3), 1–108.
67. **Cooper, I. N. and Bevan, J. W. L.** (1968) Some factors to be considered when planning the organisation of processing oil palm products. In *Oil palm developments in Malaysia*, p. 118. Incorp. Soc. of Planters, Kuala Lumpur.
68. *Stork Palmoil-review*, 1, 3 (1960).
69. **Nwanze, S. C.** (1961) W.A.I.F.O.R. Ninth Annual Report 1960–1, p. 96.
70. *Stork Palmoil-review*, 1, 4 (1960).
71. **Twitchin, J. F.** *Palm Oil machinery.* Reprinted from *The Planter*, 1955–6. Incorp. Sec. of Planters, Kuala Lumpur.
72. *Stork Palmoil-review*, 1, 5 (1960); 2, 1 (1961).
73. **Bek-Nielson, B.** (1969) Palm oil and kernel extraction plants in relation to quality. In *The quality and marketing of oil palm products*, p. 169. Incorp. Soc. of Planters, Kuala Lumpur, and *Oléagineux*, 26, 483 and 635 (1971).
74. **Wyer, G. van de.** (1927) Procédés d'extraction de l'huile de palme. *Bull. Ass. pour le Perfectionnement de Materiel Colonial*, Mai 1927.
75. **Olie, J. J.** The automatic hydraulic press and its field of application. Stork-Amsterdam, mimeograph, undated.
76. *Stork Palmoil-review*, 2, 2 (1961).
77. **Olie, J. J.** (1973) The Stork twin screw press. *Oléagineux*, 28, 33.
78. **Georgi, C. D. V.** (1932) The centrifugal extraction of palm oil at Serdang. *Malay. agric. J.*, 20, 446.
79. **Georgi, C. D. V.** (1933) Comparison of the press and centrifugal methods for treatment of palm oil fruit. *Malay. agric. J.*, 21, 103.
80. **Wolversperges, A.** (1963) The extraction of palm oil by means of screw presses. *Planter, Kuala Lumpur*, 39 (1), (2) & (3), pp. 11, 68, 111.
81. **Olie, J. J.** (1969) Total N.O.S., N.F.P.Q. and efficiency guarantees, *Oléagineux*, 24, 557.
82. **Chayen, I. H. and Ashworth, D. R.** (1953) The application of impulse rendering to the animal fat industry. *J. appl. Chem.*, 3, 529.

83. **Nwanze, S. C.** (1962) W.A.I.F.O.R. Tenth Annual Report 1961—2, p. 89.
84. **Zachariassen, B.** (1969) Moisture removal from palm oil. In *The quality and marketing of oil palm products*, p. 139. Incorp. Soc. of Planters, Kuala Lumpur, and *Oléagineux*, 25, 543.
85. **Wohlfahrt, N.** (1969) Clarification de l'huile de palme. *Oléagineux*, 24, 699.
86. **Olie, J. J. and Tjeng, T. D.** (1972) Traitment et évacuation des eaux résiduaires d'une huilerie de palme. *Oléagineux*, 27, 215.
87. **Singh, Kirat and Ng Siew Hoong** (1968) Treatment and disposal of palm oil mill effluent. *Malaysian Agric. J.*, 46, 316.
88. **Stanton, W. R.** (1974) Treatment of effluent from palm oil factories. *Planter, Kuala Lumpur*, 50, 382.
89. **Webb, B. H., Rajagopalan, K., Cheam Soon Tee and Dhiauddin Bin Noor Jantan** (1975) Palm oil mill waste recovery as a by-product industry. *Planter, Kuala Lumpur*, 51, 86 and 126.
90. **Olie, J. J.** (1969) The active development of process technology in the recovery of palm oil and kernels. *Oléagineux*, 24, 293.
91. *Stork Palmoil-review*, 3, 2 (1962).
92. **Plantation Engineering Division, Unilever.** (1967) Current Unilever developments in palm oil extraction equipment and machinery. Sabah Planters' Association. Oil Palm Technical Seminar. Mimeograph.
93. *Stork Palmoil-review*, 3, 3 (1962).
94. **Weko, B. H. and Sutiardjo,** (1969) A comparison of the hydrocyclone and clay bath separation methods for oil palm kernels. In *The quality and marketing of oil palm products*, p. 154. Incorp. Soc. of Planters.
95. **Nwanze, S. C.** (1961) The economics of the Pioneer mill. *J. W. Afr. Inst. Oil Palm Res.*, 3, 233.
96. **Natta, L. F.** (1969) The assessment of oil and kernel content of oil palm bunches. In *Progress in oil palm*, p. 126. Incorp. Soc. of Planters, Kuala Lumpur.
97. **I.R.H.O.** (1967) *Manuel de l'huilerie de palme*. Série Sci. No. 12. Paris, new edition.
98. **Roslan Abdullah** (1970) The establishment and operating costs of a 40 ton per hour palm oil mill in West Malaysia. *Review of Agric. Econ. Malaysia*, 4 (1), 11.
99. **Cornelius, J. A.** (1976) Progress towards international standards for crude palm oil. Int. Symposium on Oil Palm Processing and Marketing, Kuala Lumpur, 1976.
100. **Olie, J. J. and Tjeng, T. D.** (1974) The extraction of palm oil. Stork-Amsterdam. Mimeograph.
101. **Webb, B. H., Hutagalung, R. I. and Cheam, S. T.** (1976) Palm oil mill waste as animal feed — processing and utilization. Int. Symposium on Oil Palm Processing and Marketing, Kuala Lumpur, 1976.
102. **Hemming, M. L.** (1976) The treatment of effluents from the production of palm oil. Int. Symposium on Oil Palm Processing and Marketing, Kuala Lumpur, 1976.
103. **Southworth, A.** (1976) Process Control. Int. Symposium on Oil Palm Processing and Marketing, Kuala Lumpur, 1976.
104. **Chan, K. S.** (1976) Sampling of oil palm bunch stalk refuse to determine oil loss. Int. Symposium on Oil Palm Processing and Marketing. Kuala Lumpur, 1976.
105. **Ng, B. H., Corley, R. H. V. and Clegg, A. J.** (1976) Variation in the fatty acid composition of palm oil. *Oléagineux*, 31, 1.
106. **Babatunde, G. M., Fetuga, B. L., Odumosu, O. and Oyenuga, A.** (1975) Palm kernel meal as the major protein concentrate in the diets of pigs in the tropics. *J. Sci. Fd. Agric.*, 26, 1279.
107. **Quraishi, A. and Macfarlane, N.** (1975) The nitrogen content and amino acid composition of palm kernel meals. *Oil Palm News*, 20, 3.

Index of authors and references

Reference numbers in each chapter bibliography are given in (parentheses) after the page number. *Italic numbers* indicate a mention in the text.

Ollagnier, M. – *continued*
    579, 584(34, 41, 42, 43, 54, 55),
    585(56), 587(114, 131), 686(38),
    687(69), 690(152)
Ong Hean Tatt, 310(131)
Ooi, S. C., *259, 303*, 308(69, 72, 73,
    80, 83), 310(123)
Opsomer, J. E., *38–9*, 74(6), 194(70)
Opute, F. I., 310(131)
Owen, G., 134(32)

Pandit, S. V., 690(154)
Panton, W. P., *115, 119*, 134(38, 42),
    375, 389(26)
Park, D., 687(57)
Parnata, J., 687(68)
Patiño, V. M., 76(55), *293*, 309(97,98)
Paton, T. R., *119–20*, 134(43),
    135(49), *537*, 585(77)
Paulmier, 731
Pee, W. van, 604(22)
Perrier de la Bathie, H., 35(15), 74(5)
Perry, J. H., 776(1)
Persmark, U., 778(36)
Phang, S., 308(69)
Phillips, T. A., 36(50)
Pichel, R., 308(75)
Piggott, C. J., 587(109)
Pillai, K. R., 493(77)
Plan Décennal Congo Belge, 35(21)
Poels, G., 309(114), *337*, 359(10)
Poerck, R. De, *248*, 307(50)
Poisson, *731*
Poncelet, M., 426(18)
Ponniah, D., 690(133)
Poon Yew Chin, 584(44, 47), 585(75),
    587(120)
Portières, R., 35(16)
Pratt, N. S., 492(50)
Prendergast, A. G., 310(116), *628,
    631–2*, 685(2), 686(53), 687(58)
Preuss, L., *41*, 74(7)
Prevot, P., 191(1), 426(22), *401–10,
    515–18, 525, 533, 540*, 584(41,
    42, 43), 585(61, 62), 586(87, 103)
Price, D. T., *13*
Price, J. G. M., 493(59)
Pritchard, J. L. R., 777(24), 778(30),
    36(55)
Proceedings of Conf. OPRS, Nigeria,
    307(47)
Pronk, F., 305(4, 5)
Purba, A. Y. L., 585(57)
Purvis, C., *56, 58–9, 65*, 75(39),
    76(52), 305(21), 777(17)
Purvis, M. J., 779(63)

Quemener, J., 586(86)
Quencez, P., 360(22), 691(159)
Quraishi, A., 780(107)

Rabéchault, H., 76(65), *139*, 192(12,
    14, 16), 309(105), *605*, 684(1)
Rajagopalan, K., 360(30), 685(12),
    686(29, 48), 780(89)
Rajanaidu, N., 308(78), 310(124, 125)
Rajaratnam, J. A., *511, 524, 572*,
    584(32, 33, 39), 585(58),
    588(141), *621*, 686(37)
Ramachandram, P., 426(25)
Rao, A. P., 690(154)
Rashid bin Ahmad, M., 603(11)
Rasmussen, A. N., 493(71), *636*,
    687(67)
Raymond, W. D., 34(1), 777(18)
Rees, A. R., 34(7), 75(19, 24, 32),
    138, 141–3, 145, 191(2, 3, 7, 8,
    9, 10, 17, 26, 28, 29, 30),
    193(32, 35, 36, 38, 43), 194(61),
    327(2, 3, 4, 9), 328(19), 585(73)
Renard, J. L., 310(117), 588(145),
    629, 631–2, 685(13), 686(55),
    687(59), 690(152), 691(159)
Renault, P., 492(49)
Richards, F. J., *50*
Richardson, D. L., 76(70), 310(130)
Ringoet, M., 16, 211
Rion, G., 135(57)
Rissognol, J., 306(26)
Rivero, M. J., *293*
Robertson, G. W., 194(79)
Robertson, J. S., 310(119, 120),
    360(26), *510*, 584(30, 31), *606,
    608, 613–14, 625, 642–3*, 685(2,
    6, 10, 14, 15, 18, 25, 26),
    686(45, 46, 47, 49), 688(86, 87)
Roels, O., *70*, 76(54, 61), 307(51)
Rojas Cruz, L. A., 691(162)
Rombaut, D., 603(14)
Rosenquist, E. A., *504, 508*, 583(18),
    585(79)
Roslan Abdullah, 780(98)
Rots, O., 133(12)
Rubber Research Institute, Malaysia,
    389(8, 9), 491(20)
Ruer, P., 75(41, 42, 45), 193(40),
    194(77), 359(1), *532*, 585(72),
    587(115), 689(122)
Russell, E. W., 585(78)
Rutjers, A. A. L., *15*, 35(24)

Salisbury, E., 76(58)
Sanchez Potes, A., 687(73)

General index

Tandjong Slamat, 200
Tanganyika, Lake, 6
Tanzania, 7
Taperoa, 91, 131, 638
Tedion, 655
Tela, 104, 109, 292
Temnoschoita sp., 658, 670–2, 679
Temperature, effect on stomatal
closure, 154–5; in oil palm
regions, 96, 107, 108–9
Temperature for germination, 140–1,
316
Tenera, see fruit forms
Tennamaram Estate, 15, 204, 732
Tephrosia sp., 443
Termites, 486, 656
Terraces, 361, 370–1, 375
Terrastichus sp., 670
Terre de barre, 127, 580
Terylene bags, 227–8
Testa, 42
Tetradifon, 655
Thatching, leaves for, 8, 459
Thermomyces sp., 714
Thibenzole, 612
Thieleviopsis paradoxa, 624–7;
basicola, 613
Thinned grove, 89
Thinning, a plantation, 417, 419–23;
groves, 92, 94
Thiodan, 658
Thiram, 312, 327
Thornthwaite's potential evapo-
transpiration estimation, 102–4
Thosea sp., 662
Thryonomys swinderianus, 404, 682
Tillage, 382–4, 427, 435–7
Tin, effect on oil, 710
Tin plate industry, 31
Tingidae, 621, 657
Tiquadra sp., 659
Tirathaba mundella, 458, 467
Tissue culture, 298–9
T.M.T.D., 324
Tocopherols, 701
Togoland, 5, 9, 77, 700, 731
Tolerance of pathogens, *see* resistance
Torula sp., 707
Totox value of palm oil, 712
Toxaphene, 669
Toxicity symptoms, 514
Trade in palm products, early, 9
Transesterification, 696
Transmission of power in a mill, 732
Transpiration, 51, 55, 152–6
Transplanted palm, growth of, 152–6

Transplanting, two-leaf bare root
seedlings, 331, 342; 4–5 leaf
prenursery seedlings, 346–8; to
the field, 395–407; from a field
nursery, 395–400; from poly-
thene bags, 400; methods of,
395–400; time of, 400–2;
cultural practices at, 402–4;
mulching at, 402–3; manuring at,
403–4; preparation of site of,
372, 400, 402; bare-root
seedlings, 404–6
Tree crusher, 376
Trema sp., 430
Trentepohlia sp., 623
Triangular passage, 9
Thrichlorfon, *see* Dipterex
Trichoderma sp., 704
Trichogyia sp., 662
Trinidad, 653
Tripanosomiasis, 596
Tripsacum laxum, 597
Tropical Products Institute, 709
Trunk, *see* stem
Tsingilo, 7
Tsiribihina river, 7
Tumaco, 293
Turnaca sp., 665

Ubangi river, 4
Ufuma, 88, 265–6
Uganda, 6, 659
Ultisols, 112, 114
Umuahia, 166
Umudiki, 504, 548
Underbrushing, 368
Underplanting, 481–2
Unified activity ratio, 538
Unilever Plantations Ltd., 26, 613, 732
United Fruit Co., 28, 292
United Kingdom, 9–11, 30–2
United States of America, 30–1
Upper Stem Rot, 640–1
Urabá Gulf, 293
Uranomys ruddi, 683
Urea, as a fertilizer, 563; for hastening
trunk decay, 636, 677
Urena repens, 453
Uribe, 430
Usage of palm oil, 11, 31, 699, 708
U.S.D.A. 7th approximation, 112, 114
Ustulina sp., 641

Vacuum driers, 754–5
Valanga nigricornis, 656
Vanaspathi, 32